AF335082

# Landslide Hazard and Risk

# Landslide Hazard and Risk

**Editors**

THOMAS GLADE

*University of Bonn*

MALCOLM ANDERSON

*University of Bristol*

MICHAEL J. CROZIER

*Victoria University of Wellington*

John Wiley & Sons, Ltd

***Other Wiley Editorial Offices***

John Wiley & Sons Inc., 111 River Street, Hoboken, NJ 07030, USA

Jossey-Bass, 989 Market Street, San Francisco, CA 94103-1741, USA

Wiley-VCH Verlag GmbH, Boschstr. 12, D-69469 Weinheim, Germany

John Wiley & Sons Australia Ltd, 33 Park Road, Milton, Queensland 4064, Australia

John Wiley & Sons (Asia) Pte Ltd, 2 Clementi Loop #02-01, Jin Xing Distripark, Singapore 129809

John Wiley & Sons Canada Ltd, 22 Worcester Road, Etobicoke, Ontario, Canada M9W 1L1

Wiley also publishes its books in a variety of electronic formats. Some content that
appears in print may not be available in electronic books.

***Library of Congress Cataloging-in-Publication Data***

Landslide hazard and risk / editors, Thomas Glade, Malcolm Anderson, Michael Crozier.
    p.  cm.
Includes index.
ISBN 0-471-48663-9
1. Landslides—Risk assessment.  2. Landslides—Social aspects.  3. Slopes (Soil mechanics)
I. Glade, Thomas.  II. Anderson, Malcolm.  III. Crozier, Michael J.
QE599.A2L36 2004
551.3′07—dc22

                                                                        2004016991

***British Library Cataloguing in Publication Data***

A catalogue record for this book is available from the British Library

ISBN 0-471-48663-9

Typeset in 10/12pt Times by Integra Software Services Pvt. Ltd, Pondicherry, India
Printed and bound in Great Britain by Antony Rowe Ltd, Chippenham, Wiltshire
This book is printed on acid-free paper responsibly manufactured from sustainable
forestry in which at least two trees are planted for each one used for paper production.

# Contents

# List of Contributors

**David Alexander**   Department of Defence Management & Security Analysis, Cranfield University, Royal Military College of Science, Shrivenham, Swindon, SN6 8LA, UK

**Joan Altimir**   Euroconsult, S.A., C/Na Maria Pla, 33, 3er 2a, Edifici Illa, Andorra la Vella, Principality of Andorra

**Jordi Amigó**   Eurogeotcnica, S.A., Centre Tecnològic Europroject, Parc Tecnològic del vallès, 08290 Cerdanyola, Spain

**Malcolm G. Anderson**   School of Geographical Sciences, University of Bristol, University Road, Bristol, BS8 1SS, UK

**Francesca Ardizzone**   Consiglio Nazionale delle Ricerche, Istituto di Ricerca per la Protezione Idrogeologica, via della Madonna Alta 126, 06128 Perugia, Italy

**Fred Baynes**   Consulting Engineering Geologist, Perth, Australia

**Edward Nicholas Bromhead**   School of Engineering, Kingston University, Penrhyn Road, Kingston upon Thames, KT1 2EE, UK

**David R. Butler**   The James and Marilyn Lovell Center for Environmental Geography & Hazards Research, Department of Geography, Texas State University, San Marcos, San Marcos, TX 78666-4616, USA

**Mauro Cardinali**   Consiglio Nazionale delle Ricerche, Istituto di Ricerca per la Protezione Idrogeologica, via della Madonna Alta 126, 06128 Perugia, Italy

**Chang-Jo F. Chung**   Geological Survey of Canada, Ottawa, K1A 0E8, Canada

**Ramon Copons**   Euroconsult, S.A., C/Na Maria Pla, 33, 3er 2a, Edifici Illa, Andorra la Vella, Principality of Andorra

**Jordi Corominas**   Departament of Geotechnical Engineering and Geosciences, Civil Engineering School, Universitat Politècnica de Catalunya, Jordi Girona 1-3, 08034 Barcelona, Spain

**Michael J. Crozier**  Institute of Geography, School of Earth Sciences, Victoria University of Wellington, PO Box 600, Wellington, New Zealand

**Lisa M. DeChano**  Department of Geography, Western Michigan University, Kalamazoo, MI 49008, USA

**Andrea G. Fabbri**  DISAT, University of Milano-Bicocca, 20126 Milan, Italy

**Mirco Galli**  Consiglio Nazionale delle Ricerche, Istituto di Ricerca per la Protezione Idrogeologica, via della Madonna Alta 126, 06128 Perugia, Italy

**Thomas Glade**  Department of Geography, University of Bonn, Meckenheimer Allee 166, D-53115 Bonn, Germany

**Paula L. Gori**  US Geological Survey, 12201 Sunrise Valley Drive, Reston, VA 20192, USA

**Ken Granger**  Buderim, Australia

**Fausto Guzzetti**  Consiglio Nazionale delle Ricerche, Istituto di Ricerca per la Protezione Idrogeologica, via della Madonna Alta 126, 06128 Perugia, Italy

**Garth Harmsworth**  Landcare Research Private Bag 11052, Palmerston North, New Zealand

**A. Hart**  Scott Wilson Kirkpatrick & Co. Ltd, UK

**G.J. Hearn**  Scott Wilson Kirkpatrick & Co. Ltd, UK

**L.M. Highland**  US Geological Survey, Box 25046 Denver Fed. Ctr, Denver, CO 80225, USA

**E.A. Holcombe**  School of Geographical Sciences, University of Bristol, University Road, Bristol, BS8 1SS, UK

**Kurt Hollenstein**  ETH Zurich Forest Engineering, ETH Zentrum HG G 21.5, CH-8092 Zürich, Switzerland

**I. Ibriam**  School of Geographical Sciences, University of Bristol, University Road, Bristol, BS8 1SS, UK

**D. Karnawati**  Department of Civil Engineering, Jogyagarta University, Indonesia

**A.W. Malone**  Department of Earth Sciences, University of Hong Kong, Pokfulam Road, Hong Kong

**Michael Marden**   Landcare Research Box 445, Gisborne, New Zealand

**R. McInnes**   Isle of Wight Council, County Hall, Newport, Isle of Wight, PO30 1UD, UK

**Marion Michael-Leiba**   26 Fimister Circuit, Kambah, ACT 2902, Australia

**Sarah Michaels**   School of Planning, University of Waterloo, 200 University Avenue West, Waterloo, Ontario, N2L 3G1, Canada

**G.T. Mummery**   School of Geographical Sciences, University of Bristol, University Road, Bristol, BS8 1SS, UK

**Hans-Leo Paus**   Gerling Consulting GmbH, Frankfurter Str. 720–726, D-51145 Köln, Germany

**D.N. Petley**   Department of Geography, University of Durham, UK

**Walter Pflügner**   PlanEVAL, Nusselstrasse 2, D-81245 München, Germany

**Chris Phillips**   Landcare Research, PO Box 69, Lincoln, Christchurch, New Zealand

**Bill Raynor**   Nature Conservancy, PO Box 216, Pohnpei, FM 96941, Federated States of Micronesia

**Paola Reichenbach**   Consiglio Nazionale delle Ricerche, Istituto di Ricerca per la Protezione Idrogeologica, via della Madonna Alta 126, 06128 Perugia, Italy

**J.-P. Renaud**   School of Geographical Sciences, University of Bristol, University Road, Bristol, BS8 1SS, UK

**Greg Scott**   Geoscience Australia, Canberra, Australia

**Vern Singhroy**   Canada Centre for Remote Sensing, Ottawa, Canada

**Marino Sorriso-Valvo**   Consiglio Nazionale delle Ricerche, Istituto di Ricerca per la Protezione Idrogeologica, Sezionedi Cosenze, via Cavour, 87030 Roges di Rende (Cosenza), Italy

**Joan Manuel Vilaplana**   RISKNAT Group (SGR-01-81), Departament de Geodinàmica I Geofisica, Facultat de Geologia, Universitat de Barcelona, C/Marti Franqueès s.n., 08034 Barcelona, Spain

**Y. Wang**   Department of Civil Engineering, Imperial College, London, UK

**G.F. Wieczorek**   US Geological Survey, MS-910, 345 Middlefield Road, Meulo Park, CA 94025, USA

**Raymond C. Wilson**   US Geological Survey, MS-977, 345 Middlefield Road, Menlo Park, CA 94025, USA

# Preface

Investigation of landslide hazard and risk has been a major research focus for the international community over the last decade. During this period, efforts were directed towards two scales of investigation: site-specific analysis and regional assessment. Site-specific analysis may involve a wide spectrum of slope instability and all types of landslide activity covering areas ranging from just a few square metres to whole mountainsides. The objectives for this scale of investigation generally include a range of activities, involving: mapping the geometry and extent of the failure, mapping the environmental setting, determining the degree of activity by either surface measurements or subsurface observations, collecting soil or rock samples, analysing geophysical and geotechnical properties, assessing slope hydrology and porewater pressures, constructing geomorphic and geotechnical models of the site, and performing slope stability calculations. In particular, the last activity allows definition of site sensitivity to various changes in stability factors, enabling the modelling of future behaviour, either with or without protection measures. These investigations are commonly prompted by the existence of a specific or anticipated problem, for example real or expected failure of a road segment, movement of valley slopes along dammed lakes, cracks in buildings, displacements of lifelines such as railways, roads, sewerage systems, transmission lines and so on. Site-specific investigations are based on methods and concepts developed within engineering disciplines such as geotechnics and soil mechanics, but they are also informed by natural sciences, including engineering geology, geomorphology, geography and geophysics.

In contrast, regional assessments cover areas ranging from a few hectares to thousands of square kilometres. These assessments rarely involve the direct assessment of stress conditions; they are rather based on heuristic models, identification of parameters of indirect but theoretical significance to stability, or on statistical treatment of empirical data. A regional study may constitute the initial scoping stage of a wider landslide hazard assessment programme and may be the precursor to more detailed and expensive site investigation. Commonly, the main aim is to characterize both spatial and temporal conditions that have determined the occurrence of past events and to use these characteristics to locate future landslides in time and space. These assessments are generally carried out by natural scientists from fields such as engineering geology, geomorphology, soil science, forestry and geography.

Both scales of investigation are addressed in numerous monographs, edited books, publications and reports. Some of the best-known books covering these approaches include: Turner and Schuster (1996), Dikau *et al.* (1996), Crozier (1989), Selby (1993), Veder and Hilbert (1981), Berry and Reid (1987), Craig (1992), Terzaghi and Peck (1948), Wu (1976), Záruba and Mencl (1969), Brunsden and Prior (1984), and Anderson and Richards (1987). In addition to a wide range of scientific journals, the proceedings of international conferences on debris flows (e.g. Chen, 1997; Wieczorek and Naeser, 2000), landslides (e.g. Bonnard, 1988; Bell, 1992; Senneset, 1996; Anderson and Brooks, 1996; Bromhead *et al.*, 2000; Rybár *et al.*, 2002) and on general aspects of natural hazards and risks (e.g. INTERPRAEVENT, 2000, 2002) provide a wide coverage of recent developments.

Over the last decade, the focus of landslide research has moved beyond process investigations and stability assessment towards consequence analysis. Thus the integrated assessment of both hazard and risk is becoming accepted and expected practice in risk reduction management. Varnes (1984) was one of the early advocates of this approach to landslide research and engineering practice. The requirements for integrated hazard and risk studies are beginning to be systematized in such standard references as Turner and Schuster (1996) and Cruden and Fell (1997).

Despite the valuable research on landslide hazard and risk from an engineering and natural science perspective, there is also a need to address other important issues associated with the threat from landslides. Many of these have yet to be researched and fully understood. Some are addressed by the following questions:

- Besides immediate impacts, what is the full suite and duration of indirect and delayed effects associated with landslide hazard?
- What is the distribution of costs associated with the landslide hazard and how are they met; costs of mitigation, costs assumed by regional government, insurance, individuals, and so on?
- How do societies react to a given event in different social and cultural settings?
- What are the perceptions of landslide risk held by the different actors involved; how can they be measured and managed?
- How can levels of intolerable, tolerable and acceptable risk be measured; are they culturally specific?
- What are the most appropriate coping strategies for local authorities, communities, families and individuals?
- What levels of aesthetic, emotional and psychological impact are experienced as a result of landslide events?
- Where do responsibilities lie for: reducing risk, provision of risk information, education, communication, mitigation, remediation and rehabilitation? To what extent should risk be internalized or externalized by affected parties?

While the physical problems associated with risk assessment need continual science and engineering attention, some of the foregoing questions represent important areas of future research. Disciplines addressing these issues include social and economic science, psychology, civil and public law, planning and politics, to name only a few. Surprisingly, very few publications are available that attempt to bridge this gap between physical investigations and the social implications of hazard and risk. In this book we try to meet that challenge. In our view this can only be achieved by bringing together authors who approach the same ultimate goal of risk reduction but from widely different discipline perspectives. Rather than interpreting, paraphrasing and inevitably diluting these different perspectives, we have placed them largely unmodified in juxtaposition in broadly related sections of the book. We have striven to represent not only different disciplines but also different professional roles: university researchers, engineers, government scientists, private consultants and insurance industry officers.

This book is divided into five major parts dealing with (1) conceptual models in approaching landslide risk, (2) evaluation of landslide risk, (3) management of landslide risk, (4) end-to-end solutions of landslide risk, and finally (5) a synopsis. In the first chapter, Crozier and Glade position landslide hazard and risk within the broader generic concepts of risk, health and safety. The need for integrated multidisciplinary and holistic approaches is stressed. They identify and explain the concepts, issues, language and stakeholders in the study of landslide hazard and risk. Attention is given to the requirements, attributes, drawbacks and different levels of sophistication of the various approaches to hazard assessment, risk analysis, risk assessment and risk management.

Part 1 presents contributions largely at a conceptual level on the fundamental aspects and approaches to landslide hazard and risk. Initially, the impacts of landslide events are characterized. The means to assess the hazard are explored, along with ways of representing the spatial aspects of risk. And finally vulnerability, one of the most critical factors in explaining spatial difference of risk, is explored.

In particular, Chapter 2 on the nature of the hazard and impact, by Glade and Crozier, outlines the range of landslide hazards and the typical situations where impact occurs. It discusses the impact characteristics of different types of slope movement, from those that produce ongoing chronic problems to catastrophic failure that may bring about disaster. Standard terminology is introduced to discuss the range of hazardous situations, from debris flows in settled alpine regions, reservoir/dam shoreline failures, through a range of landslide types in urban and rural settings that are life-threatening or have the potential to degrade resources and environmental quality. The social and physical reasons for increased risk in these situations are outlined.

Chapter 3 covers the assessment of landslide hazard at various scales. Glade and Crozier discuss the range of methodologies employed to assess landslide hazard. It is acknowledged that the calculation of landslide and slope stability is relatively well established for various temporal and spatial scales. Different approaches are reviewed, from statistical to physically-based models that couple not only geotechnical and hydrological components but also significant surface elements such as vegetation. The prediction of the damaging characteristics of movement is important but less well established. This chapter also defines the different elements at risk and introduces various approaches to estimating the potential damage. Each threatened element is differently exposed to the risk. The exposure of a given risk element might change in time and/or in space. Different approaches in determining the vulnerability of risk elements are reviewed. For more than a decade, different landslide risk assessments have been undertaken, ranging from site-specific to regional scales. Different examples demonstrate the potential use of these approaches in context of their physiographic, economic and social controls.

In the following chapter, Chung and Fabbri present systematic procedures for landslide hazard mapping of risk assessments using spatial prediction models. They state that there is almost an infinite number of ways to construct prediction maps, from simple heuristic opinion-based procedures with little data, to sophisticated mathematical models supported by complex databases and using advanced software and hardware technologies. Existing quantitative techniques are reviewed and some of their common deficiencies are identified. This contribution provides guidance for avoiding the following deficiencies:

(1) simplification of input data; (2) poor handling of mixed categorical/continuous data; (3) lack of awareness of the assumptions implicit in prediction models; (4) lack of validations of prediction results; and (5) the need to use conditional probabilities of future landslides to perform risk assessment via vulnerability analysis. Most important, the authors state that, as in any prediction, the different methods of prediction do not have any scientific significance without accompanying measures of the validity of the prediction results.

Alexander devotes the fifth chapter to vulnerability to landslides and reviews the concept of vulnerability and its role in determining landslide risk. Landslide hazard processes are discussed in terms of their impact on human settlement, activities and land use. Four major sources of vulnerability to landslides are identified: expanding tropical cities, peri-urban slums, inhabited mountain areas, and densely settled steep volcanic terrains. Methods for assessing vulnerability to landslides are reviewed, including financial and category-based approaches.

Within Part 2, the focus is on the evaluation of risk. Social science research, for example sociology and psychology, explores the perception of risk held by different affected groups. Research issues include first of all the identification of the actors, defined as all parties involved (private land owners, consultants, governmental agencies, research institutions, insurance, etc.). The determination of the perception of risk within each group, but also between these groups, is of major concern. Of specific interest in this respect is the role of communication, because knowledge about the respective hazards influences the risk evaluation strongly. Personal risk evaluation is also highly dependent on the distance to the respective natural hazard as well as on the period since its last occurrence. This section addresses the sociological and psychological issues as well as highlighting the risk perception of the different actors involved in landslide management.

In Chapter 6, Butler and DeChano analyse landslide risk perception, knowledge and associated risk management, and relate the theoretical background to activities in Glacier National Park, Montana, USA. These issues are examined among visitors and employees through questionnaires. Results show that the employees have accurate perceptions of debris flow and landslide risk zones in the Park but visitors, on the other hand, are completely unaware of the likelihood of mass movements in the Park, and have poor perceptions of actual landslide risk zones.

The cultural factor in landslide risk perception is examined by Harmsworth and Raynor, using case studies from New Zealand and Micronesia in Chapter 7. Cultural groups are often attached to natural environments in ways not usually reflected or addressed by risk assessment. To understand the importance of culture within landslide risk perception, factors need to be identified that constitute cultural characteristics and differences. These factors include traditional beliefs, values, religion, social structure, historical occupation, historical and modern experiences, land tenure, learning and environmental perspectives. The landslide risk perception of cultural groups is also based on interaction with and dependence on the natural environment for economic and social benefit. Landslide risk perception is investigated for two examples: from an indigenous Maori perspective in Aotearoa–New Zealand, and from an indigenous and community perspective in Pohnpei, Micronesia.

The insurance industry plays a significant role in landslide management. Paus examines, in Chapter 8, the response of insurance companies to landslide risk in detail. Natural disasters put an increasing burden on the global economy. Although the share of mass movements is still a rather small component of the total cost, it is increasing at a disproportionate rate. Globally acting insurance companies are starting to feel the consequences of such trends and are trying to adapt their business strategies. In particular, the primary insurance sector urgently needs techniques to identify small-scale local risks. One method is presented which is based on the use of numerical earthquake simulations and on digital elevation models to identify zones of elevated landslide hazard. This method is applied to study sites in Taiwan and Switzerland and shows how the insurance sector itself gathers more detailed information on hazard and risks posed by landslides.

In Chapter 9, Hollenstein highlights the potential role of administrative bodies in the assessment of landslide risk. Administrative bodies can perform assessments themselves, define procedural guidelines, provide subsidies, require such assessments as a precondition for subsidizing their activities, approve assessments as a part of land use planning, or provide data for assessment purposes. The degree to which the administrative body is engaged in landslide risk assessments depends on the legal basis, the acceptance of the risk concept, the resource availability, the type of involvement and the organizational setting. Three case studies: the Vaiont Reservoir disaster of 1963 in Italy; a slow-moving landslide; and numerous fast-moving landslides (the latter two cases from Switzerland) are examined with respect to the role of administrative bodies during the events.

Part 3 is devoted to the management of landslide hazard and risk. Hazard and risk management involves having effective systems and procedures in place for identifying, calculating and evaluating the risks, assessing and implementing risk reduction options, and balancing the different components of associated cost in an acceptable way. The options for reducing risk, the individual and political will, and the resources vary greatly throughout the world. This part of the book describes systems and protocols for effective planning, information management, and risk reduction procedures. Procedures for arriving at appropriate solutions and putting policy into practice are discussed. Separate contributions examine detailed systems for handling spatial data and various specific risk reduction options, including reforestation, geotechnical structures and warning systems.

Michaels's contribution deals with the application of knowledge management to landslide risk in Chapter 10. Knowledge management means managing how knowledge is created, acquired, represented, disseminated and applied. Utilizing knowledge management to address landslide hazards is in its infancy. Selected explanations are provided as to why scientific knowledge of landslide hazards is not more of a consideration in decision making. However, current successes in dissemination are illustrated by examples in California. Michaels concludes that in order to move beyond the current successes in disseminating landslide hazard information, employing specific knowledge management concepts and practices is essential.

Crozier analyses issues and options of management frameworks in Chapter 11. Factors driving the awareness of risk and the desire to react are discussed and a rationale for apportioning the costs of risk management is presented. The difficult question of assessing who is responsible for creating risk (and perhaps who should pay) is discussed. Procedures and frameworks for planning and management are reviewed and examples of effective risk management legislation are offered, together with an evaluation of the fundamental human resources that are needed to make them work.

Wieczorek *et al.* describe the role of the US Geological Survey (USGS) in reducing landslide hazards and risk in the United States in Chapter 12. Since 1879 the USGS has served as the primary governmental agency in the United States, with the responsibility for conducting scientific studies of geologic hazards, in particular, of assessing landslide hazards and risks. The specifically designed USGS Landslide Hazards Program (LHP) has adopted the primary role of directing examination of post-landslide events and of developing improved methods for assessing the future hazard and risk of landslides on local, regional and national scales. With these improved methodologies, the LHP has developed methods of evaluating landslide susceptibility and probability, and of issuing landslide warnings. Information on landslide hazards is conveyed to the public in a variety of ways through the National Landslide Information Centre, which maintains a website with access to published documents and descriptions of recent landslide events. Underlying arguments, decisions and protocols for effectively addressing these issues and options are presented.

A conceptual framework for basic data and decision support for landslide management is given by Pflügner in Chapter 13. This contribution focuses on decision support, and presents a conceptual framework in terms of damage potential determination and socio-economic values for establishing an appropriate decision support system (DSS). Pflügner demands a DSS specifically designed for landslide issues, which must enhance a 'normal' GIS to provide more sophisticated integrative and dialogue-based systems.

McInnes gives an example of instability management that shows how to move from policy to practice in Chapter 14. Many parts of the world suffer from an inheritance of unplanned communities and developments built in unsustainable locations on, for example, eroding cliff tops and unstable slopes. Many problems can be reduced if there is a long-term programme of active landslide management in place. Local communities need to come to terms with the situation and learn to 'live with landslides'. In addition to an improved understanding of instability issues, dissemination avenues must be used to communicate the understanding derived from research to policy makers, key agencies and to local agencies. McInnes highlights these issues for the Ventnor community, Isle of Wight, UK.

Concepts, methods and applications of geomorphological mapping to assess landslide risk are presented by Reichenbach *et al.* in Chapter 15. The methods they use for evaluating landslide hazards and risks at the site scale involve geomorphological approaches and are based on the recognition of existing and past events, on the scrutiny of the local geological and morphological setting, and on the study of site-specific and historical information of past landslide occurrences. For each study area, a multitemporal landslide inventory map is prepared through the interpretation of various sets of stereoscopic aerial photographs. In addition, information from field mapping and the critical review of site-specific investigations enhance the approach. Distribution of vulnerable elements is ascertained and, by combining both information sources, specific landslide risk is estimated.

How earth observation systems can be used for landslide management is explained by Singhroy in Chapter 16. Over the last decade, earth observation systems have become increasingly important for landslide management. Recent studies have shown that more use can be made of current high-resolution stereo Synthetic Aperture Radar (SAR) and optical images to produce better standardized landslide inventory maps, which will assist hazard planning. In particular, Interferometric SAR (InSAR) methods could be viewed as most promising, and give spatial distribution of slope and motion maps. When conditions are suitable, InSAR is a useful tool for detecting and monitoring mass movements.

Early warning systems are becoming increasingly important in landslide management. Wilson from the USGS describes in Chapter 17 the rise and fall of a regional debris-flow warning system. Real-time warning systems can play a significant role in debris-flow hazard mitigation by alerting the public when rainfall conditions reach critical levels for hazardous debris-flow activity. Wilson describes a system that was operated for nearly a decade in the San Francisco Bay region. Unfortunately, organizational changes and decreases in funding and staffing forced the termination of the debris-flow warning system in 1995. Despite its political failings, the warning system produced several technical accomplishments, and valuable operational experience that might be useful for the development of similar systems elsewhere.

In contrast to alert systems, reforestation schemes are an alternative option for managing regional landslide risk, described by Phillips and Marden in Chapter 18. Besides engineering structures and planning tools for reducing landslide risk, 'semi-natural' protection schemes have been developed. These reforestation schemes are known to enhance slope stability and reduce the incidence of landslides. On-site and off-site benefits, but also disadvantages, are reviewed. However, the success of the reforestation scheme is hard to gauge, largely because erosion control benefits of this type are long-term in nature and there is a significant lag between tree planting and the accrual of off-site benefits, in particular. These issues are reviewed with the example of the East Coast Forestry Scheme in New Zealand and demonstrate the potential application options.

How geotechnical structures can reduce landslide risk is explored by Bromhead in Chapter 19. He highlights various possibilities of structural design for different landslide types, considering also the accrued changes in landslide risk. Within this chapter, the cost–benefit issue of such structures is discussed. Various examples give potential applications of proposed geotechnical structures.

Part 4 addresses the end-to-end solutions for landslide risk assessment. By demonstrating an integrative approach to risk reduction, various examples of integrated (end-to-end) landslide risk assessments are given by each contribution. An end-to-end methodology is formulated from specific (case-study) origins that illustrate the range of integrated approaches required in different settings.

In Chapter 20, Petley *et al.* present developments towards a landslide risk assessment for rural roads in Nepal. In the Himalayas, the occurrence of landslides has become increasingly acute. This is partly a result of increased vulnerability of people, partly due to the impact of land use change, and partly due to the initiation of infrastructure development, notably roads, which are both vulnerable to the effects of landslides and play a role in their initiation. Numerous landslide hazard and risk assessments schemes have been developed for the Himalayas. However, the use of these schemes in low-cost road projects has been essentially prevented because of their complexity and the need for high levels of technical knowledge. Therefore, Petley *et al.* suggest a simple susceptibility analysis that uses only geology and slope angle as input. Using an example from the Baglung district in West Nepal, the effectiveness of this approach is shown.

Leiba *et al.* carried out a quantitative landslide risk assessment for Cairns, Australia to provide information to the Cairns City Council on landslide hazard, vulnerability and risk, for planning and emergency management purposes. This project is fully discussed in Chapter 21, along with methods of quantitative risk analysis and their place in policy. Input requirements in terms of field measurements and digital data as well as the analytical capacities are described in detail. Due to the scale of the approach, however, detailed analyses have to be carried out to evaluate the specific risk for a given property. This chapter highlights the advances and limitations of regional landslide risk assessments.

The history of risk quantification and its place in slope safety policy in Hong Kong is reviewed in Chapter 22 by Malone. He traces the emergence in the 1960s of the idea of putting numbers to landslide risk and examines chronologically the application of the process of risk management in Hong Kong. Malone shows that risk concepts help to answer difficult questions faced by hazard-prone communities: Complete safety being unattainable, how safe is safe enough and what is an appropriate level of effort and expenditure on slope safety? How should the effectiveness of effort and expenditure on slope safety be measured?

Copons *et al.* examine rockfall risk management in high-density urban areas with an example from Andorra in Chapter 23. The rockfall risk management has been developed from three work plans: the "Rockfall Master Plan", the "Risk Mitigation Plan" and the "Surveillance Plan". The first task to be completed was the Master Plan, which included zoning of land in accordance with hazard and guidelines for urban development. Following the Master Plan, the Mitigation Plan and the Surveillance Plan have been developed simultaneously. The Mitigation Plan makes provision for the design and installation of permanent passive defences. The Surveillance Plan seeks to document the rockfalls with the aim of verifying the results obtained in the Master and Mitigation plans, and to carry out trail predictions of large rockfalls. The stepped policy of rockfall risk management seeks to protect buildings in areas of high hazard, control urban growth and to raise public awareness to rockfall problems.

A review of landslide risk assessments in Italy is given by Sorriso-Valvo in Chapter 24. Nationwide research projects on landslides were initiated in the late 1970s and still operate within the framework of an agreement between the National Civil Protection Agency and the National Research Council (CNR). Established in 1998, a specifically designed law for landslides requires deadlines for the mapping of landslides and the assessment of the related hazard and risk throughout the country. By 2002, nearly all Basin Authorities had fulfilled this requirement. However, Sorriso-Valvo concludes that while Italy has a unique, standardized ranking system of landslide hazard, assessment procedures differ significantly and, therefore, the five hazard and risk classes are not comparable from one region to the other.

Rain-induced landslides represent a major hazard in some of the poorest regions of the world. Thus community-based slope management and warning systems are being developed, the results of which can be readily communicated to those communities lacking easy access to education and social welfare. The first part of Chapter 25 by Karnawati *et al.* briefly assesses Indonesian landslide conditions. A numerical hydrology–slope stability model, parameterized by the slope conditions reviewed, is then developed and applied to typical slope conditions. From these numerical experiments a series of thresholds for rainfall and porewater conditions is identified which are formulated and presented in a sufficiently simplified manner appropriate to local community needs.

Within Part 5, an overall conclusion is drawn. Glade and Crozier integrate the diverse strands that have been presented thus far for landslide hazard and risk assessments. Suggestions are developed for appropriate models for integration which provide specifications and contexts for the next generation of process models. Such models should incorporate economic, social, legal and related components as part of their framework. The questions of the pivotal role of management and risk reduction are addressed, together with how such approaches can be implemented through different types of relevant organizations, consultant engineers, planners, landowners, emergency services and so on.

This book offers concepts and methods for landslide hazard and risk research and management that extend beyond well-established engineering and natural sciences approaches. These concepts and methods will have value for research institutions, governmental agencies, consultancies and private individuals. The variety of issues and concepts covered by this book arises from different discipline perspectives and from authors with a wide range of professional experience within universities, research institutions, governmental agencies and private consultancies.

The editors have designed this book to provide valuable insights, guidance and advice to research scientists, engineers, policy makers, planners, managers and all those who share the common interest of effective landslide risk reduction. We hope the multidisciplinary approach extends their vision, adds to their understanding and facilitates their work.

The book is rich with material that will support the teaching of the subject in educational institutions, seminars and short courses.

We wish to express our gratitude to all those authors who have made this volume possible, and to the staff of Wiley for all their support.

Thomas Glade
Malcolm Anderson
Michael Crozier

## References


Anderson, M.G. and Brooks, S.M. (eds), 1996, *Advances in Hillslope Processes*, Symposia Series, 2 vols (Chichester: John Wiley & Sons Ltd).

Anderson, M.G. and Richards, K.S., 1987, *Slope Stability* (Chichester: John Wiley & Sons Ltd).

Bell, D.H., 1992, *Landslides, Proceedings of the Sixth International Symposium*, 10–14 February 1992, Christchurch, New Zealand, 3 vols (Rotterdam: A.A. Balkema).

Berry, P.L. and Reid, D., 1987, *An Introduction to Soil Mechanics* (London: McGraw-Hill).

Bonnard, C., 1988, *Landslides, Proceedings of the fifth International Symposium on Landslides*, 10–15 July 1988, Lausanne, Switzerland, 3 vols (Rotterdam: A.A. Balkema).

Bromhead, E., Dixon, N. and Ibsen, M.-L., 2000, *Landslides, Proceedings of the Eighth International Symposium*, 26–30 June 2000, 3 vols (London: Thomas Telford).

Brunsden, D. and Prior, D.B., 1984, *Slope Instability* (Chichester: John Wiley & Sons Ltd).

Chen, C.-L., 1997, *Debris-flow Hazards Mitigation: Mechanics, Prediction, and Assessment* (San Francisco: American Society of Civil Engineers).

Craig, R.F., 1992, *Soil Mechanics* (London: Chapman and Hall).

Crozier, M.J., 1989, *Landslides: Causes, Consequences and Environment* (London: Routledge).

Cruden, D.M. and Fell, R., 1997, *Landslide Risk Assessment, Proceedings of the Workshop on Landslide Risk Assessment*, Honolulu, Hawaii, USA, 19–21 February 1997 (Rotterdam: A.A. Balkema).

Dikau, R., Brunsden, D., Schrott, L. and Ibsen, M.-L., 1996, *Landslide Recognition: Identification, Movement and Causes* (Chichester: John Wiley & Sons Ltd).

INTERPRAEVENT, 2000, *Changes within the natural and cultural habitat and consequences: Nineth International Symposium*, 3 vols (Villach, Austria: Krainer Druck).

INTERPRAEVENT, 2002, *Protection of habitat against floods, debris flows and avalanches: Tenth International Symposium*, 14–18 October 2002, Matsumoto, Japan, 2 vols (Japan: Nissei Eblo Co.).

Rybár, J., Stemberk, J. and Wagner, P., 2002, *Landslides: Proceedings of the First European Conference on Landslides*, Prague, June 2002 (Lisse: Balkema).

Selby, M.J., 1993, *Hillslope Materials and Processes* (Oxford: Oxford University Press).

Senneset, K., 1996, *Landslides – Glissements de Terrain, Seventh International Symposium on Landslides*, 3 vols (Rotterdam: A.A. Balkema).

Terzaghi, K. and Peck, R.B., 1948, *Soil Mechanics in Engineering Practice* (New York: Wiley).

Turner, A.K. and Schuster, R.L., 1996, *Landslides: Investigation and Mitigation*, Transportation Research Board Special Report 247 (Washington, DC: National Academy Press).

Varnes, D.J., 1984, *Landslides Hazard Zonation: A Review of Principles and Practice: Natural Hazards* (Paris: UNESCO).

Veder, C. and Hilbert, F., 1981, *Landslides and Their Stabilization* (New York: Springer-Verlag).

Wieczorek, G.F. and Naeser, N.D., 2000, *Proceedings of the Second International Conference on Debris-flow Hazards Mitigation*, 16–18 August 2000 Debris-flow Hazards Mitigation: Mechanics, Prediction, and Assessment (Taipei, Taiwan: Balkema).

Wu, T.H., 1976, *Soil Mechanics* (Boston: Allyn & Bacon).

Záruba, Q. and Mencl, V.E., 1969, *Landslides and Their Control* (Amsterdam, New York and Prague: Elsevier).

# 1

# Landslide Hazard and Risk:
# Issues, Concepts and Approach

*Michael J. Crozier and Thomas Glade*

## 1.1 Underpinning Issues

In the broad non-technical sense 'hazards' are defined as those processes and situations, actions or non-actions that have the potential to bring about damage, loss or other adverse effects to those attributes valued by mankind. The concept is thus applicable to all walks of life. In industry a hazard might be a power failure or a computer malfunction, in business it might be a breach of security or a poor investment decision, and in the environment it might be a spill of toxic substances or even a damaging landslide. Although the potential for something adverse to occur is appreciated, there is uncertainty as to when the hazard will realize its potential, and thus the threat is generally expressed as a likelihood or probability of occurrence of a given event magnitude in a specified period of time. Technically, we refer to this adverse condition as 'the hazard'. Thus, in common usage, the term 'hazard' has two different meanings: first, the physical process or activity that is potentially damaging; and second, the threatening state or condition, indicated by likelihood of occurrence. Generally the meanings are obvious from the context within which they are used.

The consequences of hazard occurrence can be great or small, as well as direct or indirect; the latter linked to the primary impact by a chain of dependent reactions that may be manifest at some distance in time and space from the initial occurrence. Clearly the consequences depend on the context in which they occur, the particular elements and attributes affected, and their value and level of importance.

In simple generic terms, the important concept of 'risk' can thus be seen as having two components: the likelihood of something adverse happening and the consequences if it

*Landslide Hazard and Risk*   Edited by Thomas Glade, Malcolm Anderson and Michael J. Crozier
© 2004 John Wiley & Sons, Ltd   ISBN: 0-471-48663-9

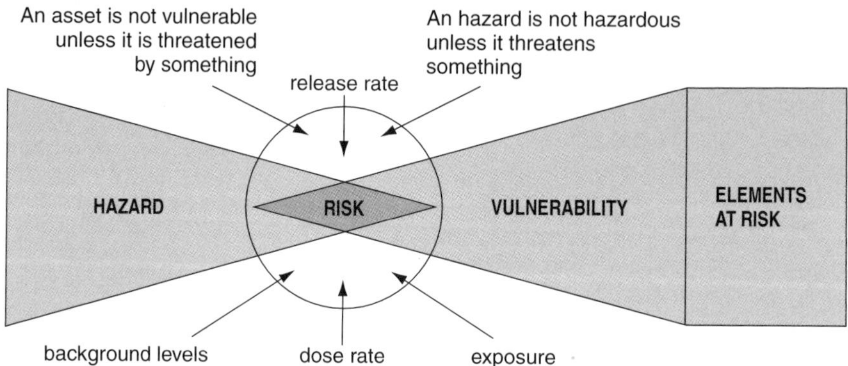


**Figure 1.1**  *Conceptual relationship between hazard, elements at risk, vulnerability and risk (Alexander, 2002)*

does. The level of risk, then, is the combination of the likelihood of something adverse occurring and the consequences if it does. The level of risk thus results from the intersection of hazard with the value of the elements at risk by way of their vulnerability (Figure 1.1).

Traditionally hazard and risk studies have been developed separately for industrial, financial, human, environmental and natural systems. For example, industrial systems may focus on operational malfunctions and consequent economic losses; financial systems on investment risks; human systems on crime, health or conflict; and environmental systems on pollution and resource quality. Within natural systems, where landslides are recognized as one of the 'natural hazards', the focus is on potentially dangerous events and situations arising from the behaviour of the atmosphere, biosphere, lithosphere and hydrosphere. The fact that natural forces are responsible for generating the threatening conditions distinguishes natural hazards from those of other systems, although there are many situations where the distinction between systems is not clear-cut.

The generic concepts pertaining to hazard and risk outlined above are equally applicable to landslides although they may be expressed in more process-specific terms. For landslides, the 'adverse something' might be a large rockslide and its 'likelihood' expressed as the probability of its occurrence. Similarly, the consequences will depend on what is affected by the landslide, the degree of damage it causes and the costs incurred.

In global terms, landslides generate a small but important component of the spectrum of hazard and increasing risk that faces mankind (Figure 1.2) (Alcántara-Ayala, 2002). If there were a choice, people would inhabit and rely for their well-being on the safe places of the earth – away from the threat of landslide. But even then that would presume there was sufficient knowledge of hazard and risk to allow an informed decision. However, mankind has been placed progressively at the mercy of nature through population pressure, increasing demands for resources, urbanization and environmental change. It is the intersection of humanity with landslide activity that has recast a natural land-forming process into a potential hazard (Figure 1.3a). Furthermore, economic globalization has enhanced reliance on communication and utility corridors. Fuel lines, water and sewage reticulation, telecommunication, energy and transport corridors, collectively referred to as 'lifelines' in hazard studies, are highly vulnerable to landslide disruption (Figure 1.3b).

Landslides present a threat to life and livelihood throughout the world, ranging from minor disruption to social and economic catastrophe. Spatial and temporal trends in

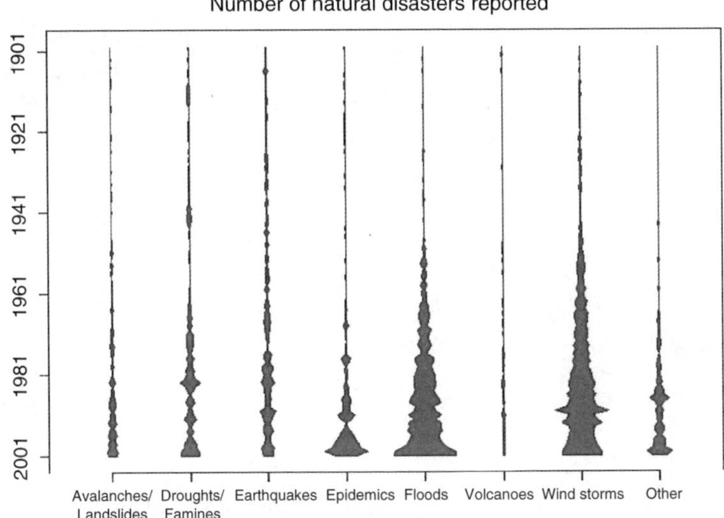



EM-DAT: The OFDA/CRED International Disaster Database
(http://www.cred.be)

***Figure 1.2*** *Comparison of the number of natural disasters from different natural hazards reported within the International Disaster Database maintained by OFDA/CRED (refer to http://www.cred.be for most recent numbers)*

the level of this threat (Figure 1.4) have driven the current international and national concerns about the issue of hazard and risk reduction. However, these trends are difficult to determine accurately because of the variable quality and consistency of record keeping. These problems arise from a range of factors, including: variability and improvements in observational techniques; changes in population density; the mix of different agencies involved and the variability of recording protocols; as well as heightened economic and social awareness. One source for economic data of damage caused by natural hazards is the statistics regularly published by the re-insurance company Munich-RE (Münchner Rückversicherung, 2000). Although one has to be cautious with interpretations based on these figures, a trend is visible of increased economic costs for the insurance companies resulting from natural events. As well as economic loss, land slides have also caused numerous humanitarian disasters throughout history. A selection of major landslide disasters of more than 1000 deaths is given in Table 1.1.

Two hundred years or so of science and practice related to slope stability problems have transformed the landslide from an 'act of God' into a comprehensible geophysical process. Society demands that such knowledge carries a responsibility, a 'duty of care' and, in some instances, an obligation to act. The formalization and apportioning of this responsibility is in its infancy in many parts of the world. Nevertheless, whether driven by legal, moral or economic concerns, there is a continuing need to seek out and refine tools for risk reduction, be they scientific, engineering, legislative, economic or educational.

In the simplest terms, landslide hazard can be depicted as the physical potential of the process to produce damage because of its particular impact characteristics and the magnitude and frequency with which it occurs (or is encountered). Landslide risk, on the

**Figure 1.3a**   *Examples of a society exposed to natural processes from Bíldudalur, Iceland. A petrol station on a bridge that crosses a drainage line susceptible to snow avalanches, debris flows and slush flows (photo by T. Glade)*

**Figure 1.3b**  *A pole of the main power supply, a freshwater tank, and a school in the background close to the same drainage line (photo by T. Glade)*

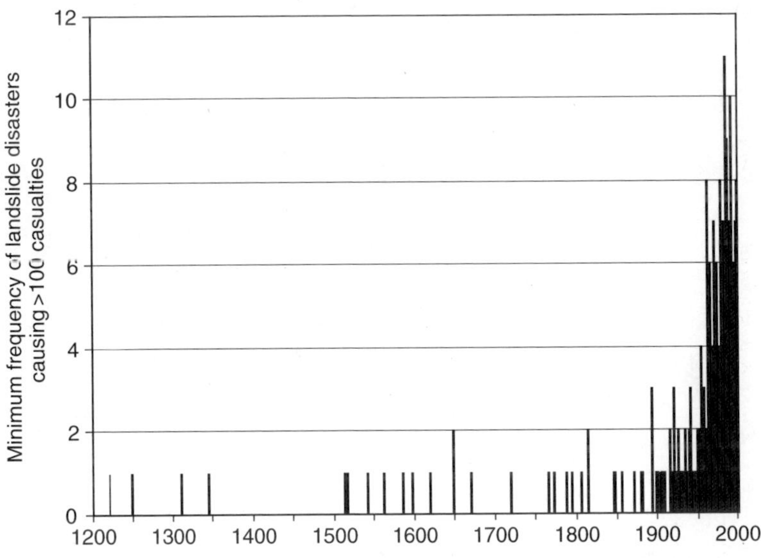

**Figure 1.4**  *Minimum frequency of recorded landslide disasters for the world, causing more than 100 deaths (Glade and Dikau, 2001). Refer to Table 1.1 for a selection of data*

**Table 1.1** Selection of major natural disasters due to landslides causing more than 1000 deaths. All data are based on respective sources. Type of process named where identifiable from the source (based on Glade and Dikau, 2001)

| Event | Trigger | Date | Location, region, country | Consequences | Source |
| --- | --- | --- | --- | --- | --- |
| Bergsturz | n.i. | 1219 | Plaine d'Oisans, Isère, France | >1000 casualties | Flageollet (1989) |
| Bergsturz | n.i. | 24.11.1248 | Mont Granier, Savoie, France | 1500 to 5000 casualties | Flageollet (1989) |
| Landslide | n.i. | 1310 | Zhigui, Hubei, China | 3466 casualties | Tianchi (1989) |
| Landslide | n.i. | 1561 | Xitan, Zhigui, Hubei, China | >1000 casualties | Tianchi (1989) |
| Bergstürze | Earthquake | 24.11.1604 | Arica, Chile | >1000 casualties | Nussbaumer (1998) |
| Bergsturz | n.i. | 25.08.1618 | Plurs, Bergell, Switzerland | >2000 casualties | Nussbaumer (1998) |
| Landslide | Earthquake | 19.06.1718 | Gansu Provonz, China | 40 000 families buried | Tianchi (1989) |
| Lahar | Volcanic erup. | 12.08.1772 | Vulkan Papandayan, Java | 2000–3000 casualties | Nussbaumer (1998) |
| Landslide | Earthquake | 10.10.1786 | Kangding-Louding, Sechuan, China | 100 000 casualties | Tianchi (1989) |
| Lahar | Volcanic erup. | 10.02.1792 | Vulkan Unzendake, Japan | 10 000 casualties | Nussbaumer (1998) |
| Lahar | Volcanic erup. | 19.02.1845 | Vulkan Nevada del Ruiz, Colombia | 1000 casualties | Nussbaumer (1998) |
| Landslide | n.i. | 09.1857 | Montem./Basilic., Italy | 5000 casualties | Nussbaumer (1998) |
| Lahar | Volcanic erup. | 1919 | Java, Indonesia | 5110 casualties, 104 villages dest. | Brand (1989) |
| Landslide | Earthquake | 16.12.1920 | Kansu, China | >200 000 casualties | Bell (1999); Nussbaumer (1998) |
| Landslide | Earthquake | 1920 | Haiyuan, China | 100 000 casualties | Tianchi (1989) |
| Landslide | Earthquake | 25.08.1933 | Sichuan, Diexi, China | 6800 casualties | Tianchi (1989) |
| Debris flow | n.i. | 14.12.1941 | Huaraz, Peru | 4000–6000 casualties, 1/4 of Huaraz dest. | Erickson *et al.* (1989); Nussbaumer (1998) |

| Type | Trigger | Date | Location | Effects | Reference |
|---|---|---|---|---|---|
| Landslide | n.i. | 1945 | Japan | 1200 casualties | Oyagi (1989) |
| Landslide | Earthquake | 1949 | Khait, Tajikistan | 12 000 casualties | Alexander (1995) |
| Landslide | Earthquake | 15.08.1950 | Assam, India | Approx. 30 000 casualties | Nussbaumer (1998) |
| Landslide | Rainfall | 1958 | Shizuoka, Japan | 1094 casualties, 19 754 buildings dest. | Oyagi (1989) |
| Landslide | n.i. | 29.10.1959 | Minatilan, Mexico | 5000 casualties | Nussbaumer (1998) |
| Debris flow | n.i. | 10.01.1962 | Nevados, Mt Huascaran, Peru | 4000 casualties, village Ranrahirca dest. | Erickson et al. (1989); Nussbaumer (1998) |
| Landslide | n.i. | 10.10.1963 | Vaiont Dam in the Piave Valley, Italy | 1189 casualties, some villages dest. | Müller (1964); Nussbaumer (1998); Petley (1996); Smith et al. (1996); Soldati (1999) |
| Landslide | Rainfall | 1966 | Rio de Janeiro, Brazil | 1000 casualties | Smith et al. (1996) |
| Landslide | Rainfall | 1967 | Sierra des Araras, Brazil | 1700 casualties | Erickson et al. (1989) |
| Lahar | Earthquake | 31.05.1970 | Ancash, Yungaytal, Peru | 66 794 casualties | Alexander (1995); Nussbaumer (1998) |
| Landslide | n.i. | 20.09.1973 | Choloma, Honduras | 2800 casualties | Nussbaumer (1998) |
| Lahar | Volcanic erup. | 13.11.1985 | Nevado del Ruiz, Colombia | >25 000 casualties | Nussbaumer (1998) |
| Landslide | n.i. | 03.04.1987 | Cochancay, Ecuador | 1000 casualties | Nussbaumer (1998) |
| Landslide | Earthquake | 23.01.1989 | Tajikistan | Up to 10 000 casualties; 2 villages dest. | Nussbaumer (1998) |
| Landslide | n.i. | 07.06.1993 | Nepal | 3000 casualties | Nussbaumer (1998) |
| Debris flow | Rainfall | 12.1999 | Venezuela | 30 000 casualties, 400 000 homeless | Larsen et al. (2001) |

other hand, is the anticipated impact or damage, loss or costs associated with that hazard. Ideally hazard can be characterized by statements of 'what', 'where', 'when', 'how strong' and 'how often', demanding knowledge of variation in both spatial conditions and temporal behaviour. The ultimate test of landslide hazard prediction would be the forecast, that is, the ability to state for particular places 'where' and 'when' something will happen and what it will be like. Our ability to forecast landslide hazard with this precision is limited, and consequently landslide hazard and risk predictions are generally couched in terms of likelihoods and probabilities. However, in global terms, even this level of landslide hazard and risk assessment is rarely achieved. For example, assessments of regional landslide 'hazard', if they exist at all, are more likely to rank components of the terrain in terms of their potential for landslide occurrence (susceptibility) or simply indicate the presence or absence of existing landslides (Crozier, 1995).

Most work on landslide 'hazard' assessment has been site-based and driven by development projects and engineering concerns. Conventionally this has been approached by stability analysis of the site, generally determined from the balance of shear stress and strength and expressed as a factor of safety. Recognition of the natural variability of factors controlling stress also suggests that the factor of safety is more realistically evaluated in probabilistic terms. The major challenge for site-based stability analysis is the conversion of the factor of safety or equivalent stability assessment into a useful expression of hazard that can then be used as a component of risk assessment. This would involve employing the factor of safety along with temporal variability in triggering factors to determine the probability of failure per unit of time. Probability of occurrence, in turn, needs to be qualified by a statement of expected behaviour of the failure in terms of its impact characteristics. Predicting the nature of the landslide, particularly for first-time failures, is yet another challenge for landslide hazard science. For example, there are sufficient studies available to allow a reasonable prediction of runout length based on landslide volume (e.g. Crosta *et al.*, 2003; Crozier, 1996; Hsu, 1975; Hungr, 1995); it is the prediction of volume, as well as other impact characteristics, of first-time failures that remains the problem.

Whereas there is a generally accepted well-defined pathway for research required to refine our understanding of hazard, there is much less unanimity on what constitutes the causes of risk, particularly the underlying causes. The wide discrepancy in losses experienced between rich countries and poor countries has focused attention on the role and causes of vulnerability. The dominant view, referred to by social scientists as the 'behaviourist' paradigm (Alexander, 2000; Smith, 2001), attributes vulnerability to a lack of knowledge, insufficient preparedness and inappropriate adjustment to specific hazards. The 'structuralist' paradigm, on the other hand, attributes vulnerability to disempowerment of the victims through political-economic structures that favour the elite at the expense of the mass of population. The denial of resources by either national or transnational concerns means that the affected populations can do little to improve their level of vulnerability. This view sees 'underdevelopment' in particular countries and regions as a product of 'development' in others. High levels of vulnerability in developing countries have also been attributed to dependence on external assistance either as disaster relief or in risk management programmes. It has been argued that these measures can override traditional coping mechanisms, suppress indigenous mitigation practices and reduce the ability or incentive to take independent measures to mitigate risk.

Landslide hazard and risk studies are clearly positioned at the nexus of social and scientific concerns – two cultures that have not always been perceived as having compatible agendas. The effective research and management of risk requires the integration of a wide range of interests. There are many stakeholders that are directly or indirectly affected by the identification of risk or the promulgation of measures to reduce risk. These include:

- The decision makers and managers, and officers responsible for executing and monitoring policy
- Directly affected property owners and those who rely on the property for income and livelihood
- Indirectly affected institutions, companies or private personnel affected by disrupted lifelines
- Financial institutions and insurance agencies
- Regulators and other government organizations that have authority over activities, including issuing consents and permits, and responsibilities for emergency management
- Politicians with electoral or portfolio interest
- Suppliers and service providers: scientists, technicians, consultants, engineers, valuers, contractors
- Public interest groups and non-government organizations such as aid agencies and environmental groups
- Media.

Balancing the interests of all the affected parties when evaluating risk or choosing risk treatment options is a fundamental consideration in risk management.

## 1.2   Landslide Risk Assessment and Risk Management

In most societies, the ultimate goal of landslide *hazard* (the definitions of terms in italics are given in Section 1.8 and are summarized in the glossary) and *risk* studies is an accurate assessment of the level of threat from landslides: an objective, reproducible, justifiable and meaningful measure of risk. The process of establishing such a measure of risk is referred to as *risk estimation*. The estimated level of risk can then be evaluated (*risk evaluation*) in the light of the benefits accrued from being exposed to that risk (*risk–benefit analysis*) and, as a result, decisions can be made on whether that level of risk is *intolerable*, *tolerable* or *acceptable*. Comparison of risks from sources other than landslides, if estimated in the same way, can be made, and priorities for *risk treatment* can be rationally established. An objective measure of risk can also be employed in terms of cost–benefit (or cost–risk reduction) analysis of proposed risk treatment measures. The full range of procedures and tasks that ultimately lead to the implementation of rational policies and appropriate measures for risk reduction are collectively referred to as *risk management*.

Figure 1.5 summarizes the components that constitute risk management and their hierarchical relationships. Each of these components is examined here in a logical sequence in order to identify some of the important issues. First, if a project were established to assess the risk from landslides, a number of fundamental questions would need to be

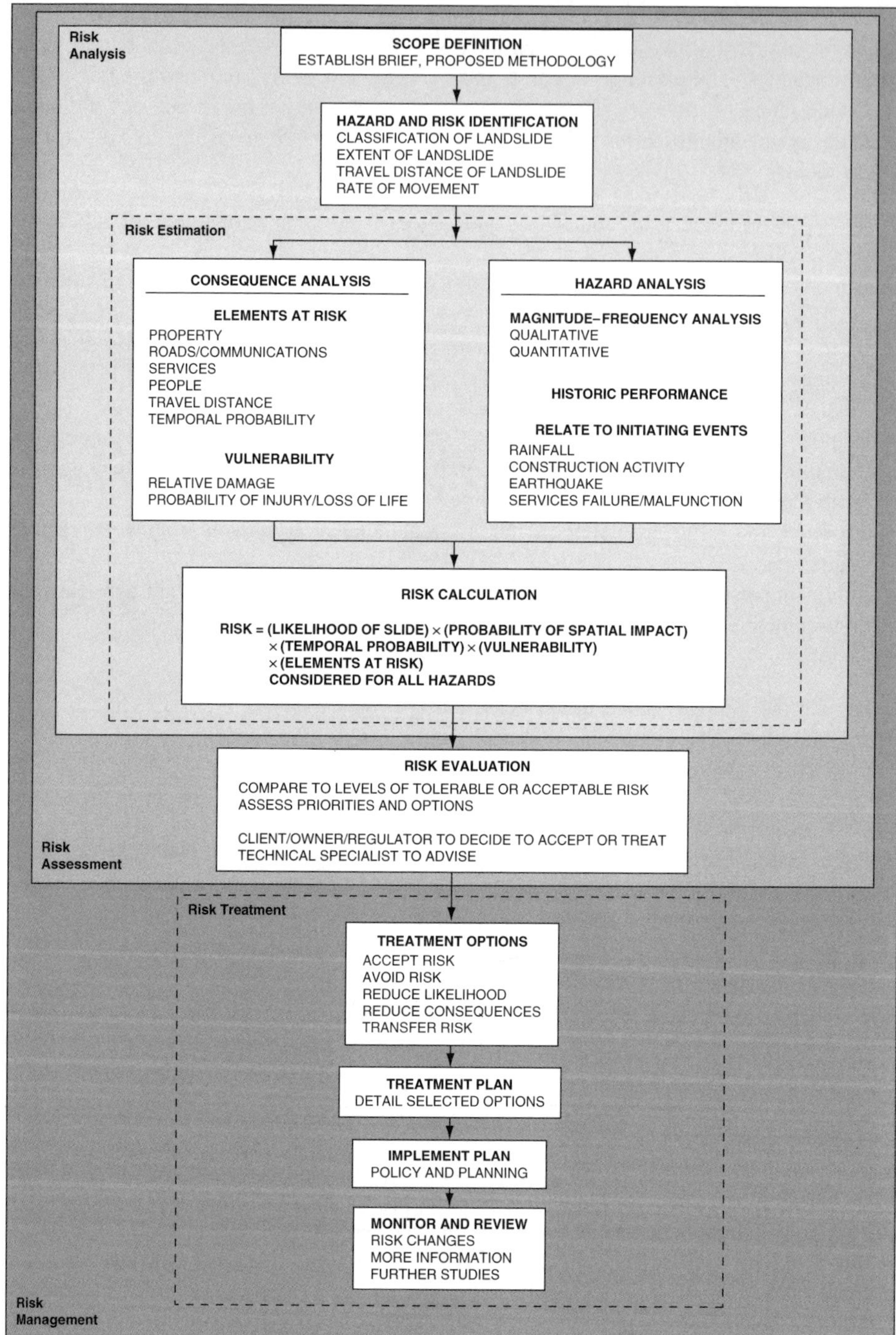


**Figure 1.5**   *Flow chart showing all the stages involved in landslide risk management (based on Australian Geomechanics Society, 2000)*

addressed from the outset. For example, what sort of information is required? What type of methodology should be employed? What resources are required, and, most importantly, what is the assessment to be used for? The resolution of these questions constitutes the scoping phase of the project. In practice, the scope is often dictated by scientific, economic, legal or social imperatives. It is likely to dictate the area or objects of concern; it may identify the time frame of interest, the resources available, and the degree of detail required by the assessment. In some respects, the success of an investigation may be measured against how it satisfies the identified scope or the prepared brief for the study. However, from a scientific and professional perspective, satisfaction of a brief may only partially address the question of risk (Dai *et al.*, 2002). If this is the case, it is incumbent on practioners to identify the shortcomings and point to ways of providing a more comprehensive and valuable answer.

Having identified the scope of the study, the next stage is *hazard and risk identification*. In one sense, this part of risk assessment is still very much a scoping exercise and should not be confused with the subsequent more detailed analyses of hazards and consequences. It should answer the question as to what types of physical processes exist. What might their impact characteristics be? At the same time, in order to identify potential risks it is essential to identify the possible *elements at risk*, their spatial and temporal relationship with the hazard, how they may be affected, as well as their possible levels of *vulnerability*.

The hazard and risk identification stage essentially identifies those factors that should be further investigated and taken into consideration in risk estimation. The process of *risk estimation* integrates the behaviour of the hazard (*hazard analysis*) with the elements at risk and their vulnerability (*consequence analysis*) in order to allow *risk calculation*, usually in the form of the generic hazard–risk equation:

$$\text{Risk} = \text{hazard} \times \text{vulnerability} \times \text{elements at risk} \tag{1}$$

This is a simple but very powerful equation that identifies separately the principal factors contributing to risk. These include the probability of occurrence of a damaging landslide of a given magnitude (hazard), the valued attributes at risk (elements at risk) and the amount of damage expected from the specified *landslide magnitude*, expressed as the ratio of the value of damage to the total value of the element (vulnerability).

Although 'risk', identified as the expected loss in a unit of time, could in some instances be determined without accounting for the other factors in the equation (by simply analysing the history of loss), it would obfuscate the dynamic role of the factors causing the problem. Because of the limited length and quality of landslide impact records, consequence analysis alone would underestimate the risk emanating particularly from *high-magnitude–low-frequency* events (e.g. Figure 1.6). Calculation of risk by multiplication of the terms is also significant. It implies that if any one of the independent factors (terms on the right of the equation) is zero, the risk will be zero. Consequently, if a natural process occurs in an unpopulated area or if the structural vulnerability is very low, the risk is zero.

In some instances, it can be useful to carry out hazard analysis as a separate exercise. By doing this it is possible to identify not only the impact on existing elements but also the potential impact related to any future development, in other words potential hazard. Despite the rubric that 'there is no hazard if there is nothing at risk', the estimation of

*Figure 1.6*  Rockfall near Randa in 1991, Matter Valley, Valais, Switzerland (photo by T. Glade), described by Schindler et al. (1993) in detail. The rockfall blocked the valley, hit lifelines and a lake was formed as a consequence. A large effort has been undertaken to reduce secondary effects, such as a flash flood resulting from a potential burst of the rockfall-dammed lake

hazard, independent of existing human constructs, can in some instances be a powerful guide to future development decisions.

Hazard analysis (at a regional scale) requires three steps: first, the analysis of all identified landslides to determine their types and potential behaviour; second, the determination of those members of the landslide population that are capable of producing damage on the basis of an analysis of impact characteristics; and third, the determination of the location, magnitude, frequency and spatial extent of the potentially damaging landslides.

As well as a term representing hazard, the hazard–risk equation (1) requires identification and valuation of the elements at risk. The elements at risk are difficult to identify and even more difficult to quantify. In the broadest sense, they are all those attributes valued by mankind. They range from life, human well-being, through monetary values of property, lifelines, resources and means of production, to ways of life, religious, aesthetic and other values. The remaining term required for the calculation of risk in the hazard–risk equation is vulnerability. This factor is probably the most difficult to represent quantitatively (see Chapter 2). Very few data exist on the degree of damage to various elements at risk from landslides in general and even fewer for vulnerability with respect to different landslide types.

Calculated values of risk are of course the product of objective scientific investigation and are of little value until their significance can be determined by those affected. The evaluation of risk is a cultural exercise. Until an estimated level of risk can be evaluated, compared to other risks, and placed in the context of the benefits derived as a function of facing that risk, few rational decisions can be made. It is on the basis of the evaluated level of risk that risk treatment options can be exercised. These may range from the simple acceptance of the risk, education and avoidance, to various measures of control, encompassing legislative and engineering solutions affecting the hazard, vulnerability or the elements at risk.

Management of risk thus involves actions that instigate, regulate and control the identification and monitoring of hazard and risk, the estimation and evaluation of risk, as well as the options for reducing risk. In turn, the reduction of risk may involve the acceptance, avoidance, prevention, mitigation or sharing of risk. The means may include engineering solutions, legislative or regulatory edicts, education, common sense, insurance, aid, preparedness and planning (see Chapter 11).

## 1.3   Approaches to Landslide Hazard Analysis

The approaches taken to analyse landslide hazard depend greatly on the scope of the problem and the physical and social context. A regional assessment, for example, is likely to differ markedly from the assessment of a high-value or potentially high-risk site. The nature of information used to assess hazard may also differ depending on whether the site involves a natural slope, artificially cut slope or constructed earthworks. The presence of existing landslides also allows the application of a different set of methodologies compared to those employed in areas where no landslides exist and the hazard from first-time landslides is being assessed. The nature of existing databases and the ability to obtain relevant surface and subsurface information also dictate the approach. Whichever approach is adopted, the initial concern is to characterize the problem by determining what physical hazards exist (hazard identification) and how they are likely to behave with respect to elements at risk.

### 1.3.1   Landslide Hazard Identification

The first step in hazard assessment, then, is the identification of the nature of hazard likely to be encountered in the area of concern. Initially, the type of landslide has to be determined, based on internationally accepted terminology of landslides such as that proposed by Cruden and Varnes (1996) or by Dikau *et al.* (1996). A detailed review of landslide types is given in Chapter 2. After having assessed landslide type, the extent of landslide activity has to be investigated. Spatial scales of investigation can range from a few $m^2$ for a single landslide to several $km^2$ for a large number of failures (Figures 1.7 and 1.8).

The hazard identification stage also needs to investigate the potential velocities of movement. At one extreme, velocities may accelerate up to m/s for rock falls and slides. On the other hand, extremely slow, creeping landslides may move at rates of only cm/year (refer to Chapter 2 for details). Appropriate investigation techniques need to be chosen with respect to potential velocities. However, most landslides do not move continuously; instead they tend to move episodically, commonly as a response to changed environmental conditions such as elevated porewater pressures. If long-term movements need to be analysed or if early warnings need to be issued, for example in the case of debris flows, measuring devices are indispensable. However, such devices generally provide information for only short periods; thus information on long-term behaviour needs to be obtained by using other techniques.

Investigations generally cover one of two general scales: site-specific analysis or regional-based investigations. Analysis of single landslides at a particular site has a long tradition and includes field mapping, soil sampling and testing, and slope stability

**Figure 1.7**  *A large, deep-seated rockslide that occurred in 1965, Hope, British Columbia (photo by M.J. Crozier)*

modelling by a wide range of techniques. In contrast, regional analysis tends to be less precise and more indicative in nature. While the first approximations of hazard at the regional scale generally involve inventory maps of old landslides, recent failures, or a combination of both, new approaches include statistical analysis and process-based methods. Depending on the sophistication of the database and methods employed, 'hazard'

**Figure 1.8**   *Shallow earth slides and earth flows, triggered by the rainstorm of February 2004, Manawatu-Wanganui, New Zealand. Note the stabilizing effect of forest plantation (photo by G. Hancox)*

may be ultimately represented by spatial distributions, derived landslide susceptibility and, in some instances, probability of occurrence. These different types of assessment are summarized by Turner and Schuster (1996).

The overall aim of landslide hazard identification (as opposed to the larger task of hazard estimation) is to scope the nature of the potential threat. Hazard identification is the initial requirement of any programme designed to estimate the risk from landslides. This initial stage of an investigation should set out to determine the physical nature of the threatening process. Is it continuous or episodic? Is it fast or slow? Is it a localized site problem or is it a regional condition? With respect to the type of movement, will displacement be slow or catastrophic; will the displacement be in the form of disrupted blocks or single units; and over what distance will the material travel (Figure 1.9)? This stage of assessment should also anticipate any likely impacts to elements at risk. For example, are the landslides likely to affect the drainage system (Figure 1.10), infrastructure, or human settlements? It is on this basis that a programme of hazard analysis can be appropriately designed.

Identification of the hazard is a small but important step towards the overall goal of hazard analysis and risk estimation. Those responsible for making decisions with respect to the landslide risk will ultimately need a much more refined statement of the problem. They need to know not only the nature of the physical threat, but also how frequently it

**Figure 1.9**   *A long-runout, rapid earthflow triggered by the rainstorm of February 2004, Manawatu-Wanganui, New Zealand (photo by G. Hancox)*

will be manifest in a given location. The scientists involved in providing such information need to be aware of the range of methodolgies currently employed to determine hazard and risk. The following sections of this chapter cover the various approaches used to solve these problems.

### 1.3.2   The Geomorphological and Geotechnical Context

A large amount of information of value to hazard and risk assessment can be gained if the existing and potential landslide hazards are considered within their geomorphological and geotechnical context. Landslides are a geomorphological process intricately linked to the landform, material, structural, hydrological, climatic and vegetative conditions within which they occur. Careful study of these relationships can reveal patterns and thresholds that differentiate stable from unstable conditions. On the assumption that the combination of factors that has led to landsliding in the past will operate in the future, the analysis of pre-existing landslides (referred to as the precedence approach) provides a useful means of assessing not only the degree of future stability but also landslide behaviour. Essential to this approach is the ability to recognize evidence of past geomorphological processes and to distinguish the landslide signature from those of other less hazardous processes. Slope form, microtopography, lineaments and depositional form and fabric, particularly when placed in the spatial context by geomorphological mapping, can provide evidence

**Figure 1.10**  *Deep-seated rockslide triggered by the rainstorm of February 2004, Manawatu-Wanganui, New Zealand (photo by G. Hancox)*

of current activity, velocity, age, type and extent of past landslides (Crozier, 1984). The correlation of dated landslides with changes in environmental conditions, as indicated by other geomorphological evidence, can also reveal the nature and intensity of conditions leading to landslide initiation (Hutchinson, 2001). The advantage of the precedence approach over a lab and desktop assessment is that geomorphological evidence has the potential to record the influence of a wide range of factors (including climate and hydrology) experienced by the slope (Crozier, 1997). In some instances, however, the conditions that prevailed in the past may be significantly altered by factors such as climate change, earthworks and land use activity (e.g. Figure 1.11). In particular, landslide events themselves can change the susceptibility of the terrain to future events, commonly by removing susceptible material and thereby increasing the resistance of the terrain (Crozier and Preston, 1999; Bovis and Jakob, 1999; Dykes, 2002; Glade, 2004a; Zimmermann and Haeberli, 1992; Zimmermann *et al.*, 1997). These changes strongly influence hazard, and thus a comprehensive geomorphological and geotechnical assessment of their influence is an indispensable requirement for any sustainable landslide hazard scheme.

Understanding the interrelationships between the hillslope system and associated fluvial or coastal systems can provide valuable information not only on the magnitude and frequency of landslides but also on the implications of slope protection and stabilization measures. In one example from the unstable coastal cliffs of North Yorkshire, England,

***Figure 1.11*** *Landslide occurrence following a rainstorm in 2002, Gisborne area, New Zealand. Note the differences in landslide coverage, which can be attributed to land use (photo by M.J. Crozier)*

Moore and McInnes (2002) demonstrate the close links not only between marine erosion, landsliding, cliff and coastline movements (both seaward and landward), but also the relationship between landsliding and the supply of beach sediment. Stabilization measures to protect cliff top property need to be viewed alongside their potential to affect the maintenance of the beach, which in itself is a valuable recreational asset (Figure 1.12).

Careful reading of the ground can indicate the stress history of the slope. Evidence of past erosion may indicate the presence of overconsolidated material, indications of former movement can reveal that material strength has been reduced below its original peak strength, while fault evidence may point to the presence of crushed and myolinite zones. Groundwater conditions may be anticipated by spring seeps or soil precipitates, and the potential for perched water tables or artesian pressure can be signalled by stratigraphic conditions.

### 1.3.3   Determining Landslide Susceptibility and Hazard

Methodologies for analysing hazard range from theoretical determinism based on slope physics, through empiricism, to historical description. As mentioned before, each of these approaches may be treated quantitatively or qualitatively and many can be validated and explored through computer or lab-based physical simulation. Deterministic methods may

**Figure 1.12**  *Stabilized landslide site at a coast near Ventnor, Isle of Wight. Stabilization is integrated in recreational usage of the coast (photo by T. Glade)*

be argued through geomechanical principles and mathematical solutions, while stochastic attributes and statistical association validate other methods.

Theoretical deterministic methods depend on understanding the causes of landslides. Causes, or factors conducive to instability, can be recognized at various levels of abstraction from the slope itself. There are those factors such as cohesion or porewater pressure that directly control the magnitude of stress within the slope. These direct factors can be influenced by other factors recognized at successively more remote levels of abstraction. For example, porewater pressure may be related to the rate of rainfall infiltration through the ground surface, which in turn may be related to the density of the vegetation cover. Similarly, the vegetation cover may be subject to change as a result of climate conditions or land use activity. Such chains of relationships may be critical in reducing the stability of a slope over time to a point where triggering of movement may occur.

*Landslide susceptibility* is thus a function of the degree of the inherent stability of the slope (as indicated by the factor of safety or excess strength) together with the presence and activity of causative factors capable of reducing the excess strength and ultimately triggering movement. The identification of causative factors is the basis of many methods of susceptibility/stability assessment. The factors may be dynamic (e.g. porewater pressure), or passive (e.g. rock structure), and may also be considered in terms of the roles they perform in destabilizing a slope (Crozier, 1989). In this sense, the factors recognized are preconditioning factors (e.g. slope steepness), preparatory factors (e.g. deforestation) and triggering factors (e.g. seismic shaking). For a full discussion see Chapter 2.

### 1.3.3.1   Susceptibility assessment

An initial stage of hazard analysis based on the deterministic approach is susceptibility assessment and mapping. This provides a measure of the propensity of a site or area to produce landslides based on the presence of known causative factors, the history of slope behaviour (precedence approach) or the comparison of shear and resisting stresses (factor of safety).

*Factor (parameter) mapping.* Factor mapping is commonly employed as an initial stage of regional stability assessments. It involves identifying the spatial distribution of one or more of the causative factors or their combined effect and the subsequent ranking of unit areas on the anticipated interrelated influence these may have on susceptibility (e.g. rock type and slope angle). A comprehensive list of stability factors commonly employed in this approach is given by Crozier (1989), Turner and Schuster (1996) and Guzzetti *et al.* (1999). Factor mapping can be carried out both in areas with landslides and in areas with no previous landslide history. However, if landslides are present, they can be used subsequently to determine the relative importance of factors employed in this form of susceptibility assessment.

A good example of this approach has been produced for assessing debris-flow risk (Moon *et al.*, 1991). The range of techniques that can be applied is discussed by Hansen (1984) and Gee (1992).

*Precedence approach.* If landslides or evidence of former landslides are present in an area, a useful first approach to the determination of susceptibility can be gained by discriminating between those factors associated with landslides and those factors associated with stable ground (Rice *et al.*, 1969). For example, it may be possible to identify slope angle and height combinations above which landslides are always found and below which conditions appear stable. The application of such spatial thresholds beyond the area within which they were established needs to done with great care, as critical conditions may vary between sites. In addition, as with all assessments based on historical conditions, subsequent changes in conditions that may affect stability need to be taken into account.

*Factor of safety.* A more sophisticated approach represents the terrain in terms of differences in inherent stability based on the factor of safety (FoS). The FoS is the ratio of shear strength to the shear stress mobilized. In simple terms, the value of the FoS is assumed to be 1.0 at the moment of failure and values successively greater than 1.0 represent increasing stability and hence lower susceptibility to failure. This approach can be pursued with a wide range of available methods (Duncan, 1996), depending on the nature of the anticipated failure. Simple methods such as the infinite approach require little information on the geometry of the potential displaced mass, while others partition the mass into components and may involve two-dimensional or three-dimensional stress analysis. Recently, finite element models that predict deformations within the slope are becoming increasingly used to inform engineering solutions.

In order to account for the stresses operating in a slope, a considerable amount of information is required, relating to such factors as the shape of the potential failure surface, the geometry of the slope, porewater pressure conditions, and material properties including friction and cohesion (determined for the appropriate drainage conditions), stress and strain history and likely rate of failure. Determination of the FoS permits limiting equilibrium analysis of a slope and is particularly useful in the design of the type and magnitude of remedial measures required to achieve an acceptable FoS. Because of the detailed data requirements, generally obtainable only by subsurface exploration and rigorous laboratory testing, limiting equilibrium analysis is generally only employed on a site-by-site basis, and then only where potential risk is high. If limiting equilibrium analysis is linked to the behaviour of potential triggering factors, it has the potential to convert a static FoS into a statement of hazard. For example, if a critical factor, such as rise in porewater pressure, is successfully correlated with rainfall conditions, it may be possible (with reference to the rainfall record) to determine the probability with which porewater conditions exceed a critical threshold and initiate failure.

*Physically based simulation models.* Dynamic modelling of hydrology and resultant slope strength conditions can be achieved using sophisticated computer simulation programs. One such model, the Combined Hydrology Slope Stability Model (CHASM™) can simulate the changes in slope stability during the course of a rainstorm and anticipate the factors of safety during the course of the event (Anderson *et al.*, 1988). In critical situations, this can lead to a prediction of the time and size of failure during a rainstorm. Such models require information on rainfall intensity, antecedent hydrological conditions, soil properties, and saturated and unsaturated hydraulic conductivities. The CHASM model employs both unsaturated flow and groundwater flow as factors determining porewater pressure throughout the slope. Factors of safety are iteratively determined while conventional slope stability methods are used to isolate the most likely failure surface. Similar approaches have been used for distributed catchment hydrology modelling and slope stability (Burton and Bathurst, 1998; Montgomery and Dietrich, 1994) and have the advantage of being able to simulate areas of drainage convergence and divergence based on upslope surface morphological conditions derived from a digital elevation models.

*Deterministic physical lab modelling.* There have been a number of attempts to set up hardware, scale models of slopes and landslides in the laboratory environment and to determine empirically the conditions that control initiation and behaviour of landslides. This approach has been used successfully for debris flows (Tognacca *et al.*, 2000). As with all scale modelling there are difficulties in scaling down all field parameters. Furthermore, in controlled laboratory conditions, it is difficult to account for dynamic changes in the geomorphic environment that occur in reality at initiation sites.

Susceptibility and stability assessments, while useful contributions to hazard analysis, do not generally provide a direct assessment of magnitude and frequency of occurrence. Increasingly, stability analyses that can link critical dynamic changes of geotechnical properties to behaviour of external triggering conditions are capable of providing an estimate of probability of occurrence and magnitude of movement (van Asch *et al.*, 1999) and therefore a representation of hazard.

## 1.3.3.2　Historically determined frequency and magnitude

Sources for historically determining frequency–magnitude relationships are based either on natural archives from the field (e.g. hillslope evidence – morphology, deposits; dendrogeomorphology; varved lake sediments) or on human archives (e.g. church chronicles, postcards, newspaper, letters).

*Field evidence.* Evidence of former landsliding can be determined from slope morphology, sedimentary deposits, or impact features (e.g. deformed trees). As this type of evidence deteriorates or is obliterated progressively with time, care has to be taken in establishing long-term trends in occurrence. Off-site landslide deposits may be better preserved in the lacustrine sedimentary record. While careful dating of landslide-derived strata can provide an accurate measure of frequency, the actual magnitude and character of the formative landslide activity is less easy to establish. An excellent illustration of the use of lacustrine records to establish the frequency and magnitude of landsliding is provided by Page *et al.* (1994). A wide range of both relative and absolute methods has been employed for dating of field evidence (Lang *et al.*, 1999; Bull, 1996). A number of papers dealing with the determination of frequency and magnitude of occurrence from field evidence can be found in Matthews *et al.* (1997).

*Historic archives.* Another important source of past landslide activity is historical information. In this context, the term 'historical' refers to information recorded either intentionally or unintentionally by humans. Such sources may include maps, newspaper articles, church chronicles, and even postcards, drawings and personal letters. When using these data, it has to be remembered that not all events will have been recorded. Despite the fact that the quality of historical evidence is strongly dependent on recording procedures and available records, this approach provides an indication of at least the minimum level of landslide activity in an area. The issue of using historical data in natural hazard assessments is discussed by Guzzetti *et al.* (1994), Glade *et al.* (2001), and Petrucci and Polemio (2002) and is specifically addressed to landslides by Glade (2001), Bozzano *et al.* (1996) and Guzzetti (2000).

There are problems associated with all historically based approaches if they are to be used to estimate existing or future hazard. First, the historically based frequency–magnitude record may be a response to environmental conditions that no longer pertain in the area. Second, longevity of evidence is a function of time and magnitude of event. This means that the record may give a false impression, indicating, for example, that in the distant past there were fewer but bigger landslides compared with more recent periods of the record. However, this approach has the capability of including the influence of critical slope stability factors that may be missed by the inaccuracies of sampling and laboratory analyses.

### 1.3.3.3   Triggering threshold analysis

Analysis of the influence and behaviour of landslide-triggering agents can be used to assess the frequency and sometimes frequency–magnitude behaviour of landslides. This can be a useful approach because, in some cases (e.g. climate records), the triggering agent record is much longer and more reliable than the record of actual landslide occurrence. Despite the stability status of a given slope or catchment, a trigger (usually extrinsic) is needed to initiate the movement. In nature, these triggers can be identified as: rainfall, earthquakes, volcanic eruptions, or the undercutting of slopes by fluvial, coastal or weathering processes. Human-induced triggers may include explosions, slope cutting, slope loading (with buildings, material or water) or drainage systems that lead to a change of soil moisture regime. While human-induced triggers are difficult to assess – in particular with respect to determination of the probability of occurrence and a consequent triggering threshold – investigation of natural triggers has been successfully used to estimate landslide hazard. Triggering threshold analysis involves identifying the critical conditions associated with the initiation of landslides in the past and the comparison of these with the conditions that did *not* initiate movement. Based on the assumption that the triggering conditions and environmental setting are constant, a threshold analysis can be carried out. Such analyses are strongly dependent on a number of factors, including: the quality of the database, the quantity of items in the database, and standardization of record-keeping techniques. Another issue is the representation of a region by point-source data. For example, landslide locations and rainfall recording sites are commonly some distance apart. Thus care has to be taken when extrapolating rainfall conditions over wide areas.

Rainfall as trigger has been extensively investigated by various authors (e.g. Wieczorek and Glade, in press; Wieczorek and Guzzetti, 2000; Polemio and Petrucci, 2000; Toll, 2001; Zêzere, 2000). While some studies focus on specific locations (e.g. Finlay *et al.*, 1997), regional assessments have been performed for the USA (e.g. Larson, 1995; Wilson *et al.*, 1993; Wilson and Wieczorek, 1995), for Italy (e.g. Polloni *et al.*, 1996), and for New Zealand (e.g. Glade, 2000), to name only a few. In areas where there is a comprehensive database, it has been possible to develop triggering models to provide early warning schemes (e.g. Crozier, 1999; Wilson *et al.*, 1993). Some authors have developed rainfall thresholds for landslides and floods (Aleotti *et al.*, 1996; Reichenbach *et al.*, 1998). Inherent in all of these methods is their empirical, somewhat 'black-box' approach, leading to some uncertainty as to which stress conditions are actually critical in the triggering process (Chowdhury and Flentje, 2002).

In contrast, studies on earthquakes as landslide triggers are not as extensive (e.g. Bommer and Rodriguez, 2002). Wilson and Keefer (1985) suggested a method to predict spatial limits of earthquake-induced landslides, based on earthquake magnitude and intensity. More recently, Jibson and Keefer (1993) investigated earthquakes and related thresholds for the Madrid region, while the Northridge earthquake in 1994, which triggered numerous landslides, has been studied by Harp and Jibson (1995). A regional method for relating landslide occurrence and earthquake activity has been proposed by Jibson *et al.* (1998). As with rainfall triggers, these studies involve empirical methods and thus numerous data are required. The basis of the relationships established by these methods can be location-specific and therefore application of derived models to other regions is limited.

## 1.4    Techniques Employed in Obtaining Information of Value to Hazard Estimation

Landslide occurrence is a complex, multivariate problem. The accuracy with which landslide hazard can be represented varies depending on the quality and quantity of data available. The quality of the database in turn can often be a function of the availability of time, money and other resources. At the scoping stage of any landslide hazard and risk investigation, important decisions need to be made on the nature of the solution sought and the consequent detail and techniques required with respect to data acquisition. In general, investigation types can be differentiated, including:

- Surface investigations
- Subsurface investigations

- Laboratory analysis
- Modelling approaches
- Dating techniques
- GIS techniques.

Commonly, each field study starts with a detailed mapping campaign. Depending on the study design, this may incorporate topographic characteristics, geotechnical information, geomorphological features, lithological and structural information, hydrological conditions and so on. These can be registered on a base map, or dominant positions can be determined by tachymetry or GPS. Both techniques have the advantage that fixed ground-control points can be regularly revisited to provide information on surface movement (e.g. Malet *et al.*, 2002). Present-day movement can also be monitored by a range of recording devices (e.g. extensiometers as described in Angeli *et al.*, 1999). In addition, remote sensing techniques allow spatial information to be accessed for even remote areas. These techniques include airborne-derived data such as aerial photography and oblique photography, and satellite imagery (e.g. Zhou *et al.*, 2001). Where a suitable vegetation cover exists, dendrogeomorphological investigations can be used to determine the record of surface movements (e.g. Gers *et al.*, 2001).

Subsurface movement is generally monitored using inclinometer tubes. These measurements are often used in conjunction with borehole drillings, drop-penetration tests and geophysical investigations (e.g. seismic reflection and refraction, georadar and geo-electric sensing; refer to Mauritsch *et al.* (2000) for a typical case study). In addition, soil hydrology can be monitored using tensiometers, piezometers and pressure cells installed at different depths.

Age of events can also be assessed using a range of methods. Generally, these refer to either indirect techniques such as relative dating (e.g. by stratigraphic position of landslide sediment) or to direct approaches such as lichenometry, radiocarbon dating ($C^{14}$), luminescence (TL or OSL), or nucleide dating of exposed surfaces (Lang *et al.*, 1999). Recently, spatial analysis using GIS techniques has become common. Point-derived data can be coupled with spatial data sets (e.g. Digital Terrain Model, geology, soil, vegetation, etc.) in order to gain additional information of relevance to landslide distribution and movement. An introduction to GIS techniques of different complexities is given by Soeters and van Westen (1996) and by Carrara and Guzzetti (1995). It has to be emphasized, however, that spatial analysis may involve large errors due to data uncertainty. Therefore, any spatial analysis should be verified by a range of independent validation techniques (e.g. Chung and Fabbri, 1999). A valuable source of information on techniques and methods available for landslide hazard assessment has been compiled by Turner and Schuster (1996).

## 1.5  Approaches to Risk Estimation

A simple approach to risk estimation might involve just a frequency analysis of past *consequences*. For example, the number of deaths resulting from aircraft incidents per unit distance travelled could provide a measure of risk from air travel. However, because aircraft are becoming safer and patterns of air travel change, the measure of risk produced in this way has limited temporal significance. While safer aircraft mean that there are fewer fatalities per air mile travelled and thus on average *individual risk* may be less, *societal risk* has increased because there are more people flying and more air miles being flown. Clearly, both the dynamics of the source of the problem (hazard) and elements at risk also need to be assessed. Similarly, landslide risk cannot be adequately represented by consequence analysis alone. The main reason for this is that the record of landslide impact is often too short or too obscure to have captured the very high-magnitude–low-frequency events that constitute a major component of landslide risk. Thus a comprehensive risk assessment should involve the sequential identification and analysis of a number of components that influence risk.

There are two sources of uncertainty in risk estimation: first, the uncertainty attached to both the hazard and consequence components of risk; and second, the accuracy (margin of error) of the estimate itself. Estimates of risk can be arrived at and expressed both qualitatively and quantitatively. No matter which approach is taken, the value of the estimate depends on the accuracy of initial hazard and risk identification. That initial stage of investigation should be widely scoped to include not just direct and immediate impacts but also consequential hazards, indirect and delayed impacts. For example, episodes of landsliding in parts of New Zealand have ultimately led to downstream aggradation, loss of channel capacity and severe flooding. In some cases this has resulted in abandonment of farming operations or the installation of expensive flood protection works (Page *et al.*, 2001). Thus landslide impacts can be direct or indirect, immediate or delayed, and in some instances generate consequential hazards.

### 1.5.1   Vulnerability Assessment

Vulnerability relates to the Latin verb *vulnerare*, 'to wound' or 'to be susceptible', and is explained in the dictionary as 'liability to be damaged or wounded; not protected against attack'. Hence the vulnerability relates to the consequences, or the results of an impact of a natural force, and not to the natural process or force itself (Lewis, 1999). In practice, vulnerability and consequences are irrevocably linked. Two fundamentally different perspectives for examining vulnerability exist: investigations based on natural science and those based on social science methods and assumptions.

Unfortunately, there exists no uniform definition of 'vulnerability' in social sciences. Numerous definitions are reviewed and listed by Weichselgartner (2001). Wilches-Chaux (1992) summarized different views of vulnerability and differentiates between natural, physical, ecological, technical, economical, social, political, institutional, ideological, cultural and educative vulnerability. Also Cutter (1996) states that there are no unique definitions of vulnerability in social sciences. Chambers (1989) refers to both internal and external dimensions affecting vulnerability. While the internal dimensions include defencelessness and insecurity of threatened people, the external dimension refers to exposure to risk, shock and stress (Bohle, 2001). Hence vulnerability is determined by factors closely related to both external conditions, and to whether humans and their resources are able to withstand or cope with a natural disaster, or not (Hewitt, 1997; Smith, 2001).

Commonly, vulnerability assessments in landslide risk research are based on natural science approaches such as Liu *et al.* (2002). In contrast to other natural processes such as flooding and earthquakes, it is very difficult to assess vulnerability to landslides due to the complexity and the wide range of variety of landslide processes (Leroi, 1996). As Glade (2004b) summarizes, diverse effects have to be considered:

- Vulnerability of different elements at risk varies for similar processes.
  Fell (1994: 263) states that 'a house may have similar vulnerability to a slow- and a fast-moving landslide, but persons living in the house may have a low vulnerability to the slow-moving landslide (they can move out of the way) but a higher vulnerability to the fast-moving landslide . . .' because they cannot escape. If the scale of investigation is increased, there are also differences within a single house. For example, rooms facing towards the slope are more vulnerable to debris flows than valley-facing rooms. Furthermore, the larger the windows, the more vulnerable is the room and the respective content. Even people sleeping in this room will have a higher probability of death than other occupants of the house (Fell, 1994; Fell and Hartford, 1997).
- Temporal probability for a person of being present during the landslide event is variable.
  While a house is fixed to the ground, a car or inhabitants are mobile and might not be present during the event. For example, at night, a family is sleeping in the house whereas during the day, children are at school and the parents are working; hence the house would be empty. In contrast, fewer people are in commercial buildings at night; hence the potential consequences would be less severe, although property damage might be extensive.
- Different groups of humans have different coping potentials.
  In contrast to most adults, children might not be able to react adequately to endangering processes. Similarly, elderly or handicapped people might not have the possibility to escape, although the endangering process is correctly judged. This is one example of different coping potentials that has been addressed for landslide risk analysis by Liu *et al.* (2002).
- Early warning systems affect the vulnerability of people.
  If a warning system is installed, people might be able to escape (Smith, 2001), or at least reach safe places (Fell and Hartford, 1997) and thus change their vulnerability to given event magnitudes.
- Spatial probability of landslide occurrence varies.
  The spatial probability of the occurrence of a potentially damaging event at a given location has to be considered. For example, although a landslide occurs in the predicted zone, the probability that a small item or an individual human is affected is significantly different for a single rockfall compared with a widespread debris flow. Hence it is important to differentiate landslides by type, such as rockfall, debris flow, or translational earth slides, to name a few only (Fell, 1994).

Although this list could be extended, it gives an indication of potential factors that have to be considered in vulnerability assessment within landslide risk analysis. Despite all these limitations and complex, sometimes even unsolved problems, it is an economic and political necessity to assess vulnerability to landslides. Various attempts have been made. For preliminary studies, vulnerability is commonly set to 1, referring to a total damage as soon as the element at risk is hit by a landslide (e.g. Carrara, 1993; Glade *et al.*, submitted). More detailed investigations apply damage matrices (Leone *et al.*, 1996) based on either qualitative (e.g. Cardinali *et al.*, 2001) or quantitative approaches (e.g. Fell,

1994; Finlay and Fell, 1997; Heinimann, 1999; Leone *et al.*, 1996; Michael-Leiba *et al.*, 2000; Ragozin, 1996). Vulnerability assessment is a complex issue, which is regularly not considered in an appropriate and thoughtful manner.

### 1.5.2 Qualitative Risk Estimation

While the ultimate aim of risk estimation is the derivation of some reproducible standard measure of risk that can be compared and evaluated along with other similarly estimated risks, this is not always achievable. Resource and data constraints or preconceived notions of risk may dictate that quantitative estimations are neither warranted nor achievable. In such cases, risk may be determined by judgement and experience and expressed in qualitative terms. Intuition and professional judgement have long been defended as a legitimate approach to risk assessment, particularly among the engineering fraternity. If this approach is adopted, it needs to be explained and supported by ample reasons and statement of significance. Some examples of qualitative assessment of frequency, consequences and risk with respect to property are given in Tables 1.2, 1.3 and 1.4.

The subjectivity, latitude and cultural specificity of the terms used in qualitative estimation of risk lend themselves to a diversity of interpretation. Whereas intuitive estimates of risk, calling on judgement and experience, may be entirely appropriate

**Table 1.2**  *Qualitative measures of likelihood (Australian Geomechanics Society, 2000)*

| Level | Descriptor | Description | Indicative annual probability |
|---|---|---|---|
| A | Almost certain | The event is expected to occur | $>=10^{-1}$ |
| B | Likely | The event will probably occur under adverse conditions | $=10^{-2}$ |
| C | Possible | The event could occur under adverse conditions | $=10^{-3}$ |
| D | Unlikely | The event might occur under very adverse circumstances | $=10^{-4}$ |
| E | Rare | The event is conceivable but only under very exceptional circumstances | $=10^{-5}$ |
| F | Not credible | The event is inconceivable or fanciful | $<=10^{-6}$ |

**Table 1.3**  *Qualitative measures of consequences to property (Australian Geomechanics Society, 2000)*

| Level | Descriptor | Description |
|---|---|---|
| 1 | Catastrophic | Structure completely destroyed or large-scale damage requiring major works for stabilization |
| 2 | Major | Extensive damage to most of the structure, or extending beyond site boundaries requiring significant stabilization works |
| 3 | Medium | Moderate damage to some structure, or significant part of the site requiring large stabilization works |
| 4 | Minor | Limited damage to part of structure, or part of site requiring some reinstatement/stabilization works |
| 5 | Insignificant | Little damage |

**Table 1.4**   *Qualitative risk-level implications (Australian Geomechanics Society, 2000)*

| Risk level | General guide to implications |
| --- | --- |
| Very high risk | Extensive detailed investigation and research planning and implementation of treatment options essential to reduce risk to acceptable levels: may be too expensive and not practicable |
| High risk | Detailed investigation, planning and implementation of treatment options required to reduce risk to acceptable levels |
| Moderate risk | Tolerable provided plan is implemented to maintain or reduce risks. May be accepted. May require investigation and planning of treatment options |
| Low risk | Usually accepted. Treatment requirements and responsibility to be defined to maintain or reduce risk |
| Very low risk | Acceptable. Manage by normal slope maintenance procedures |

and even the best approach in some circumstances, they can sometimes be difficult to reproduce and substantiate by other parties. Where possible, a universal estimate of hazard and risk is best addressed in standard, objective, quantitative terms. The derived values can then be appropriately placed and evaluated within the relevant social context.

### 1.5.3   Quantitative Risk Calculation

Quantitative risk calculation is carried out by expressing hazard frequency and consequences in measured, numerical terms and determining their product. For example, the risk from property can be calculated (Australian Geomechanics Society, 2000) from:

$$R_{(Prop)} = P_{(H)} \times P_{(S:H)} \times V_{(Prop:S)} \times E \qquad (2)$$

where:
$R_{(Prop)}$   is the risk to property (annual loss of property value)
$P_{(H)}$   is the annual probability of the hazardous event (the landslide)
$P_{(S:H)}$   is the probability of spatial impact by the hazard (i.e. of the landslide affecting the property and, for vehicles, for example, the temporal probability)
$V_{(Prop:S)}$   is the vulnerability of the property to spatial impact (proportion of the property value lost)
$E$   is the element at risk (e.g. the value or net present value of the property)

Because the areal unit used in assessing hazard and risk is not always identical to the area specifically affected by the landslide, the term 'probability of spatial impact' $[P_{(S:H)}]$ is included in the above equation. Spatial probability is the ratio of the area affected by the landslide to the assessment area multiplied by the ratio of the area of the element of interest to the assessment area. Similarly, some elements at risk are mobile and have only temporary presence in the area affected by the landslide. The probability of presence can be taken into account by including the term temporal probability $[T_{(P:S)}]$. For example, a person may occupy a threatened building for only part of the time or a vehicle may be in the location only for a proportion of the time. It should be stressed that a quantitative approach such as indicated in equation (2) provides only a very limited estimate of risk, dealing with only one component, essentially direct damage to property in economic terms. There are likely to be many other indirect consequences associated with property damage. For example, in the case of damage to an industrial plant, this may involve loss of profit, loss of clients, loss of employment and earnings, as well as the adverse effects experienced by retailers and suppliers of raw materials associated with that industrial plant.

## 1.6 Approaches to Risk Evaluation

Risk evaluation is the processes of determining the significance of a risk to the individual, organization or community. Only after significance of risk is assessed can an appropriate response be determined. Essentially the risk needs to be judged as *acceptable*, *tolerable*, or *intolerable* (Fell, 1994; Finlay and Fell, 1997). These judgements are, however, hugely influenced by psychological, social and cultural values (Fischhoff *et al.*, 1981). Therefore it is important that risk is understood, evaluated and response options determined by those that live with the risk. Perception of risk involves an intuitive evaluation by an individual or group and perceptions can vary widely between individuals even within the same community (refer to Chapter 6 for more details). Perception is influenced by a multitude of factors, including: education, acquired knowledge, experience of previous hazards, gender, age and so on, and has been the subject of extensive psychological and sociological research (Garrick and Willard, 1991). From a management perspective, it is important that the variability of perception is reduced and that, through education and communication, the margin between reality and perception is narrowed. The terminology used to express risk is also difficult for the non-expert to understand. Difficulty may be experienced in interpreting and differentiating between expressions such as 'highly likely', or a 'probability of $10^{-2}$', or a chance of occurrence of '10% in fifty years'. One of the ways of improving the understanding of risk is by risk comparison; that is, comparing risk estimates with those of more familiar, easily understood risks. Risk comparison and ranking is also a useful means of prioritizing response (Table 1.5).

The nature of response to risk depends on the judgements of whether the risk is acceptable, tolerable or intolerable. As with perception, these judgements are highly subjective and influenced by psychological, sociological and cultural perspectives. Judgements can also be influenced by the nature of the risk. For example, people are more likely to accept a given level of risk emanating from a natural hazard as opposed to risk associated with

**Table 1.5** *Comparison of individual risk of death from hazards in New Zealand (population ~3.5 million). Annual average between 1840 and 1990 (Tephra, 1994)*

| Hazard | Deaths per year | Probability of death per person per year |
|---|---|---|
| Smoking | 4 000 | $1.1 \times 10^{-3}$ |
| Road accident | 600 | $1.7 \times 10^{-4}$ |
| Suicide | 380 | $1.1 \times 10^{-4}$ |
| Falls | 300 | $8.6 \times 10^{-5}$ |
| Drowning | 120 | $3.5 \times 10^{-5}$ |
| Homicide | 50 | $1.4 \times 10^{-5}$ |
| Fire | 32 | $9.0 \times 10^{-6}$ |
| Natural hazards | 6 | $1.6 \times 10^{-6}$ |

an artificial source such as a nuclear power plant. Similarly, attitudes toward risk vary, depending on whether the risk is localized or widespread throughout the community or whether it is voluntary (such as rock climbing) or involuntary (such as earthquakes).

Despite the subjectivity involved in risk judgement, some authorities and industries have set standards used to determine the acceptability of risk. For example, interim guidelines set by the Geotechnical Engineering Office of the Hong Kong Government for landslides and boulder falls on natural terrain indicate, in terms of annual loss of life, an 'acceptable' individual risk level at $1 \times 10^{-5}$ for new developments and $1 \times 10^{-4}$ for existing developments (Moore *et al.*, 2001).

Acceptability of societal risk, on the other hand, is sometimes based on the use of the FN diagram, which shows the ratio of frequency of events to the severity of the consequences of those events, expressed in terms of fatalities (Figure 1.13).

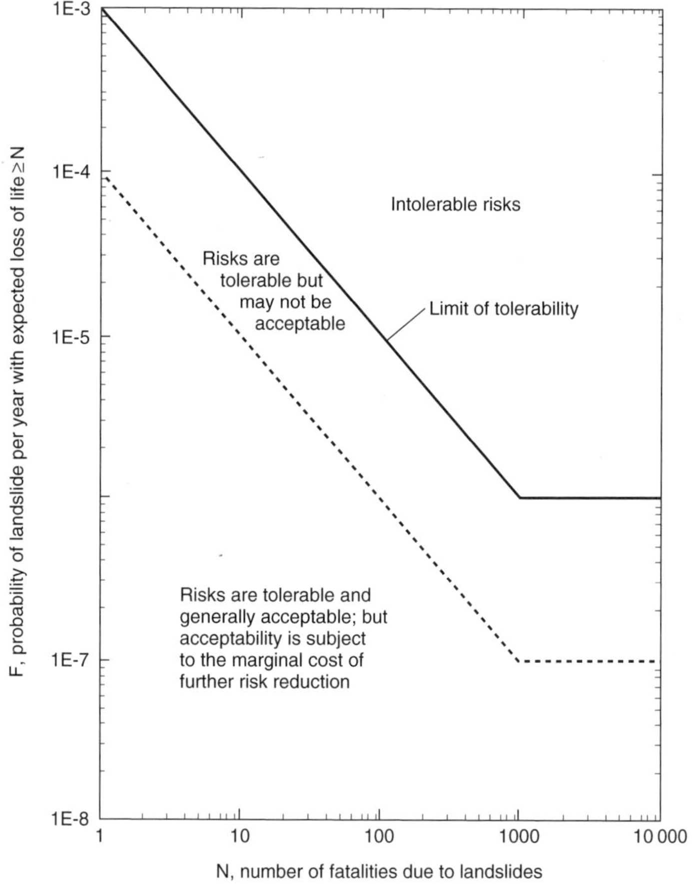


**Figure 1.13**   *The frequency of events of given magnitude (number of deaths) plotted against the number of deaths represented by those events (adapted from Australian Geomechanics Society, 2000). These diagrams are referred to as FN diagrams*

Implicit or explicit in any decision on acceptability of risk, whether as an individual or government authority, is the exercise of risk–benefit analysis. This is the comparison of the level of risk with benefits associated with being exposed to that risk. Even though there might be a relatively high risk, the associated benefits are sufficient to accept or tolerate the risk. For example, living on the top of a coastal cliff may expose the inhabitants to landslide risk but the view and other attributes may be considered to outweigh that risk (see Chapter 11 for discussion of this issue).

## 1.7   Risk Treatment

The risk management cycle (Figure 1.14) provides a generic ideal representation of the range and relationships of all the components of management aimed at managing and reducing risk and responding to emergencies (refer to Chapter 11 for further details). Landslide risk management is fully discussed in Part 3 of this book and mentioned here are only those aspects of management that are directly hazard-specific, namely mitigation and prevention (treatment options).

The risk/benefit ratio and the absolute level of risk strongly influence not only the acceptability of risk but also the nature of the response. High levels of risk may warrant

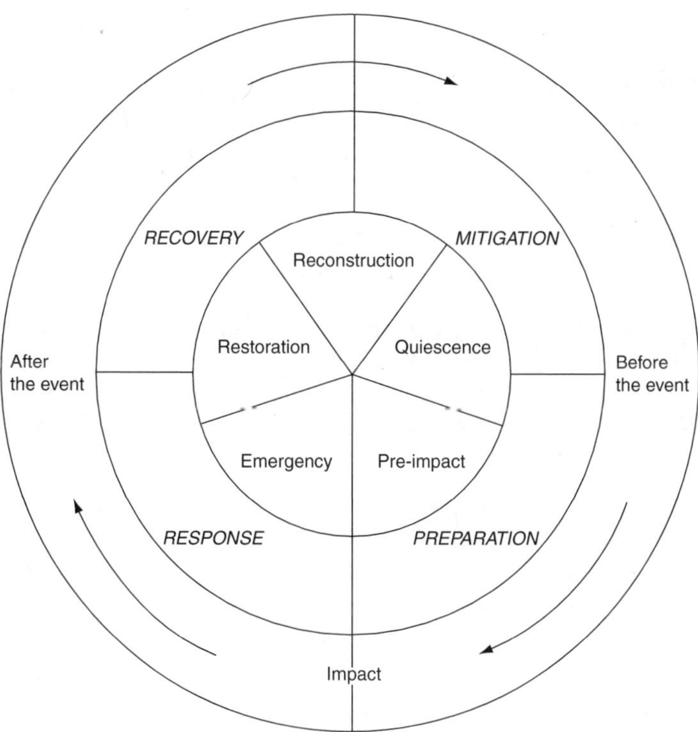


**Figure 1.14**   *The general risk management cycle as described by Alexander (2000, 2002)*

a legislative or sophisticated engineering response, while low levels of risk may be accepted or treated by common sense and education. The physical characteristics of the hazard also dictate the type of treatment measures adopted. For example, the hazard from shallow soil landslides may be prevented or reduced by tree planting, whereas with deep-seated landslides trees may produce very little benefit. If landslides are of the type that recur in specific locations, for example debris-flow tracks, alarm systems, warning information or zoning may be employed. The type of material, size, rate, type and depth of movement can all be matched with a range of appropriate engineering slope control measures. Determination of the factors critical to the onset of movement can also point to appropriate remedial solutions. For example, if movement relates to high groundwater levels, slope drainage becomes an option. In areas where toe erosion instigates movement, buttressing and toe reinforcement are appropriate measures.

The range of options available for reducing landslide risk can be grouped as follows:

- Hazard modification: usually engineering solutions aimed at modifying the impact characteristics and reducing the frequency – in other words, keeping the hazard away from people.
- Behaviour modification: reducing the consequences by options such as avoidance, warning systems, reduction of vulnerability, development planning, education, regulations and economic incentives.
- Loss sharing: including systems for insurance, disaster relief, development aid and compensation.


## 1.8   Definitions

Because hazard and risk studies in general have been approached from so many different discipline perspectives there is inevitably a level of confusion in the terminology employed. This lack of standardization and consistency of use of the terminology has also found its way into the study of landslide hazard and risk. Listed below are definitions for many important terms that are fundamental to communication and understanding of landslide hazard and risk, although many of the definitions offered are sufficiently generic to apply to other forms of hazard. Wherever possible the definitions have been expressed initially in the most broadly applicable generic sense and where necessary subsequently explained with reference to landslides. Most but not all of the definitions are in accord with those proposed by Australian Geomechanics Society (2000) and Fell and Hartford (1997).

**Acceptable risk**: level of risk that a given society is prepared to accept because of the marginal cost of any further risk reduction. Risk management may aim to reduce all risks to this level.

**Consequences (impacts)**: the effects, usually but not always negative or adverse, resulting from hazard. Negative consequences may be referred to as losses or costs involving both economic and non-economic values.

**Consequence analysis**: identification and analysis of adverse effects or potential adverse effects arising from landslides, including immediate and delayed effects from direct landslide impact or indirectly through the disruption of other systems.

**Consequential landslide hazard**: a hazard (type I or type II) resulting not directly from the landslide itself but as a result of a consequential process set in train by the landslide. For example, a wave set up by a catastrophic landslide entering a water body would be considered a consequential hazard in this context.

**Dynamic hazard**: hazard resulting from the active, generally episodic behaviour of the natural process.

**Elements at risk**: all valued attributes threatened by the hazard (the landslide); may include structures, land, resources, social and physical infrastructure productive and non-productive activities, environmental qualities, life and physical and mental well-being. Some of these attributes are quantifiable, some can be expressed in economic terms and others defy ready quantification.

**Frequency**: a measure of the likelihood expressed as the number of occurrences of an event in a given time. For many natural hazards, including landslides, the basic unit of time used in frequency analysis is the year. For example a frequency of five events ($n$) recorded in a 100-year period ($t$) can also be expressed as an average frequency ($n/t$) of one event every 20 years on average. The term $n/t$ (in this example 20 years) is referred to as the **recurrence interval** or return period. The reciprocal of the return period expressed in years provides the annual *probability*; in this example 1/20 yields an annual probability of 0.05. In other words, there is a 5% chance of the event occurring in any one year, on average.

**Hazard**: in natural hazard usage, there are two accepted definitions of 'hazard'. The first (Hazard I) refers to an actual physical entity (process or situation) that has the potential to cause damage (e.g. a large rockslide or a long runout debris flow). This is the common non-technical understanding of hazard. However, this use of the term hazard is also found in some legal and statutory documents, with statements of the form: 'It is council policy to record the date and location of hazards. These include landslide, debris flow, surface flooding, subsidence etc. ...'. The second definition (Hazard II) is more technical and refers not to a process but rather to a threatening condition resulting from the behaviour of that process, expressed as the probability of occurrence of a damaging landslide.

**Hazard I**: a hazard is a potentially damaging process or situation (the landslide), for example an earthquake above a certain intensity or a landslide of sufficient size, depth, or displacement to cause damage or disruption or, as an example of a situation, the presence of weak foundation material.

**Hazard II**: the probability of a potentially damaging event (a landslide) occurring in a unit of time. This probability varies with the magnitude of the event (generally small landslides occur more frequently than large landslides). Consequently hazard is often expressed as the probability of occurrence of a given magnitude of event (see *magnitude–frequency relationship*). Defined in this way, hazard represents a state or condition and is assessed and applied to a particular place, for example site, unit area of land surface, region or object, lifelines, hydro dams and so on.

**Hazard analysis**: the process of identifying the probability of occurrence of a damaging event.

**Hazard identification**: the process of recognizing and accounting for all possible hazards that might occur within the place and time period of interest. For landslides this involves identifying landslide type, *landslide impact characteristics* and *consequential hazards*. The process needs to consider the types of element at risk as well as the relationship in time and space between landslides and *elements at risk*.

**Individual risk**: total risk divided by the population at risk. For example, if a region with a population of one million people experiences on average 5 deaths from landslides per year, the individual risk of being killed by a landslide in that region is 5/1 000 000, usually expressed in orders of magnitude as $5 \times 10^{-6}$.

**Intolerable risk**: level of risk that society is not prepared to live with and which must be reduced, removed, or avoided.

**Landslide impact characteristics**: characteristics of the landslide that may control the potential impact, including: degree of disruption of the displaced mass, areal extent and distance of runout, depth, area affected, velocity, discharge per unit width, and kinetic energy per unit area. Collectively these have been considered to represent **'landslide intensity'**.

**Landslide magnitude**: a measure of landslide size generally taken as the mass or volume of material displaced. With many other natural hazards the standard magnitude parameter (e.g. Richter magnitude for earthquakes or peak discharge for floods) is directly related to potential impact. Landslide magnitude, however, is a less reliable index of impact potential – other characteristics of the movement may be more important in determining impact (see *landslide impact characteristics*).

**Landslide susceptibility**: the propensity of an area to undergo landsliding. It is a function of the degree of inherent stability of the slope (as indicated by the factor of safety or excess strength) together with the presence of factors capable of reducing the excess strength and ultimately triggering movement.

**Magnitude–frequency relationship**: the relationship between the size of landslides and the frequency with which they occur in time or space. Essentially, big events are rare and small events are common. Some form of declining exponential or power-law function generally represents the relationship between magnitude and frequency.

**Probability**: the likelihood of a specific outcome, measured by the ratio of specific outcomes to the total number of possible outcomes. Probability is generally expressed as a number between 0 and 1, with 0 indicating an impossible outcome, and 1 indicating that an outcome is certain.

**Risk**: a measure of the probability and severity of loss to the *elements at risk*, usually expressed for a unit area, object, or activity, over a specified period of time.

**Risk analysis**: the overall process involving scoping, *hazard and risk identification* and *risk estimation*.

**Risk assessment**: the combined processes of risk analysis and risk evaluation, leading to the stage where personal judgements and treatment decisions can be rationally made.

**Risk estimation**: the process of deriving a measure of the probability and severity of loss to the *elements at risk* by the integration of hazard and consequence analysis. This can be carried out quantitatively (involving risk calculation, sometimes referred to as Quantitative Risk Analysis [QRA]) or qualitatively.

**Risk evaluation**: the process of determining the importance and relevance (significance) of the results of *risk analysis* with reference to the social and physical context within which they occur. This process determines whether risk is tolerable or acceptable. Risk evaluation may involve considerations of risk perception, risk communication and risk comparison with the aim of developing some appropriate level or form of response. It generally implicitly or explicitly balances the risk with the benefits associated with exposure to that risk.

**Risk management**: the process of developing and applying policies, procedures and practices to the tasks of assessment, monitoring, communication and *treatment of risk*.

**Risk treatment**: actions taken to address the results of risk; may involve acceptance, avoidance, reduction of frequency or intensity of hazard, reduction of consequences or the transferral of risk.

**Risk–benefit analysis**: the process of relating the level of risk to the level of benefits associated with exposure to that risk.

**Societal risk**: the total risk attributed to the society responsible for bearing that risk.

**Specific risk**: hazard probability × vulnerability for a given element at risk and/or for a given type of process.

**Static hazard**: hazard arising not through episodic behaviour of the natural agent but by human actions leading to the encounter of static hazardous conditions, for example building on weak foundation material. Hazard in this case is determined from the probability of human encounter or the number of damaging incidents per unit of time associated with the deposit.

**Tolerable risk**: level of risk that a society is prepared to live with because there are net benefits in doing so, as long as that risk is monitored and controlled and action is taken to reduce it.

**Total risk**: the expected consequences (loss) resulting from the level of hazard in a place, over a specified time period. It depends not only on the different hazardous process involved but also on elements at risk and their vulnerability.

**Vulnerability**: the expected degree of loss experienced by the elements at risk for a given magnitude of hazard.


# References

Alcántara-Ayala, I., 2002, Geomorphology, natural hazards, vulnerability and prevention of natural disasters in devloping countries, *Geomorphology*, **47**, 107–124.
Aleotti, P., Baldelli, P. and Polloni, G., 1996, Landsliding and flooding event triggered by heavy rains in the Tanaro Basin (Italy) Internationales Symposium Interpraevent 1996, Garmisch-Partenkirchen, VHB, **1**, 435–446.

Alexander, D., 1995, *Natural Disasters* (London: UCL Press).

Alexander, D.E., 2000, *Confronting Catastrophe* (New York: Oxford University Press).

Alexander, D.E., 2002, *Principles of Emergency Planning and Management* (New York: Oxford University Press).

Anderson, M.G., Kemp, M.J. and Lloyd, D.M., 1988, Applications of soil water finite difference models to slope stability problems, in C. Bonnard (ed.), *Proceedings of the 5th International Symposium on Landslides*, 10–15 July 1988, Lausanne, Switzerland (Rotterdam: A.A. Balkema), 525–530.

Angeli, M.G., Pasuto, A. and Silvano, S., 1999, Towards the definition of slope instability behaviour in the Alvera mudslide (Cortina d'Ampezzo, Italy), *Geomorphology*, **30**, 201–211.

Australian Geomechanics Society, 2000, Landslide risk management concepts and guidelines, *Australian Geomechanics*, March, **35**(1), 49–92.

Bell, F.G., 1999, *Geological Hazards – Their Assessment, Avoidance and Mitigation* (London and New York: E & FN Spon).

Bohle, H.-G., 2001, Vulnerability and criticality: perspectives from social geography. *International Human Dimensions Programme on Global Environmental Change Update*, **2**, 1–5.

Bommer, J.J. and Rodriguez, C.E., 2002, Earthquake-induced landslides in central America, *Engineering Geology*, **63**, 189–220.

Bovis, M.J. and Jakob, M., 1999, The role of debris supply conditions in predicting debris-flow activity, *Earth Surface Processes and Landforms*, **24**, 1039–1054.

Bozzano, F., De Pari, P. and Mugnozza, G.S., 1996, Historical data in evaluating landslide hazard in some villages in Southern Italy, in K. Senneset (ed.), *Landslides – Glissements de Terrain* (Rotterdam: A.A. Balkema), vol. 1, 159–164.

Brand, E.W., 1989, Occurrence and significance of landslides in Southeast Asia, in E.E. Brabb and B.L. Harrod (eds), *Landslides – Extent and Economic Significance* (Rotterdam: A.A. Balkema) 303–324.

Bull, W.H., 1996, Prehistorical earthquakes on the Alpine fault, New Zealand, *Journal of Geophysical Research*, **101**, 6037–6050.

Burton, A. and Bathurst, J.C., 1998, Physically based modelling of shallow sediment yield at catchment scale, *Environmental Geology*, **35**, 89–99.

Cardinali, M., Reichenbach, P., Guzetti, F., Adrizzone, F., Antonini, G., Galli, M., Cacciano, M., Castellani, M. and Salvati, P., 2001, A geomorphological approach to the estimation of landslide hazards and risks in Umbria, Central Italy, *Natural Hazards and Earth System Sciences*, **2**, 1–16.

Carrara, A., 1993, Uncertainty in evaluating landslide hazard and risk, in J. Nemec, J.M. Nigg and F. Siccardi (eds), *Prediction and Perception of Natural Hazards, Proceedings Symposium*, 22–26 October 1990, Perugia, Italy. (Dordrecht: Kluwer Academic Publishers), vol. 2, 101–109.

Carrara, A. and Guzzetti, F. (eds), 1995, *Geographical Information Systems in Assessing Natural Hazards: Advances in Natural and Technological Hazards Research* (Dordrecht: Kluwer Academic Publishers).

Chambers, R., 1989, Vulnerability, coping and policy, *Institute of Development Studies Bulletin*, **20**, 1–7.

Chowdhury, R. and Flentje, P., 2002, Uncertainties in rainfall-induced landslide hazard, *Quarterly Journal of Engineering Geology and Hydrogeology*, **35**, 61–69.

Chung, C.-J.F. and Fabbri, A.G., 1999, Probabilistic prediction models for landslide hazard mapping, *Photogrammetric Engineering & Remote Sensing*, **65**, 1389–1399.

Crosta, G.B., Imposimato, S. and Roddeman, D.G., 2003, Numerical modelling of large landslide stability and runout, *Natural Hazard and Earth System Science*, **3**, 523–538.

Crozier, M.J., 1984, Field assessment of slope instability, in D. Brunsden and D.B. Prior (eds), *Slope Instability* (London: John Wiley & Sons Ltd), 103–142.

Crozier, M.J., 1989, *Landslides: Causes, Consequences and Environment* (London: Routledge).

Crozier, M.J., 1995, Landslide hazard assessment: a review of papers submitted to theme G4, in D.H. Bell (ed.), *Proceedings of the Sixth International Symposium*, 10–14 February 1992, Christchurch, New Zealand (Rotterdam: A.A. Balkema), 1843–1848.

Crozier, M.J., 1996, Runout behaviour of shallow, rapid earthflows, *Zeitschrift für Geomorphologie (Supplementband)*, **105**, 35–48.

Crozier, M.J., 1997, The climate–landslide couple: a southern hemisphere perspective, in J.A. Matthews, D. Brunsden, B. Frenzel, B. Gläser and M.M. Weiß (eds), *Rapid Mass Movement As a Source of Climatic Evidence for the Holocene* (Stuttgart: Gustav Fischer), vol. 19, 333–354.

Crozier, M.J., 1999, Prediction of rainfall-triggered landslides: a test of the antecedent water status model, *Earth Surface Processes and Landforms*, **24**, 825–833.

Crozier, M.J., and Preston, N.J., 1999. Modelling changes in terrain resistance as a component of landform evolution in unstable hill country, in S.Hergarten and H.J. Neugebauer (eds), Process Modelling and Landform Evolution. Lecture Notes in Earth Sciences. Spring-Verlag, Heidelberg, pp. 267–284.

Cruden, D.M. and Varnes, D.J., 1996, Landslide types and processes, in A.K. Turner and R.L. Schuster (eds), *Landslides: Investigation and Mitigation* (Washington, DC: National Academy Press), Special Report 247, 36–75.

Cutter, S.L., 1996, Vulnerability to environmental hazards, *Progress in Human Geography*, **20**, 529–539.

Dai, F.C., Lee, C.F. and Ngai, Y.Y., 2002, Landslide risk assessment and management: an overview, *Engineering Geology*, **64**, 65–87.

Dikau, R., Brunsden, D., Schrott, L. and Ibsen, M. (eds), 1996, *Landslide Recognition: Identification, Movement and Causes* (Chichester: John Wiley & Sons Ltd).

Duncan, J.M., 1996, Soil slope stability analysis, in A.K. Turner and R.L. Schuster (eds), *Landslides: Investigation and Mitigation* (Washington, DC: National Academy Press), Special Report 247, 337–371.

Dykes, A.P., 2002, Weathering-limited rainfall-triggered shallow mass movements in undisturbed steepland tropical rainforest, *Geomorphology*, **46**, 73–93.

Erickson, G.E., Ramirez, C.F., Concha, J.F., Tisnado, G.M. and Urquidi, F.B., 1989, Landslide hazards in the central and southern Andes, in E.E. Brabb and B.L. Harrod (eds), *Landslides: Extent and Economic Significance* (Rotterdam: A.A. Balkema), 111–118.

Fell, R., 1994, Landslide risk assessment and acceptable risk, *Canadian Geotechnical Journal*, **31**, 261–272.

Fell, R. and Hartford, D., 1997, Landslide risk management, in D.M. Cruden and R. Fell (eds), *Landslide Risk Assessment – Proceedings of the Workshop on Landslide Risk Assessment*, Honolulu, Hawaii, USA, 19–21 February 1997 (Rotterdam: A.A. Balkema), 51–109.

Finlay, P.J. and Fell, R., 1997, Landslides: risk perception and acceptance, *Canadian Geotechnical Journal*, **34**, 169–188.

Finlay, P.J., Fell, R. and Maguire, P.K., 1997, The relationship between the probability of landslide occurrence and rainfall, *Canadian Geotechnical Journal*, **34**, 811–824.

Fischhoff, B., Lichtensetin, B.S., Slovic, P., Derby, S.L. and Kenney, R.L., 1981, *Acceptable Risk* (Cambridge: Cambridge University Press).

Flageollet, J.C., 1989, Landslides in France: a risk reduced by recent legal provisions, in E.E. Brabb and B.L. Harrod (eds), *Landslides: Extent and Economic Significance* (Rotterdam: A.A. Balkema), 157–168.

Garrick, B.J. and Willard, C.G. (eds), 1991, *The Analysis, Communication and Perception of Risk* (New York: Plenum Press).

Gee, M.D., 1992, Classification of landslide hazard zonation methods, in D.H. Bell (ed.) *Proceedings of the Sixth International Symposium*, 10–14 February 1992, Christchurch, New Zealand (Rotterdam: Balkema), 947–952.

Gers, E., Florin, N., Gärtner, H., Glade, T., Dikau, R. and Schweingruber, F.H., 2001, The application of shrubs for dendrogeomorphological analysis to reconstruct spatial and temporal landslide patterns – A preliminary study, *Zeitschrift für Geomorphologie, Supplementband*, **125**, 163–175.

Glade, T., 2000, Modelling landslide triggering rainfall thresholds at a range of complexities, in E. Bromhead, N. Dixon and M.-L. Ibsen (eds), *Landslides in Research, Theory and Practice* (Cardiff: Thomas Telford), 633–640.

Glade, T., 2001, Landslide hazard assessment and historical landslide data – an inseparable couple? in T. Glade, F. Frances and P. Albini (eds), *The Use of Historical Data in Natural Hazard Assessments* (Dordrecht: Kluwer Academic Publishers), vol. 7, 153–168.

Glade, T., 2004a, Linking natural hazard and risk analysis with geomorphology assessments, *Geomorphology*, in press.

Glade, T., 2004b, Vulnerability assessment in landslide risk analysis, *Die Erde*, **134**, 121–138.

Glade, T. and Dikau, R., 2001, Gravitative Massenbewegungen – vom Naturereignis zur Naturkatastrophe, *Petermanns Geographische Mitteilungen*, **145**, 42–55.

Glade, T., Frances, F. and Albini, P. (eds), 2001, *The Use of Historical Data in Natural Hazard Assessments: Advances in Natural and Technological Hazards Research* (Dordrecht: Kluwer Academic Publishers).

Glade, T., von Davertzhofen, U. and Dikau, R. (submitted) GIS-based landslide risk analysis in Rheinhessen, Germany *Natural Hazards.*

Guzzetti, F., 2000, Landslide fatalities and the evaluation of landslide risk in Italy, *Engineering Geology*, **58**, 89–107.

Guzzetti, F., Cardinali, M. and Reichenbach, P., 1994, The AVI Project: A Bibliographical and Archive Inventory of Landslides and Floods in Italy, *Environmental Management*, **18**, 623–633.

Guzzetti, F., Carrara, A., Cardinali, M. and Reichenbach, P., 1999, Landslide hazard evaluation: a review of current techniques and their application in a multi-scale study, Central Italy, *Geomorphology*, **31**, 181–216.

Hansen, A., 1984, Landslide hazard analysis, in D. Brunsden and D.B. Prior (eds) *Slope instability* (Chichester: John Wiley & Sons Ltd), 523–602.

Harp, E.L. and Jibson, R.W., 1995, Inventory of landslides triggered by the 1994 Northridge, California earthquake Denver, *US Geological Survey*, **17**.

Heinimann, H.R., 1999, *Risikoanalyse bei gravitativen Naturgefahren – Methode.* Bern.

Hewitt, K., 1997, *Regions of risk: A geographical introduction to disasters* (Harlow: Addison Wesley Longman Limited).

Hsu, K.J., 1975, Catastrophic debris streams (sturtzstrom) generated by rockfalls. *Geological Society of America Bulletin*, **86**, 128–140.

Hungr, O., 1995, A model for the runout analysis of rapid flow slides, debris flows, and avalanches, *Canadian Geotechnical Journal*, **32**, 610–623.

Hutchinson, J.N., 2001. Reading the ground: morphology and geology in site appraisal. The Fourth Glossop Lecture. *Quarterly Journal of Engineering Geology and Hydrogeology*, **34**, 7–50.

Jibson, R.W. and Keefer, D.K., 1993, Analysis of the seismic origin of landslides: examples from the New Madrid seismic zone, *Geological Society of America Bulletin*, **105**, 521–536.

Jibson, R.W., Harp, E.L. and Michael, J.A., 1998, A method for producing digital probabilistic seismic landslide hazard maps: an example from Los Angeles, California, area Los Angeles, *US Geological Survey*, **21**.

Lang, A., Moya, J., Corominas, J., Schrott, L. and Dikau, R., 1999, Classic and new dating methods for assessing the temporal occurrence of mass movements, *Geomorphology*, **30**, 33–52.

Larsen, M.C., Wieczorek, G.F., Eaton, S. and Sierra, H.T., 2001, The Venezuela landslide and flash flood disaster of December 1999, in Mugnai, A. (ed.), *2nd Plinius Conference on Mediterranean Storms*, 16–18 October 2000. Siena, Italy, EGS.

Larson, R.A., 1995, Slope failures in southern California: rainfall threshold, prediction, and human causes, *Environmental and Engineering Geoscience*, **1**, 393–401.

Leone, F., Asté, J.P. and Leroi, E., 1996, Vulnerability assessment of elements exposed to mass-movement: working toward a better risk perception, in K. Senneset (ed.), *Landslides – Glissements de Terrain* (Rotterdam: A.A. Balkema), vol. 1, 263–270.

Leroi, E., 1996, Landslide hazard – Risk maps at different scales: objectives, tools and development, in Senneset, K. (ed.), *Landslides – Glissements de Terrain, 7th International Symposium on Landslides* (Rotterdam: A.A. Balkema), 35–51.

Lewis, J., 1999, *Development in disaster-prone places – studies of vulnerability* (London: Intermediate Technology Publications Ltd).

Liu, X.L., Yue, Z.Q., Tham, L.G. and Lee, C.F., 2002, Empirical assessment of debris-flow risk on a regional scale in Yunnan province, southwestern China, *Environmental Management*, **30**, 249–264.

Malet, J.P., Maquaire, O. and Calais, E., 2002, The use of Global Positioning System techniques for the continuous monitoring of landslides: application to the Super-Sauze earthflow (Alpes-de-Haute-Provence, France), *Geomorphology*, **43**, 33–54.

Matthews, J.A., Brunsden, D., Frenzel, B., Gläser, B. and Weiß, M.M. (eds), 1997, *Rapid mass movement as a source of climatic evidence for the Holocene: Paläoklimaforschung Paleoclimate Research* (Stuttgart, Jena, Lübeck and Ulm: Gustav Fischer Verlag), **19**.

Mauritsch, H.J., Seiberl, W., Arndt, R., Römer, A., Schneiderbauer, K. and Sendlhofer, G.P., 2000, Geophysical investigations of large landslides in the Carnic region of southern Austria, *Engineering Geology*, **56**, 373–388.

Michael-Leiba, M., Baynes, F. and Scott, G., 2000, Quantitative landslide risk assessment of Cairns, Australia, in Bromhead, E., Dixon, N. and Ibsen, M.-L. (eds), *Landslides in Research, Theory and Practice* (Cardiff: Thomas Telford), 1059–1064.

Montgomery, D.R. and Dietrich, W.E., 1994, A physically based model for the Topographic control on shallow landsliding, *Water Resources Research*, **30**, 1153–1171.

Moon, A.T., Olds, R.J., Wilson, R.A. and Burman, B.C., 1991, Debris-flow risk zoning at Montrose, Victoria, in D.H. Bell (ed.), *Landslides* (Rotterdam: Balkema), vol. 2, 1015–1022.

Moore, R., Hencher, S.R. and Evans, N.C., 2001, An approach for area and site-specific natural terrain hazard and risk assessment, Hong Kong Geotechnical Engineering Meeting Society's Needs, *Proceedings of the 14th Southeast Asia Geotechnical Conference*, Hong Kong, December 2001, 155–160.

Moore, R. and McInnes, R.G., 2002. Cowes to Gurnard costal stability: providing the tools and information for effective planning and management of unstable land, in R.G. McInnes and J. Jakeways (eds), *Instability Planning and Management. Proceedings of the International Conference*. Thomas Telford, Isle of Wight, pp. 109–116.

Müller, L., 1964, The rock slide in the Vaiont valley, *Felsmechanik und Ingenieurgeologie*, **2**, 148–212.

Münchner Rückversicherung, 2000, *Topics 2000: Naturkatastrophen – Stand der Dinge München* (Münchner Rückversicherung), 126.

Nussbaumer, J., 1998, *Die Gewalt der Natur. Eine Chronik der Naturkatastrophen von 1500 bis heute*, ed. Grünbach Sandkorn.

Oyagi, N., 1989, Geological and economic extent of landslides in Japan and Korea, in E.E. Brabb and B.L. Harrod (eds), *Landslides: Extent and Economic Significance* (Rotterdam: A.A. Balkema), 289–302.

Page, M.J., Trustrum, N.A. and DeRose, R.C., 1994, A high-resolution record of storm induced erosion from lake sediments, New Zealand, *Journal of Paleolimnology*, **11**, 333–348.

Page, M.J., Trustrum, N. and Gomez, B., 2001, Implications of a century of anthropogenic erosion for future land use in the Gisborne – East Coast region of New Zealand, *New Zealand Geographer*, **56**, 13–24.

Petley, D., 1996, The Mechanics and Landforms of Deep-Seated Landslides, in M.G. Anderson and S.M. Brooks (eds), *Advances in Hillslope Processes* (Chichester: John Wiley & Sons Ltd), vol. 2, 823–835.

Petrucci, O. and Polemio, M., 2002, Hydrogeological multiple hazard: a characterisation based on the use of historical data, in J. Rybár, J. Stemberk and P. Wagner (eds), *Landslides: Proceedings of the First European Conference on Landslides, Prague*, June 2002 (Lisse: Balkema), 269–274.

Polemio, M. and Petrucci, O., 2000, Rainfall as a landslide triggering factor: an overview of recent international research, in E. Bromhead, N. Dixon and M.-L. Ibsen (eds), *Landslides in Research, Theory and Practice, Proceedings of the 8th International Symposium on Landslides* (Cardiff: Thomas Telford), 1219–1226.

Polloni, G., Aleotti, P., Baldelli, P., Nosetto, A. and Casavecchia, K., 1996, Heavy rain triggered landslides in the Alba area during November 1994 flooding event in the Piemonte Region (Italy), in Senneset (ed.), *Landslides – Glissements de Terrain* (Rotterdam: Balkema), vol. 3, 1955–1960.

Ragozin, A.L., 1996, Modern problems and quantitative methods of landslide risk assessment, in K. Senneset (ed.), *Landslides – Glissements de Terrain* (Rotterdam: A.A. Balkema), vol. 1, 339–344.

Reichenbach, P., Cardinali, M., De Vita, P. and Guzzetti, F., 1998, Regional hydrological thresholds for landslides and floods in the Tiber River Basin (central Italy), *Environmental Geology*, **35**, 146–159.

Rice, R.M., Corbett, E.S. and Bailey, R.G., 1969, Soil slips related to vegetation, topography, and soil in Southern California, *Water Resources Research*, **7**, 647–659.

Schindler, C., Cuenod, Y., Eisenlohr, T. and Joris, C.L., 1993, The events of randa, April 18th and May 9th, 1991 – an uncommon type of rockfall, *Eclogae Geologicae Helvetiae*, **86**, 643–665.

Smith, K., 2001, *Environmental Hazards: Assessing Risk and Reducing Disaster* (London: Routledge).

Smith, R.P., Jackson, S.M. and Hackett, W.R., 1996, Paleoseismology and seismic hazards evaluations in extensional volcanic terrains, *Journal of Geophysical Research – Solid Earth*, **101**, 6277–6292.

Soeters, R. and van Westen, C.J., 1996, Slope instability recognition, analysis, and zonation, in A.K. Turner and R.L. Schuster (eds), *Landslides: Investigation and Mitigation* (Washington, DC: National Academy Press), Special Report 247, 129–177.

Soldati, M., 1999, Landslide hazard investigation in the Dolomites (Italy): the case study of Cortina d'Ampezzo, in R. Casale and C. Margottini (eds), *Floods and Landslides: Integrated Risk Assessment* (Berlin: Springer-Verlag), 281–294.

Tephra, 1994. National Report of New Zealand. Ministry of Civil Defence, **13**(1), 7–29.

Tianchi, L.C., 1989. Landslides: Extend and economic significance in China, in E.E. Brabb and B.L. Harrod (eds), *Landslides: Extent and Economic Significance*. (Rotterdam: A.A. Balkema) pp. 271–288.

Tognacca, C., Bezzola, G.R. and Minor, H.-E., 2000, Threshold criterion for debris-flow initiation due to channel bed failure, in G.F. Wieczorek and N.D. Naeser (eds), *Debris-flow Hazards Mitigation: Mechanics, Prediction, and Assessment*, 16–18 August 2000, Taipei, Taiwan (Rotterdam: A.A. Balkema), 89–98.

Toll, D.G., 2001, Rainfall-induced landslides in Singapore. *Proceedings of the Institution of Civil Engineers – Geotechnical Engineering*, **149**, 211–216.

Turner, A.K. and Schuster, R.L. (eds), 1996, *Landslides: Investigation and Mitigation*, Transportation Research Board (Washington, DC: National Academy Press), Special Report 247, 675.

van Asch, T.W.J., Buma, J. and Van Beek, L.P.H., 1999, A view on some hydrological triggering systems in landslides, *Geomorphology*, **30**, 25–32.

Weichselgartner, J., 2001, Disaster mitigation: the concept of vulnerability revisited, *Disaster Prevention and Management*, **10**, 85–94.

Wieczorek, G. and Glade, T., 2005, Climatic factors influencing occurrence of debris flows, in M. Jakob and O. Hungr (eds), *Debris-Flow Hazards and Related Phenomena* (Heidelberg: Springer).

Wieczorek, G.F. and Guzzetti, F., 2000, A review of rainfall thresholds for triggering landslides, *Mediterranean Storms – Proceedings of the EGS Plinius Conference*, Maratea, Italy, October 1999, 407–414.

Wilches-Chaux, G., 1992, The global vulnerability, in Y. Aysan and I. Davis (eds), *Disasters and the Small Dwelling*, 30–35.

Wilson, R.C. and Keefer, D.K., 1985, Predicting areal limits of earthquake-induced landsliding, *US Geological Survey*.

Wilson, R.C. and Wieczorek, G.F., 1995, Rainfall thresholds for the initiation of debris flows at La Honda, California, *Environmental and Engineering Geoscience*, **1**, 11–27.

Wilson, R.C., Mark, R.K. and Barbato, G., 1993, Operation of a real-time warning system for debris flows in the San Francisco Bay Area, California Hydraulic Engineering, ASCE (San Francisco, CA: Hydraulics Division, ASCE), 1908–1913.

Zêzere, J.L., 2000, Rainfall triggering of landslides in the area north of Lisbon (Portugal), in E. Bromhead, N. Dixon and M.-L. Ibsen (eds), *Landslides in Research, Theory and Practice, Proceedings of the 8th International Symposium on Landslides* (Cardiff: Thomas Telford), 1629–1634.

Zhou, C.H., Yue, Z.Q., Lee, C.F., Zhu, B.Q. and Wang, Z.H., 2001, Satellite image analysis of a huge landslide at Yi Gong, Tibet, China, *Quarterly Journal of Engineering Geology and Hydrogeology*, **34**, 325–332.

Zimmermann, M. and Haeberli, W., 1992, Climatic change and debris-flow activity in high-mountain areas – a case study in the Swiss Alps, *Catena Supplement*, **22**, 59–72.

Zimmermann, M., Mani, P. and Romang, H., 1997, Magnitude–frequency aspects of alpine debris flows, *Eclogae Geologicae Helvetiae*, **90**, 415–420.

# PART 1
## CONCEPTUAL MODELS IN APPROACHING LANDSLIDE RISK

# 2

# The Nature of Landslide Hazard Impact

*Thomas Glade and Michael J. Crozier*

## 2.1  Introduction

Landsliding is one of the many natural processes that shape the surface of the earth. It is only when landslides threaten mankind that they represent a hazard. Landslides belong to a much broader group of slope processes referred to as mass movement. The definition of mass movement includes all those processes that involve the outward or downward movement of slope-forming material under the influence of gravity. Some mass movement processes, such as soil creep, are almost imperceptibly slow and diffuse while others, such as landslides, are capable of moving at high velocity, are discrete, and have clearly identifiable boundaries, often in the form of shear surfaces (Crozier, 1999a). Landslides are a manifestation of slope instability. This chapter discusses the stability of slopes, the factors that promote instability and the adverse effects that landslides can have on human well-being, land and livelihood. In particular, it identifies those aspects of landslides that make them hazardous and analyses the vulnerability of elements at risk in the face of landslide activity.

## 2.2  Slope Stability Considerations

Because of the destructive potential of landslides, scientists and engineers have long tried to identify the conditions of a slope that give rise to landsliding and in particular to determine how readily the slope may fail, that is, the 'stability' of the slope. Thus 'slope stability' and its corollary 'slope instability' are defined as the propensity for a slope to undergo morphologically and structurally disruptive landslide processes.

*Landslide Hazard and Risk*   Edited by Thomas Glade, Malcolm Anderson and Michael J. Crozier
© 2004 John Wiley & Sons, Ltd   ISBN: 0-471-48663-9

Slow, distributed forms of mass movement such as soil creep are generally considered not sufficiently disruptive to be included in this definition. From a hazard and engineering perspective, assessments of slope stability are generally intended to apply to periods ranging from days to decades, or in some cases to specified periods relating to the design-life of a potentially affected structure. However, slope stability may also be treated as a factor in landform evolution and therefore its significance in this role has to be measured over much longer time scales (Cendrero and Dramis, 1996; Schmidt and Preston, 1999).

In every slope, there are stresses that tend to promote downslope movement of material (shear stress) and opposing stresses that tend to resist movement (shear strength). In order to assess the degree of stability, these stresses can be calculated for a known or assumed failure surface within the slope and compared to provide a factor of safety (defined as the ratio of shear strength to shear stress). In a static slope, shear strength exceeds shear stress and the factor of safety is greater than 1.0, whereas, for slopes on the point of movement, shear strength is just balanced by shear stress and the factor of safety is assumed to be 1.0 (Selby, 1993).

While engineering codes of practice may specify a particular factor of safety to be achieved for completed earthworks, there are limitations to this measure of stability. Consider for example two slopes (A and B) that have the same factor of safety but that have large absolute differences in excess strength (i.e. strength minus shear stress). Let us assume that the strength to stress ratio, in unspecified stress units for a slope (A) is 400/200 and for a slope (B) is 200/100; thus both slopes yield a factor of safety of 2.0. However, slope (A) has an excess strength of 200 units while slope (B) has an excess strength of only 100 units. As excess strength is the quantity that must be entirely reduced (by reduction in strength or increase in shear stress) in order to produce failure, it represents the inherent stability of the slope or, in other words, the 'margin of stability' against failure. Thus spatial differences in inherent stability are better represented by excess strength than by the factor of safety. Instability in its broadest sense, however, is determined not only by the margin of stability of the existing slope but also by the magnitude and frequency of (external) destabilizing forces acting on the slope that are capable of reducing that margin and initiating landslides. Defined in this way, slope stability/instability is akin to the concept of 'susceptibility' (see Chapter 1).

Slopes can therefore be viewed as existing at various points along a stability spectrum ranging from high margins of stability with low probabilities of failure at one end, to actively failing slopes, with no margin of stability, at the other (Figure 2.1). It useful to define three theoretical stability states along this spectrum, based on the ability of dynamic external forces to produce failure (Crozier, 1989). First is the 'stable state', defined as slopes with a margin of stability which is sufficiently high to withstand the action of all natural dynamic destabilizing forces likely to be imposed under the current environmental/geomorphic regime. Second is the 'marginally stable state', represented by static slopes, not currently undergoing failure, but susceptible to failure at any time that dynamic external forces exceed a certain threshold. Third is the 'actively unstable state', represented by slopes with a margin of stability close to zero and which undergo continuous or intermittent movement (Figure 2.2).

The margin of stability is thus a measure of slope sensitivity to destabilizing factors and, together with an assessment of the potential effect of destabilizing factors affecting

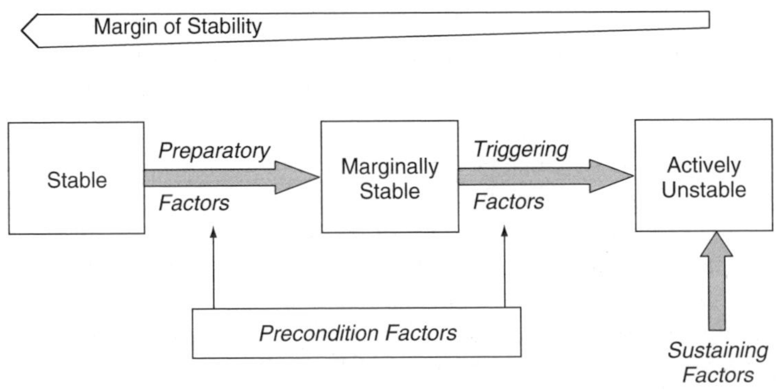


**Figure 2.1**   *Stability states and destabilizing factors (based on Crozier, 1989)*

**Figure 2.2**   *Actively unstable slopes, subject to deep-seated earthflows, Poverty Bay, New Zealand (photo by Ministry of Works, New Zealand)*

the slope, provides a measure of susceptibility/instability. In turn, an understanding and quantification of the relationship between the margin of stability and the frequency and magnitude of dynamic destabilizing factors provides one way of determining the probability of landslide occurrence. Ultimately, the probability of failure together with its magnitude provides a measure of landslide hazard.

Factors promoting slope instability are important to consider. The concept of three stability states offers a useful framework for understanding the causes and development of instability. In this context four groups of factors promoting instability ('destabilizing factors') can be identified on the basis of function (Figure 2.1).

**Precondition (predisposing) factors** are static, inherent factors which not only influence the margin of stability but more importantly in this context act as catalysts to allow other dynamic destabilizing factors to operate more effectively. For example, slope materials that lose strength more readily than others in the presence of water predispose the slope to failure during a rainstorm; or a particular orientation of rock structure may enhance the destabilizing effects of undercutting.

**Preparatory factors** are dynamic factors that by definition reduce the margin of stability in a slope over time without actually initiating movement. Hence, facilitated by preconditions, they are responsible for shifting a slope from a 'stable' to a 'marginally stable' state. Some factors, such as reduction in strength by weathering (Chandler, 1972), climate change (Dehn *et al.*, 2000), and tectonic uplift (Shroder and Bishop, 1998), operate over long periods of geomorphic time whereas others may be effective in shorter time periods, for example slope oversteepening by erosional activity (Preston, 2000), deforestation (Schmidt *et al.*, 2001), or slope disturbance by human activity (Rybár, 1997).

**Triggering factors** are those factors that initiate movement, that is, shift the slope from a 'marginally stable' to an 'actively unstable' state. The most common triggering factors are intense rainstorms, prolonged periods of wet weather or rapid snowmelt, seismic shaking and slope undercutting. Thus, if a slope is in a state of marginal stability it is possible to recognize a threshold value for the triggering factor that is responsible for initiating movement. The common triggering factors are usually external forces imposed on the slope and the initiating thresholds are thus referred to as extrinsic thresholds (Schumm, 1979). In certain instances, however, movement may be initiated in the absence of an identifiable external triggering force, and therefore it is assumed that some intrinsic threshold has been surpassed within the slope. For example, the Mount Cook rock avalanche from New Zealand's highest mountain in 1991 appears to have been triggered in this way (McSaveney, 2002). For this event it is suggested that gradual weakening of the rock mass, perhaps by mechanical weathering or dilation from unloading by continual erosion, lowered the rock mass strength below the prevailing gravitationally induced stress, allowing failure to occur.

In most cases, however, an extrinsic triggering threshold for landslide occurrence is identifiable and presents two useful opportunities for hazard estimation. The first opportunity recognizes that the triggering threshold varies with the inherent stability of the terrain. Thus spatial differences in the value of triggering thresholds can provide a relative measure of the geographical distribution of terrain susceptibility to landslide occurrence (Crozier, 1989; Glade, 1998). The second opportunity is that, having identified the triggering threshold for a given terrain, the triggering value may be used to determine the frequency of occurrence of landslide-generating conditions by reference, for example, to the seismic or climatic record for the region (Glade *et al.*, 2000; Brooks *et al.*, 2004). The advantage of this approach over determination of frequency from the historical inventory of landsliding is that climate records are usually much longer and more reliable than historical landslide records (Barnikel *et al.*, 2003). In addition, these thresholds can be used for warning systems and forecasting of landslide activity (Crozier, 1999b).

While triggering threshold analysis has many advantages over other approaches for determining probability of occurrence for hazard estimation, there are two components of the analysis which need particular attention. First, it is essential that the threshold analysis is not based solely on values of the initiating agent that occur during landslide initiation. These may be in excess of the minimum triggering value and the computed frequencies would thus underestimate the true frequencies. Second, it is clear that in some situations the triggering threshold for a given terrain is not a constant but varies temporally as a result of landslide occurrence. As susceptible material is successively removed from hillslopes there is a residual strengthening of the terrain and the triggering threshold rises. This phenomenon is referred to as 'event resistance' (Crozier and Preston, 1999). A similar phenomenon can be observed with debris-flow occurrence. The activation of debris flows depends not only on the magnitude of the triggering event but also on the availability of transportable material. For example, if all source material is removed by a rainfall-triggered debris flow, further rainstorms of the same magnitude are unlikely to generate flows. Triggering thresholds can thus also be seen as a function of the time required to establish a critical volume of rock debris in the source area (Glade, 2004). The implication of sediment availability/removal and event resistance for hazard estimation is that historically derived magnitude–frequency relationships may not always be a reliable measure of future activity.

**Sustaining factors** are those that dictate the behaviour of 'actively unstable' slopes, for example duration, rate and form of movement. While some of these may be dynamic external factors such as rainfall, others may relate to the progressive state of landslide movement or the terrain encountered in the landslide path.

## 2.3  Landslide Types

The range of landslide types identified by most classifications also provides an approximation of the range of potential impacts. Although the impact of a given landslide type is not always predictable, the class of landslide does present an indication of the type of movement and its destructive potential. Within the field of landslide research, many different landslide classifications can be found. The most commonly used landslide classifications are based on material type (e.g. rock, debris, earth), mechanisms of movement (e.g. fall, topple, slide, flow, creep) and degree of disruption of the displaced mass. Landslide classifications are discussed by Hutchinson (1988), Crozier (1989), Cruden and Varnes (1996), and Dikau *et al.* (1996). Landslide types are classified as shown in Table 2.1.

In practice, it is difficult to assign a landslide to a particular class. Commonly, landslides are complex processes, for example with rotational shear planes in the upper part and

**Table 2.1**  *Landslide classification based on Dikau* et al. *(1996)*

| Process | Material | | |
|---|---|---|---|
| | Rock | Debris | Earth |
| Fall | Rockfall | Debris fall | Earthfall |
| Topple | Rock topple | Debris topple | Earth topple |
| Rotational slide | Single (slump) | Single | Single |
| | Multiple | Multiple | Multiple |
| | Successive | Successive | Successive |
| Translational slide | Block slide | Block slide | Slab slide |
| Planar | Rockslide | Debris slide | Mudslide |
| Lateral spreading | Rock spreading | Debris spread | Earth spreading |
| Flow | Rockflow (Sackung) | Debris flow | Earthflow |
| Complex | e.g. Rock avalanche, Bergsturz | e.g. flow slide | e.g. slump – earthflow |

**Figure 2.3**   *Earth slides on the slopes converted to mudflows in the valley during a rainstorm in 1977 in Wairarapa, New Zealand (photo by M.J. Crozier)*

flow structures in the lower reach. It is even more complex when several types of slope material are present in the one landslide. Also, external factors determine landslide types. While a given slope segment might fail as a translational debris slide under moderate moisture conditions, the same slide might convert to a debris avalanche or debris flow under wet conditions, thus increasing runout (Figure 2.3). Similarly, an earthslide may change to a mudflow as a result of slope morphological and hydrological conditions. In addition, vegetation cover can also influence the type of movement.

## 2.4   Impact

The juxtaposition of landslides and human presence exacts a cost. That cost can arise from the damage resulting from landslide impact or from the expense required to sustain measures to mitigate the impact. In a sense there is no escaping the cost; it can be transferred and transformed but, nevertheless, one way or another there still remains a price for living within a hazardous environment.

If landslide hazard is defined as the probability of occurrence of a potentially damaging landslide, the following questions become fundamental:

- What constitutes a damaging landslide?
- Which attributes of the landslide are capable of producing what kind of damage?
- What is the recurrence frequency for landslides either on specific sites, or somewhere in the region?

The following sections set out to address these questions.

Landslide impact is discussed in terms of the physical impact *mechanism* exhibited by the landslide (the destructive behaviour of material as it moves downslope) and the *type* of impact. The type of impact refers to how these slope movements can create damage in time and space. Landslide impacts can be direct or indirect, immediate or delayed, and in some instances generate consequential hazards. Not all landslides are equally hazardous.

### 2.4.1   Impact Mechanisms

Landslides directly affect physical elements at risk by a range of impact mechanisms, including: burial, collision impact, earth pressures, differential shearing in tension, compression or torque, plastic deformation (flow), by object displacement and by removal or deformation of valued ground, such as productive soil and foundation substrate. The degree to which these mechanisms are manifest is generally reflected by the type of landslide. However, many landslides exhibit complex behaviour and a variety of impact mechanisms may be represented in the one landslide type. For example, an earth slide may change to a mudflow as a result of slope morphological and hydrological conditions. This increases difficulties of assigning structural damage to specific landslide types. Despite this problem, a classification scheme has been suggested by Flageolett (1999) in Figure 2.4.

### 2.4.2   Physical Impact Type

The elements subject to these impact mechanisms show several types of physical impact. The impacts may be direct or indirect, acute (immediate) or chronic (delayed), or may lead to the development of consequential hazards. *Direct impacts* are those consequences incurred by direct physical contact with the landslide itself. *Indirect impacts*, on the other hand, are changes brought about in the properties and behaviour of other natural systems as a result of landslide activity. Some of these induced changes may give rise to consequential hazards, for example a wave generated by a landslide entering a reservoir. Indirect impacts can be immediate or delayed, occur in the proximity of the landslide or at some distance from the landslide site. *Acute impacts* are short-lived, while *chronic impacts* may be manifest over a longer period of time.

### 2.4.3   Direct Impacts

Direct impacts arising from landslide activity upslope of a site can affect structures by: collapse or damage by crushing from burial, collision impact, associated air blast, or distortion by gradual earth pressure (Casale and Margottini, 1999). The impact on humans and animals from these mechanisms may include loss of life or injury by trauma from collision impact, crushing or asphyxiation, whether directly affected by the landslide or indirectly through structural collapse. Vegetation, including large trees, may be root-wrenched, uprooted or buried. Landslide deposits can also extensively inundate productive agricultural land, at least temporarily reducing productivity (Figure 2.5).

Landslides occurring underneath or downslope of structures cause removal of basal support, leading to collapse, deformation and displacement (Figure 2.6). If a structure intersects a shear or tensional rupture zone, damage can result in simple relative displacement (e.g. rupture of a pipeline) or distortion and collapse (Figure 2.7). Where landslide

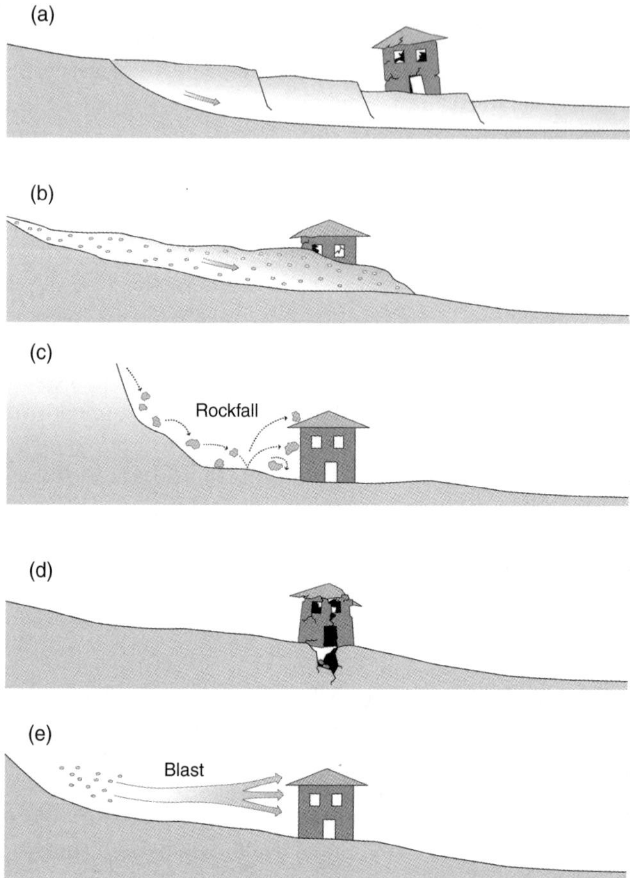


**Figure 2.4** *Schematic representation of structural damage to buildings for different landslide types (according to Flageolett, 1999). (a) Damage is assigned to slide and flow processes, (b) to flows, (c) to falls and topples, (d) to subsidence and (e) to rock avalanches or large rock failures, such as a Bergsturz*

displacement occurs by relatively undeformed blocks, the physical impact to structures may result in their translocation rather than destruction. A schematic diagram of a compound landslide showing typical destructive components such as crown scarp, tensional zones, lateral shears and compressive zones is shown in Figure 2.8. Acute impacts that occur instantaneously or take place over a short period of time are usually much more life-threatening than chronic impacts, which nevertheless can create expensive ongoing problems (Figure 2.9).

### 2.4.4 Indirect Impact

Indirect impacts may involve the interaction of landslides with other systems or processes, for example fluvial systems, artificial or natural lakes, and they may be responsible for

**Figure 2.5**   *Removal of soil from the slope and burial of soil on valley floor in the rainstorm of 6 August 2002, Gisborne, New Zealand (photo by M.J. Crozier)*

tsunami, coastal erosion, soil depletion and increased storm runoff. These impacts are described in more detail below.

Many of the most serious indirect impacts arise from the coupling of landslides with the fluvial system. The way in which landslides interact with the fluvial system can have important implications for the resultant level of hazard. Korup (2003), in his study of the unpopulated southwest Southern Alps of New Zealand, attributes potential impact to the orientation of the landslide track with respect to the fluvial receiving system (Table 2.2). The range of on-site impacts resulting from landslide/fluvial coupling is given in Table 2.3.

There is a range of long-term, long-range effects associated with the coupling modes and direct impacts described in Tables 2.2 and 2.3, which include consequential hazards such as channel avulsions at the landslide site, or upstream and downstream of the landslide body as well as aggradation and subsequent potential for landslide dam-burst events. Costa and Schuster (1988) observe that 85% of landslide dams fail within one year of emplacement. Dam failures may take place as catastrophic events causing widespread damage and destruction downstream (Korup, 2002; Evan & DeGraff 2002). Some landslide dams, however, may last for thousands of years and affect the fluvial system by entrapment of bedload and downstream starvation of sediment (Figures 2.10, 2.11 and 2.12) (Riley and Read, 1992).

**Figure 2.6**  *House destruction in the graben of Abbotsford landslide, 8 August 1979, Dunedin, New Zealand (photo by Allied Press Ltd)*

**Figure 2.7a**  *Left lateral shear surface of mudslide, Otago Peninsula, South Island, New Zealand (photo by M.J. Crozier)*

***Figure 2.7b*** *Right lateral shear zone of mudslide, Biferno River Valley, South Italy (photo by M.C. Salvatore)*

**Figure 2.7c**  *Tree split by left lateral shear zone, Gisborne, New Zealand (photo by M.J. Crozier)*

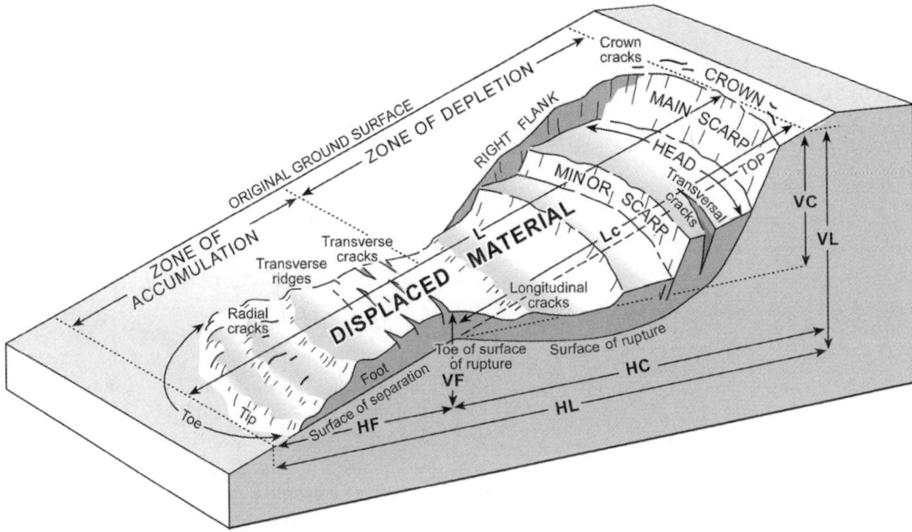



**Figure 2.8** *A compound landslide showing typical destructive components such as crown scarp, tensional zones, lateral shears, compressive zones. Note: H is the horizontal distance and V is the vertical distance for various parts of the landslide indicated (based on Varnes, 1978)*

**Figure 2.9** *Instantaneous failure, resulting from rainstorm of December 1976, Central Terrace Wellington, New Zealand (photo by M.J. Crozier)*

**Table 2.2**    *The coupling mode: the geometric relationship between landslides and the fluvial system (The ALPIN classification), after Korup (2003)*

| Geomorphic coupling interface class | Subclass | Diagnostic characteristics |
|---|---|---|
| (A) Area | | Very large landslide bodies in excess of $100\,km^2$ that obliterate low-order drainage divides and reorientate drainage systems |
| (L) Linear | | More than 50% of the runout direction of landslides is oriented in the direction of the fluvial drainage system |
| (P) Point | | Emplacement of the landslide deposit is normal or near normal to the planform direction of the channel. This mode of coupling favours the development of landslide dams |
| (I) Impounded | | Landslides terminate in a standing water body, e.g. fjords, moraine/landslide or floodplain lakes |
| (N) Nil | | Landslide are in a state of geomorphic decoupling, with no physical contact between the toe of the deposit and the channel system |
| | Nh | Landslide deposits are stored in morphological or structural depressions (colluvial storage) |
| | Nv | Landslide deposits are stored on valley fill, e.g. floodplain alluvium, terraces, fans, moraines or older landslide deposits |
| | Ni | Landslides buffered on ice or snow |

**Table 2.3**    *On-site impact on fluvial systems resulting from the coupling mode, after Korup (2003)*

| Geomorphic impact class | Diagnostic landform |
|---|---|
| Buffered | Landslide makes no physical, direct contact with the fluvial system |
| Riparian | Direct contact with the channel, where fluvial erosion dominates, controlling landslide initiation and removal |
| Occlusion | Landslide diverts river channel around toe of landslide, with up- and downstream influence |
| Blockage | Occurrence of a landslide dammed lake |
| Obliteration | Complete burial of extensive valley-floor section with drainage reversals, landslide ponds and dams |

Impounded coupling (Table 2.2) may produce some of the most intractable problems for hazard estimation. Landslides on the margins of reservoirs, depending on the velocity of emplacement and volume of material involved, have the potential to create large waves that can overtop or destroy dams and create serious catastrophic inundation downstream. A tragic example of this type of consequential hazard emptied the artificial lake in Vaiont in 1963, causing the deaths of over 2500 people in the Italian town of Longarone and surrounding villages (Petley, 1996; Voight and Faust, 1992). A description of the

**Figure 2.10**  *Occlusion impact, Shotover River, Otago, New Zealand (photo by Allied Press Ltd)*

administrative response to the Vaiont disaster is given in Chapter 9 by Hollenstein. Similar catastrophes have occurred in Peru, when snow and rock avalanches have entered moraine-dammed lakes, causing overtopping or dam breach. The major objective for hazard estimation in these cases is the determination of likely landslide volumes and velocities (Gillon and Hancox, 1992). If first-time failures are being assessed, the initial problem is the estimation of landslide volume. In the case where existing landslides occur on reservoir margins, it may be possible to estimate volume by locating boundary shear surfaces, but the question of velocity is much less readily resolved. It is generally assumed that in an existing landslide, brittle failure, often associated with rapid movement, will have already taken place and that subsequent movement will mobilize residual strength and result in more gradual displacement. However, instances have been recorded where the reactivation of existing landslides has resulted in high-velocity surges of movement (Prior and Stephens, 1972).

Subaerial coastal landslides and submarine landslides, in some cases of huge dimensions, are capable of generating high-magnitude tsunami (Dawson, 1999; Driscoll *et al.*, 2000; Hampton *et al.*, 1996). For example, different ages of Holocene mass movements are known from Norway fjords (Boe *et al.*, 2003). One of the best known is the Storegga submarine landslide, which occurred between 7300 and 6400 C$^{14}$ yr BP

***Figure 2.11***   *Blockage impact: lakes formed by earthquake-triggered landslides circa 1300 years BP, Waverley, New Zealand (photo by Lands and Survey, New Zealand)*

(Grauert *et al.*, 2001) and from the resulting tsunami caused considerable impact along the Norwegian coast (Bondevik *et al.*, 1997), and also in the Faeroe Islands (Grauert *et al.*, 2001) and Scotland (Dawson and Smith, 2000). Similarly, the Sissano Papua New Guinea tsunami disaster of 1998 is thought to have been caused by a seismically triggered submarine landslide (Tappin *et al.*, 2001).

More subtle, but none the less important, are the impacts of landslides which remove or destroy the pedological soil, particularly in areas relying on primary production from those soils. A number of studies in New Zealand (e.g. Crozier *et al.*, 1980; Page *et al.*, 1994) have shown that, in one event, multiple landslides can remove soil from up to 10% of areas involving hundreds of square kilometres (Figure 2.13). The cumulative effect of a series of these events in New Zealand hill country (40% of NZ land area) has seen soil depleted from 20–50% of the area in the hundred or so years since forest clearance. Each landslide usually removes the entire soil mantle from the underlying bedrock. Although these sites regain a soil cover with time, 20-year-old landslides have been shown to yield only 70–80% of the productivity on undisturbed slopes and even after 80 years productivity is still only 80% (Lambert *et al.*, 1984). The limiting factor to growth appears to be not so much nutrient availability as soil moisture availability in the thin recovering soil.

A further indirect impact, resulting from the removal of soil by shallow landslides and the consequent reduction of slope water storage capacity, is increased storm runoff (e.g. Dietrich *et al.*, 1993). This effect, combined with the reduction in channel capacity from landslide-derived sediment, increases the frequency and magnitude of overbank flooding (Figure 2.14).

**Figure 2.12**   *Linear coupling: deposits from the Vancouver Ridge landslide totalled 170 million tonnes and travelled 3.5 km downstream, August 1989, Ok Tedi, Papua New Guinea (photo by M.J. Crozier)*

***Figure 2.13***   *Shallow earthflows, Kiwi Valley, Wairoa, New Zealand, triggered by a rainstorm of 965 mm in 72 hours in 1977 (photo by Hawke's Bay Catchment Board)*

Slow-moving deep-seated mudslides and earthflows may have the opposite effect on slope hydrology compared to the impact of rapid shallow soil failures. The progressive surface deformation within and upslope of the displaced mass tends to disrupt and obliterate existing drainage lines and channels and impound water on the slope (Figure 2.15). This gives rise to surface ponding and saturated hollows, leading to a die-off of the usual slope vegetation and the ultimate replacement with more water-tolerant species.

### 2.4.5   Impact Characteristics (Intensity) of Landslides

The mechanism and severity of impact depends on the type of landslide, its impact characteristics, and the location of elements at risk with respect to the particular morphological components of the landslide. In their review of 23 case histories of catastrophic landslides in South America, Schuster *et al.* (2002) observed that most casualties were caused by high-velocity debris avalanches and high- to medium-velocity, highly mobile, long-runout debris flows. The impact potential or power of a landslide is primarily a function of its mass and velocity. At the most dangerous end of the power spectrum are rock avalanches that can attain volumes of tens of millions of cubic metres and travel at velocities up to 60–80 m/s (McSaveney, 2002).

**Figure 2.14** *Channel instability induced by reduction of storage capacity on slopes by regolith landslides, January 1990, Waitotora, New Zealand (photo by M.J. Crozier)*

The range of landslide velocities is shown in Table 2.4. Although a given landslide type may carry out most of its movement at a characteristic velocity, it may be also be capable of moving at a wide range of velocities. For example, a rock slide can creep at a rate of cm/year, but most of its displacement will be at rates of cm/s to m/s.

The appropriate management response to a landslide hazard depends on the expected velocity of movement. For example, a large rotational landslide creeping at rates of mm/year can still be used for settlements or infrastructure lines. Appropriate counter-measures such as flexible sewage lines, moveable basements for railway tracks, or strong house foundations might allow intensive usage (refer to Chapter 19 for such geotechnical applications). The decision on whether a usage is still economically viable is mostly based on cost/benefit analysis. In contrast, if the similar block moves with a speed of cm/day, safe, economic use of the site may not be possible. Figure 2.16 gives schematic examples of the role of velocity of movement to the consequences.

In general, major factors controlling the speed of movement are the mass in motion, the horizontal and vertical travel distances and thus the slope angle, the moisture of the transported material and, for lower-magnitude events, the vegetation cover. These factors also influence the runout distances (Hungr, 1995). Physical models are regularly used to calculate runout distances for different landslide types (e.g. Miao *et al.*, 2001). If the landslide size/magnitude increases, movement patterns become too complex to be accurately modelled (Hutter *et al.*, 1996). Another approach involves empirical models, relating for example landslide dimensions or/and topographic conditions to volume in order to

***Figure 2.15*** *Landslide dam four days after initiation, August 2002, Gisborne, New Zealand (photo by M.J. Crozier)*

approximate runout length (Rickenmann, 1999). Various authors have applied these models to assess potential runout zones (e.g. Corominas, 1996; Crozier, 1996; Fannin and Wise, 2001; McClung, 2001). If linked to frequency of landslides events, either established by historical information or by investigating the recurrence intervals of the trigger, the runout zones can be transferred to hazard zones (Glade, 2002). Extreme runout zones of some kilometres have been observed for the Parinacota debris avalanche, northern Chile by Clavero *et al.* (2002) and for Mt Cook, New Zealand by McSaveney (2002).

With slow-moving landslides, impact potential is also related to the amount of displacement per unit of time providing destructive earth pressure and differential shearing rather than collision impact. Next to volume and velocity, the degree of disruption of the displaced mass influences the type of impact and the degree of destruction of elements at risk. The depth of movement is also an important impact characteristic and dictates not only the type of impact but also the type of remedial measures than can be successfully applied.

## 2.5    Frequency–Magnitude Issues

As already indicated, the frequency and magnitude of landslides are of particular concern for any hazard and risk analysis. There are two approaches to assessing frequency and magnitude: first, temporal investigations that may include stability analysis of a site

**Table 2.4**  *Classification of velocity of movement according to Cruden and Varnes (1996) and Australian Geomechanics Society (2002)*

| Speed class | Description | Velocity (mm/s) | Typ. velocity | Probable destructive significance |
|---|---|---|---|---|
| 7 | Extremely fast | | | Disaster of major violence, buildings destroyed by impact of displaced material, many deaths, escape unlikely |
| | | $5 \times 10^3$ | 5 m/s | |
| 6 | Very fast | | | Some lives lost; velocity too great to permit all persons to escape |
| | | $5 \times 10^1$ | 3 m/min | |
| 5 | Fast | | | Escape evacuation possible; structures, possessions and equipment destroyed |
| | | $5 \times 10^{-1}$ | 1.8 m/hr | |
| 4 | Moderate | | | Some temporary and insensitive structures can be temporarily maintained |
| | | $5 \times 10^{-3}$ | 13 m/month | |
| 3 | Slow | | | Remedial construction can be undertaken during movement; insensitive structures can be maintained with frequent maintenance work if total movement is not large during a particular acceleration phase |
| | | $5 \times 10^{-5}$ | 1.6 m/year | |
| 2 | Very slow | | | Some permanent structures undamaged by movement |
| | | $5 \times 10^{-7}$ | 16 mm/year | |
| 1 | Extremely slow | | | Imperceptible without instruments, construction possible with precautions |

or analysis of external triggers or landslide occurrence (e.g. Glade, 1998; Hungr *et al.*, 1999), and second, the spatial analysis of frequency distributions of landslide size in a given area (e.g. Guzzetti *et al.*, 2002; Hovius *et al.*, 1997).

Temporal studies can either investigate the response of a single landslide to climatic inputs or relate the occurrence of landslides within a larger region to climatic conditions that characterize the region; in this case occurrence within a region, rather than the specific landslide location, is the parameter of interest. The simplest approach is to characterize the behaviour of the triggering agent at the time of landslide occurrence. For example, this may result in the establishment of a threshold rainstorm value, above which landslides can be expected to occur (Glade, 1998). A more advanced technique is to associate other temporal information with the triggering threshold. For example, the antecedent climate or slope hydrological conditions may also be taken into account (Glade *et al.*, 2000). By including information on physical characteristics of the soil, these models can be refined for specific environmental conditions (Glade, 2000). The established thresholds can then be used to calculate the probability of exceedence of this climatic threshold within different periods of time. Such an analysis has been undertaken by Crozier and

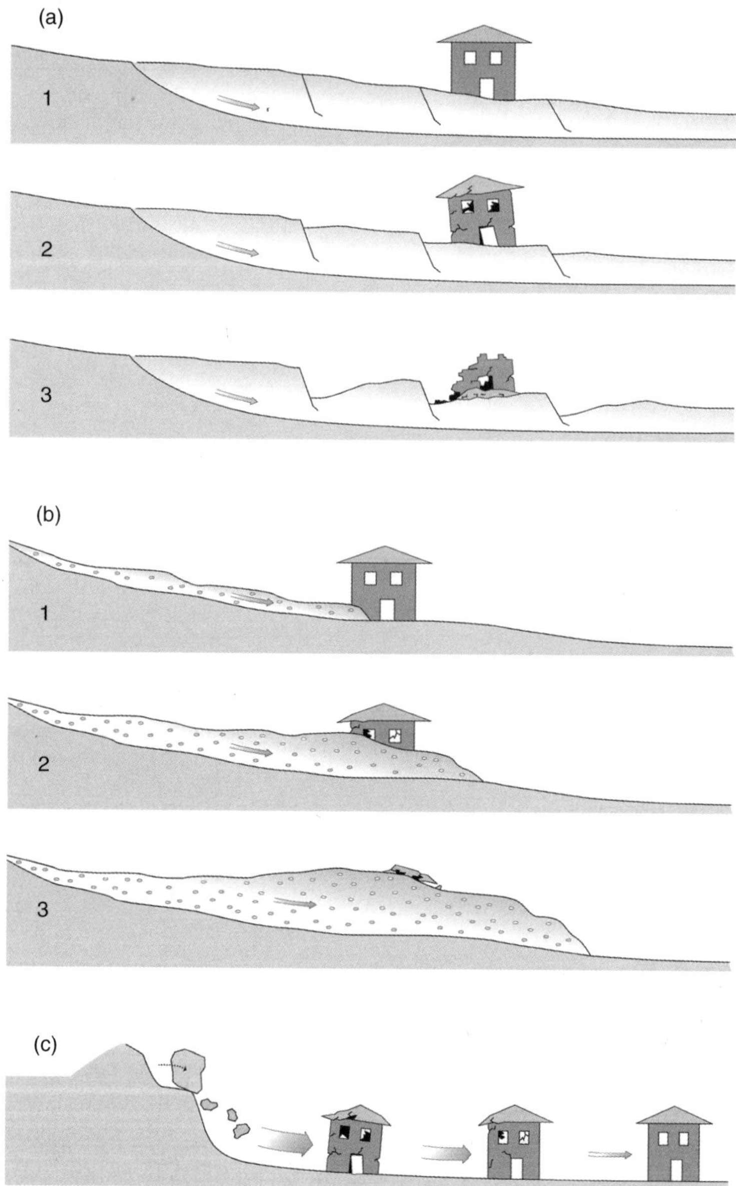

**Figure 2.16** *Schematic consequences of different velocities of movement for different land-slide types (adopted from Flageolett, 1999). (a1) A slide creeps or (a3) fails suddenly. (b1) A debris flow progresses in low or (b3) high velocities with respective changes in flow height. (c) A slow- or fast-moving rockfall damages, depending on the size and consequent momentum, elements at risk to a different degree. The degree depends on the distance between the process and the location of the element at risk*

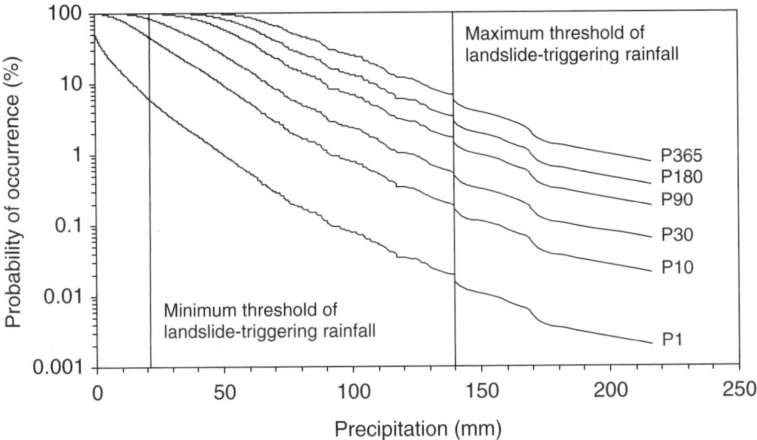



**Figure 2.17** *Probability of occurrence of daily precipitation equalling or exceeding given values in Wellington, New Zealand. (Note: different lines refer to a probability of occurrence of a specific rainfall magnitude at each single day (P1), within a period of 10 days (P10), within a month (P30), etc. The empirically established minimum and maximum thresholds of landslide-triggering rainfall (140 mm) are shown by the thin vertical lines. Method is described by Crozier and Glade, 1999.)*

Glade (1999) and one result is shown in Figure 2.17. By adding the threshold lines, respective values can be used for estimation of the frequency of a landslide-triggering rainstorm event for different time periods.

The determination of the frequency of occurrence based on triggering thresholds has so far been developed for regional scale analysis based on the history of landslide occurrence. Although the models have the capacity to be run for different landslide types, this has not been performed yet due to data limitations.

In addition to temporal analysis, spatial impacts of widespread landsliding following an intense triggering event have also been investigated in a number of localities. The sort of event that is suited to this type of analysis is shown in Figure 2.18. Spatial analysis uses frequency–area statistics of landslides. Results of analysis by different authors show that these distributions follow commonly a power-law relation with a negative exponent (e.g. Czirok *et al.*, 1997; Guzzetti *et al.*, 2002; Hovius *et al.*, 1997). Essentially these show that small landslides are common while large landslides are relatively rare. This relationship appears to remain constant irrespective of the size of the data set, over a population range from 100 to more than 10 000 landslides. Also, the power–law distributions seem to be independent on the type of trigger. For example, Guzzetti *et al.* (2002) found a comparable distribution for both earthquake- and snowmelt-triggered landslides (Figure 2.19). This is a particularly interesting result with the potential to be used in hazard and risk assessments in the future.

Whether spatial or temporal approaches are used, they both require a reliable database. It is clear that further use of these methods depends on a standardized system for collecting and archiving data on landslide occurrence. For temporal studies, particularly

**Figure 2.18**   *Widespread landsliding as a result of Cyclone Bola in March 1988, East Coast, North Island, New Zealand (photo by N. Trustrum)*

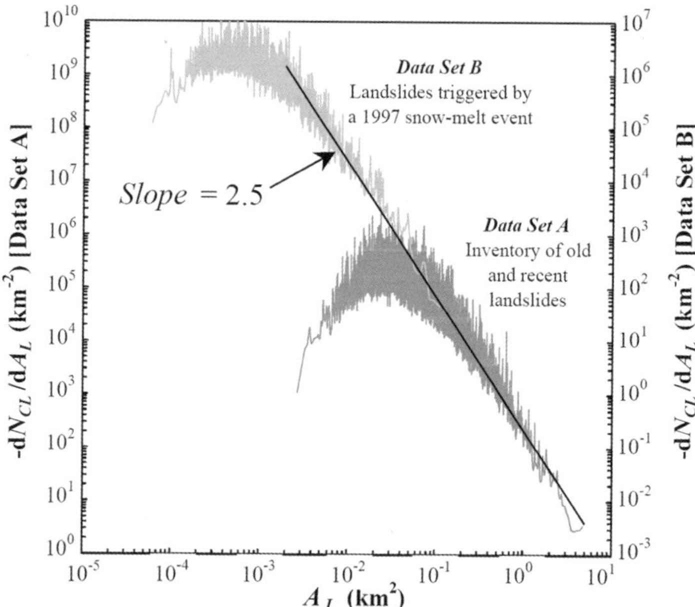


**Figure 2.19**   *Non-cumulative frequency-area distribution of central Italian landslides (Guzzetti et al., 2002). (Note: $A_L$ = landslide area; d = derivative; $N_{CL}$ = cumulative number of landslides with areas greater than $A_L$.)*

those requiring correlation with climatic conditions, it is essential to have accurate information on the date of occurrence. An approximation of period of occurrence may be obtained through the comparison of time-slice air photography. In some cases, however, no information on either the trigger or the time of landslide occurrence is available. In such cases, temporal information can only be given in relative terms. A reference of features indicating activity or relative age of landslides was proposed by Crozier (1984). Relative age assessments can be used to determine whether landslides belong to the same age cohort and therefore indicate triggering by a single event (Table 2.5).

More robust age information for longer time periods, for example the Holocene, can be obtained using absolute dating techniques (e.g. Lang *et al.*, 1999). The material dated is generally either sediment, in particular quartz grains (TL, OSL), or buried organic material ($C^{14}$). Both materials can be taken from the basins or ponds that have developed within the previously moved mass, from sediments in dammed lakes, or from fossil surfaces buried by the landslide. A number of studies on the temporal occurrence of landslides in the Holocene with worldwide examples is given by Matthews *et al.* (1997).

Databases on landslides may have a wide range of accuracy. As one might expect, the more recent the landslides, the greater the level of available detail. Also, in assessing older events, evidence of some of the smaller landslides may have become obliterated.

**Table 2.5**   *Features indicating activity or relative age of landslides (after Crozier, 1984)*

| Active/Recent | Inactive/old |
|---|---|
| Scarps, blocks and crevices with sharp edges (Fig. 2.20a) | Scarps, blocks and crevices with rounded edges (Fig. 2.20b) |
| Crevice and depressions without secondary depositional infilling | Crevice and depressions with secondary depositional infilling |
| Secondary mass movement on scarp | No secondary mass movement on scarp |
| Surface-of-rupture and marginal shear surfaces show fresh slikensides and striations | Surface-of-rupture and marginal shear surfaces show no or subdued slikensides and striations |
| Fresh fractured surfaces on blocks, little lichen cover | Weathering on fractured surfaces of blocks, established lichen cover |
| Disarranged or non-integrated drainage system; many ponds and undrained depressions | Integrated drainage system |
| Pressure ridges in contact with slide margin | Deflated lobes and abandoned levees |
| No soil development or airfall deposits on exposed failure surfaces | Soil development on exposed failure surfaces, mantle of airfall deposits |
| Presence of fast-growing, colonizing vegetation species on disrupted surfaces | Presence of slow-growing, climax vegetation species on disrupted surfaces |
| Distinct vegetation differences 'on' and 'off' slide | No distinction between vegetation 'on' and 'off' slide |
| Tilted trees with no new vertical growth | Tilted trees with subsequent vertical growth |
| No new supportive, secondary tissue on trunks | New supportive, secondary tissue on trunks |

**Figure 2.20**   (a) Landslide displaying features of relative youth – disrupted blocks with sharp distinctive form, dated at circa 1.3 thousand years BP. (b) Landslide displaying features of greater age – subdued, smooth surface with soil mantle of airfall deposits, dated at circa 31 thousand years BP

Therefore, in any frequency–magnitude investigation, it is most important to consider the limitations of the database, particularly with respect to the age of the landslides examined.

Although general frequency–magnitude assessments are at the core of landslide hazard estimation, magnitude itself (as measured by volume or area) may not be the most important parameter in producing impact. In many cases, other impact characteristics such as velocity and degree of disruption or runout distance may be more important. There have yet to be any comprehensive assessments of the frequency of occurrence of some of these higher-impact parameters.

## 2.6  Vulnerability with Respect to Landslide Types

Various approaches can be used to assess vulnerability to different landslide types (reviewed by Glade, 2004). These approaches vary significantly in the detail of analysis and the final vulnerability values. In contrast to Heinimann (1999b), most approaches do not distinguish between landslide types (e.g. Leone *et al.*, 1996; Ragozin and Tikhvinsky, 2000; Wong *et al.*, 1997) or landslide magnitudes (e.g. Leone *et al.*, 1996; Michael-Leiba *et al.*, 2000; Ragozin and Tikhvinsky, 2000; Wong *et al.*, 1997). Also vulnerability estimates for elements at risk vary. Although the vulnerability of buildings is assessed in terms of degree of loss (e.g. Leone *et al.*, 1996), absolute values of vulnerability differ significantly. Similarly, vulnerability of people is treated in a variety of ways. Some authors distinguish between different levels of injury and the 'final' loss of life (e.g. Ragozin and Tikhvinsky, 2000), while others just define the probability of loss of life (e.g. Michael-Leiba *et al.*, 2000; Wong *et al.*, 1997). In addition, the resultant absolute values for vulnerability are spread over a wide range and make consequent comparisons of approaches very difficult (Glade, 2004; refer also to Chapter 5 Alexander).

Various reasons might explain these large differences:

- Not all authors explicitly state in detail how the values of vulnerability for different landslide types were derived. No uniform methodology exists. It is suspected that most of the values have been assumed.
- Most studies are based on empirical data, for example Wong *et al.* (1997) used such an approach for Hong Kong.
- Local historical databases have been reviewed; for example Michael-Leiba *et al.* (1999) assessed the vulnerability of buildings and people by using the Australian Land-slide Database and of roads by information provided by the Cairns City Council. Derived results are thus heavily dependent on such databases containing socio-economic indicators of community vulnerability to natural hazards (e.g. King, 2001).
- Back-analysis of specific past events; for example Ragozin and Tikhvinsky (2000) examined past landslide and earthquake events and Heinimann (1999a, 1999b) investigated past events and derived estimates, but assumed missing values.

Indeed, uncertainty is inherent in all different vulnerability studies, but the margin of error remains unknown in detail. It can be concluded that – although Heinimann (1999a, 1999b) introduces a very detailed approach in determining risk to gravitational mass movements – a general strategy in determining vulnerability of elements at risk to specific landslide types and magnitudes is missing. This is a major drawback for any landslide

**Table 2.6**   *Vulnerability of a person being affected by a landslide in open space, in a vehicle and in a building (modified by Glade, 2004 after Wong et al., 1997)*

| Location | Description | Vulnerability of a person | | |
|---|---|---|---|---|
| | | Data range | Recommended value | Comment |
| Open space | Struck by rockfall | 0.1–0.7 | 0.5 | May be injured but death unlikely |
| | Buried by debris | 0.8–1 | 1 | Death by asphyxia |
| | Not buried, but hit by debris | 0.1–0.5 | 0.1 | High chance of survival |
| Vehicle | Vehicle is buried/crushed | 0.9–1 | 1 | Death almost certain |
| | Vehicle is damaged only | 0–0.3 | 0.3 | High chance of survival |
| Building | Building collapse | 0.9–1 | 1 | Death almost certain |
| | Building inundated with debris and person is buried | 0.8–1 | 1 | Death highly likely |
| | Building inundated with debris, but person is not buried | 0–0.5 | 0.2 | High chance of survival |
| | Debris strikes the building only | 0–0.1 | 0.05 | Virtually no danger |

risk analysis. Most values adopted in landslide risk analysis are based on experience of previous events and on common sense. One example of such a classification for different landslides and associated vulnerability of a person in different situations is given by Wong *et al.* (1997) (Table 2.6).

## 2.7   Conclusion

Landslides are natural events occurring worldwide and pose a threat to affected communities. Conditions promoting slope instability include predisposing factors, preparatory factors, triggering factors and sustaining factors. The importance of each factor varies from place to place and differs for each landslide type. Some of these causative factors are readily affected by human activity, some are controllable for mitigation purposes while others we must simply learn to live with.

The physical impact potential of landslides is a function of the mass of displaced material, depth, degree of disruption, and velocity. It is clear that no uniform impact condition can be unequivocally related to a specific landslide type. In response to external and internal factors, similar landslide types can behave differently; thus a careful assessment of movement patterns is essential. Landslide impacts are described in terms of their impact mechanisms, the physical impact type, and the immediacy of their effect over time and space. Frequency–magnitude issues of landsliding are discussed for both temporal and spatial analysis. Finally, the difficulty of establishing vulnerability to the landslide threat is discussed. The following conclusions on the nature of landslide hazard impact can be derived:

- No unique and simple method is currently available for the prediction of impact within landslide risk analysis.
- Impact estimates are heavily dependent on historical data for the region and the landslide type respectively, and therefore may not have direct relevance to the estimation of future risk.
- Even when information on past events is available, details of landslide impact to elements at risk with respect to specific type and magnitude of process are frequently missing.
- If none of the information sources is available, impacts to elements at risk have to be estimated based on examples from other regions, or even other processes (e.g. earthquakes, floods).


# References


Australian Geomechanics Society, 2002, Landslide risk management concepts and guidelines, in Australian Geomechanics Society Sub-committee on Landslide Risk Management (ed.), *Australian Geomechanics*, 51–70.

Barnikel,publication details F., Frank, C. and Becht, M., 2003, Natural hazard maps in the Alps derived from historical data on a local scale – results from the Tegernsee Valley (Bavaria, Germany), in V.R. Thorndycraft, G. Benito, M.C. Llasat and M. Barriendos (eds), *Paleofloods, Historical Data & Climatic Variability: Applications in Flood Risk Assessment, Proceedings of the PHEFRA Workshop*, 16–19 october 2002, Barcelona, 155–162.

Boe, R., Rise, L., Blikra, L.H., Longva, O. and Eide, A., 2003, Holocene mass-movement processes in Trondheimsfjorden, Central Norway, *Norwegian Journal of Geology*, **83**, 3–22.

Bondevik, S., Svendsen, J.I., Johnsen, G., Mangerud, J. and Kaland, P.E., 1997, The Storegga tsunami along the Norwegian coast, its age and runup, *Boreas*, **26**, 29–53.

Brooks, S.M., Crozier, M.J., Glade, T. and Anderson, M.G., 2004, Towards establishing climatic thresholds for slope instability: use of a physically-based combined soild hydrology-slope stability model, *Pure and Applied Geophysics*, **161**.

Casale, R. and Margottini, C. (eds), 1999, Floods and landslides, *Environmental Science* (Berlin and Heidelberg: Springer).

Cendrero, A. and Dramis, F., 1996, The contribution of landslides to landscape evolution in Europe, *Geomorphology*, **15**, 191–211.

Chandler, R.J., 1972, Lias clay: weathering processes and their effect on shear strength, *Geotechnique*, **22**, 403–431.

Clavero, J.E., Sparks, R.S.J. and Huppert, H.E., 2002, Geological constraints on the emplacement mechanism of the Parinacota debris avalanche, northern Chile, *Bulletin of Volcanology*, **64**, 40–54.

Corominas, J., 1996, The angle of reach as a mobility index for small and large landslides, *Canadian Geotechnical Journal*, **33**, 260–271.

Costa, J.E. and Schuster, R.L., 1988, The formation and failure of natural dams, *Geological Society of America Bulletin*, **100**, 1054–1068.

Crozier, M.J., 1984, Field assessment of slope instability, in D. Brunsden and D.B. Prior (eds), *Slope Instability* (Chichester: John Wiley & Sons Ltd), 103–142.

Crozier, M.J., 1989, *Landslides: Causes, Consequences and Environment* (London: Routledge).

Crozier, M.J., 1996, Runout behaviour of shallow, rapid earthflows, *Zeitschrift für Geomorphologie (Supplementband)*, **105**, 35–48.

Crozier, M.J., 1999a, Landslides, in D.E. Alexander and R.W. Fairbridge (eds), *Encyclopedia of Environmental Science* (Dordrecht: Kluwer), 371–375.

Crozier, M.J., 1999b, Prediction of rainfall-triggered landslides: a test of the antecedent water status model, *Earth Surface Processes and Landforms*, **24**, 825–833.

Crozier, M.J. and Glade, T., 1999, Frequency and magnitude of landsliding: fundamental research issues, *Zeitschrift für Geomorphologie (Supplementband)*, **115**, 141–155.

Crozier, M.J. and Preston, N.J., 1999, Modelling changes in terrain resistance as a component of landform evolution in unstable hill country, in S. Hergarten and H.J. Neugebauer (eds), *Process Modelling and Landform Evolution* (Heidelberg: Springer), **78**, 267–284.

Crozier, M.J., Eyles, R.J., Marx, S.L., McConchie, J.A. and Owen, R.C., 1980, Distribution of landslips in the Wairarapa hill country, *New Zealand Journal of Geology and Geophysics*, **23**, 575–586.

Cruden, D.M. and Varnes, D.J., 1996, Landslide types and processes, in A.K. Turner and R.L. Schuster (eds), *Landslides: Investigation and Mitigation* (Washington, DC: National Academy Press), Special Report, **247**, 36–75.

Czirok, A., Somfai, E. and Vicsek, T., 1997, Fractal scaling and power–law landslide distribution in a micromodel of geomorphological evolution, *Geologische Rundschau*, **86**, 525–530.

Dawson, A.G., 1999, Linking tsunami deposits, submarine slides and offshore earthquakes, *Quaternary International*, **60**, 119–126.

Dawson, S. and Smith, D.E., 2000, The sedimentology of Middle Holocene tsunami facies in northern Sutherland, Scotland, UK, *Marine Geology*, **170**, 69–79.

Dehn, M., Bürger, G., Buma, J. and Gasparetto, P., 2000, Impact of climate change on slope stability using expanded downscaling, *Engineering Geology*, **55**, 193–204.

Dietrich, W.E., Wilson, C.J., Montgomery, D.R. and McKean, J., 1993, Analysis of Erosion Thresholds, Channel Networks, and Landscape Morphology using a Digital Terrain Model, *Journal of Geology*, **101**, 259–278.

Dikau, R., Brunsden, D., Schrott, L. and Ibsen, M. (eds), 1996, *Landslide Recognition: Identification, movement and causes* (Chichester: John Wiley & Sons Ltd).

Driscoll, N.W., Weissel, J.K. and Goff, J.A., 2000, Potential for large-scale submarine slope failure and tsunami generation along the US mid-Atlantic coast, *Geology*, **28**, 407–410.

Evans, S.G. and DeGraff, J.V. (eds), 2002, *Catastrophic landslides: Reviews in Engineering Geology* (Boulder, Co: Geological Society of America).

Fannin, R.J. and Wise, M.P., 2001, An empirical–statistical model for debris flow travel distance, *Canadian Geotechnical Journal*, **38**, 982–994.

Flageolett, J.-C., 1999, Landslide hazard – a conceptual approach in risk viewpoint, in R. Casale and C. Margottini (eds), *Floods and Landslides: Integrated Risk Assessment* (Berlin: Springer-Verlag), 3–18.

Gillon, M.D. and Hancox, G.T., 1992, Cromwell Gorge Landslides – A general overview, in D.H. Bell (ed.), *Proceedings of the sixth International Symposium*, 10–14 February 1992, Christchurch, New Zealand (Rotterdam: A.A. Balkema), 83–120.

Glade, T., 1998, Establishing the frequency and magnitude of landslide-triggering rainstorm events in New Zealand, *Environmental Geology*, **35**, 160–174.

Glade, T., 2000, Modelling landslide-triggering rainfalls in different regions in New Zealand – the soil water status model, *Zeitschrift für Geomorphologie*, **122**, 63–84.

Glade, T., 2002, Ranging scales in spatial landslide hazard and risk analysis, in C.A. Brebbia (ed.), *Third International Conference on Risk Analysis*, 19–21 June 2002, Sintra, Portugal, 719–729.

Glade, T., 2004, Vulnerability assessment in landslide risk analysis, *Die Erde*, **134**, 121–138.

Glade, T., Crozier, M.J. and Smith, P., 2000, Applying probability determination to refine landslide-triggering rainfall thresholds using an empirical 'Antecedent Daily Rainfall Model', *Pure and Applied Geophysics*, **157**, 1059–1079.

Grauert, M., Bjorck, S. and Bondevik, S., 2001, Storegga tsunami deposits in a coastal lake on Suouroy, the Faroe Islands, *Boreas*, **30**, 263–271.

Guzzetti, F., Malamud, B.D., Turcotte, D.L. and Reichenbach, P., 2002, Power–law correlations of landslide areas in central Italy, *Earth and Planetary Science Letters*, **195**, 169–183.

Hampton, M.A., Lee, H.J. and Locat, J., 1996, Submarine landslides, *Reviews of Geophysics*, **34**, 33–59.

Heinimann, H.R., 1999a, *Risikoanalyse bei gravitativen Naturgefahren – Fallbeispiele und Daten* (Bern).

Heinimann, H.R., 1999b, *Risikoanalyse bei gravitativen Naturgefahren – Methode* (Bern).

Hovius, N., Stark, C.P. and Allen, P.A., 1997, Sediment flux from a mountain belt derived by landslide mapping, *Geology*, **25**, 231–234.

Hungr, O., 1995, A model for the runout analysis of rapid flow slides, debris flows, and avalanches, *Canadian Geotechnical Journal*, **32**, 610–623.

Hungr, O., Evans, S.G. and Harzard, J., 1999, Magnitude and frequency of rock falls and rock slides along the main transportation corridors of southwestern British Columbia, *Canadian Geotechnical Journal*, **36**, 224–238.

Hutchinson, J., 1988, General Report: morphological and geotechnical parameters of landslides in relation to geology and hydrogeology, in C. Bonnard (ed.), *Proceedings of the 5th International Symposium on Landslides*, 10–15 July 1988. Lausanne, Switzerland (Rotterdam: A.A. Balkema), 3–35.

Hutter, K., Svendsen, B. and Rickenmann, D., 1996, Debris flow modeling: a review, *Continuum Mechanics and Thermodynamics*, **8**, 1–35.

King, D., 2001, Uses and limitations of socioeconomic indicators of community vulnerability to natural hazards: data and disasters in Northern Australia, *Natural Hazards*, **24**, 147–156.

Korup, O., 2002, Recent research on landslide dams – a literature review with special attention to New Zealand, *Progress in Physical Geography*, **26**, 206–235.

Korup, O., 2003, Landslide-induced river disruption: geomorphic imprints and scaling effects in Alpine catchments of South Westland and Fiordland, New Zealand School of Earth Sciences, Wellington, Victoria University of Wellington.

Lambert, M.G., Trustrum, N.A. and Costall, D.A., 1984, Effect of soil slip erosion on seasonally dry Wairarapa hill pastures, *New Zealand Journal of Agricultural Research*, **27**, 57–64.

Lang, A., Moya, J., Corominas, J., Schrott, L. and Dikau, R., 1999, Classic and new dating methods for assessing the temporal occurrence of mass movements, *Geomorphology*, **30**, 33–52.

Leone, F., Asté, J.P. and Leroi, E., 1996, Vulnerability assessment of elements exposed to mass-movement: working toward a better risk perception, in K. Senneset (ed.), *Landslides – Glissements de Terrain* (Rotterdam: A.A. Balkema), vol. 1, 263–270.

Matthews, J.A., Brunsden, D., Frenzel, B., Gläser, B. and Weiß, M.M. (eds), 1997, *Rapid Mass Movement as a Source of Climatic Evidence for the Holocene. Paläoklimaforschung Paleoclimate Research* (Stuttgart, Jena, Lübeck and Ulm: Gustav Fischer Verlag).

McClung, D.M., 2001, Extreme avalanche runout: a comparison of empirical models, *Canadian Geotechnical Journal*, **38**, 1254–1265.

McSaveney, M.J., 2002, Recent rockfalls and rock avalanches in Mount Cook National Park, New Zealand, in S.G. Evans and J.V. DeGraff (eds), *Catastrophic Landslides: Effects, Occurrence, and Mechanisms*, **15**, 35–70.

Miao, T.D., Liu, Z.Y., Niu, Y.H. and Ma, C.W., 2001, A sliding block model for the runout prediction of high-speed landslides, *Canadian Geotechnical Journal*, **38**, 217–226.

Michael-Leiba, M., Baynes, F. and Scott, G., 1999, Quantitative landslide risk assessment of Cairns, *Australian Geological Survey Organisation*, **36**, 51.

Michael-Leiba, M., Baynes, F. and Scott, G., 2000, Quantitative landslide risk assessment of Cairns, Australia, in E. Bromhead, N. Dixon and M.-L. Ibsen (eds), *Landslides in research, theory and practice* (Cardiff: Thomas Telford), 1059–1064.

Page, M.J., Trustrum, N.A. and Dymond, J.R., 1994, Sediment budget to assess the geomorphic effect of a cyclonic storm, New Zealand, *Geomorphology*, **9**, 169–188.

Petley, D., 1996, The Mechanics and Landforms of Deep-Seated Landslides, in Anderson, M.G. and Brooks, S.M. (eds), *Advances in Hillslope Processes* (Chichester, John Wiley & Sons Ltd), vol. 2, 823–835.

Preston, N.J., 2000, Feedback effects of rainfall-triggered shallow landsliding, in E. Bromhead, N. Dixon and M.-L. Ibsen (eds), *Landslides in Research, Theory and Practice* (Cardiff: Thomas Telford), 1239–1244.

Prior, D.B. and Stephens, N., 1972, Some movement patterns of temperate mudflows: examples from northeastern Ireland, *Geological Society of America Bulletin*, **83**, 2533–2544.

Ragozin, A.L. and Tikhvinsky, I.O., 2000, Landslide hazard, vulnerability and risk assessment, in E. Bromhead, N. Dixon and M.-L. Ibsen (eds), *Landslides in Research, Theory and Practice* (Cardiff: Thomas Telford), 1257–1262.

Rickenmann, D., 1999, Empirical relationships for debris flows, *Natural Hazards*, **19**, 47–77.

Riley, P.B. and Read, S.A.L., 1992, Lake Waikaremoana – present day stability of landslide barrier, in D.H. Bell (ed.), *Proceedings of the sixth International Symposium*, 10–14 February 1992, Christchurch, New Zealand (Rotterdam: A.A. Balkema), 1249–1256.

Rybár, J., 1997, Increasing impact of anthropogenic activities upon natural slope stability, in P.G. Marinos, G.C. Koukis, G.C. Tsiambaos and G.C. Stournaras (eds), *Proceedings of the International Symposium on Engineering Geology and the Environment*, 23–27 June 1997, Athens, Greece (Rotterdam: A.A. Balkema), 1015–1020.

Schmidt, J. and Preston, N., 1999, Landform evolution on regional scales – a conceptional modelling approach.

Schmidt, K.M., Roering, J.J., Stock, J.D., Dietrich, W.E., Montgomery, D.R. and Schaub, T., 2001, The variability of root cohesion as an influence on shallow landslide susceptibility in the Oregon Coast Range, *Canadian Geotechnical Journal*, **38**, 995–1024.

Schumm, S.A., 1979, Geomorphic thresholds: the concept and its applications, *Transactions Institute of British Geographers* (New Series), **4**, 485–515.

Schuster, R.L., Salcedo, D.A. and Valenzuela, L., 2002, Overview of catastrophic landslides of South America in the twentieth century, in S.G. Evans and J.V. DeGraff (eds), *Catastrophic Landslides: Effects, Occurrence, and Mechanisms* (Boulder, CO: Geological Society of America, Reviews in Engineering Geology), **15**, 1–34.

Selby, M.J., 1993, *Hillslope materials and processes* (Oxford: Oxford University Press).

Shroder, J.F. and Bishop, M.P., 1998, Mass movement in the Himalaya: new insights and research directions. *Geomorphology*, **26**, 13–35.

Tappin, D.R., Watts, P., McMurtry, G.M., Lafoy, Y. and Matsumoto, T., 2001, The Sissano, Papua New Guinea tsunami of July 1998 – offshore evidence on the source mechanism, *Marine Geology*, **175**, 1–23.

Varnes, D.J., 1978, Slope movement: types and processes, in R.L. Schuster and Krizek, R.J. (eds), *Landslides: Analysis and Control* (Washington, DC: Transportation Research Board National Academy Press), Special Report, **176**, 11–33.

Voight, B. and Faust, C., 1992, Frictional heat and strength loss in some rapid landslides – error correction and affirmation of mechanism for the vaiont landslide, *Géotechnique*, **42**, 641–643.

Wong, H.N., Ho, K.K.S. and Chan, Y.C., 1997, Assessment of consequences of landslides, in Cruden, D.M. and Fell, R. (eds), *Landslide Risk Assessment – Proceedings of the Workshop on Landslide Risk Assessment*, Honolulu, Hawaii, USA, 19–21 February 1997 (Rotterdam: A.A. Balkema), 111–149.

# 3

# A Review of Scale Dependency in Landslide Hazard and Risk Analysis

*Thomas Glade and Michael J. Crozier*

## 3.1  Introduction

Landslides occur at various spatial and temporal scales. They are a natural part of landscape evolution, and differ greatly in their contribution to slope-forming processes in different environmental settings. When landslides occur, they can move quickly downslope at rates of several m/s, or they can creep slowly at rates of only a few mm/year. On the one hand, they can move instantaneously following a specific trigger such as an earthquake, an intense rainfall event, an explosion, or undercutting event. On the other hand, they may show a delayed response to critical triggering conditions, for example after a prolonged rainfall event with a gradual rise in porewater pressures. The range of spatial and temporal scales covered by different landslide types is shown schematically in Figure 3.1.

The wide range of both spatial and temporal scales distinguishes landslide processes significantly from other natural processes such as floods, earthquake shaking or tsunamis. Some examples of the range of landslide occurrences are given in Figure 3.2. The relative spatial and temporal coverage of these examples is indicated in Figure 3.1. Despite these extreme variations, some general patterns of occurrences can be recognized.

The spatial and temporal behaviour of landslides and the occurrence of specific types of landslide can be linked to particular environmental domains, but only in the most general terms (Figure 3.2). For example, all types and scales of movement can be found in mountainous terrain. However, rock avalanches (Bergsturz) and instantaneous rock

*Landslide Hazard and Risk*  Edited by Thomas Glade, Malcolm Anderson and Michael J. Crozier
© 2004 John Wiley & Sons, Ltd  ISBN: 0-471-48663-9

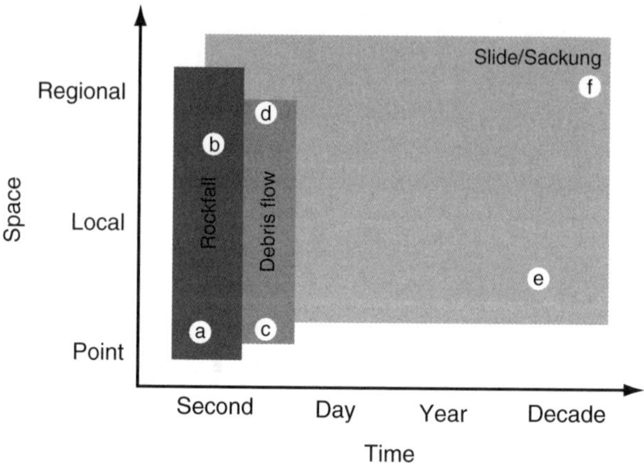


**Figure 3.1**  *Schematic diagram of scales of landslide occurrence. Letters refer to examples shown in Figures 3.2(a–f)*

and debris falls, slides, and flows with long runouts are generally restricted to these steep mountainous areas. Such areas provide the potential and kinetic energy requirements by having high relief and steep slopes, as well as providing large rock-dominated slopes, and sources of mechanically weathered debris. Nevertheless, rotational failures (slumps) require rock and soil conditions that are massive and free from structural control in order to achieve their full development. Typically, such conditions are met in softer rock in more gentle terrain. Large failures, however, are not restricted to any terrain. Block slides, rotational failures (slumps) and lateral spreading of large dimensions have been recorded in areas of very low slope angle as well as low relative relief. Critical in these instances is the presence of weak or failure-prone material. The slump flows in the quick clay of Scandanavia and North America are prime examples (e.g. Larsen *et al.*, 1999). Other problem situations are commonly found in areas where slopes have been actively and recently destabilized, usually by active undercutting such as on river banks and coasts or where human construction has taken place. All forms of movement are possible in these locations and their magnitude and behaviour are largely dictated by the available relief and slope angle. Regolith and soil failures are, by definition, characteristic of areas that are of sufficiently low angle or sufficiently susceptible to weathering to have produced and retained a regolith mantle (for example rolling hill country). The soil and debris slides and flows that eminate from these areas are supply-constrained. Their frequency of occurrence is dependent not only on triggering forces but also on the availability of material. These failures become most threatening in areas where they can become channelized. Thus moderate to steep terrain, retaining a regolith and drained by high-angle valleys, provides the potential for high-velocity, high-magnitude events.

The character of magnitude and frequency distributions can also be related to the nature of the triggering event. Earthquake shaking and extreme climatic conditions (including intense rainfall) can trigger movements over areas of many square kilometres in extent (Crozier and Preston, 1999; Eyles *et al.*, 1978; Keefer, 2002). These situations commonly produce multiple-occurrence events with up to thousands of landslides occurring over hundreds of square kilometres in the range of a few minutes or hours. Their impact can

**Figure 3.2a** *Examples of landslides occurring at different temporal and spatial scales. Rockfall in the Ahr Valley, Germany (photo by T. Glade)*

**Figure  3.2b**   Vaimont rockslide/Bergsturz (photo by E. Bromhead)

**Figure  3.2c**   Debris flow in the Matter Valley, Switzerland (photo by H. Gärtner)

**Figure 3.2d** *Debris and earthslides flows in Makahoni, New Zealand (photo by M.J. Crozier)*

**Figure 3.2e** *Coastal landslides in the south of the Isle of Wight, UK (photo by T. Glade)*

**Figure 3.2f**  *Large rock slumps in King Country, New Zealand (photo by M.J. Crozier)*

be registered in all types of terrain, from gentle relief to mountainous terrain. Analysis of such events has indicated difficulties in differentiating the landslide signature arising from earthquake- and rainfall-triggered events. Crozier (1997) suggests that climatically triggered events have a predominance of small to medium-size landslides, with only the rare large event, whereas he concludes that earthquake-triggered events are capable of producing a high proportion of large failures. Alternatively, Guzzetti *et al.* (2002b) suggest that the magnitude–frequency distribution of events triggered by rainfall and earthquakes are indistinguishable.

Irrespective of the triggering mechanism, on most slopes, landslides will occur where inherent susceptibility (excess strength) is lowest. However, failure sites for climatically triggered events will normally occur where surface and groundwater concentrate or where sufficient depth of susceptible material occurs (e.g. hillslope hollows, Crozier *et al.*, 1990). In some situations, prevailing antecedent soil-water conditions may be related to slope aspect, consequently dictating the distribution of landslide occurrence during an event (Crozier *et al.*, 1980). In contrast, seismically triggered failures may occur preferentially on ridge crests where topographic enhancement of earthquake waves occurs or within material susceptible to liquefaction. Other triggering mechanisms such as undercutting by geomorphic process occur in predictable locations such as the outside bends of stream channels and exposed coastal cliffs. Triggering by human action is indiscriminate (Baroni *et al.*, 2000), generally confined to areas of undercutting, mining or oversteepening or to areas that have been loaded by material or excess drainage. However, human action as a preparatory factor (see Chapter 2) can exert its influence over wide areas, such as in the case of deforestation (e.g. Glade, 2003a; Guthrie, 2002;

Marden and Rowan, 1993; Montgomery *et al.*, 2000; Vanacker *et al.*, 2003; Wu and Swanston, 1980).

## 3.2   Philosophy of Spatial Modelling

The temporal and spatial behaviour of landslides dictates the method of hazard and risk analysis as well as the treatment of the problem. While individual landslides may be treated with site investigations and possibly advanced numerical stability models, spatial distributions require other techniques. Generally, it can be assumed that with increasing spatial resolution more data are available to analyse the phenomenon, hence the system complexity also increases. Accordingly, the model generalizing reality expands in its complexity. Consequently, the more data available, the higher is the model complexity, and the predictive potential of the result is more robust. This dependency has been described to determine spatial patterns of catchment hydrology by Grayson and Blöschl (2000) as well as Grayson *et al.* (2002) and is transferred to spatial landslide observations in Figure 3.3.

The conceptual relation between data availability, model complexity and predictive capacity shows that there is an 'optimum model complexity' (bold dashed line in Figure 3.3) best describing the relation of these three variables. The following example demonstrates this dependency for a given data set (bold line in Figure 3.4). Analysis of a medium-sized data set shows a decreasing model performance after having passed the line of 'optimum model complexity'. Even better and more advanced models describe the data set with less predictive capacity. This relation can be attributed to the fact that a specific data set allows only the application of specific models; more advanced models do not necessarily increase the accuracy of prediction. Similarly, a model with a given complexity can only be used to predict a data set of a given quality. Even better data sets (in quality, quantity, resolution, etc.) do not significantly enhance the prediction given by the similar model complexity. Although the general trend shown in this figure is reasonable, some problems are inherent in details. The line of 'optimum model complexity' is not necessarily as straight as shown in Figure 3.3. Another possible relation is a step-wise increase in prediction accuracy, which is given when the data availability increases, but model complexity and the predictive surface stay constant. In contrast,

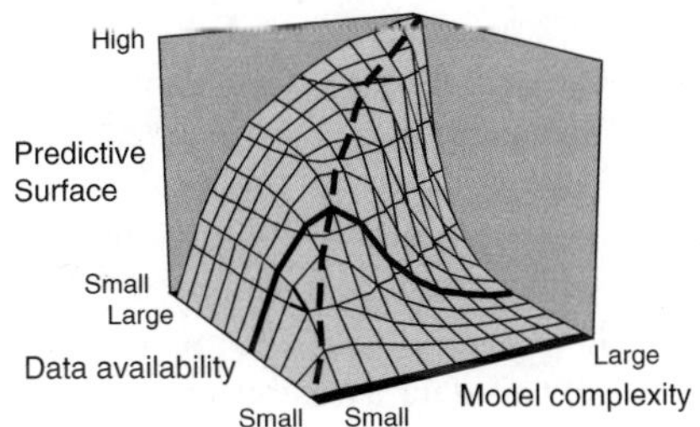


*Figure 3.3*   *Schematic diagram showing one relation between data availability, model complexity and predictive capacity of the result (based on Grayson and Blöschl, 2000 and Grayson et al., 2002). The 'optimum model complexity' (Grayson et al., 2002) is marked as a bold dashed line and described in the text*

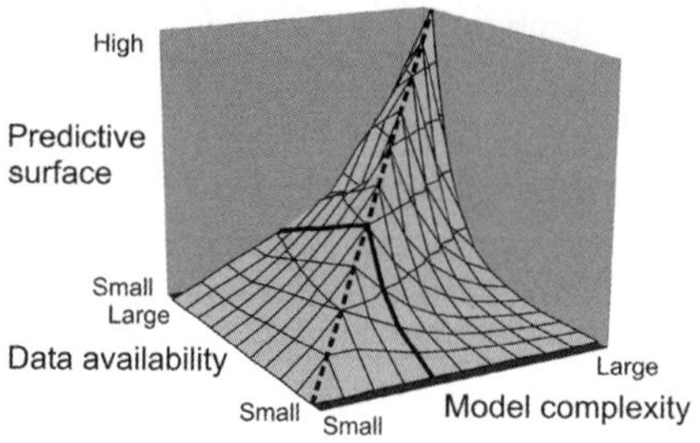


***Figure 3.4*** *Schematic relation between data availability, model complexity and predictive capacity of the result. This figure gives a probable relation of current reality. The bold line indicates a decrease of the predictive surface after exceeding the line of 'optimum model performance' (bold dashed line) although the data availability increases*

models with increasing complexity applied to the same data set do not necessarily change the predictive surface. In addition, often the accuracy of results (or predictive surface) from different models with increasing complexity *decreases* after exceeding the line of 'optimum model complexity' (bold line in Figure 3.4). This can be related to the fact that the more variables included in a data set, the more uncertain are the interrelationships, and positive or negative feedback loops between the variables exist. Larger model complexities cannot address these interrelations and loops adequately. Consequently, more data do not necessarily allow better predictive surfaces, as indicated in Figure 3.3. Therefore, the minimum set of information that best explains the system behaviour with current methods and techniques has to be determined. It must be asked whether the most accurate predictive result is better modelled using smaller data sets than applying additional data to a similar model. In addition, the simpler black-box models (i.e. models where only input and output are known and no knowledge on internal links is available) are often more robust than advanced numerical models. Hence, after exceeding the line of 'optimum model complexity' there is no constant prediction surface associated with increasing data sets for similar models as shown in Figure 3.3; rather the prediction surface decreases again with larger data sets. This dependency is given schematically in Figure 3.4.

To summarize, Figure 3.3 gives theoretical relations, but these relations can often not be verified by analysis for the reasons explained in the previous paragraph. Independent of the previously mentioned constraints, however, research should aim to move towards a relation such as given in Figure 3.3.

For practical application (e.g. planning purposes), it is most important to consider the cost of the analysis, and the benefit of the proposed measures. Such cost–benefit considerations are often the driving force of practical solutions and thus the line of 'optimum model complexity' helps to define which method describes the available data set with highest precision for which resolution. Consequently, it is most important to be sure that the result of the applied method meets the requirements of the study aim.

Having this schematic concept of data availability, model complexity and prediction capacity in mind, the following sections review approaches for local investigations and spatial analysis. Three distinct different landslide types have been selected: rockfall, debris flow and translational/rotational earth- and soil slides. These three groups are the most common landslide types and are thus briefly reviewed with respect to susceptibility, hazard and risk.

## 3.3   Landslide Susceptibility and Hazard Analysis

### 3.3.1   Site-based Stability and Hazard Analysis

#### 3.3.1.1   Rock slope analysis

Rock slope failures, rockfalls and rock topples can occur in any size. Worldwide examples are summarized by Evans and DeGraff (2002). Analytical techniques for rockfalls and rock slopes have been developed since the beginning of the twentieth century. Albert Heim (1932) analysed rockfalls systematically in the European Alps. This work has been extended by Abele (1974) to produce one of the most comprehensive monographs on rockfalls in the European Alps. On the basis of this type of information various other authors have continued to develop the empirical methods. Many of these methods use, in particular, fall height and rock volume to establish empirical estimates of runout distance (e.g. Scheidegger, 1973, 1984). Mapping of field evidence and the characterization of different terrain along with landslide attributes are common to many empirical models of this type (e.g. Li Tianchi, 1983). In contrast, detailed rock slope monitoring is often required in order to give predictions for rockfall occurrences (e.g. Monma *et al.*, 2000). These studies require more advanced models.

A summary of those various analytical methods and techniques is given in Giani (1992) and Erismann and Abele (2001). Following Coggan *et al.* (1998), Moser (2002) differentiates between conventional techniques and numerical methods for rock slope analysis. The conventional methods include stereographic and kinematic analysis, limit equilibrium analysis and physical modelling, including the use of rockfall simulators. Stereographic and kinematic analyses aim at determining critical slope, discontinuity geometry and approximate shear strength characteristics. Limiting equilibrium analysis focuses on determining the degree of stability of a slope and requires information on slope geometric and material characteristics, rock mass shear strength parameters (cohesion and friction), as well as groundwater conditions (Stead *et al.*, 2001). Physical models use material characteristics at appropriate scaling factors. Rockfall simulators are based on slope geometry, rock block sizes, shapes along with density, and on the coefficients of restitution (Moser, 2002). Examples of such models widely applied for practical use include the Colorado Rockfall Simulation Program CRSP (Jones *et al.*, 2000) and the 'Rockfall' model developed by Spang and Sönser (1995), which additionally considers the influence of vegetation characteristics (Ploner and Sönser, 1999). A similar ROCKFALL model, developed by Evans and Hungr (1993), is based on a random collision lumped mass modelling approach. The ROCKFALL model uses two restitution coefficients and a transition to rolling criterion (Evans and Hungr, 1993). A comparison of some rockfall models is given by Guzzetti *et al.* (2002a).

Numerical models may include continuum modelling (e.g. finite-elements, finite-difference), discontinuum modelling (e.g. distinct-elements, discrete-elements) (e.g. Yamagami *et al.*, 2001), and hybrid/coupled modelling (Moser, 2002). In general, advantages of these numerical approaches are: a basis on general physical laws, a deformation and stability consideration performed within one model only, any kind of support or construction is incorporated, and dynamic impacts such as vibrations or earthquakes can be modelled. These models are mainly used in mining and civil engineering situations. Specific applications include tunnel constructions, foundations, and surface excavations

(Kliche, 1999). The main disadvantage is, however, the high demand for precise data, which are often not available in view of the cost involved or the high complexity of the slopes.

### 3.3.1.2 Debris-flow analysis

Debris flows are complex mass movement processes determined by hydraulic flow behaviour, which is strongly dependent on the composition of the solids (Hungr, 2005; Hungr *et al.*, 2001). One of the first monographs specifically devoted to debris flows was published by Stiny (1910). The most recent textbook on debris-flows and debris avalanches is edited by Jakob and Hungr (2005). The methods used to assess debris flows on a site-specific scale range from general geometric relations to advanced numerical modelling. Current research on debris flows is summarized in Chen (1997), Wieczorek and Naeser (2000), Rickenmann and Chen (2003) as well as within the proceedings of the International Symposium INTERPRAEVENT (proceedings of last symposia are INTERPRAEVENT, 2000a, b, c, 2002a, b).

Relatively simple empirical and semi-empirical methods commonly relate geometric parameters to debris-flow characteristics. Due to practical demands, one of the most common debris-flow characteristics to be modelled is the runout distance (e.g. Rickenmann, 1999; Wieczorek *et al.*, 2000). Although originally developed for rockfalls (as suggested by Heim, 1932 and further developed by Scheidegger, 1975, Li Tianchi, 1983 and others), the empirical model describing the relationship between volume and travel distance, and in some cases relief (height difference between the starting and deposition point) has also been widely applied to debris flows (e.g. Cannon, 1989; Corominas, 1996; Mark and Ellen, 1995; Rickenmann, 1999; Wong and Ho, 1996; Zimmermann *et al.*, 1997). Other studies using statistical analysis of slope geometry to predict landslide travel distances are limited to cut slopes, fill slopes, retaining walls and boulder falls (e.g. Finlay *et al.*, 1999). However, there are some drawbacks in these empirical approaches. First, some models do not consider slope breaks within the longitudinal channel profile (e.g. Cannon, 1993; Fannin *et al.*, 1997). Second, some models give statistical relationships between various factors which have been calculated for specific regions only, and are therefore not easily applicable to other regions. Additionally, it is impossible to model or include complex flow mechanisms involved in the equations. Despite all of these limitations, Rickenmann (1999) has shown a surprisingly good fit of general and global trends for these empirical models.

Rheological and physical-based modelling of debris flows needs detailed information on rheologic, hydrologic and hydraulic properties (e.g. Coussot *et al.*, 1998). For example, Hungr (2000) analysed debris-flow surges using the theory of uniformly progressive flow. Numerous authors are working with such physical models (e.g. Costa and Wieczorek, 1987; Iverson, 1997a, b; Major and Iverson, 1999; Revellino *et al.*, 2002). A recent review of different approaches is given by Hutter *et al.* (1996), Jan and Shen (1997), Chen and Lee (2000) and within Rickenmann and Chen (2003).

### 3.3.1.3 Slide stability investigations

Slide stability analysis have a long history going back to Terzaghi (1925), Terzaghi and Peck (1948), Skempton and Northey (1952), and Skempton (1953). Besides modelling the stability of unfailed slopes, it is also of interest to get more information on the importance of certain stability factors of previous events, which can be verified by back-analysis. For example, a large event which interests researchers until today is the Vaiont slide (e.g. Kiersch, 1980; Müller, 1964; Petley, 1996; Skempton, 1966; Voight and Faust, 1992). Most recently, Vardoulakis (2002) performed a dynamic analysis and presented two early stages of the earth slide considering two mechanically coupled substructures: (a) the rapidly deforming shear band at the base of the slide, and (b) the accelerating (rotating) rigid body.

Most recent reviews of slope stability concepts and techniques have been reviewed by Bromhead (1996, 1997). Applications of numerical modelling tools to slope stability assessments for single landslides are given in Bandis (1999). Additionally, the use of neural networks for slope stability modelling is becoming popular (e.g. Mayoraz *et al.*, 1996), and some authors also used this method to predict slope movements (e.g. Fernández-Steeger and Czurda, 2001). Collections of most recent approaches of slope stability modelling are within the conference proceedings edited by Anderson and Brooks (1996), Li *et al.* (1998), especially by Ho and Li (2003).

Actual research tries to extend sophisticated models originally developed for two-dimensional approaches to the third dimension (e.g. Bromhead *et al.*, 2002; Wang *et al.*, 2001). One example is CHASM in its latest version 4.0. Within this Combined Hydrology And Stability Model, geometrical characteristics, geotechnical properties, hydrologic conditions and vegetation-related information are defined for squares with three dimensions. In combination with triggering conditions, both rainfall events and earthquakes, slope stability calculations give most likely failure surfaces with respective factor-of-safety values, and runout distance can be obtained (Lloyd *et al.*, in press). Another recent method is the Energy Approach (EA) developed by Ekanayake and Phillips (1999). The newly proposed approach incorporates, within the stability analysis, the ability of soil with roots to withstand strain, based on a consideration of the energy consumed during the shearing process of the soil-root system (Ekanayake and Phillips, 1999). All these new promising approaches cannot be used at larger spatial scales, because neither data are available in the required detail nor does the computational capacity exist. However, with further development of computer technology, these approaches have the potential to be applied within the next years.

### 3.3.1.4  Conclusion

Rock slope analyses are commonly based on empirical estimates, conventional stability analysis techniques, and more sophisticated numerical methods. The more advanced the models, the higher the input data requirement and thus, the more complex the assessment. Hence empirical and conventional techniques are applied either for back-analysis or for preliminary assessments. Detailed site-specific investigations require numerical models based on continuum modelling, discontinuum modelling, or hybrid/coupled modelling. The last models, in particular, are used in mining and civil engineering applications.

Debris-flow analysis is strongly determined by hydraulic-flow behaviour. Empirical and semi-empirical methods relate geometric parameters to debris-flow characteristics. Despite restrictive assumptions these relatively simple methods have proven their potential in practical applications. Rheological and physical-based modelling approaches have been further developed over the last decades. Although these approaches allow a detailed modelling of debris flows, data requirements are very high and thus such models applied to practical applications are limited.

Slide stability analysis usually provides a statement of site susceptibility in terms of a factor of safety (FoS). In the case of first-time failures, the magnitude of event is largely unknown. However, modelling multiple potential failure surface permits some estimate, usually in a two-dimensional sense, of the likely magnitude involved. Moreover, the magnitude of movement associated with pre-existing failures can be addressed by locating the boundary shear surfaces within the slope. In addition, if the significance of dynamic stability factors (such as porewater pressure) can be determined through sensitivity analysis, then the behaviour of such critical factors may be linked to external triggering factors (such as rainfall). An examination of the climatic record may then reveal the frequency with which critical conditions may be reached within the slope. In some instances, the importance of certain stability factors can be verified by back-analysis of previous events.

For site-based analysis, irrespective of the process types and the applied method, the main objective should be the determination of both the magnitude and frequency of landslide occurrence, in order to properly estimate the hazard. By definition, the general location and, in some cases, the actual landslide itself is predetermined in site-based analyses. If no information on frequency is available, then it is only possible to determine the susceptibility of a given location towards the respective process. In some cases, frequency–magnitude information may be obtained by using historical archives or field evidences to approximate temporal landslide occurrence (e.g. Glade *et al.*, 2001a).

### 3.3.2   Spatial Susceptibility and Hazard Analysis

Investigations of numerous landslides extending over large regions have been performed for decades. Many of the first regional assessments carried out were based on mapping techniques as part of extensive field survey campaigns (e.g. Brabb and Pampeyan, 1972). With the development of new computer technologies, particularly GIS techniques (e.g. Carrara and Guzzetti, 1995), controlled automated mapping procedures are becoming more popular (e.g. McKean and Roering, 2004). These techniques are commonly based on remote sensing data and use either aerial photography or satellite images to obtain spatial information on landslide occurrence and movement (e.g. Hervás *et al.*, 2003). These automated procedures are constantly being developed with new computer generations, along with the availability of remote sensing imagery with increased resolution and accuracy. The main advantage of any GIS technique is its capacity for spatial analysis of large data sets. Different spatial information can be linked and coupled, new data sets can be created, and additional information can be obtained. Thus these recent advances provide a powerful tool for spatial landslide assessment.

Within the last decade, techniques of spatial landslide analysis have been greatly improved (e.g. summarized in Carrara and Guzzetti, 1995). Based on the scale classification for engineering geology maps (International Association of Engineering Geology, 1976), Soeters and van Westen (1996) have carried out extensive assessments of spatial landslide hazard. They slightly modified the original classification to produce the following classes ranging from large scales (<1:10 000), medium scales (1:15 000–1:100 000), regional scales (1:125 000–1:500 000), to national scales (>1:750 000). A typical method of analysis can be assigned to each investigation scale. This classification is summarized in Table 3.1.

**Table 3.1**  *Recommended scales for different spatial landslide analysis (extended from Soeters and van Westen, 1996)*

| Scale | Qualitative methods | | Quantitative methods | | |
|---|---|---|---|---|---|
| | Inventory | Heuristic analysis | Statistical analysis | Probabilistic prediction analysis | Process-based and numerical analysis |
| <1:10 000 | Yes | Yes | Yes | Yes | Yes |
| 1:15 000–1:100 000 | Yes | Yes | Yes | Yes | Probable |
| 1:125 000–1:500 000 | Yes | Yes | Probable | Probable | No |
| >1:750 000 | Yes | Yes | No | No | No |

Two main types of investigation can be differentiated on the basis of methodology: qualitative and quantitative. Landslide inventories plus heuristic approaches are grouped within the qualitative methods. In nearly all spatial investigations, landslide inventories are the basis for developing and/or verifying the method. Even if the chosen method does not use landslide locations for model development (e.g. numerical models), information on locations is needed for verification and validation of the results (e.g. Santacana *et al.*, 2003). These inventories are thus of great importance, and provide a potential source of information for future developments in spatial analysis (Guzzetti *et al.*, 1999). Consequently, a high proportion of project resources should be allocated for the development of inventories, because only high-quality inventories allow a reliable proof for spatial analysis. A second qualitative method is the heuristic approach. Based on *a priori* knowledge, local experiences, as well as expert judgement, are included. The heuristic approach also uses spatial information in explaining landslide occurrence. Commonly, such information includes topographic, hydrological, geologic, geotechnical, or geomorphic factors, and often vegetation coverage along with land use is considered, too. These factors are determined by either field campaigns or aerial photograph interpretation. In particular, spatial geomorphic factor maps offer a first approximation of the activity degree regarding the respective landslide processes (e.g. Cardinali *et al.*, 2002). In addition to inventories and other factor maps, this geomorphic information is an important basis for any further assessment (e.g. Glade and Jensen, 2004). Experts weight the importance of different environmental factors based on personal knowledge and experience, thus providing an initial assessment of landslide susceptibility. Indeed, qualitative weightings are heavily dependent on the experience of the person or expert group responsible for the analysis. Criteria for the assessments are not always identifiable by others, which is a major limitation of the heuristic approach. Thus the objectivity is not measurable, and consequently the reproducibility is often difficult. However, if the expert has a profound understanding of the processes involved and knows the study region in detail, such assessments can also be accurate and applicable, in particular for first approximations of landslide susceptibility.

In contrast, approaches using quantitative methods are generally based on objective criteria and are thus, in theory at least, repeatable, producing identical results for similar data sets. The quantitative methods include statistical, probabilistic prediction, process-based, or numerical approaches. The statistical methods are the most popular ones. Factor maps such as geology, soils, or topographic conditions (e.g. slope angle, horizontal and vertical curvature, aspect, distance to divide, etc.) are compared with landslide distribution from inventory maps and landslide density is calculated. Initially bivariate statistical analysis may be used to compare each factor separately with landslide locations, and weighting factors are computed on this basis for each factor. However, using multivariate statistics, any combination of factor maps can be related to landslide locations and the resulting matrix is then analysed using statistical tests, such as multiple regression or discriminant analysis (e.g. Chung *et al.*, 1995). The statistical tests then provide information on which factor or which combination of factors best explains landslide occurrence. The areas with factor scores equivalent to those for areas associated with landslides, but without former landslide occurrence, are thus considered prone to future landslides. Resulting maps give only spatial landslide susceptibility, because they do not contain any direct information on the hazard, that is, temporal variation of magnitude and frequency of landslides. Other statistical methods providing probabilistic prediction models (e.g. Bayesian probability, fuzzy logic) can also be used to produce landslide susceptibility maps (e.g. Binaghi *et al.*, 1998; Chung and Fabbri, 1999; Fabbri *et al.*, 2002; Fernández-Steeger *et al.*, 2002; Pistocchi *et al.*, 2002). For example, the fuzzy method simply applies 'if-then' rules to the different factor sets, and is thus based on a decision tree approach (e.g. Ercanoglu and Gokceoglu, 2002; Mackay *et al.*, 2003). The result is still a susceptibility map. Basic assumptions in both statistical approaches are static environmental and triggering boundary conditions. Considering the ongoing debate on the effects of climate change on landslide occurrence (e.g. Dehn, 1999; Schmidt and Glade, 2003), on changes of catchment conditions following each landslide event (e.g. Crozier and Preston, 1999), and on human impact on environmental conditions through, for example, land use change (e.g. Frattini and Crosta, 2002), it is obvious that these assumptions strongly influence the interpretation of the result.

The use of different data sets for spatial analysis requires a good deal of caution. First, large data sets are required which are difficult to assess for some remote regions. Second, the input data need to be of identical quality and resolution. For example, generating a 10 m raster resolution from a 1:2 750 000 soil map using downscaling techniques provided in any GIS is very easy. This downscaled high-resolution raster can be used for large-scale analysis, for example at a scale of 1:25 000. However, the information stored with the 10 m raster pixel still relates to the original scale, and is thus of little value for comparison with more detailed data sets, for example landslide locations. Although this pitfall is obvious, one might be tempted to apply this procedure in order to gain a result; but when analysing data sets with two different resolutions, the result can lead to an incorrect conclusion. As a general rule-of-thumb, spatial analysis can only be carried out at the scale of the data set with the coarsest resolution. Nevertheless, despite all these potential pitfalls and limitations, the beauty of this approach is its simplicity and reproducibility. And for numerous applications, the derived information on landslide susceptibility is sufficient.

The second group of quantitative methods includes the empirical and deterministic, process-based methods. Within this set of methods, topographic attributes (e.g. slope angle, vertical and horizontal curvatures, slope aspect, distance to divide or channel, contributing area, etc.) are coupled with hydrological conditions (e.g. soil saturation, permeability, hydraulic conductivity) and generalized geotechnical information on soil properties (e.g. cohesion, angle of internal friction, specific weight) in order to perform a stability analysis. Most of the available models are based on the infinite slope approach (e.g. Vanacker *et al.*, 2003).

Verification of modelled results, however, is an important task which is not always carried out (Chowdhury and Flentje, 2003). One example of a spatial application of the infinite slope approach is the SHALSTAB model, which has been developed by Montgomery and Dietrich (1994) and Dietrich *et al.* (1995) and was applied to various sites in the United States (e.g. Dietrich and Sitar, 1997; Montgomery *et al.*, 2000; Montgomery *et al.*, 1998) and in Rio de Janeiro (e.g. Fernandes *et al.*, 2004). A recent development is the application of numerical cinematic approaches to spatial analysis (e.g. Günther *et al.*, 2002a, b).

After having addressed major issues in site-specific and spatial landslide analysis, the final part of this chapter focuses on spatial landslide assessments. Due to the numerous demands from agencies responsible for spatial planning and to the increasing numbers of studies published in recent years, it is important to give an overview of spatial assessments. Consequently, the following sections give some examples of different kinds of spatial landslide susceptibility and hazard, but also risk investigations.

## 3.4 A Review of Spatial Landslide Susceptibility and Hazard Investigations

Qualitative methods and approaches are popular for providing a preliminary estimation of landslide susceptibility and hazard. While some investigations do not distinguish between the different types of landslide, others treat specific types separately. To illustrate different types of analysis, some examples of the many studies that have been carried out are given below. Whenever possible, the studies have been classed in the two groups of 'catchment and regional scale' and 'national scale' analysis.

### 3.4.1 General Landslide Information

Table 3.2 lists sources providing information on the spatial distribution of landslides. These sources treat landslides collectively and do not provide an analysis on the basis of landslide type. The nature of the data provided (whether in the form of general information, landslide distribution, or inventory) is noted for each entry. For some sources, it was difficult to determine which form of spatial information was used. If no details on the spatial data set were available, the label 'information' was added. Table 3.2 shows that numerous spatial landslide studies have been carried out. These data sets provide a rich information base for future detailed analysis.

Table 3.3 includes references to those spatial data sets providing estimations of landslide susceptibility and hazard. None of these, however, differentiates between different types of landslide. These sources of information have been classified in the table as susceptibility, hazard, zonation, or qualitative assessment. This table demonstrates the performance of numerous spatial analyses throughout the world and the availability of spatial landslide susceptibility and hazard estimates for numerous catchments and regions.

**Table 3.2** *Sources of information on spatial landslide distribution and inventories for different regions worldwide*

| Continent | Country | Region | Type | Reference(s) |
|---|---|---|---|---|
| | | | Catchment and regional scale | |
| Africa | Nigeria | General | Information | Schoeneich and Bouzou (1996) |
| | | Southern Nigeria | Distribution | Okagbue (1994) |
| | Southern Africa | General | Distribution | Paige-Green (1989) |
| Asia | China | Yunnan Province | Inventory | Tang and Grunert (1999) |
| | | Gansu region | Distribution | Derbyshire *et al.* (1991) |
| | | Hong Kong | Inventory | Brand *et al.* (1984); Chan *et al.* (2003); King (1999); Pun *et al.* (2003); Wong and Hanson (1995) |
| | India | Darjeeling | Inventory | Basu (2001); Jana (2000); Sarkar (1999) |
| | | Northeastern India | Susceptibility | Gupta (2000) |
| | Japan | Hokkaido | Inventory | Yamagishi *et al.* (2002) |
| | | Kobe | Distribution | Sassa *et al.* (1999) |
| | Jordan | Northern & Central | Distribution | Farhan (1999) |
| | Korea | Gyeonggi Province | Susceptibility | Kim *et al.* (2001) |
| | Taiwan | Western Foothills | Frequency and spatial distribution | Chang and Slaymaker (2002) |
| | | Central Range | Inventory | Hovius *et al.* (2000) |
| Europe | Croatia | Medvednica Range | Distribution | Jurak *et al.* (1998) |
| | Czech Republic | Vizovická vrchovina Highland | Distribution | Kirchner (2002) |
| | France | Mercantour Massif, French Riviera | Inventory | Julian and Anthony (1996) |
| | Germany | Bonn Region | Inventory | Grunert and Schmanke (1997); Hardenbicker (1994) |
| | | Rheinhessen | Inventory | Dikau and Jäger (1995) |
| | | Hessen, Thüringen | Inventory | Schmidt and Beyer (2001) (2002) |
| | | Schwäbische Alb | Inventory | Bibus and Terhorst (1999); Schädel and Stober (1988); Thein (2000) |
| | | Fränkische Alb | Inventory | Moser and Rentschler (1999); Streit (1991) |
| | | Bavarian Alps | Inventory | Mayer *et al.* (2002); von Poschinger and Haas (1997) |

| Country | Region | Type | Reference |
| --- | --- | --- | --- |
| Great Britain | Isle of Wight | Distribution | Hutchinson and Bromhead (2002) |
| | Scotland | Distribution | Ballantyne (1997) |
| | South Coast | Temporal and spatial distribution | Brunsden and Ibsen (1994) |
| | South Kent | Distribution | Bromhead *et al.* (1998) |
| Hungary | Danubian Bluffs | Inventory | Kertész and Schweitzer (1991) |
| | Hernád Valley | Distribution | Szabó (1999) |
| Italy | Calabria | Distribution | Sorriso-Valvo (1997) |
| | Cardoso T. basin | Inventory | D'Amato Avanzi *et al.* (2000) |
| | Cortina d'Ampezzo | Inventory | Panizza *et al.* (1996) (1997); Pasuto and Soldati (1999) |
| | Naples | Distribution | Calcaterra *et al.* (2002) |
| | Northern Calabria | Inventory | Carrara and Merenda (1976) |
| | Central Calabria | Distribution | Antronico and Gullà (2000) |
| | Pizzo d'Alvano | Distribution | Gudagno and Zampelli (2000) |
| | Sicilia & Southestern | Distribution | Nicoletti *et al.* (2000); Pantano *et al.* (2002) |
| | Umbria region | Inventory | Guzzetti and Cardinali (1990); Guzzetti *et al.* (1994) (2002b) |
| Poland | Carpathians | Inventory | Alexandrowicz (1993); Alexandrowicz (1997); Margielewski (2002); Ostaficzuk (1999); Starkel (1997) |
| Portugal | | Distribution | Zêzere *et al.* (1999) |
| Romania | General | Inventory | Ielenicz *et al.* (1999); Rosenbaum and Popescu (1996) |
| Slovakia | Orava region | Distribution | Janova (2000) |
| Spain | Asturias, Meredela Valley | Distribution | Cuesta *et al.* (1999); Sánchez *et al.* (1999) |
| | Barranco de Tirajana Basin, Gran Canaria | Distribution | Lomoschitz (1999) |
| | Izbor basin, Granada | Inventory | El Hamdouni *et al.* (2000) |
| | Los Guajares Mountains, Granada | Inventory | Fernandez *et al.* (1996); Irigaray *et al.* (1996) |
| | La Pobla de Lillet area | Inventory | Santacana *et al.* (2003); Santacana and Corominas (2002) |

**Table 3.2** *(Continued)*

| Continent | Country | Region | Type | Reference(s) |
|---|---|---|---|---|
| | | Catchment and regional scale | | |
| | | Río Serpis basin | Inventory | van Beek (2002) |
| | | Sorbas | Inventory & Distribution | Hart and Griffiths (1999) |
| | | Southeastern Pyrenees | Distribution | Moya *et al.* (1997) |
| | Sweden | Kärkevagge | Distribution | Jonasson *et al.* (1997) |
| | UK | Broadway area | Distribution | Whitworth *et al.* (2000) |
| | | Scarborough coast | Distribution | Lee and Clark (2000) |
| Northern America | Canada | Alberta | Inventory | Cruden (1996) |
| | | Capilano Watershed, British Columbia | Inventory | Brardinoni *et al.* (2003) |
| | | Queen Charlotte Islands | Inventory | Martin *et al.* (2002) |
| | | Vancouver Island | Inventory | Guthrie (2002) |
| | Puerto Rico | Tropical region | Inventory | Larsen and Torres-Sanchez (1998) |
| | USA | New Mexico | Inventory | Brabb *et al.* (1989); Dikau and Jäger (1995); Reneau and Dethier (1996) |
| | | Northridge, California | Inventory | Harp and Jibson (1995) |
| | | San Fransisco Bay | Inventory | Ellen and Wieczorek (1988); Wieczorek (1984) |
| | | Utah | Inventory | Hylland and Lowe (1997) |
| | | Lewis County, Washington | Inventory | Dragovich *et al.* (1993) |
| Southern America | Brazil | Rio de Janeiro | Inventory | Amaral and Palmeiro (1997); Amaral *et al.* (1996); Jones (1973) |
| | Chile | Antofagasta | Distribution | Van Sint Jan (1994) |
| | | Rinihue | Distribution | Erickson *et al.* (1989) |
| | Colombia | Cudinamarca | Inventory | Forero-Duenas and Caro-Pena (1996) |
| | | Paez region | Distribution | Martinez *et al.* (1995) |
| | | Different regions | Inventory | van Westen (1994) |
| | Ecuador | | Distribution | Schuster *et al.* (1996); Tibaldi *et al.* (1995) |

| Region | Country | Location | Type | References |
| --- | --- | --- | --- | --- |
| Pacific | El Salvador | Corillera Costera | Distribution | Agnesi *et al.* (2002a) |
| | Peru | Nevados Huascaran | Distribution | Keefer (1984); Plafker *et al.* (1971) |
| | Fiji | Viti Levu, Wainitubatolu Catchment | Distribution | Crozier *et al.* (1981) |
| South Pacific | Philippines | Luzon | Distribution | Arboleda and Punongbayan (1999) |
| | Salomon Island | MISSING | Distribution | Trustrum *et al.* (1990) |
| | Australia | Bumbunga Hill | Distribution | Twidale (2000) |
| | New Zealand | Hawke Bay | Inventory | Glade (1997); Harmsworth *et al.* (1987); Page *et al.* (1994) |
| | | Taranaki | Distribution | Crozier and Pillans (1991); DeRose *et al.* (1993) |
| | | Waipaoa | Distribution | Page *et al.* (1999) |
| | | Wairarapa | Inventory | Crozier *et al.* (1980); Glade (1997); Trustrum and Stephens (1981) |
| | | Wairoa | Distribution | Douglas *et al.* (1986) |
| | | Wellington | Inventory | Brabhaharan *et al.* (1994); Crozier *et al.* (1978); Eyles *et al.* (1974) (1978); Glade (1997) |

National scale

| Region | Country | | Type | References |
| --- | --- | --- | --- | --- |
| Asia | Armenia | | Distribution | Boynagryan *et al.* (2000) |
| | China | | Inventory | Yin *et al.* (2002) |
| Europe | Austria | | Inventory of large landslides | Moser (2002) |
| | France | | Inventory | Asté *et al.* (1995) |
| | Hungary | | Distribution | Juhász (1997) |
| | Italy | | Inventory | Guzzetti *et al.* (1994) |
| | Spain | | Distribution | Ferrer and Ayala-Carcedo (1997) |
| | United Kingdom | | Inventory | Jones and Lee (1994); Lee *et al.* (2000) |
| North America | USA | | Inventory | Brabb *et al.* (1999); Eldredge (1988); Wieczorek (1984) |
| South Pacific | New Zealand | | Inventory | Glade (1996); Harmsworth and Page (1991) |

**Table 3.3** *Sources of information on spatial landslide susceptibility and hazard for different region of the world*

| Continent | Country | Region | Type of analysis | Reference(s) |
|---|---|---|---|---|
| | | | Catchment and regional scale | |
| Africa | Ethiopia | Blue Nile Basin | Susceptibility | Ayalew (2000) |
| Asia | China | Gansu Province | Hazard | Meng *et al.* (2000) |
| | | Hong Kong | Susceptibility | Dai and Lee (2001)(2002); Lee *et al.* (2001) |
| | | Lawngthlai, Southern Miziram | Susceptibility | Khullar *et al.* (2000) |
| | India | Darjiling, Himalaya | Susceptibility | Basu (2000) |
| | | Garhwal Himalaya | Susceptibility | Anbalagan *et al.* (2000) |
| | | Munipur River basin | Susceptibility | Nagarajan (2002) |
| | | Rakti Basin | Susceptibility | Bhattacharya (1999) |
| | Iran | Jiroft watershed | Susceptibility | Uromeihy (2000) |
| | | Khorshrostam area | Susceptibility | Mahdavifar (2000) |
| | | Shahrood drainage basin | Susceptibility | Feiznia and Bodaghi (2000) |
| | Japan | Amahata River basin | Susceptibility | Aniya (1985) |
| | | Fukushima Pref. | Susceptibility | Sasaki *et al.* (2002) |
| | | Hanshin district | Susceptibility | Kamai *et al.* (2000) |
| | | Higashikubiki region | Susceptibility | Iwahashi *et al.* (2003) |
| | | MISSING | Susceptibility | Kubota (1994) |
| | | MISSING | Susceptibility | Massari and Atkinson (1999) |
| | Jordan | Wadi Mujib Canyon | Susceptibility | De Jaeger (2000) |
| | Korea | Yanghung area | Susceptibility | Lee *et al.* (2002) |
| | | Yongin | Susceptibility | Lee and Min (2001) |
| | Nepal | Kulekhani watershed | Susceptibility | Dhakal *et al.* (2000) |
| | Ukraine | Southern region | Susceptibility | Cherkez *et al.* (2000) |
| Europe | Austria | Bad Ischl | Susceptibility | Fernández-Steeger *et al.* (2002) |
| | Belgium | Manaihant | Susceptibility | Demoulin and Chung submitted |
| | Czech Republic | North Bohemia | Susceptibility | Hroch *et al.* (2002) |
| | Germany | Bonn region | Susceptibility | Schmanke (2001) |
| | | Rheinhessen | Susceptibility/ hazard | GLA (1989); Jäger (1997) |
| | | Schwäbische Alb | Susceptibility | Thein (2000) |
| | | Hessen and Thüringen | Susceptibility | Schmidt and Beyer (2001) (2002) |

| | | | |
|---|---|---|---|
| Great Britain | Barbados, Scotland | Susceptibility | Hodgson *et al.* (2002) |
| | Starkholmes, Derbyshire | Susceptibility | Thurston and Degg (2000) |
| Italy | Bratica T. Basin | Susceptibility | Clerici (2002) |
| | Calabria Region | Susceptibility | Carrara *et al.* (1977b) |
| | Centenora catchment, Northern Apennines | Susceptibility | Casadei and Farabegoli (2003) |
| | Corniglio | Hazard | Froldi and Bonini (2000) |
| | MISSING | Susceptibility | Carrara (1989) |
| | Mignone basin | Hazard | Del Monte *et al.* (2003) |
| | Forlì-Cesena, Emilia Romagna | Susceptibility | Pistocchi *et al.* (2002) |
| | Lecco province, Lombardy region | Susceptibility | Frattini and Crosta (2002) |
| | Messina Straits Crossing site | Susceptibility | Baldelli *et al.* (1996) |
| | Orcia drainage basin | Hazard | Del Monte *et al.* (2003) |
| | Potenza region | Susceptibility | Ferrigno and Spilotro (2002) |
| | Trionto basin | Hazard | Del Monte *et al.* (2003) |
| | Umbria, Marche Regions | Susceptibility and hazard | Carrara *et al.* (1995); Guzzetti *et al.* (1999) |
| Portugal | Coimbra region | Susceptibility | Tavares and Soares (2002) |
| | Fanhoes-Trancao Region | Susceptibility | Fabbri *et al.* (2002); Zêzere *et al.* (2000) |
| Slovak Republic | Handlovská kotlina Basin | Susceptibility | Paudits and Bednárik (2002) |
| | Kosice region | Susceptibility | Petro *et al.* (2002) |
| Spain | Deba Valley, Province Guipuzoa | Susceptibility | Fabbri *et al.* (2002) |
| | Rio Aguas | Susceptibility | Griffiths *et al.* (2002) |
| | Malorca Island | Susceptibility | Mateos Ruiz (2002) |
| | La Pobla de Lillet area | Susceptibility | Baeza and Corominas (1996); Santacana *et al.* (2003) |
| | Jerte Valley | Susceptibility | Carrasco *et al.* (2000) |
| Turkey | Mengen | Susceptibility | Gökceoglu and Aksoy (1996) |
| | Yenice | Susceptibility | Ercanoglu and Gokceoglu (2002) |
| North America | Canada | Hull-Gatineau region, Quebec | Susceptibility | Clouatre *et al.* (1996) |
| | | Lilloet River watershed, British Columbia | Susceptibility | Holm *et al.* (2004) |

**Table 3.3** (Continued)

| Continent | Country | Region | Type of analysis | Reference(s) |
|---|---|---|---|---|
| | | | Catchment and regional scale | |
| | USA | Anchorage | Susceptibility | Dobrovolny (1971) |
| | | Cincinnati, Ohio | Hazard | Bernknopf *et al.* (1988) |
| | | Coos Bay, Oregon | Susceptibility | Casadei and Dietrich (2003) |
| | | Oregon Coast Range | Susceptibility | Schmidt *et al.* (2001) |
| | | San Mateo County | Susceptibility | Brabb (1993); Brabb *et al.* (1978); Roth (1983) #2940 |
| | | Travis County, Texas | Susceptibility | Wachal and Hudak (2000) |
| | | Washington State | Hazard | Harp *et al.* (1997) |
| | Jamaika | St Andrew | Susceptibility | Maharaj (1993) |
| | Argentina | Mendoza province | Susceptibility | Moreiras (2004) |
| South America | Brazil | Rio de Janeiro | Susceptibility | Barros *et al.* (1991); Fernandes *et al.* (2004) |
| | Colombia | Chinchina region | Hazard | Chung *et al.* (1995); Chung *et al.* (2003); van Asch *et al.* (1992) |
| | El Salvador | Corillera del Balsamo | Susceptibility | Agnesi *et al.* (2002b) |
| Pacific Islands | Fiji | Viti Levu | Susceptibility | Crozier (1989); Greenbaum *et al.* (1995) |
| | Papa New Guinea | Ok Tedi | Susceptibility | Crozier (1991) |
| South Pacific | Australia | Southeast Queensland | Susceptibility | Hayne and Gordon (2001) |
| | NewZealand | Hawke Bay | Susceptibility | Glade (2001) |
| | | Wairarapa | Hazard | Wilson and Crozier (2000) |
| | | | National scale | |
| Africa | South Africa | | Zonation based on expert judgement | Paige-Green (1985) |
| Asia | China | | Hazard mapping and management | Tianchi (1996) |
| Europe | Germany | | Qualitative assessment | Dikau and Glade (2003); Glade *et al.* in prep.a |

### 3.4.2   Rock Slope Analysis

Spatial rock slope analysis focuses mainly on rockfalls and rock slides, the latter mostly of large dimension. Information on spatial studies on rockfalls, topples, slides and avalanches is summarized in Table 3.4. Inventories give spatial distributions (e.g. Gardner, 1983; Luckman, 1972; McSaveney, 2002). Other inventories have been further analysed using statistical approaches (e.g. Bartsch *et al.*, 2002) and apply empirical models to spatial rockfall analysis (e.g. Dorren and Seijmonsberegen, 2003; Meißl, 2001; Wieczorek *et al.*, 1998). Most recently, numerical models have been developed to calculate spatial movement patterns (e.g. Guzzetti *et al.*, 2002a). Although a few general national inventories provide information on rockfalls and topples (e.g. Guzzetti *et al.*, 1994), no nationwide inventory has been carried out specifically for rock slope events.

### 3.4.3   Debris-flow Analysis

In contrast, debris flows have been investigated at catchment, regional and national scales (Table 3.5). Such investigations have been focused on general inventories of spatial debris-flow occurrence (e.g. Calcaterra *et al.*, 1996a) or on distributions following distinct triggering events (e.g. Del Prete *et al.*, 1998; Pareschi *et al.*, 2000; Rickenmann, 1990; Villi and Dal Pra', 2002). Statistical techniques along with numerical approaches to assess debris-flow susceptibility and hazard have been applied in various regions worldwide (e.g. D'Ambrosio *et al.*, 2003a; D'Ambrosio *et al.*, 2003b; Lorente *et al.*, 2002; Mark and Ellen, 1995). Besides the catchment and regional analysis, national scale investigations have also been carried out. For example, maps showing the reported debris flows, debris avalanches and mudflows (Bert, 1980), as well as inventory and regional susceptibility for Holocene debris flows and related fast-moving landslides (Brabb *et al.*, 1999), are available for the USA or for Switzerland (Zimmermann *et al.*, 1997).

### 3.4.4   Slide Analysis

References related to spatial assessments of soil and earth flows and slides are summarized in Table 3.6. While some authors record deep-seated landslides only (e.g. Yamagishi *et al.*, 2002), others focus on shallow translational slides. Several papers employ infinite limiting equilibrium slope stability analysis. This method has been applied in particular to shallow landsliding (e.g. Dietrich *et al.*, 1995; Montgomery and Dietrich, 1994; Montgomery *et al.*, 2000; Wu and Abdel-Latif, 2000) to estimate the factor of safety (FoS) and probability of failure. Derived from hydrological response units, soil mechanical response units have been suggested by Möller *et al.* (2001) for application to the infinite slope model. Some authors also include soil root strength (e.g. Ekanayake and Phillips, 1999). Simple heuristic techniques are also applied to national scale investigations (e.g. Fallsvik and Viberg, 1998; Viberg *et al.*, 2002). In addition, Perov *et al.* (1997) presented a global distribution of mudflows. Although this analysis is based on expert judgement, it gives a first approximation of mudflow distributions, thus providing a starting point for further, more detailed analysis applying more advanced models.

### 3.4.5   Summary

Tables 3.2 to 3.6 demonstrate the wide application of spatial landslide analysis over the last thirty years. Types of information range from landslide distributions and inventories to

**Table 3.4** *Selections of spatial assessments of rock topple, fall, slide and avalanche for different regions of the world*

| Continent | Country | Region | Type of analysis | Reference(s) |
| --- | --- | --- | --- | --- |
| Asia | Pakistan | Karakoram Himalaya | Distribution of large failures | Hewitt (2002) |
| Europe | European Alps | | Distribution of large failures | Abele (1974); Heim (1932); von Poschinger (2002) |
| | Austria | Innsbruck | Process-based modelling | Meißl (2001) |
| | | Gaschurn, Montafon | Numerical modelling of rock falls | Dorren *et al.* (2004); Dorren and Seijmonsberegen (2003) |
| | | Vorarlberg | Process-based modelling | Ruff *et al.* (2002) |
| | Germany | Oker basin | Rock slide modelling | Günther *et al.* (2002a) |
| | Ireland | Co. Antrim | Rock fall distribution | Douglas (1980) |
| | Italy | Camonica Valley, Lombardi region | Numerical modelling of rock falls | Guzzetti *et al.* (2002a) |
| | | Valle San Giacmo | Rock fall modelling | Mazzoccola and Sciesa (2000) |
| | Spain | Northern Spain | Rock fall susceptibility | Duarte and Marquinez (2002) |
| | Sweden | Kärevagge | Rock fall – statistical analysis | Bartsch *et al.* (2002) |
| North America | Canada | Canadian Cordillera | Distribution of large failures | Cruden (1985) |
| | | Surprise Valley, Jasper National Park | Rock fall inventory | Luckman (1972) |
| | | Highwood Pass Area, Alberta | Rock fall and slide inventory | Gardner (1983) |
| | USA | Yosemite Valley | Rock fall hazard | Guzzetti *et al.* (2003); Wieczorek *et al.* (1998) |
| | | Tully Valley Area, Finger Lakes Region, New York | Rock fall susceptibility | Jäger and Wieczorek (1994) |
| South America | Argentinia | Puna Plateau | Inventory of rock avalanches | Hermanns *et al.* (2002) |
| South Pacific | New Zealand | Mount Cook National Park | Rock fall and avalanches | McSaveney (2002) |

**Table 3.5** *Selections of spatial susceptibility and hazard analysis of debris flow for different regions of the world*

| Continent | Country | Region | Type of analysis | Reference(s) |
|---|---|---|---|---|
| | | | Regional and catchment scale | |
| Asia | Japan | Miyakejima volcano | Distribution | Yamakoshi *et al.* (2003) |
| | Kazakstan | Southeast | Susceptibility | Medeuov and Beisenbinova (1997) |
| | Nepal | Kulekhani watershed | Distribution | Dhital (2003) |
| | Taiwan | Chen-You-Lan River basin | Hazard | Lin *et al.* (2000) |
| | | Central taiwan | Distribution | Cheng *et al.* (2003) |
| Europe | Austria | Salvensen Valley | Distribution | Becht and Rieger (1997) |
| | Germany | Faltenbach Valley | Susceptibility | Becht and Rieger (1997) |
| | Iceland | Gleidarhjalli area | Hazard | Decaulne and Saemundsson (2003) |
| | | Northwestfjords region | Susceptibility | Glade and Jensen (2004) |
| | Italy | Serre Massif – Calabria | Distribution | Calcaterra *et al.* (1996a) |
| | | Circum-Vesuvian areas & Sarno Mountains | Distribution/Hazard | Calcaterra *et al.* (2000); Cinque *et al.* (2000); D'Ambrosio *et al.* (2003a) (2003b); Del Prete *et al.* (1998); Fiorillo *et al.* (2001); Pareschi *et al.* (2000) |
| | | Isarco Valley | Distribution | Villi and Dal Pra (2002) |
| | | Lecco area, Lombardy | Susceptibility | Bathurst *et al.* (2003) |
| | | Versilia, Garfagnana | Susceptibility | Martello *et al.* (2000) |
| | Spain | Upper Aragón and Gállego Valley, Central Pyrenees | Bivariate Statistics | Lorente *et al.* (2002) |
| | Switzerland | Mattertal, Wallis | Distribution | Dikau *et al.* (1996) |
| North America | USA | Honolulu of Oahu, Hawaii | Hazard | Ellen and Mark (1993); Ellen *et al.* (1993); Reid *et al.* (1991) |
| | | Madison County, Virginia | Hazard | Wieczorek *et al.* (2003) |
| | | Mount Rainier, Washington | Hazard | Hoblitt *et al.* (1995); Iverson *et al.* (1998); Schilling and Iverson (1997); Scott *et al.* (1995) |

**Table 3.5**  (Continued)

| Continent | Country | Region | Type of analysis | Reference(s) |
| --- | --- | --- | --- | --- |
| | | Regional and catchment scale | | |
| | | Northwestern California | Distribution | Reid *et al.* (2003) |
| | | Noyo watershed, California | Susceptibility | Dietrich and Sitar (1997) |
| | | Oakland, California | Susceptibility | Campbell and Bernkopf (1997); Campbell *et al.* (1994) |
| | | Oregon | Inventory & Susceptibility | Hofmeister (2000); Hofmeister and Miller (2003) |
| | | San Mateo County, California | Susceptibility | Mark (1992) |
| | | Santa Cruz Mountain, California | Inventory | Wieczorek (1984) |
| | | Blue Ridge of Central Virginia | Hazard | Wieczorek *et al.* (2000) |
| | | Wasatch Front, Utah | Distribution | Wieczorek *et al.* (1989) |
| South America | Ecuador | Pichincha massif | Hazard | Canuti *et al.* (2002) |
| | El Salvador | San Salvador, San Vicente & San Miguel volcanoes | Hazard | Major *et al.* (2003) |
| | Venezuela | Northern region | Distribution | Lopez *et al.* (2003) |
| South Pacific | Australia | Montrose, Victoria | Hazard | Fell and Hartford (1997) |
| | | Wollongong | Distribution | Flentje *et al.* (2000) |
| | | National scale | | |
| Europe | Austria | | Distribution | Andrecs (1995) |
| | Switzerland | | Distribution | Rickenmann (1990); Zimmermann *et al.* (1997) |
| North America | USA | | Inventory of debris flow, avalanches, and mud flows | Bert (1980); Brabb *et al.* (1999) |

**Table 3.6**  Selections of spatial assessment of shallow translational and rotational earth and soil slides for different regions of the world

| Continent | Country | Region | Type of analysis | Reference(s) |
| --- | --- | --- | --- | --- |
| | | | Catchment and regional scale | |
| Asia | Japan | Hokkaido | Inventory | Yamagishi *et al.* (2002) |
| | | Taiyo-no Kuni | Inventory | Chigira (2002) |
| | | Sasebo district | Numerical 3d modelling | Xie *et al.* (2001) |
| Europe | Bulgaria | Baltchik area | Distribution | Koleva-Rekalowa *et al.* (1996) |
| | Germany | Rheinhessen | Physically based modelling | Möller *et al.* (2001) |
| | Italy | MISSING | Physically based modelling | Ekanayake and Phillips (1999) |
| | | Lemezzo basin, Piemonte region | Physically based modelling | Campus *et al.* (2001) |
| | | Serre Massif, Calabria | Distribution | Calcaterra *et al.* (1996b) |
| North America | Canada | Grondines and Trois Rivieres areas, Quebec | Distribution | Karrow (1972); Mollard and Hughes (1973) |
| | | Lemieux, Ontario | Distribution | Evans and Brooks (1999) |
| | USA | Northern California | Physically based modelling | Dietrich *et al.* (1995) |
| | | Van Duzen River basin, California | Distribution | Kelsey (1978) |
| | | Seattle, Washington | Hazard | Savage *et al.* (2003) |
| | | South Fork of Tilton River, Cascade Mountains, Washington State | Mechanics based approach | Wu and Abdel-Latif (2000) |
| | | MISSING | Physically based modelling | Montgomery and Dietrich (1994); Montgomery *et al.* (2000) |
| South America | Ecuador | Gordeleg catchment | Physically based modelling | Vanacker *et al.* (2003) |
| South Pacific | New Zealand | Otago | Distribution | Crozier (1968) (1969) (1996) |
| | | | National scale | |
| Asia | USSR | | Qualitative assessment | Perov and Budarina (2000); Sidorova (1997) |
| | Central & Southeast | | Qualitative assessment | Belaia *et al.* (2000) |
| Europe | Sweden | | Qualitative assessment | Fallsvik and Viberg (1998); Viberg *et al.* (2002) |
| | | | World | |
| | | | Qualitative assessment | Perov *et al.* (1997) |

advanced mathematical modelling of spatial data sets at catchment, regional and national scales. At regional scales, statistical models have been widely applied to assess landslide susceptibility (e.g. Baeza and Corominas, 2001; Carrara, 1983, 1989; Carrara *et al.*, 1977a; Fernandes *et al.*, 2004; Griffiths *et al.*, 2002; Jäger, 1997) and hazard (e.g. Guzzetti *et al.*, 1999; van Asch *et al.*, 1992). Also statistical techniques such as the fuzzy approach (e.g. Ercanoglu and Gokceoglu, 2002; Pistocchi *et al.*, 2002) as well as different probabilistic prediction models (e.g. Pistocchi *et al.*, 2002 or most recently Chung, Chapter 4 in this book) have been applied recently to assess landslide susceptibility. At national scale, Paige-Green (1985) has produced a classification of different susceptibility classes based on expert judgement. Information on landsliding in Great Britain was summarized by Jones and Lee (1994) and a comprehensive landslide inventory is provided by Guzzetti *et al.* (1994) for Italy. For Germany, a national landslide susceptibility map was estimated based on lithology and slope geometry (Dikau and Glade, 2003). The latter examples show, despite the fact of landslide occurrence at distinct locations or within restricted regions, the large potential for analysis at the national scale. Any available local or regional landslide information can be used to validate and verify the results gained at national scale analysis. Although major differences in the resolution and quality of basic data sets and in the type of analysis appear, spatial landslide information is available and provides a valuable source for further analysis, for example to estimate regional landslide risk by combination with elements at risk and respective socio-economic attributes. For some regions, such regional landslide risk estimates have already been carried out. Some examples are given in the following section.

## 3.5  Landslide Risk Assessments

The history and basic concepts of landslide risk assessments and analysis are explained in Chapters 1 and 2 of this book (refer also to Chowdhury, 1988; Evans, 1997; Kong, 2002). The following section summarizes regional examples of landslide risk assessment. Due to limited information, Table 3.7 does not distinguish between different landslide types, nor between different methods used to assess the elements at risk and the respective consequences. Methods may involve different spatial resolution of elements at risk (e.g. single houses versus 'urban settlement') and different depth of quality and quantity of socio-economic data (e.g. monetary value of a building including its content or of an industrial site including goods, number of persons of different ages in a house versus 'population density', population per km$^2$). Such socio-economic data are fundamental to an accurate assessment of vulnerability (Romang *et al.*, 2003). Comprehensive expressions of vulnerability involve not only structural measures (e.g. the degree of damage to a building hit by a given magnitude debris flow), but have also a social dimension (e.g. coping capacity (resilience) of the affected person/family/community) as described by Solana and Kilburn (2003).

Once landslide hazard maps have been produced and further spatial information on potential consequences is available, landslide risk can be estimated (e.g. Wu *et al.*, 1996). Thus the consequences of the natural hazard occurring are the product of the elements at risk and the vulnerability. A measure of vulnerability is essential for the determination of consequences and is defined as the degree of loss for a given element at risk, or set

**Table 3.7** References on spatial landslide risk assessments for different regions of the world

| Continent | Country | Region | Type | Reference(s) |
| --- | --- | --- | --- | --- |
| Catchment and regional scale | | | | |
| Asia | China | Yunnan Province | Debris-flow risk | Liu *et al.* 2002 |
| | | Hong Kong | Analysis; Quantitative risk assessments | Hardingham *et al.* (1998); Ho and Wong (2001); Moore *et al.* (2001); Papin *et al.* (2001); Pinches *et al.* (2001); Reeves *et al.* (1998); Smallwood *et al.* (1997) |
| | India | Kumaun Himalaya | Assessment | Anbalagan and Singh (1996) |
| | Taiwan | Fong-Chui area | Debris-flow risk | Lin (2003) |
| Europe | Germany | Rheinhessen | Analysis | Glade *et al.* in prep.b |
| | Iceland | Bíldudalur | Analysis | Bell and Glade (2004) |
| | Italy | Italian Alps | Assessment | Eusebio *et al.* (1996) |
| | | Northern Calabria | | Ragozin (1996) |
| | | Piecmont region | Risk assessment | Aleotti *et al.* (2000) |
| | | Sarno region | Debris-flow risk | Toyos *et al.* (2003) |
| | | Umbria region | Qualitative risk assessment | Cardinali *et al.* (2002) |
| | Switzerland | La Veveyse and Veveyse de Figre valleys | Risk assessment | Sarkar *et al.* (2000) |
| Northern America | Canada | Vancover | Debris-flow risk | Morgan *et al.* (1992) |
| | USA | Alameda County, California | Landslide damage | Godt *et al.* (2000); Godt and Savage (1999) |
| | | Montrose, Victoria | Debris-flow risk zoning | Moon *et al.* (1991) |
| | | Seattle, Washington | Debris-flow risk | Gori *et al.* (2003) |
| Southern America | Argentinia | Rio Grande Basin | Zonation mapping | Espizua and Bengochea (2002) |
| | Ecuador | Precupa | Hazard and vulnerability map | Basabe and Bonnard (2002) |
| Pacific | Indonesia | Yogyakarta | Lahar risk assessment | Lavigne (1999) |
| South Pacific | Australia | Cairns | Quantitative landslide risk assessment | Michael-Leiba *et al.* (2000); Michael-Leiba *et al.* (2003) |
| | | Wollongong | Risk assessment | Flentje *et al.* (2000) |
| National scale | | | | |
| Europe | Italy | | Assessment | Guzzetti (2000) |

of elements at risk, resulting from event occurrence of a given magnitude (Newman and Strojan, 1998). Vulnerability is commonly expressed on a scale of 0 (no loss) to 1 (total loss) and is expressed either in monetary terms, such as the loss experienced by a given property, or to loss of life. The vulnerability concept has been reviewed for landslide risk assessments by Alexander (Chapter 5 in this book) and Glade (2004).

The risk concept (*hazard × elements at risk × vulnerability*) (UNDRO, 1982) has been transferred to landslides issues by various authors (Brabb, 1984; Einstein, 1988; Fell, 1994; Gill, 1974; Hearn and Griffiths, 2001; Hicks and Smith, 1981; Leone *et al.*, 1996; Leroi, 1996; Stevenson, 1977; Stevenson and Sloane, 1980; Wu and Swanston, 1980). One comprehensive publication summarizing various attempts to address landslide risk is the proceedings of a workshop on landslide risk assessment edited by Cruden and Fell (1997). Since then, various case studies have been published on landslide risk (e.g. Cardinali *et al.*, 2002; Dai *et al.*, 2002; Finlay *et al.*, 1999; Guzzetti, 2000; Hardingham *et al.*, 1998; Hearn and Griffiths, 2001; Michael-Leiba *et al.*, 2000). A comprehensive and generalized definition of landslide risk has been proposed by the Australian Geomechanics Society by Fell (2000) and adopted by the IUGS Working Group on Landslides – Committee on Risk Assessment (1997). This report refers not only to the definitions given in Chapter 1 and in the glossary of this book, but also focuses on the notions of 'acceptable', 'tolerable', 'single' (individual) and 'collective' (societal) risk. As a conclusion, however, the majority of landslide hazard and risk literature is based on natural science approaches to assess landslide risk (Aleotti and Chowdhury, 1999). Social science studies looking at coping strategies or resilience capacities of affected communities for landslide occurrence are rather limited in contrast to those available for other natural processes such as floods or earthquakes.

Table 3.7 gives an overview of various spatial landslide risk assessments for different regions worldwide. While some authors present landslide hazard and risk zonation based on mapping procedures (e.g. Espizua and Bengochea, 2002), others propose empirical assessments for specific landslide types, for example debris flows (Liu *et al.*, 2002), or use probabilistic methods to analyse landslide risk (e.g. Chung and Fabbri, 2002; Rezig *et al.*, 1996). Common to all approaches is the attempt to relate socio-economic data to spatial landslide hazard information in order to gain more informative data on the potential consequences of landslide occurrence. Numerous publications are available which use 'risk' in their title and text, but do not cover the risk concept as previously defined. Such studies have not been included in the presented tables. In order to demonstrate the different depth of analysis, the following section gives examples of local and spatial landslide risk assessments at varying levels of generalization.

## 3.6   Examples of Landslide Risk Analysis

Spatial landslide risk analysis provides a valuable tool for gaining risk estimates at the regional scale. As with any spatial assessment, the choice of model type and the performance of the model are strongly dependent on the data sets available for analysis. Two examples of varying depth of analysis and data sets of different resolution give some idea on the variety of details in spatial landslide risk analysis. Hence the focus of the following examples is not on the calculation of the hazard using advanced methods (e.g. Guzzetti *et al.*, 2003); rather it aims to demonstrate the application of different information on elements at risk and potential consequences for spatial landslide risk analysis.

### 3.6.1   A Quantitative Rockfall Risk Analysis in Bíldudalur, Iceland

A comprehensive, object-oriented assessment of landslide risk has been carried out by Glade and Jensen (2004) for Bíldudalur in the northwest fjord region of Iceland (Figure 3.5). To illustrate the result of the applied methodology of risk analysis, the following description focuses on rockfalls. A detailed report of environmental settings of Bíldudalur, local rockfall history along with the method and results of calculating runout zones for rockfalls are described in detail in Glade and Jensen (2004).

Based on this report, Bell and Glade (2004) developed a methodology for landslide risk analysis as part of a general landslide risk assessment. For this methodology, the approach of Heinimann (1999) was applied, which determines the vulnerability of buildings according to building structure and their resistance to rockfalls of different magnitude. Historical data could not be used to prove the reliability of vulnerability values because suitable information was not available. Within the whole historical record, no fatalities have been caused by rockfall events (Glade and Jensen, 2004). Although there is no previous evidence of serious consequences, there still is an inherent risk to life which needs to be calculated to support responsible administration to take appropriate countermeasures. Therefore the probability of loss of life in a building for both individuals (individual risk of life) and all people living or working inside a house (object risk to life, thus a risk to life considering all the people staying inside one building) has been calculated.

Rockfall runout zones determined by Glade and Jensen (2004) have been transformed into hazard zones by attributing a return period to each rock size used within the runout calculations. Rockfall risk was calculated using these hazard zones in combination with potential damage values and respective vulnerabilities of the elements at risk. The spatial distribution of one set of elements at risk (number of residents and employees per building) are shown in Figure 3.6. The consequence analysis was carried out considering the vulnerability, the probability of spatial and temporal impact, as well as the probability of seasonal impact of the rockfall at any given location in the study area. Resulting risk maps include individual risk to life and object risk to life, which are given in Figure 3.7. On these maps, areas with different probabilities of loss of life can be identified (refer to Bell and Glade, 2004 for a comprehensive description).

The individual risk to life due to rockfalls ranges between $1.1 \times 10^{-5}$/year and $5.6 \times 10^{-5}$/year and is thus relatively low (Figure 3.7a). Of the total area, 92% belong to low risk and 8% to very low risk. Taking the total number of people in a building into account (object risk to life), the risk increases (Figure 3.7b) and ranges between $1.6 \times 10^{-3}$/year and $2.1 \times 10^{-5}$/year. For the total region, 4% relate to very low risk, 27% to low risk, 58% to medium risk, and 11% to high risk. The calculated total risk to life is 0.009 deaths per year.

Similar procedures can be used to calculate the monetary risk of the community. One of the main advantages of such an approach is that this type of analysis can be performed for just about any natural processes (e.g. rockfall, debris flow, snow avalanches, tsunami) and a combined multi-risk analysis can be derived (Bell and Glade, 2004). Whether appropriate countermeasures have to be organized is the decision of the responsible

**Figure 3.5** (a) Northwards view to Bíldudalur, Northwest Iceland. Relief difference is approx. 400 m. (b) Rock with diametres up to 1.7 m above a house in Bíldudalur (photos by T. Glade)

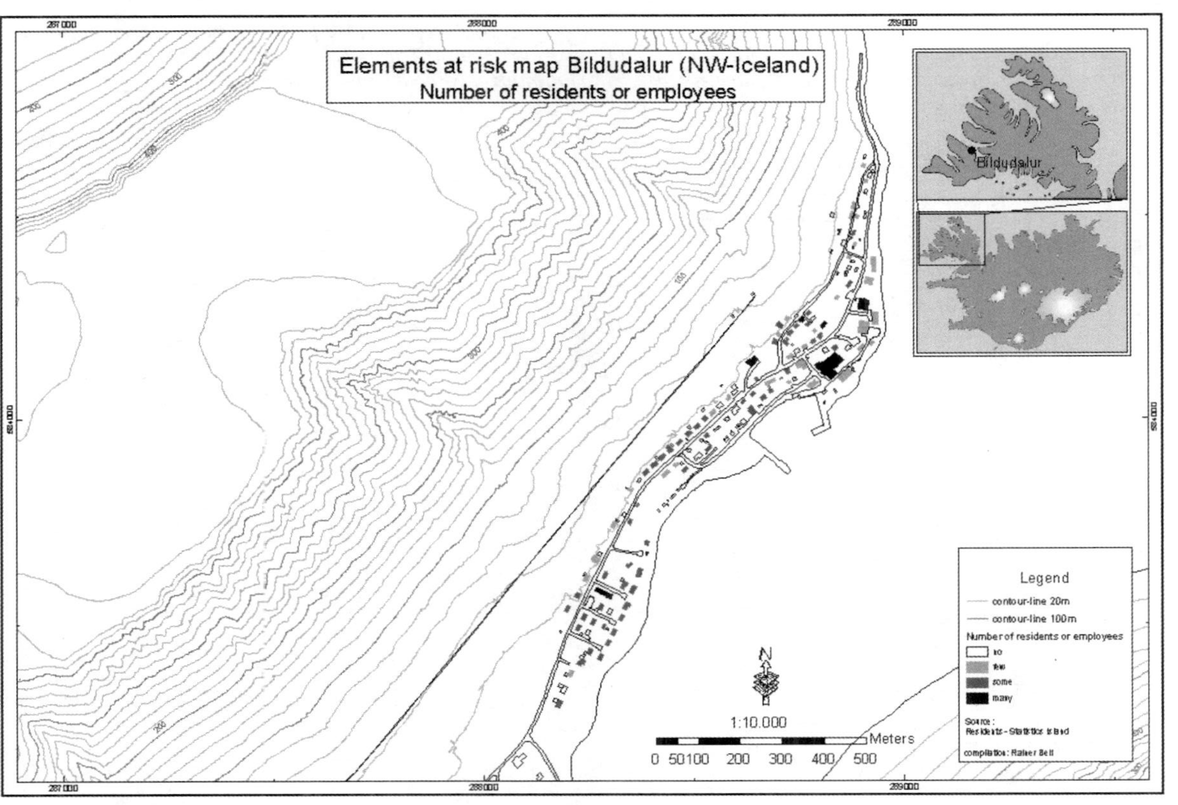


**Figure 3.6** *From all elements at risk, the number of residents and employees per building are given using the four classes of residents: 'no', 'few' (1–2 persons), 'some' (3–6 persons), and 'many' (>7 persons). Eighty-nine buildings are garages and barns and are grouped as 'no' persons, 'few' persons reside in 26 buildings, 46 buildings accommodate 'some' persons, and only two buildings belong to the largest class (Bell and Glade, 2004)*

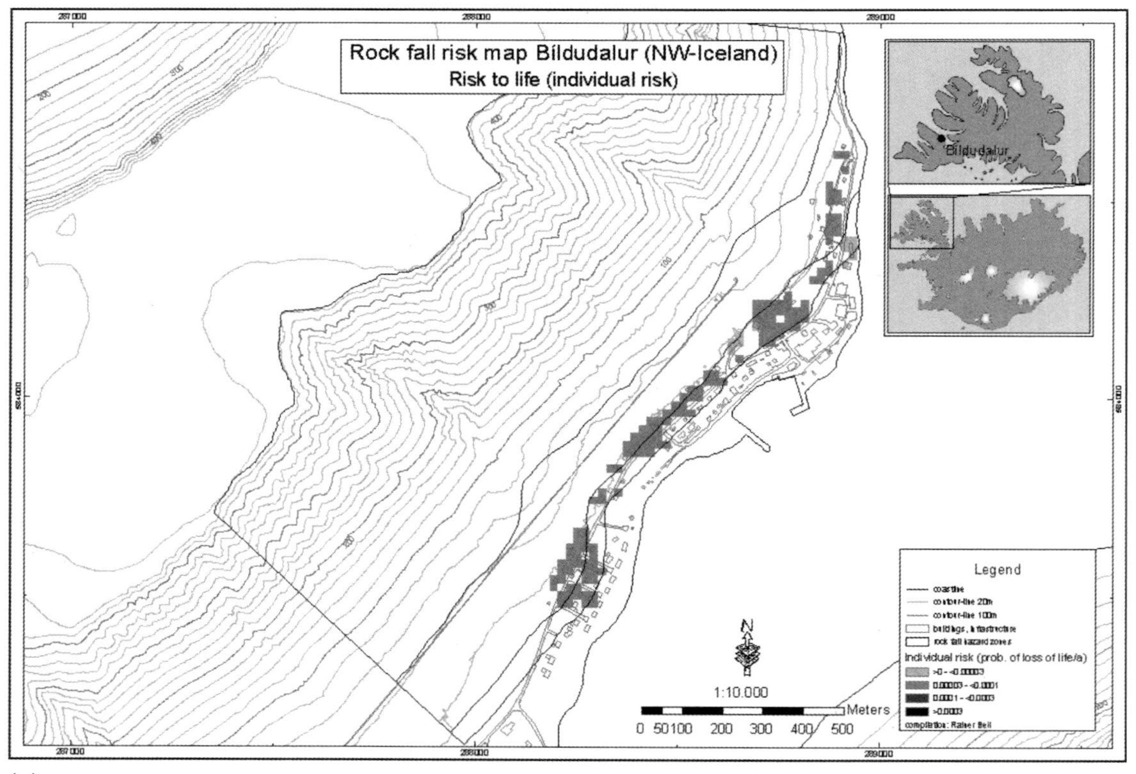


(a)

**Figure 3.7** *The rockfall risk map gives two different types of risks in buildings. (a) refers to the individual risk to life for each person and (b) gives the object risk to life considering all people in a building, and hence is an average risk to life (Bell and Glade, 2004)*

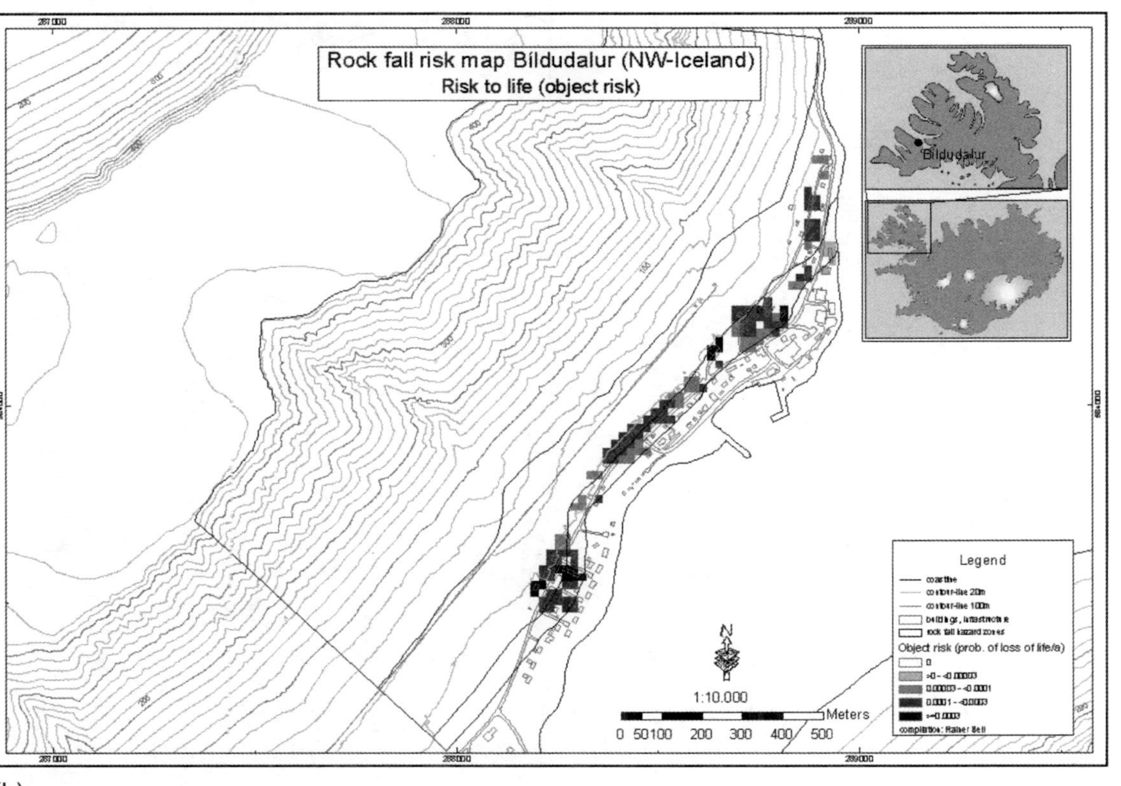


*Figure 3.7* (Continued)

administration. This type of analysis, however, provides the local administrations with important information.

### 3.6.2   A Regional Approach to Address Regional Landslide Risk

The Rheinhessen study was designed to provide a landslide risk analysis by applying simplified vulnerability values and generalized monetary values based on regional mean values. Regional details and the general background of slope instability in Rheinhessen are given in Glade *et al.* (2001b) and Glade *et al.* (in prep.b). Dominant landslide types are shallow translational failures and rotational slides (Figure 3.8).

First, landslide risk analysis is based on landslide hazard map derived by Jäger (1997), but extended resolution using a 20 m DTM instead of the original 40 m. Second, elements at risk have been determined for different land use groups and digitized from official land use plans. Afterwards, for each element at risk, a damage potential has been defined based on literature review and on data from national statistics yearbooks (Table 3.8). For this region, no information on vulnerability of elements at risk from landslide initiation was available. Therefore it was assumed that if an element at risk is affected by a landslide, it is totally destroyed. Consequently, vulnerability has been assigned as 1 to all elements at risk. Due to the low probability that a person will be injured or even killed from a landslide event, risk to life has been excluded from the analysis. Details on methods, analysis and results are given by Glade *et al.* (in prep.b).

The classified elements at risk are summarized in Table 3.8. Respective damage potentials have been assigned to enable a calculation of economic value for each class.

**Figure 3.8**   *Example of the rotational landslide OCK3 in northwest Rheinhessen, view to east (photo by T. Glade)*

***Table 3.8***   *Elements at risk with attributed damage potential in (€/m² ) (refer to Glade et al. (in prep.b) for details of sources and calculations)*

| Risk element | Monetary value (€/m²) | Risk element | Monetary value (€/m²) |
| --- | --- | --- | --- |
| Residential Area | 255 | Pasture | 0.5–0.7 |
| Mixed usage | 255–410 | Agricultural areas | 0.3 |
| Industrial Region | 205–255 | Viniculture | 10 |
| Specialized Region | 205 | Forest | 2 |
| Road | 13–15 | Highway | 85–128 |

These classes have been combined with natural hazard information and the elements at risks. A qualitative matrix of the combination of these parameters resulted in different landslide risk classes, which are shown in the landslide risk map (Figure 3.9 – see also Colour Plate section Plate 1).

The landslide risk map includes 'low', 'medium', 'high' and 'very high' risk classes. Of the total area, 90% has been classified as 'low', 8% as 'medium', 2% as 'high', and 0.2% as 'very high' landslide risk. In general, 'low' risk areas refer to flat or moderately steep slopes with pasture. In contrast, 'high' and 'very high' risk classes represent the steep slope segments with either buildings or vineyards. This result highlights the importance of the potential effects of landslides in the study area, which is representative for the whole Rheinhessen area. Due to its generalized input data, the resulting risk map cannot be used by local administration for detailed planning, but it is of great value for both local and regional governments to locate areas prone to landslide risk and to organize more detailed analysis in the identified 'hot spot' areas.

### 3.6.3   Summary

Both examples demonstrate the potential of landslide risk assessments at various scales and with different levels of analysis. While detailed risk assessments are indispensable for site-specific problems, more generalized risk analysis is also of major importance to gain an overview of a large area. Besides the scale of interest of the administrative authorities, detail of analysis is also highly dependent on numerous other factors such as financial resources, time constraints, data availability and quality. However, it is important to use the resources in the most profitable way to provide methods and concepts which can be applied to gain the most benefit from lowest costs.

## 3.7   Influence of the Triggering Agent

The previous discussion on local and spatial landslide investigations gave no details of the respective landslide triggering agents. Nearly all reviewed landslide investigations are related either to rainfall and subsequent soil moisture regimes or to earthquake triggers. In terms of establishing an inventory or a susceptibility map, the landslide trigger is of minor importance. Irrespective of the cause, the principal interest of these investigations is the landslide location and the environmental factors, which give some indication of landslide susceptibility. Indeed, some environmental factors are more important for earthquakes

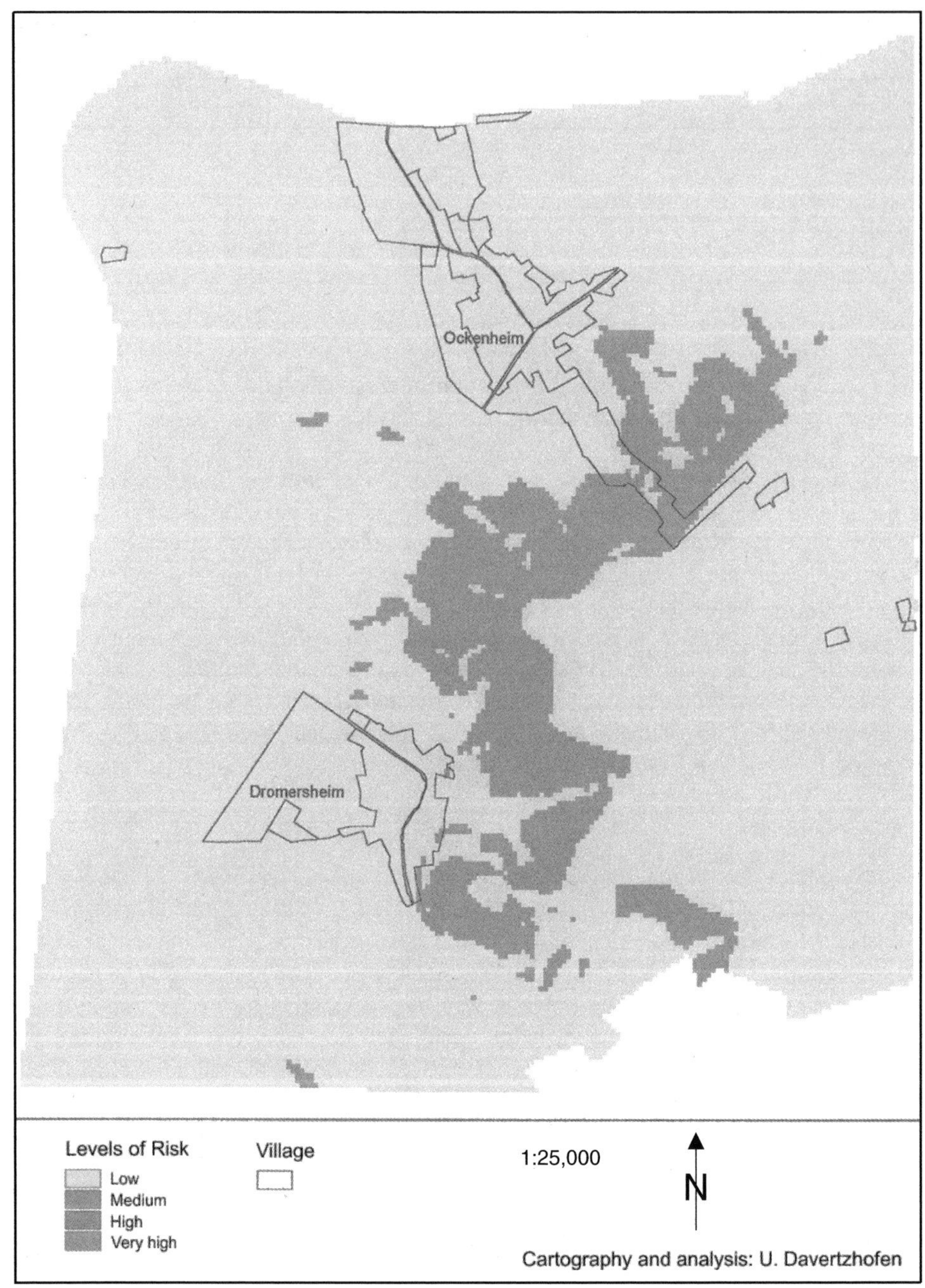


**Figure 3.9**  *Regional landslide risk in Rheinhessen, Germany (Glade et al., in prep.b). Vulnerability to elements at risk is assumed to be 1, referring to total loss if an element is affected by a landslide. (See also Plate 1)*

than for rainfall (e.g. orientation of geologic structure and landforms, distance to tectonic lineaments). But most other factors are important for both triggers (e.g. slope geometry, soils, vegetation). In any case, if the analysis extends further to address hazard, for a specific landslide type, information on the triggering agent can be extremely valuable as a component of the analysis.

Generally, it is easier to establish a temporal record of rainfall-triggered landslides than of earthquake-triggered failures. Rainfall records coupled with historical landslide information allow the calculation of the temporal probability of rainfall-triggered landslides. In contrast, information on landslide occurrence related to recurrence intervals of different-sized earthquakes is more difficult to assess due to the low return periods of these events. Despite these constraints, attempts to model the spatial extent of both triggers using empirical and/or numerical approaches are in progress. These scenarios of probable future triggers have the potential to be linked with empirical or numerical models of landslide movement. This procedure allows an approximation of the change of landslide hazard for different trigger magnitudes. Thus it enables a shift from static to dynamic conditions. This scenario modelling is a powerful tool for any landslide hazard assessment.

The consequences of a landslide event are also not dependent on the nature of the trigger. Structural damage of elements at risk results purely from the landslide types and expected magnitudes and intensities. Direct damage from earthquakes is not within the scope of this work. Possibly, some elements at risk may already have been weakened by foreshocks or an earlier earthquake (e.g. cracks in foundations, etc.) and are thus more vulnerable to the subsequent landslides, while other elements at risk might become less vulnerable. Foreshocks or the first few seconds of an earthquake might allow people to be better prepared for the subsequent landslides, for example by moving into other rooms in the case of debris flows, leaving the house in the case of large rotational slides, or seeking shelter in the case of small rockfalls. In general, it is rather difficult to forecast the consequences of a trigger and thus their consideration within the landslide risk analysis is complex.

## 3.8  Summary and Conclusion

The review of inventory, susceptibility and hazard analysis has shown the wide range of studies and applications. Despite the numerous studies from worldwide examples, many other regions are also affected by landslides. These also need to be examined in detail. It is demonstrated that landslide inventories are of major value for any susceptibility, hazard and risk analysis. Such inventories can be used as input data for the direct calculation of susceptibility. Moreover, if there is temporal and magnitude information available in the inventory, the probability of landslide occurrence of a given magnitude in a specific time period and a predefined location can also be estimated, and thus landslide hazard estimates delineated. Another application of landslide inventories is their use for verification and validation of calculated susceptibility or hazard. If inventories need to be used for both analysis and validation of results, the data sets can be split in two groups, one for analysis and one for validation (Chung and Fabbri, 1999). This is a major and fundamental issue which is often ignored.

Independent of scale, the concepts and approaches to landslide hazard and risk analysis outlined in this chapter allow a standardized and, in some cases, objective assessment of potential consequences of an assumed triggering event. As well as the ultimate determination of a level of risk, decision makers and planners should also be aware of the concepts, assumptions, methods or limitations involved in its computation. As with any modelling procedure, limitations of the approach have to be appreciated when using the information for making subsequent decisions on policy and management:

- Any spatial landslide information contains uncertainties that are difficult to evaluate (e.g. Ardizzone *et al.*, 2002; Carrara *et al.*, 1992).
- The resolution and quality of the socio-economic data influence the accuracy of the resulting risk.
- In most cases, the vulnerability of structures and of societies can only be roughly estimated or approximated (e.g. Glade, 2003b).
- The risk model is always a generalization of reality, and the model performance is strongly dependent on data constraints.
- The calculated landslide risk is a stationary expression of reality at the time of analysis.

Alternatively, there are many advantages of landslide risk assessments (e.g. Petrascheck and Kienholz, 2003). These are, in particular:

- Risk values and information are transparent and comprehensible.
- Scenarios allow assessment of the consequences of future developments.
- Reliability of the model performance is strongly dependent on data quantity and quality; thus with increasing data availability, the reliability of the risk estimate increases.
- Most models of landslide risk can be adapted to significant changes in the environment, such as vegetation changes or changes in land use or suburban developments. Therefore the potential exists to regularly update the static risk information.
- The conceptual approach and established methods allow a comparison not only of risk from different landslide types, but also from other natural hazards.

These advantages can be used to trace the evolution of landslide risk. Change of landslide risk is not only dependent on the change of the underlying landslide processes. Even while the level of landslide hazard remains constant, the risk may change as a result of human activity. Landslide risk is consequently not only an expression of the natural environment, but is also related to human interference with nature.

# References

Abele, G., 1974, *Bergstürze in den Alpen: Ihre Verbreitung, Morphologie und Folgeerscheinungen* (München: Deutscher und Österreichischer Alpenverein).

Agnesi, V., Di Maggio, C., Fiorito, S. and Rotigliano, E., 2002a, Analytical approach for landslide hazard assessment in El Salvador (C.A.), in Delahaye, D., Levoy, F. and Maquaire, O. (eds), *Proceedings of the Symposium 'Geomorphology: from Expert Opinion to Modelling'*, 26–27 April 2002, Strasbourg, France (Rouen: SODIMPAL), 263–270.

Agnesi, V., Di Maggio, C. and Rotigliano, E., 2002b, Earthquake-induced landslides in El Salvador, in Delahaye, D., Levoy, F. and Maquaire, O. (eds), *Proceedings of the Symposium 'Geomorphology: From Expert Opinion to Modelling'*, 26–27 April 2002, Strasbourg, France (Rouen: SODIMPAL), 283–286.

Aleotti, P. and Chowdhury, R.N., 1999, Landslide hazard assessment: summary review and new perspectives, *Bulletin of Engineering Geology and Environment*, **58**, 21–44.

Aleotti, P., Baldelli, P. and Polloni, G., 2000, Hydrogeological risk assessment of the Po River basin (Italy), in Bromhead, E.N., Dixon, N. and Ibsen, M.-L. (eds), *Landslides in Research, Theory and Practice, Proceedings of the 8th International Symposium on Landslides*, 26–30 June 2000 (Cardiff: Thomas Telford), 13–18.

Alexandrowicz, S.W., 1993, Late Quaternary landslides at eastern periphery of the National Park of Pieniny Mountains, Carpathians, Poland, *Studia Geol. Pol.*, **102**, 209–225.

Alexandrowicz, S.W., 1997, Holocene dated landslides in the Polish Carpathians, in J.A. Matthews, D. Brunsden, B. Frenzel, B. Gläser and M.M. Weiß (eds), *Rapid Mass Movement as a Source of Climatic Evidence for the Holocene* (Stuttgart, Jena, Lübeck and Ulm: Gustav Fischer Verlag), **12**, 75–84.

Amaral, C. and Palmeiro, F., 1997, Local landslide inventory of Rio de Janeiro: state of the art and access, in ABMS ABGE & ISSMGE (ed.), *2nd Pan-American Symposium on Landslides* (II PSL/2a COBRAE), Rio de Janeiro, 195–200.

Amaral, C., Vargas, E. and Krauter, E., 1996, Analysis of Rio de Janeiro landslide inventory data, in K. Senneset (ed.), *Landslides – Glissements de Terrain* (Rotterdam: A.A. Balkema), vol. 3, 1843–1846.

Anbalagan, R. and Singh, B., 1996, Landslide hazard and risk assessment mapping of mountainous terrains – a case study from Kumaun Himalaya, India, *Engineering Geology*, **43**, 237–246.

Anbalagan, R., Srivastava, N.C.N. and Jain, V., 2000, Slope stability studies of Vyasi dma reservoir area, Garhwal Himalaya, U.P. India, in E.N. Bromhead, N. Dixon and M.-L. Ibsen (eds), *Landslides in Research, Theory and Practice, Proceedings of the 8th International Symposium on Landslides*, 26–30 June 2000 (Cardiff: Thomas Telford), 51–56.

Anderson, M.G. and Brooks, S.M. (eds), 1996, *Advances in Hillslope Processes*, Symposia Series (Chichester: John Wiley & Sons Ltd).

Andrecs, P., 1995, Einige Aspekte der Murenereignisse in Österreich 1972–1992, *Wildbach- und Lawinenverbauung*, **59**, 75–91.

Aniya, M., 1985, Landslide-susceptibility mapping in the Amahata River basin, Japan, *Annals of the Association of American Geographers*, **75**, 102–114.

Antronico, L. and Gullà, G., 2000, Slopes affected by soil slips: validation of an evolutive model, in E.N. Bromhead, N. Dixon and M.-L. Ibsen (eds), *Landslides in Research, Theory and Practice, Proceedings of the 8th International Symposium on Landslides*, 26–30 June 2000 (Cardiff: Thomas Telford), 77–84.

Arboleda, R.A. and Punongbayan, R.S., 1999, Landslides induced by the 16 July 1990 Luzon, Philippines, earthquake, in K. Sassa (ed.), *Landslides of the World* (Kyoto: Kyoto University Press), 230–234.

Ardizzone, F., Cardinali, M., Carrara, A., Guzzetti, F. and Reichenbach, P., 2002, Uncertainty and errors in landslide mapping and landslide hazard assessment, *Natural Hazard and Earth System Science*, **2**, 3–14.

Asté, J.P., Gouisset, Y. and Leroi, E., 1995, The French 'INVI' project: national inventory of unstable slopes, in D.H. Bell (ed.), *Proceedings of the Sixth International Symposium*, 10–14 February 1992, Christchurch, New Zealand (Rotterdam: A.A. Balkema), 1547–1552.

Ayalew, L., 2000, Factors affecting slope stability in the Blue Nile Basin, in E.N. Bromhead, N. Dixon and M.-L. Ibsen (eds), *Landslides in Research, Theory and Practice, Proceedings of the 8th International Symposium on Landslides*, 26–30 June 2000 (Cardiff: Thomas Telford), 101–106.

Baeza, C. and Corominas, J., 1996, Assessment of shallow landslide susceptibility by means of statistical techniques, in K. Senneset (ed.), *Landslides – Glissements de Terrain* (Rotterdam: A.A. Balkema), vol. 1, 147–152.

Baeza, C. and Corominas, J., 2001, Assessment of shallow landslide susceptibility by means of multivariate statistical techniques, *Earth Surface Processes and Landforms*, **26**, 1251–1263.

Baldelli, P., Aleotti, P. and Polloni, G., 1996, Landslide-susceptibility numerical mapping at the Messina Straits Crossing site, Italy, in K. Senneset (ed.), *Landslides – Glissements de Terrain* (Rotterdam: A.A. Balkema), vol. 1, 153–158.

Ballantyne, C.K., 1997, Holocene rock-slope failures in the Scottish Highlands, in J.A. Matthews, D. Brunsden, B. Frenzel, B. Gläser and M.M. Weiß (eds), *Rapid Mass Movement as a Source of Climatic Evidence for the Holocene* (Stuttgart, Jena, Lübeck and Ulm: Gustav Fischer Verlag), **12**, 197–206.

Bandis, S.C., 1999, Numerical modelling techniques as predictive tools of ground instability, in R. Casale and C. Margottini (eds), *Floods and Landslides: Integrated Risk Assessment* (Berlin: Springer-Verlag), 101–118.

Baroni, C., Bruschi, G. and Ribolini, A., 2000, Human-induced hazardous debris flows in Carrara marble basins (Tuscany, Italy), *Earth Surface Processes and Landforms*, **25**, 93–103.

Barros, W.T., Amaral, C. and D'Orsi, R.N., 1991, Landslide susceptibility map of Rio de Janeiro, in D.H. Bell (ed.), *Landslides – Proceedings of the Sixth International Symposium*, 10–14 February 1992 (Rotterdam: A.A. Balkema), 869–871.

Bartsch, A., Gude, M., Jonasson, C. and Scherer, D., 2002, Identification of geomorphic process units in Kärevagge, Northern Sweden, by remote sensing and digital terrain analysis, *Geografiska Annaler*, **84A**, 171–178.

Basabe, P. and Bonnard, C., 2002, Instability management in Ecuador – from policy to practice, in R.G. McInnes and J. Jakeways (eds), *Instability Planning and Management* (Isle of Wight: Thomas Telford), 659–670.

Basu, S.R., 2000, Causes of landslides in the Darjiling town of the Eastern Himalaya, in E.N. Bromhead, N. Dixon and M.-L. Ibsen (eds), *Landslides in Research, Theory and Practice, Proceedings of the 8th International Symposium on Landslides*, 26–30 June 2000 (Cardiff: Thomas Telford), 133–138.

Basu, S.R., 2001, A systematic study of landslides along the arterial routes to Darjeeling and their control, in K.K.S. Ho and K.S. Li (eds), *Geotechnical Engineering – Meeting Society's Needs, Proceedings of the 14th Southeast Asian Geotechnical Conference*, 10–14 December 2001, Hong Kong (Rotterdam: A.A. Balkema), 691–695.

Bathurst, J.C., Crosta, G.B., Dfarcía-Ruiz, J.M., Guzzetti, F., Lenzi, M.A. and Ríos Aragüés, S., 2003, DAMOCLES: Debris-fall assessment in mountain catchments for local end-users, in D. Rickenmann and C.-L. Chen (eds), *Debris-Flow Hazards Mitigation: Mechanics, Prediction, and Assessment*, 10–12 September 2003, Davos, Switzerland (Rotterdam: Millpress), 1073–1083.

Becht, M. and Rieger, D., 1997, Spatial and temporal distribution of Debris-Flow occurrence on slopes in the Eastern Alps, in C.-L. Chen (ed.), *Proceedings, First International Conference on Debris-Flow Hazards Mitigation: Mechanics, Prediction, and Assessment*, 7–9 August 1997 (San Francisco, CA: ASCE), 516–529.

Belaia, N.L., Perov, V.F. and Sidorova, T.L., 2000, Distribution and regime of mudflow phenomena in Asia, not including the FSU territory, in G.F. Wieczorek and N.D. Naeser (eds), *Debris-Flow Hazards Mitigation: Mechanics, Prediction, and Assessment*, 16–18 August 2000, Taipei, Taiwan (Rotterdam: A.A. Balkema), 495–502.

Bell, R. and Glade, T., 2004, Landslide risk analysis for Bíldudalur, NW-Iceland, *Natural Hazard and Earth System Science*, **4**, 1–15.

Bernknopf, R.L., Campbell, R.H., Brookshire, D.A. and Shapiro, C.D., 1988, A probabilistic approach to landslide hazard mapping in Cincinnati, Ohio, with applications for economic evaluation, *Bulletin of the Association of Engineering Geologists*, **25**, 39–56.

Bert, T., 1980, Map showing the reported debris flows, debris avalanches, and mud flows in the United States, *US Geological Survey*, **46**.

Bhattacharya, S.K., 1999, A Constructive Approach to Landslides through Susceptibility Zoning and Case Study in the Rakti Basin of Eastern Himalaya, *Transactions, Japanese Geomorphological Union*, **20**, 317–333.

Bibus, E. and Terhorst, B. (eds) 1999, in Bibus E. and Terhorst B. (eds), *Tübinger Geowissenschaftliche Arbeiten*, Tübinger, 241. *Angewandte Studien zu Massenbewegungen.*

Binaghi, E., Luzi, L., Madella, P., Pergalani, F. and Rampini, A., 1998, Slope instability zonation: a comparison between certainty factor and fuzzy Dempster–Shafer approaches, *Natural Hazards*, **17**, 77–97.

Boynagryan, V.R., Boynagryan, B.V. and Boynagyran, A.V., 2000, Regularities of spreading and forming of landslides in Armenia, in E.N. Bromhead, N. Dixon and M.-L. Ibsen (eds),

*Landslides in Research, Theory and Practice, Proceedings of the 8th International Symposium on Landslides*, 26–30 June 2000 (Cardiff: Thomas Telford), 163–166.

Brabb, E.E., 1984, Innovative approaches to landslide hazard and risk mapping, *The 4th International Symposium on Landslides* (Toronto), 307–324.

Brabb, E.E., 1993, The San Mateo County GIS project for predicting the consequences of hazardous geologic processes, *Meeting on Geographical Information Systems for Assessing Natural Hazards* (Perugia, Italy, International Landslide Research Group), 1–5.

Brabb, E.E. and Pampeyan, E.H., 1972, Preliminary map of landslide deposits in San Mateo County, California, *US Geological Survey*.

Brabb, E.E., Pampeyan, E.H. and Bonilla, M.G., 1978, Landslide susceptibility in San Mateo County, California Reston, Virginia, *US Geological Survey*.

Brabb, E.E., Colgan, J.P. and Best, T.C., 1999, Map showing inventory and regional susceptibility for Holocene debris flows and related fast-moving landslides in the conterminous United States, *US Geological Survey*.

Brabb, E.E., Guzzetti, F., Mark, R. and Simpson, R.W., 1989, The extent of landsliding in Northern New Mexico and similar semi-arid and arid regions, in P.M. Sadler and D.M. Morton (eds), *Landslides in a Semi-Arid Environment: Studies from the Inland Valleys of Southern California* (Riverside, Publications of the Inland Geological Society), **2**, 163–173.

Brabhaharan, P., Hancox, G.T., Perrin, N.D. and Dellow, G.D., 1994, *Earthquake induced slope failure hazard study*, Wellington Region, Study Area 1 – Wellington City, Wellington Regional Council Works Consultancy Services Ltd, Institute of Geological and Nuclear Sciences.

Brand, E.W., Premchitt, J. and Phillipson, H.B., 1984, Relationship between rainfall and landslides in Hong Kong, in Canadian Geotechnical Society (ed.), *Proceedings of the 4th International Symposium on Landslides*, Toronto, Canada (Rotterdam: A.A. Balkema), 377–384.

Brardinoni, F., Hassan, M.A. and Slaymaker, H.O., 2003, Landslide inventroy in arugged forest watershed: a comparison between air-photo and field-survey data, *Geomorphology*, **54**, 179–196.

Bromhead, E., 1996, Slope stability modelling: an overview, in R. Dikau, D. Brunsden, L. Schrott and M. Ibsen (eds), *Landslide Recognition: Identification, Movement and Causes* (Chichester: John Wiley & Sons Ltd), 231–235 (Appendix 3).

Bromhead, E.N., 1997, The treatment of landslides, *Geotechnical Engineering*, **125**, 85–96.

Bromhead, E.N., Hopper, A.C. and Ibsen, M.-L., 1998, Landslides in the Lower Greensand escarpment in south Kent. *Bulletin Engineering Geological Environment*, **57**, 131–144.

Bromhead, E.N., Ibsen, M.L., Papanastassiou, X. and Zemichael, A.A., 2002, Three-dimensional stability analysis of a coastal landslide at Hanover Point, Isle of Wight, *Quarterly Journal of Engineering Geology and Hydrogeology*, **35**, 79–88.

Brunsden, D. and Ibsen, M., 1994, The spatial and temporal distribution of landslides on the south coast of Great Britain, in R. Cascale, R. Fantechi and J.C. Flagelollet (eds), *Temporal Occurrenece and Forecasting of Landslides in the European Community* (European Community), **1**, 385–426.

Calcaterra, D., Parise, M. and Dattola, L., 1996a, Debris flows in deeply weathered granitoids (Serre Massif – Calabria, Southern Italy), in K. Senneset (ed.), *Landslides – Glissements de Terrain* (Rotterdam: A.A. Balkema), vol. 1, 171–176.

Calcaterra, D., Parise, M. and Dattola, L., 1996b, Debris flows in deeply weathered granitoids (Serre Massif – Calabria, Southern Italy), in K. Senneset (ed.), *Proceedings of the 7th International Symposium on Landslides*, 17–21 June 1996, Trondheim (Rotterdam: A.A. Balkema), 171–176.

Calcaterra, D., de Riso, R., Nave, A. and Sgambati, D., 2002, The role of historical information in landlside hazard assessment of urban areas: the case of Naples (Italy), in J. Rybár, J. Stemberk and P. Wagner (eds), *Landslides*, 24–26 June 2002, Prague, Czech Republic (Rotterdam: A.A. Balkema), 129–135.

Calcaterra, D., Parise, M., Palma, B. and Pelella, L., 2000, Multiple debris flows in volcaniclastic materials mantling carbonate slopes, in G.F. Wieczorek and N.D. Naeser (eds), *Debris-Flow Hazards Mitigation: Mechanics, Prediction, and Assessment*, 16–18 August 2000, Taipei, Taiwan (Rotterdam: A.A. Balkema), 99–107.

Campbell, R.H. and Bernkopf, R.L., 1997, Debris-flow hazard map units from gridded probabilities, in C.-L. Chen (ed.), *Proceedings, First International Conference on Debris-Flow Hazards*

*Mitigation: Mechanics, Prediction, and Assessment*, Hydraulics Division, American Society of Civil Engineers, 7–9, August 1997 (San Francisco, CA: ASCE), 165–175.

Campbell, R.H., Bernkopf, R.L. and Soller, D.R., 1994, Mapping time-dependent changes in soil slip – debris flow probability Denver, *US Geological Survey*, **74**.

Campus, S., Forlati, F., Sarri, H. and Scavia, C., 2001, Shallow landslides hazard assessment based on multidisciplinary studies, in K.K.S. Ho and K.S. Li (eds), *Geotechnical engineering – Meeting society's needs, Proceedings of the 14th Southeast Asian Geotechnical Conference*, 10–14 December 2001, Hong Kong (Rotterdam: A.A. Balkema), 703–708.

Cannon, S.H., 1989, An evaluation of the travel-distance potential of debris flows Salt Lake City, *Utah Geological and Mineral Survey*, Utah Department of Natural Resources, **35**.

Cannon, S.H., 1993, An empirical model for the volume change behaviour of debris flows, in H.W. Shen and F. Wen (eds), *Proceedings of the Hydraulic Engineering* (ASCE), 1678–1773.

Canuti, P., Casagli, N., Catani, F., Falorni, G. and Fanti, R., 2002, Lahar modelling at Guaga Pichincha volcano, in J. Rybár, J. Stemberk and P. Wagner (eds), *Landslides*, 24–26 June 2002, Prague, Czech Republic (Rotterdam: A.A. Balkema), 517–522.

Cardinali, M., Reichenbach, P., Guzzetti, F., Adrizzone, F., Antonini, G., Galli, M., Cacciano, M., Castellani, M. and Salvati, P., 2002, A geomorphological approach to estimate landslide hazards and risk in urban and rural areas in Umbria, Central Italy, *Natural Hazards and Earth System Sciences*, **2**, 57–72.

Carrara, A., 1983, Multivariate models for landslide hazard evaluation, *Mathematical Geology*, **15**, 403–426.

Carrara, A., 1989, Landslide hazard mapping by statistical methods: a 'black-box' model approach, *Proceedings International Workshop on Natural Disasters in European–Mediterranian Countries*, Perugia, 27 June–1 July 1988 (CNR-ESNSF), 205–224.

Carrara, A. and Guzzetti, F. (eds), 1995, Geographical Information Systems in Assessing Natural Hazards, *Advances in Natural and Technological Hazards Research* (Dordrecht: Kluwer Academic Publishers).

Carrara, A. and Merenda, L., 1976, Landslide inventory in northern Calabria, southern Italy, *Geological Society of America Bulletin*, **87**, 1153–1162.

Carrara, A., Carratelli, E.P. and Merenda, L., 1977a, Computer-based data bank and statistical analysis of slope instability phenomena, *Zeitschrift für Geomorphologie*, **21**, 187–222.

Carrara, A., Cardinali, M. and Guzzetti, F., 1992, Uncertainty in assessing landslide hazard and risk, *ITC Journal*, The Netherlands, vol. 2, 172–183.

Carrara, A., Cardinali, M., Guzzetti, F. and Reichenbach, P., 1995, GIS technology in mapping landslide hazard, in Carrara, A. and Guzzetti, F. (eds), *Geographical Information Systems in Assessing Natural Hazards* (Dordrecht: Kluwer Academic Publishers), vol. 5, 135–175.

Carrara, A., Catalano, E., Sorriso-Valvo, M., Reali, C., Merenda, L. and Rizzo, V., 1977b, Landslide morphometry and typology in two zones, Calabria, Italy, *Bulletin of the International Association of Engineering Geology*, **16**, 8–13.

Carrasco, R.M., Pedraza, J., Martín-Duque, Mattera, M. and Sanz, M.A., 2000, Landslide succeptibility zoning for risk analysis using a geographical information system (GIS) in the Jerte Valley (Spanish Central System), in C.A. Brebbia (ed.), *Risk Analysis II – Second International Conference on Computer Simulation in Risk Analysis and Hazard Mitigation*, 329–344.

Casadei, M. and Dietrich, W.E., 2003, Controls on shallow landslide size, in D. Rickenmann and C.-L. Chen (eds), *Debris-Flow Hazards Mitigation: Mechanics, Prediction, and Assessment*, 10–12 September 2003, Davos, Switzerland (Rotterdam: Millpress), 91–101.

Casadei, M. and Farabegoli, E., 2003, Estimation of the effects of slope map computing on shallow landslide hazard zonation: a case history in the Northern Apennines (Italy), in R.J. Allison (ed.), *Applied Geomorphology: Theory and Practice* (Chichester: John Wiley & Sons Ltd.), vol. 10, 283–302.

Chan, R.K.S., Pang, P.L.R. and Pun, W.K., 2003, Recent developments in the Landslip Warning System in Hong Kong, in K.K.S. Ho and K.S. Li (eds), *Geotechnical Engineering – Meeting Society's Needs, Proceedings of the 14th Southeast Asian Geotechnical Conference*, 10–14 December 2001, Hong Kong (Rotterdam: A.A. Balkema), 219–224.

Chang, J.C. and Slaymaker, O., 2002, Frequency and spatial distribution of landslides in a mountainous drainage basin: Western Foothills, Taiwan, *Catena*, **46**, 285–307.

Chen, C.-I., 1997, *Debris-Flow Hazards Mitigation: Mechanics, Prediction, and Assessment* (San Francisco, CA: American Society of Civil Engineers).

Chen, H. and Lee, C.F., 2000, Numerical simulation of debris flows, *Canadian Geotechnical Journal*, **37**, 146–160.

Cheng, J.D., Yeh, J.L., Deng, Y.H., Wu, H.L. and Hsei, C.D., 2003, Landslides and debris flows induced by typhoon Traji, 29–30 July, 2001 in central Taiwan, in D. Rickenmann and C.-L. Chen (eds), *Debris-Flow Hazards Mitigation: Mechanics, Prediction, and Assessment*, 10–12 September 2003, Davos, Switzerland (Rotterdam: Millpress), 919–929.

Cherkez, Y., Kozlova, V., Shmourakto, V., Kharitonov, V. and Karavan, A., 2000, Landslide in the North-Western Black sea region, in E.N. Bromhead, N. Dixon and M.-L. Ibsen (eds), *Landslides in Research, Theory and Practice, Proceedings of the 8th International Symposium on Landslides*, 26–30 June 2000 (Cardiff: Thomas Telford), 251–254.

Chigira, M., 2002, Geologic factors contributing to landslide generation in a pyroclastic area: August 1998 Nishigo Village, Japan, *Geomorphology*, **46**, 117–128.

Chowdhury, R.N., 1988, Special lecture: analysis methods of assessing landslide risk – recent developments, in C. Bonnard (ed.), *Proceedings of the 5th International Symposium on Landslides* (Rotterdam: A.A. Balkema), 515–524.

Chowdhury, R.N., and Flentje, P.N., 2003, Role of slope reliability analysis in landslide risk management, *Bulletin of Engineering Geology and the Environment*, **61**, 41–46.

Chung, C.-J.F. and Fabbri, A.G., 1999, Probabilistic prediction models for landslide hazard mapping, *Photogrammetric Engineering & Remote Sensing*, **65**, 1389–1399.

Chung, C.-J.F. and Fabbri, A.G., 2002, Landslide risk analysis from prediction of future occurrences based on geomorphology-related spatial data, *IAMG*, 15–20 September 2002 (Berlin).

Chung, C.-J.F., Fabbri, A.G. and van Westen, C.J., 1995, Multivariate regression analysis for landslide hazard zonation, in A. Carrara and F. Guzzetti (eds), *Geographical Information Systems in Assessing Natural Hazards* (Dordrecht: Kluwer Academic Publishers), vol. 5, 107–134.

Chung, C.-J.F., Kojima, H. and Fabbri, A.G., 2003, Stability analysis of prediction models for landslide hazard mapping, in R.J. Allison (ed.), *Applied Geomorphology: Theory and Practice* (Chichester: John Wiley & Sons Ltd), vol. 10, 3–20.

Cinque, A., Robustelli, G., Scarciglia, F. and Terribile, F., 2000, The dramatic cluster of pyroclastic debris flows which occured on 5th and 6th May 1998 in the Sarno Mountains (Vesuvius region, Southern Italy): a geomorphological perspective, in E.N. Bromhead, N. Dixon and M.-L. Ibsen (eds), *Landslides in Research, Theory and Practice. Proceedings of the 8th International Symposium on Landslides*, 26–30 June 2000. (Cardiff: Thomas Telford), 273–278.

Clerici, A., 2002, A GRASS GIS based Shell script for landslide susceptibility zonation by the conditional analysis method Open Source Free Software, *GIS–GRASS users conference 2002*, Trento, Italy, 11–13 September 2002, 1–17.

Clouatre, E., Dubois, J.M.M. and Poulin, A., 1996, The geographic information system and regional delimitation of zones at risk for landslides: Hull-Gatineau region, Quebec. *Canadian Geographer – Geographe Canadien*, **40**, 367–386.

Coggan, J.S., Stead, D. and Exre, J.M., 1998, Evaluation of techniques for quarry slope stability assessment, *Transactions of the Institution of Mining and Metallurgy*, **107**, 139–147.

Corominas, J., 1996, The angle of reach as a mobility index for small and large landslides, *Canadian Geotechnical Journal*, **33**, 260–271.

Costa, J.E. and Wieczorek, G.F., 1987, *Debris flows/avalanches: process, recognition, and mitigation* (Boulder, CO: The Geological Society of America).

Coussot, P., Laigle, D., Arattano, M., Deganutti, A.M. and Marchi, L., 1998, Direct determination of rheological characteristics of debris flow, *Journal of Hydraulic Engineering*, **124**, 865–868.

Crozier, M.J., 1968, Earthflows and related environmental factors of Eastern Otago, *Journal of Hydrology* (New Zealand), **7**, 4–12.

Crozier, M.J., 1969, Earthflow occurrence during high intensity rainfall in Eastern Otago, New Zealand, *New Zealand Journal of Engineering Geology*, **3**, 325–334.

Crozier, M.J., 1989, Landslide hazard in the Pacific Islands, in E.E. Brabb and B.L. Harrod (eds), *Landslides: Extent and Economic Significance* (Rotterdam: A.A. Balkema), 357–366.

Crozier, M.J., 1991, Landslide problems at Ok Tedi, Papua New Guinea, *N.Z. Geomechanics News*, **42**, 26–27.

Crozier, M.J., 1996, Runout behaviour of shallow, rapid earthflows, *Zeitschrift für Geomorphologie (Supplementband)*, **105**, 35–48.

Crozier, M.J., 1997, The climate–landslide couple: a southern hemisphere perspective, in J.A. Matthews, D. Brunsden, B. Frenzel, B. Gläser and M.M. Weiß (eds), *Rapid Mass Movement as a Source of Climatic Evidence for the Holocene* (Stuttgart: Gustav Fischer), **19**, 333–354.

Crozier, M.J. and Pillans, B.J., 1991, Geomorphic events and landform response in south-eastern Taranaki, New Zealand, in M.J. Crozier (ed.), *Geomorphology in Unstable Regions, Catena*, **18**, 471–487.

Crozier, M.J. and Preston, N.J., 1999, Modelling changes in terrain resistance as a component of landform evolution in unstable hill country, in S. Hergarten and H.J. Neugebauer (eds), *Process Modelling and Landform Evolution* (Heidelberg: Springer-Verlag), **78**, 267–284.

Crozier, M.J., Eyles, R.J. and Wheeler, R.H., 1978, Landslips in Wellington City, *NZ Geographer*, **34**, 58–74.

Crozier, M.J., Howrth, R. and Grant, I.J., 1981, Landslide activity during Cyclone Wally, Fiji: A case study of Wainitubatolu Catchment, *Pacific Viewpoint*, **22**, 69–80.

Crozier, M.J., Vaughan, E.E. and Tippet, J.M., 1990, Relative instability in colluvium filled bedrock depressions, *Earth Surface Processes and Landforms*, **15**, 329–339.

Crozier, M.J., Eyles, R.J., Marx, S.L., McConchie, J.A. and Owen, R.C., 1980, Distribution of landslips in the Wairarapa hill country, *New Zealand Journal of Geology and Geophysics*, **23**, 575–586.

Cruden, D.M., 1985, Rock slope movements in the Canadian Cordillera, *Canadian Geotechnical Journal*, **22**, 528–540.

Cruden, D.M., 1996, An inventory of landslides in Alberta, Canada, in K. Senneset (ed.), *Proceedings of the 7th Intenational Symposium on Landslides*, 17–21 June 1996, Trondheim (Rotterdam: A.A. Balkema), vol. 3, 1877–1882.

Cruden, D.M. and Fell, R. (eds), 1997, *Landslide Risk Assessment – Proceedings of the Workshop on Landslide Risk Assessment*, Honolulu, Hawaii, USA, 19–21 February 1997 (Rotterdam and Brookfield: A.A. Balkema).

Cuesta, M.J.D., Sánchez, M.J. and García, A.R., 1999, Press archives as temporal records of landslides in the North of Spain: relationships between rainfall and instability slope events, *Geomorphology*, **30**, 125–132.

D'Ambrosio, D., Di Gregorio, S. and Iovine, G., 2003a, Simulating debris-flows through a hexagonal cellular automata model: SCIDDICA S3-hex, *Natural Hazard and Earth System Science*, **3**, 545–559.

D'Ambrosio, D., Di Gregorio, S., Iovine, G., Lupiano, V., Rongo, R. and Spartaro, W., 2003b, First simulations of the Sarno debris flows through Cellular Automata modelling, *Geomorphology*, **54**, 91–117.

Dai, F.C. and Lee, C.F., 2001, Terrain-based mapping of landslide susceptibility using a geographical information system: a case study, *Canadian Geotechnical Journal*, **38**, 911–923.

Dai, F.C. and Lee, C.F., 2002, Landslides on natural terrain – Physical characteristics and susceptibility mapping in Hong Kong, *Mountain Research and Development*, **22**, 40–47.

Dai, F.C., Lee, C.F. and Ngai, Y.Y., 2002, Landslide risk assessment and management: an overview, *Engineering Geology*, **64**, 65–87.

D'Amato Avanzi, G., Giannecchini, R. and Puccinelli, A., 2000, Geologic and geomorphic factors of the landslides triggered in the Cardoso T. basin (Tuscany, Italy) by the 19th June 1996 intense rainstorm, in E.N. Bromhead, N. Dixon and M.-L. Ibsen (eds), *Landslides in Research, Theory and Practice. Proceedings of the 8th International Symposium on Landslides*, 26–30 June 2000 (Cardiff: Thomas Telford), 381–386.

Decaulne, A. and Saemundsson, T., 2003, Debris-flow characteristics in the Gleidarhjalli area, northwestern Iceland, in D. Rickenmann and C.-L. Chen (eds), *Debris-Flow Hazards Mitigation: Mechanics, Prediction, and Assessment*, 10–12 September 2003, Davos, Switzerland (Rotterdam: Millpress), 1107–1118.

Dehn, M., 1999, Application of an analog downscaling technique to the assessment of future landslide activity – a case study in the Italian Alps, *Climate Research*, **13**, 103–113.

De Jaeger, C., 2000, Influence on landsliding and slope development of the particular environment of the Dead Sea region: a case study for the Wadi Mujib Canyon (Jordan), in E.N. Bromhead,

N. Dixon and M.-L. Ibsen (eds), *Landslides in Research, Theory and Practice, Proceedings of the 8th International Symposium on Landslides*, 26–30 June 2000 (Cardiff: Thomas Telford), 421–426.

Del Monte, M., Fredi, P., Palmieri, E.L. and Marini, R., 2003, Contribution of quantitative geomorphic analysis to the evaluation of geomorphological hazards: case study in Italy, in R.J. Allison (ed.), *Applied Geomorphology: Theory and Practice* (Chichester: John Wiley & Sons Ltd), **10**, 335–358.

Del Prete, M., Guadagno, F.M. and Hawkins, A.B., 1998, Preliminary report on the landslides of 5 May 1998, Campania, southern Italy, *Bulletin Engineering Geological Environment*, **57**, 113–129.

Demoulin, A. and Chung, C.-J.F. (submitted for publication) Landslide hazard prediction in the Pays de Herve (Belgium): a quantitative susceptibility map, *Geomorphology*.

Derbyshire, E., Jingtai, W., Zexian, J., Billard, A., Egels, Y., Kasser, M., Jones, D.K.C., Muxart, T. and Owen, L.A., 1991, Landslides in the Gansu loess of China, *Catena*, **20**, 119–145.

DeRose, R.C., Trustrum, N.A. and Blaschke, P.M., 1993, Post-deforestation soil loss from steepland hillslopes in Taranaki, New Zealand, *Earth Surface Processes and Landforms*, **18**, 131–144.

Dhakal, S., Amada, T. and Aniya, M., 2000, Databases and Geographical Information Systems for medium scale landslide hazard evaluation: an example from typical mountain watershed in Nepal, in E.N. Bromhead, N. Dixon and M.-L. Ibsen (eds), *Landslides in Research, Theory and Practice, Proceedings of the 8th International Symposium on Landslides*, 26–30 June 2000 (Cardiff: Thomas Telford), 457–462.

Dhital, M.R., 2003, Causes and consequences of the 1993 debris flows and landslides in Kulekhani watershed, central Nepal, in D. Rickenmann and C.-L. Chen (eds), *Debris-Flow Hazards Mitigation: Mechanics, Prediction, and Assessment*, 10–12 September 2003, Davos, Switzerland (Rotterdam: Millpress), 931–942.

Dietrich, W.E. and Sitar, N., 1997, Geoscience and geotechnical engineering aspects of debris-flow hazard assessment, in C.-L. Chen (ed.), *Proceedings, First International Conference on Debris-Flow Hazards Mitigation: Mechanics, Prediction, and Assessment*, Hydraulics Division, American Society of Civil Engineers, 7–9 August 1997 (San Francisco, CA: ASCE), 656–676.

Dietrich, W.E., Reiss, R., Hsu, M.-L. and Montgomery, D.R., 1995, A process-based model for colluvial soil depth and shallow landsliding using digital elevation data, *Hydrological Processes*, **9**, 383–400.

Dikau, R. and Jäger, S., 1995, Landslide hazard modelling in New Mexico and Germany, in D.F.M. McGregor and D.A. Thompson (eds), *Geomorphology and Land Management in a Changing Environment* (Chichester: John Wiley & Sons Ltd), 51–68.

Dikau, R. and Glade, T., 2003, Nationale Gefahrenhinweiskarte gravitativer Massenbewegungen, in H. Liedtke, R. Mäusbacher and K.-H. Schmidt (eds), *Relief, Boden und Wasser* (Heidelberg: Spektrum Akademischer Verlag), vol. 2, 98–99.

Dikau, R., Gärtner, H., Holl, B., Kienholz, H., Mani, P. and Zimmermann, M., 1996, Untersuchungen zur Murgangaktivität im Mattertal, Wallis, Schweiz. INTERPRAEVENT (ed.), INTERPRAEVENT 1996, Garmisch-Partenkirchen, 397–408.

Dobrovolny, E., 1971, Landslide susceptibility in and near Anchorage as interpreted from topographic and geologic maps, *The Great Alaska Earthquake of 1964: Biology*, Washington, DC: National Academy of Science.

Dorren, L.K.A. and Seijmonsbergen, A.C., 2003, Comparison of three GIS-based models for predicting rockfall runout zones at a regional scale, *Geomorphology*, **56**, 49–64.

Dorren, L.K.A., Maier, B., Putters, U.S. and Seijmonsberegen, A.C., 2004, Combining field and modelling techniques to assess rockfall dynamics on a protection forest hillslope in the European Alps, *Geomorphology*, **57**, 151–167.

Douglas, D.J., 1980, Magnitude and frequency study of rockfall in Co. Antrim, N-Ireland, *Earth Surface Processes and Landforms*, **5**, 123–129.

Douglas, G.B., Trustrum, N.A. and Brown, I.C., 1986, Effect of soil slip erosion on Wairoa hill pasture production and composition, *N.Z. Journal of Agricultural Research*, **29**, 183–192.

Dragovich, J.D., Brunengo, M.J. and Gerstel, W.J., 1993, Landslide Inventory and Analysis of the Tilton River–Mineral Creek Area, Lewis County, Washington, *Washington Geology*, **21**, 19–30.

Duarte, R.S. and Marquinez, J., 2002, The influence of environmental and lithologic factors on rockfall at a regional scale: an evaluation using GIS, *Geomorphology*, **43**, 117–136.

Einstein, H.H., 1988, Special lecture: landslide risk assessment procedure, in C. Bonnard (ed.), *Proceedings of the 5th International Symposium on Landslides*, 10–15 July 1988, Lausanne, Switzerland (Rotterdam: A.A. Balkema), 1075–1090.

Ekanayake, J.C. and Phillips, C.J., 1999, A method for stability analysis of vegetated hillslopes: an energy approach, *Canadian Geotechnical Journal*, **36**, 1172–1184.

Eldredge, S.N., 1988, An overview of landslide inventories predominantly of North America, *Utah Geological and Mineral Survey*, **67**.

El Hamdouni, R., Irigaray, C., Fernández, T., Sanz de Galdeano, C. and Chacón, J., 2000, Slope movements and active tectonics in the Izbor River Basin (Granada, Spain), in E.N. Bromhead, N. Dixon and M.-L. Ibsen (eds), *Landslides in Research, Theory and Practice, Proceedings of the 8th International Symposium on Landslides*, 26–30 June 2000 (Cardiff: Thomas Telford), 501–506.

Ellen, S.D. and Mark, R.K., 1993, Mapping debris-flow hazard in Honolulu using a DEM, *Hydraulic Engineering '93*, **2**, 1774–1779.

Ellen, S.D. and Wieczorek, G.F. (eds), 1988, *Landslides, floods, and marine effects of the storm of January 3–5, 1982, in the San Francisco Bay region, California*, US Geological Survey Professional Paper (Washington: United States Goverment Printing Office).

Ellen, S.D., Mark, R.K., Cannon, S.H. and Knifong, D.L., 1993, Map of debris-flow hazard in the Honolulu District of Oahu, Hawaii, US Department of the Interior, *US Geological Survey*, **25**.

Ercanoglu, M. and Gokceoglu, C., 2002, Assessment of landslide susceptibility for a landslide-prone area (north of Yenice, NW Turkey) by fuzzy approach, *Environmental Geology*, **41**, 720–730.

Erickson, G.E., Ramirez, C.F., Concha, J.F., Tisnado, G.M. and Urquidi, F.B., 1989, Landslide hazards in the central and southern Andes, in E.E. Brabb and B.L. Harrod (eds), *Landslides: Extent and Economic Significance* (Rotterdam: A.A. Balkema), 111–118.

Erismann, T.H. and Abele, G., 2001, *Dynamics of rockslides and rockfalls* (Heidelberg: Springer).

Espizua, L.E. and Bengochea, J.D., 2002, Landslide hazard and risk zonation mapping in the Rio Grande Basin, Central Andes of Mendoza, Argentina, *Mountain Research and Development*, **22**, 177–185.

Eusebio, A., Grasso, P., Mahtab, A. and Morino, A., 1996, Assessment of risk and prevention of landslides in urban areas of the Italian Alps, in K. Senneset (ed.), *Landslides – Glissements de Terrain* (Rotterdam: A.A. Balkema), vol. 1, 190–194.

Evans, S.G., 1997, Fatal landslides and landslide risk in Canada, in D.M. Cruden and R. Fell (eds), *Landslide Risk Assessment – Proceedings of the Workshop on Landslide Risk Assessment*, Honolulu, Hawaii, USA, 19–21 February 1997 (Rotterdam: A.A. Balkema), 185–196.

Evans, S.G. and Brooks, G.R., 1999, A large retrogressive earthflow at Lemieux, Ontario, Canada, 20 June 1993, in K. Sassa (ed.), *Landslides of the World* (Kyoto: Kyoto University Press), 293–296.

Evans, S.G. and DeGraff, J.V. (eds), 2002, *Catastrophic Landslides, Reviews in Engineering Geology* (Boulder, CO: Geological Society of America).

Evans, S.G. and Hungr, O., 1993, The Assessment of Rockfall Hazard at the Base of Talus Slopes, *Canadian Geotechnical Journal*, **30**, 620–636.

Eyles, R.J., Crozier, M.J. and Wheeler, R.H., 1974, Landslides in Wellington City, *Soil & Water*, **11**, 17–20.

Eyles, R.J., Crozier, M.J. and Wheeler, R.H., 1978, Landslips in Wellington City, *New Zealand Geographer*, **34**, 58–74.

Fabbri, A.G., Chung, C.-J., Napolitano, P., Remondo, J. and Zêzere, J.L., 2002, Prediction rate functions of landslide susceptibility applied in the Iberian Peninsula, in C.A. Brebbia (ed.), *Third International Conference on Risk Analysis*, 19–21 June 2002, Sintra, Portugal, 703–718.

Fallsvik, J. and Viberg, L., 1998, Early stage landslide and erosion risk assessment – a method for a national survey in Sweden, *Erdwissenschaftliche Aspekte des Umweltschutzes. 4. Arbeitstagung des Bereiches Umwelt*, April, Wien 151–153.

Fannin, R.J., Wise, M.P., Wilkinson, J.M.T., Thomson, B. and Hetherington, E.D., 1997, Debris-flow hazard assessment in British Columbia, in C.-L. Chen (ed.), *Proceedings, First International Conference on Debris-Flow Hazards Mitigation: Mechanics, Prediction, and Assessment*, Hydraulics Division, American Society of Civil Engineers, 7–9 August 1997 (San Francisco, CA: ASCE), 197–206.

Farhan, Y., 1999, Landslide hazards and highway engineering in Central and Northern Jordan, in J.S. Griffiths, M.R. Stokes and R.G. Thomas (eds), *Landslides – Proceedings of the Ninth International Conference and Field Trip on Landslides*, Bristol, United Kingdom, 5–16 September 1999 (Rotterdam: A.A. Balkema), 37–45.

Feiznia, S. and Bodaghi, B., 2000, A statistcal approach for logical modelling of a landslide hazard zonation in Shahrood Drainage Basin, in E.N. Bromhead, N. Dixon and M.-L. Ibsen (eds), *Landslides in Research, Theory and Practice, Proceedings of the 8th International Symposium on Landslides*, 26–30 June 2000 (Cardiff, UK: Thomas Telford) 549–552.

Fell, R., 1994, Landslide risk assessment and acceptable risk. *Canadian Geotechnical Journal*, **31**, 261–272.

Fell, R., 2000, Landslide risk management concepts and guidelines – Australian geomechanics society sub-committee on landslide risk management, in International Union of Geological Sciences (ed.), *Landslides* (Cardiff, UK: International Union of Geological Sciences), 51–93.

Fell, R. and Hartford, D., 1997, Landslide risk management, in D.M. Cruden and R. Fell (eds), *Landslide Risk Assessment-Proceedings of the Workshop on Landslide Risk Assessment*, Honolulu, Hawaii, USA, 19–21 February 1997 (Rotterdam: A.A. Balkema), 51–109.

Fernandez, T., Irigary, C. and Chacón, J., 1996, Inventory and analysis of landslide determinant factors in Los Guajares Mountains, Granada (Southern Spain), in K. Senneset (ed.), *Landslides – Glissements de Terrain* (Rotterdam: A.A. Balkema), vol. 3, 1891–1896.

Fernandes, N.F., Guimaraes, R.F., Gomes, R.A.T., Vieira, B.C., Montgomery, D.R. and Greenberg, H., 2004, Topographic controls of landslides in Rio de Janeiro: field evidence and modeling, *Catena* **55**, 163–181.

Fernández-Steeger, T. and Czurda, K., 2001, Erkennung von Rutschungen mit neuronalen Netzen. *Geotechnik – Sonderband zur 13. Nat. Tagung f. Ingenieurgeologie Karlsruhe*, 61–66.

Fernández-Steeger, T.M., Rohn, J. and Czurda, K., 2002, Identification of landslide areas with neural nets for hazard analysis, in J. Rybár, J. Stemberk and P. Wagner (eds.), *Landslides*, 24–26 June 2002, Prague, Czech Republic (Rotterdam: A.A. Balkema), 163–168.

Ferrer, M. and Ayala-Carcedo, F., 1997, Landslides in Spain: extent and assessment of the climatic susceptibility, in P.G. Marinos, G.C. Koukis, G.C. Tsiambaos and G.C. Stournaras (eds.), *Engineering Geology and the Environment. Proceedings International Symposium on Engineering Geology and the Environment*, 23–27 June 1997 (Rotterdam: A.A. Balkema), 625–631.

Ferrigno, L. and Spilotro, G., 2002, Analysis of land sensitivity through the use of remote sensing derived thematic maps: landslide hazard, in R.G. McInnes and J. Jakeways (eds.), *Instability Planning and Management* (Isle of Wight: Thomas Telford), 229–236.

Finlay, P.J., Mostyn, G.R. and Fell, R., 1999, Landslide risk assessment: prediction of travel distance. *Canadian Geotechnical Journal*, **36**, 556–562.

Fiorillo, F., Guadagno, F.M., Aquino, S. and De Blasio, A., 2001, The December 1999 Cervinara landslides: further debris flows in the pyroclastic deposits of Campania (southern Italy). *Bulletin of Engineering Geology and the Environment*, **60**, 171–184.

Flentje, P.N., Chowdhury, R.N. and Tobin, P., 2000, Management of landslides triggered by a major storm event in Wollongong, Australia, in G.F. Wieczorek and N.D. Naeser (eds.), *Debris-Flow Hazards Mitigation: Mechanics, Prediction, and Assessment*, 16–18 August 2000, Taipei, Taiwan (Rotterdam: A.A. Balkema), 479–487.

Forero-Duenas C.A. and Caro-Pena, P.E., 1996, Inventory and study of landslide hazards in Cudinamarca, in K. Senneset (ed.), *Landslides – Glissements de Terrain* (Rotterdam: A.A. Balkema), vol. 3, 1897–1902.

Frattini, P. and Crosta, G.B., 2002, Modelling the impact of forest management changes on landslide occurrence, in R.G. McInnes and J. Jakeways (eds), *Instability Planning and Management* (Isle of Wight: Thomas Telford), 257–264.

Froldi, P. and Bonini, G., 2000, A new method to assess the landslide hazard in Argillitic terrains: Corniglio case-history, in E.N. Bromhead, N. Dixon and M.-L. Ibsen (eds), *Landslides in*

*Research, Theory and Practice, Proceedings of the 8th International Symposium on Landslides*, 26–30 June 2000 (Cardiff, UK: Thomas Telford), 585–590.

Gardner, J.S., 1983, Rockfall frequency and distribution in the Highwood Pass area, Canadian Rock Mountains, *Zeitschrift für Geomorphologie*, N.F. **27**, 311–324.

Giani, G.P., 1992, *Rock Slope Stability Analysis* (Aldershot: Ashgate Publishing Ltd).

Gill, J.L., 1974, Risks, legalities and insurance of slope stability, *Symposium on Stability of Slopes in Natural Ground. New Zealand*, 2.1–2.6.

GLA, 1989, *Rutschungs-Kataster Mainz*, Geologisches Landesamt Rheinland-Pfalz.

Glade, T., 1996, The temporal and spatial occurrence of landslide-triggering rainstorms in New Zealand, in R. Mäusbacher and A. Schulte (eds), *Beiträge zur Physiogeographie – Festschrift für Dietrich Barsch* (Heidelberg, Selbstverlag des Geographischen Instituts der Universität Heidelberg), **104**, 237–250.

Glade, T., 1997, The temporal and spatial occurrence of rainstorm-triggered landslide events in New Zealand, *School of Earth Science, Institute of Geography*, Wellington, Victoria University of Welington, 380.

Glade, T., 2001, Landslide hazard assessment and historical landslide data – an inseparable couple? in T. Glade, F. Frances and P. Albini (eds), *The Use of Historical Data in Natural Hazard Assessments* (Dordrecht: Kluwer Academic Publishers), **7**, 153–168.

Glade, T., 2003a, Landslide occurrence as a response to land use change: a review of evidence from New Zealand. *Catena*, **51**, 297–314.

Glade, T., 2003b, Vulnerability assessment in landslide risk analysis. *Die Erde*, **134**, 121–138.

Glade, T., 2004, Linking natural hazard and risk analysis with geomorphology assessments. Geomorphology: in press.

Glade, T. and Jensen, E.H., 2004, *Landslide Hazard Assessments for Bolungarvík and vesturbyggð, NW-Iceland* (Reykjavik: Icelandic Metereological office).

Glade, T., Frances, F. and Albini, P. (eds), 2001a, The use of historical data in natural hazard assessments. *Advances in Natural and Technological Hazards Research* (Dordrecht: Kluwer Academic Publishers).

Glade, T., Kadereit, A. and Dikau, R., 2001b, Landslides at the Tertiary escarpment of Rheinhessen, Southwest Germany, *Zeitschrift für Geomorphologie, Supplementband*, **125**, 65–92.

Glade, T., Dikau, R. and Bell, R. (in preparation a) National landslide susceptibility analysis for Germany, *Natural Hazard and Earth System Science*.

Glade, T., von Davertzhofen, U. and Dikau, R. (in preparation b) GIS-based landslide risk analysis in Rheinhessen, Germany, *Natural Hazards*.

Godt, J.W. and Savage, W.Z., 1999, El Niño 1997–1998: direct costs of damaging landslides in the San Francisco Bay region, in J.S. Griffiths, M.R. Stokes and R.G. Thomas (eds), *Landslides – Proceedings of the Ninth International Conference and Field Trip on Landslides*, Bristol, United Kingdom, 5–16 September 1999 (Rotterdam: A.A. Balkema), 47–55.

Godt, J.W., Coe, J.A. and Savage, W.Z., 2000, Relation between cost of damaging landslides and construction age, Alameda County, California, USA, El Niño winter storm season, 1997–98, in E.N. Bromhead, N. Dixon and M.-L. Ibsen (eds), *Landslides in Research, Theory and Practice, Proceedings of the 8th International Symposium on Landslides*, 26–30 June 2000 (Cardiff, UK: Thomas Telford) 641–646.

Gökceoglu, C. and Aksoy, H., 1996, Landslide susceptibility mapping of the slopes in the residual soils of the Mengen region (Turkey) by deterministic stability analysis and image processing techniques, *Engineering Geology*, **44**, 147–161.

Gori, P.L., Jeer, S.P. and Highland, L.M., 2003, Enlisting the support of land-use planners to reduce debris-flow hazards in the United States, in D. Rickenmann and C.-L. Chen (eds), *Debris-Flow Hazards Mitigation: Mechanics, Prediction, and Assessment*, 10–12 September 2003. Davos, Switzerland (Rotterdam: Millpress), 1119–1127.

Grayson, R.B. and Blöschl, G., 2000, Spatial modelling of catchment dynamics, in R.B. Grayson and G. Blöschl (eds), *Spatial Patterns in Catchment Hydrology: Observations and Modelling* (Cambridge: Cambridge University Press), 51–81.

Grayson, R.B., Blöschl, G., Western, A.W. and McMahon, T.A., 2002, Advances in the use of observed spatial pattern of catchment hydrological response. *Advances in Water Resources*, **25**, 1313–1334.

Greenbaum, D., Bowker, M.R., Dau, I., Dropsy, H., Greally, K.B., McDonald, A.J.W., Marsh, S.H., Northmore, K.J., O'Connor, E.A., Prasad, S. and Tragheim, D.G., 1995, Rapid methods of landslide hazard mapping: Fiji case study Keyworth, Nottingham, *British Geological Survey*, **107**.

Griffiths, J.S., Mather, A.E. and Hart, A.B., 2002, Landslide susceptibility in the Rio Aguas catchment, SE Spain, *Quarterly Journal of Engineering Geology and Hydrogeology*, **35**, 9–17.

Grunert, J. and Schmanke, V., 1997, Hangstabilität im Südwesten Bonns, *Geographische Rundschau*, **49**, 584–590.

Gudagno, F.M. and Zampelli, S.P., 2000, Triggering mechanisms of the landslides that inundated Sarno, Quindici, Siano, and Bracigliano (S. Italy) on May 5–6, 1998, in E.N. Bromhead, N. Dixon and M.-L. Ibsen (eds), *Landslides in Research, Theory and Practice, Proceedings of the 8th International Symposium on Landslides*, 26–30 June 2000 (Cardiff, UK: Thomas Telford) 671–676.

Günther, A.F., Carstensen, A. and Pohl, W., 2002a, GIS-application in slope stability assessments, in J. Rybár, J. Stemberk and P. Wagner (eds), *Landslides*, 24–26 June 2002, Prague, Czech Republic (Rotterdam: A.A. Balkema), 175–184.

Günther, A.F., Carstensen, A. and Pohl, W., 2002b, Slope stability management using GIS, in R.G. McInnes and J. Jakeways (eds), *Instability Planning and Management* (Isle of Wight, Thomas Telford) 265–272.

Gupta, S.K., 2000, 1897 Great Assam earthquake-generated landslides: distribution, pattern and correlation with landslide potentiality of Northeastern India, in E.N. Bromhead, N. Dixon and M.-L. Ibsen (eds), *Landslides in Research, Theory and Practice, Proceedings of the 8th International Symposium on Landslides*, 26–30 June 2000, Cardiff, UK, Thomas Telford, **2**, 677–682.

Guthrie, R.H., 2002, The effects of logging on frequency and distribution of landslides in three watersheds on Vancouver Island, British Columbia, *Geomorphology*, **43**, 273–292.

Guzzetti, F., 2000, Landslide fatalities and the evaluation of landslide risk in Italy. *Engineering Geology*, **58**, 89–107.

Guzzetti, F. and Cardinali, M., 1990, Landslide inventory map of the Umbria Region, Central Italy, in ALPS 90 ALPS (ed.), *Sixth International Conference And Field Workshop On Landslides*, 273–284.

Guzzetti, F., Cardinali, M. and Reichenbach, P., 1994, The AVI project: a bibliographical and archive inventory of landslides and floods in Italy, *Environmental Management*, **18**, 623–633.

Guzzetti, F., Reichenbach, P. and Wieczorek, G.F., 2003, Rockfall hazard and risk assessment in the Yosemite Valley, California, USA, *Natural Hazard and Earth System Science*, **3**, 491–503.

Guzzetti, F., Carrara, A., Cardinali, M. and Reichenbach, P., 1999, Landslide hazard evaluation: a review of current techniques and their application in a multi-scale study, Central Italy, *Geomorphology*, **31**, 181–216.

Guzzetti, F., Crosta, G., Detti, R. and Agliardi, F., 2002a, STONE: a computer program for the three-dimensional simulation of rock-falls, *Computers and Geosciences*, **28**, 1079–1093.

Guzzetti, F., Malamud, B.D., Turcotte, D.L. and Reichenbach, P., 2002b, Power–law correlations of landslide areas in central Italy, *Earth and Planetary Science Letters*, **195**, 169–183.

Hardenbicker, U., 1994, *Hangrutschungen im Bonner Raum – Naturräumliche Einordnung und ihre anthropogenen Ursachen* (Bonn, Ferd. Duemmlers Verlag).

Hardingham, A.D., Ditchfield, C.S., Ho, K.K.S. and Smallwood, A.R.H., 1998, Quantitative risk assessment of landslides – a case history from Hong Kong, in K.S. Li, J.N. Kay and K.K.S. Ho (eds), *Slope Engineering in Hong Kong* (Hong Kong: A.A. Balkema), 145–151.

Harmsworth, G.R., Hope, G.D., Page, M.J. and Manson, P.A., 1987, *An assessment of storm damage at Otoi in northern Hawke's Bay*, Soil Conservation Centre, Aokautere, Palmerston North.

Harmsworth, G.R. and Page, M.J., 1991, A review of selected storm damage assessments in New Zealand, *DSIR Land Resources*.

Harp, E.L. and Jibson, R.W., 1995, Inventory of landslides triggered by the 1994 Northridge, California earthquake Denver, *US Geological Survey*, 17.

Harp, E.L., Chleborad, A.F., Schuster, R.L., Cannon, S.H., Reid, M.E. and Wilson, R.C., 1997, *Landslides and Landslide Hazards in Washington State due to February 5–9, 1996 Storm Denver*, US Geological Administrative Report.

Hart, A.B. and Griffiths, J.S., 1999, Mass movement features in the vicinity of the town of Sorbas, South-east Spain, in J.S. Griffiths, M.R. Stokes and R.G. Thomas (eds), *Landslides – Proceedings of the Ninth International Conference and Field Trip on Landslides*, Bristol, United Kingdom, 5–16 September 1999 (Rotterdam: A.A. Balkema), 57–63.

Hayne, M.C. and Gordon, D., 2001, Regional landslide hazard estimation, a GIS/decision tree analysis: Southeast Queensland, Australia, in K.K.S. Ho and K.S. Li (eds), *Geotechnical Engineering – Meeting Society's Needs, Proceedings of the 14th Southeast Asian Geotechnical Conference*, 10–14 December 2001 (Hong Kong: A.A. Balkema), 115–121.

Hearn, G.J. and Griffiths, J.S., 2001, Landslide hazard mapping and risk assessment, in Griffiths, J.S. (ed.), *Land Surface Evaluation for Engineering Practice* (London: Geological Society), **18**, 43–52.

Heim, A., 1932, *Bergsturz und Menschenleben* (Zürich).

Heinimann, H.R., 1999, *Risikoanalyse bei gravitativen Naturgefahren – Methode* (Bern).

Hermanns, R.L., Niedermann, S., Strecker, M.R., Trauth, M.H., Alonso, R. and Fauque, L., 2002, Prehistoric rock avalanches in the NW-Argentine Andes: boundary conditions and hazard assessment, in J. Rybár, J. Stemberk and P. Wagner (eds), *Landslides*, 24–26 June 2002, Prague, Czech Republic (Rotterdam: A.A. Balkema) 199–205.

Hervás, J., Barredo, J.I., Rosin, P.L., Pasuto, A., Mantovani, F. and Silvano, S., 2003, Monitoring landslides from optical remotely sensed imagery: the case history of Tessina landslide, Italy, *Geomorphology*, **54**, 63–75.

Hewitt, K., 2002, Styles of rock-avalanche depositional complexes conditioned by very rugged terrain, Karakoram Himalaya, Pakistan, in S.G. Evans and J.V. DeGraff (eds), *Catastrophic Landslides: Effects, Occurrence, and Mechanisms*, **15**, 345–377.

Hicks, B.G. and Smith, R.D., 1981, Management of steeplands impacts by landslide hazard zonation and risk evaluation, *Journal of Hydrology* (NZ), **20**, 63–70.

Ho, K.K.S. and Li, K.S. (eds), 2003, *Geotechnical Engineering – Meeting Society's Needs* (Lisse: A.A. Balkema).

Ho, K.K.S. and Wong, H.N., 2001, Application of quantitative risk assessment in landslide risk management in Hong Kong, in K.K.S. Ho and K.S. Li (eds), *Geotechnical Engineering – Meeting society's needs, Proceedings of the 14th Southeast Asian Geotechnical Conference*, 10–14 December 2001 (Hong Kong: A.A. Balkema) 123–128.

Hoblitt, R.P., Walder, J.S., Driedger, C.L., Scott, K.M., Pringle, P.T. and Vallance, J.W., 1995, *Volcano Hazards from Mount Rainier* (Washington: US Geological Survey).

Hodgson, I.F., Hearn, G.J. and Lucas, G., 2002, Landslide hazard assessment for land management and development planning, Scotland District, Barbados, in R.G. McInnes and J. Jakeways (eds), *Instability Planning and Management* (Isle of Wight, Thomas Telford) 281–290.

Hofmeister, R.J., 2000, *Slope failures in Oregon: GIS inventory for three 1996/97 storm events*, Oregon Department of Geology and Mineral Industries.

Hofmeister, R.J. and Miller, D.J., 2003, GIS-based modeling of debris-flow initiation, transport and deposition zones for regional hazard assessments in western Oregon, USA, in D. Rickenmann and C.-L. Chen (eds), *Debris-Flow Hazards Mitigation: Mechanics, Prediction, and Assessment*, 10–12 September 2003, Davos, Switzerland (Rotterdam: Millpress) 1141–1149.

Holm, K., Bovis, M. and Jakob, M., 2004, The landslide response of alpine basins to post-Lille Ice Age glacial thinning and retreat in sothwestern British Columbia, *Geomorphology*, **57**, 201–216.

Hovius, N., Stark, C.P., Chu, H.T. and Lin, J.C., 2000, Supply and removal of sediment in a landslide-dominated mountain belt: Central Range, Taiwan, *Journal of Geology*, **108**, 73–89.

Hroch, Z., Kycl, P. and Sebesta, J., 2002, Landslide hazards in North Bohemia, in J. Rybár, J. Stemberk and P. Wagner (eds), *Landslides*, 24–26 June 2002 (Prague, Czech Republic: A.A. Balkema) 207–212.

Hungr, O., 2000, Analysis of debris flow surges using the theory of uniformly progressive flow, *Earth Surface Processes and Landforms*, **25**, 483–495.

Hungr, O., 2005, Classification and terminology, in M. Jakob and O. Hungr (eds), *Debris Flow Hazards and Related Phenomena* (Heidelberg: Springer).

Hungr, O., Evans, S.G., Bovis, M.M. and Hutchinson, J.N., 2001, A review of the classification of landslides of the flow type, *Environmental and Engineering Geoscience*, **7**, 221–238.

Hutchinson, J.N. and Bromhead, E.N., 2002, Isle of Wight landslides, in R.G. McInnes and J. Jakeways (eds), *Instability Planning and Management* (Isle of Wight: Thomas Telford), 3–72.

Hutter, K., Svendsen, B. and Rickenmann, D., 1996, Debris flow modeling: a review, *Continuum Mechanics and Thermodynamics*, vol. 8, 1–35.

Hylland, M.D. and Lowe, M., 1997, Regional landslide-hazard evaluation using landslide slopes, Western Wasatch County, Utah, *Environmental and Engineering Geoscience*, **III**, 31–43.

Ielenicz, M., Patru, I. and Mihai, B., 1999, Some Geomorphic Types of Landslides in Romania, *Transactions, Japanese Geomorphological Union*, **20**, 287–297.

International Association of Engineering Geology, 1976, *Engineering Geological Maps: A Guide to their Preparation* (Paris: UNESCO Press), 79.

INTERPRAEVENT, 2000a, Changes within the natural and cultural habitat and consequences, in INTERPRAEVENT (ed.) *9. Internationales Symposium* (Villach, Austria: Krainer Druck), 480.

INTERPRAEVENT, 2000b, Changes within the natural and cultural habitat and consequences, INTERPRAEVENT (ed.) *9. Internationales Symposium* (Villach, Austria: Krainer Druck), 374.

INTERPRAEVENT, 2000c, Changes within the natural and cultural habitat and consequences, INTERPRAEVENT (ed.) *9. Internationales Symposium* (Villach, Austria: Krainer Druck), 374.

INTERPRAEVENT, 2002a, Protection of habitat against floods, debris flows and avalanches, INTERPRAEVENT (ed.) *10. Internationales Symposium*, 14–18 October 2002 (Matsumoto, Japan: Nissei Eblo Co.), 492.

INTERPRAEVENT, 2002b, Protection of habitat against floods, debris flows and avalanches, INTERPRAEVENT (ed.), *10. Internationales Symposium*, 14–18 October 2002 (Matsumoto, Japan: Nissei Eblo Co.), 513.

Irigaray, C., Fernandez, T. and Chacón, J., 1996, Inventory and analysis of determining factors by a GIS in the northern edge of the Granada Basin, in K. Senneset (ed.), *Landslides – Glissements de Terrain* (Rotterdam: A.A. Balkema), vol. **3**, 1915–1921.

IUGS Working Group on Landslides – Committee on Risk Assessment, 1997, Quantitative assessment for slopes and landslides – The state of the art, in D.M. Cruden and R. Fell (eds), *Landslide Risk Assessment – Proceedings of the Workshop on Landslide Risk Assessment*, Honolulu, Hawaii, USA, 19–21 February 1997 (Rotterdam: A.A. Balkema), 3–12.

Iverson, R.M., 1997a, Hydraulic modeling of unsteady debris-flow surges with soild-fluid interactions, in C.-L. Chen (ed.), *Proceedings, First International Conference on Debris-Flow Hazards Mitigation: Mechanics, Prediction, and Assessment*, Hydraulics Division, American Society of Civil Engineers, 7–9 August 1997 (San Francisco, CA: ASCE), 550–560.

Iverson, R.M., 1997b, The physics of debris flows, *Reviews of Geophysics*, **35**, 245–296.

Iverson, R.M., Schilling, S.P. and Vallance, J.W., 1998, Objective delineation of lahar-inundation hazard zones. *Geological Society of America Bulletin*, **110**, 972–984.

Iwahashi, J., Watanabe, S. and Furuya, T., 2003, Mean slope-angle distribution and size frequency distribution of landslide masses in Higashikubiki area, Japan. *Geomorphology*, **50**, 349–364.

Jäger, S., 1997, *Fallstudien zur Bewertung von Massenbewegungen als geomorphologische Naturgefahr*. (Heidelberg: Selbstverlag des Geographischen Instituts).

Jäger, S. and Wieczorek, G.F., 1994, *Landslide susceptibility in the Tully Valley Area, Finger Lakes Region* (New York: US Geological Survey).

Jakob, M. and Hungr, O. (eds), 2005, *Debris-Flow Hazards and Related Phenomena* (Heidelberg: Springer).

Jan, C.-D. and Shen, H.W., 1997, Review dynamic modeling of debris flows, *Armanini & Michiue*, 93–116.

Jana, M.M., 2000, Landslides: their causes and measures in Darjiling, Himalaya, India, in E.N. Bromhead, N. Dixon and M.-L. Ibsen (eds), *Landslides in Research, Theory and Practice, Proceedings of the 8th International Symposium on Landslides*, 26–30 June 2000 (Cardiff, UK: Thomas Telford) 775–782.

Janova, V., 2000, Analyses of slope failurs in the Orava region, in E.N. Bromhead, N. Dixon and M.-L. Ibsen (eds), *Landslides in Research, Theory and Practice, Proceedings of the 8th International Symposium on Landslides*, 26–30 June 2000 (Cardiff, UK: Thomas Telford) 783–788.

Jonasson, C., Nyberg, R. and Rapp, A., 1997, Dating of rapid mass movements in Scandinavia: talus rockfalls, large rockslides, debris flows and slush avalanches, in J.A. Matthews, D. Brunsden, B. Frenzel, B. Gläser and M.M. Weiß (eds), *Rapid Mass Movement as a Source of Climatic Evidence for the Holocene* (Stuttgart, Jena, Lübeck and Ulm: Gustav Fischer Verlag), vol. **12**, 267–282.

Jones, C.L., Higgins, J.D. and Andrew, R.D., 2000, *Colorado Rockfall Simulation Program* (Denver: Colorado Department of Transportation).

Jones, D.K.C. and Lee, M., 1994, *Landsliding in Great Britain* (London: HSMO).

Jones, F., 1973, *Landslides of Rio de Janeiro and the Serra das Araras Escarpment, Brazil*, Geological Survey Professional Paper, 697, 42.

Juhász, Á., 1997, Landslides and climate in Hungary, in J.A. Matthews, D. Brunsden, B. Frenzel, B. Gläser and M.M. Weiß (eds), *Rapid Mass Movement as a Source of Climatic Evidence for the Holocene* (Stuttgart, Jena, Lübeck and Ulm: Gustav Fischer Verlag), **12**, 109–126.

Julian, M. and Anthony, E., 1996, Aspects of landslide activity in the Mercantour Massif and the French Riviera, southeastern France, *Geomorphology*, **15**, 275–289.

Jurak, V., Matkovic, I., Miklin, Z. and Cvijanovic, D., 1998, Landslide hazard in the Medvednica submountain area under dynamic conditions, in B. Maric, Z. Lisac and A. Szavits-Nossan (eds), *Proceedings of the 10th Danube – European Conference on Soil Mechanics and Geotechnical Engineering*, 25–29 May 1998, Porec, Croatia (Rotterdam: A.A. Balkema), 827–834.

Kamai, T., Kobayashi, Y., Jinbo, C. and Shuzui, H., 2000, Earthquake risk assessments of fill-slope instability in urban residential areas in Japan, in E.N. Bromhead, N. Dixon and M.-L. Ibsen (eds), *Landslides in Research, Theory and Practice, Proceedings of the 8th International Symposium on Landslides*, 26–30 June 2000 (Cardiff, UK: Thomas Telford), 801–806.

Karrow, P.F., 1972, Earthflows in the Grondines and Trois Rivieres areas, Quebec, *Canadian Journal of Earth Science*, **9**, 561–573.

Keefer, D.K., 1984, Landslides caused by earthquakes, *Geological Society of America Bulletin*, **95**, 406–421.

Keefer, D.K., 2002, Investigating landslides caused by earthquakes – a historical review, *Surveys in Geophysics*, **23**, 473–510.

Kelsey, H.M., 1978, Earthflows in Franciscan melange, Van Duzen River basin, California, *Geology*, **6**, 361–364.

Kertész, A. and Schweitzer, F., 1991, Geomorphological mapping of Landslides in Hungary with a case study on mapping Danubian Bluffs, *Catena*, **18**, 529–536.

Khullar, V.K., Sharma, R.P. and Pramanik, K., 2000, A GIS approach in the landslide zone of Lawngthlai in southern Mizoram, North East India, in E.N. Bromhead, N. Dixon and M.-L. Ibsen (eds), *Landslides in Research, Theory and Practice, Proceedings of the 8th International Symposium on Landslides*, 26–30 June 2000 (Cardiff, UK: Thomas Telford), 825–830.

Kiersch, G.A., 1980, Vaiont reservoir disaster, *Civil Engineering*, **34**, 32–39.

Kim, W.Y., Chae, B.G., Kim, K.S. and Cho, Y.C., 2001, Approach to quantitative prediction of landslides on natural mountain slopes, Korea, in K.K.S. Ho and K.S. Li (eds), *Geotechnical Engineering – Meeting Society's Needs, Proceedings of the 14th Southeast Asian Geotechnical Conference*, 10–14 December 2001 (Hong Kong: A.A. Balkema), 795–799.

King, J.P., 1999, *Natural Terrain Landslide Study: Natural Terrain Landslide Inventory Hong Kong*, Geotechnical Engineering Office (GEO).

Kirchner, K., 2002, To the distribution of slope deformations in the northeastern vicinty of Zlín town, Vizovická vrchovina Highland (Outer Western Carpathians), in J. Rybár, J. Stemberk and P. Wagner (eds), Landslides, 24–26 June 2002 (Prague, Czech Republic: A.A. Balkema), 363–366.

Kliche, C.A., 1999, *Rock Slope Stability*, Society for Mining Metallurgy & Exploration.

Koleva-Rekalowa, E., Dobrey, N. and Ivanov, P., 1996, Earthflows in the Baltchik landslide area, North-eastern Bulgaria, in K. Senneset (ed.), *Landslides – Glissements de Terrain* (Rotterdam: A.A. Balkema), vol. **1**, 473–478.

Kong, W.K., 2002, Risk assessment of slopes, *Quarterly Journal Engineering Geology and Hydrogeology*, **35**, 213–222.

Kubota, T., 1994, A study of fractal dimension of landslides – the feasibility for susceptibility index, *Journal of Japan Landslide Society*, **31**, 9–15.

Larsen, J.O., Grande, L., Matsuura, S., Okamoto, T., Asano, S. and Gyu Park, S., 1999, Slide activity in quick clay related to pore water pressure and weather parameters, in J.S. Griffiths, M.R. Stokes and R.G. Thomas (eds), *Proceedings of the ninth International Conference and Field Trip on Landslides*, Bristol, United Kingdom, 5–16 September 1999 (Rotterdam: A.A. Balkema), 81–88.

Larsen, M.C. and Torres-Sanchez, A.J., 1998, The frequency and distribution of recent landslides in three montane tropical regions of Puerto Rico, *Geomorphology*, **24**, 309–331.

Lavigne, F., 1999, Lahar hazard micro-zonation and risk assessment in Yogyakarta city, Indonesia, *GeoJournal*, **49**, 173–183.

Lee, C.F., Ye, H., Yeung, M.R., Shan, X. and Chen, G., 2001, AIGIS-based methodology for natural terrain landslide susceptibility mapping in Hong Kong, *Episodes*, **24(3)**, 150–159.

Lee, E.M. and Clark, A.R., 2000, The use of archive records in landslide risk assessment: historical landslide events on the Scarborough coast, UK, in E.N. Bromhead, N. Dixon and M.-L. Ibsen (eds), *Landslides in Research, Theory and Practice, Proceedings of the 8th International Symposium on Landslides*, 26–30 June 2000 (Cardiff, UK: Thomas Telford), 905–910.

Lee, E.M., Jones, D.K.C. and Brunsden, D., 2000, The landslide environment of Great Britain, in E.N. Bromhead, N. Dixon and M.-L. Ibsen (eds), *Landslides in Research, Theory and Practice, Proceedings of the 8th International Symposium on Landslides*, 26–30 June 2000 (Cardiff, UK: Thomas Telford), 911–916.

Lee, S. and Min, K., 2001, Statistical analysis of landslide susceptibility at Yongin, Korea, *Environmental Geology*, **40**, 1095–1113.

Lee, S., Chwae, U. and Min, K.D., 2002, Landslide susceptibility mapping by correlation between topography and geological structure: the Janghung area, Korea, *Geomorphology*, **46**, 149–162.

Leone, F., Asté, J.P. and Leroi, E., 1996, Vulnerability assessment of elements exposed to mass-movement: working toward a better risk perception, in K. Senneset (ed.), *Landslides – Glissements de Terrain* (Rotterdam: A.A. Balkema), vol. **1**, 263–270.

Leroi, E., 1996, Landslide hazard – risk maps at different scales: objectives, tools and development, in K. Senneset (ed.), *Landslides – Glissements de Terrain, 7th International Symposium on Landslides*, Trondheim, Norway (Rotterdam: Balkema), 35–51.

Li, K.S., Kay, J.N. and Ho, K.K.S., 1998, Slope Engineering in Hong Kong, *Proceedings of the Annual Seminar on Slope Engineering in Hong Kong*, 2 May 1997 (Hong Kong: A.A. Balkema), 342.

Li Tianchi, C., 1983, A mathematical model for predicting the extent of a major rockfall, *Zeitschrift für Geomorphologie, Supplementband*, **24**, 473–482.

Lin, M.-L., 2003, Management of debris flow in Taiwan, in K.K.S. Ho and K.S. Li (eds), *Geotechnical Engineering – Meeting society's needs, Proceedings of the 14th Southeast Asian Geotechnical Conference*, 10–14 December 2001 (Hong Kong: A.A. Balkema), 181–185.

Lin, P.S., Hung, J.C., Lin, J.Y. and Yang, M.D., 2000, Risk assessment of potential debris-flows using GIS, in G.F. Wieczorek and N.D. Naeser (eds), *Debris-Flow Hazards Mitigation: Mechanics, Prediction, and Assessment*, 16–18 August 2000, Taipei, Taiwan (Rotterdam: A.A. Balkema), 431–440.

Liu, X.L., Yue, Z.Q., Tham, L.G. and Lee, C.F., 2002, Empirical assessment of debris flow risk on a regional scale in Yunnan province, southwestern China, *Environmental Management*, **30**, 249–264.

Lloyd, D.M., Anderson, M.G., Renaud, J.P., Wilkinson, P.W. and Brooks, S.M., 2004, On the need to determine appropriate model domains for hydrology slope stability models. *Advances in Environmental Research*, **8(3–4)**, 379–386.

Lomoschitz, A., 1999, Old and recent landslides of the Barranco de Tirajana basin, Gran Canaria, Spain, in J.S. Griffiths, M.R. Stokes and R.G. Thomas (eds), *Landslides – Proceedings of the Ninth International Conference and Field Trip on Landslides*, Bristol, United Kingdom, 5–16 September 1999 (Rotterdam: A.A. Balkema), 89–95.

Lopez, J.L., Perez, D. and Garcia, R., 2003, Hydrologic and geomorphologic evaluation of the 1999 debris-flow event in Venezuela, in D. Rickenmann and C.-L. Chen (eds), *Debris-Flow Hazards Mitigation: Mechanics, Prediction, and Assessment*, 10–12 September 2003. Davos, Switzerland (Rotterdam: Millpress), 989–1000.

Lorente, A., Garcia-Ruiz, J.M., Beguería, S. and Arnáez, J., 2002, Factors explaining the spatial distribution of hillslope debris flows: a case study in the Flysch sector of the central Spanish pyrenees, *Mountain Research and Development*, **22**, 32–39.

Luckman, B.H., 1972, Some observations on the erosion of talus slopes by snow avalanches in Surprise Valley, Jasper National Park, Alberta, in H.O. Slaymaker and H.J. McPherson (eds), *Mountain Geomorphology: Geomorphological Processes in the Canadian Cordillera* (Vancouver: Tantalus Research Ltd), 85–92.

Mackay, D.S., Samanta, S., Ahl, D.E., Ewers, B.E., Gower, S.T. and Burrows, S.N., 2003, Automated Parameterization of Land Surface Process Models Using Fuzzy Logic, *Transactions in GIS*, **7**, 139–153.

Maharaj, R.J., 1993, Landslide processes and landslide susceptibility analysis from an upland watershed: a case study from St. Andrew, Jamaica, West Indies, *Engineering Geology*, **34**, 53–79.

Mahdavifar, M.R., 2000, Fuzzy information processing in landslide hazard zonation and preparing the computer system, in E.N. Bromhead, N. Dixon and M.-L. Ibsen (eds), *Landslides in Research, Theory and Practice, Proceedings of the 8th International Symposium on Landslides*, 26–30 June 2000 (Cardiff, UK: Thomas Telford), 993–998.

Major, J.J., Chilling, S.P. and Pullinger, C.R., 2003, Volcanic debris flow in developing countries – the extreme need for public education and awareness of debris-flow hazards, in D. Rickenmann and C.-L. Chen (eds), *Debris-flow hazards mitigation: mechanics, prediction, and assessment*, 10–12 September 2003, Davos, Switzerland (Rotterdam: Millpress), 1185–1196.

Major, J.J. and Iverson, R.M., 1999, Debris-flow deposition: effects of pore-fluid pressure and friction concentrated at flow margins, *Geological Society of America Bulletin*, **111**, 1424–1434.

Marden, M. and Rowan, D., 1993, Protective value of vegetation on tertiary terrain before and during Cyclone Bola, East Coast, North Island, New Zealand, *NZ Journal of Forestry Science*, **23**, 255–263.

Margielewski, W., 2002, Late Glacial and Holocene climatic changes registered in landslide forms and their deposits in the Polish Flysch Carpathians, in J. Rybár, J. Stemberk and P. Wagner (eds), *Landslides*, 24–26 June 2002, Prague, Czech Republic (Rotterdam: A.A. Balkema), 399–404.

Mark, R.K., 1992, Map of debris-flow probability, San Mateo County, California, USGS *Miscellaneous Investigation Series*, US Department of the Interior.

Mark, R.K. and Ellen, S.D., 1995, Statistical and simulation models for mapping debris-flow hazard, in A. Carrara and F. Guzzetti (eds), *Geographical Information Systems in Assessing Natural Hazards* (Dordrecht: Kluwer Academic Publishers), vol. **5**, 93–106.

Martello, S., Catani, F. and Casagli, N., 2000, The role of geomorphological settings and triggering factors in debris flow initiation during 19th June 1996 meteorological event in verslia and Garfagnana (Tuscany, Italy), in E.N. Bromhead, N. Dixon and M.-L. Ibsen (eds), *Landslides in Research, Theory and Practice, Proceedings of the 8th International Symposium on Landslides*, 26–30 June 2000 (Cardiff, UK: Thomas Telford), 1017–1022.

Martin, Y., Rood, K., Schwab, J.W. and Church, M., 2002, Sediment transfer by shallow landsliding in the Queen Charlotte Islands, British Columbia, *Canadian Journal of Earth Sciences*, **39**, 189–205.

Martinez, J.M., Avila, G., Agudelo, A., Schuster, R.L., Casadevall, T.J. and Scott, K.M., 1995, Landslides and debris flows triggered by the 6 June 1994 Paez earthquake, southwestern Columbia, *Landslide News*, **9**, 13–15.

Massari, R. and Atkinson, P.M., 1999, Modelling susceptibility to landsliding: an approach based on individual landslide type, *Transactions, Japanese Geomorphological Union*, **20**, 151–168.

Mateos Ruiz, R.M., 2002, Slope movements in the Majorca Island (Spain). Hazard analysis, in R.G. McInnes and J. Jakeways (eds), *Instability Planning and Management* (Isle of Wight: Thomas Telford), 339–346.

Mayer, K., Müller-Koch, K. and von Poschinger, A., 2002, Dealing with landslide hazards in the Bavarian Alps, in J. Rybár, J. Stemberk and P. Wagner (eds), *Landslides*, 24–26 June 2002, Prague, Czech Republic (Rotterdam: A.A. Balkema), 417–421.

Mayoraz, F., Cornu, T. and Vulliet, L., 1996, Using neural networks to predict slope movements, in K. Senneset (ed.), *Landslides – Glissements de Terrain* (Rotterdam: A.A. Balkema), vol. 1, 295–300.

Mazzoccola, D. and Sciesa, E., 2000, Implementation and comparison of different methods for rockfall hazard assessment in the Italian Alps, in E.N. Bromhead, N. Dixon and M.-L. Ibsen (eds), *Landslides in Research, Theory and Practice, Proceedings of the 8th International Symposium on Landslides*, 26–30 June 2000 (Cardiff, UK: Thomas Telford), 1035–1040.

McKean, J. and Roering, J., 2004, Objective landslide detection and surface morphology mapping using high-resolution airborne laser altimetry, *Geomorphology*, **57**, 331–351.

McSaveney, M.J., 2002, Recent rockfalls and rock avalanches in Mount Cook National Park, New Zealand, in S.G. Evans and J.V. DeGraff (eds), *Catastrophic Landslides: Effects, Occurrence, and Mechanisms* (Boulder, CO: Geological Society of America), **15**, 35–70.

Medeuov, A. and Beisenbinova, A., 1997, Regional debris-flow hazard assessment of mountain territories of the Republic of Kazakstan, in C.-L. Chen (ed.), *Proceedings, First International Conference on Debris-Flow Hazards Mitigation: Mechanics, Prediction, and Assessment*, Hydraulics Division, American Society of Civil Engineers, 7–9 August 1997 (San Francisco, CA: ASCE), 415–424.

Meißl, G., 2001, Modelling the runout distances of rockfalls using a geographic infomation system, *Zeitschrift für Geomorphologie, Supplementband*, **125**, 129–137.

Meng, X., Derbyshire, E. and Du, Y., 2000, Landslide hazard in the eastern part of Gansu Province, China, *Zeitschrift der Geologischen Gesellschaft*, **151**, 31–47.

Michael-Leiba, M., Baynes, F. and Scott, G., 2000, Quantitative landslide risk assessment of Cairns, Australia, in E.N. Bromhead, N. Dixon and M.-L. Ibsen (eds), *Landslides in Reserach, Theory and Practice, Proceedings of the 8th International Symposium on Landslides*, 26–30 June 2000 (Cardiff, UK: Thomas Telford), 1059–1064.

Michael-Leiba, M., Baynes, F., Scott, G. and Granger, K., 2003, Regional landslide risk to the Cairns community, *Natural Hazards*, **30**, 233–249.

Mollard, J.D. and Hughes, G.T., 1973, Earthflows in the Grondines and Trois Rivieres areas, Quebec: discussion, *Canadian Journal of Earth Science*, **10**, 324–326.

Möller, R., Glade, T. and Dikau, R., 2001, Determining and applying soil-mechanical response units in regional landslide hazard assessments, *Zeitschrift für Geomorphologie, Supplementband*, **125**, 139–151.

Monma, K., Kojima, S. and Kobayashi, T., 2000, Rock slope monitoring and rock fall prediction, *Landslide News*, **13**, 33–35.

Montgomery, D.R. and Dietrich, W.E., 1994, A physically based model for the topographic control on shallow landsliding, *Water Resources Research*, **30**, 1153–1171.

Montgomery, D.R., Schmidt, K.M., Greenberg, H.M. and Dietrich, W.E., 2000, Forest clearing and regional landsliding, *Geology*, **28**, 311–314.

Montgomery, D.R., Sullivan, K. and Greenberg, H.M., 1998, Regional test of a model for shallow landsliding, *Hydrological Processes*, **12**, 943–955.

Moon, A.T., Olds, R.J., Wilson, R.A. and Burman, B.C., 1991, Debris flow risk zoning at Montrose, Victoria, in D.H. Bell (ed.), *Landslides* (Rotterdam: Balkema), vol. 2, 1015–1022.

Moore, R., Hencher, S.R. and Evans, N.C., 2001, An approach for area and site-specific natural terrain hazard and risk assessment, Hong Kong, in K.K.S. Ho and K.S. Li (eds), *Geotechnical Engineering – Meeting Society's Needs, Proceedings of the 14th Southeast Asian Geotechnical Conference*, 10–14 December 2001 (Hong Kong: A.A. Balkema), 155–160.

Moreiras, S., 2004, Landslide incidence zonation in the Rio Mendoza valley, Mendoza Province, Argentina, *Earth Surface Processes and Landforms*, **29**, 255–266.

Morgan, G.C., Rawlings, G.E. and Sobkowicz, J.C., 1992, Evaluating total risk to communities from large debris flows, *Geotechnique and Natural Hazards*, 6–9 May 1992, Vancouver, Canada, 225–236.

Moser, M., 2002, Geotechnical aspects of landslides in the Alps, in J. Rybár, J. Stemberk and P. Wagner (eds), *Landslides*, 24–26 June 2002, Prague, Czech Republic (Rotterdam: A.A. Balkema), 23–44.

Moser, M. and Rentschler, K., 1999, Geotechnik der Kriech- und Gleitprozesse im Bereich des Juras der Frankenalb, in E. Bibus and B. Terhorst (eds), *Angewandte Studien zu Massenbewegungen. Reihe D: Geoökologie und Quartärforschung* (D2), 193–212.

Moya, J., Vilaplana, J.M. and Corominas, J., 1997, Late Quaternary and historical landslides in the south-eastern Pyrenees, in J.A. Matthews, D. Brunsden, B. Frenzel, B. Gläser and

M.M. Weiß (eds), *Rapid Mass Movement as a Source of Climatic Evidence for the Holocene* (Stuttgart, Jena, Lübeck and Ulm: Gustav Fischer Verlag), vol. 12, 55–74.

Müller, L., 1964, The rock slide in the Vaiont valley, *Felsmechanik und Ingenieurgeologie*, **2**, 148–212.

Nagarajan, R., 2002, Rapird assessment procedure to demarcate areas susceptible to earthquake-induced ground failures for environment management – a case study from parts of northeast India, *Bulletin of Engineering Geology and the Environment*, **61**, 99–119.

Newman, M.C. and Strojan, C.L., 1998, *Risk assessment: Logic and Measurement*. (Chelsea, MI: Ann Arbor Press).

Nicoletti, P.G., Iovine, G. and Catalano, E., 2000, Earthquake-triggered landsliding and historical seismicity in Southeastern Sicily: a discrepancy, in E.N. Bromhead, N. Dixon and M.-L. Ibsen (eds), *Landslides in Research, Theory and Practice, Proceedings of the 8th International Symposium on Landslides*, 26–30 June 2000 (Cardiff, UK: Thomas Telford) 1111–1116.

Okagbue, C.O., 1994, Classification and Distribution of Recent Historic Landslides in Southern Nigeria, *Engineering Geology*, **37**, 263–270.

Ostaficzuk, S.R., 1999, Variety of landslides in Poland, and their possible dependence on neo-geodynamics, in J.S. Griffiths, M.R. Stokes and R.G. Thomas (eds), *Landslides – Proceedings of the Ninth International Conference and Field Trip on Landslides*, Bristol, United Kingdom, 5–16 September 1999 (Rotterdam: A.A. Balkema) 111–127.

Page, M.J., Reid, L.M. and Lynn, I.H., 1999, Sediment production from Cyclone Bola landslides, Waipaoa catchment, *Journal of Hydrology (New Zealand)*, **38**, 289–308.

Page, M.J., Trustrum, N.A. and Dymond, J.R., 1994, Sediment budget to assess the geomorphic effect of a cyclonic storm, New Zealand, *Geomorphology*, **9**, 169–188.

Paige-Green, P., 1985, The development of a regional landslide susceptibility map for Southern Africa annual transportation convention, *Südafrika*, **8**.

Paige-Green, P., 1989, Landslides: extent and economic significance in southern Africa, in Brabb and Harrad (eds), *Landslides*, 261–269.

Panizza, M., Pasuto, A., Silvano, S. and Soldati, M., 1996, Temporal occurrence and activity of landslides in the area of Cortina d'Ampezzo (Dolomites, Italy), *Geomorphology*, **15**, 311–326.

Panizza, M., Pasuto, A., Silvano, S. and Soldati, M., 1997, Landsliding during the Holocene in the Cortina d'Ampezzo region, Italian Dolomites, in J.A. Matthews, D. Brunsden, B. Frenzel, B. Gläser and M.M. Weiß (eds), *Rapid Mass Movement as a Source of Climatic Evidence for the Holocene* (Stuttgart, Jena, Lübeck and Ulm: Gustav Fischer Verlag), **12**, 17–32.

Pantano, F.G., Nicoletti, P.G. and Parise, M., 2002, Historical and geological evidence for seismic origin of newly recognized landslides in southeastern Sicily, and its significance in terms of hazard, *Environmental Management*, **29**, 116–131.

Papin, J.W., Ng, M., Venkatesh, S. and Au-Yeung, Y.S., 2001, Quantitative risk assessment of collapses of deep excavations in Hong Kong, in K.K.S. Ho and K.S. Li (eds), *Geotechnical Engineering – Meeting Society's Needs, Proceedings of the 14th Southeast Asian Geotechnical Conference*, 10–14 December 2001 (Hong Kong: A.A. Balkema) 161–166.

Pareschi, M.T., Favalli, M., Giannini, F., Sulpizio, R., Zanchetta, G. and Santacroce, R., 2000, May 5, 1998, debris flows in circum-Vesuvian areas (southern Italy): insights for hazard assessment, *Geology*, **28**, 639–642.

Pasuto, A. and Soldati, M., 1999, The use of landslide units in geomorphological mapping: an example in the Italian Dolomites, *Geomorphology*, **30**, 53–64.

Paudits, P. and Bednárik, M., 2002, Using Grass in evaluation of landslide susceptibility in Handlovská Kotlina Basin Open Source Free Software GIS–GRASS users' conference 2002, Trento, Italy, 11–13 September 2002, 1–11.

Perov, V.F., Artyukhova, I.S., Budarina, O.I., Glazovskaya, T.G. and Sidorov, T.L., 1997, Map of the world mudflow phenomena, in C.-I. Chen (ed.), *Proceedings, First International Conference on Debris-Flow Hazards Mitigation: Mechanics, Prediction, and Assessment*, Hydraulics Division, American Society of Civil Engineers, 7–9 August 1997 (San Francisco, CA: ASCE) 322–331.

Perov, V.F. and Budarina, O.I., 2000, Mudflow hazard assessment for the Russian Federation, in G.F. Wieczorek and N.D. Naeser (eds), *Debris-Flow Hazards Mitigation: Mechanics, Prediction, and Assessment*, 16–18 August 2000, Taipei, Taiwan (Rotterdam: A.A. Balkema), 489–494.

Petley, D.N., 1996, The Mechanics and Landforms of Deep-Seated Landslides, in M.G. Anderson and S.M. Brooks (eds), *Advances in Hillslope Processes* (Chichester: John Wiley & Sons), vol. 2, 823–835.

Petrascheck, A. and Kienholz, H., 2003, Hazard assessment and mapping of mountain risks in Switzerland, in D. Rickenmann and C.-L. Chen (eds), *Debris-Flow Hazards Mitigation: Mechanics, Prediction, and Assessment*, 10–12 September 2003, Davos, Switzerland (Rotterdam: Millpress), 25–38.

Petro, L., Polascinová, E. and Höger, A., 2002, Landslides in the Kosice region, in J. Rybár, J. Stemberk and P. Wagner (eds), *Landslides*, 24–26 June 2002, Prague, Czech Republic (Rotterdam: A.A. Balkema), 443–447.

Pinches, G.M., Hardingham, A.D. and Smallwood, A.R.H., 2001, The natural terrain hazard study of Yam O Tuk: a case study, in K.K.S. Ho and K.S. Li (eds), *Geotechnical Engineering – Meeting Society's Needs, Proceedings of the 14th Southeast Asian Geotechnical Conference*, 10–14 December 2001 (Hong Kong: A.A. Balkema), 887–892.

Pistocchi, A., Luzi, L. and Napolitano, P., 2002, The use of predictive modeling techniques for optimal exploitation of spatial databases: a case study in landslide hazard mapping with expert system-like methods, *Environmental Geology*, **41**, 765–775.

Plafker, G., Ericksen, G.E. and Fernandez Concha, J., 1971, Geological aspects of the May 31, 1970, Peru earthquake, *Seismological Society of America Bulletin*, **61**, 543–578.

Ploner, A. and Sönser, T., 1999, Natural Hazard Zoning as an important basement of the evaluation of forest protection functions in Austria, in F. Gillet and F. Zanolini (eds), *Risques naturels en montagne* (Grenoble), 431–436.

Pun, W.K., Wong, A.C.W. and Pang, P.L.R., 2003, A review of the relationship between rainfall and landslides in Hong Kong, in K.K.S. Ho and K.S. Li (eds), *Geotechnical Engineering – Meeting Society's Needs, Proceedings of the 14th Southeast Asian Geotechnical Conference*, 10–14 December 2001 (Hong Kong: A.A. Balkema), 211–216.

Ragozin, A.L., 1996, Modern problems and quantitative methods of landslide risk assessment, in K. Senneset (ed.), *Landslides – Glissements de Terrain* (Rotterdam: A.A. Balkema), vol. 1, 339–344.

Reeves, A., Chan, H.C. and Lam, K.C., 1998, Preliminary quantitative risk assessment of Boulder falls in Hong Kong, in K.S. Li, J.N. Kay and K.K.S. Ho (eds), *Slope Engineering in Hong Kong* (Hong Kong: A.A. Balkema), 185–191.

Reid, M.E., Ellen, S.D., Baum, R.L., Fleming, R.W. and Wilson, R.C., 1991, Landslides and debris flows near Honolulu, Hawaii, USA. *Landslide News*, **5**, 28–30.

Reid, M.E., Brien, D.L., LaHusen, R.G., Roering, J.J., de la Fuente, J. and Ellen, S.D., 2003, Debris-flow initiation from large, slow-moving landslides, in D. Rickenmann and C.-L. Chen (eds), *Debris Flow Hazards Mitigation: Mechanics, Prediction, and Assessment*, 10–12 September 2003, Davos, Switzerland (Rotterdam: Millpress), 155–166.

Reneau, S.L. and Dethier, D.P., 1996, Late Pleistocene landslide-dammed lakes along the Rio Grande, White Rock canyon, New Mexico, *Geological Society of America Bulletin*, **108**, 1492–1507.

Revellino, P., Hungr, O., Gudagno, F.M. and Evans, S.G., 2002, Dynamic analysis of recent destructive debris flows and debris avalanches in pyroclastic deposits, Campania region, Italy, in R.G. McInnes and J. Jakeways (eds), *Instability Planning and Management* (Isle of Wight: Thomas Telford), 363–372.

Rezig, S., Favre, J.L. and Leroi, E., 1996, The probalistic evaluation of landslide risk, in K. Senneset (ed.), *Landslides – Glissements de Terrain* (Rotterdam: A.A. Balkema), vol. 1, 351–356.

Rickenmann, D., 1990, *The 1987 Debris Flows in Switzerland: Modelling and Fluvial Sediment Transport* (Wallingford: Oxfordshire), vol. 194, 371–378.

Rickenmann, D., 1999, Empirical relationships for debris flows, *Natural Hazards*, **19**, 47–77.

Rickenmann, D. and Chen, C.-L., 2003, *Debris-Flow Hazards Mitigation: Mechanics, Prediction, and Assessment*, 10–12 September 2003, Davos, Switzerland (Rotterdam: Millpress), 1335.

Romang, H., Kienholz, H., Kimmerle, R. and Böll, A., 2003, Control structures, vulnerability, cost-effectiveness – a contribution to the management of risks from debris torrents, in D. Rickenmann and C.-L. Chen (eds), *Debris-Flow Hazards Mitigation: Mechanics, Prediction, and Assessment*, 10–12 September 2003, Netherlands, Davos, Switzerland (Rotterdam: Millpress), pp. 1303–1313.

Rosenbaum, M.S. and Popescu, M.E., 1996, Usind a geographical information system to record and assess landslide-related risks in Romania, in K. Senneset (ed.), *Landslides – Glissements de Terrain* (Rotterdam: A.A. Balkema), vol. **1**, 363–370.

Roth, R.A., 1983, Factors affecting landslide-susceptibility in San Mateo County, California, *Bulletin of the Association of Engineering Geologists*, **20**(4): 353–372.

Ruff, M., Schanz, C. and Czurda, K., 2002, Geological hazard assessment and rockfall modelling in the Northern Calcareous Alps, Vorarlberg/Austria, in J. Rybár, J. Stemberk and P. Wagner (eds), *Landslides*, 24–26 June 2002, Prague, Czech Republic (Rotterdam: A.A. Balkema), 461–464.

Sánchez, M.J., Farias, P., Rodriguez, A. and Duarte, R.A.M., 1999, Landslide development in a coastal valley in Northern Spain: conditioning factors and temporal occurrence, *Geomorphology*, **30**, 115–123.

Santacana, N., Baeza, C., Corominas, J., de Paz, A. and Marturià, J., 2003, A GIS-based multivariate statistical analysis for shallow landslide stability mapping in La Pobla de Llet area (Eastern Pyrenees, Spain), *Natural Hazards*, **30**(3), 281–295.

Santacana, N. and Corominas, J., 2002, Example of validation of landslide suceptibility maps, in J. Rybár, J. Stemberk and P. Wagner (eds), *Landslides*, 24–26 June 2002, Prague, Czech Republic (Rotterdam: A.A. Balkema), 305–310.

Sarkar, S., 1999, Landslides in Darjeeling Himalayas, India, *Transactions, Japanese Geomorphological Union*, **20**, 299–315.

Sarkar, S., Bonnard, C. and Noverraz, F., 2000, Risk Assessment of potential landslide dams in the valleys of La Veveyse and Veveyse de Figre, Switzerland, in E. Bromhead, N. Dixon and M.-L. Ibsen (eds), *Landslides in Research, Theory and Practice* (Cardiff: Thomas Telford), 1309–1314.

Sasaki, H., Dobrev, N. and Wakizaka, Y., 2002, The detailed hazard map of road slopes in Japan, in R.G. McInnes and J. Jakeways (eds), *Instability Planning and Management* (Isle of Wight: Thomas Telford), 381–388.

Sassa, K., Fukuoka, H., Scarascia-Mugnozza, G., Irikura, K. and Okimura, T., 1999, The Hyogoken-Nanbu earthquake and the distribution of triggered landslides, in K. Sassa (ed.), *Landslides of the world* (Kyoto: Kyoto University Press), 21–26.

Savage, W.Z., Godt, J.W. and Baum, R.L., 2003, A model for spatially and temporally distributed shallow landslide initiation by rainfall infiltration, in D. Rickenmann and C.-L. Chen (eds), *Debris-Flow Hazards Mitigation: Mechanics, Prediction, and Assessment*, 10–12 September 2003, Davos, Switzerland, (Rotterdam: Millpress), 179–187.

Schädel, K. and Stober, I., 1988, Rezente Großrutschungen an der Schwäbischen Alb, *Jahreshefte des Geologischen Landesamtes Baden-Württemberg*, **30**, 431–439.

Scheidegger, A., 1975, *Physical Aspects of Natural Catastrophes* (Amsterdam: Elsevier).

Scheidegger, A.E., 1973, On the prediction of the reach and velocity of catastrophic landslides, *Rock Mechanics*, **5**, 231–236.

Scheidegger, A.E., 1984, A review of recent work on mass movements on slopes and rock falls, *Earth Science Reviews*, **21**, 225–249.

Schilling, S.P. and Iverson, R.M., 1997, Automated, reproducible delinieation of zones at risk from inundation by large volcanic debris flows, in C.-L. Chen (ed.), *Proceedings, First International Conference on Debris-Flow Hazards Mitigation: Mechanics, Prediction, and Assessment*, Hydraulics Division, American Society of Civil Engineers, 7–9 August 1997 (San Francisco, CA: ASCE), 176–186.

Schmanke, V., 2001, *Hangrutschungen im Bonner Raum* (Mainz).

Schmidt, K.-H. and Beyer, I., 2001, Factors controlling mass movement susceptibility on the Wellenkalk-scarp in Hesse and Thuringia, *Zeitschrift für Geomorphologie, Supplementband*, **125**, 43–63.

Schmidt, K.-H. and Beyer, I., 2002, High-magnitude landslide events on a limestone-scarp in central Germany: morphometric characteristics and climate controls, *Geomorphology*, **49**, 323–342.

Schmidt, K.M., Roering, J.J., Stock, J.D., Dietrich, W.E., Montgomery, D.R. and Schaub, T., 2001, The variability of root cohesion as an influence on shallow landslide susceptibility in the Oregon Coast Range, *Canadian Geotechnical Journal*, **38**, 995–1024.

Schmidt, M. and Glade, T., 2003, Linking global circulation model outputs to regional geomorphic models: a case study of landslide activity in New Zealand, *Climate Research*, **25**, 135–150.

Schoeneich, P. and Bouzou, I., 1996, Landslides in Niger (West Africa), in K. Senneset (ed.), *Landslides* (Rotterdam: Balkema), 1967–1972.

Schuster, R.L., Nieto, A.S., O'Rourke, T.D., Crespo, E. and Plaza-Nieto, G., 1996, Mass wasting triggered by the 5 March 1987 Ecuador earthquakes, *Engineering Geology*, **42**, 1–23.

Scott, K.M., Vallance, J.W. and Pringle, P.T., 1995, *Sedimentology, behaviour, and hazards of debris flows at Mount Rainier* (Washington: US Geological Survey).

Sidorova, T.L., 1997, Potential changes of mudflow phenomena due to global warming, in C.-L. Chen (ed.), *Proceedings, First International Conference on Debris-Flow Hazards Mitigation: Mechanics, Prediction, and Assessment*, Hydraulics Division, American Society of Civil Engineers, 7–9 August 1997 (San Francisco, CA: ASCE), 540–549.

Skempton, A.W., 1953, Soil mechanics in relation to geology, *Proceedings of the Yorkshire Geological Society*, **29**, 33–62.

Skempton, A.W., 1966, Bedding-plane slip, residual strength and the Vaiont landslide, *Géotechnique*, **16**, 82–84.

Skempton, A.W. and Northey, R.D., 1952, The sensitivity of clays, *Géotechnique*, **3**, 30–53.

Smallwood, A.R.H., Morley, R.S., Hardingham, A.D., Ditchfield, C. and Castleman, J., 1997, Quantitative risk assessment of landslides: case histories from Hong Kong, in P.G. Marinos, G.C. Koukis, G.C. Tsiambaos and G.C. Stournaras (eds), *Proceedings of the International Symposium on Engineering Geology and the Environment*, 23–27 June 1997 (Athens, Greece: A.A. Balkema), 1055–1060.

Soeters, R. and van Westen, C.J., 1996, Slope instability recognition, analysis, and zonation, in A.K. Turner and R.L. Schuster (eds), *Landslides: Investigation and Mitigation* (Washington, DC: National Academy Press), vol. **247**, 129–177.

Solana, M.C. and Kilburn, C.R.J., 2003, Public awareness of landslide hazards: the Barranco de Tirjana, Gran Canaria, Spain, *Geomorphology*, **54**, 39–48.

Sorriso-Valvo, M., 1997, Landsliding during the Holocene in Calabria, Italy, in J.A. Matthews, D. Brunsden, B. Frenzel, B. Gläser and M.M. Weiß (eds), *Rapid Mass Movement as a Source of Climatic Evidence for the Holocene* (Stuttgart, Jena, Lübeck and Ulm: Gustav Fischer Verlag), **12**, 97–108.

Spang, R.M. and Sönser, T., 1995, Optimized rockfall protection by 'Rockfall', in T. Fuji (ed.), *8. International Congress on Rock Mechanics* (Tokyo, Japan), 1233–1242.

Starkel, L., 1997, Mass movements during the Holocene: a Carpathian example and the European perspective, in J.A. Matthews, D. Brunsden, B. Frenzel, B. Gläser and M.M. Weiß (eds), *Rapid Mass Movement as a Source of Climatic Evidence for the Holocene* (Stuttgart, Jena, Lübeck and Ulm: Gustav Fischer), vol. 12, 385–400.

Stead, D.S., Eberhardt, E., Coggan, J. and Benko, B., 2001, Advanced numerical techniques in rock slope stability analysis – applications and limitations, in M. Kühne, H.H. Einstein, E. Krauter, H. Klapperich and R. Pöttler (eds), *Landslides – Causes, Impacts and Countermeasures*, 17–21 June 2001 (Davos, Switzerland: Verlag Glückauf Essen).

Stevenson, P.C., 1977, An empirical method for the evaluation of relative landslide risk, *International Association of Engineering Geology Bulletin*, **16**, 69–72.

Stevenson, P.C. and Sloane, D.J., 1980, The evolution of a risk-zoning system for landslide areas in Tasmania, Australia, *Proceedings of the 3rd Australia and New Zealand Geomechanics Conference* (Wellington), 7.

Stiny, J., 1910, *Die Muren* (Innsbruck).

Streit, R., 1991, *Eiszeitliche und postglaziale Rutschmassen bei Plüttlach, Nördliche Frankenalb* (Köln).

Szabó, J., 1999, Landslide activity and land utilisation at the high river bank zones, in J.S. Griffiths, M.R. Stokes and R.G. Thomas (eds), *Landslides – Proceedings of the Ninth International Conference and Field Trip on Landslides*, Bristol, United Kingdom, 5–16 September 1999 (Rotterdam: A.A. Balkema), 147–154.

Tang, C. and Grunert, J., 1999, Inventory of landslides triggered by the 1996 Lijiang earthquake, Yunnan Province, China, *Transactions, Japanese Geomorphological Union*, **20**, 335–349.

Tavares, A.O. and Soares, A.F., 2002, Instability relevance on land use planning in Coimbra municipality (Portugal), in R.G. McInnes and J. Jakeways (eds), *Instability Planning and Management* (Isle of Wight: Thomas Telford), 177–184.

Terzaghi, K., 1925, Principles of soil mechanics, *Engineering News Record*, **95**, 742.

Terzaghi, K. and Peck, R.B., 1948, *Soil Mechanics in Engineering Practice* (New York: Wiley).

Thein, S., 2000, Massenverlagerungen an der Schwäbischen Alb – Statistische Vorhersagemodelle und regionale Gefährdungskarten unter Anwendung eines Geographischen Informationssystems.

Thurston, N. and Degg, M., 2000, Transferability and terrain reconstruction within a GIS landslide hazard mapping model: Derbyshire Peak District, in E.N. Bromhead, N. Dixon and M.-L. Ibsen (eds), *Landslides in Research, Theory and Practice, Proceedings of the 8th International Symposium on Landslides*, 26–30 June 2000 (Cardiff, UK: Thomas Telford), 1467–1472.

Tianchi, L., 1996, *Landslide Hazard Mapping and Management in China Kathmandu, Nepal*, International Centre for Integrated Mountain Development (ICIMOD), 36.

Tibaldi, A., Ferrari, L. and Pasquare, G., 1995, Landslides triggered by earthquakes and their relations with faults and mountain slope geometry – an example from ecuador, *Geomorphology*, **11**, 215–226.

Toyos, G., Oppenheimer, C., Pareschi, M.T., Sulpizio, R., Zanchetta, G. and Zuccaro, G., 2003, Building damage by debris flows in the Sarno area, southern Italy, in D. Rickenmann and C.-L. Chen (eds), *Debris-Flow Hazards Mitigation: Mechanics, Prediction, and Assessment*, 10–12 September 2003. Davos, Switzerland (Rotterdam: Millpress), 1209–1220.

Trustrum, N.A. and Stephens, P.R., 1981, Selection of hill country pasture measurement sites by interpretation of sequential aerial photographs, *New Zealand Journal Experimental Agriculture*, **9**, 31–34.

Trustrum, N.A., Whitehouse, I.E., Blaschle, P.M. and Stephens, P.R., 1990, Flood and landslide hazard mapping, Solomon Islands, in R.R. Ziemer, C.L. O'Loughlin and L.S. Hamilton (eds), *Research Needs and Applications to Reduce Erosion and Sedimentation in Tropical Steeplands*, International Association of Hydrological Sciences, **192**, 138–146.

Twidale, C.R., 2000, The Lochiel landslip, a mass movement developing in 1974 but orginating 600–700 million years earlier, in E.N. Bromhead, N. Dixon and M.-L. Ibsen (eds), *Landslides in Research, Theory and Practice, Proceedings of the 8th International Symposium on Landslides*, 26–30 June 2000 (Cardiff, UK: Thomas Telford), 1489–1494.

UNDRO, 1982, *Natural Disasters and Vulnerability Analysis* (Geneva: United Nations Disaster Relief Organisation).

Uromeihy, A., 2000, Use of landslide hazard zonation map in the evaluation of slope instability in the Jiroft-dam watershed, in E.N. Bromhead, N. Dixon and M.-L. Ibsen (eds), *Landslides in Research, Theory and Practice, Proceedings of the 8th International Symposium on Landslides*, 26–30 June 2000 (Cardiff, UK: Thomas Telford), 1501–1508.

van Asch, T.W.J., Van Westen, C., Blijenberg, H. and Terlien, M.I.J., 1992, Quantitative landslide hazard analyses in volcanic ashes of the chinchina area, *Colombia Primer Simposio Internacional Sobre Sensores Remotos y Sistemas de Informacion Geografica (SIG) Para el Estudio de Riesgos Naturales*, 8 a 15 de marzo de 1992 (Santafé de Bogotá, Colombia), 433–443.

van Beek, R., 2002, *Assessment of the Influence of Changes in Land Use and Climate on Landslide Activity in a Mediterranean Environment*. (Utrecht.).

Van Sint Jan, M.L., 1994, Characteristcs and cost evaluation of landslides in Chile, in International Association of Engineering Geology (ed.), *Proceedings of the 7th International Congress* (Lisbon, Portugal), 1713–1720.

van Westen, C.J., 1994, GIS in landslide hazard zonation: a review, with examples from the Colombian Andes, in M.F. Price and D.I. Heywood (eds), *Mountain Environments and Geographical Information Systems* (London: Taylor & Francis), 135–166.

Vanacker, V., Vanderschaeghe, M., Govers, G., Willems, E., Poesen, J., Deckers, J. and De Bievre, B., 2003, Linking hydrological, infinite slope stability and land-use change models through GIS for assessing the impact of deforestation on slope stability in high Andean watersheds, *Geomorphology*, **52**, 299–315.

Vardoulakis, I., 2002, Dynamic thermo-poro-mechanical analysis of catastrophic landslides, *Géotechnique*, **52**, 157–171.

Viberg, L., Fallsvik, J. and Johansson, K., 2002, National map data base on landslide prerequisites in clay and silt areas – development of prototype, in R.G. McInnes and J. Jakeways (eds), *Instability Planning and Management* (Isle of Wight: Thomas Telford), 407–414.

Villi, V. and Dal Pra, A., 2002, Debris flow in the upper Isarco valley, Italy – 14 August 1998, *Bulletin of Engineering Geology and Environment*, **61**, 49–57.

Voight, B. and Faust, C., 1992, Frictional heat and strength loss in some rapid landslides – error correction and affirmation of mechanism for the vaiont landslide, *Géotechnique*, **42**, 641–643.

von Poschinger, A., 2002, Large rockslides in the Alps: a commentary on the contribution of G. Abele (1937–1994) and a review of some recent developments, in S.G. Evans and J.V. DeGraff (eds), *Catastrophic Landslides: Effects, Occurrence, and Mechanisms* (Boulder, CO: Geological Society of America), **15**, 237–255.

von Poschinger, A. and Haas, U., 1997, Risiken durch Hangbewegungen in den bayerischen Alpen, Erfahrungen mit dem Projekt GEORISK nach 10 Jahren Praxis, in B. Schwaighofer, H.W. Müller and J.F. Schneider (eds), *Angewandte Geowissenschaften* (Wien: Institut für Angewandte Geologie), **8**, 43–52.

Wachal, D.J. and Hudak, P.F., 2000, Mapping landslide susceptibility in Travis County, Texas, USA, *GeoJournal*, **51**, 245–253.

Wang, Y.-J., Chen, Z., Lee, C.F. and Yin, J.-H., 2001, Three-dimensional stability analysis of slopes using a plastic limit method, in K.K.S. Ho and K.S. Li (eds), *Geotechnical Engineering – Meeting Society's Needs, Proceedings of the 14th Southeast Asian Geotechnical Conference*, 10–14 December 2001 (Hong Kong: A.A. Balkema), 923–928.

Whitworth, M.C.Z., Giles, D.P., Murphy, W. and Petley, D.N., 2000, Spectral properties of active, suspended and relict landslides derived from airborne thematic mapper imagery, in E.N. Bromhead, N. Dixon and M.-L. Ibsen (eds), *Landslides in Research, Theory and Practice, Proceedings of the 8th International Symposium on Landslides*, 26–30 June 2000 (Cardiff, UK: Thomas Telford), 1569–1574.

Wieczorek, G.F., 1984, Preparing a detailed landslide-inventory map for hazard evaluation and reduction, *Bulletin of the Association of Engineering Geologists*, **21**, 337–342.

Wieczorek, G.F., Coe, J.A. and Godt, J.W., 2003, Remote sensing of rainfall for debris-flow hazard assessment, D. Rickenmann and C.-L. Chen (eds), *Debris-Flow Hazards Mitigation: Mechanics, Prediction, and Assessment*, 10–12 September 2003, Davos, Switzerland (Rotterdam: Millpress), 1257–1268.

Wieczorek, G.F., Lips, E.W. and Ellen, S.D., 1989, Debris flows and hyperconentrated floods along the Wasatch Front, Utah, 1983 and 1984, *Bulletin of the Association of Engineering Geologists*, **26**, 191–208.

Wieczorek, G.F., Morgan, B.A. and Campbell, R.H., 2000, Debris-flow hazards in the Blue Ridge of Central Virginia, *Environmental and Engineering Geoscience*, **6**, 3–23.

Wieczorek, G.F., Morrissey, M.M., Iovine, G. and Godt, J., 1998, *Rock-Fall Hazards in the Yosemite Valley*, U.S. Geological Survey, 8.

Wieczorek, G.F. and Naeser, N.D., 2000, *Proceedings of the Second International Conference on Debris-Flow Hazards Mitigation*, 16–18 August 2000 *Debris-flow Hazards Mitigation: Mechanics, Prediction, and Assessment*, Taipei, Taiwan, (Rotterdam: Balkema), 608.

Wilson, H. and Crozier, M.J., 2000, Quantitative hazard assessment: rainfall-triggered landslides, in E.N. Bromhead, N. Dixon and M.-L. Ibsen (eds), *Landslides in Research, Theory and Practice, Proceedings of the 8th International Symposium on Landslides*, 26–30 June 2000 (Cardiff, UK: Thomas Telford), 1575–1580.

Wong, H.N. and Ho, K.K.S., 1996, Travel distance of landslide debris, in K. Senneset (ed.), *Landslides – Glissements de Terrain* (Rotterdam: A.A. Balkema), vol. 1, 417–422.

Wong, T.T. and Hanson, A., 1995, The use of GIS for inventory and assessment of natural landslides, in *Hong Kong International Symposium on Remote Sensing, Geographic Information Systems and Global Positioning Systems for sustainable development and environmental monitoring*, 26–28 May 1995 (Hong Kong), 930–937.

Wu, T.H. and Abdel-Latif, M.A., 2000, Prediction and mapping of landslide hazard, *Canadian Geotechnical Journal*, **37**, 781–795.

Wu, T.H. and Swanston, D.N., 1980, Risk of landslides in shallow soils and its relation to clearcutting in southeastern Alaska, *Forest Science*, **26**, 495–510.

Wu, T.H., Tang, W.H. and Einstein, H.E., 1996, Landslide hazard and risk assessment, in A.K. Turner and R.L. Schuster (eds), *Landslides, Investigation and Mitigation*, Special Report 247 (Washington, DC: National Academy Press), 106–118.

Xie, M., Esaki, T., Zhou, G. and Mitani, Y., 2001, Effective data management for GIS-based 3D critical slip searching, in K.K.S. Ho and K.S. Li (eds), *Geotechnical Engineering – Meeting Society's Needs, Proceedings of the 14th Southeast Asian Geotechnical Conference*, 10–14 December 2001 (Hong Kong: A.A. Balkema), 947–952.

Yamagami, T., Jiang, J.-C. and Yokino, K., 2001, An identification of DEM parameters for rockfall analysis, in K.K.S. Ho and K.S. Li (eds), *Geotechnical Engineering – Meeting Society's Needs, Proceedings of the 14th Southeast Asian Geotechnical Conference*, 10–14 December 2001 (Hong Kong: A.A. Balkema), 953–958.

Yamagishi, H., Ito, Y. and Kawamura, M., 2002, Characteristics of deep-seated landslides of Hokkaido: analyses of a database of landslides of Hokkaido, Japan, *Environmental and Engineering Geoscience*, **8**, 35–46.

Yamakoshi, T., Nakano, M., Watari, M., Mizuyama, T. and Chiba, T., 2003, Debris-flow occurrence after the 2000 eruption of Miyakejima volcano, in D. Rickenmann and C.-L. Chen (eds), *Debris-Flow Hazards Mitigation: Mechanics, Prediction, and Assessment*, 10–12 September 2003. Davos, Switzerland (Rotterdam: Millpress), 1049–1057.

Yin, K., Liu, Y., Zhang, L., Wu, Y. and Zhu, L., 2002, Overview geo-hazard assessment of China by GIS, in R.G. McInnes and J. Jakeways (eds), *Instability Planning and Management* (Isle of Wight: Thomas Telford), 423–432.

Zêzere, J.L., Ferreira, A.B., Vieira, G., Reis, E. and Rodrigues, M.L., 2000, The use of Baysian probability for landslide susceptibility evaluation: A case study in the area of Lisbon (Portugal), in E.N. Bromhead, N. Dixon and M.-L. Ibsen (eds), *Landslides in Research, Theory and Practice, Proceedings of the 8th International Symposium on Landslides*, 26–30 June 2000 (Cardiff, UK: Thomas Telford), 1635–1640.

Zêzere, J.L., Ferreira, A.D. and Rodrigues, M.L., 1999, The role of conditioning and triggering factors in the occurrence of landslides: a case study in the area north of Lisbon (Portugal), *Geomorphology*, **30**, 133–146.

Zimmermann, M., Mani, P., Gamma, P., Gsteiger, P., Heiniger, O. and Hunziker, G., 1997, *Murganggefahr und Klimaänderung – ein GIS-basierter Ansatz (NFP 31 Schlussbericht)* (Zürich: vdf Hochschulverlag).

# 4

# Systematic Procedures of Landslide Hazard Mapping for Risk Assessment Using Spatial Prediction Models

*Chang-Jo F. Chung and Andrea G. Fabbri*

## 4.1 Introduction

This chapter reviews advances in thematic map derivation by the application of mathematical models to stacks of map layers in digital form. Such spatially distributed data layers must describe, not only visually but also structurally, indicators of processes that capture spatial variations and relationships. For instance, the variations in elevation indicated by contour lines approximate the topographic surface that is the product of orogenetic, lithogenetic and erosion processes. Other examples are a land-cover map that describes the distribution of biological, erosion and anthropogenic processes, or the map of the drainage network that describes aspects of the hydrologic processes. The crucial layer is the distribution of landslide scars such as the shallow translational landslides that document the past surficial land degradation processes.

While such maps have until the 1970s been produced in the analog form of paper products, they are now mostly in digital form, accessible in a variety of media beside hard copies on paper. Furthermore, the digital output of spatial data processing is not limited to static visual renderings but can provide means of interpreting spatial variability and the relationships between the processes.

Given compatible spatial and 'conceptual' granularities (sampling densities and mapping units), several representations can be integrated into a thematic map to express the

*Landslide Hazard and Risk*   Edited by Thomas Glade, Malcolm Anderson and Michael J. Crozier
© 2004 John Wiley & Sons, Ltd   ISBN: 0-471-48663-9

typical physical settings of events generated by the processes. Integration work has been the task of earth scientists, who have generated cartographic products such as surficial geology, engineering geology, and landslide hazard maps. They represent qualitative characterizations to assist in understanding the thematic distribution of classes, intervals, or mapping units to better plan land uses or to locate undesired features such as landscape instability areas.

Spatial understanding and planning have become increasingly important with the densification of urban agglomerates and human activities, so that spatial representations are now sought for daily decision making over continuous stretches of land at operational levels of detail. The term operational level implies a hierarchy of relative resolution increments as a function of evaluating thematic maps of lesser resolution, that is, greater generality, thus optimizing the data requirements and the uniformity of significance.

For instance, spatial thematic representations can now be generated owing to three facts: (1) the desired digital data are often easily obtainable or even downloadable from the Internet; (2) powerful spatial data processing systems are common; and (3) the theoretical background of hazardous processes has been developed in many cases.

The decision makers, however, prefer 'predictive' representations over 'static' inventories of past hazardous events, whenever the specific purpose of the decision to be taken is loss avoidance or damage prevention/impact mitigation. To clarify what needs to be mapped to assist in decision making, Varnes *et al.* (1984: 10) have provided the definitions in Table 4.1.

The damages and casualties due to landslides are extensive in both the developed and the developing countries. In the 1970s alone, thousands of lives have been lost and examples of direct and indirect costs in the USA have been estimated in excess of $1000 million. In Italy alone, annual losses have been reported as over $1000 million during the 1970s (Varnes *et al.*, 1984). At present, such damages and casualties seem to be on the increase, mainly because of an increase in population growth and urbanization in hazardous locations (Terlien, 1996; Terlien *et al.*, 1995) so that hazard zonation should support not only the prevention of measures for disasters but also hazard mitigation programmes.

This means that decision makers in institutes with mandates such as civil protection and land planning must anticipate the occurrences and location of future disasters and provide documentation to support the measures to be taken. In practice, for instance, the general public needs to know the locations to be affected by future landslides in areas of concern within, say, the next 30 years. What is traditionally provided instead are the locations of past landslides, their description and characteristics, and some qualitative geomorphological models that explain what is contained in a geomorphological map or in a hazard map compiled from it. Seldom is a representation provided that predicts the location or distribution in time and in space of future landslides. In addition, when such prediction maps are produced, no measure of reliability and effectiveness accompanies them to provide support for the responsibilities that would be undertaken by using them in the decision process. To quote Varnes *et al.* (1984: 10): 'Many hundreds of maps of landslides or of their deposits old or new or active, have been made throughout the world . . . ' but there are a 'far fewer number of studies that go further and attempt to assign degrees of hazard to mapped areas'. We can add to that by stating that, at present, the production of probability maps expressing hazard, vulnerability and risk has yet to become a common practice even in the light of the extensive and repeated damage to society and the recent research efforts by earth scientists and engineers.

**Table 4.1**   *Terms related to hazard and risk after (Varnes et al., 1984: 10, reproduced by permission of UNESCO)*

| Term | Symbol | Definition |
|---|---|---|
| Landslide | | A term that comprises almost all varieties of mass movements on slope, including some, such as rockfalls, topples, and debris flows, that involve little or no true sliding |
| Zonation | | A term that applies in a general sense to division of the land surface into areas and the ranking of these according to degrees of actual or potential hazard for landslides or other mass movements on slopes |
| Natural hazard | $H$ | The probability of occurrence within a specified period of time and within a given area of a potentially damaging phenomenon |
| Vulnerability | $V$ | The degree of loss to a given element or set of elements at risk (see below) resulting from the occurrence of a natural phenomenon of a given magnitude. It is expressed on a scale from 0 (no damage) to 1 (total loss) |
| Specific risk | $R_s$ | The expected degree of loss due to a particular natural phenomenon. It may be expressed by a product of $H$ times $V$ |
| Element at risk | $E$ | The population, properties, economic activities, including public services, etc. at risk in a given area |
| Total risk | $R_t$ | The expected number of lives lost, persons injured, damage to property, or disruption of economic activity due to a particular natural phenomenon, and is therefore the product of specific risk ($R_s$) and elements at risk ($E$). Thus $R_t = E \cdot R_s = E \cdot H \cdot V$ |

This chapter deals with an analytical strategy to estimate the conditional probability of occurrence of future landslides within a specified time period within each hazard prediction class. The reconstruction of the typical settings in which individual dynamic types of landslides tend to occur is obtained by processing a spatial database of geomorphological and related data such as remotely sensed land cover and land use images, digital elevation models (DEM) and of features extracted from them, maps of the distribution of landforms identifying surficial deposits and mass movement areas, including dynamic type, scars, time and magnitude of the landslides. The latter two characteristics are seldom part of the database, however, due to the practical difficulties in recording the time of failure and in measuring the extent of the individual movements. In the databases used in this chapter, the magnitude of the landslides was not available and was therefore not included in the prediction models. In order to include the magnitude in the spatial models proposed here, that of past landslides is required at each pixel in the study area. We have dealt with the magnitude of the landslides indirectly by considering the areas occupied by the scars (or by the scarps) of the landslides. The assumption made is that the landslides with large magnitude tend to have large scars (or scarps) and consequently they strongly influence the construction of the spatial prediction models proposed. Another possible way to infer their magnitude is to assume that the magnitude of a landslide at a pixel is inversely proportional to the minimum distance between the pixel and the boundary of the scar. This assumption was not made in this study.

A cross-validation procedure is used to evaluate the reliability of the prediction classes using a spatial/temporal partition of the distribution of past landslides. From the statistics of the spatial cross-validation, measures to estimate the conditional probability of the damaging events are obtained via the simulation of vulnerability scenarios and landslide scar dynamic simulations from the prediction pattern of landslide trigger area distribution.

A review of past traditional hazard zonation work and of key recent mathematical developments sets the background for the strategy proposed here. Several application examples provide support to the strategy and indicate what type of information at which operational levels of detail spatial data infrastructures have to provide locally or for a region or for a country, to enable a vulnerability analysis for landslide risk mapping. Considerations on the future of spatial prediction models are offered to conclude the chapter.

## 4.2   Some Examples of Hazard Zonation

The following review provides some procedures for landslide prediction modelling that are important to understand the earlier studies on the complexities of quantitative hazard zonation and to identify promising grounds for spatial predictions.

Examples of the application of quantitative techniques to landslide hazard mapping were presented by Carrara (1989), where multivariate or 'black-box' models were portrayed as capable of successfully predicting actual and potential slope failures in each of several zones studied in Italy. As 'warning devices' or, better, 'tools' for the selection of sites for further investigation, that author listed three procedures: (1) direct estimation by expert surveyors; (2) 'index' thematic maps obtained by overlaying maps representing encoded slope instability factors that are rated and weighted; and (3) statistical approach to actual/potential instability assessment using multivariate models to build hazard probability levels. The fundamental assumption in the procedures was that 'the factors which caused and can cause the failures in the training areas are the same as those generating landslides' throughout the whole study area. One of the basic units used for the analysis was 'geomorphologically meaningful' slope units of variable size, over which geomorphic processes had been observed. The various factors were analysed and aggregated into hazard levels using a grid mesh of convenient size. Stepwise discriminant analysis was then used to separate stable from unstable slopes into a classification using scores that the author considered easily convertible into probabilities. The resulting maps were then compared with the distribution of the actual/mapped landslide zones to obtain a 'degree of reliability'.

Unfortunately, such comparisons happen to be completely meaningless for measuring the reliability of the prediction because in practice the same landslides that guided in the construction of the prediction analysis were used as confirmation of the four classes of hazard arbitrary selected.

More recently, the same author and his collaborators (Carrara *et al.*, 1995 and Guzzetti *et al.*, 1999) discussed three types of terrain units used in the application of multivariate models for mapping landslide hazard: grid-cell, unique-condition and slope units. They attempted a rough comparison of three hazard models using an arbitrary selection of hazard levels: I – discriminant analysis on slope units, II – 'conditional analysis' on unique-condition units, and III – discriminant analysis on unique-condition units. In particular, hazard model I was used to classify 266 slope units consisting of two groups: those considered as landslide-free because containing less than 2% of landslide areas, and the remainder that were considered as landslide-bearing. Forty factors were used for constructing the discriminating function to classify the units as either stable or unstable. The function was then tested in three experiments in which randomly selected subsets of 65% of the units were used as training data and the remaining 35% were used as test data. The results led to 83.8%, 82% and 75% of 'correctly' classified units.

Several problems are implicit in the above analyses: (1) no clear statistical relationship had been established or mentioned between the location of individual mass movements and of the corresponding factors, but only between the preclassified slope units and the factors; (2) no subdivision was made of the mass movements in time slices corresponding to the time of the aerial photo coverage; (3) the assumption that the factors were 'the same' through time is very strict and does not lead to modelling degrees of similarities in factors that could consider changes in the rates of geomorphological processes; (4) the probabilities of occurrences of landslides cannot be obtained without several specific assumptions on the occurrences of the future events. The assumptions were never specified, but were hidden. The assumptions required for the estimation of the probabilities include the number of future landslides expected and the sizes of the landslides within a time period. For example, if we expect one future landslide in the study area, the estimated probability of the occurrence of the future landslide should be much smaller than the estimator under an assumption of 1000 expected future landslides. If the assumptions were not specified, the probabilities estimated were completely meaningless; (5) the division of the units into 65% as training data and the remaining 35% as test data enabled us to check the degree of reliability of a classification; (6) a quantitative comparison between relative classifications should have been done using the same sizes of predicted hazard areas from the three prediction studies and the statistics from the distribution of the unused mass movements within identical area proportions of hazard levels. It was not clear whether the authors had done so for their comparisons. The approaches used by these authors were not only unsystematic but led to neither explainable hazard levels nor to validated slope units.

Liener *et al.* (1996) proposed a rule-based procedure, termed SLIDISP, to locate landslide prone areas using simplified safety factors obtained from geotechnical investigations (slope topography, soil strength parameters, depth and shape of potential shear plane, and hydraulic behaviour). In a Swiss study area, one type of hazard map, termed Simple Landslide Map or SLM, classified areas where the slope angle was steeper than the lowest critical slope angle as potentially landslide prone. Different slope angles were estimated for different combinations of landslide types and soils. To confirm the SLM maps, the limits of the slope angle class were compared with the slope angles of the occurred landslides: 86% of the pixels were correctly classified. In the method proposed by the authors, while the derivation of safety factors of slope and soil combinations extended the use of deterministic factors, the rule-based classification did not generate a prediction map in which the relationships between the location of the landslide events and the several map layers or parameter combinations were established so that a validation could be performed of the hazard index obtained.

Mejía-Navarro and Garcia (1996) presented IPDSS, an Integrated Planning Decision Support System, for hazard, vulnerability and risk assessment based on a GIS platform and a graphic user interface. Landslide hazard (susceptibility) was computed as a weighted summation of ratings assigned to natural physical factors such as: topography, aspect, bedrock, surficial and structural geology, geomorphology, soils, land cover, land use, hydrology, precipitation, floodway maps and historical data on previous hazards. Relative ratings of the units of each factor map were assigned and overlaid and cross-tabulated with the map of geomorphic processes. The units in which failures have been occurring most intensively were assigned a rating of 10 and relatively lesser frequencies were assigned lower values within a 0–10 interval. In addition, relative 0–10 weights were assigned to each factor map based on either statistical analysis using multivariate criteria or on expert knowledge. Tables of commented ratings and weights were stored in their system for interaction and could be modified by the experts. Weighted summation overlays of such factors generated the hazard representations assuming a 50-year return period as a function of an estimated probability of maximum rainfall. While such a weighted summation aggregation of predefined ratings and weights of the different factor maps allowed the DSS to process the data and generate the hazard maps, no systematic attempt was described to subdivide the mass movements into time intervals and to characterize their presence and location in terms of association of factors that led to prediction patterns that could be validated. The results of the hazard maps, generated by weighted summation, are tailored to approximate as much as possible the settings that were considered as typical by expert surveyors. Although IPDSS proceeded further to represent vulnerabilities and risks, the procedure was affected by a variety of arbitrary characterizations and choices. No mathematical foundation was discussed that led to predictions and to their interpretations.

Many environmental factors have the potential to affect landsliding, as discussed by Soeters and van Westen (1996), but only a few can be effectively used to generate a variety of landslide hazard maps. No systematic prediction strategy, however, can be extracted from their review and discussion of existing approaches.

Guzzetti *et al.* (1999) have extensively reviewed 'current techniques' on landslide hazard evaluation and stated: 'The reliability of those maps [of landslide hazard prediction maps by current techniques] and the criteria behind those hazard evaluations are ill-formulated or poorly documented.' In their review Guzzetti *et al.* (1999) did not provide any possible suggestion to avoid the problem, but observed that 'predictive models of landslide hazard can not be readily tested by traditional scientific methods. Indeed, the only way a landslide predictive map can be validated is through time.' Then they concluded that 'Solutions to these challenging problems may come from a new scientific practice enabling to cope with large uncertainties.' In short, although they have recognized the problem, which is significant, they did not see how to tackle it, and the only possible solution is 'wait and see'. We have tried to provide a possible solution to the problem of the validation in Chung and Fabbri (1999, 2003).

Cardinali *et al.* (2002) produced and analysed multitemporal landslide inventory maps obtained through aerial photo interpretations and field verifications. They restricted their investigations to the immediate neighbourhoods of the mapped existing and past mass movements observed during 60 years to focus on the evolutional changes and distribution of the various types of failures. Hazard, vulnerability and risk indices of the landslide neighbourhoods were obtained using the following strategy: 1 – define extent of study area; 2 – produce multitemporal landslide inventory and classification map; 3 – define landslide hazard zone 'around existing' simple and multiple landslides; 4 – perform landslide hazard assessment; 5 – identify and map elements at risk and assess their vulnerabilities to the different landslide types; and 6 – evaluate landslide risk. It was assumed that the landslides in the study area tended to occur in time and in space within and in the vicinity of other landslides, or in the same slope or watershed. By overlaying and merging, an aggregated map was obtained from the multitemporal landslide inventory maps of deep-seated and shallow mass movements. Arbitrary tables were generated and used for the following landslide aspects: expected intensity, frequency, velocity, volume and type. To economize in data acquisition, the authors opted for the definition of local landslide hazard instead of a regional one, that is, not for an entire area or drainage basin, municipality or province, but for only the areas of evolution of existing (mapped) landslides. Their landslide hazard zonation was limited to the areas of possible evolution of the mass movements. The definition of hazard levels was obtained by cross-tabulation of classes of estimated landslide frequency and of the observed intensity of the landslides into $4 \times 4$ classes of low/light, medium, high and very high values that, however, did not provide absolute rankings of hazard levels. According to them, some of the rankings may be 'a matter of opinion'. A generalized map of 11 elements at risk in an example of a study area was prepared at 1:10 000 and a three-level vulnerability table was constructed for each landslide intensity and type based on an inferred relationship (from literature), historical data of the region, and expert judgement. Also, the landslide risk index that they have obtained did not provide an absolute ranking of risk levels. For cost-effectiveness in their study, 79 study areas were considered in which 21.4% was covered, corresponding to a subset of 980 landslide hazard zones of 210, that is, $20 \, \text{km}^2$. As stated by the authors, the method says nothing about the hazards outside a landslide hazard zone, and 'at present it is not possible to judge quantitatively how good the proposed method is'. This work is of particular interest for its efforts to introduce time through a 60-year aerial photo interpretation of mass movements in their evolutional occurrence; however, it has three severe limitations in that: (1) it does not provide quantitative absolute ranking of the hazard levels; (2) it does not allow validation of the hazard assessment/prediction results; and (3) it does not use consistent spatial units nor spatial coverage.

Clerici *et al.* (2002) proposed a procedure for landslide susceptibility zonation based on conditional probabilities using both a GIS platform and a scripting language for iterative processing. They considered five environmental factors probably related to landslide occurrences (the respective number of units used within parentheses): geology (12), land use (5), slope (5), rainfall (5), and bedding/topographic slope relationships (5). By overlaying the five maps, the whole study area was divided into a number of 'unique-condition polygons'. Each polygon was homogeneous and contained the same geology, land use, slope, rainfall and bedding/topographic slope relationships unit. As an example of application of their procedure, they used a spatial database of $5 \, \text{m} \times 5 \, \text{m}$ resolution of over 13 million cells, in which the frequency of landslide events (proportion of 5 m cells containing at least one landslide event) was assumed to be the estimator of the probability that that event would occur within the unique-condition polygon. That meant that landslide density equals landslide susceptibility. Such densities were computed for unique-condition areas or areas with unique combinations of factor units. The computed density classes were grouped into five final susceptibility classes defined with the mean density value for the entire area in the middle point of the middle class. Arbitrary very low and low classes were set below the latter and high and very high classes above it. In that way different study area percentages were obtained for the susceptibility map that consisted of 2131 unique-condition areas, of which 1542 were affected by at least one landslide. Six different dynamic types of landslides were considered without differentiation, so that the susceptibilities generated referred only to a 'generic' failure; however, the author mentioned that it would be advisable to carry out landslide susceptibility zonation for each type. No particular theory was presented for predictive statistical analysis and no technique was employed to validate the resulting susceptibility map – only an interpretation by general geomorphological considerations.

Those authors attempted to provide a useful procedure; however, the approach appears to ignore to some extent the theoretical background available in predictive modelling. In addition, no separation was made in the analysis between the crown and the accumulation zones of the landslides. Moreover, their approach provided no means of evaluating the degree of success of the final susceptibility map. In fact, Chung *et al.* (1995) had discussed that very procedure (later used by Clerici *et al.*, 2002), a procedure that had been termed 'direct method', and that, according to them, should not be used as a prediction model but just as a benchmark of input data layers as the causal factors of the landslides.

Dai and Lee (2002) have used a logistic analysis to predict slope instability in Lantau Island, Hong Kong, where they have also studied the runout behaviour of the landslide masses, but again they failed to evaluate the prediction results.

Let us now consider some contributions that provided approaches that seem more in line with the one proposed in this chapter. Leroi (1996) reviewed several landslide-related map productions in France since 1970 that, for instance, have led to the Plans for the Exposure to predictable natural Risks, or PERs, instituted by law on 13 July 1982. These maps are drawn at 1:5000 to 1:10 000, are legally binding, and have created a link between prevention and compensation. Levoi observed that the evaluation of risk of slope instability to be based on explicit and rigorous mapping of the hazard implies answers to the following six questions of which, in most cases, risk hazard maps are based on answering to only the first two: (1) Which type of movement is involved? (2) Where are the potentially unstable areas? (3) At which moment can the identified phenomenon be triggered? (4) How far can the phenomenon be propagated? (5) What are the interactions with the environments, whether natural or modified by man? (6) What is the cost of the caused damage? He discussed three main mapping methods available: (i) expert evaluation (subjective and needing explanations and statistical techniques); (ii) back-analysis through shape recognition (using the mass movement patterns as learning areas and groups of causal factors constructed from reliable spatial databases, that is, a method that 'is particularly interesting and should be used as often as possible'; and (iii) mechanical analysis (similar to the previous method but based on more deterministic stability models, when they are available for evaluating landslides). Therefore he envisaged two stages in the making of risk maps: (1) evaluation of hazard at a scale of a risk based on 1:25 000 and identification and analysis of stakes; and (2) modification of the hazard map into a 1:5000 risk map of a deterministic protection level. He concluded that hardly any of the mapping methods considered to date correctly integrate the notion of time, and called for 'modulating the preventive capital investments in terms of stakes', and for historical analyses in which exhaustive information allows proposing realistic hazard models.

Wu *et al.* (1996) discussed the common conditions of uncertainty that are implicit when dealing with future events that may trigger landslides, with insufficient information about site conditions or understanding of landslide mechanism, and with the inclusion of probability of success or failure of hazard reduction measures. When dealing with hazard, they discuss the evaluation of the performance reliability of geotechnical systems as a performance function (for instance of a safety factor). In the discussion of a decision-making process for landslide mitigation, they envisage the following two stages within a multicriteria decisional framework: 1 – site characterization; 2 – identification of failure modes; 3 – evaluation of hazard for each failure mode, which needs: 3.1 – the computation of prior probabilities, 3.2 – observations of the study area, and 3.3 – the estimation of posterior probabilities; 4 – evaluation of the consequences of each failure mode (vulnerability estimation); 5 – evaluation of the risk of each management option (scenarios); and 6 – selection of management option (it could require more observations from step 3.3 and further iterations). In addition they discussed how historic failure rates can be of help to hazard predictions because, ideally, landslide hazard should be expressed as the probability of failure per time per area. None of the applications examples mentioned by those authors, however, seemed to provide such a representation in a systematic manner.

Glade (2001) proposed an approach for a comprehensive natural risk assessment in which methods based on engineering and natural sciences were combined with risk evaluation and perception techniques used in the social sciences. He dealt with two applications on differently scaled landslide risk analyses. In Iceland he identified on 1:5000 maps landslide sources, travel path and deposit area and constructed four hazard classes: very high, high, medium and low, for which he estimated the corresponding rock and debris volume and specific travel and runout distance. Next, he transferred the landslide runout map into a hazard map, based on a process-based rockfall and deterministic debris-flow runout scenarios. In Germany, using a 40 m grid resolution and a 20 m DTM from 1:25 000 maps, he reported the derivation of landslide hazard from multivariate statistical analyses of a study area in which a worst-case scenario of high vulnerability was used due to missing information. Risk evaluation was obtained from aerial photos and land use plans. Historical landslide records were used together with more recent records on the local geological setting and the recurrence interval of rainfall triggering events, so that a return period of 50 years was selected. An average potential damage value was assigned to each risk element to identify each combination of landslide hazard and risk element into low, medium and high risk. The highest landslide risk zone in the study area was indicated on a map for risk communication and minimization.

Mark and Ellen (1995) described the use of logistic regression on a database of thousands of debris flows and five physical attribute factor maps assuming as trigger a storm event identical to the one that affected a Californian study area during a 1982 storm. Their study of historical landslides led further to use a simulation technique with a 10 m DEM (digital elevation model). They used the distribution and frequency of shallow landslides to model the initiation sites, the estimated volume, and the volume-change behaviour of a random subset of 200 historical landslide scars. Categories of high, moderate and low hazard that took into consideration the zones of deposition; that is, the maximum likely extension of the hazard or runout distance was obtained and compared with the known landslide extensions.

In addition to a number of different approaches to landslide hazard evaluation reviewed by Soeters and van Westen (1996), Disperati *et al.* (2002) preferred to use a statistical technique proposed by Chung and Fabbri (1993). They used a 1:10 000 spatial database in which crown, scarp and displaced material of the extent of landslides were stored along with landslide scarp lithology, slope steepness and orientation, site elevation and land use, that is, the mapped factors considered by experts to be related to the mass movement initiation. In their predictive experiments they used a randomly selected half of the available scarp areas (as training data set) and different combinations of the causal factors mentioned to obtain hazard predictions maps, and the remainder of the scarp areas to validate the results. They computed prediction-rate curves following a calculation proposed by Chung and Fabbri (1999). Although the prediction rates obtained for their study area in central Italy were marginally acceptable, they have taken an important step in using a systematic procedure for spatial prediction generation and interpretation of the prediction results.

Another interesting development towards the analysis and use of spatial uncertainties in landslide hazard mapping was proposed by Park *et al.* (2002), who characterized the fuzziness of the boundaries of categorical maps used to represent the causal factors in spatial prediction models. Prediction-rate curves were used to interpret and validate the results of the different imposed levels of spatial uncertainty in the map data characterized by different spread parameters and combinations of factors.

Gritzner *et al.* (2001) have attempted to validate the prediction results by partitioning the past landslides into two groups by a random division and have used the landslides in one group to develop their model. The resulting predictions of the model were evaluated by using the landslides from the other group, as discussed in Chung and Fabbri (1999), but not using time-related landslide groups as strongly recommended earlier by Chung *et al.* (1995). Their approach by partitioning the landslides into two groups, however, is one right step toward evaluating the hazard models.

Leone *et al.* (1996) proposed damage functions as part of an explicit framework for structuring the concept of vulnerability to generate acceptable risk representations and perceptions. For these, the collection and comparison of historical data and the return analysis of the past events are vital. The essential representation is the spatial subdivision of a study area into zones with different probabilities of landslide occurrence.

The eight promising approaches above point to various aspects that need to·be considered in predictive spatial data analysis for hazard zonation: official hazard maps integrating the notion of time, representation of uncertainty for hazard reduction measures, the importance of historical records on landslide occurrence and runout patterns, the simulation of runout distances from predictions, prediction validation by data set partitioning, and the integration of hazard classes and vulnerability estimations into risk maps. The discussion at the end of this chapter will connect the proposed approach and applications with those aspects.

## 4.3 Deficiencies of Existing Quantitative Prediction Models

We have identified five common unfortunate deficiencies in the existing prediction models that we have reviewed:

1. Simplification of input data
2. Handling of mixture of categorical and continuous data layers
3. No statement on the assumptions made in the prediction models
4. Validation of the prediction results
5. Estimation of the conditional probabilities of future landslides given geomorphological characterizations of an area within the study area.

We shall look at each deficiency separately.

### 4.3.1 Simplification of Input Data

By simplifying input data, we lose much detailed information. Very often, the continuous data such as slope angles and elevations are first categorized into four to ten classes, not because the simplification is desirable but because the proposed models or the associated computer programs cannot handle continuous measurements. Clerici *et al.* (2002) have proposed the use of the unique-condition areas for their prediction model, which required that all continuous data layers including the slope angle map be converted into categorized data layers, although the original DEM were obtained from 1:10 000 topographic base maps. Likewise, in most of the landslide hazard mapping studies encountered, very high-resolution DEM data (5 m or 10 m pixels) were used to describe the geomorphological characteristics of landscapes such as 'concavity/convexity' or the scars of the landslides. However, those high-resolution continuous original data were rarely used in the prediction analysis. Carrara and his coworkers (Carrara *et al.*, 1995; Guzzetti *et al.*, 1999) have proposed 'morphological units or slope angle units' for their prediction analysis. The sizes of the units ranged from a few square metres to several square kilometres, but only one slope angle or slope angle class in each unit was assigned, although their original DEM had often 10 m resolution pixels. Up until the early 1990s, because of the limits of computer capacities, some simplifications were a necessity and that limitation forced the scientists to simplify the data and to adapt the quantitative models to the modified data for predictive modelling, as was done by Chung *et al.* (1995). However, such a simplification is no longer required and must be considered as a 'thing of the past' based on current computer technology.

### 4.3.2 Handling of a Mixture of Categorical and Continuous Data Layers

The geomorphological data layers used as causal factors for landslide hazard zonation in all published quantitative models consist of both continuous data layers such as the slope angle map and categorized data layers such as the surficial geologic map. Instead of keeping the two types of data layers as they are, the predictive models proposed opted for the conversion of either all categorized data layers into sets of binary representations for each class, or all continuous data layers into categorized data layers. Whenever the data layers are converted into other types of data layers, the conversions lose much of the original meaning of the data, which is not a desirable step to take in developing predictive models for landslide hazard mapping

### 4.3.3   No Statement on the Assumptions Made in the Prediction Models

All quantitative prediction models from simple 'conditional analysis' by Clerici *et al.* (2002) to the 'multivariate statistical techniques' by Carrara and his coworkers (Carrara, 1989; Carrara *et al.*, 1995, Guzzetti *et al.*, 1999) and by Chung *et al.* (1995), Chung and Fabbri (1999, 2001) are based on many sets of assumptions. For example, Carrara *et al.* (1995) have expressed their landslide hazard assessment maps in terms of four groups according to three 'probabilities' based on the discriminant analysis. However, the probability was not clearly defined although it supposedly represented 'probability of correct membership classification' and somehow it was also linked to the 'probability of the occurrences of future landslides'. Throughout the paper there was no discussion on what kinds of assumptions were made to come up with their probabilities.

### 4.3.4   Validation of the Prediction Results

Strictly speaking, the validation of the prediction of future events is not possible without a 'wait-and-see' validation. However, as in any predictions, the methods of prediction have no scientific significance without measuring the validity of the prediction results. Because landslide hazard zonation maps supposedly show the locations of the future landslides, attempts to measure the validity of the prediction results must be made and we should address what kinds of assumptions are required. Chung and Fabbri (2003) discussed how to provide empirical measures of significance of the prediction results through space/time partitioning of the spatial databases. That work expanded the validation techniques earlier used by Chung *et al.* (1995) and Chung and Fabbri (1999) to interpret the predictions. Fabbri *et al.* (2003) have used similar validation techniques to resolve several misconceptions about the quality of spatial databases used for landslide hazard zonation and the significance of the predictions obtained from them. Fabbri and Chung (2001) have used validation techniques to measure the spatial support of individual map layers, of combinations of them, and to resolve the problem of how to count the number of landslides falling within each hazard class.

### 4.3.5   Estimation of Probabilities of Future Landslides

A landslide risk map is usually generated from the landslide hazard map, at a greater and more expensive level of detail, by integrating the hazard map with socio-economic spatial data including the spatial distributions of populations, infrastructures and related economic parameters. For the integration for socio-economic analysis including an expected 'cost–benefit analysis', the levels of hazard in the map should be expressed in terms of the probabilities of the occurrences of future landslides, so that acceptable (operational) probability levels can be selected for the successive vulnerability and risk analyses. Fabbri *et al.* (2002) proposed an example of such an integration and pointed out how validation techniques can be used to integrate hazard levels and vulnerability scenarios for risk representations. When we partition the past landslides into two groups, it is almost essential to try to partition the landslides into two time periods, not into two random divisions as discussed Gritzner *et al.* (2001), because the random divisions tend to overestimate the 'certainty' of the predictions. Further discussion of the extended use of validation techniques to compute the probabilities of occurrence of landslides will be made later for the two application examples.

## 4.4   Mathematical Background

This section reviews the mathematical background to spatial prediction models in general, but for brevity and clarity uses only one example of models, the likelihood ratio function. The background is essential to the formulation of prediction strategies that allow risk representations for decision making.

The term favourability function, or FF, was used by Chung and Fabbri (1993) to indicate a unified mathematical framework for the modelling of spatial predictions of: (i) mineral deposit discoveries; (ii) natural hazard occurrences; or (iii) environmental impact of human activities. What the three application areas have in common is that the predictions are based on spatial databases that represent the distribution of mapping units (either discrete categorized data or continuous data or both) and the corresponding distribution of future events such as resource discoveries, or hazardous occurrences, or negative environmental impacts.

Properties of the databases that are important for predictive modelling are:

1. the data layers sufficiently represent the typical conditions of the events;
2. the condition and rates of occurrence are similar through time; and
3. the databases can be partitioned in time and/or in space to allow validating the prediction results, thus providing a measure of confidence.

Such properties are additional to the obvious congruence of scale, density of detail, geo-referencing, co-registration, and consistency of the spatial and the associated non-spatial information.

### 4.4.1   Favourability Function

Suppose that we have $m$ data layers as the causal factors of the landslides, with the landslide layer describing the distribution of the past landslides.

Consider a unit area, $a$, with $m$ causal factors, $(a_1, \ldots, a_m)$, one value in each layer. As discussed in Chung (2003), a favourability function $g$ at a unit $a$ with $(a_1, \ldots, a_m)$ must satisfy three basic conditions:

1. $g(a_1, \ldots, a_m)$ represents (or measures) a relative level of hazard of the unit $a$.
2. Considering two unit areas, $a$ and $b$ with $(a_1, \ldots, a_m)$ and $(b_1, \ldots, b_m)$, respectively: if $a$ is a more hazardous geomorphological environment for the landslides than $b$, then $g(a_1, \ldots, a_m) > g(b_1, \ldots, b_m)$.
3. Based on the $m$ data layers with the landslide layer, we should be able to estimate the favourability function, $\hat{g}(c_1, \ldots, c_m)$ for any unit $c$ in the study area.

As discussed in Chung and Fabbri (1993, 1998, 1999, 2001, 2002) and Chung (2003), several representations, within well-established mathematical frameworks, can be used as a favourability function, such as the conditional probability function, the likelihood ratio function (also the certainty factor function or the weights of evidence function as special cases of the likelihood ratio function), the Dempster–Shafer belief functions, and the fuzzy set membership functions. In this contribution, we will use only one representation, namely the likelihood ratio, as a favourability function.

The use of the broad term favourability function was due to the need to simultaneously cover many different interpretations of models and problem applications in a unified approach. For instance, Bayesian probability, certainty factors, likelihood ratios, Dempster–Shafer belief functions, and fuzzy set membership functions have been used to obtain predictions of hazards and of mineral discovery potential.

### 4.4.2   The Likelihood Ratio Function Model

To assess the role played by an individual data layer with respect to the occurrences of landslides, let us consider an example of a slope angle map from a DEM. Suppose that a study area is divided into two non-overlapping sub-areas, the scarps (or trigger zones) of the landslides and the remaining areas. Suppose further that the slope angles provide useful information to identify the scarps, and that the slope angle data of the scarps should have unique characteristics that are different from the slope angle data in the remaining areas. This suggests that the frequency distribution function (black curve in Figure 4.2) of the scarps (solid polygons in Figure 4.1 – see also colour plate section,

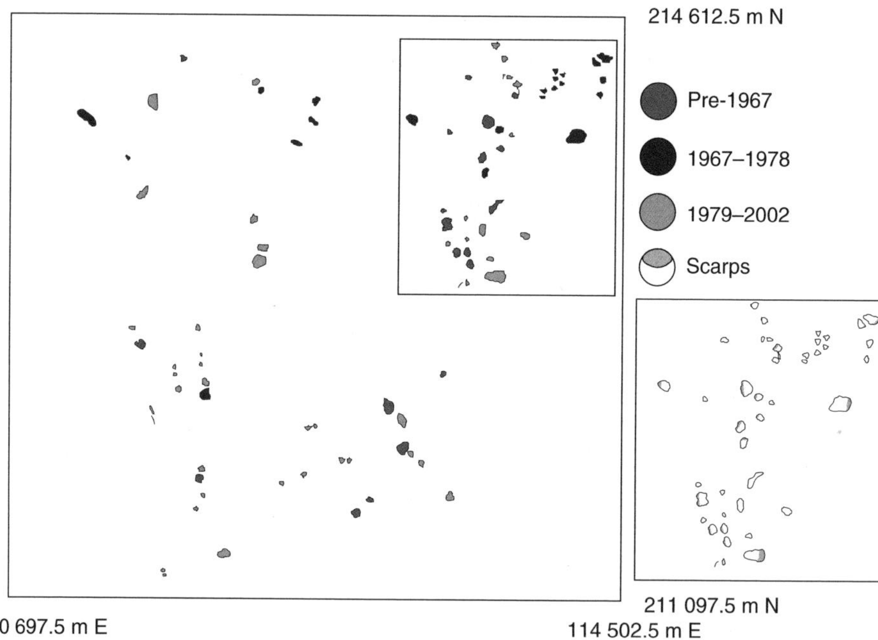


***Figure 4.1*** *Distribution of three time periods of shallow translational landslides in the Fanhões–Trancão area, north of Lisbon in Portugal. The different shades indicate the time periods for the 92 landslides. The inset to the lower right shows the separation of the scarps in solid shades. (See also Plate 2)*

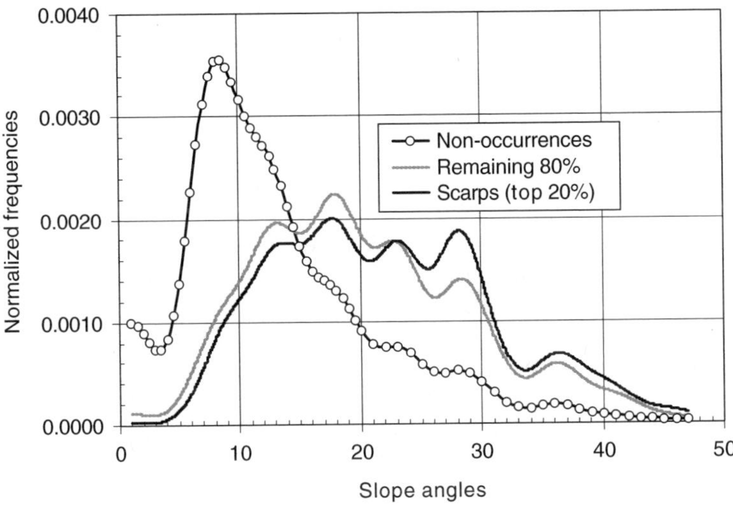


**Figure 4.2**  *Three empirical distribution functions of the slope angles of the pixels from the scarps (black curve) whole scars–scarps (grey curve), and of the remaining areas (curve with circles) from the area in Figure 4.1*

plate 2) and that (curve with circles in Figure 4.2) of the remaining areas should be distinctly different. The difference is shown in Figure 4.2. The likelihood function, which is the ratio of two frequency distribution functions, not only highlights this difference but can also be the FF satisfying the three conditions discussed.

Consider a number of map layers (images, fields or variables) contained in the database of the study area. Based on the likelihood ratio function of the multivariate frequency distribution functions calculated for these layers, a prediction model identifying the areas (trigger zones) of future landslide occurrence can be generated. To formalize the idea, let us consider a point $c$ with $m$ pixel values $(c_1, \ldots, c_m)$ in the whole study area $A$ consisting of two sub-areas, the area affected by landslides $\mathbf{M}$ and the remaining landslide-free area $\overline{\mathbf{M}}$. Let $f\{c_1, \ldots, c_m|\mathbf{M}\}$ and $f\{c_1, \ldots, c_m|\overline{\mathbf{M}}\}$ be the multivariate frequency distribution functions assuming that the pixels are from $\mathbf{M}$ and from $\overline{\mathbf{M}}$, respectively. Then the 'likelihood ratio' (Kshirsagar, 1972; Press, 1972; Cacoullos, 1973) at $c$ is defined as:

$$\lambda(c_1, \ldots, c_m) = \frac{f\{c_1, \ldots, c_m|\mathbf{M}\}}{f\{c_1, \ldots, c_m|\overline{\mathbf{M}}\}}. \tag{1}$$

We can estimate $\lambda(c_1, \ldots, c_m)$ in (1), for instance, using discriminant analysis or Bayesian methods (see Chung, 2003; Chung and Keating, 2002; and Chung *et al.*, 2002a for a discussion and application of the likelihood ratio functions in mineral potential mapping). For every pixel, we estimate $\lambda(c_1, \ldots, c_m)$. According to this model, the pixel with the largest estimate is considered as the most likely unit area containing the future landslides. The model can include the time and magnitude, if such information is available in the databases. While the dynamic types of the mass movements are commonly recorded, the time and the magnitude may not be part of the database. In such a common situation, the prediction model proposed is restricted to the spatial probabilities only.

Using the FF modelling, we generate a predicted hazard map showing a relative hazard level in a continuous scale at every point on the map. This is similar to what is done for hazard maps by geomorphologists or for safety factor maps by civil engineers.

### 4.4.3   First Two Deficiencies and the Likelihood Ratio Function Model

The likelihood ratio function not only satisfies the three conditions of the favourability function but also overcomes the first two deficiencies mentioned in Section 4.3, namely simplification of data and mixture of continuous and categorized data, of the earlier prediction models. The causal factor map layers usually consist of both categorical data layers and continuous data layers. As in equation (1), because the likelihood ratio function is based on two multivariate distribution functions, we need neither to simplify the data nor to unify two data types into one data type. Assuming that the first $k$ layers correspond to categorical data and the remaining $h$ layers represent continuous data, let $f\{x_1, \ldots, x_k, y_1, \ldots, y_h | \mathbf{M}\}$ and $f\{x_1, \ldots, x_k, y_1, \ldots, y_h | \overline{\mathbf{M}}\}$ be the multivariate frequency distribution functions from $\mathbf{M}$ and from $\overline{\mathbf{M}}$, respectively, where the first $k$ values, $x_1, \ldots, x_k$ correspond to the categorical data layers and subsequent $h$ values, $y_1, \ldots, y_h$ represent the continuous data layers. Then the likelihood ratio, the multivariate generalization of the likelihood ratio function in (1), is:

$$\lambda(x_1, \ldots, x_k, y_1, \ldots, y_h) = \frac{f\{x_1, \ldots, x_k, y_1, \ldots, y_h | \mathbf{M}\}}{f\{x_1, \ldots, x_k, y_1, \ldots, y_h | \overline{\mathbf{M}}\}}. \tag{2}$$

To handle the multivariate frequency distribution functions of two different types of data layers, we have made the following assumption of conditional independence:

$$f\{x_1, \ldots, x_k, y_1, \ldots, y_h | \mathbf{M}\} = f\{x_1, \ldots, x_k | \mathbf{M}\} f\{y_1, \ldots, y_h | \mathbf{M}\},$$

$$\text{and} \tag{3}$$

$$f\{x_1, \ldots, x_k, y_1, \ldots, y_h | \overline{\mathbf{M}}\} = f\{x_1, \ldots, x_k | \overline{\mathbf{M}}\} f\{y_1, \ldots, y_h | \overline{\mathbf{M}}\}.$$

Under the assumption of (3), the $k + h(=m)$-dimensional multivariate frequency distribution function, $f\{x_1, \ldots, x_k, y_1, \ldots, y_h | \mathbf{M}\}$ is expressed as a multiple of the $k$-dimensional multivariate discrete distribution function for the categorical data and the $h$-dimensional multivariate continuous distribution function for the continuous data. The same argument is applied to $f\{x_1, \ldots, x_k, y_1, \ldots, y_h | \overline{\mathbf{M}}\}$. Under (3), the likelihood ratio function is simply a multiple of two likelihood ratio functions:

$$\lambda(x_1, \ldots, x_k, y_1, \ldots, y_h) = \lambda(x_1, \ldots, x_k) \cdot \lambda(y_1, \ldots, y_h), \tag{4}$$

and hence we will estimate the likelihood ratio function as a multiple of two estimated likelihood ratio functions, one for categorical data layers and the other for continuous data layers.

### 4.4.4   The Assumptions Required for the Likelihood Ratio
###          Function Model

To handle the mixture of the categorical data layers and the continuous data layers, we have already made the assumption of the conditional independence in (3). Furthermore, we need several additional assumptions to estimate the two likelihood ratio functions, one for categorical data layers and the other for continuous data layers, as shown in (4).

### 4.4.4.1 Estimation of the likelihood ratio function for categorical data layers

There are two procedures to estimate the likelihood ratio function for categorical data layers, $\lambda(x_1, \ldots, x_k)$ for a point $x$ with $k$ pixel values $(x_1, \ldots, x_k)$, each $x_i$ representing a categorical class where the point $x$ belongs to the $i$th data layer, and each procedure requiring a different assumption. The first procedure requires the conditional independence as in (3). We assume that the $k$ categorical layers are conditionally independent and hence, we have:

$$\lambda(x_1, \ldots, x_k) = \lambda(x_1) \cdots \lambda(x_k). \tag{5}$$

Instead of considering all the $k$ layers at the same time, we may estimate the likelihood ratio function as a multiple of $k$ separate univariate likelihood ratio functions, where each univariate ratio function is estimated by each single categorical data layer in conjunction with the distribution of the occurrences of the landslides:

$$\check{\lambda}(x_i) = \frac{\text{No. of landslide pixels within } x_i \text{ category within the } i\text{th layer}}{\text{No. of non-landslide pixels within } x_i \text{ category within the } i\text{th layer}} \tag{6}$$

Using (5) and (6) under the conditional independence assumption, we obtain an estimate:

$$\check{\lambda}(x_1, \ldots, x_k) = \check{\lambda}(x_1) \cdots \check{\lambda}(x_k). \tag{7}$$

The next procedure requires the assumption that two distributions, $f\{x_1, \ldots, x_k | \mathbf{M}\}$ and $f\{x_1, \ldots, x_k | \overline{\mathbf{M}}\}$ for $\lambda(x_1, \ldots, x_k)$ are the multivariate multinomial frequency distribution functions (Johnson and Kotz, 1969) from $\mathbf{M}$ and from $\overline{\mathbf{M}}$, respectively. Then, as discussed by Chung (2003), the maximum likelihood estimates of these two multinomial distribution functions are two corresponding $m_1 \times m_2 \times \cdots \times m_k$ cross-classified contingency tables (Johnson and Kotz, 1969), one for $\mathbf{M}$ and the other for $\overline{\mathbf{M}}$. By taking the ratios of two corresponding cells in two contingency tables, we obtain the second estimate:

$$\tilde{\lambda}(x_1, \ldots, x_k) = \frac{\text{No. of lanslide pixels within } U_{x_1 \cdots x_k}}{\text{No. of non-lanslide pixels within } U_{x_1 \cdots x_k}}, \tag{8}$$

where $U_{x_1 \cdots x_k}$ is defined as the unique condition sub-area (Chung *et al.*, 1995; Clerici *et al.*, 2002) with the classes, $x_1, \ldots, x_k$, and belongs to the $x_1$ categorical class in the first layer, the $x_2$ class in the second layer and the $x_k$ class in the $k$th layer. The estimate in (8) is identical to the estimators used in the 'Conditional analysis' by Clerici *et al.* (2002), when all layers were converted into categorical data.

We strongly recommend the use of equation (7) rather than (8) when the number of categorized data layers is more than two. When the number is greater than two and the number of the classes in each layer is more than five, then the sizes of many of the unique condition sub-areas become very small and consequently many of the $\tilde{\lambda}(x_1, \ldots, x_k)$ are equal to zero. When we integrate the zero value with the likelihood ratio functions from the continuous data layers, they generate an undesirable negative impact on the prediction maps.

### 4.4.4.2   Estimation of the likelihood ratio function for continuous data layers

We will present here only two procedures to estimate the likelihood ratio function for continuous data layers, $\lambda(y_1, \ldots, y_h)$ for a point $y$ with $k$ pixel values $(y_1, \ldots, y_h)$, where each $y_j$ represents a real value at point $y$ within the $j$th continuous data layer, and each procedure requires a different assumption. The first procedure requires conditional independence, as in (3). We assume that the $h$ continuous layers are conditionally independent and hence we have:

$$\lambda(y_1, \ldots, y_h) = \lambda(y_1) \cdots \lambda(y_h). \tag{9}$$

Instead of considering $k$-dimensional multivariate distribution functions, we may estimate the likelihood ratio function as a multiple of $k$ separate univariate likelihood ratio functions, where each univariate ratio function is estimated by:

$$\ddot{\lambda}(y_j) = \frac{\ddot{f}\{y_j|\mathbf{M}\}}{\ddot{f}\{y_j|\overline{\mathbf{M}}\}}, \tag{10}$$

where $\ddot{f}\{y_j|\mathbf{M}\}$ and $\ddot{f}\{y_j|\overline{\mathbf{M}}\}$ are two empirical distribution functions estimated from the $j$th continuous data layer in conjunction with the distribution of the occurrences of the landslides. Using (9) and (10) under the conditional independence assumption, we obtain an estimate:

$$\ddot{\lambda}(x_1, \ldots, x_k) = \ddot{\lambda}(x_1) \cdots \ddot{\lambda}(x_k). \tag{11}$$

The second estimate of $\lambda(y_1, \ldots, y_h)$ is obtained by assuming that the two corresponding distribution functions, $f\{y_1, \ldots, y_h|\mathbf{M}\}$ and $f\{y_1, \ldots, y_h|\overline{\mathbf{M}}\}$ are the multivariate discrete distribution functions, $N(\mu_\mathbf{M}, \Sigma_\mathbf{M})$ and $N(\mu_{\overline{\mathbf{M}}}, \Sigma_{\overline{\mathbf{M}}})$, where $\mu_\mathbf{M}, \Sigma_\mathbf{M}, \mu_{\overline{\mathbf{M}}}$ and $\Sigma_{\overline{\mathbf{M}}}$ are the $k$-dimensional mean vectors for $\mathbf{M}$, the $k \times k$ dimensional covariance matrix for $\mathbf{M}$, the $k$-dimensional mean vectors for $\overline{\mathbf{M}}$, and the $k \times k$ dimensional covariance matrix for $\overline{\mathbf{M}}$, respectively. Under the normality assumption, we obtain the second estimate:

$$\dddot{\lambda}(y_1, \ldots, y_h) = \frac{\sqrt{|\hat{\Sigma}_{\overline{\mathbf{M}}}|}\, \exp\left\{-0.5(y - \hat{\mu}_\mathbf{M})' \hat{\Sigma}_\mathbf{M}(y - \hat{\mu}_\mathbf{M})\right\}}{\sqrt{|\hat{\Sigma}_\mathbf{M}|}\, \exp\left\{-0.5(y - \hat{\mu}_{\overline{\mathbf{M}}})' \hat{\Sigma}_{\overline{\mathbf{M}}}(y - \hat{\mu}_{\overline{\mathbf{M}}})\right\}}, \tag{12}$$

where $y = (y_1, \ldots, y_h)$ is an $h$-dimensional vector containing $h$ observed values, one for each layer, of point $y$, $\hat{\mu}_\mathbf{M}, \hat{\Sigma}_\mathbf{M}, \hat{\mu}_{\overline{\mathbf{M}}}$ and $\hat{\Sigma}_{\overline{\mathbf{M}}}$ are the sample mean vectors and the sample covariance matrices for $\mu_\mathbf{M}, \Sigma_\mathbf{M}, \mu_{\overline{\mathbf{M}}}$ and $\Sigma_{\overline{\mathbf{M}}}$ and the computation operations in (12) are vector and matrix calculations. The estimated likelihood ratio function is also termed linear discriminant function with two different covariance functions (see Press, 1972). If the normality assumptions are reasonably realistic conditions, then we recommend the estimate shown in (12). However, if the empirical distribution function appears to be far from normality, then we recommend the estimate in (11) under the conditional independence assumption. As discussed in Chung (1978), and Dai and Lee (2002), we can assume that $f\{y_1, \ldots, y_h | \mathbf{M}\}$ is the logistic distribution function instead of the normal distribution function. Then it can be shown that the log-likelihood ratio function is reduced to a simple linear function instead of the one in equation (12). The parameters in the logistic model can be estimated by the maximum likelihood method as discussed in Cox (1970) and Chung (1978), and we obtain the estimate:

$$\lambda'''(y_1, \ldots, y_h) = \exp\{y_1\hat{\beta}_1 + \cdots + y_h\hat{\beta}_h\} \cdot \hat{C}, \qquad (13)$$

where $(\hat{\beta}_1, \ldots, \hat{\beta}_h)$ are the maximum likelihood estimators and $\hat{C}$ is a constant.

The final estimate for $\lambda(x_1, \ldots, x_k, y_1, \ldots, y_h)$ is obtained by:

$$\hat{\lambda}(x_1, \ldots, x_k, y_1, \ldots, y_h) = \hat{\lambda}(x_1, \ldots, x_k) \cdot \hat{\lambda}(y_1, \ldots, y_h), \qquad (14)$$

where $\hat{\lambda}(x_1, \ldots, x_k)$ is obtained from either (7) or (8) and $\hat{\lambda}(y_1, \ldots, y_h)$ is obtained from (11), (12) or (13).

The mathematical background so far exposed for the favourability function can now be used to develop prediction strategies. The next section introduces a two-stage analytical strategy.


## 4.5    The Approach Proposed: The Two-stage Strategy

The two-stage procedure, I and II, is being proposed as a response to compensate for and eradicate the latter two deficiencies, (4) and (5), identified earlier. Landslide hazard zonation maps produced by FF models represent a relative hazard level for future landslides in a continuous scale at every point in a map, and commonly the hazard levels are sliced into a fixed or convenient number of hazard classes for visualization, as a prediction pattern. Table 4.2 illustrates a general strategy for landslide hazard models from the data layer preparation in Step I.1 of Stage I, to the estimation of probability of the occurrence of future landslide (probability table) from the prediction-rate curve and the integration in Steps II.3 and II.4 of Stage II.

A critical issue in predictive modelling is the interpretation of the hazard levels in a predicted map (Step I.2). For this, prediction-rate curves have been generated. One way in which the curves are obtained is the following. It requires partitioning the distribution of the past landslides into two groups, *Estimation* and *Validation* groups (Step II.1). Whenever possible, time partition, where the past landslides are divided into two time periods, is strongly recommended. We may partition the past landslides into two randomly selected subgroups. Other partitioning criteria can also be used to generate two groups of past landslides. We can generate a new predicted hazard map using the landslides in the *Estimation* group. The next step is to overlay the landslides in the *Validation* group over the new hazard map and then count the number of landslides at each hazard level. An example of a prediction-rate curve is shown in Figure 4.4. The horizontal axis of the curve in Figure 4.4 is for the first ratio described in Step II.2 and indicates the proportion of the study area classified as hazardous. The vertical axis is for the ratio described in Step II.2, and indicates the proportion of the *Validation* group landslides within the selected hazardous pixels.

**Table 4.2**  *General two-stage strategy based on favourability function (FF) modelling for landslide hazard mapping*

| Stage.Step | Task | Description |
| --- | --- | --- |
| I.1 | Preparation | Prepare the past landslide layer containing the locations (or the scars) of the occurrences of the past landslides in study area $A$, and delineate the scarps (trigger areas) of the scars, if required. Also prepare $m$ data layers of causal (or correlated) factors of the occurrences. Each causal map layer consists of a number of mapping units and continuous field values that provide evidence for finding the future landslides. For each data layer, compare two empirical distribution functions from the scars and the scarps. Decide whether the scars or the scarps are to be used for the analysis.<br>Co-register all the layers (landslide and data) using the geo-reference and select an FF model for analysis. |
| I.2 | Estimation and Prediction | Construct FF models (Chung and Fabbri, 1993, 1998, 1999, 2001). Compute an FF value at each pixel: it shows the relative hazard level for that pixel being a part of the scarp (or scar) of a future landslide. These computed FF values are in a continuous scale and constitute the predicted hazard map. The values can be divided into a number of hazard classes for visualization. |
| II.1 | Partition | To evaluate the reliability of the prediction hazard map and for the interpretation of the prediction, we divide the scarps (or scars) of the past landslides into two groups, termed *Estimation group* and *Validation group*. Either time partition or random partition is generally obtained. Whenever possible, time partition, where past landslides are divided into two time periods, is strongly recommended. |
| II.2 | Cross-validation | Perform the Estimation and Prediction described in Step I.2. Generate a second predicted hazard map using only the *Estimation group* landslides described in the Partition in Step II.1. This second prediction map based on the *Estimation group* landslides is compared with the distribution of the *Validation group* landslides.<br>For any given hazard level in the second predicted hazard map, select all the pixels whose hazard levels are greater than the given level. Within the selected pixels, count the scarps (or scars) in the *Validation group* landslides. At each hazard level, compute two ratios: the first is for the number of selected pixels and the total number of pixels in $A$, and the second is for the counted scarps within the selected pixels and the total number of scarps in the *Validation group* landslides. The sets of two ratios constitute the prediction-rate curve of the first predicted hazard map based on all landslides.<br>As the hazard level decreases, the number of selected pixels will increase, and both the ratios will increase to 1. Examples of prediction-rate curves are shown in Figures 4.4 and 4.6. |

| | | |
|---|---|---|
| II.3 | Probability of occurrence | For a decisional scenario containing appropriate assumptions, obtain the estimated probability of the occurrence of a future landslide from the prediction-rate curve for any given hazard class. Using equation (15), estimate the corresponding probability that any given sub-area will be damaged. An example of such probabilities is in Table 4.3 |
| II.4 | Integration | To construct the landslide risk map, the prediction hazard map generated in Step I.2 is now being integrated with the probability table discussed in Step II.3. |

Although the prediction-rate curve is obtained by comparing the new hazard map with the distribution of the *Validation* group landslides, it will be used to interpret the original predicted hazard map using all past landslides in Step I.2. The partitioning criteria of the past landslides determine the interpretation of the prediction-rate curve and consequently the interpretation of the predicted hazard map.

Using the prediction-rate curves, we can also measure the degree of support for combinations of causal factors, the effectiveness of different FF models, the time or space robustness of the predictions and so on (Chung and Fabbri, 1999; Kojima *et al.*, 1998; Fabbri and Chung, 2001), depending on the partition criteria. In practice, we can use the prediction-rate curve for the cost–benefit analysis of the corresponding predicted hazard maps. For instance, a decisional issue could be to identify areas of predicted highest hazard value that cover no more than 10% of the study area. Such areas could be a first priority for direct inspection on the field or for investing resources in prevention works and in the risk analysis to follow. In addition, as we shall see, we can estimate similarly the probabilities of the occurrences of future landslides within any predicted hazard level in the predicted map that is shown in Step II.4 of Table 4.2.

To further proceed in the assessment of risk, we will have to provide a spatial inventory of all the vulnerable elements in the study area to all types of landslides hazards, and finally we will aggregate all hazards and all vulnerable elements. According to Varnes *et al.* (1984), vulnerability is the degree of loss for a given element or a set of vulnerable elements resulting from the occurrence of a natural phenomenon of a given magnitude. The specific risk is the expected degree of loss due to a specific natural phenomenon, and then the total risk is a comprehensive aggregation of expected number of lives lost, persons injured, damage to properties and disruption of economic activities due to natural phenomena.

It seems evident that the critical part of risk analysis is the identification of hazard classes in a study area. They have been represented here as the values of probability of occurrence associated with the different classes of relative levels of predicted hazard. The computation of the probabilities can be obtained by setting up assumptions for the individual vulnerable elements and by analysing the prediction-rate curve obtained from the validation of the prediction results (with the time partitioning of past landslide distribution).

In our approach to spatial predictions, the reason for separating the two stages is that, while predictions are a rather common product of geomorphological or engineering practice (e.g. hazard zonation maps and engineering slope safety factor maps are also forms of predictions, without a validation strategy), it is impossible to establish their significance or prove their robustness. The validation process allows an empirical measure of the spatial support so that the relationships between the prediction-rate curve and the prediction pattern can be transformed into probabilities of occurrences. The latter are essential to link hazard and risk via appropriate scenarios. We believe that this two-stage procedure provides the basis for guiding the decision makers in taking more informed and justifiable decisions.

## 4.6   Application Examples of Hazard Zonations

To illustrate the proposed two-stage systematic procedure, we present two case studies in different geomorphological settings, one from a study area in Portugal and the other from Canada.

### 4.6.1   The Portuguese Case Study

A spatial database for the Fanhões–Trancão area, north of Lisbon in Portugal, was documented by Zêzere (1996a, b, 1997) of the University of Lisbon as part of two European Union Projects, NEWTECH (Corominas *et al.*, 1998) and ALARM (http://www.spinlab.vu.nl/alarm). The study area is 13.37 km$^2$ and is part of the dip-downstream slope of the Lousa–Bucelas *cuesta*, a substructural slope defined by a general concordance between topographic surface and south and southwest dip of the strata with angular values of 12°. Geologically the region is part of the Portuguese Meso-Cenozoic sedimentary basin and is located close to the contact between that morphostructural unit and the Tagus River alluvional plain. The maximum elevation does not exceed 350 m a.s.l. The yearly average precipitation is only 700 mm; however, the area is characterized by a great irregularity of rainfall regime considered as a failure-triggering factor (Zêzere, 1996a). Detailed geological–geomorphological mapping at 1:2000 identified 92 shallow translational landslides but 1:10 000 maps were compiled and digitized into a 5 m × 5 m resolution spatial database consisting of digital images of 761 × 702 pixels. The causal factors are: elevation, slope, aspect maps forming the digital elevation model (DEM), geology map and land use map. The past landslides consisted of the 92 shallow trans-lational slides. The analysis described in the next section will use the following causal factors: the DEM set, geology (six map units), surficial deposits (seven units) and land use (five units). The 92 shallow translational landslides were divided into three time periods, pre-1967 (21 landslides with 485 pixels in total), 1968–1979 (19 landslides with 380 pixels) and 1980–2002 (52 landslides with 761 pixels). A detailed description of the database and of its statistical analysis is in Corominas *et al.* (1998). A morphologic synthesis of the study region has been provided by Zêzere (1996a, b, 1997). Figure 4.1 (see also Colour Plate section, Plate 2) shows the distribution of the three time periods of landslides used to illustrate the systematic procedures proposed. As shown in the inset in the illustration, each scar was divided into two sub-areas, the scarp as the trigger area and the remaining sub-area (scar–scarp).

The first preliminary step (Step I.1 of Stage I) in the predictive modelling is to determine whether the whole scars or the scarps are to be used for the analysis. Using the empirical distribution functions, each of the input data layers is analysed with respect to the scarps and the whole scars. For example, let us look at the two distribution functions of the pixels from the scarps and the remaining sub-areas (whole scars–scarps). Figure 4.2 shows two empirical distribution functions of the slope angles of the pixels from the scarps (black curve); and the remaining sub-areas (whole scars–scarps) (grey curve). There is no significant difference between two curves in this case study. This implies that there is no reason to separate the scarp from the scar in each landslide for the predictive modelling based on the slope angle data layer. Similarly, all the other data layers were studied through the empirical distribution functions.

When the signatures from these two populations, the scarps and the scars, are not significantly different, the scars should be used for the modelling. Based on our experiences, for certain types of landslides such as the 'debris-flow' type, where there can be a relatively extensive horizontal movement, it is desirable to use the scarps instead of scars. However, for the 'slip' type or the 'translational' type landslides, it may be sufficient to use the scars themselves for the modelling.

The next step (Step I.2) is to determine which type of FF model and what type of estimation formula are to be used for the analysis. In this example, we will use the likelihood ratio function in equation (1). Based on equation (14) under the conditional independence assumption, we will estimate the two ratio functions, one for the three continuous data layers (slope, elevation and aspect) and the other for the three categorical data layers (geology, surficial material and land use) separately (see Chung, 2003). For the estimation of the likelihood ratio function for the categorical data layers, we have used equation (7). For the continuous data layers, we have used equation (12). Using all the 92 scars of the shallow translational landslides, we have obtained the prediction map shown in Figure 4.3 (see also Colour Plate section, Plate 3) which, it is hoped, indicates the possible locations of future shallow translational landslides in the study area. For the prediction-rate curves in Figure 4.4, the estimated ratio function $\hat{\lambda}(x_1, \ldots, x_k, y_1, \ldots, y_h)$ was obtained for each of the 534 222 pixels in the area. According to the rank order of $\hat{\lambda}(x_1, \ldots, x_k, y_1, \ldots, y_h)$, a new prediction index was assigned to each pixel. The pixel with the largest $\hat{\lambda}(x_1, \ldots, x_k, y_1, \ldots, y_h)$ was assigned a new index value 1/534 222, and the pixel with the smallest $\hat{\lambda}(x_1, \ldots, x_k, y_1, \ldots, y_h)$ was given the value 1. For instance, a pixel with a new index value 0.01 indicates that the estimated ratio function of the pixel has approximately the 528 880th largest $\hat{\lambda}(x_1, \ldots, x_k, y_1, \ldots, y_h)$ among all the 534 222 estimates. The class of the pixels with the indices smaller than 0.01 consists of 5342 pixels and covers approximately 133 550 m$^2$ or 1% of the study area. Based on these new indices, a simplified set of 40 classes with an equal number of pixels is shown in Figure 4.3 (see also Colour Plate section, Plate 3). The most hazardous class (13 350 pixels, which covers approximately 333 750 m$^2$ or 2.5% of the study area) is displayed as purple and the subsequent most hazardous 13 350 pixels are pink in Figure 4.3. The pseudo-colour legend consists of 40 coloured bars, each bar representing 13 350 pixels.

In Stage II, we try to evaluate the reliability of the prediction results shown in Figure 4.3. For Step II.1, we have first partitioned the 92 landslides into two time periods, the pre-1967 group (*Estimation group*: 21 landslides) and the 1968–2002 group (*Validation group*: 71 landslides). In Step II.2, as in Step I.2, we have again used equation (14) with (7) and (12) but using only 21 landslides in the *Estimation group* to generate a prediction map. As in Step I.2, the prediction map consisted of the new indices ranging from 1/534 222 to 1, based on the 21 landslides. The 71 landslides in the *Validation*

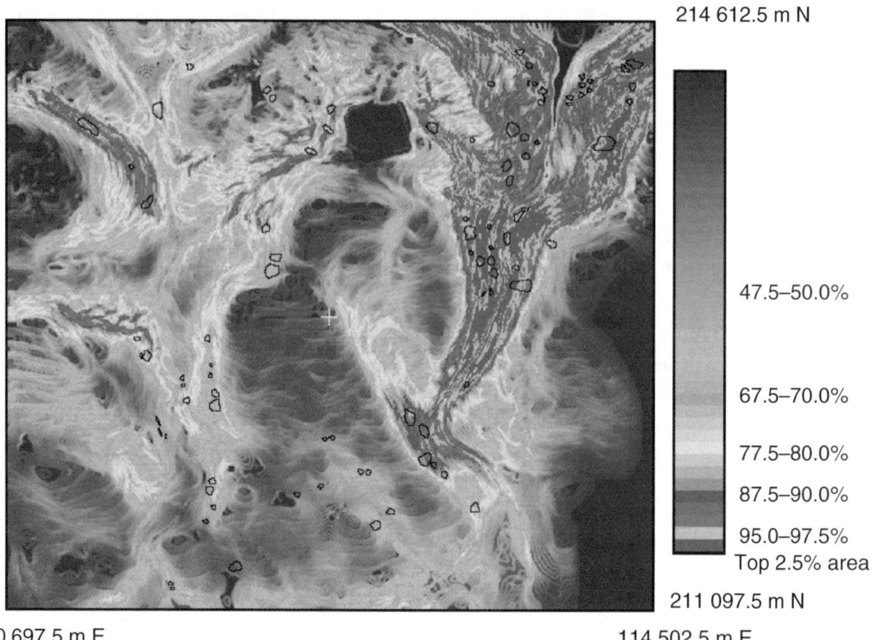



**Figure 4.3**  *Landslide hazard prediction map of the Fanhões–Trancão area, Portugal, based on the 92 landslides (shown in Figure 4.1) and six layers (bedrock geology, land use, surficial material, elevation, aspect angle and slope angle maps) of geomorphological map information using the linear discriminant analysis model. The 92 black polygons represent the 92 landslides. The pseudo-colours and the associated numbers in the legend refer to the predicted hazard class area percentages. (See also Plate 3)*

*group* were overlaid on the map. At each index value starting from 0, the number of the overlaid 71 landslides was counted within the class of the pixels with the corresponding indices larger than the index. When a landslide was only partially overlapped by the class, the landslide was counted only if at least 50% of the scar was covered by the class. Cumulative counts with respect to the indices are shown in Figure 4.4 as a grey 'prediction-rate curve'.

Theoretically speaking, the prediction-rate curve should have the following three properties: (i) the slopes of the curve are monotonically decreasing with respect to the indices; (ii) the steeper the slope angle in the prediction-rate curve, the higher the prediction power of the map, and it also indicates its reliability; and (iii) although the curve is constructed by comparing a prediction map based on the *Estimation group* landslides to the distribution of the *Validation group* landslides, the prediction-rate curve is used for the interpretation of the prediction map constructed based on the landslides from both the *Estimation and Validation groups*. The curve for a 'randomly' constructed prediction map is a straight line with slope = 1. The black broken line in Figure 4.4 is a prediction-rate curve for a 'randomly' constructed prediction map. That is, if we randomly take 1%

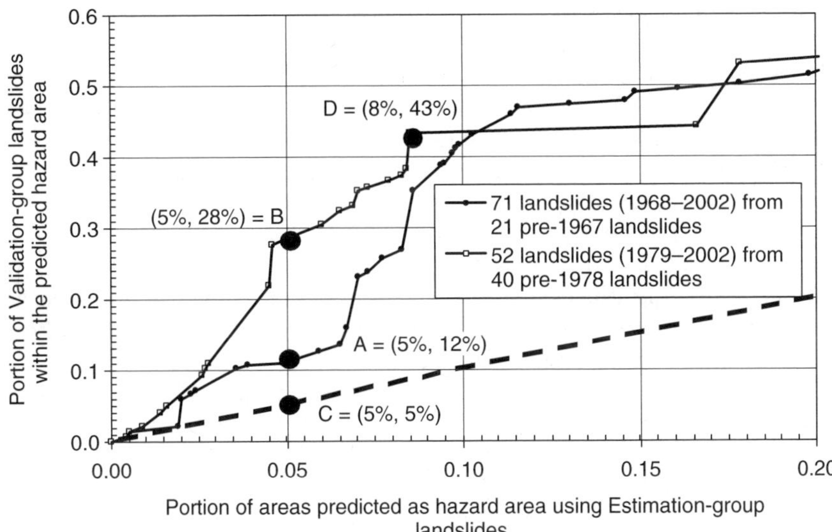


**Figure 4.4** *Prediction-rate curves for the Fanhões–Trancão area, Portugal. They have been obtained by comparing the hazard classes generated from 21 pre-1967 landslides with the distribution of 71 1968–2002 validation landslides (light grey curve with solid circles), and from 40 pre-1978 landslides with the distribution of 52 1979–2002 validation landslides (dark grey curve with open circles). The broken line has slope = 1 and represents the curve for a 'randomly' constructed prediction map. See text for explanation*

of the study area as hazardous, then we expect that 1% of the future landslides will be located within that 1% hazardous area selected.

The empirical prediction-rate curves do not always have the first monotonic property. A severe violation of the monotonic property indicates that such prediction-rate curves should not be used for interpretation of the prediction map and that the reliability of the prediction map is questionable. From our experience, the grey prediction-rate curve based on the pre-1967 and the 1968–2002 groups indicates a problematic violation of that property. For that reason it would be difficult to use it for the interpretation of the prediction map, shown in Figure 4.3 (Plate 3) that was constructed using the 92 landslides in Step I.2. If the prediction-rate curve were certain, then it would have been functional to interpret the prediction in Figure 4.3 (Plate 3) for the next future 30–40 years.

Consider the point A = (0.05, 0.12) on the grey curve in Figure 4.4. It implies that when we take the most hazardous 26 700 pixels (the largest indices that cover 5% of the study area) in the prediction map based on the 21 *Estimation group* landslides, they contain 8 of the 71 (12%) *Validation group* landslides.

Because the grey prediction-rate curve was not satisfactory, we have also partitioned the 92 landslides into other two time periods, the pre-1978 group (*Estimation group*: 40 landslides) and the 1979–2002 group (*Validation group*: 52 landslides). We repeated Step II.2 using that new partition. The 52 landslides in the *Validation group* were overlaid on the prediction map based on the 40 *Estimation group* landslides. As before, cumulative counts with respect to the indices are shown in Figure 4.4 as a solid black 'prediction-rate curve'.

This second prediction-rate curve is still problematic, but it may be acceptable until the point D = (0.08, 0.35) in Figure 4.4. After point D, the slopes of the curve are less than the black broken line (with slope = 1) and hence it has no significance. This prediction-rate curve is not usable for the interpretation of all the classes beyond the most hazardous class with the index value of 8% of the prediction map in Figure 4.3 (Plate 3). Similar to point A, point B suggests that when we take the most hazardous 26 700 pixels (the largest indices that cover 5% of the study area) of the prediction map based on the 42 *Estimation group* landslides, they contain 14 of the 50 *Validation group* landslides. This is significantly better than the results from the first partition. It could be effectively used to interpret the prediction in Figure 4.3 (Plate 3) for the next 20–23 years.

In our earlier study (Fabbri *et al.*, 2002), we divided these 92 landslides into two random groups because the dates of the occurrences of the landslides were unknown to us. The resulting prediction-rate curve (shown in black in Figure 4.3 of Fabbri *et al.*, 2002) was drastically better than the prediction-rate curves shown in Figure 4.4 of this chapter. Unlike the prediction-rate curves in Figure 4.4 that were not satisfactory, the prediction-rate curve based on two random divisions had a satisfactory monotonic property. Using the prediction-rate curve of the random division, we have said that the most hazardous (the largest indices) 26 700 pixels in Figure 4.3 of Fabbri *et al.* (2002) expected to contain 43% of the future landslides, a much higher proportion than the 28% obtained from the black prediction-rate curve in Figure 4.4, that was based on two time partitions of the pre-1978 and the 1979–2002 landslides. Obviously, the earlier conclusion was not correct and if we had used the earlier prediction-rate curve based on random division for the interpretation of the prediction map in Figure 4.3, such an interpretation would have been wrong. Chung *et al.* (2002b) used validation procedures to separate influent from non-influent factor layers. In this study, all layers appear to provide strong support for the prediction results, so that their separation was not necessary.

### 4.6.2   Estimation of Conditional Probability for Stage II

In Step II.3 of Stage II, we estimate the probability of landslide occurrence for each class of predicted hazard using a scenario based on a set of assumptions, such as those in the following scenario.

Suppose that we build a house of size $10\,m \times 25\,m\,(250\,m^2)$ within the most hazardous class (that covers approximately $667\,750\,m^2$ or 26 700 pixels or 5% of the study area) in the prediction map in Figure 4.3 constructed in Step I.2. The next logical task is to estimate the conditional probability that a future landslide will affect the house within the next 20 years. The reason that we are discussing the next 20 years rather than any other period is that we are going to estimate the probability empirically using the prediction-rate curve (black curve) in Figure 4.4 based on the partition of pre-1978 and 1979–2002 (23 years, i.e., 20~25).

To estimate the probability, we need some more assumptions on the future landslides that are to occur within the next 20 years. We need to have the 'expected' number of future landslides in the area within that time interval, and the 'expected' average size of the landslides. Owing to the fact that in the study area we had 52 landslides covering 761 pixels for the past 23 years, we can additionally assume that:

(i)  50 landslides will occur in the study area in the next 20 years;
(ii)  50 'future' landslides will affect 800 pixels.

If we were to build a house of size $10\,\text{m} \times 25\,\text{m}$ ($250\,\text{m}^2$ or 10 pixels) in the most hazardous 5% area (26 700 pixels), then the probability that the house will be part of the affected area can be estimated as follows:

$$\text{Estimate of probability} = 1 - \left[1 - p_\gamma\right]^{\frac{n_\alpha}{n_\gamma}\kappa}, \tag{15}$$

where $n_\alpha$ is the number of pixels in the affected area expected as future landslides, $n_\gamma$ is the number of pixels in the hazard class, $p_\gamma$ is the corresponding probability of the hazard class in the prediction-rate curve and $\kappa$ is the number of pixels occupied by the house. Using equation (15), we have that the estimate is $0.094 = 1 - [1 - 0.28]^{8000/26\,700}$. Similarly, using (15) and the prediction-rate curve in Figure 4.4, the probabilities given other scenarios can be estimated.

As discussed, because the prediction-rate curve may not be applicable to interpret the prediction map beyond the classes with index higher than 8%, we can only discuss the probabilities of possible damage to a house built within the most hazardous 8% area (i.e. 42 720 pixels). This means that the database provides sufficient support only for the 8% area with the highest predicted values. This is a very critical piece of knowledge about the data significance. Not knowing that, we would provide entirely meaningless or non-interpretable prediction patterns, like some of those in qualitative hazard maps commonly produced.

### 4.6.3   The Canadian Case Study

La Baie, a study area in Québec, Canada, covers $10\,\text{km} \times 6\,\text{km}$. The digital images in its database consist of $2000 \times 1200$ pixels and each pixel covers $5\,\text{m} \times 5\,\text{m}$ on the ground. Five layers of geomorphological information could be related to the landslides in the area: (1) bedrock geology (12 lithologic categorical units); (2) forest coverage (binary); (3) elevations; (4) aspect angles; and (5) slope angles (the last three are continuous values). The locations were available of the 22 landslides that occurred in 1964, and the 51 landslides that occurred in 1976 and 1996, but no scars of the landslides were available in the database. Seven of the latter 51 landslides occurred at the same locations as the 22 that occurred in 1964. The average size of the 73 landslides is approximately $15\,\text{m} \times 15\,\text{m}$, thus covering 9 pixels. Among the 2 400 000 pixels, 445 164 pixels corresponded to a lake and rivers. The pixels corresponding to lake and rivers were excluded in the study, and the remaining 1 954 836 pixels were analysed.

As can be seen from this description of the data, the input to the modelling again consisted of two different types of spatial data: (i) categorical data layers, bedrock geology and forest coverage; and (ii) continuous data layers, elevations, aspects and slopes. In Stage I, with Steps I.1 and I.2 of Table 4.2, we constructed a hazard map using the 66 locations of all the 73 landslides that have occurred during the past 38 years (1964–2002) and the five data layers. We computed an estimate of $\lambda$ assuming conditional independence of the five layers, $\hat{\lambda}(x_1, \ldots, x_k, y_1, \ldots, y_h)$ in equation (14) with (7) and (11) for each of the 1 954 836 pixels in the study area. According to the rank order of $\hat{\lambda}(x_1, \ldots, x_k, y_1, \ldots, y_h)$ for all those pixels, we replaced the $\hat{\lambda}(x_1, \ldots, x_k, y_1, \ldots, y_h)$ by the new indices discussed in the Portuguese case study.

### 4.6.4   Estimation of Conditional Probability for Stage II

In Stage II, Steps II.1 to II.3 of Table 4.2, we estimated the probability of landslide occurrence for each class of predicted hazard using a scenario based on a set of assumptions similar to the Portuguese case study.

For Step II.1, we first partitioned the 73 landslides into two time periods, a pre-1964 group (*Estimation group*: 22 landslides) and a 1965–2002 group (*Validation group*: 51 landslides). In Step II.2, as in Step I.2, we have again used equation (14) with (7) and (11), but used only the 22 landslides in the *Estimation group* to generate a prediction map. As in Step I.2, the prediction map consisted of the new indices ranging from 1/1 954 836 to 1, based on the 22 landslides. Each class contained 1955 pixels (covering approximately 0.05 km$^2$). These 1000 classes are not shown individually, but are displayed in Figure 4.5 (see also Colour Plate section, Plate 4) as 40 groups. The pseudo-colour legend consists of 40 colour bars, each bar representing 25 classes of the 1000 original ones. The most hazardous 25 classes (48 871 pixels that cover 1.22 km$^2$ or 2.5% of the study area) are displayed in purple and the subsequent most hazardous 48 871 pixels are pink. The 51 landslides in the *Validation group* were overlaid on the map as shown in Figure 4.5 (Plate 4). The number of the contained 51 landslides was

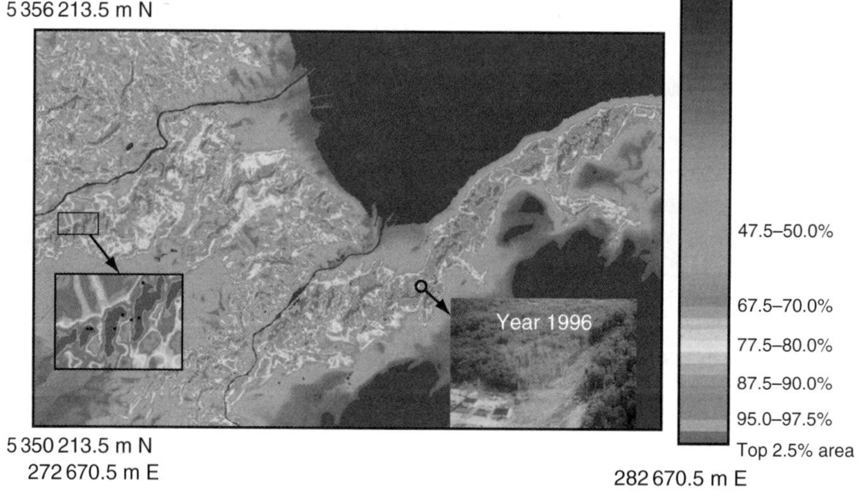


*Figure 4.5*  Landslide hazard prediction map for the La Baie study area, Québec, Canada, based on 22 landslides that occurred in 1967 and five layers (bedrock geology, forest coverage, elevation, aspect angle and slope angle maps) of geomorphological map information using the likelihood ratio function model. The 51 black dots represent the 51 landslides that occurred in 1976 and 1996. The left-side inset is an enlargement of a small area in the black rectangle in the middle left-side. The right-side inset with 'Year 1996' is a photograph of a landslide that occurred in 1996 at the site of the black circle in the middle of the illustration. The pseudo-colours and the associated numbers in the legend refer to the predicted hazard class area percentages. The classes are regrouped in different percentages from those in Table 4.3. (See also Plate 4)

counted within each of the 1000 classes. Cumulative counts with respect to the new indices are shown in Figure 4.6 as a 'prediction-rate curve', a black curve with circles.

Unlike in the previous Portuguese case study, the prediction-rate curve in Figure 4.6 satisfies the monotonic property reasonably well up to about the 20% index. Hence the prediction-rate curve would be used for the interpretation of the prediction map and the assessment of the reliability of the prediction map up to the classes with indices of less than 20%. Because of the time partition of 37 years (2002–1965), it can be used to interpret the prediction for the next 30–37 years.

Consider a house of size $10\,\text{m} \times 25\,\text{m}\,(250\,\text{m}^2)$ within the most hazardous class (that covers approximately $0.5\,\text{km}^2$) of the 1000 classes calculated. To estimate the conditional probability that a future landslide will affect the house within the next 30 years, we will follow the identical steps as in the Portuguese case study.

The first column in Table 4.3 represents the portion of the whole study area classified as 'hazardous' for future landslides. The first row, 'Top 1%', in the column is for the group of the most hazardous 10 classes of the original 1000 classes. The subsequent '1–2%' group is for the next 10 classes. To generate the second column in Table 4.3, in each of the 1000 classes, we first made a cumulative count of the 51 landslides. For the classes without the landslides, instead of the cumulative counts, we used interpolated values. Among the 1000 pairs, we selected the 20 pairs shown in the second column of

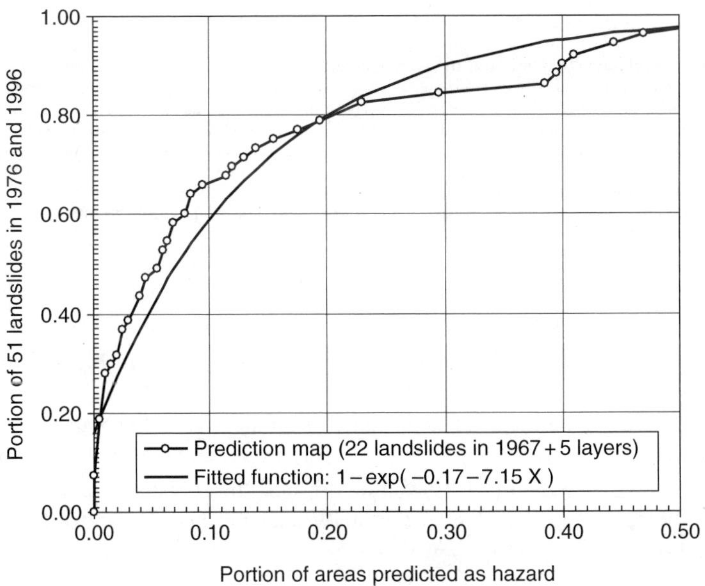


**Figure 4.6** *Prediction-rate curves for the map in Figure 4.5. They were obtained by comparing the 1000 hazard classes generated for Figure 4.5 and the 51 landslides that occurred in 1976 and 1996 as discussed in the text. The 20 pairs in the second column of Table 4.3 constitute the 2/5th of the dark grey curve with circles. The fitted function is shown as a light grey curve*

**Table 4.3**   *Prediction results for the La Baie study area*

| Portion of the study area assigned as hazard area | Cumulative portion of 51 landslides within the class | Empirical estimation based on the cumulative portion | Fitted function $1 - e^{-0.17 - 7.15area}$ | Estimated from the fitted exponential function |
|---|---|---|---|---|
| Top 1% | 0.2800 | 0.5307 | 0.2177 | 0.4319 |
| 1–2% | 0.0373 | 0.0838 | 0.0540 | 0.1200 |
| 2–3% | 0.0704 | 0.1547 | 0.0503 | 0.1121 |
| 3–4% | 0.0476 | 0.1063 | 0.0468 | 0.1045 |
| 4–5% | 0.0452 | 0.1010 | 0.0436 | 0.0976 |
| 5–6% | 0.0480 | 0.1071 | 0.0406 | 0.0910 |
| 6–7% | 0.0559 | 0.1241 | 0.0378 | 0.0849 |
| 7–8% | 0.0186 | 0.0423 | 0.0352 | 0.0792 |
| 8–9% | 0.0455 | 0.1017 | 0.0327 | 0.0737 |
| 9–10% | 0.0120 | 0.0274 | 0.0305 | 0.0688 |
| 10–11% | 0.0150 | 0.0342 | 0.0284 | 0.0642 |
| 11–12% | 0.0207 | 0.0470 | 0.0264 | 0.0598 |
| 12–13% | 0.0186 | 0.0423 | 0.0246 | 0.0557 |
| 13–14% | 0.0186 | 0.0423 | 0.0229 | 0.0519 |
| 14–15% | 0.0171 | 0.0389 | 0.0213 | 0.0484 |
| 16–17% | 0.0100 | 0.0229 | 0.0198 | 0.0450 |
| 17–18% | 0.0100 | 0.0229 | 0.0185 | 0.0421 |
| 18–19% | 0.0090 | 0.0261 | 0.0172 | 0.0392 |
| 19–20% | 0.0090 | 0.0261 | 0.0160 | 0.0365 |

*Note:*
The first column represents the portion of the whole study area classified as 'hazardous' for future landslides. As discussed in the text, the second column was generated comparing the 1000 classes for Figure 4.5 and the distribution of the 51 landslides that occurred in 1976 and 1996. The fourth column was based on a fitted function shown in (17) for the empirical values in the second column. The third and fifth columns show, under the assumptions in (16), the estimated probabilities that a house of size $10\,m \times 25\,m$ ($250\,m^2$ or 10 pixels) in the corresponding 1% areas will be affected by a future landslides within the next 30 years using the probabilities shown in the second and fourth, respectively. Obviously while the third column is based in empirical estimates, the fifth is based on the grey fitted prediction-rate curve shown in Figure 4.6. The corresponding plots of the third and fifth column are shown in Figure 4.7.

Table 4.3 that constitute the 2/5th of the prediction-rate curve in black with circles in Figure 4.6.

To estimate the probability, we need more assumptions on the future landslides that will occur within the next 30 years. We need to have the 'expected' number of future landslides in the area within that time interval, and the 'expected' average size of the landslides. Owing to the fact that in the study area we had 51 landslides for the past 35 years and of average size of approximately $15\,m \times 15\,m$, we can additionally assume that:

(i)  50 landslides will occur in the study area in the next 30 years;                         (16)
(ii) the average size of the 50 'future' landslides is $15\,m \times 15\,m$.

From the assumptions in (16), the whole area affected by the 50 landslides expected within the next 30 years is $50 \times 15\,m \times 15\,m$ or 450 pixels (of resolution $5\,m \times 5\,m$). If we were to build a house of size $10\,\text{m} \times 25\,\text{m}$ ($250\,\text{m}^2$ or 10 pixels) in the most hazardous 1% area ('Top 1%' area in column 1 of Table 4.3), then the probability that the house will be part of the affected area can be estimated by equation (15).

The probability estimated is $0.5307 = 1 - [1 - 0.28]^{4500/1954}$ and is shown in the corresponding row for the 'Top 1%' area of the second column of Table 4.3. The estimate is 53.07%, shown in the corresponding row for 'Top 1%' area of the third column. Similarly the numbers in the third column are generated from (15) using the corresponding probabilities of the second column of Table 4.3.

Based on the three properties of the prediction-rate curve, we have fitted a linear exponential function to the empirical prediction-rate curve (black curve with circles) in Figure 4.6 that is shown as a grey curve. The equation is:

$$\text{Fitted function} = 1 - e^{-0.17 - 7.15\text{area}} \tag{17}$$

The 'area' in equation (17) represents the portion of the whole study area as shown in the first column of Table 4.3. The corresponding fitted values are shown in the fourth column. Using the probabilities in the fourth column, the numbers in the fifth column were generated from equation (15). Here the estimate is 4.90% in the corresponding row of 'Top 1%' area of the fifth column.

The first 20 values in the third and fifth columns were plotted as histograms in Figure 4.7. Under the assumptions in (16), they are the estimated probabilities that a house of size $10\,\text{m} \times 25\,\text{m}$ ($250\,\text{m}^2$ or 10 pixels) in the corresponding 1% area will be affected by future landslides within the next 30 years. Obviously while the third column is based on empirical estimates, the fifth column is based on the fitted prediction-rate curve shown in grey in Figure 4.6.

In the third and fifth columns of Table 4.3, we can see that the probability of landslide occurrence for subsequent 1% portions of the study area changes from 53.07 to 8.38, 15.5, 10.6, 10.1, etc., and from 43.2 to 12.1, 11.2, 10.5, 9.0, etc., respectively. Evidently, the most interesting probability value is that for the 'Top 1%' area, from a cost–benefit point of view. It means that subsequent hazard classes have rather similar probabilities associated with the respective areas, so that the additional costs of taking care of additional 1% areas will have to be balanced by the additional benefit of only about 1% probability of occurrence.

The subdivision of the study area into classes of different probability of occurrence of a particular type of landslide satisfies the representation of hazard susceptibility, that is, the susceptibility of occurrence of a damaging phenomenon within a given area controlled by the combination of several physical factors (Meja-Navarro and Garcia, 1996). Furthermore, in the example described here, the time partitioning of the landslides into two groups 37 years apart has been used to relate the probability to a specific time interval of human concern. We could then assume that, besides the uniformity of physical settings during the 30 years, also we have a similarity of triggering factors, that is, the probability **P** that during the 30 years an event of a magnitude equal to or greater than those observed 37 years earlier will occur over the study area. Then we can consider our probability classes as a simple form of natural hazard map, that is, representing the probability that a potentially damaging phenomenon of an intensity equal to or greater than 'i' occurs during a period of exposure 't' (according to the definition of Varnes *et al.*

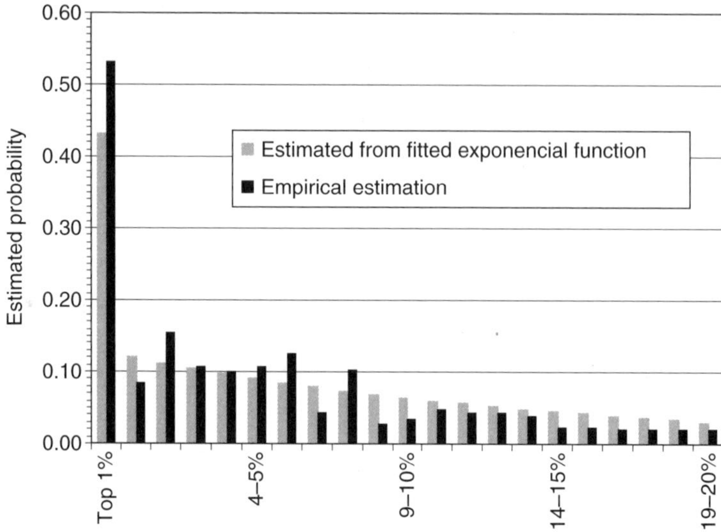



***Figure 4.7*** *Probability of occurrence curves for the map in Figure 4.5. The two histograms show, under the assumptions in (16), the estimated probabilities that a house of size 10 m × 25 m (250 m² or 10 pixels) in the corresponding 1% areas will be affected by a future landslide within the next 35 years and the prediction-rate curves in Figure 4.6. Obviously, while the dark grey histogram is based on empirical estimates, the light grey histogram is based on the fitted prediction-rate function using (17) shown as a light grey curve in Figure 4.6. The corresponding table values are shown in the third and fifth columns of Table 4.3*

(1984). Had we not been able to time-partition the landslides, we could, for instance, have resorted to a random division of the landslides into two groups. In that case, the validation would only relate to the 'next landslides' in the second group, without a time connotation but with just a spatial one, that is, predicting only where the next group of landslides is likely to occur. In such a situation, we would only obtain the susceptibility. However, as shown in the Portuguese case study, a random division may not provide a required prediction-rate curve for a proper interpretation of the prediction map.

## 4.7   Towards Risk Assessment

The conditional probabilities estimated in subsections 4.6.2 and 4.6.4 for the Portuguese and the Canadian case studies are the estimates of $H$ in Table 4.1. When we obtain the vulnerability $V$ of the house, the risk $R_s$ can be calculated by $R_s = H \cdot V$. Therefore, with the probabilities in Table 4.3, we are now ready to tackle the problem of risk assessment from landslide hazard. One of the essential components is the estimation of $H$. Obviously, without the magnitude value for each pixel of a landslide in the database, it is not possible here to perform predictions for the magnitude of the predicted future landslides. The procedure exposed is not restricted to spatial predictions only and can easily handle time partitions and magnitudes, therefore satisfying a more complete expression of risk as not only a product of hazard and vulnerability but of $H$, $V$ and $D$ ($D$ is the total damage of an element at risk).

To illustrate how to use the prediction-rate curve to estimate the probabilities of the occurrences of future landslides for risk analysis, let us summarize Section 4.6.2 for the Portuguese study area as an example. The first step is to generate a predicted hazard map using the same FF model (as shown in Figure 4.3) based on all six layers of the causal factors and the 92 past landslides. The 92 landslides are divided into two time periods, pre-1978 as *Estimation group* and 1979–2002 as *Validation group*. Using all six layers and only the 40 *Estimation group* landslides, we construct the second prediction map that is then compared with the distribution of the 52 *Validation group* landslides. The comparison generates the black prediction-rate curve shown in Figure 4.4. Consider the most hazardous $667\,750\,\text{m}^2$ ($26\,700$ pixels of $5\,\text{m} \times 5\,\text{m}$) from the predicted hazard map based on all the 92 landslides shown in Figure 4.3. Let us look at the prediction-rate curve (black curve in Figure 4.4). When the X-axis is at the value 0.05 (5%), the corresponding Y-axis value on the prediction-rate curve is 0.28 (or 28%). Suppose that we now assume that we expect 50 landslides covering 800 pixels over the next 25 years. If we build a house of the size of $10\,\text{m} \times 15\,\text{m}$ within the most hazardous $667\,750\,\text{m}^2$, the estimate the probability that the house will be part of the 50 future landslides is approximately 9%. Obviously as we change the expected number of future landslides, the average size of future landslides, and/or the size of house, the estimate of the corresponding probability will also be changing accordingly.

Based on the estimates of such probabilities, once an acceptable high-hazard area is identified as a priority for further analysis or for prevention follow-up, a second stage in the analysis can be initiated that requires an inventory of all human activities and assets that are within the its reach, or spatial domain, or zone of influence.

We are proposing a new way to express hazard spatially and statistically. We generate numerical predictions maps in which we try to estimate the probability of occurrence of each class. We produce a prediction-rate curve by using a validation technique. If our database allows temporal validation, we can predict within the time interval provided by the temporal partitioning.

This procedure would extend the general strategy described in Table 4.2, from a first stage in which we generate hazard predictions, to a second stage in which we estimate the probability for each hazard class and the corresponding spatial zone of influence.

We believe that the two-stage procedure can be developed to obtain the robustness and transparency needed for decision makers and for the dissemination of official risk maps. In particular, it can be expanded to represent the uncertainty levels of the input data and of the predictive models. In addition, regional predictions can be refined by the simulation of runout distances whenever appropriate historical data are compiled and made available through institutional geoinformation infrastructures. Further research work by the authors is directed to this target.

## Acknowledgements

Chang-Jo F. Chung carried out a part of the research while he was a visiting professor at the University of Tokyo, Japan in 2001. He is grateful for financial support from the University and the host Professor T. Shoji for his kind hospitality and for many stimulating discussions at his cottage. The authors acknowledge that Dr Didier Parret of the Geological Survey of Canada and Dr Zêzere of the University of Lisbon have kindly provided the spatial data for the La Baie case study and the data for the Fanhões–Trancão case study, respectively. This research is also being supported by a research network project on the 'Assessment of Landslide Risk and Mitigation in Mountain Areas, ALARM' (Contract EVG1-CT-2001-00038) of the European Commission's Fifth Framework Programme (http://www.spinlab.vu.nl/alarm). We acknowledge that all computations in this chapter were carried out by using Spatial Prediction Modeling System by Spatial Models Inc. (http://www.spatialmodels.com).

# References


Cacoullos, T., 1973, *Discriminant Analysis and Applications* (New York: Academic Press).

Cardinali, M., Reichenbach, P., Guzzetti, F., Ardizzone, F., Antonini, G., Galli, M., Cacciano, M., Castellani, M. and Salvati, P., 2002, A geomorphological approach to the estimation of landslide hazards and risks in Umbria, Central Italy, *Natural Hazards and Earth System Sciences*, **2**, 57–72.

Carrara, A., 1989, Landslide hazard mapping by statistical methods: a 'black-box' model approach, *Proceedings of the International Workshop on Natural Disasters in European–Mediterranean Countries*, Perugia, Italy, 27 June–1 July 1988, CNR-USNSF, 205–224.

Carrara, A., Cardinali, M., Guzzetti, F. and Reichenbach, P., 1995, GIS technology in mapping landslide hazard, in A. Carrara and F. Guzzetti (eds), *Geographical Information Systems in Assessing Natural Hazards* (Dordrecht: Kluwer Academic Publishers), 135–175.

Chung, C.F., 1978, *Computer program for the logistic model to estimate the probability of occurrence of discrete events*, GSC Paper 78–11, Geological Survey of Canada, Ottawa, Canada.

Chung, C.F., 2003, Modeling the conditional probability of the occurrences of future landslides for risk assessment, Manuscript submitted for a publication to *Computer and Geosciences*.

Chung, C.F. and Fabbri, A.G., 1993, The representation of geoscience information for data integration, *Nonrenewable Resources*, **2**(2), 122–139.

Chung, C.F. and Fabbri, A.G., 1998, Three Bayesian prediction models for landslide hazard, in *Proceedings of the International Association for Mathematical Geology Annual Meeting IAMG 1998*, Ischia, Italy, October 1998, 204–211.

Chung, C.F. and Fabbri, A.G., 1999, Probabilistic prediction models for landslide hazard mapping, *Photogrammetric Engineering & Remote Sensing*, **65**(12), 1389–1399.

Chung, C.F. and Fabbri, A.G., 2001, Prediction models for landslide hazard using a fuzzy set approach, in M. Marchetti and V. Rivas (eds), *Geomorphology and Environmental Impact Assessment* (Rotterdam: Balkema), 31–47.

Chung, C.F. and Fabbri, A.G., 2002, Modeling the conditional probability of the occurrences of future landslides in a study area characterized by spatial data, in *Proceedings of ISPRS 2002*, Ottawa, Canada, 8–12 July 2002, CD.

Chung, C.F. and Fabbri, A.G., 2003, Validation of spatial prediction models for landslide hazard mapping, *Natural Hazards*, **30**, 451–472.

Chung, C.F. and Keating, P., 2002, Mineral potential evaluation based on airborne geophysical data, *Geophysical Exploration*, **33**(1), 28–34.

Chung, C.F., Fabbri, A.G. and Van Westen, C.J., 1995, Multivariate regression analysis for landslide hazard zonation, in A. Carrara and F. Guzzetti (eds), *Geographical Information Systems in Assessing Natural Hazards* (Dordrecht: Kluwer Academic Publishers), 107–133.

Chung, C.F., Fabbri, A.G. and Chi, K.H., 2002a, A strategy for sustainable development of nonrenewable resources using spatial prediction models, in A.G. Fabbri, G. Gáal and R.B. McCammon (eds), *Geoenvironmental Deposit Models for Resource Exploitation and Environmental Security* (Kluwer: Dordrecht), 101–118.

Chung, C.F., Kojima, H. and Fabbri, A.G., 2002b, Stability analysis of prediction models for landslide hazard mapping, in R.J. Allison (ed.), *Applied Geomorphology: Theory and Practice* (New York: John Wiley & Sons Ltd), 3–19.

Clerici, A., Perego, S., Tellini, C. and Vescovi, P., 2002, A procedure for landslide susceptibility zonation by the conditional analysis method, *Geomorphology*, **48**, 349–364.

Corominas, J., Moya, J., Ledesma, A., Gili, J., Lloret, A. and Rius, J. (eds), 1998, 'New Technologies for Landslide Hazard Assessment and Management in Europe (NEWTECH). Final Report, October 1998', CEC Environment Programme, Contract ENV-CT96-0248. Barcelona, Department of Engineering and Geosciences, Civil Engineering School, Technical University of Catalunya (unpublished report).

Cox, D.R., 1970, *Analysis of Binary Data* (London: Methuen).

Dai, F.C. and Lee, C.F., 2002, Landslide characteristics and slope instability modeling using GIS, Lantau Island, Hong Kong, *Geomorphology*, **42**, 213–228.

Disperati, L., Guastaldi, E. and Carmignani, L., 2002, Landslide mapping and hazard prediction in the Pergola area (Marche, Italy), *Proceedings of the 8th Annual Conference of the International Association for Mathematical Geology*, Berlin, Germany, 15–20 September 2002 (Berlin: Alfred-Wegener-Stiftung), 507–512.

Fabbri, A.G. and Chung, C.F., 2001, Spatial support in landslide hazard prediction based on map overlays, in *Proceedings of IAMG 2001, International Association for Mathematical Geology Annual Meeting*, Cancun, Mexico, 10–12 September 2001, website:http://www.kgs.ukans.edu/conference/iamg/program.html/.

Fabbri, A.G., Chung, C.F., Napolitano, P., Remondo, J. and Zêzere, J.L., 2002, Prediction rate functions of landslide susceptibility applied in the Iberian Peninsula, in C.A. Brebbia (ed.), *Risk Analysis III* (Southampton, Boston: WIT Press), 703–718.

Fabbri, A.G., Chung, C.F., Cendrero, A. and Remondo, J., 2003, Is prediction of future landslides possible with a GIS?, *Natural Hazards*, **30**, 487–499.

Glade, T., 2001, Ranging scales in spatial landslide hazard and risk analysis, in C.A. Brebbia (ed.), *Risk Analysis III* (Southampton, Boston: WIT Press), 719–726.

Gritzner, M.L., Marcus, W.A., Aspinall, R. and Custer, S.G., 2001, Assessing landslide potential using GIS, soil wetness modeling and topographic attributes, Payette River, Idaho, *Geomorphology*, **37**, 149–165.

Guzzetti, F., Carrara, A., Cardinali, M. and Reichenbach, P., 1999, Landslide hazard evaluation: a review of current techniques and their application in a multi-scale study, Central Italy, *Geomorphology*, **31**, 181–216.

Johnson, N.L. and Kotz, S., 1969, *Discrete Distributions* (Boston: Houghton Mifflin).

Kojima, H., Chung, C.F., Obayashi, S. and Fabbri, A.G., 1998, Comparison of strategies in prediction modeling of landslide hazard zonation, in *Proceedings of IAMG 1998, International Association for Mathematical Geology Annual Meeting*, Ischia, Italy, 3–7 October 1998, eds A. Buccianti, R. Potenza and G. Nardi, 218–223.

Kshirsagar, A.M., 1972, *Multivariate Analysis* (New York: Marcel Dekker Inc.).

Leone, F., Asté, J.P. and Leroi, E., 1996, Vulnerability assessment of elements exposed to mass-movement: working toward a better risk perception, in K. Senneset (ed.), *Landslides* (Rotterdam: Balkema), 263–269.

Leroi, E., 1996, Landslide hazard-risk maps at different scales: objectives, tools and developments, in K. Senneset (ed.), *Landslides* (Rotterdam: Balkema), 35–51.

Liener, S., Kienholtz, H., Liniger, M. and Krummenacher, B., 1996, SDLISP – A procedure to locate landslide prone areas, in K. Senneset (ed.), *Landslides* (Rotterdam: Balkema), 279–284.

Mark, T.K. and Ellen, S.D., 1995, Statistical and simulation models for mapping debris-flow hazard, in A. Carrara and F. Guzzetti (eds), *Geographical Information Systems in Assessing Natural Hazards* (Dordrecht: Kluwer Academic Publishers), 93–106.

Mejía-Navarro, M. and Garcia, L.A., 1996, Natural hazard and risk assessment using decision support systems, application: Glenwood Springs, Colorado, *Environmental & Engineering Geoscience*, **II**(3), 299–324.

Park, N.W., Chi, K.H. and Chung, C.J., 2002, Effects of uncertainties of boundaries in thematic maps for spatial prediction models in landslide hazard mapping, in *Proceedings of the 8th Annual*

*Conference of the International Association for Mathematical Geology*, Berlin, Germany, 15–20, September 2002 (Berlin: Alfred-Wegener-Stiftung), 541–546.

Press, S.J., 1972, *Applied Multivariate Analysis* (New York: Holt, Rinehart and Winston).

Soeters, R. and van Westen, C.J., 1996, Slope stability recognition analysis and zonation, in A.K. Turner and R.L. Schuster (eds), *Landslides: Investigation and Mitigation*, Special Report 247, Transportation Research Board, National Research Council, Washington, DC, Chapter 8, 129–177.

Terlien, M.T.J., 1996, *Modelling Spatial and Temporal Variations in Rainfall-Triggered Land-slides*, Ph.D. Thesis, International Institute for Aerospace Survey and Earth Sciences, Publication Number 32, Enschede, The Netherlands.

Terlien, M.T.J., van Westen, C.J. and van Asch, T.W.J., 1995, Deterministic modeling in GIS-based landslide hazard assessment, in A. Carrara and F. Guzzetti (eds), *Geographical Information Systems in Assessing Natural Hazards* (Dordrecht: Kluwer Academic Publishers), 57–78.

Varnes, D.J. and International Association of Engineering Geology Commission on Landslides and Other Mass Movements on Slopes, 1984, *Landslide Hazard Zonation – a review of principles and practice*, Natural Hazard Series, no. 3 (Paris: UNESCO).

Wu, T.H., Tang, W.H. and Einstein, H., 1996, Landslide hazard and risk assessment, in A.K. Turner and R.L. Schuster (eds), *Landslides: Investigation and Mitigation*, Special Report 247, Transportation Research Board, National Research Council, Washington, DC, Chapter 6, 106–118.

Zêzere, J.L., 1996a, Landslides in the North of Lisbon region, *Fifth European Intensive Course on Applied Geomorphology – Mediterranean and Urban Areas*, eds A.B. Ferreira and G.T. Vieira, Departamento de Geografia, Universidade de Lisboa, 79–89.

Zêzere, J.L., 1996b, Mass movements and geomorphological hazard assessment in the Trancão valley between Bucelas and Tojal, *Fifth European Intensive Course on Applied Geomorphology – Mediterranean and Urban Areas*, eds A.B. Ferreira and G.T. Vieira, Departamento de Geografia, Universidade de Lisboa, 101–105.

Zêzere, J.L., 1997, *Movimentos de vertente e perigosidade geomorfológica na Região a Norte de Lisboa*, Ph.D. Thesis, University of Lisbon, Portugal.

# 5

# Vulnerability to Landslides

*David Alexander*

## 5.1  Introduction

The word 'vulnerability' comes from the Latin verb *vulnerare*, 'to wound', and signifies exposure to physical or moral harm. Thus it is a hypothetical concept that only assumes a tangible reality when impact transforms it into damage. Despite this, vulnerability to landslides is one of the most widespread components of natural hazard risk, for it reflects the ubiquity of slopes and erosional processes.

People and things are vulnerable to natural hazards in that they are susceptible to damage and losses. In many cases, vulnerability determines the losses to a greater degree than does hazard. For example, in 1964, following the Great Alaska Earthquake, 29 million m$^3$ of rock slid at 180 km/hr into the Sherman Valley, but the area was entirely uninhabited and the result was a mere geological curiosity, not a disaster. By contrast, on 21 October 1966 in the Welsh mining village of Aberfan, 42 000 m$^3$ of debris moved down a mining spoil heap at walking pace until it overwhelmed and killed 144 people. The Welsh example was nearly 700 times smaller and 30 times slower than the Alaskan one, but was very much more destructive. Difference in vulnerability determined the relative disaster potentials (Alexander, 1993: 9–10).

This chapter will first enquire into the theoretical nature of vulnerability as a concept and then consider how it applies to landslides. Forms of vulnerability will be examined in the context of rural and urban environments, and with respect to the different categories of mass movements and their settings. Next, patterns and trends in vulnerability to landslide disaster will be analysed at the world scale for an eight-year period of recent history. A methodology for assessing vulnerability to landslides is presented in an appendix.

*Landslide Hazard and Risk*   Edited by Thomas Glade, Malcolm Anderson and Michael J. Crozier
© 2004 John Wiley & Sons, Ltd   ISBN: 0-471-48663-9

## 5.2   What is Vulnerability?

This section will examine the definition and meaning of the concept of vulnerability in order to provide a basis for its application to landslide hazards.

A loose working definition of the term is 'potential for losses or other adverse impacts'. In this respect, vulnerability cannot be assessed in the absence of hazards posed by dangerously extreme phenomena (Figure 5.1). At its simplest, risk is considered to be the product of hazard, $H$, acting upon vulnerability, $V$ (UNDRO, 1982). To express total risk, $R_t$, these quantities should be summed for all vulnerable elements, $E$, that exist in a particular situation:

$$R_t = \sum[E(H.V)]$$

Factors that complicate this relationship include *exposure* and *dose rate* (or *release rate*). For example, a person who for ten minutes a day travels to and from work along a stretch of road which is threatened by rockfalls is exposed to the hazard for $10/(60 \times 24 \times 7) = 0.001$ of a week. This in physical terms is the person's exposure level (Tobin and Montz, 1997: 308–312; Hewitt, 1997: 144–145), but it assumes a constant level of risk during the periods of exposure. If rockfalls are only a significant hazard, for example, when rainfall has recently occurred, then levels of vulnerability, risk and exposure must be quantified by release rate.

Release or dose rate can also be considered as a probability function of hazard. Hence total risk, considered in the light of exposure and release rates, becomes

$$R_t = fcn\{t_e, V(E), P(H)\}$$

where $t_e$ is a temporal function describing exposure, $V(E)$ is the overall vulnerability of elements at risk, and $P(H)$ is the probability that hazardous landsliding will occur in a

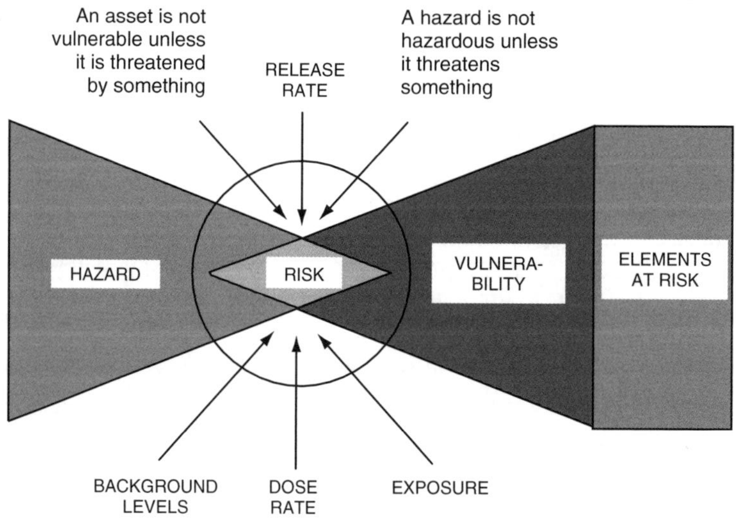


***Figure 5.1***   *Concepts in risk analysis*

specified manner. For instance, $t_e$ may describe a risk that increases over time as slopes weather and become more unstable or are increasingly destabilized by human activities. $V(E)$ may represent a certain potential value of total losses with respect to property located on the unstable slopes, and $P(H)$ may express a likelihood that slumping will take place at susceptible locations during a defined period of time.

This analysis assumes that vulnerability is a definable and essentially fixed quantity. That is not necessarily so. To begin with, there is the question of what level of losses to assume as the theoretical maximum value. Here we encounter another definition of exposure – not under threat for given periods of time, but at risk to the extent of given degrees of loss. Many vulnerability analyses of landslide hazard situations assume that total loss (death of people at risk, destruction of assets) is the maximum theoretical extent of potential losses. In practice, there are many cases in which losses are unlikely to be total, especially where slow-acting landslides cause progressive damage to buildings that can be put right as it occurs. One answer to this is to grade vulnerability levels according to the *degree* of potential loss. A methodology for vulnerability estimation which takes into account the potential for partial losses is described in the appendix to this chapter.

Apart from the question of what level of loss to expect, vulnerability tends to be a dynamic concept in relation to the perpetual duality between efforts to reduce or mitigate risks and human actions that create risks or increase their levels. Viewed in terms of risk management, vulnerability to landslides of socio-economic and environmental systems is a function of the costs (or other drawbacks) and benefits of inhabiting areas with significant landslide hazards (Alexander, 1991) mediated by decisions taken on the basis of perception of the risks:

$$\begin{array}{ccccccc} \text{total} & & \text{landslide} & & \text{landslide} & & \text{landslide risk} \\ \text{vulnerability} & \propto & \text{exacerbating} & - & \text{mitigation} & \pm & \text{perception} \\ \text{to landslides} & & \text{processes} & & \text{measures} & & \text{factors} \end{array}$$

Perception affects vulnerability, given that if a risk is perceived to be high or a hazard particularly dangerous (the usual term here is 'salient'), there exists an incentive or a social mechanism to reduce it. Moreover, if perception of vulnerability to landslide hazards is acute, the risks may be reduced much more than they would be if perception were low, irrespective of the *objective* levels of hazard, vulnerability and risk.

Hitherto, this discussion has emphasized the negative aspects of risk: factors that lead to increases in vulnerability, or at least to its persistence. However, the more positive way of considering it is as follows:

$$\text{vulnerability} = 1/\text{resilience}$$

that is, the inverse of vulnerability is composed of mechanisms for avoiding impacts or absorbing them by coping (Blaikie *et al.*, 1994). Landslide mitigation is capable of providing some of the highest cost–benefit ratios of all natural hazards (Leighton, 1976). Moreover, structural mitigation is highly developed (Veder, 1981) and non-structural defences are gaining ground, especially in response to advances in automated hazard and risk mapping (Carrara and Guzzetti, 1995). Finally, there is increasing evidence of a cultural ecology of human adaptation to landslide damage in urban areas. Examples are reviewed in Alexander (2000: 57–60).

The next section will examine the vulnerability of human systems in relation to different sorts of landslide hazard.


## 5.3    What is Landslide Vulnerability?

The investigation of landslide vulnerability first requires a brief conceptualization of hazard. Lithology, slope and climate together predispose areas to landslide activity. The weakest lithologies, such as unconsolidated sands, do not form slopes that carry a high landslide risk, but cohesion, consolidation or interlayering of materials with highly varied permeabilities can lead to high degrees of fracturing, jointing or erosional dissection. In short, nature must create long slopes that are steep relative to the strength conditions of the lithologies in which they are cut (Cruden and Varnes, 1996).

At hard rock sites, climate is not quite so fundamental, but in all other lithologies it is usually the vital determinant of landsliding through the processes of water inputs to slope systems (Enoki *et al.*, 1999). Widespread landsliding is usually stimulated by high porewater pressures as a result of intense or prolonged rainfall, ice that grows in cracks or along potential shear planes, and thermal expansion–contraction forces. In this sense, vulnerability is a temporal phenomenon in that the hazardousness which calls it into play varies with weather patterns – in ways that are only now coming to be known through detailed research (Miller, 1988; Crosta, 1998).

As human vulnerability depends primarily on patterns of activity and land use, characteristic scenarios of landsliding emerge where these coincide with susceptible terrains and high climatic inputs (Dai *et al.*, 2002).

In humid tropical mountain areas, 'feral' topography evolves in response to high rates of weathering and erosion. Harder lithologies develop rockfalls and debris flows; softer ones mudflows and slumps (Franks, 1999). Aggregate vulnerability is highest where settlement is densest or most exposed to the hazards (e.g. concentrated along the slope foot). The most vulnerable settlement involves either multi-storey buildings in well-established parts of major cities or precarious slums – informal housing – in poorer urban areas (Jiménez Díaz, 1992).

With regard to the former category, Hong Kong and Kuala Lumpur are pre-eminent examples (Lumb, 1975). Much of the urbanization in Hong Kong is very high density and multi-storey in character. It lines the coasts of the Kowloon Peninsula and surrounding islands and extends back into high-angle vegetated slopes in fractured, weathered rocks (Figure 5.2). With average population densities of 5800 persons per $km^2$, it is inevitable that major slope failures sometimes prove lethal. Torrential rains, often associated with the passage of typhoons, increase porewater pressures in slope soils to very high levels (Zhou *et al.*, 2002). Hence, in the 1960s and 1970s, some very serious and destructive failures occurred (So, 1971).

The Hong Kong authorities have adopted a very active programme of landslide hazard reduction that has had a considerable impact on levels of vulnerability. Besides monitoring, mapping and predicting the hazards (Dai and Lee, 2002), micro-management of slopes is common throughout the Special Administrative Region. Maintenance of good drainage and vegetation cover are the bases of this policy, though cement spray, gabions, retaining walls and buttresses are also widely used. This has enabled landslide risk to be

***Figure 5.2***  *Interdigitation of slopes and dense urbanization on Hong Kong Island*

abated without reduction or significant alteration of the fundamental elements at risk – the urban areas. This is therefore a straightforward case of hazard reduction without engineering a decrease in vulnerability.

Similar vulnerability, though perhaps with less risk mitigation, exists elsewhere in Southeast Asia. For example, in the 1980s and 1990s at least ten large buildings collapsed totally or partially in Malaysia, but the only case to involve an occupied structure was that of the luxury Menara Highlands apartment block in Ulu Kelang, near Kuala Lumpur, which was demolished by a landslide in December 1993. The 12-storey building contained 52 apartments, and 48 of their occupants died in the collapse. Unregulated urbanization was the principal destabilizing factor, as the area upslope of the building had been destabilized by construction work. Rainwater infiltrated the ground and the resulting mass movement destroyed the foundations of the building (Chan, 1998).

The second form of major vulnerability to landslides is that of the precarious squatter settlements that have developed on or at the foot of steep, unstable slopes in the major cities of developing countries. Caracas (Jiménez Díaz, 1992), Rio de Janeiro (Jones, 1973), Ponce (Puerto Rico) and Cuzco (Peru) all contain areas of informal housing (*barrios*, *favelas*, *bidonvilles*) that clings precariously to unstable slopes and can be washed away by debris flows and mudflows during episodes of torrential rain (Alexander, 1989). In Guatemala City, the urbanized sides of canyons are vulnerable to rockfalls caused by amplification of seismic forces during major earthquakes (Harp *et al.*, 1981), while hurricanes provide the climatic inputs to catastrophic slope failure in the *barrios* of

Tegucigalpa, Honduras. It is important to note that such risks are taken involuntarily by the inhabitants of the unstable slopes, who have little choice but to occupy such dangerous sites. An analogous form of voluntary risk-taking is the urbanization of canyons with unstable side slopes in southern California, where economics do not constitute such a barrier to risk avoidance.

A third kind of major vulnerability is found in settled mountain belts. Exposed, tectonically destabilized slopes impinge upon routeways and settlements, which leads to damage and casualties during earthquakes and periods of wet weather. Most of the world's major orogens are involved. For example, the European Alps and the Himalayas, Karakoram and Hindu Kush areas suffer considerably from this problem, as do the Andes (Figure 5.3). Vulnerability to landslides in rural terrain has much to do with the essential fragility of socio-economic systems in those areas where poverty and deprivation are common. Thus the Faizabad, Afghanistan, earthquake of 30 May 1998 (magnitude 6.9) may have involved several thousand deaths in landslides that buried entire villages. No forms of protection or hazard microzonation existed. Yet even in France, Spain and Italy, where detailed scientific knowledge of landslide hazards has been accumulated for decades, campsites have been overwhelmed by mudflows, traffic swept off roads by debris avalanches, and buildings crushed by rockslides (Figure 5.4).

However, in general the risk to human life is considerably greater in developing countries, especially in Central Asia and South America. In both regions high rates of tectonic uplift lead to steep, unstable slopes and populations are concentrated in deep valleys where rockfalls, debris slides and rock avalanches can occur suddenly and with great devastation.

A subset of vulnerability in mountain areas concerns the hazard of catastrophic breaching of landslide-dammed lakes. Costa and Schuster (1988) reviewed 65 cases of mass movements that invaded active watercourses and concluded that most natural dams fail within two weeks of formation. There have been various cases in which this led to catastrophe for downstream settlements (e.g. on Mount Huascaran in Peru, 1970 – Browning, 1973) or has threatened them with disaster (e.g. in the Italian Alps in 1987 – Alexander, 1988).

Finally, considerable vulnerability to landslides exists where the flanks of volcanoes, and the valleys that drain them, are densely settled. The supreme example is that of the lahar (volcanic mudflow) that destroyed Armero, Colombia, in 1985 during the eruption of Nevado del Ruiz Volcano, which led to the loss of 22 000 lives (Voight, 1990). However, many other cases are important. For instance, lahar damage has repeatedly occurred on the flanks of Mount Pinatubo in the Philippines (Pierson, 1992) and major loss of life occurred in the lahar provoked on Casita Volcano in Nicaragua by Hurricane Mitch in 1998 – see below (Sheridan *et al.*, 1999). In this context, it is important to note that mass movements can occur in the absence of eruptions. Only about half of the 17 different causes of lahars involve the direct action of a volcanic cataclysm (Alexander, 1993: 99). The breaching of a crater wall, for example, may involve porewater pressure increments caused by rainfall, as in the Casita example. In any case, the presence of large population agglomerations in the potential paths of lahars or other volcanic mass movements is a striking example of vulnerability to intermittent hazards that tend to shift their foci with the effects of eruptions upon the physical landscape, thus continually altering risk levels.

**Figure 5.3** *At Cuyocuyo in the Andean Eastern Cordillera of Peru, houses have been damaged by rockslides from the 42° slopes that flank this 1500 m deep valley. The valley bottom is the most suitable land for building, but is acutely vulnerable to mass movements*

**Figure 5.4**  At Passo Falzarego in the Italian Dolomites a debris slide-flow threatens a newly built hotel complex

Thus there are four major sources of vulnerability to landslides: expanding tropical cities (e.g. Cuzco, Caracas), peri-urban slums (e.g. Rio de Janeiro and the nearby Serra das Araras escarpment), inhabited mountain areas (e.g. the Karakoram Himalaya), and densely settled steep volcanic terrains (e.g. Casita Volcano, Nicaragua). To these can be added areas that have undergone major land use changes, especially those subject to deforestation or devegetation (Glade, 2003), and suburban areas where development has interfered with slope stability (e.g. Cotton and Cochrane, 1982).

## 5.4  How is Vulnerability to Landslides Assessed?

In practice it is often hard to separate vulnerability from hazard and risk, as these concepts are intertwined in complex ways (Alexander, 2000: 16–20; Figure 5.1). Hence it is difficult to design a standard, all-embracing method of assessing the vulnerability of communities and structures to landslides. The appendix to this chapter describes in detail a method of assessing vulnerability to landslides based on asset recognition and estimating potential death tolls and the cost of damage. Generally, potential losses are assessed in monetary terms and with respect to casualty totals. As this is hard to achieve in accurate quantitative terms, in many cases the methodology is based upon categories, for which values are summed (see Table 5.1). In other cases, the assessment is restricted to particular categories, for example buildings only (Carrara *et al.*, 1992). Some authors have adopted a policy base approach (Olshansky, 1990), while others have concentrated on quantifying the hazard, leaving vulnerability to take care of itself (e.g. Morgenstern, 1997).

One particular problem that complicates the evaluation of vulnerability with respect to landslides is the reactivation of pre-existing mass movements (Galadini *et al.*, 2003). Consider the case of the small town of Campomaggiore, in the southern Apennines of Italy, which was founded in classical times as a Roman fort. In 1885 it was completely ruined by a major rockfall from neighbouring limestone cliffs. Survivors transferred the entire town to an apparently safer locality some 3 km from the original site. Unfortunately, the new site proved to be on a concealed palaeo-landslide in Plio-Pleistocene clays, and the building works reactivated it. Damage was slower to occur this time, but no less profound. Reactivation of landslides is so common that most landslide hazard maps are based on the assumption that where mass movements have occurred in the past they will happen again in the future (Wieczorek, 1984). However, it cannot be assumed that reactivation is a function of vulnerability, only that it will add a further layer of complication to landslide risk.

Despite these complications there is a clear distinction between vulnerability to fast and slow landslides. Those events that fall into the 'extremely rapid' category of the Varnes classification (Cruden and Varnes, 1996) may threaten life, as there is little time to react to them. Landslide warning systems (Angeli *et al.*, 1994, Wieczorek *et al.*, 1990) can help vulnerable families and communities to evacuate or take other precautions before rockfalls or debris flows destroy buildings, but they are by no means common. In contrast, slow and extremely slow landslides rarely threaten life, but can inexorably tear buildings apart (Alexander, 1984). It is not usually that damage cannot be remedied or stopped, but that the economic cost of putting it right is exorbitant. Thus, once major

**Table 5.1**   *Elements at risk*

| Infrastructure | Buildings and rural production |
| --- | --- |
| **Roads**<br><br>– unasphalted rural roads<br>– asphalted rural roads<br>– main roads<br>– divided highways (dual carriageways)<br>– limited access freeways (motorways)<br>– urban access roads (asphalted)<br>– private drives<br><br>**Railways**<br><br>– main lines<br>– branch lines<br>– sidings<br>– buildings (stations, etc.)<br><br>**Bridges**<br><br>– major road, rail, pipeline bridges and viaducts<br>– minor bridges<br>– culverts<br><br>**Electricity transmission**<br><br>– low-tension lines, on poles<br>– high-tension lines, on pylons<br>– transformers, switching stations and substations<br><br>**Telephone**<br><br>– low-tension lines, on poles<br>– cellular telephone repeaters and their electricity supplies<br><br>**Pipelines**<br><br>– water supply: main pipelines and distribution networks<br>– sewer lines<br>– methane gas: main pipelines and distribution networks<br>– septic tanks and their feeder systems<br><br>**Other**<br><br>– canals, navigable rivers and drainage channels<br>– water towers and tanks<br>– gas and oil storage facilities<br>– airfields, airports | **Houses**<br><br>– single family homes<br>– semi-detached (duplex) and terraced (row) housing<br>– blocks of apartments (flats)<br>– urban insulae (historic or modern city block)<br>– farmhouses<br>– farm outhouses, stalls, barns, etc.<br>– villas and isolated dwellings<br>– prefabricated buildings<br><br>**Public buildings**<br><br>– town halls and public administration offices<br>– hospitals and clinics<br>– sports centres, stadia and sports fields<br>– cemeteries<br>– churches and chapels<br>– schools and other educational institutions<br>– fire and ambulance stations<br>– armed forces barracks and police stations<br><br>**Architectural heritage**<br><br>– historic buildings<br>– fortifications<br>– monuments<br><br>**Commercial buildings**<br><br>– shops and stores<br>– office blocks<br>– warehouses and storage areas<br>– factories<br>– artisans' premises and small businesses<br>– mechanics' premises, motor showrooms and engineering works<br>– heavy industrial plants and refineries<br><br>**Agriculture**<br><br>– tilled fields<br>– market gardens |

landsliding has begun to destabilize the foundations of a structure, it may inexorably be vulnerable to yet more damage. This can be measured and classified using the landslide damage intensity scales (Alexander, 1989).

The next section will consider how such situations can be managed.


## 5.5   Vulnerability and Risk Management

As landslide risk is fundamentally a product of hazard and vulnerability, these two phenomena can be managed in mutually varying proportions. Given that they are entirely interdependent (Figure 5.1), any change in one will necessarily alter the level of the other. Thus there is a continuum of possible management strategies. It varies from complete focus on hazard abatement to total reliance on vulnerability reduction, considering hazard as an immutable quantity.

A good example of the former strategy is provided by the central Italian city of Perugia. Slope instability has existed in the western part of the urban area since it was first settled by the Etruscans 2400 years ago. In the 1960s and 1970s considerable urban expansion occurred into areas formerly avoided and the new buildings began to be damaged by extremely slow but persistent landsliding, mostly roto-translational slumping. In the mid-1990s it was decided to build a large commercial centre on the site. This goal was achieved after the slope had been stabilized by constructing a linked network of 240 drainage wells and an automated system for monitoring porewater pressure and pumping out groundwater. Geographical inertia (Alexander, 1993: 5) determined that the site would be redeveloped (urban services, infrastructure, landownership criteria and investment decisions all came into play here). However, one could argue that the whole process introduced a form of secondary vulnerability by relying on a technological system that could break down and render large amounts of expensive new property vulnerable to damage. Moreover, this 'hazard aversion' strategy involved astronomical costs to implement and maintain (Alexander, 2000: 25–6).

Examples of 'landslide vulnerability aversion' by total prohibition of development are much harder to find than are those related to 'hazard aversion'. This reflects a prevailing ethos that landslides are essentially controllable phenomena, though it is not one that is necessarily based on a rational consideration of costs (Schuster and Fleming, 1986).

Suburban environments in many parts of the world are susceptible to landslide damage. The highest sustained annual costs of slope failure in the USA come from Allegheny County, Pennsylvania (suburban Pittsburgh), Hamilton County, Ohio (suburban Cincinnati), Pacific Palisades (suburban Southern California) and the San Francisco Bay area (Fleming and Taylor, 1980). The spread of single-family housing onto unstable slopes does not necessarily lead to spectacular landslide impacts, but both the high aggregate value of property exposed to risk (mainly to slumping and debris flows) and the yearly recurrence of damaging slope movements add up to potentially high costs (Fleming and Taylor, 1980). The processes that give rise to this situation were analysed in a previous work (Alexander, 2000: 13–14) and are summarized in Figure 5.5 as a vicious circle of risk-taking permissiveness. Successful control of the problem involves using planning measures and land use control to break out of the vicious circle.

Analysis of geomorphological hazards in Los Angeles led Cooke (1984) to conclude that many of the older cores of cities have abated their risks by gradually coming to terms

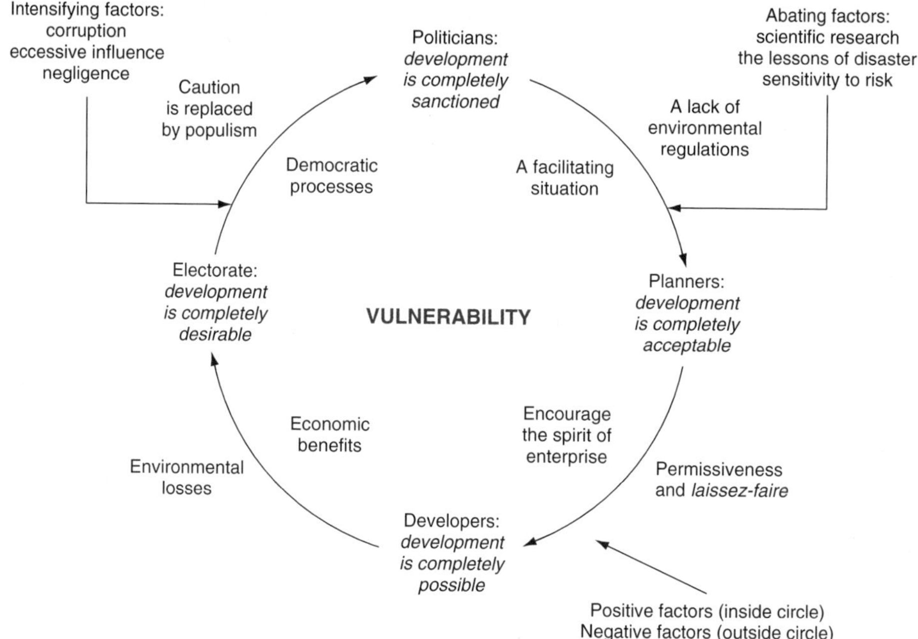



**Figure 5.5**  *The vicious circle of increases in vulnerability: a situation of positive feedback*

with the limitations on settlement and land use that hazards pose. Vulnerability may have shifted to the periphery. Thus the Los Angeles metropolitan area includes the San Gabriel Mountain Front, where mudflows spread across alluvial fans and *bajadas*, many of which have had expensive homes built on them. Likewise, the Wasach Front in Utah has taken the overspill settlement from Salt Lake City but is also highly vulnerable to debris-flow damage (Wieczorek *et al.*, 1989). The density of housing is not very high but development is space-extensive. Again, the solution involves moratoria or prohibitions on building, coupled with microzonation of slope environments.

In this context, it should be noted that the California landslide warning system (Wieczorek *et al.*, 1990) does not actually reduce the vulnerability of property, though it does enable the occupants of buildings to take evasive action.

Having examined what landslide vulnerability consists of, the next section will investigate patterns and trends in this phenomenon over a contemporary study period.

## 5.6  Landslide Vulnerability over the Period 1993–2002

In order to investigate vulnerability to landslides in the context of a sequence of actual events, the following account will focus on the period August 1993–May 2002, for which comprehensive information is available. Throughout this period I collected situation reports and news briefings on natural disasters, about 350 of which involved damage or casualties caused by mass movements. This data set, which spans 8.75 years, involved continuous accumulation of material on damaging natural hazard events from wire-service news reports, NGO and UN agency situation reports, and field reports filed by disaster workers. The data set, which is worldwide in scope, has been cross-checked and verified as much as possible with reference to journal articles and other academic sources. It is reasonably comprehensive, but has the following drawbacks:

- Not all significant landslide events were internationally reported by the news media and aid agencies. Significant underreporting of damage and casualties has undoubtably occurred. In fact, it seems that information on landslide impacts is published even less systematically than that on other forms of natural disasters, such as floods, volcanic eruptions and earthquakes.
- No clear definition emerges as to the lower threshold of damage and disruption that encourages the media and agencies to report an event. In addition, no distinction is made here between a disaster and a mere incident: all events that were reported during the study period have been analysed equally.
- The choice of a starting date was somewhat arbitrary and the ending date is simply the last event to be reported at the time of writing this chapter. It is not clear whether the period in question is representative of conditions over any other comparable span of recent history, especially as the impacts of natural disasters generally tend to be very uneven in time.
- Not only are landslide disasters underreported, they also tend to be described without differentiating them from the floods, earthquakes or volcanic eruptions which caused the slope movements. Hence it is necessary to distinguish between effects, particularly deaths and injuries, that are definitely the result of landslide impacts, and those in which some of the effects are due to floods or other agents.

Nevertheless, every effort was made to collect as comprehensive a data set as possible, and the result does furnish a comprehensive picture of landslide impacts, and by implication, of the human vulnerability patterns that gave rise to them (i.e. the aspects of vulnerability that, by interacting with hazards, facilitated disaster), over the study period. However, the statistics reported below should be considered as minimum values, as many other incidents, casualties and damages probably went unreported.

One event, the floods and debris flows of 16 December 1999 in Venezuela, dominates the picture to such an extent that the death toll, approximately 30 000, is three times as high as that of all other events in the list considered together. However, despite the skewing effect of this major catastrophe, many trends are evident.

Significant variations exist in the average numbers of deaths in landslide disasters as reported in the hazards literature. The Centre for Research on the Epidemiology of Disasters at the University of Louvain reported about 790 deaths per year, worldwide, over the last quarter of the twentieth century, rising to 955 over the last decade, 1991–2000 (ICRCRCS, 2001). Alexander (1989) reported a slightly lower figure for the 1960s to 1980s. Landslide disasters accounted for about 4.4% of deaths in all natural disasters, though in certain years it reached almost double that, either by higher than average landslide deaths, or lower than average deaths in other types of disaster. The present data set (Tables 5.2 and 5.3) gives figures of 716–5443 deaths per year, depending on the criteria used to sum and average them, whether or not the 1999 Venezuela event is included, and what proportion of flood-related deaths are caused by landslides.

***Table 5.2***   *Regional breakdown of landslide disasters, August 1993–May 2002 (author's data set)*

| Region | Sub-region | No. of events | Total no. of deaths attributed directly to landslides | Total no. of deaths attributed to landslides with other events | No. of events with no deaths reported |
|---|---|---|---|---|---|
| **Africa** | | **7** | **237** | **40** | **1** |
| | North | 2 | 40 | 40 | – |
| | Central | 3 | 20 | – | 1 |
| | South | 2 | 177 | – | – |
| **Americas** | | **126** | **36 263** | **538** | **45** |
| | North | 30 | 19 | 2 | 24 |
| | Caribbean | 10 | 787 | 132 | 2 |
| | Central | 21 | 3 214 | 264 | 6 |
| | South | 65 | 32 243 | 140 | 13 |
| **Asia** | | **153** | **3 682** | **6271** | **38** |
| | South and Central | 35 | 1 362 | 2409 | 5 |
| | East Asia | 75 | 1 710 | 1651 | 25 |
| | Southeast Asia | 43 | 610 | 2211 | 8 |
| **Australasia** | | **11** | **121** | **–** | **4** |
| | Australia & New Zealand | 3 | 19 | – | 1 |
| | Pacific Islands | 8 | 102 | | 3 |
| **Europe** | | **50** | **337** | **71** | **26** |
| | Eastern | 13 | 39 | – | 10 |
| | Western | 37 | 298 | 71 | 16 |
| **Middle East** | | **5** | **62** | **–** | **2** |
| **Totals** | | **352** | **40 702** | **6920** | **116** |

Regardless of whether the Venezuela event is taken into account, Latin America furnishes the largest death tolls, followed by Central and East Asia (Table 5.2). As Table 5.3 demonstrates, two kinds of country are particularly susceptible to landslide disasters: tropical nations subject to hurricane-force storms or torrential monsoon rains; and those countries that have seismically active orogens. Venezuela and China fall into both of these categories. Hence vulnerability reaches its peak, and landslide disasters are most recurrent, in areas where population densities are high and slope instability is provoked by torrential rain, flood conditions, earthquake shaking or vulcanism.

Overwhelmingly, deaths were caused by debris flows and mudflows. Rather less common sources of disaster were slumps, debris avalanches and rockfalls. Episodes of high landslide damage were mainly caused by intense and prolonged rainfall, in many cases associated with tropical storms and usually accompanied by floods. Earthquakes and volcanic activity were much less frequent sources of landslide hazard and were,

***Table* 5.3**    *Landslide events that caused casualties or damage over the period August 1993–May 2002 (author's data set)*

| Rank | Country | No. of events | Total no. of deaths attributed directly to landslides | Total no. of deaths attributed to landslides with other events | No. of events with no deaths reported |
|---|---|---|---|---|---|
| 1 | Venezuela | 7 | 30218 | – | 2 |
| 2 | Nicaragua | 1 | 2200 | – | – |
| 3 | Colombia | 15 | 1438 | – | 1 |
| 4 | China | 19 | 1173 | 1544 | 2 |
| 5 | Haiti | 2 | 777 | – | – |
| 6 | El Salvador | 3 | 712 | – | 1 |
| 7 | India | 17 | 624 | 2150 | 3 |
| 8 | Peru | 11 | 356 | – | 4 |
| 9 | Mexico | 11 | 276 | 202 | 4 |
| 10 | Philippines | 25 | 256 | 894 | 2 |
| 11 | Indonesia | 11 | 226 | 1225 | 5 |
| 12 | Nepal | 4 | 203 | – | – |
| 13 | Italy | 11 | 169 | 59 | 3 |
| 13 | Mozambique | 1 | 169 | – | – |
| 15 | Afghanistan | 4 | 154 | 200 | 1 |
| 16 | Japan | 25 | 119 | 10 | 8 |
| 16 | Taiwan | 19 | 119 | 86 | 10 |
| 18 | Brazil | 13 | 118 | 88 | – |
| 19 | Tajikistan | 1 | 100 | – | – |
| 20 | Malaysia | 4 | 98 | – | – |
| 21 | Spain | 3 | 77 | 6 | – |
| 22 | Papua New Guinea | 3 | 70 | – | 1 |
| 23 | Ecuador | 8 | 68 | 52 | – |
| 24 | Pakistan | 3 | 65 | – | 1 |
| 25 | Tibet | 1 | 53 | – | – |
| 26 | Kyrgyzstan | 1 | 51 | – | – |
| 27 | Sri Lanka | 1 | 50 | – | – |
| 27 | Iran | 1 | 50 | – | – |
| 27 | Bhutan | 1 | 50 | – | – |
| 30 | Bolivia | 2 | 41 | – | – |
| 31 | Ethiopia | 1 | 40 | – | – |
| 32 | South Korea | 6 | 36 | 11 | 1 |
| 33 | Thailand | 3 | 30 | 20 | – |
| 34 | Azores | 1 | 29 | – | – |
| 35 | Russia | 3 | 24 | – | 1 |
| 36 | Kenya | 1 | 20 | – | – |
| 37 | Australia | 2 | 19 | – | – |
| 38 | USA | 36 | 17 | 2 | 17 |
|  | Others | 71 | 407 | 371 | 37 |
|  | Total | 352 | 40702 | 6920 | 104 |

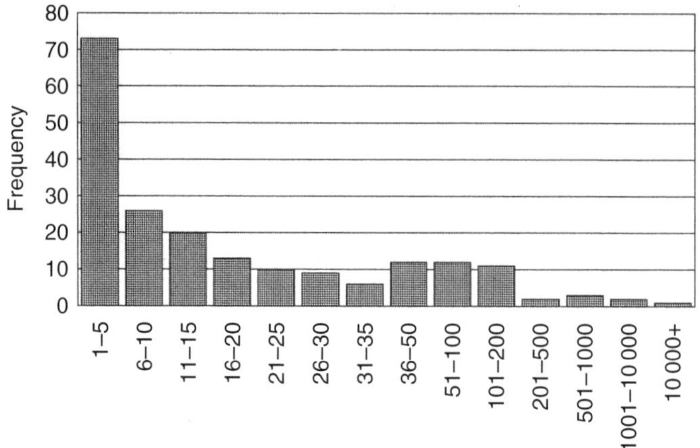

**Figure 5.6**  *Magnitude–frequency relationship for number of deaths per landslide event, August 1993–May 2002 (x axis: no. of deaths per event, y axis: no. of events)*

of course, limited to appropriate areas of the world. Deaths per landslide event were generally rather small, following the classic magnitude–frequency relationship for natural hazards: half the lethal events in the data set involved ten or fewer deaths, two-thirds, 20 or fewer, and 90% fewer than 100 deaths. Only three events that occurred during the 8.75 year study period involved more than 1000 deaths (Figure 5.6).

Awareness is one key to avoiding the threat to life posed by landslides. Hence most of the deaths involved people who, judging by the lack of prior evacuation, were unaware of the hazard at the time it struck: evacuation is usually the best means of avoiding imminent danger but can only be carried out if there is sufficient awareness and preparation. One of the most significant causes of mortality in this category is the fast-moving debris flow that strikes and demolishes a dwelling at night when its occupants are asleep. A smaller but none the less significant mortality occurs on roads when cars, buses and goods vehicles are swept away by mudflows or, less commonly, crushed by rockfalls.

During the study period there were four examples of landslide disasters in volcanic terrains. None of them involved a direct connection with an eruption, but three involved secondary lahars – that is, volcanic mudflows taking place during a period of quiescence. Both Mexico's Popocatepetl and the Philippines' Pinatubo are volcanic edifices that frequently generate lahars, many of which are highly destructive. However, the collapse of a crater lake wall on Casita Volcano in Nicaragua, which followed the torrential rains caused in October 1998 by Hurricane Mitch, generated a lahar that overwhelmed 2200 people at the base of the mountain. In contrast, most of the landslides caused by intense rain in Papua New Guinea were the result of more general saturation of open-textured volcanic soils, leading to more widespread slumping and debris flowage.

It has long been known that the precarious shelters that the poorest city dwellers construct on steep peri-urban slopes are particularly vulnerable to being swept away by landslides: slumps and rockfalls during earthquakes, debris flows and mudflows caused by storms and intense rainfall. During the study period, *favelas* were devastated in São Paulo State, Brazil, and *barrios* in Vargas State, Venezuela, and all over Haiti. Though death tolls were limited in the Brazilian case, 750 died in Haiti (in Tropical Storm Gordon during November 1994) and a sizeable proportion of the estimated 30 000 deaths in Venezuela in December 1999 occurred in the slums of the poor (ICRCRCS, 2001: 83). This followed a previous episode, dated August 1993, in which 100 died when debris flows tore apart the *barrios* of urban Venezuela.

At least 34 landslide events were triggered by earthquakes. Magnitudes varied from 4.3 to 7.6 and the most common effects reported involved blockage of roads, causing traffic accidents and hampering the supply of relief. Although the number of landslides caused was often very high, and a wide catalogue of slope failures was usually involved, most of the 34 events did not involve significantly high death tolls as a result of the mass movements themselves. There were three exceptions: the magnitude 6.4 event that occurred in Colombia in June 1994, in which landslides buried several villages and killed 1109 people; the magnitude 7.6 event in El Salvador in January 2001 in which a landslide killed 700 residents of the Las Colinas neighbourhood; and the magnitude 7.2 event of March 2002 in Afghanistan, in which 150 were killed by rockfalls and other landslides. Both Central America and the Hindu Kush have long histories of devastatingly lethal seismic landslides.

Nearly 50 events involved landsliding provoked by hurricanes (typhoons, tropical cyclones). Some 36 of these were in Asia, and the countries most affected were the Philippines (which has an average of 22 typhoons a year), Taiwan, Japan, coastal southern China (including Hong Kong), and South Korea. The tectonically disturbed terrains of the Japanese archipelago were particularly at risk, but the deforested slopes of the Philippines exhibited the greatest vulnerability, especially to debris flows, many of which were associated with flash flooding during or shortly after the passage of hurricanes. The same was true of Central America (Honduras, Nicaragua, El Salvador, Mexico, etc.) where, for instance, in October 1998 Hurricane Mitch caused enormous numbers of debris flows on denuded and unprotected slopes. Although it is not known how many of the 7000 victims of Mitch were killed by mass movements, in the Philippines, a contemporary typhoon, code-named Babs, killed more than 200 in debris flows, including 36 in a single instance. In a typical example, Danas, the typhoon of September 2001, provoked 83 landslides in Japan. However, good mitigation, warning and emergency management systems in this and 13 other typhoons that caused landsliding kept death tolls to a minimum. The same was true in Hong Kong, where only two people died in the six typhoons that caused mass movement there during the study period. Death tolls were also limited in landsliding caused by tropical cyclones passing over oceanic islands, though in September 1998 Hurricane Georges killed 140 people in three Caribbean countries: the bulk of these died in Haiti, where persistent devegetation had prepared the ground for debris flows to occur.

The lessons of these events are clear. Irrespective of the levels of hazard, tolls of death and destruction are much greater in poor, unprotected communities than they are in areas of well-funded, technologically advanced mitigation and preparedness. Poverty is not exactly synonymous with vulnerability, but it is quite close to being so (Cannon, 1994). Thus very serious landslide hazards threaten densely settled areas in Japan and Hong Kong, but without much of the devastation and loss of life that occur in Venezuela, Brazil and Nicaragua. Thus, for landslides as for other natural hazards, *risk* is determined rather more by *vulnerability* than by *hazard* (Hewitt, 1983).

## 5.7  Conclusions

The preponderance of landslide-induced mortality in some of the world's poorest countries (Afghanistan and Haiti, for example) ought to be cause for profound reflection on the meaning of vulnerability. While advanced technological systems of mapping, microzoning and mitigating landslide risk are now well developed, a large proportion of the inhabitants of the world's most hazardous areas has no access to such tools and does not benefit from their application. Many authors regard vulnerability as a concept to be the product of Western culture, and some argue that as a phenomenon it is the consequence of the rich nations' hegemony over the poor ones (Bankoff, 2001; Blaikie *et al.*, 1994, Boyce, 2000). According to the classification given in Alexander (2000: 16–17), this is a mixture of *pristine vulnerability* (of the original, unmitigated kind) and *deprived vulnerability* (i.e. resulting from lack of the economic means to achieve mitigation). It outweighs the *technological vulnerability* (resulting from the use and deployment of technological systems) and *wilful vulnerability* (caused by ignoring hazard information or protective regulations) that characterize high-income countries.

As the analyses reported above demonstrate, landslide disasters follow the magnitude–frequency relationship for natural hazards, with many small events and a few very large ones. Yet, given the widespread nature of slope failure and the inexorable rise in world population, the cumulative effect of the smaller disasters tends to negate the value of the recovery time between big events. For example, in the Philippines in 1996 there were 31 major floods, 29 earthquakes, 10 typhoons and 7 tornadoes. Population pressure has denuded large areas of Luzon and other islands of their vegetation, and this makes settlements and routeways extraordinarily vulnerable to landslides. Twelve major episodes of slope failure occurred in the archipelago during 1996.

Similar problems were highlighted in Central America in the wake of Hurricane Mitch in 1998. The disaster is estimated to have set back development by 25 years, largely as a result of the way it accelerated environmental devastation that was already in progress. From this we may conclude that the key to landslide vulnerability reduction lies not in better mapping schemes, warning systems and research on slope processes, but on spreading models of sustainable development and socio-economic stability. The answer may well lie in the pattern of global economics rather than that of natural hazards (Wisner, 2001).

Land tenure may be the critical issue here. Groups of poor people tend to live on, or under, unstable slopes because they have no choice. The safe land is owned by richer people. Land reform can be used to free slopes of vulnerable urbanization, which both protects the beneficiaries and, by enabling better environmental management, reduces the risk of slope collapse. In other cases it is a question of reducing the incentives for deforestation or other land uses that make slopes unstable (Glade, 2003). However, there is still a long way to go before the benefits of wise land uses are adequately perceived before the costs of not implementing them are incurred.

In synthesis, some of the examples discussed in this chapter suggest that the role of vulnerability in determining landslide risk levels has consistently been underestimated. Much is now known about landslide *hazards*, but *vulnerability* is a more elusive concept, as it depends on patterns of decision making and behaviour that are more or less complex. However, further reduction of landslide risks – and even to an extent hazards – will depend critically on the abatement of vulnerability levels, which will require planning, funding and, above all, community participation in sustainable development processes.


## Acknowledgements

Parts of this chapter are derived from a report on landslide risk written for the Italian National Research Council (IRPI–Perugia) and Region of Umbria. The author thanks the IPRI and Dr Fausto Guzzetti for permission to reproduce this work in the present chapter.


## Appendix: Vulnerability Estimation

This appendix provides an example of a methodology for the estimation of vulnerability using data that can be collected by field survey or other methods. With respect to elements at risk, vulnerability can be considered either as susceptibility to damage in mass movements of given types or sizes, or in terms of value, which is expressed in any of three different ways:

*Monetary value*   The price or current value of the asset, or the cost of replacing it with a similar or identical asset if it were totally lost or written off.

*Intrinsic value*   The extent to which an asset (such as an ancient monument) is considered important and irreplaceable.

*Utilitarian value*   The usefulness of a given asset, or the monetary value of its usage averaged over a specified length of time.

Human life constitutes a special case in that its intrinsic value when threatened by a hazard such as landslides is incalculable. Despite this, several measures are used in actuarial work to put a monetary value on death or injury (Linnerooth, 1979). The first, the *value of a statistical life*, simply allots a standard figure, based on lost earnings, which is, in 1990s figures, averaged for the industrialized countries, about US$1.75 million for death, $10 000 for serious injury, and $1000 for minor injury. The second, termed the *private value of a statistical life*, is based on lost earnings, medical expenses and indirect costs. The third, known as the *social value of a statistical life*, includes the private value, plus foregone taxes and general medical, emergency, legal, court and public assistance administration costs. Foregone taxes are estimated by developing an age, sex and income profile of potential victims and calculating their future tax liabilities. On this basis, the average monetary values of a human life have been variously estimated at between US$873 000 and $7 million. The wide diversity reflects not only age, social status and earning capacity, but also the value of court awards when damages are sought.

As deaths and injuries are unlikely in most slow-landsliding situations, the following analysis will be restricted to the assessment of property losses and other material costs. Three methods of characterizing assets for estimation of vulnerability and risk are appropriate here (see Figure 5.7).

*Single asset method*   For each delimited hazard area the possible maximum value of landslide damage or losses to each asset that is present should be assessed. Figures are

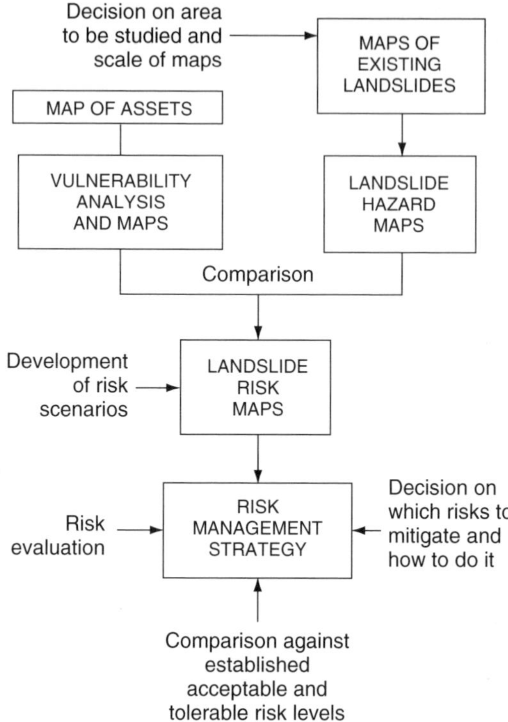



**Figure 5.7**    *Methods of characterizing assets for estimation of vulnerability and risk*

not averaged between assets. This is appropriate where assets show widely diverse vulnerability levels, especially where uses, functions and values differ substantially from one asset to another. However, it may produce an excessively detailed picture of vulnerability and possibly losses that are inflated by disaggregating them.

***Summed asset method***    For each delimited hazard area, the vulnerabilities of individual assets should be established. The data should be summed, averaged and weighted to round them to the nearest integer between 0 and 4, so as to represent the average vulnerability of assets in the given hazard area in terms of the classes described below.

***Generalized assets method***    A general level of vulnerability for all assets in a delimited hazard area should be heuristically estimated.

The assets are characterized as elements at risk and can be selected from the categories shown in Table 5.1.

For each asset, or spatial grouping of elements at risk, vulnerability classes can be assigned on the basis of a hypothesis about the degree of losses likely to be sustained when landsliding occurs (i.e. when the hazard manifests itself). As the nature of vulnerability differs between buildings and structures, human lives and socio-economic activities, these should be codified separately. It should be noted that the classes are unlikely to generate the same spreads of vulnerability. For example, high vulnerability to loss of life (V3–V4) is likely to be concentrated in few locations and only a few cases, while high vulnerability of buildings to damage will be much more common and widespread.

The classes are as follows:

*Buildings and structures*

V0 Total loss would not cause severe problems or represent a significant loss of utility or intrinsic value. This is only likely to be the case where assets are of no value (e.g. abandoned buildings).

V1 Total loss would cause minor problems or result in small and affordable costs. Damage would be easily repairable and no interruption in socio-economic activities would be necessary.

V2 Total loss would be moderately significant or result in moderate costs or hardship. Damage would be repairable, though it could be costly.

V3 Total loss would be very significant or costly relative to available resources or reserves. Damage would not necessarily be repairable, and if it were would be extremely expensive to remedy.

V4 Total loss would be extremely significant or very expensive (relative to available resources or reserves) and would be difficult or impossible to rectify.

*Human lives*

V0 No threat to life or safety exists.

V1 Casualties are highly unlikely and fatalities are virtually ruled out. If they were to occur, injuries would probably not be life-threatening.

V2 Casualties are unlikely, especially fatalities. If the worst happens, only small numbers of people are likely to be involved.

V3 Casualties might occur and death in a landslide is possible, though not likely. Non-fatal injuries could be serious.

V4 Casualties could well occur and in the worst cases might include high loss of life and very serious injuries.

*Socio-economic activities*

V0 No disruption to activities or associated losses are likely.

V1 Interruption of socio-economic activities would occur, if at all, to nuisance level only.

V2 Socio-economic activities could be disrupted significantly, with interruptions of some basic services for the duration of the emergency and costs incurred by rerouting, rescheduling or otherwise adapting other activities.

V3 Socio-economic activities would be seriously disrupted and some of the more important ones brought to a halt for the duration of the landslide emergency. High losses would probably be sustained.

V4 Socio-economic activities would be brought to a halt. Very high losses would be incurred as a result. Disruption could last for months or years.

To assess vulnerability using these scales, a means must be found to equate damage to buildings, people and activities. It is probably advisable to take the worst case from the distribution of each of the three scales, as this represents the maximum vulnerability (to structures, lives or services) at each site. The category should be reduced by one

**Table 5.4**  *Determination of vulnerability categories from proportion of losses expected (excluding vulnerability of human lives)*

| Loss 100% | 100% > loss > 50% | 50% > loss > 0% |
| --- | --- | --- |
| V0 | V0 | V0 |
| V1 | V1 | V0 |
| V2 | V1 | V0 |
| V3 | V2 | V1 |
| V4 | V3 | V2 |

grade (e.g. V4 to V3) if loss is partial but at least 50%, or by more if it is less that 50% (Table 5.4), unless the intrinsic value is still irremediably compromised by the loss, in which case the latter should be treated as total. This will be the case when loss of life is hypothesized.

Guzzetti and his colleagues (IRPI, 2000) reduced the definition of elements at risk to ten classes: high-density settlements, rural settlements, sports infrastructure, industrial areas, animal farming structures, quarries and landfills, main roads, secondary and local roads, railway lines, and cemeteries. In this context it is necessary to decide whether to lump together the single elements that go to make up these categories (e.g. consider entire settlements) or disaggregate them (i.e. consider settlements as groups of individual buildings and structures of varied vulnerability). The former is a quicker, cheaper method, but the latter is decidedly more accurate.

Determinations of vulnerability made by this method can be mapped and compared with similar determinations of hazard in order to determine and manage risk, using $H \times V = R$ or similar relationships. A flow chart for this process is given in Figure 5.8.

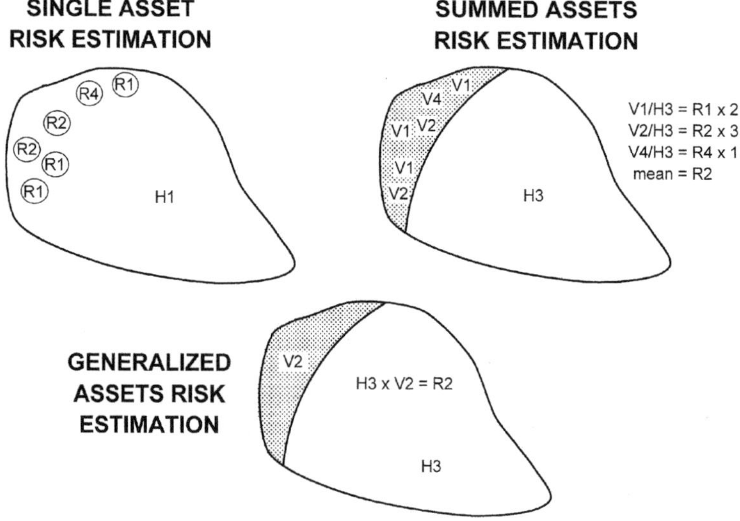


**Figure 5.8**  *Flow chart of steps in landslide estimation*

# References


Alexander, D.E., 1984, Building damage by landslide: the case of Ancona, Italy, *Ekistics*, **51**, 452–462.

Alexander, D.E., 1988, Valtellina landslide and flood emergency, northern Italy, 1987, *Disasters*, **12**, 212–222.

Alexander, D.E., 1989, Urban landslides, *Progress in Physical Geography*, **13**, 157–191.

Alexander, D.E., 1991, Natural disasters: a framework for research and teaching, *Disasters*, **15**, 209–226.

Alexander, D.E., 1993, *Natural Disasters* (London: Routledge and Boston: Kluwer Academic Publishers).

Alexander, D.E., 2000, *Confronting Catastrophe: New Perspectives on Natural Disaster* (Terra Publishing: Harpenden, UK and Oxford University Press: New York).

Angeli, M.G., Gasparetto, P., Menotti, R.M., Pasuto, A. and Silvano, S., 1994, A system of monitoring and warning in a complex landslide in northeastern Italy, *Landslide News*, **8**, 12–15.

Bankoff, G., 2001, Rendering the world unsafe: 'vulnerability' as Western discourse, *Disasters*, **25**, 19–35.

Blaikie, P., Cannon, T., Davis, I. and Wisner, B., 1994, *At Risk: Natural Hazards, People's Vulnerability and Disasters* (London: Routledge).

Boyce, J.K., 2000, Let them eat risk: wealth, rights and disaster vulnerability, *Disasters*, **24**, 254–261.

Browning, J.M., 1973, Catastrophic rock slide, Mount Huascaran, north-central Perú, May 31, 1970, *Bulletin of the American Association of Petroleum Geologists*, **57**, 1335–1341.

Cannon, T., 1994, Vulnerability analysis and the explanation of 'natural' disasters, in A. Varley (ed.), *Disasters, Development and Environment*. (Chichester: John Wiley & Sons Ltd), 13–30.

Carrara, A. and Guzzetti, F. (eds), 1995, *Geographical Information Systems in Assessing Natural Hazards* (Dordrecht: Kluwer).

Carrara, A., Cardinali, M. and Guzzetti, F., 1992, Uncertainty in assessing landslide hazard and risk, *ITC Journal*, **2**, 172–183.

Chan, N.W., 1998, Responding to landslide hazards in rapidly developing Malaysia: a case of economics versus environmental protection, *Disaster Prevention and Management*, **7**, 14–27.

Cooke, R.U., 1984, *Geomorphological Hazards in Los Angeles* (London: Unwin Hyman).

Costa, J.E. and Schuster, R.L., 1988, The formation and failure of natural dams, *Bulletin of the Geological Society of America*, **100**, 1054–1068.

Cotton, W.R. and Cochrane, D.A., 1982, Love Creek landslide disaster, January 5, 1982, Santa Cruz County, *California Geology*, **35**, 153–157.

Crosta, G., 1998, Regionalization of rainfall thresholds: an aid to landslide hazard evaluation, *Environmental Geology*, **35**, 131–145.

Cruden, D.M. and Varnes, D.J., 1996, Landslide types and processes, in *Landslides: Investigation and Mitigation*. Special Report 247, Transportation Research Board, National Research Council, Washington, DC, Chapter 3, 36 75.

Dai, F.C. and Lee, C.F., 2002, Landslide characteristics and slope instability modeling using GIS, Lantau Island, Hong Kong, *Geomorphology*, **42**, 213–228.

Dai, F.C., Lee, C.F. and Ngai, Y.Y., 2002, Landslide risk assessment and management: an overview, *Engineering Geology*, **64**, 65–87.

Enoki, E., Kokubu, A.A. and Ikeda, Y., 1999, Infiltration of rainwater and slope failure, in Griffiths, J.S., Stokes, M. and Thomas, R.G. (eds), *Landslides* (Rotterdam: Balkema).

Fleming, R.W. and Taylor, F.A., 1980, Estimating the cost of landslide damage in the United States, *US Geological Survey Circular*, **832**.

Franks, C.A.M., 1999, Characteristics of some rainfall-induced landslides on natural slopes, Lantau Island, Hong Kong, *Quarterly Journal of Engineering Geology*, **32**, 247–259.

Galadini, F., Colini, L., Giaccio, B., Messina, P., Salvi, S. and Sposato, A., 2003, Persisting effects of the Colfiorito (central Italy) Pleistocene paleo-landslide in the planning of land use: upper palaeolithic and proto-historical coexistence and antique-modern modifications, *Environmental Geology*, **43**, 621–634.

Glade, T., 2003, Landslide occurrence as a response to land use change: a review of evidence from New Zealand, *Catena*, **51**, 297–314.

Harp, E.L., Wilson, R.C. and Wieczorek, G.F., 1981, *Landslides from the February 4, 1976, Guatemala earthquake*, US Geological Survey Professional Paper 1204A.

Hewitt, K. (ed.), 1983, *Interpretations of Calamity: From the Viewpoint of Human Ecology* (London: George Allen and Unwin).

Hewitt, K., 1997, *Regions of Risk: A Geographical Introduction to Disasters* (Harlow, UK: Addison Wesley Longman).

ICRCRCS, 2001, *World Disasters Report 2001: Focus on Recovery.* International Federation of Red Cross and Red Crescent Societies, Geneva, Chapter 4.

IRPI, 2000, *L'acquisizione di nuove informazioni sui fenomeni franosi nella Regione dell'Umbria: Primo rapporto* (F Guzzetti, ed.). (Perugia, Italy: Istituto Regionale per la Protezione Idrogeologica in Italia Centrale).

Jiménez Díaz, V., 1992, Landslides in the squatter settlements of Caracas: towards a better understanding of causative factors, *Environment and Urbanization*, **4**, 80–89.

Jones, F.O., 1973, *Landslides of Rio de Janeiro and the Serra das Araras escarpment, Brazil*, US Geological Survey Professional Paper, 697.

Leighton, F.B., 1976, Urban landslides: targets for land-use planning in California, in D.R. Coates (ed.), *Urban Geomorphology*. Special Paper 174 (Boulder, CO: Geological Society of America), 37–60.

Linnerooth, J., 1979, The value of human life: a review of the models, *Economic Inquiry*, **17**, 52–74.

Lumb, P., 1975, Slope failures in Hong Kong, *Quarterly Journal of Engineering Geology*, **8**, 31–65.

Miller, S.M., 1988, A temporal model for landslide risk based on historical precipitation, *Mathematical Geology*, **20**, 529–542.

Morgenstern, N.R., 1997, Toward landslide risk assessment in practice, in D.M. Cruden and R. Fell (eds), *Landslide Risk Assessment* (Rotterdam: Balkema), 15–23.

Olshansky, R.B., 1990, *Landslide Hazard in the United States: Case Studies in Planning and Policy Development* (New York: Garland Publishing).

Pierson, T.C., 1992, Rainfall triggered lahars at Mount Pinatubo, Philippines, following the June 1991 eruption, *Landslide News*, **6**, 6–9.

Schuster, R.L. and Fleming, R.W., 1986, Economic losses and fatalities due to landslides, *Bulletin of the Association of Engineering Geologists*, **23**, 11–28.

Sheridan, M.F., Bonnard, C., Carreno, R., Siebe, C., Strauch, W., Navarro, M., Calero, J.C. and Trujilo, N.B., 1999, 30 October 1998 rock fall/avalanche and breakout flow of Casita Volcano, Nicaragua, triggered by Hurricane Mitch, *Landslide News*, **12**, 2–5.

So, C.L., 1971, Mass movements associated with the rainstorm of June 1966 in Hong Kong, *Transactions of the Institute of British Geographers*, **53**, 55–65.

Tobin, G.A. and Montz, B.E., 1997, *Natural Hazards: Explanation and Integration* (New York: Guilford).

UNDRO, 1982, *Natural Disasters and Vulnerability Analysis*, Office of the United Nations (Geneva: Disaster Relief Co-ordinator).

Veder, C., 1981, *Landslides and their Stabilization* (New York: Springer-Verlag).

Voight, B., 1990, The 1985 Nevado del Ruiz Volcano catastrophe: anatomy and retrospection, *Journal of Volcanology and Geothermal Research*, **42**, 151–188.

Wieczorek, G.F., 1984, Preparing a detailed landslide-inventory map for hazard evaluation and reduction, *Bulletin of the Association of Engineering Geologists*, **21**, 337–342.

Wieczorek, G.F., Lips, E.W. and Ellen, S.D., 1989, Debris flows and hyperconcentrated floods along the Wasatch Front, Utah, 1983 and 1984, *Bulletin of the Association of Engineering Geologists*, **26**, 191–208.

Wieczorek, G.F., Wilson, R.C., Mark, R.K., Keefer, D.K., Harp, E.L., Ellen, S.D., Brown, W.M., III and Rice, P., 1990, Landslide warning system in the San Francisco Bay Region, California, *Landslide News*, **4**, 5–8.

Wisner, B., 2001, Risk and the neoliberal state: why post-Mitch lessons didn't reduce El Salvador's earthquake losses, *Disasters*, **25**, 251–268.

Zhou, C.H., Lee, C.F., Li, J. and Xu, Z.W., 2002, On the spatial relationship between landslides and causative factors on Lantau Island, Hong Kong, *Geomorphology*, **43**, 197–207.

# PART 2
## EVALUATION OF RISK

# 6

# Landslide Risk Perception, Knowledge and Associated Risk Management: Case Studies and General Lessons from Glacier National Park, Montana, USA

*David R. Butler and Lisa M. DeChano*

## 6.1  Introduction

The history of attempts to reduce landslide risk is largely a history of management of landslide terrain, whether through engineering attempts at slope stabilization, construction of protective structures and/or monitoring and warning systems, or ever-increasingly sophisticated methods for mapping and delineating areas prone to landslides (see Dai *et al.*, 2002; Marchi *et al.*, 2002). Little attention has been paid to the *people* who live in or visit landslide-prone areas. Depressingly little is known about the knowledge base and experiences brought by residents or visitors into landslide terrain. Equally little is known about how residents and visitors perceive the dangers presented by landslide terrain. In this chapter, we examine the perceptions, experience and knowledge base of both permanent residents and visitors in a popular American National Park that is prone to a wide variety of hazardous landslide processes.

*Landslide Hazard and Risk*   Edited by Thomas Glade, Malcolm Anderson and Michael J. Crozier
© 2004 John Wiley & Sons, Ltd   ISBN: 0-471-48663-9

## 6.2   Background

The literature on landslide and mass movement risk perception is extremely limited. Studies of snow-avalanche hazard perception are probably most common, including those carried out by the senior author and described later in this chapter (Butler, 1987, 1997). Potentially large variations in avalanche frequency and magnitude, in addition to variations in human factors, produce variations in public perception of avalanche hazards (Simpson-Housley and Fitzharris, 1979). Simple warnings or dissemination of information on the nature of the problem do not necessarily lead to accurate awareness of the hazard (Butler, 1987). Non-site-specific avalanche warnings that merely state 'very high hazard throughout area' do little to improve public confidence in avalanche warnings (Heywood and Tufnell, 1985). Studies in Norway cite 'considerable psychological pressures' caused by avalanche threats, resulting in public overreaction, fear and anxiety (Ramsli, 1974), that is, mountain users reacted to the perceived avalanche threat rather than to the real avalanche danger. In a study of rockfalls in Wales, UK, Williams and Williams (1988) reported that past experience is useful in regard to people adopting adjustments to rockfall events. They also noted significant relationships between people's perception and their previous experience with rockfall hazards. Through an experimental examination of people's responses to a variety of hazard warning signs, they also determined the most effective designs for rockfall warning signs for their study area.

## 6.3   The Study Area

### 6.3.1   Location

Glacier National Park (GNP) encompasses approximately 0.4 million hectares, bisected by the Continental Divide, in northwestern Montana, USA (Figure 6.1). Together with the adjacent Waterton Lakes National Park in Alberta, Canada, it is a UN-designated World Heritage Site. Two major transportation corridors allow access around and through the Park: US Highway 2 (US 2), which borders GNP on the south and southwest; and Going-to-the-Sun Road (GSR), a narrow road providing the only trans-Park route. US 2 is open all year but subject to temporary winter closings due to snow avalanching (Butler and Malanson, 1985; Butler, 1987, 1997). GSR is open seasonally, usually early June through early October. Both roads allow visitors to drive through areas of steep cliffs, rock overhangs and rugged glacial terrain, providing the potential for interaction of visitors with a variety of forms of landslides. Additional roads that offer access into valleys used heavily by Park visitors, and which are also subject to a variety of landslide types, penetrate the Many Glacier/Swiftcurrent, and Two Medicine, areas (Figure 6.1).

### 6.3.2   Geological and Geomorphic Setting

In addition to the placement of the primary Park transportation corridors in steep terrain, with the concomitant exposure of residents and visitors to the potential for landslides, the area's geological structure and geomorphic history exacerbate the likelihood of landslides.

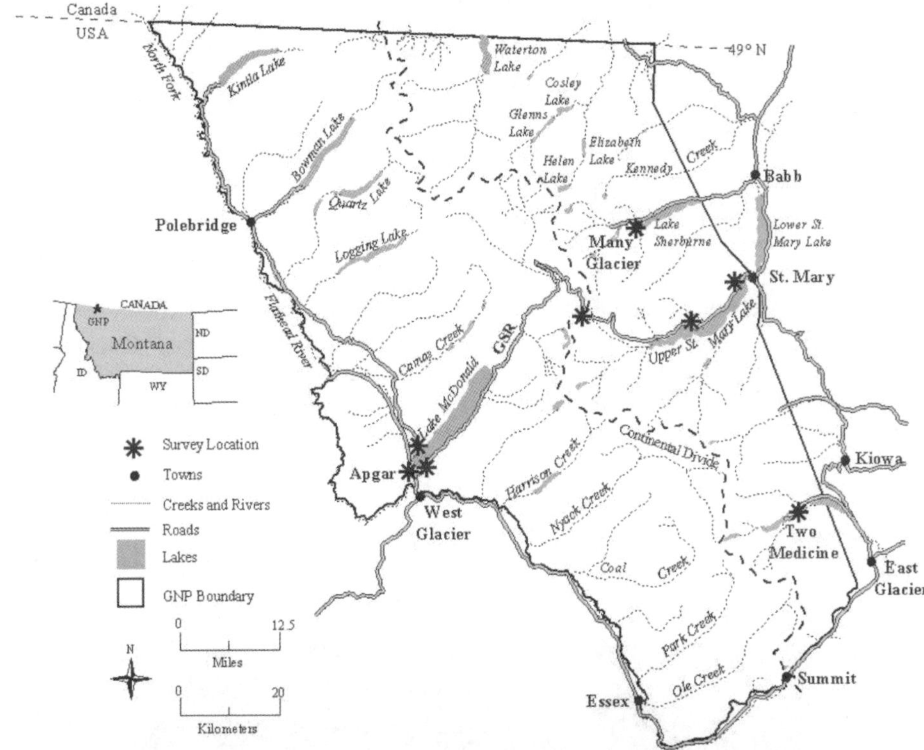


**Figure 6.1**  *Map of Glacier National Park, Montana, showing the survey locations and primary roads. GSR, Going-to-the-Sun Road*

The Lewis Overthrust, extending along the entire length of the Park near its eastern boundary, emplaced faulted and highly fractured Precambrian sedimentary rocks over folded and substantially weaker Cretaceous sedimentary units (Ross, 1959; Whipple, 1992), rendering the entire eastern section of GNP highly susceptible to landslide activity (Oelfke and Butler, 1985; Butler, *et al.*, 1986). Extensive Pleistocene glaciation produced spectacular U-shaped valleys throughout the Park, with slopes in many places in excess of 40°, further enhancing inherent slope instability.

### 6.3.3  Climate

The climate of GNP is generally continental, but the existence of the Continental Divide creates areas of maritime climate in the lowlands on the western side of the Park. Heavy precipitation during the late spring and summer months, associated either with frontal passage or with isolated convectional thunderstorms, acts as a trigger for landslides along the steep mountain slopes (Butler *et al.*, 1991; DeChano and Butler, 2001). Also during this time of year, temperatures tend to fluctuate frequently across the freezing point, which in conjunction with abundant moisture can result in freeze–thaw fracturing of the weak bedrock to provide additional material prone to landsliding (Butler, 1990).

### 6.3.4    Landslides of Glacier National Park

A wide variety of landslide types exists in GNP, ranging from slow-moving slumps in glacial till (not examined here, although they do damage Park roads and cause inconvenience to visitors as a result of necessary repair work) to potentially catastrophic, high-speed rockfall avalanches *sturzstroms*. Carrara (1990) and Dutton and Marrett (1997) provided maps of the Park that illustrate the geographical distribution of slow-moving, non-hazardous landslip areas east of the Lewis Overthrust Fault, but their maps lacked information on more potentially hazardous landslides to which Park visitors and residents may be exposed.

Hazardous landslides in GNP include rockfalls, rockfall avalanches, debris flows and snow avalanches. We do not examine snow avalanches and avalanche perception in detail here, as they have been described elsewhere (Butler and Malanson, 1985; Butler, 1987, 1997). Butler (1990) described several dozen historical incidents of rockfalls along Going-to-the-Sun Road, US 2, and the Many Glacier Road, including several that produced injuries and fatalities. Both Park visitors and employees have been injured by rockfalls. Rockfalls in the Park are triggered by freeze–thaw (Butler, 1990), as well as by frontal and convectional precipitation (Butler, 1990; DeChano, 2000). Park roads may be temporarily closed during times of high rockfall likelihood (DeChano, 2000), but the

***Figure 6.2*** *Aerial photograph of rockfall-avalanche deposits and their impounded lakes, Otatso Creek drainage, northeastern GNP. Larger lake was impounded in 1910, smaller lake in 1946 (Butler et al., 1986, 1991, 1998)*

primary response to the rockfall hazard along GSR is the simple posting of warning signs notifying motorists 'rockfall – next 12 miles'. No other information on rockfall likelihood or location is provided to Park visitors.

Rockfall avalanches (*sturzstrom*) have occurred in several locations along the Lewis Overthrust Fault during the twentieth century (Butler *et al.*, 1986, 1991, 1998); fortunately, no injuries have yet been recorded from these high-speed landslides, but their occurrence has caused temporary road and trail closures (Butler *et al.*, 1998). Rockfall avalanches have also impounded two potentially hazardous landslide-dammed lakes (in 1910 and 1946) in the Otatso Creek drainage in the northeastern corner of the Park (Butler *et al.*, 1986, 1991; Butler and Malanson, 1993) (Figure 6.2).

Debris flows in GNP are widespread phenomena (Butler and Walsh, 1994; Walsh and Butler, 1997), with literally thousands of debris-flow deposits distributed throughout the Park. Debris flows produce frequent threats to visitors and Park employees along GSR (DeChano, 2000). During the evening of 28 July 1998, a strong frontal storm brought drenching rains to GNP, and numerous debris flows were triggered throughout the Park. Three flows crossed GSR (Figure 6.3), trapping cars between the flows and exposing their occupants to the possibility of further flows. National Park Service personnel quickly rescued the trapped vehicle occupants, but it took over 24 hours for the debris-flow deposits (in excessive of 20–40 tons of sediment per deposit; DeChano and Butler, 2001) to be

**Figure 6.3** *Typical fresh debris-flow deposits emplaced overtop talus cones in GNP, Siyeh Creek drainage*

cleared from the road, thus temporarily shutting down GSR during the height of the tourist season.

In terms of number of hazardous encounters and fatalities, it is useful to compare results from studies of rockfall (Butler, 1990) and general 'landslides' (DeChano and Butler, 2001) with human/grizzly bear encounters (DeChano and Butler, 2002) in GNP. A total of 76 'landslide events' have affected Park roads as reported in Park literature and incidence reports, with no fatalities (DeChano and Butler, 2001). Rockfalls injure roughly one person per year in the Park, numerous cars and trucks have been heavily damaged, and in 1962 a visitor was killed in her car when a massive slab of rock demolished the car (Butler, 1990). By comparison, in the period 1946–88, grizzly bears (*Ursus arctos horribilis*) mauled 65 Park visitors who survived, and killed an additional 10 people (DeChano and Butler, 2002). These data are difficult to compare, however, in that virtually all the grizzly bear maulings and fatalities occurred either on backcountry trails or in campsites, where visitors are concentrated in prime grizzly habitat; whereas landslide and rockfall injuries and property damage are largely restricted to narrow road corridors.

## 6.4 Previous Hazard Perception Research in Glacier National Park

Except for the authors' past research, there have been no studies of geomorphic hazard perception in GNP. This past work is briefly described here.

### 6.4.1 Snow Avalanche Hazard Perception

In the mid-1980s, Butler (1987) distributed a postal survey to 167 residents of the US 2 region along the GNP southern boundary. He received back 58 usable completed questionnaires, a respectable 34% return rate. This questionnaire enquired as to individuals' driving habits in the hazardous US 2 region during times of high avalanche likelihood (frequency of trips, time of day or night of trips, etc.), whether their driving habits altered during such times, and from whence had they garnered their information about the likelihood of increased avalanche activity. The survey revealed that even long-time residents of the area did not alter driving habits during times of high avalanche danger, nor were they effectively gathering information about avalanche conditions in the area from the local avalanche warning system.

In February of 1996, after another temporary closure of US 2 and the local railroad due to avalanche deposition throughout the area, Butler (1997) conducted a similar survey, but in this case did so on site in the US 2 region, less than two weeks after the widespread avalanching and closures (Figures 6.4a, b. Approximately 60 individuals were directly interviewed, from which 38 full-time or seasonal residents were willing to provide answers to the survey. Personal interviews were conducted by the senior author and a graduate student. These 38 individuals provided a more representative cross-section of the community (in terms of age, educational background and gender) than had resulted from the postal survey of the 1980s. The individuals interviewed in 1996 illustrated in general a greater knowledge of, and respect for, the snow-avalanche

**Figure 6.4a, b**  *Snow-avalanche deposits at same location along US Highway 2 (in foreground), in 1996 (above) and 2002 (below). Snowploughs have removed snow from the surface of the highway. Senior author for scale in both photos*

hazard in the region, but were still shockingly unaware of impending danger at any given time. Furthermore, very little improvement in reaching the public via the local avalanche warning system had occurred (the local avalanche warning system is comprised of twice-weekly radio recordings of snow conditions and avalanche hazard level. These recordings are also posted on the Internet, an additional dissemination method unavailable in the 1980s). As with the responses to the 1980s postal survey, local 'word of mouth' was still the primary method by which residents derived their information about avalanche conditions in the area. Although snow-avalanche perception is not discussed in further detail here, this background information is useful for comparison with that described below gathered for rockfall and landslide events, and because the snow-avalanche questionnaire used by Butler (1987, 1997) served as a basis for the subsequent questionnaire employed by DeChano (2000) and described in detail also in DeChano and Butler (2001). This subsequent questionnaire is reproduced here in Appendix 1.

### 6.4.2   Rockfall Hazard Perception

In spite of the numerous rockfall near misses, injuries and fatalities from rockfalls in GNP documented by Butler (1990), no information on the distribution or likelihood of rockfall is presented to visitors to the Park. Although tour bus drivers pass through several zones of high rockfall hazard (often driving more than 10 000 km [>6000 miles] per summer), they are not provided information about the likelihood of rockfall occurrence, where rockfalls occur, nor what to do in the possible event of rockfall blockage of the roads. From this absence of information provided to tour bus drivers, Butler (1990) concluded that it was extremely unlikely that visitors had adequate knowledge of rockfall hazard distribution and past occurrence in the Park. Continued rockfall hazard disruption to traffic in the area since that paper was published (DeChano, 2000) illustrates an ongoing lack of adequate information dissemination to the visiting public.

### 6.4.3   Debris-flow Hazard Perception

Earlier we briefly described the debris flows of July 1998 that blocked and temporarily closed GSR, trapping several cars and drivers between flow deposits (Figures 6.5a, 6.5b and 6.5c). Immediately before this debris-flow event, DeChano had directly interviewed 71 park visitors via the use of questionnaires at a nearby visitor centre on GSR as to their knowledge of the occurrence and distribution of past landslides along the road. Those results illustrated that visitors were essentially completely unaware of the landslide and debris-flow history of GSR. Two days after the debris flows, 56 additional visitors were willing to be interviewed and to fill out questionnaires at the same site. Although this timing resulted in a pool of interviewees different from those previously encountered, we assumed that the recent debris-flow events and road closure would have been noticed by the visiting public (DeChano and Butler, 2001). Such was not the case, however; our results showed no significant changes in public perception of danger to self from landslides (including debris flows), or in the perceived locations of where landslides may occur. We also interviewed several NPS entrance station personnel, and those individuals were providing no information to visitors about the recent road closures or debris-flow events.

(a)

***Figure 6.5*** *(a) View downslope to GSR from debris-flow deposit, emplaced during the rainstorm of July 1998; (b) View upslope from GSR to debris-flow deposit, emplaced during the rainstorm of July 1998; (c) View of GSR where it was blocked by a multi-ton debris-flow deposit, emplaced during the rainstorm of July 1998*

## 6.5  Experiences with, and Knowledge of, Landslides in Glacier National Park

Our past research and work experiences in GNP have illustrated the sad lack of knowledge that many visitors possess concerning the location and likelihood of landslide occurrences there. However, our admittedly anecdotal evidence of lack of knowledge of rockfalls, and the fact that respondents from the pre- and post-debris-flow surveys were surveyed at only

(b)

(c)

***Figure 6.5***    *(Continued)*

one site in GNP, motivated us to survey a much broader cross-section of Park visitors, from sites throughout the Park (Figure 6.1). We also surveyed permanent and seasonal GNP National Park Service (NPS) employees to ascertain whether they possessed a greater perception and knowledge base of, and experience with, landslides than did the visiting public. Although we initially included a survey of seasonal, non-Park Service employees, we determined that the variability in the training that those individuals received before taking up their employment positions in the Park was too broad to allow for meaningful comparisons with the standardized training received by NPS employees, or with the total lack of information provided to Park visitors. The non-NPS employees are not, therefore, discussed further here.

## 6.6   Mapping of Historical Distribution of Hazardous Mass Movements

In order to assess whether Park employees and visitors have accurate perceptions of where mass movements have occurred that posed threats to them, that is, along roads and popular trails, such occurrences needed to be mapped. Our data sources included NPS incident reports, filed when a visitor or Park employee was involved with a mass movement such that economic loss, injury or death occurred; files and photographs of the GNP archives, located in the Park headquarters complex in West Glacier; and the *Hungry Horse News*, a weekly newspaper published in nearby Columbia Falls, Montana, that provides coverage of Park news and events of note. Every issue of the newspaper, from its inception in 1946 through 1999, was scanned for information about landslide events. When encountered, the type of landslide (if available) and data on its time and location (as accurately as possible; however, in some cases phrases such as 'on the west side of GSR' or 'sometime last evening' precluded accurate mapping or specific data categorization) were recorded. From these data sources, a total of 40 rockfall events (25 west of the Continental Divide, and 15 east of the divide) were recorded and mapped, with the greatest number occurring along GSR. Twenty-eight unspecified 'landslides' were recorded, with 14 west, and 12 east, of the divide, and two where the location could not be pinpointed. Our experience with the 1998 debris flows, and examination of photographs of many of the 'landslides' described or depicted in our data sources, lead us to believe that the vast majority of these unspecified events would be more accurately categorized as debris flows.

## 6.7   NPS Employees' Survey and Results

A total of 28 full-time and 22 seasonal NPS employees were surveyed (Appendix 1) at a variety of locations throughout GNP. They answered questions in categories including demographics (age, gender, educational background, etc.), past experiences with landslides in GNP, perception of where landslides occur in the Park, and from where (if anywhere) they gathered information about current landslide conditions in the Park before venturing into it (see Appendix 1). They were also presented with a list of possible hazards affecting the Park, ranging from hazards that actually do occur there to several that are rare or non-existent (such as hurricanes), and asked to rank the level of concern engendered by those hazards using a standard 5-point Likert scale (where 1 = very serious concern, 3 = moderately serious, and 5 = not of any concern or consequence). Hazards listed as possibly affecting Park employees included: landslides, grizzly bears, strong damaging winds, rockfalls, floods, wildfires, snow avalanches, earthquakes, hail, tornadoes and hurricanes.

NPS employees surveyed were well educated (43 of the 50 total respondents had some form of college or university education), in early middle age (mean age for permanent employees, 43; for seasonal employees, 38). Over one-half of those surveyed (29 of 50) were women. Over 75% of the respondents stated that they enquired about hazardous conditions before venturing into the Park, although their primary source of information about such conditions was other NPS employees. Radio reports and newspapers were the other major sources of information used by NPS employees. Thirteen employees reported having 'direct experience' with landslides (type unspecified), and more than half the total number of employees surveyed (30) reported direct experience with rockfalls.

Those hazards that presented the highest level of concern to NPS employees varied slightly between permanent and seasonal employees. Permanent employees targeted grizzly bears, landslides and snow avalanches as the most serious hazards (in descending order, with mean Likert scores of 2.3, 2.4 and 2.5, respectively), whereas seasonal NPS employees identified strong damaging winds of most concern, with snow avalanches, rockfalls and grizzly bears tied for second (mean Likert scores of 2.2 for damaging winds, and 2.5 for the other three hazards). Using a T-test, we found no significant differences between male and female NPS employees in their self-concern perceptions. We also found no significant differences among the employees resulting from age differences.

Given their high degree of experience with past rockfall and landslide events, one could expect that NPS employees possess fairly accurate perceptions as to where hazardous landslides and rockfalls occur in GNP. Employees were asked to map locations where they believed there exists a significant rockfall or a significant landslide hazard in the Park. The maps they generated were compared with actual maps created from the historical database described above; the specific methodology for this comparison is available in DeChano (2000). NPS employees had precise perceptions as to the locations of rockfalls along GSR and the road into the Many Glacier region, where the most actual hazardous rockfalls have occurred; they were less precise about actual rockfall hazard zones along US 2. Park employees also had accurate perceptions as to the geographical locations of landslides along GSR, the Many Glacier road, and the US 2 region (unlike their knowledge of rockfalls there).

## 6.8   Visitors Survey and Results

The same survey (Appendix 1) given to NPS employees was given to a total of 454 Park visitors, at Park Service Visitor Centers and campgrounds throughout the Park (Figure 6.1). Visitor Centers provide the most diverse demographic cross-section of Park visitors, ranging over all ages, education levels and incomes, whereas campgrounds provide data gathered more generally from middle-aged and younger segments of the population. The questionnaire was employed at the Apgar and Logan Pass (twice, before and after the debris flows of 28 July 1998) Visitor Centers, and at the Swiftcurrent, St Mary, Rising Sun, Apgar, Fish Creek and Two Medicine campgrounds.

The mean age of visitor respondents was 41 years, similar to that of NPS employees, but with a distinctly older modal age of 50. Gender of respondents was nearly equal, 51% male and 49% female. Like the NPS employees in general, the visitors sampled were well educated, with all but 15% of respondents possessing some form of post-high-school education. Over 64% of those responding noted that this was their first trip to GNP, suggesting that nearly two-thirds of the visiting public may be assumed to have little experience of the Park.

Only 18 visitors (4% of the sample total) stated that they had had experience with rockfalls in GNP, and only nine visitors (2%) had had experience with landslides. Of those nine, however, six listed the 28 July 1998 debris flows as their event of experience. Only slightly over one-half of the total sample group (54%) enquired about hazardous conditions in the Park. The degree of enquiry was essentially similar parkwide, ranging from 50% at St Mary campground to 61% at the Fish Creek, Rising Sun and Two Medicine campgrounds. The top two sources of information concerning hazardous conditions in the Park were GNP literature, typically provided upon entry to the Park at the West Glacier, St Mary, Many Glacier and Two Medicine entrance stations (and also available at visitor centres at Apgar, Logan Pass and St Mary); and the Internet. Individual sample sites followed this general trend throughout the Park, except at Swiftcurrent/Many Glacier, where books and the Internet were the top two choices for such information.

Grizzly bears topped the list of hazards identified by visitors as causing them self-concern (a mean of 2.5 on the 5-point Likert scale). This finding is not surprising, given the NPS efforts in publicizing the negative effects of human–bear encounters, and the literature concerning such interactions that every visitor is given upon entry into the Park. However, as stated earlier, virtually all grizzly bear encounters, maulings and fatalities are restricted to campsites and backcountry trails. Grizzly bears do *not* cause injury or death to people at visitor centres or along Park roads! Rockfalls and wildfires were tied (Likert scale mean of 2.7) for second among hazards causing self-concern, and landslides was fourth (but at only a mild level of concern of 3.0). It is notable, however, that the average level of concern of visitors (2.7 and 3.0) for rockfalls and landslides was significantly less than that of NPS employees described above, illustrating in general what we believe is an unrealistic under-appreciation for the possibility of such events (especially in light of such events as 28 July 1998).

Visitors' perceptions as to where rockfalls and landslides might occur in the Park were relatively similar to those of Park employees, with steep slopes along GSR identified as prone to both. Perhaps more notable was the identification of virtually any steep slope or peak as prone to landslide and rockfall occurrence, illustrative of an over-generalized view of where such processes might occur. By identifying basically every steep slope and peak in the Park as prone to such hazards, visitors in effect become 'numb' to the increased likelihood of mass movement occurrence in particularly hazardous areas such as those along the upper segments of GSR.

We employed T-tests to further examine our data on the basis of key demographic variables. No significant differences emerged among different categories of age, occupation, or education for rockfalls, landslides and snow avalanches. On the basis of gender, significant differences in perceived threat to self from landslides and rockfalls emerged, but not for snow avalanches. This result is not surprising given that our survey was carried out during mid-summer, and that summer visitors have no experience with snow avalanches, unlike NPS employees, who view snow avalanches as a strong secondary hazard and are present in the Park during avalanche season. In the case of both rockfalls and landslides, females perceived a greater self-concern than did their male counterparts. These results are consistent with other published studies that have investigated risk perception differences between genders (see Butler, 1987; Westmoreland, 1995; Riechard and Peterson, 1998).

## 6.9  Concluding Remarks

Few differences or inaccuracies associated with hazard awareness or perception were found to exist among GNP employees, suggesting that the NPS does a good job in educating those individuals as to the nature of the hazardous environment in which they are employed. In contrast, visitors view virtually the entire Park landscape as potentially prone to rockfall or landslides, yet do not rank either hazard as more significant than the threat of attack by grizzly bears. This finding emphasizes two aspects of the NPS literature provided to Park visitors – whereas it does a thorough (and perhaps excessive) job of warning about the dangers presented to visitors by bear–human contacts, it does a very poor job of explaining the nature of mass movement hazards that may be faced. The complete lack of information presented to the public immediately following the potentially disastrous debris flows of 28 July 1998 (See Figures 6.5a, 6.5b and 6.5c) clearly illustrates this fact, as does the absence of information on landslides and rockfalls in the NPS literature provided to visitors.

We should note that the GNP Visitor Center website provides warnings to visitors in the early summer season about the possibility of snow avalanches and rockfalls. Also recall that the Internet was the second most utilized source by visitors for information on hazardous conditions in the Park. Opportunities exist for expanding public awareness of rockfall and landslide hazards in GNP via the Internet, and the NPS should continue its policy of providing this information. It must also be noted, however, that many people still do not use the Internet, so do not receive NPS information on hazardous conditions. We also have no information on *which* websites were being viewed by Park visitors. There are plenty of websites that provide information about GNP, many of which are operated by commercial firms that offer lodging and services in the area in and around the Park. It is in their best financial interests to underplay or obfuscate the likelihood of a hazardous encounter of any sort during a visitor's stay in GNP, and thus these websites cannot be relied upon for the transfer of accurate hazard information to visitors. Until such time as all Park visitors use the GNP Visitor Center website before entering the Park, the NPS should provide written literature to all Park visitors on the nature of mass movements hazards there. It would take only a small amount of effort to incorporate this information into currently existing pamphlets and newsletters provided at entrance stations to all Park visitors, and it would greatly enhance the visiting public's awareness of mass movement hazards in Glacier National Park.

No monitoring or measuring devices currently exist in the study area for the detection of mass movement hazards (landslide, debris flow, rockfall, snow avalanche). When debris flows or other hazards occur above Park highways and roads, it is sheer chance as to whether they will affect the road, and cause property damage or harm to people. GNP has been relatively lucky in this regard, but should not hope that such good fortune will continue. Sometime in the near future, Park employees or visitors may be seriously harmed or killed by debris flows, snow avalanches, rockfalls, or landslides along Park roads. When, not if, this occurs, perhaps then there will be sufficient public outcry to force the National Park Service to recognize the hazardous nature of road travel in GNP, and to install monitoring and detection devices along those portions of the road with the highest potential mass movement hazard. Until that happens, road travel in GNP will continue to be a high-risk undertaking, and a high-risk undertaking by visitors who are totally unaware of the dangers from mass movements to which they are being exposed.


## Acknowledgements

Funding for risk perception data collection in Glacier National Park came from a Quick Response Grant to DRB from the Natural Hazards Research and Applications Information Center of the University of Colorado (1996), and from a Research Enhancement Grant to DRB from Texas State University. The cooperation of NPS personnel in GNP is gratefully acknowledged and appreciated. This chapter is a contribution from the Mountain GeoDynamics Research Group.


## References

Butler, D.R., 1987, Snow-avalanche hazards, southern Glacier National Park, Montana: the nature of local knowledge and individual responses, *Disasters*, **11**(3), 214–220.

Butler, D.R., 1990, The geography of rockfall hazards in Glacier National Park, Montana, *The Geographical Bulletin*, **32**(2), 81–88.

Butler, D.R., 1997, A major snow-avalanche episode in northwest Montana, February, 1996. Quick Response Report 100, Natural Hazards Research and Applications Information Center, University of Colorado at Boulder. Available at http://www.colorado.edu/hazards/qr/qr100.html, last accessed 29 October 2003.

Butler, D.R. and Malanson, G.P., 1985, A history of high-magnitude snow avalanches, southern Glacier National Park, Montana, U.S.A., *Mountain Research and Development*, **5**(2), 175–182.

Butler, D.R. and Malanson, G.P., 1993, Characteristics of two landslide-dammed lakes in a glaciated alpine environment, *Limnology and Oceanography*, **38**(2), 441–445.

Butler, D.R. and Walsh, S.J., 1994, Site characteristics of debris flows and their relationship to alpine treeline, *Physical Geography*, **15**(2), 181–199.

Butler, D.R., Oelfke, J.G. and Oelfke, L.A., 1986, Historic rockfall avalanches, northeastern Glacier National Park, Montana, U.S.A., *Mountain Research and Development*, **6**(3), 261–271.

Butler, D.R., Malanson, G.P. and Oelfke, J.G., 1991, Potential catastrophic flooding from landslide-dammed lakes, Glacier National Park, Montana, USA, *Zeitschrift für Geomorphologie, Supplementband*, **83**, 195–209.

Butler, D.R., Malanson, G.P., Wilkerson, F.D. and Schmid, G.L., 1998, Late Holocene sturzstroms in Glacier National Park, Montana, U.S.A., in J. Kalvoda and C. Rosenfeld (eds), *Geomorphological Hazards in High Mountain Areas*, GeoJournal Library (Dordrecht: Kluwer Academic Publishers), 149–166.

Carrara, P.E., 1990, Surficial Geologic Map of Glacier National Park, Montana, US Geological Survey Miscellaneous Investigations, Map I-1508D, Washington, DC.

Dai, F.C., Lee, C.F. and Ngai, Y.Y., 2002, Landslide risk assessment and management: an overview, *Engineering Geology*, **64**(1), 65–87.

DeChano, L.M., 2000, 'Geohazard Perception in Glacier National Park, Montana, USA', unpublished Doctoral Dissertation, Department of Geography, Southwest Texas State University, San Marcos, TX.

DeChano, L.M. and Butler, D.R., 2001, Analysis of public perception of debris flow hazard, *Disaster Prevention and Management*, **10**(4), 261–269.

DeChano, L.M. and Butler, D.R., 2002, An analysis of attacks by grizzly bears (*Ursus arctos horribilis*) in Glacier National Park, Montana, *The Geographical Bulletin*, **44**(1), 30–41.

Dutton, B.L. and Marrett, D.J., 1997, *Soils of Glacier National Park East of the Continental Divide*, Land & Water Consulting Inc., Missoula, MT.

Heywood, D.I. and Tufnell, L., 1985, Snow avalanche hazards in the Glen Nevis and Glen Coe areas of Scotland, *Disasters*, **9**(1), 51–56.

Marchi, L., Arattano, M. and Deganutti, A.M., 2002, Ten years of debris-flow monitoring in the Moscardo Torrent (Italian Alps), *Geomorphology*, **46**(1–2), 1–17.

Oelfke, J.G. and Butler, D.R., 1985, Landslides along the Lewis Overthrust Fault, Glacier National Park, Montana, *The Geographical Bulletin*, **27**, 7–15.

Ramsli, G., 1974, Avalanche problems in Norway, in G.F. White (ed.), *Natural Hazards – Local, National, Global* (Oxford: Oxford University Press), 175–180.

Riechard, D.E. and Peterson, S.J., 1998, Perception of environmental risk related to gender, community socioeconomic setting, age, and locus of control, *The Journal of Environmental Education*, **30**(1), 11–19.

Ross, C.P., 1959, *Geology of Glacier National Park and the Flathead Region, Northwestern Montana*, US Geological Survey Professional Paper 296, Washington, DC.

Simpson-Housley, P. and Fitzharris, B.B., 1979, Perception of the avalanche hazard, in I.F. Owens and C.L. O'Loughlin (eds), *Snow Avalanches – A Review with Special References to New Zealand*, New Zealand Mountain Safety Council, 74–81.

Walsh, S.J. and Butler, D.R., 1997, Morphometric and multispectral image analysis of debris flows for natural hazard assessment, *Geocarto International*, **12**(1), 59–70.

Westmoreland, G., 1995, Perception of Risk from Environmental Hazards: Relationships among Nation, Gender, and Locus-of-Control, unpublished Ph.D. dissertation, Emory University, Atlanta, GA.

Whipple, J.W., 1992, Geologic Map of Glacier National Park, Montana, US Geological Survey Miscellaneous Investigations Series Map I-1508-F, Washington, DC.

Williams, M.J. and Williams, A.T., 1988, The perception of, and adjustment to rockfall hazards along the Glamorgan Heritage Coast, Wales, United Kingdom, *Ocean and Shoreline Management*, **11**, 319–339.

## Appendix 1   Survey Used in Obtaining Respondent Data

**Natural Hazard Perception in Glacier National Park, Montana**

1. City, State/Province, County of residence: _____________________

2. ( ) Female ( ) Male

3. Age: _____________________

4. Education:
   ( ) Primary School              ( ) Some high school
   ( ) High School Diploma         ( ) Bachelors Degree
   ( ) Some College                ( ) Associate/Vocational Degree
   ( ) Masters Degree              ( ) Ph.D.
   ( ) Other, specify _____________________

5. Occupation:
   ( ) Education (teacher, educational administrator, college professor, etc.)
   ( ) Professional (doctor, pharmacist, engineer, CEO, lawyer, etc.)
   ( ) General/Technical (construction, driver, cook, etc.)
   ( ) National Park Service Employee          ( ) Homemaker
   ( ) Self-employed                           ( ) Student
   ( ) Retired                                 ( ) Unemployed
   ( ) Other, specify _____________________

6. What natural hazards do you perceive to exist in Glacier National Park, if any? How
   serious do you perceive each of them to be to you?

| | Very serious | | Moderately serious | | Not of any consequence | No opinion |
|---|---|---|---|---|---|---|
| Snow avalanches | 1 | 2 | 3 | 4 | 5 | A |
| Grizzly bears | 1 | 2 | 3 | 4 | 5 | A |
| Rockfalls | 1 | 2 | 3 | 4 | 5 | A |
| Tornadoes | 1 | 2 | 3 | 4 | 5 | A |
| Floods | 1 | 2 | 3 | 4 | 5 | A |
| Hail | 1 | 2 | 3 | 4 | 5 | A |
| Strong damaging winds | 1 | 2 | 3 | 4 | 5 | A |
| Wildfires | 1 | 2 | 3 | 4 | 5 | A |
| Hurricanes | 1 | 2 | 3 | 4 | 5 | A |
| Landslides | 1 | 2 | 3 | 4 | 5 | A |
| Earthquakes | 1 | 2 | 3 | 4 | 5 | A |
| Other, specify | 1 | 2 | 3 | 4 | 5 | A |

7. Each of the following 5 pages has a map of Glacier National Park numbered and labelled corresponding to the list below. On each map please draw a boundary around the area where you believe that specified hazard exists in Glacier National Park.

   1. Rockfalls       4. Snow avalanches
   2. Landslides      5. Strong damaging winds
   3. Floods

# 7
# Cultural Consideration in Landslide Risk Perception

*Garth Harmsworth and Bill Raynor*

## 7.1  Introduction

Some cultural groups have a unique dependence on or intimate relationship with their natural environment, distinctive from other cultural groups, which often results in cultural differences and perspectives. This difference in the way people understand, interpret, perceive, assess and manage risk within a contemporary environment is further attributed to a cultural world-view, commonly derived from combinations of traditional beliefs and values, knowledge, custom, religion, social structure, land tenure, length of time in coexistence with or occupation of a particular geographical location, and historical and modern experiences. Landslide risk perception is also based on human interaction and dependence on the natural environment, and can be heightened by a close interdependence. To explain these differences in a cultural context, and the cultural issues that arise in risk assessment, we must first examine the cultural characteristics of a group that distinguish them from other groups. Understanding cultural differences and group dynamics is fundamental for understanding differences in risk perception, and can help guide risk assessment and risk mitigation, and determine future actions. This provides another dimension to consider in risk assessment and management, requiring in-depth understanding of cultural values, learning and human behaviour. All cultures differ in some way in their understanding, perception, reaction, coping mechanisms and solutions to risk. Culture and impacts on culture are also significant determinants for understanding behaviours and causal factors that lead to risk.

In this chapter we illustrate the importance of cultural consideration in landslide risk assessment by selecting two indigenous group case studies, one from

---

*Landslide Hazard and Risk*   Edited by Thomas Glade, Malcolm Anderson and Michael J. Crozier
© 2004 John Wiley & Sons, Ltd   ISBN: 0-471-48663-9

Aotearoa–New Zealand and one from Pohnpei, Micronesia. For many indigenous groups that have a long association and intimate relationship with natural resources in one area, risk perception often evolves through long-term cumulative impact, usually a combination of experience and effects on human and social values. The following case studies demonstrate that sustainable management solutions need to be holistic, and that risk reduction measures can be improved by understanding cultural traits and differences, including local authority, values, kinship relationships, environmental relationships, land tenure and changes to or evolution of culture.

In both case studies, indigenous beliefs and values resonate strongly within a contemporary world, and for many indigenous cultures there is an increasing tension between the modern and the traditional. The challenge for many cultural groups, particularly those that are religious or indigenous based, is to try to blend traditional perspectives and values with more modern-day perspectives to cope with the pressures and stresses of westernization and modern living. This tension usually has its origins in colonization and, more recently, westernization, and many indigenous cultures and communities have had to adjust within a relatively short time to major social and cultural upheaval and change. This has commonly resulted in the undermining of cultural values, practice, custom, principles and lore affecting the communities' relationship with the natural environment. Around the world, groups have adjusted to change in different ways: some have become marginalized from the larger general population, some have changed markedly from their original culture in order to adapt, while others have readjusted by finding a balance between the traditional and the modern.

Risk perception, assessment and management, therefore, result from a combination of and balance between the traditional and the contemporary provided in these two case studies. In Aotearoa–New Zealand many indigenous Maori struggle in the general population, which is reflected in socio-economic and constitutional disparities, ongoing distrust and conflict, and a general lack of understanding by the large non-Maori population of indigenous cultural identity, historical grievance and cultural values. Maori have undergone huge social and cultural changes that have undermined every part of their society. Three general stages are evident: an early colonial period in the nineteenth and early twentieth century with much conflict between Maori and non-Maori, marginalization and oppression of Maori, along with rapid transformation of the natural environment; the mid- to late twentieth century, with growing economic and social dependence by Maori on the state; the late twentieth to early twenty-first century, with further assimilation into the general population, a cultural renaissance within the context of westernization and globalization, an increasing move to self-suffiency and sustainable development within a free market economy, and with increasing development pressures on the natural and cultural environment. In all these stages, Maori culture has suffered detrimentally and significantly, but through resilience, action and a strong belief in cultural identity, values and custom, the culture has remained strong and is vitally important to most Maori. In Pohnpei, communities are vulnerable, undergoing large and rapid social transformation, and struggling to cope with transition from traditional ways of life, based on traditional values and practices, to more western values and practices, which are accompanied by a breakdown in tribal and leadership structure. The Pohnpeian society has moved from being traditionally based on subsistence agriculture and fishing, through being an economy dependent on external funding, such as that from the USA, to a promoted self-sufficiency based on a cash economy, particularly cash agriculture in highland areas that is greatly increasing the landslide risk on the island. As Pohnpei shifts to a modern independent society, its traditional values and social structure are undergoing much change, with increasing pressure on the local community to generate economic wealth from an environmentally fragile island.

Landslide risk is a complex global issue. Management solutions to reducing risk require an integrated approach that encourages inter-agency and community cooperation, improved learning and uptake of knowledge, and an increasing need to understand the role of human behaviour and human activity in landscape modification. To understand how cultural groups can be active participants in assessing and developing solutions for landslide risk, it is important to incorporate cultural understanding into all environmental planning and policy, and find solutions that use a holistic approach to achieve cultural, social, economic and environmental goals. It is hoped that the case studies that follow will help advance our understanding and thinking of managed landscapes and highlight the need for cultural consideration in landslide risk.

## 7.2 Case study: Aotearoa–New Zealand

### 7.2.1 Introduction

People from northern Polynesia migrated to Aotearoa–New Zealand (Figure 7.1) about 1000 years ago. It was in this new country that Maori culture developed and flourished, drawing on the early Polynesian cultural beliefs, customs, language and philosophies. At present, the indigenous Maori make up around 15% of the total population of 4 million in a largely homogeneous multicultural society. This society is very different from when Europeans first colonized New Zealand in the early nineteenth century: then there were two distinct and separate cultures, one Maori, one English. Maori culture since the arrival of the European has gone from being strong and vibrant, through a long period of being at risk from the pressures of colonization, to a new-found Maori cultural renaissance that has progressively grown from the latter half of the twentieth century to the present. Very early traditional beliefs, values and cultural perspectives still resonate strongly in this contemporary world, and have taken on new importance through a resurgence of interest in cultural identity, cultural philosophies and preservation of indigenous language. Furthermore, an increasing worldwide shift to sustaining natural resources, reducing environmental impacts, integrating social, economic, and environmental planning and policy, and protecting biodiversity has often mirrored traditional, indigenous environmental philosophies.

Much of the recognition of indigenous rights in New Zealand is based on, and can be attributed to, the signing of the Treaty of Waitangi in 1840. This document – in two versions, one English and one Maori – provided a basis for bicultural development and partnership. However, that the two language versions do not entirely agree has resulted in arduous debate and interpretation. The Treaty has become a baseline document for most legislation in New Zealand and, as such, most laws and statutes highlight the responsibility and obligation to include a cultural component or approach within all economic, social and environmental planning and policy. Cultural consideration addressing a wide range of issues and initiatives is therefore set in law (Durie, 1998).

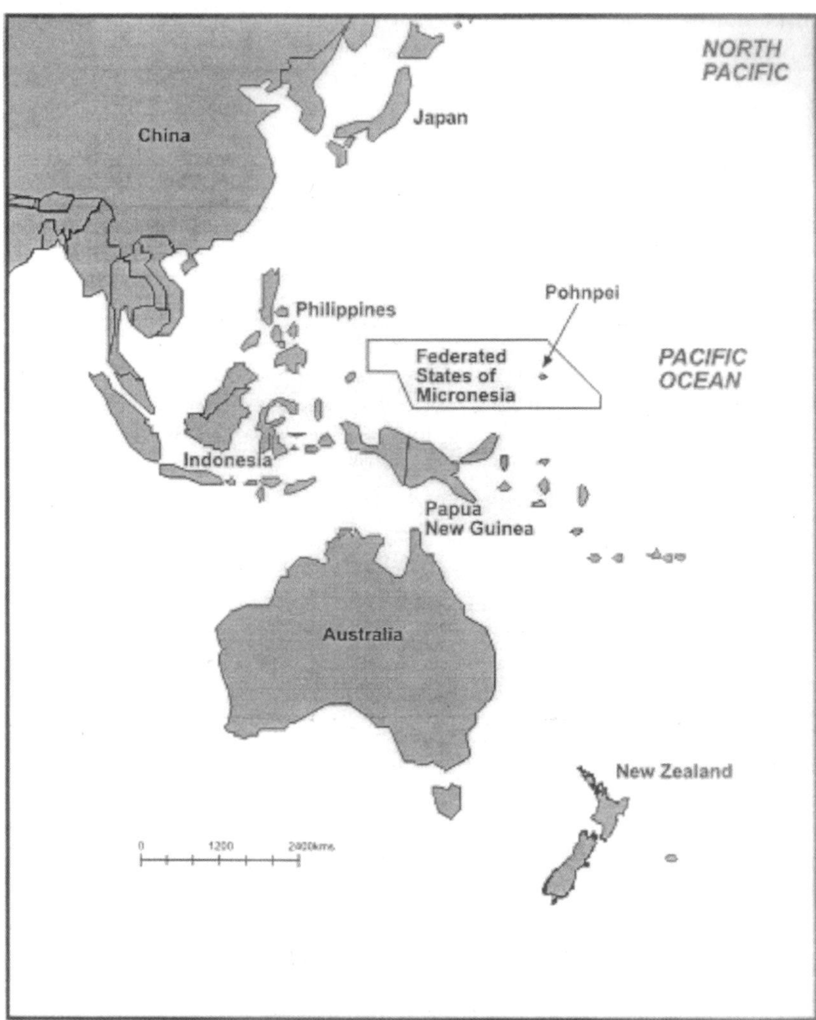


**Figure 7.1**   *Location map of New Zealand and Pohnpei, Micronesia*

### 7.2.2   Maori Beliefs

Maori beliefs, customs and values are derived from a mixture of cosmogony, cosmology, mythology, religion and anthropology. Within this complex and evolutionary belief system are the stories of the origin of the universe and of Maori people, the sources of knowledge and wisdom that have fashioned the concepts and relationship Maori have with the environment today.

From a Maori perspective, the origin of the universe and the world can be traced through a series of ordered genealogical webs that go back hundreds of generations to the beginning. This genealogical sequence is referred to as *whakapapa*, and places Maori in an environmental context with all other flora and fauna and natural resources as part of a hierarchical genetic assemblage with identifiable and established bonds.

*Whakapapa* follows a sequence beginning with the nothingness, the void, the darkness, a supreme god, emerging light, through to the creation of the tangible world, the creation of two primeval parents, *Ranginui* – the sky father, and *Papatuanuku* – the earth mother, the birth of their children, such as the forest, the sea, the rivers, the animals, through to the creation of mankind. The two primeval parents, once inseparable, had many children, often termed departmental *atua* or Maori gods, each with supernatural powers. In a plan carried out by the children to create light and flourish, the parents were prised apart. The separation of the parents led to *Ranginui* forming the sky, resulting in the rain as he continued to weep for his separated wife *Papatuanuku*, and *Papatuanuku* forming the land providing the sustained nourishment for all her children. As part of this ancestry, a large number of responsibilities and obligations were conferred on Maori to sustain and maintain the well-being of people, communities and natural resources.

### 7.2.3  Maori Values

Maori values are instruments through which Maori make sense of, experience and interpret their environment (Marsden, 1988). They form the basis for explaining the Maori world-view, and provide the concepts, principles and lore Maori use in everyday life to varying degrees, affecting their interaction with others, and their responsibilities and relationship with the environment. *Tikanga* denotes the Maori body of rules and values used to govern or shape people's behaviour. Some important Maori values include: *Tino Rangatiratanga* and *Mana Motuhake* – self-determination, independence or interdependence; *Mana Whenua* – rights of self-governance, rights to authority over traditional tribal land and resources; *Whanaugatanga* – family connections and family relationships; *Kaitiakitanga* – guardianship of the environment; *Manaakitanga* – reciprocal and unqualified acts of giving, caring and hospitality; *Arohatanga* – the notion of care, respect, love and compassion; *Awhinatanga* – to assist for or care for; *Whakakoha* – the act of giving; *Whakapono* – trust, honesty, integrity; *Whakakotahitanga* – respect for individual differences and participatory inclusion for decision making; *Wairua* – the spiritual dimension to life.

Maori values can be represented in many forms: they may be represented in the environment as places of significance, or sites; they form the basis for recognizing Maori treasures (*taonga*), or significant biodiversity and environmental concerns; they may be represented in language; they may be represented through people or organizations in terms of relationships; and they may be the intrinsic cultural basis for controlling or modifying human behaviour.

### 7.2.4  Maori Social Structure

Before colonization, Maori lived together in small, geographically distinct groups and settlements as part of a larger hierarchical tribal structure. They often identified themselves and their ancestors (*tupuna*) with landmarks and physiographic features in their settlement area. Today, Maori social structure is highly fragmented, with more complex subgroupings, and distinct areas of urban and rural Maori. However, they are very proud of their heritage and continue to affiliate with hierarchical groups such as *iwi*, *hapu* and *whanau* based on *whakapapa* (Figure 7.2). The largest socio-political group is the *iwi*, a distinct

**Figure 7.2**  *The communal gathering place for many contemporary Maori is the* marae, *which includes the meeting house* (Whare Whakairo). *Location: Waiapu catchment (photo Tui Aroha Warmenhoven)*

tribe or nation belonging to a large geographical region or district. The next sub-level is the sub-tribe, termed *hapu*, confined to a smaller geographical area, for example, around coastal areas, bays, rivers and mountains. *Hapu* are made up of *whanau* or extended families. The basic unit of Maori society is the *whanau*, which includes an extended family of parents, brothers, sisters, cousins, uncles, aunts, grandparents, grandchildren, and siblings once, twice or three times removed. Traditionally, *whanau* was the residential unit with designated areas of land where every individual had a right to share resources equally. Today, the *hapu* or *iwi* are the main groupings involved in pooling resources for health and economic service delivery, education, economic development, and planning and policy for environmental and resource management. The *whanau* provides the basic unit for decision making, administering specific blocks of land, and utilizing specific natural and human resources. Maori also have customary rights for using and managing natural resources within distinct tribal areas.

Traditionally, Maori beliefs about *tikanga* gave rise to a communal or team-based society where Maori lived and worked together, made decisions together based on the common good, worked together towards common goals (which further reinforced the importance of community), cared for each other, developed land together, sustained and managed natural resources collectively, and adapted to change. These are still very important concepts within Maori society, although colonization and western laws and economics have placed huge stresses on Maori social structure, which has greatly affected and altered Maori collectivism, resource ownership and abilities for decision making.

## 7.2.5   Maori Land Tenure

Traditionally, Maori had a well-developed communal land tenure system, where resources were shared within the tribe. Today, they have only a fraction of the land and natural resources to which they once had rights or title and live in a more fragmented, modern economic society. In 1840 most land in New Zealand was under Maori control and ownership; that figure currently stands at about 6% of New Zealand's total land base (Durie, 1998). However, Maori have indigenous rights over all areas where they have a recognized cultural tribal relationship, so are included in decision making for lands, water and sea under private or Crown ownership. Where land is under Maori ownership, it comes under a Maori Land Act that focuses on retaining Maori land, acknowledging rights of *whakapapa*, allowing multiple ownership, and facilitating and promoting effective use, management and development. Most Maori-owned land in New Zealand is held under different types of Maori trust to protect and encourage owners' rights and to promote participatory decision making.

## 7.2.6   Maori Environmental Perspectives

Maori, connected to the natural environment through *whakapapa*, perceive themselves therefore as an integral part of nature, and expect to relate to it in a responsible and meaningful way. Traditionally, they relied totally on the environment for food, medicines, implements, shelter, clothing and identity. In a modern society, the way Maori source and access food and other natural resources from the natural environment, and interact with that environment, has changed substantially over the last 150 years, but the bond with the natural environment remains. Many traditional practices are still continued, albeit to a much lesser extent, and Maori environmental perspectives are very much based on the traditional Maori belief and values system, and on action and association.

The notion of how central land was, and still is to Maori, captured by Asher and Naulls (1987): 'To the early Maori, land was everything. Bound up with it was survival, politics, myth, and religion. It was not part of life but life itself.' Taking culture in its widest context, there was no part of early Maori culture that was not touched by the physical environment, and land in particular.

Maori concepts for environmental management and the perception of risk are still very much based on traditional beliefs, values and philosophies, such as:

- *Whakapapa* – genealogical descent, ancestral lineage;
- Maori knowledge including *matauranga* (traditional knowledge);
- *Mauri* – denoting health and spirit, a sustaining life force, an essential essence of being, an energy or element that permeates all living things;
- *Ritenga* – the area of customs, protocols, laws that regulate actions and behaviours related to the physical environment and people. Includes concepts such as *tapu*, *rahui* and *noa*, which were practical rules to sustain the well-being of people, communities and natural resources. Everything was balanced between regulated and deregulated states where *tapu* was sacred, *rahui* was restricted and *noa* was relaxed access or unrestricted.
- *Mana* – a sense of prestige and authority.

Traditional Maori values from *tupuna* or ancestral Maori emphasize the importance of everything, where everything has a *whakapapa* showing connection, and where *mauri*, the sustaining life force or spirit, is based on *whakapapa*. All plants, animals and water have *mauri*, cultural heritage sites, bodies of water and mountains are often sacred or *tapu*, and all natural resources are precious and referred to as *taonga*. Natural and cultural resources and cultural sites are interconnected, and the environment is seen holistically, where concepts of cumulative effects, biodiversity and integrated catchment management fit easily with Maori environmental philosophies.

In terms of cultural perception of environmental risk, therefore, Maori base their perception of risk on the total environment, not on a portion of it, and seek to understand the relationship between different parts of the environment, including the part people play. Damage or contamination to the environment is therefore damage to or loss of *mauri*. Maori view the health of the environment in terms of environmental and cultural changes through time, of what they remember as children, what an area was like, what it is like now, what it was like when their ancestors were alive, what natural and cultural resources were once plentiful and in good condition and what has been irretrievably lost or degraded. Perceptions are also based on the relationship between people and the environment, and the attitudes and beliefs of different groups of Maori people, particularly at the community level. Important concepts like *mauri* and *whakapapa* permeate the whole environment.

In respect to landsliding, therefore, Maori look at the whole picture (i.e. holistically) in terms of cumulative effects by describing that what happens on steep hillslopes will inevitably have effects downstream, and will ultimately affect the coastal and marine environment. An increased perception of landslide risk is also based on memory and experience, such as living through a severe storm event, and being aware of damage from erosion, flooding and sedimentation. All damage impacts on people's lives and well-being: they are affected by the degree to which their livelihoods are disrupted, or the modification or destruction of culturally significant resources (e.g. fishing grounds – *mahinga mataitai*, and food resource areas – *mahinga kai*).

In examples around New Zealand, such as on the East Coast of the North Island, Maori demonstrate acute awareness of the relationship between the wholesale forest clearance about 100 years ago (Harmsworth *et al.*, 2002) and the subsequent increased risk of landsliding, gully erosion and flooding. They are equally aware of the link between high-intensity cyclonic rainstorms and landsliding that has greatly affected their lives and cultural resources.

### 7.2.7   Erosion in New Zealand

About 60% of the New Zealand land area is classified as hill or mountainous terrain, with slopes above 22°. New Zealand is also vulnerable to regular high-intensity rainstorms, sometimes localized, sometimes widespread (Glade, 1996, 1998). With rapid deforestation and the transformation from indigenous forest to grassland under pastoralism, particularly between 1860 and 1930, New Zealand's fragile landscape became increasingly prone and at risk to landslide erosion (Figure 7.3). Today, landslide erosion is evident on about 7.7 million ha, about 30% of New Zealand's total land area (Eyles, 1983, 1985), with most landsliding predominating on slopes above 28°. Since free market economic reforms in the 1980s, some of the more marginal or unproductive hill country has been allowed to revert to scrub, but the landslide risk on large areas of hill and mountainous country still remains very high across about 30% of the total New Zealand land area

**Figure 7.3** *The New Zealand landscape is susceptible to widespread landslide erosion, as seen during Cyclone Bola 1988. Devils Elbow, northern Hawkes Bay looking west (photo reproduced by permission of Noel Trustrum)*

(Glade, 1996, 1998). Some of this risk may be further heightened by the onset of climatic warming, with predictions of more frequent high-intensity rainstorms (Harmsworth and Page, 1991).

Maori knowledge together with paleo-environmental information indicate an awareness of the New Zealand landscape from about 1000 years ago when Polynesians first arrived and New Zealand's indigenous forest was estimated to cover 86% of the country, through to 1770 when 51% of the total land area was covered, and on to 1840 when 50% of the land area was under indigenous forest cover, with an additional 50% in alpine, scrub, fern, swamp and grassland. At present about 23% of New Zealand is in indigenous forest (Wards, 1976; McGlone, 1983, 1989; Newsome, 1987; Ministry for the Environment, 1997; Landcare Research (2001) GIS tables). With successive reductions in indigenous forest and scrub area there has been an associated increased erosion risk through time and a reduction in indigenous biodiversity and environmental health. Before 1800, Polynesian burning was instrumental in changing large areas of indigenous forest to lowland scrub and fernland, as fernland was a prized Maori food resource that represented a period of shifting cultivation by Maori. However, the change up to 1840 of about 35% of the total indigenous forest area to scrub and fernland occurred over a period of about 400 years. Based on today's knowledge of erosion under scrub and fernland, the landslide risk, although increased, was probably still relatively small compared with what occurred as a result of conversion to grassland following European settlement in 1840. This assumption is further supported by historic records such as sediment production studies by Trustrum *et al.* (1999) and Page and Trustrum (1997), which show a ten-fold or 1000%

increase in sediment production on a representative East Coast area of the North Island, with the conversion of indigenous scrub, fernland and forest to grassland. Many New Zealand studies show that landslide risk was greatly heightened by an order of magnitude after European settlement, following wholesale forest and scrub clearance that was part of widespread changes to pastoralism. Although Maori were players in this large-scale forest clearance in the late nineteenth and early twentieth centuries, it was under the auspices of a European, not a Maori, vision for New Zealand.

Extensive pastoralism now covers 44% of New Zealand, while intensive pastoralism covers 7% of the total land area (Ministry for the Environment, 1997). This transformation of the landscape from native forest to grassland within a period of only 100 to 130 years resulted in a dramatic increase in landslide risk and increased sediment production. Grassland now represents about 51% of the New Zealand land area, much being on hill country prone to landslide.

Maori have therefore been affected in numerous ways both directly and indirectly by areas with high landslide risk. Removal of indigenous forest and scrub has led to widespread loss of cultural resources, such as flora and fauna, with around 70% of all indigenous native forest being destroyed since Polynesians first arrived in New Zealand. Following land clearance for pasture, increased landslide and gully erosion have often been translated off site through environmental systems in terms of greater downstream flooding risk, increased sediment production, decreasing water quality, loss of habitat, and sediment deposition in coastal and marine environments. Although landslides have their origins in hill and mountainous terrain, the off-site impacts from debris, sediment deposition and flooding are often huge in downriver and stream areas such as the floodplain, and on low terraces near the floodplain (Figure 7.4). Many of the areas previously heavily populated by Maori were on the floodplains, which occupy about 9% of New Zealand's total land area, with the floodplain and extensive coastline (approx. 14 500 km) including some of the most significant cultural food source areas. Today these areas represent some of the most highly productive agricultural land in New Zealand and are still of high cultural value, although many areas have become depleted of resources, or degraded or polluted close to urban environments. The increased landslide risk due to deforestation and changes to pastoralism has undoubtedly had a great impact on the lives of Maori through erosion, flooding, sediment generation, impacts on coastal environments and destruction and modification of indigenous forest and scrub. Many culturally significant areas, and the flora and faunal species that inhabited them, have been greatly modified or lost forever.

### 7.2.8 Landslide Risk Perception

Contemporary Maori perceive landslide risk within a wider context of social, cultural, environmental and economic issues. A large number of social and economic disparities have existed, and still exist, between Maori and non-Maori in New Zealand (Te Puni Kokiri, 2000; Statistics NZ, 1996). Although Maori are often referred to as 'culturally rich', a large part of the Maori population are socially and economically 'poor' in contrast to their non-Maori counterparts, and issues such as health, housing, employment, education, household incomes, land ownership and crime are shown in more negative statistics (Te Puni Kokiri, 2000; Statistics NZ, 1996) and often take precedence over environmental issues. These issues and disparities obviously affect the way Maori see

**Figure 7.4**  *During Cyclone Bola 1988, landsliding caused widespread damage to lowland and floodplain areas. Tologa Bay, looking west (photo reproduced by permission of Noel Trustrum)*

their lives, their families, their environment and their perceptions, and are pivotal in constructing the modern Maori world-view and the priorities and issues they either struggle with or take advantage of.

However, where landslide erosion is an issue, Maori understand the effects as having both direct and indirect ramifications for their lives and well-being, for their cultural resources, and for their relationship with the environment.

### 7.2.9   Example: East Coast, North Island

Maori population represents 42% of the total regional population in the Gisborne–East Coast of New Zealand. The region has many social and economic issues, and includes some of the lowest Maori household incomes in New Zealand (Statistics NZ, 1996). It also comprises relatively large areas of Maori-owned land, mainly multiple owned, predominantly in pastoral farming, forestry and large tracts of undeveloped land covered in shrubland (scrub) and indigenous forest. Many Maori in the district are pastoral farmers, mainly beef and sheep, and grow maize and other fodder crops on and adjacent to floodplains. About 70% of farms are located in hill country but most Maori live near or on the floodplains.

The East Coast of the North Island has always experienced periodic destructive storm and flood events (Page *et al.*, 2001a, b), as well as other natural phenomena such as earthquakes and erosion. However, with the advent of pastoral farming between 1880 and 1920, following clearance of large areas of erosion-prone land from indigenous forest, scrub and fernland to pasture, the erosion risk was greatly heightened; perhaps a hundred-fold in many areas. Maori have had to adjust to this rapid transformation of their landscape, not only losing large tracts of cultural resource such as native forest comprising culturally significant flora and fauna, but also losing communication and roading networks, suffering damage to housing and productive farmland, and then slowly adapting to an environment with a greatly heightened erosion and flooding risk. Many Maori on the East Coast remember people drowned in rivers and streams, particularly during floods, heightening their awareness of erosion and floods. A number of interviews carried out as part of research in the Waiapu catchment from 1998 to 2002 (Harmsworth *et al.*, 2002) indicated that many people believed the occurrence and intensity of floods and landslide erosion became progressively worse from 1970 onwards. This belief is most probably a result of experiences of the much-publicized 1988 storm event, Cyclone Bola, and of a number of other high-intensity damaging rain and cyclonic storms at the end of the twentieth century. One response (Harmsworth *et al.*, 2002) was: 'My thoughts go back about the Waiapu. In that time there was only one big flood, 1938. That is the flood I remember in that time. . . . That was the only flood, not like the floods of today where there is a flood every year . . .'

Most Maori living on the East Coast have a very good recollection of Cyclone Bola. It is a relatively recent storm event that provides a strong basis for contemporary thinking about landslide risk. However, a few of the more elderly Maori also remember preceding storm and flood damaging events in the region, such as the 1938 storm that caused extensive landsliding and flooding, and others in 1916 and 1918 that severely damaged bridges in the region. 'The storm of 1938 unleashed the biggest problem to farming for years to come. The bush had been felled off the hills long enough for the root systems to have rotted and there was nothing to hold those steep hills together.' 'The amount of slipping and movement in the hills and the consequent build up of debris in the river beds had to be seen to be believed' (Rau, 1993). Some recovery of scars under pasture occurred over six to eight years (Rau, 1993), and most farmers believed a storm of such magnitude would not affect the region for another 100 years (Rau, 1993).

Cyclone Bola was widespread, affecting more than 20 000 km$^2$ across New Zealand, with estimated total damage of US$72 million (Glade, 1998). The storm had a major impact on the East Coast region of the North Island, with rainfall between 500 and 900 mm over four days (Page *et al.*, 1999), the main damaging erosion types being landslide and gully. These were the highest rainfall figures since records began in the region in 1876 (Rau, 1993).

The cyclone had a devastating effect on the lives of people living on the East Coast, Maori and non-Maori alike, and it was seen as the worst storm in living memory. It was fortunate that no lives were lost, but the storm has had long-term repercussions, both on the local and regional economy and on the social, cultural and psychological well-being of people who live in the region. Harmsworth *et al.* (2002) recorded a number of excerpts, as part of Maori research in the Waiapu catchment, that highlight the impact Cyclone Bola had on people's lives.

> The catastrophic reality of Cyclone Bola was an event that etched its memory into the hearts and minds of all Ngati Porou residents and landowners. The storm struck with such force that both the landscape and people were never to fully recover.

The face of the Waiapu had changed forever and its whole community could only stand aghast in absolute disbelief and actual fear as the whole scene unfolded and the clouds and mists cleared to reveal what had actually happened. (Rau, 1993)

The tribe that had always had a proud history of habitation for hundreds and hundreds of years . . . . And now it was laid bare by a cyclone . . . that crashed its way down the hillsides . . . created new rivers, new river beds, new paths, Destroyed things in its path . . .

When Cyclone Bola hit us, that's when I saw the fury of the Waiapu river. Anaru, my son, and I drove down to the Waiapu. We drove up from the bottom of the road. And there were huge waves coming towards the Waiapu. When we saw this 100-foot-long pine tree, there was one time when I thought the bridge was going to go under, in terms that the pine tree hit the bridge pylon straight on. And there was this huge thing – it was like an explosion. The whole pine tree splattered into millions and millions of matchsticks. Just like a bomb that exploded, such as was the force of the river. I wept . . .

The damage was unbelievable! All the streams and rivers had risen to heights never before imagined, carrying millions upon millions of tonnes of silt and debris, that slid off the hills, down to lie metres thick of silt and rubble on every piece of flat in their path. (Rau, 1993)

Excerpts about the need to replant trees to stop landslide erosion include: we need to start and replant our whenua [land], because the health of the river depends on the health of the mountains, hills, its tributaries. And if it is taken away from us, what hope do we have?

I went from a farmer with a steady job to someone with nothing, not even a house to live in, all because of Cyclone Bola. My house burnt down because of that. The excess rain, the ceiling sagged and the wires split . . . electrical wire split and, the whole house went up. And then I built a shack with a dirt floor. I was living in a tin shack with a dirt floor after Bola.

Well, there's too much of those big slips lying down in the creeks. It will take a long time before it all levels out. After Bola, the slips up there just come down and block the creeks, and it's going to take a long time before it gets washed away. It just keeps rising. You can hear the stuff rolling down when it floods. If you went up the road back here, you'd just see the roads falling away . . .

When they cut down the trees and then the flood came and the silt and wood came down the river, the hills slipped away. They came down and were left all over things . . .

It is when you see that the earth is slipping away, that is when you worry! You know when it is slipping, you can hear the earth crashing into the water . . .

When the big storm came – Bola – this catchment (Maraehara) was the only place that wasn't hit by erosion . . . because we had all our trees on the land . . . that is why it didn't slip . . .

Very few people could recall exactly what the Waiapu catchment was like back in the 1930s but this interview excerpt indicates the great changes that must have taken place:

No! No . . . Back then the river was not as wide. I'm talking about the 1930s, the late 30s and 1940s, the time of the War, World War II, it was narrow . . . There were no floods in that time, things were good . . .

It also indicates that people remember flood and storm events in different ways, given that a large storm and flood affected the catchment in 1938.

Many Maori interviewed in the Waiapu research Harmsworth *et al.*, 2002) also believed the environment was paying people back for showing lack of respect through actions such as widespread removal of forest cover:

> it's the whole process of that clearing, it's just gone gradually year in and year out. Like I said, it's just all take and no give – well, something's got to give! There's no bush cover, it's silt – it rises and chokes the springs and that... The river is always seen as a living entity just like the plants or human life... The river is our taonga [significant cultural treasure], and our life essence. Land erosion reflects how we are becoming as people. We are losing our mana [status and relationship with the land]. The river is eating away at our land. Without this land we are nothing.

### 7.2.10   Managing Landslide Erosion

Reduction of landslide risk in the East Coast has required both regional and central government initiatives (NWASCO, 1970). Since the 1960s a number of schemes have been adopted to promote sustainable land use and retire erosion-prone land, with widespread planting of exotic forest (mainly *Pinus radiata*) seen as one of the solutions along with extensive soil conservation plantings such as spaced tree planting on large tracts of pastoral land. The erosion problem on the East Coast continues to be significant and ongoing (Jessen *et al.*, 1999). The East Coast forestry programme (East Coast Project, 1978) became operative in the 1980s to target and subsidize exotic forest plantings on marginal land. Efforts to reduce the East Coast erosion risk continue, with local government playing a major coordinating, regulatory and educational role, and planning and implementing sustainable land management. Maori also have a major role in sustainable land management on the East Coast but often feel helpless and isolated because of factors such as the magnitude of the problem; the lack of resources; a lack of human capacity, such as skills to engage or participate fully in planning and policy; adjustment to a western paradigm for solutions; difficulties in coordinating large numbers of landowners; and difficulties in developing partnerships and participatory projects between Maori and the Crown (Government). Many Maori tribes in New Zealand have prepared or are preparing environmental management plans, based on Maori values and *whakapapa*, to articulate Maori aspirations, identify environmental problems and issues, and plan practical solutions for themselves and others (Harmsworth, 1995, 1997, 1998). These plans are often essential cultural platforms to address and prioritize cultural and environmental issues but are not always recognized or adequately resourced by local and central government, and some documents are too generalized to address specific issues. Much of this work between Maori and non-Maori continues to lack a participatory commitment by key players and an acknowledgement of Maori values, and seldom offers a culturally based set of actions.

The East Coast environment, along with many other parts of New Zealand, continues to have a significant landslide and flooding risk, where landscapes are still evolving and readjusting to the earlier wholesale clearance of indigenous forest and scrub. Many areas continue to be vulnerable to erosion and storms even after significant forest plantings, soil conservation and scrub reversion. Maori have had to adjust to this new, modified physical environment in many ways, while remembering that the environment is so different from the one their ancestors walked and associated with. Their relationship with this changed environment, however, is still culturally based and as strong as ever.

More recent storm events and the effects of these on their lives have changed Maori perception and heightened awareness of erosion, landslide and flooding risk within whole catchments, particularly those susceptible to erosion with limited protective vegetative cover. This heightened awareness and understanding of cumulative effects at a catchment scale has led to advocacy for participatory planning from the headwaters to the sea.

## 7.3   Case Study: Pohnpei, Micronesia

### 7.3.1   Introduction

Pohnpei Island is one of the four states of the Federated States of Micronesia (Figure 7.1). Population in 1996 was about 34 000, with about 25% living in the single urban centre of Kolonia and its environs. Although the third largest island in Micronesia, Pohnpei is small – 355 km$^2$. The centre of the island is mountainous and forested, with the highest peak, *Ngihneni* (Spirit's Tooth), rising to 795 m a.s.l. Pohnpei, along with nearby atolls Ahnd and Pakin, is the remnant of a giant shield volcano that began its growth about 10 million years ago. Although sporadic eruptions continued to as recently as 1 million years ago, the forces of subsidence and erosion have substantially diminished the island's earlier size (Spengler *et al.*, 1992). The climate is humid tropical with annual rainfall averaging 3090 mm on the coast and up to 9000 mm in the interior (Spengler, 1990). The island lies at the eastern edge of the typhoon belt, and thus damaging storms are rarer than on other Micronesian islands located to the west. Because of the wet climate, the island is dissected by numerous streams and rivers.

Archaeological evidence suggests that the first humans arrived on Pohnpei more than 2500 years ago (Haun, 1984). Before their arrival, the entire island and basaltic islets of the lagoon were covered with rainforest (Glassman, 1952). In coastal areas and coastal valleys vegetation was extensively modified by humans, mainly through the historic conversion to traditional agroforestry, which has maintained a forest cover. Increasingly through the twentieth century species composition has been altered in favour of plants with a social or economic value, and coastal vegetation now is primarily agroforest or secondary forest, with areas of grassland. The shoreline remains fringed by thick mangrove forests, and an offshore barrier reef forms a lagoon. Inland, mountain slopes are still covered with dense rainforest, mainly characterized by broadleaf forest dominated by *Campnosperma brevipetiolata*, *Elaeocarpus* spp., and other tree species. Pure stands of the native palm *Clinostigma ponapensis* are found on upper elevation ridges. Swamp forest also occurs in small patches in the upland forest, signified by the presence of the endemic ivory nut palm, *Metroxylon amicarum*. The relative age and isolation of the island make the flora of Pohnpei's upland forests some of the most diverse in Micronesia, with high levels of endemicity. A combination of strong traditional respect for the forest and heavy human depopulation due to introduced diseases during the twentieth century has, at least until recently, spared the upland interior forests from much of the disturbance and destruction that has occurred on other Pacific islands.

### 7.3.2   Traditional Beliefs and Values

As with Maori, Pohnpei's legends and beliefs tie the people strongly to their land. The island's cultural history begins with the arrival of *Sapkini*, a master canoe builder from a land to the south called 'Eihr'. According to Pohnpei's creation legend, when *Sapkini* arrived at the island's location, he found only a tiny bit of coral rubble on a reef. With supernatural assistance and seven further canoe loads of materials, the island was created on the reef. As the island was being created, the waves and rain kept knocking down what the people built. In response, the various gods helped by becoming the reef, mangrove forests and upland forest that finally held the island together and secured the humans' work. This legend explains the origin of the island's name Pohnpei, which literally means 'upon (*pohn*) a stone altar (*pei*)', and demonstrates the basic Pohnpeian belief in the sacredness of nature.

Pohnpeian's rich history and deep social, political and spiritual connection with the land are also evident through the very high density of place-names on the island. Hanlon (1988) reports 'every bit of Pohnpei is named, and each name bears a history'. The names connect the people of modern Pohnpei with the creation of the island. Naming is the concrete act that unites collective environmental experience with history (Dahl, 1993). However, knowledge of the names of hills, streams, channels and so on remains esoteric and closely guarded, a result of attempts to safeguard the power inherent in these names. Residence and lineage are intertwined, thus place and social definition of self are highly integrated. In turn, the political hierarchy is recognized through its ability to intensify agricultural production and direct its output through tribute and redistribution. Agricultural production and redistribution, the central theme of both traditional and modern Pohnpeian culture, is firmly based on the potential of the land to produce.

### 7.3.3   Social Structure

With the first canoes that arrived on Pohnpei came the founders of the matrilineal clans that continue to be the basis for Pohnpeian social life today. All Pohnpeians are members of named, totemic, exogamous, matrilineal descent clans (*sou*). While the clans, or their subclans (*keimw*) were localized at one time, all originating from a single woman ancestor, with intermarriage and association they eventually evolved into wider island chiefdoms. The essential social–political unit on Pohnpei – a group of people living and working in the same area, claiming close kin ties, and having a leader with a formal title – gradually shifted from the kin group, the *sou*, to a territorial group, the *kousapw*. The *kousapw* is modern Pohnpei's true community, and is made up of a localized group of farmsteads (*peliensapw*) organized under the leadership of a local chief (*soumas*). Above the *kousapws* are five traditional autonomous kingdoms (*wehi*) overseen by the paramount chief, the *Nahnmwahrki*. While the traditional kingdom and their sub-units were characterized by fission and fusion, the constant reorganizing of geopolitical entities ground to a halt with the arrival of foreigners and the imposition of colonial rule (Petersen, 1990).

Because the *kousapw* are relatively small (100–200 members), they are numerous, and it is possible for a person to live in one and participate in another. Adult men participate either in a *kousapw* where they have matrilineal ties with the ruling clan, or in a *kousapw* where they have patrilineal ties to the complementary line of titles reserved for the sons of men in the ruling line. Pohnpeian social activity is intricately dependent on the *kousapw*'s continual feasting, and the matrilineages ensure the viability of the *kousapw* they rule through promoting an active round of feasting and ritual and producing goods that are presented to the chiefs and then redistributed in the course of these events (Petersen, 1982). Men prove their skill and worthiness by competing to produce the largest, the most, and the finest of the prestige goods – yams, kava and pigs. The more they contribute, the higher up the political hierarchy they advance.

### 7.3.4  Traditional Resource Management and Land Tenure on Pohnpei

Based mainly on anecdotal evidence, it appears that the native population developed a complex system of resource management early in their exploitation of the island (Haun, 1984). According to Pohnpei beliefs, common people are interposed between the chiefs as representatives of mankind and nature. People exploit or transform nature in the service of society, represented by the chiefs. Untransformed nature is considered *sarawi* (sacred), and is contrasted with the social domain by *wahu* (respect, honour) derived from proper behaviour in relation to the social hierarchy (Dahl, 1999). This dualism is also expressed spatially, and Pohnpeians divide their island into several concentric domains (Figure 7.5). An inner core – the upland forest (*nanwel*), and the outer rings in marine space – mangrove forest (*naniak*), lagoon (*nansed*), and ocean (*nanmadau*), were believed to be controlled by spirits, or *eni*, by virtue of their location outside the sphere of human influence. The middle concentric ring in the Pohnpei 'world' was made up of settled coastal areas (*nansapw*). The *nansapw* were considered to be land wrestled from the *eni* through the human activities of clearing and planting (*sapwasapw*). Conversely, abandoned lands that reverted to forest could be considered as returning to the stewardship of the *eni*. In addition, the political boundaries of the village (*kousapw*) and kingdom (*wehi*) formed contrasting radial divisions that encompassed the island's entire marine and terrestrial environmental diversity (Dahl and Raynor, 1996). In both the forest and marine areas it was believed (and to some extent the belief persists) that lack of respect for the *eni* or spirit guardians of these areas, either through not following proper etiquette

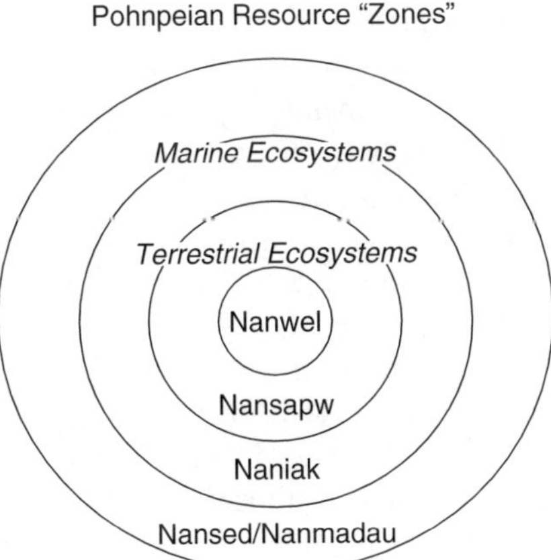


**Figure 7.5**  *Pohnpeians understand their environment by dividing their island into several traditional concentric 'resource zones', or 'spheres on influence', based on dominant vegetation, ecosystems, and traditional activity and use*

in these zones or through improper use of resources, was punished supernaturally by severe illness or even death.

Two broad property classes apparently existed within prehistoric Pohnpei. Settled lands were within the human domain and were under the trusteeship (*kohwa*) of the paramount chief. Surrounding these humanized areas were the *luhwen wehi*, common property open to a variety of semi-secret and temporary uses. Ranked titles reflected the political structure within each *kousapw* and *wehi*, and among these, titles connoting a resource regulation function were common, for example *Sou Madau*, master of the ocean, *Souwel Lapalap*, great master of the forest. The function of these titles, however, has been lost for at least several generations – no contemporary evidence exists for the exercise of such functions (Dahl and Raynor, 1996).

Management of land and waters, and thus effectively all natural resources, was carried out through the traditional leadership system. In this complex dual lineage system, which still exists today, people are divided into two lines with titles attributed to either the *Nanmwahrki* or the *Nahnken*. Each member of Pohnpeian society had a station, and held a unique rank within that station. Each station or level of leadership had clear responsibilities and powers, understood by all Pohnpeians. Promotion within the system was based on a combination of blood or clan, passed on through the mother's side, and on achievements, especially in warfare and to a lesser extent in special skills and/or exceptional agricultural and marine production.

### 7.3.5    Contemporary Changes in Resource Management and Land Tenure

Europeans began to frequent Pohnpei in the mid-1800s, but the Spanish were the first to colonize the island (1886–99). Largely ineffectual as colonizers, they were followed by the Germans (1899–1914). Between 1912 and 1914, the German Administration instituted individual ownership of land by deed and inheritance by primogeniture. The German code also assigned the *luhwen wehi* to a specific district (*wehi*), which was to be administered jointly by the paramount chief and the German governor. The Japanese, who assumed control of Pohnpei shortly after the German land reforms were instituted (1914–45), considered unused land as belonging to the Administration. In addition to the upland forest, these public lands included swamp-lands and marine areas. The area under administrative control was additionally increased by forced sale. Some of these lands were made available to Japanese settlers; other tracts were appropriated for military use. When the USA took over the island at the end of World War II, all this land remained under administrative control as public land.

Demographic change during the colonial period strongly influenced the gradual expropriation of land from traditional management and control. Like many other Pacific islands ravaged by foreign diseases to which they had no immunity, the population declined sharply after European contact, from an estimated 15 000 in 1840 to only 1705 in 1891. Subsequently, the population recovered slowly until the mid-1960s. During this period, much previously occupied land reverted to forest, and then came under administrative jurisdiction. At the same time, the traditional form of land allocation nominally regulated by the paramount chiefs was replaced by bureaucratic forms of land administration (Dahl and Raynor, 1996). In the process, traditional resource management was undermined by loss, at least legally, of local authority. Since the early 1960s, population growth and an expanding economy have worked together to intensify resource exploitation.

Besides population growth, discontinuities in the Pohnpei land tenure system have also pushed settlement into previously unoccupied upland areas. The government-sponsored bureaucratic apparatus that regulates and certifies landownership has been combined with beliefs about rights of use formed through tradition and history (Dahl and Raynor, 1996). Population pressure on coastal lands has been exacerbated as a result of insecure tenure. Traditionally, the right of occupation was derived from bringing land under cultivation, thereby 'humanizing' it. As *kohwa*, land was not owned in a western sense, but a century of colonialism has increased the desire for secure title, which represents the conveyance of rights by the state. However, occupation is still considered as the primary basis for assuming rights over land use.

### 7.3.6   Erosion in Pohnpei

About 60% of Pohnpei island is classified as steep and mountainous (Laird, 1982), and much of the mountainous interior is composed of slopes of 27° or greater. The current steep relief can be attributed to a long period of sustained geomorphic incision and sculpturing by landslide and fluvial processes, along with high-average rainfall that is reflected today in the deeply weathered soils, regolith and vegetation. Under natural conditions of healthy rainforest cover, sheet-wash erosion rates are estimated to be from 0.2 tons/acre for 7° slopes to 1.5 tons per acre for 27° slopes[1] (USDA, 1995), but this does not include sediment generated from landsliding. Typical of many Western Pacific islands, upper watersheds generally have very high landslide susceptibility under certain conditions (Trustrum *et al.*, 1989, 1990). The probability of landslide occurrence is usually low under healthy forest but can be greatly increased during cyclonic storms (especially with return periods of 50 years or greater) or where natural vegetation has been modified; both conditions can result in extensive landsliding (Figure 7.6). Sediment production on these islands is usually related to landslide frequency coupled with the magnitude of the event. The downstream or off-site sediment impacts caused by landslides are often most damaging to environments used for food gathering and cultivation, such as 'downstream' terraces, rivers and coastal margins, frequently where villages and towns are located (Figure 7.7). After large magnitude storms extensive sediment plumes along coastlines, harbours, and estuaries are common.

The danger of soil erosion was one of the factors that prompted the development of complex resource management systems throughout the Pacific, and human activity has been directed at the most effective use of each habitat and natural phenomena (OTA, 1987). Traditional agroforestry on Pohnpei, which made up nearly one-third of the island's area (13 090 ha) in 1995, is an example of this type of adaptation. Pohnpei agroforestry generally conserves soil well. At all stages of development, a multi-storey vegetation layer and substantial ground cover and plant litter are maintained. During agroforestry establishment, vegetation is altered to some extent through slashing and some girdling of larger trees, but much of the forest canopy is maintained, all the slash remains on site, and burning is not practised. Root systems, important in holding soil, remain relatively intact. Within a short time, underplanted crops and regeneration of selected components of the original vegetation re-establish the

**Figure 7.6** *Extensive landsliding can occur in Pohnpei and other western Pacific islands during cyclonic storms or where vegetation is modified. Many old scars and debris tails are evident in steep re-vegetated terrain. Near Pilen Pahnigin, looking northeast (photo reproduced by permission of Noel Trustrum)*

***Figure 7.7***   *Sediment from landslides can greatly impact on inhabitated lowland areas and fragile coastal ecosystems. Near Sakartik, looking west (photo reproduced by permission of Noel Trustrum)*

understorey and lower canopy. Scientists estimate that erosion under Pohnpei agro-forestry is only slightly higher than under native forest, perhaps 0.5 to 4.6 tons/acre/year (USDA, 1995).

However, infrastructure development, moves towards a cash economy and changes in agricultural systems over the last few decades have greatly increased erosion. The greatest single cause has been the construction of roads. Until early in the twentieth century, Pohnpeians had no means of transportation besides walking, and low-impact trails were the main route for getting around the island. Under the German Administration, road construction became a priority and roads were constructed in the vicinity of the main German settlement at Kolonia, out to Sokehs and around US municipalities in the north of the island. Efforts were interrupted by a rebellion in 1910 and later by the outbreak of World War I. Under the Japanese, roads were expanded around the island to transport commodities to the harbour in Kolonia. Many of these roads subsequently reduced to trails through neglect by the early US Adminstration that replaced the Japanese in 1945. Over the last three decades, however, the US and local administration have made a concerted effort to improve infrastructure, and a circumferential road was finally completed in 1986. Since then numerous access roads have been constructed off the main road. In general, these roads are not hard surfaced and road gradients are often extreme (Zeimer and Megahan, 1991). These roads alter drainage patterns, increase runoff, and contribute substantially to sediment load in streams from erosion from the roadbed, reducing water quality and negatively affecting downstream ecosystems, including the mangrove forests, seagrass beds and the coral reefs of the lagoon.

Roads and population growth encourage even more people to move into previously undeveloped areas, thus leading to the second major cause of erosion on Pohnpei – the cultivation and gradual conversion of forest by people from coastal areas. This upland cultivation both supplements existing lowland areas under cultivation and prepares the area for permanent homesteading. *Piper methysticum*, locally known as *sakau*, has emerged as the foremost crop leading to forest conversion. The roots of this plant are pounded to make a narcotic beverage that has long been of central cultural importance on Pohnpei. It was traditionally consumed only by the higher-ranking members of society, but prohibitions against consumption by the general populace have been relaxed since World War II. *Sakau* (or kava) has since emerged as the premier cash crop for many of the island's population, who have little prospect of finding wage employment. Commercial *sakau* production involves clearing forests for the richer soil and moist environment found there. Since commercially grown *sakau* requires direct sunlight, the forest canopy must be opened by felling or ring-barking overstorey trees. Because *sakau* is shallow-rooted, planting on steep slopes can lead to soil erosion and mass wasting during major storm events. Loss of forest habitat also negatively affects biodiversity.

The magnitude of recent forest conversion has only now been appreciated. Aerial photography and vegetation mapping efforts in 1995 revealed that intact native forest on Pohnpei had been reduced from 15 008 ha (42% of the island's land area) to 5169 ha (15%) during the 20-year period between 1975 and 1995 (Trustrum, 1996). The ongoing cycle of cultivation, settlement and road building has three broad impacts. First, land clearance increases erosion, which exacerbates the downstream impacts of sediment on mangroves, lagoons and coral reefs. Second, more intensive resource exploitation is becoming unsustainable. This is already the case with avidly hunted bird species like the Micronesian Pigeon (*Ducula oceanica*) and the Purple-capped Fruit Dove (*Ptilinopus porphyraceus*), which have experienced drastic population reductions in recent years (Buden, 2000). Finally, forest conversion results in a loss of species diversity. Since terrestrial endemism is relatively high, the local extinction of a species could be equivalent to its complete loss. Since the upland forest is relatively small anyway, it may already be close to a critical threshold in terms of habitat value.

Due to a number of factors the risk of erosion on many Western Pacific islands is becoming more serious. Over thousands of years, Pohnpei's steep mountain slopes have adjusted to high rainfall, and landslides do not occur in any great numbers except when slopes are subjected to abnormally high rainfall duration and/or intensities (Harp and Savage, 1997). Landslide events generally occur in groups and are linked to tropical storms and typhoons, which are difficult to predict but can be very destructive. The most serious recent landslide event on Pohnpei occurred in April 1997, and was the first event to result in human fatalities. It was caused by the combined effects of Super Typhoon Isa and Tropical Storm Jimmy, which passed the island approximately one week apart. Over 30 landslides were triggered during the night of 20–21 April, which resulted in 19 fatalities and the destruction of 14 dwellings in the villages of Oumoar and Iohl on the northwest side of the island. Precipitation from Isa on 11–13 April saturated the slopes on the island, and one week later over 250 mm (10 in) of rain from Tropical Storm Jimmy fell during a four-hour period, triggering the landslides. Although data are lacking, it appears that an intense precipitation cell was centred, for several hours, over the area of maximum landslide concentration. Residents in the area reported that they had never seen rainfall so intense, even on Pohnpei. A similar landslide event, centred in the mountains of southeast Pohnpei, occurred in 1991, and another from Cyclone Axel in 1992 caused landsliding and downstream damage. In 1991 several *sakau* farms were destroyed and two rivers were severely affected by large sediment loads caused by damming from landslides. People still attribute the destruction of several rich shellfish areas in the southeast part of the island to the extreme sediment deposited in mangrove areas as a result of the 1991 landslides.

### 7.3.7   Landslide Risk Perception

Up until the 1980s, most landslides occurred in the interior forest, far from human settlement, and most were natural in origin. As such, landslide risk was not a serious consideration in Pohnpeian communities. However, with increased population and establishment of homesteads inland on steeper more landslide-prone slopes, contemporary perceptions of landslide risk are becoming more acute as landslide occurrences become more common and their consequences more serious. The government responded in 1983, based on evidence that the island interior was being rapidly deforested (MacLean *et al.*, 1986). The Pohnpei State Division of Forestry requested assistance from the Pacific Islands Forester Office (USDA Forest Service Institute of Pacific Islands Forestry, Honolulu) and, using the 1975 aerial photography of the island, the soils survey (Laird, 1982) and aerial reconnaissance, the two agencies closely cooperated in legislative efforts that resulted in the passing of The Pohnpei Watershed Forest Reserve and Mangrove Protection Act of 1987. The Act designated some 5100 ha of the central upland forest area and 5525 ha of coastal mangrove forests of Pohnpei Island as a protected area, to be managed and enforced by the Pohnpei Department of Resource Management and Development. The legislative intent was that all use of the upland and mangrove forests within the reserves would have to be coordinated with state officials so that further upland settlement and other perceived unsustainable activities could be restrained.

Community involvement in the development of the law, however, was virtually non-existent, and the proposed rules and regulations, failing to recognize traditional Pohnpei resource use and authority, were universally rejected. As a result, around the island government boundary survey teams were turned back by angry villagers. These setbacks led in 1990 to the formation of the Watershed Steering Committee (WSC), an inter-agency task force made up of representatives from several Pohnpei State Government agencies, community leaders and NGOs. With funding from the US Forest Service and subsequently from SPREP, the WSC initiated a watershed education and negotiation programme, which was extended round the entire island over two years. The programme had two parts: an overview of why Pohnpei's forested watersheds must be protected, and a critical review of the 1987 law. Two major changes were unanimously insisted upon by the local communities in over 200 meetings:

1. Paramount chiefs and their village representatives (*Soumas*) need to be partners in the management process;
2. Environmentally sustainable management should be extended beyond the WFR to encompass the entire island, from the mountains to the reefs.

Over the last 10 years, intensive efforts have been made to reinvigorate traditional forest management on Pohnpei. A number of obstacles have been encountered, including continued unsustainable population growth and the resulting growing shortage of available land. Also, the role of the traditional system in resource management appears to have eroded considerably over the past century. Further, while some of the former responsibilities of traditional managers have been taken over by the government, government edicts and programmes have decreasing influence outside the main towns because they are poorly integrated with customary structures. The result is a confusion between traditional and government systems in some areas, with a consequent reduction in the effectiveness of resource management.

In 1996, the Pohnpei Community Planning Program (PCRMP) was launched by The Nature Conservancy and local partners as an innovative attempt to support the island's communities as the primary managers of their biological resources (Raynor, 1996). The programme aims to develop coordinated management within and between communities (Figure 7.8) that will maintain subsistence and cash resources while also protecting the island's natural ecosystems and remarkable biodiversity. It also hopes to develop a legal and administrative framework for equitable co-management between government and customary authorities on the island. The PCRMP was developed under the framework established by the Pohnpei Watershed Management Strategy (The Nature Conservancy, 1996). This strategy recognizes the central role of communities in determining resource use and managing natural environments. The Strategy seeks to ensure the sustainable management of Pohnpei's natural resources; to help communities develop strategies

**Figure 7.8**  *The Pohnpeian community are often engaged in resource management and watershed planning (photo reproduced by permission of Noel Trustrum)*

for ecologically sustainable business; to strengthen community ties; and to maintain management of cultural and sacred sites.

Landslide events occur infrequently (once or twice per decade) in conjunction with major storm events, and are usually localized. However, while the damage resulting from the most recent storm events has no historical precedent, Pohnpeians are increasingly connecting increased settlement, agricultural activity and forest clearing with an increase in hazard and landscape risk. This heightened awareness and concern has led to new partnerships with government and other agencies being formed to improve landscape risk management at the ground level.

## 7.4   Discussion

Cultural perception of landslide risk is often related to the association with and dependence certain groups have on a particular environment. It is strongly based on the type of physical environment, the cultural system, the values and beliefs on which perception is based, and the degree to which landslide erosion is understood to affect people's lives and resources. This perception can be heightened by close cultural, social and spiritual links, environmental knowledge, economic dependence, and experience and learning over time. However, in many parts of the world, the close relationship with the natural environment based on traditional cultural values is being undermined through a progressive shift in cultural world-view and behaviour. This change in values, largely because of economic and lifestyle pressures, is often towards a world-view more dependent on exploiting resources and damaging natural ecosystems for short-term economic gain. For many indigenous groups the shift in cultural relationship, attachment and interdependence is increasingly exhibited in land use and land management examples detrimental to the environment.

The perception of landslide risk in most situations, as seen in the two case studies, is most often increased, improving understanding and chances for behaviour modification, if people have experienced a damaging storm or flood event. This may be an event that has directly or indirectly affected their lives, destroyed or diminished economic assets, affected the infrastructure on which a community relies, or degraded significant cultural or natural resources. Risk perception can also be learned over time through education and discussion, where an awareness of environmental risks, relationships, cumulative effects from landslide, deforestation and flooding is increased through activities such as collaborative learning, tribal or community-led projects, and participatory projects and research.

These two case studies show that from a cultural perspective, risk perception can be based on cumulative traditional and historic events, often passed on through several generations (see glossary of terms, Table 7.1) through song (*waiata*), quotations (*whakatauki, pepeha, lepin kahs*), chants (*ngis*), dances (*dokia*), folklore, knowledgeable people (*kaumatua, sou*) and stories (*te reo, kupu, poadoapoad*) that describe a damaging event or events such as landslides or floods. However, with most people the recall is usually recent, such as a catastrophic or sudden event they physically experienced in their lifetime. Both ways of recalling events provide a record of loss of life, injury, destruction or modification to natural resources, such as loss of soil resources, impacts on values, and effects and damage on community infrastructure.

**Table 7.1**    *Glossary of terms*

| Maori–New Zealand | Pohnpei |
| --- | --- |
| Iwi (tribe) | Wehi (tribe or kingdom) |
| Hapuu (sub-tribe) | Kousapw (community) |
| Whaanau (extended family) | Sou, Keimw (family) |
| Tangata (people) | Aramas (people) |
| Kaumaatua (knowledgeable person, respected person, Maori elder) | Sou (older expert) |
| Te Reo, Kupu (language, words, stories) | Poadoapoad (stories, legends) |
| Whakatauki, Pepeha (sayings, proverbs, quotes) | Lepin kahs (sayings, quotes) |
| Waiata (song) | Ngis (chants) |
| Haka (dance) | Dokia (singing dances) |
| Maatauranga (traditional knowledge) | Tiahk en sahpw (culture of the land) |

Management solutions for reducing landslide risk and achieving sustainable land use increasingly require an integrated and participatory approach to identify the cause and effect of activities leading to risk, and to find local and regional solutions to, in most cases, quite complex problems. The case studies in this chapter highlight the need to understand culture as a determinant of human behaviour. To find solutions we must therefore have some fundamental understanding of cultural perspectives and background. These studies present examples of the types of information required to understand complex issues in many areas around the world, and the role culture has in advancing our knowledge and thinking for the development of effective planning and policy to reduce risk and achieve good environmental outcomes. Five major knowledge strands, derived from the two case studies, are recognized as increasing or heightening the perception of landslide risk. The strands described below work in combination to advance our understanding of risk perception, and form an integral part of collaborative learning.

1. **Perception of risk formed from loss, destruction, depletion or degradation of natural resources**    Landslide risk perception is increased when associated with loss, damage, or modification to culturally significant natural resources, for example, damage or destruction to forest and other ecosystems, destruction or degradation of ecological habitats, loss of culturally significant flora and fauna species, loss of *mauri* (life force or life essence) of water, loss or damage to fishing grounds, etc. On the East Coast of New Zealand, extensive areas of native fish and bird habitat were destroyed during Cyclone Bola in 1988, and water quality seriously degraded by sediment. Anecdotal evidence was also given on impacts of coastal and marine fishing grounds. In Pohnpei, landslide events and high-sediment loads resulting from tropical storms and typhoons have seriously damaged culturally significant forest zones, and stream, river and coastal habitats. In 1991, several important shellfish harvest areas were destroyed around the coast as a result of landslide events and excessive sedimentation. With indigenous cultures, the loss of a cultural resource often has long-term ramifications and a great effect on human and social well-being

and health, culminating, for Maori, in loss of tribal status (*mana*), well-being (*ora*), identity (*whakapapa*) and spirituality (*wairua*).

2. **Perception of risk from impacts on economic assets and production**   Landslide risk perception is greatly heightened through experience of and learning from an event that damages economic assets, property, production, and the infrastructure of people and communities, affecting individual, community, regional and national economies. This increased perception of risk can result from damage to individual property or on-farm damage, with lost productivity, loss of soil resource, crop damage, livestock loss, roading and fence damage, cost of repair, reduced cash flows, or can be to the wider community in terms of property damage, schools, buildings, roads, bridges and other infrastructure such as telecommunications, electricity, water, etc. These impacts can result in an overall reduction in the productive capacity of land, damaged economic assets, and degradation of natural resources on which economic activity and production are based. Examples of these types of impacts and experiences on people's lives were clearly evident following Cyclone Bola in 1988, when thousands of hectares of land, productive grassland and crops in the East Coast region were lost through erosion or ruined by sediment deposition (Figure 7.4), resulting in millions of dollars of damage. In Pohnpei in 1991 storms caused landslides and high-sediment loads, destroying several *sakau* farms and extensively modifying several rivers. In 1992 Cyclone Axel caused landsliding, with a resulting landslide dam becoming breached within days of formation, causing significant downstream damage to crops and infrastructure (roads, bridges and buildings). In 1997, Super Typhoon Isa and Tropical Storm Jimmy again caused widespread economic and infrastructure damage to the island, with 19 people killed, 14 dwellings destroyed, and roads and rivers badly damaged. The effects on local and regional economies can take decades to repair. Many Maori remember the socio-economic impacts on their lives from Cyclone Bola through lost work, lost production, costs of repair, rebuilding the infrastructure of roads, communication and electricity for their communities, the time taken to repair property damage and regain production. Many people never fully recovered from the damaging storm and flood events of 1988, which have been etched in their minds for ever and have heightened their awareness both of the vulnerability of the landscape they live in and the huge costs associated with storm damage.

3. **Perception of risk formed from effects on human well-being, health, and mortality**   Landslide risk perception is again heightened through experience, memory or stories of an actual damaging event such as a storm or major flood that has affected people's lives. This may result in cultural, social and psychological effects ranging from stress, psychological and physical illness, through to loss of life associated with a major event. In New Zealand Maori families recalled drownings from floods associated with storm events and cyclones since the 1950s, while others recalled very close escapes from injury or even death during storms and floods. A few Maori believed that loss of life and resources was a form of punishment for disrespect and lack of care for the environment, which demonstrated the close spiritual links people have with their environment. In Pohnpei in 1997, landslides associated with Super Typhoon Isa killed 19 people. This storm is well remembered and greatly heightens awareness of the ferocity of such events, the associated hazard and risk, the precautions that must be taken, and the amount of damage inflicted in such a short time.

4. **Perception of risk formed from impact on cultural features and icons**   The perception of landslide risk can be greatly increased when associated with damage, modification, or removal of traditional sites, landmarks, features and icons. This perception based on cultural and historical relationships is significant for both Maori and Pohnpeians and helps them understand their environment in a cultural context. One such example on the East Coast, New Zealand, is of a famous ancestral rock – believed to be derived from an ancestral mountain – and residing in a river channel for hundreds of years. The rock was either removed or buried by sediment during Cyclone Bola. Many other cultural sites, such as rocks, river islands, sacred ground, terraces and forest stands, disappeared or were greatly modified during Cyclone Bola. Traditional stories, quotations, or songs are important for recording the history of damaging events such as storms, floods, erosion and other natural phenomena. Typically such songs and stories record when and where lives were lost, cultural features or landmarks modified, damaged or destroyed, or cultural resources (e.g. flora and fauna, forest, habitats, fish) degraded or depleted. Maori and Pohnpeian history, tradition and myth typically record historic environmental change through cultural association with ancestors, oral stories and adventures.

5. **Perception of risk formed from education, collaborative learning and participation**   Landslide risk perception is often increased through collaborative learning, improved understanding of environmental change through both cultural and mainstream knowledge, and improved understanding of science such as ecological and catchment/watershed processes. Such understanding results in improved recognition and awareness of cumulative effects, the interconnectedness of ecological systems, and the nature of environmental change. This can be carried out in workshops, through community-based projects, and through encouraging participatory approaches to address specific environmental issues such as erosion, biodiversity and how best to sustain and develop natural resources.

These five strands provide both a framework to advance our knowledge on risk perception and guidance on how cultural information can be acquired, collated and incorporated into planning and policy. In a world with an increasing array of complex environmental, social and economic problems, it is essential to understand cultural differences, to acknowledge that different peoples around the world see their environments differently, and to realize that universal human solutions cannot always be applied to every situation. Understanding the cultural basis for landslide risk perception is essential. Once the specific way communities and individuals perceive risk has been understood, it is easier to facilitate and plan remedies and solutions to lessen risk. This should involve engaging groups to develop, own and implement planning and policies that are culturally, socially, environmentally and economically based.

## Note

1. These are well below the estimated soil loss tolerance (level of soil erosion that can occur on a per area basis without reducing soil quality or crop yield potential) of 4.9–11.21 tons/ha/year (2–5 tons/acre/year) (USDA, 2001).

# References


Asher, G. and Naulls, D., 1987, *Maori Land*, Planning Paper 29, New Zealand Planning Council, Wellington.

Buden, D., 2000, A Comparison of 1983 and 1994 Bird Surveys of Pohnpei, Federated States of Micronesia, *The Wilson Bulletin*, **112**(3), 403–410.

Dahl, C., 1993, Recommendations for community based watershed planning and management for Pohnpei Island (FSM) based on an investigation of local nouns and place names, unpublished manuscript. Pacific Island Network–University of Hawaii Sea Grant Program.

Dahl, C., 1999, *The State and Tradition: Conceptions of Land Tenure on the Island of Pohnpei*, Ph.D. Dissertation, University of Hawaii.

Dahl, C. and Raynor, B., 1996, Watershed Planning and Management: Pohnpei, Federated States of Micronesia, *Asia Pacific Viewpoint*, **37**(3), 235–253.

Durie, M., 1998, *Te Mana, Te Kaawanatanga: The Politics of Maori Self-Determination* (Oxford: Oxford University Press).

East Coast Project, 1978, Report of land use planning and development study for erosion prone land of the East Cape region, Section 1, The East Coast, May 1978. A report by the Poverty Bay Catchment Board, The 'Red Report'.

Eyles, G.O., 1983, The distribution and severity of present soil erosion in New Zealand, *New Zealand* Geographer, **39**(1), 12–28.

Eyles, G.O., 1985, *The New Zealand Land Resource Inventory Erosion Classification*, Water and Soil Miscellaneous Publication 85.

Glade, T., 1996, The temporal and spatial occurrence of landslide triggering rainstorms in New Zealand, *Heidelberger Geographische Arbeiten*, **104**, 237–250.

Glade, T., 1998, Establishing the frequency and magnitude of landslide triggering rainstorm events in New Zealand, *Environmental Geology*, **35**(2–3), August, 160–174.

Glassman, S., 1952, The flora of Ponape, *Bulletin 209*, Bernice P. Bishop Museum, Honolulu.

Hanlon, D., 1988, *Upon a Stone Altar: A History of Pohnpei Island to 1890* (Hondulu: University of Hawaii Press).

Harmsworth, G.R., 1995, Maori values for land use planning. Discussion document, unpublished Manaaki Whenua – Landcare Research Report.

Harmsworth, G.R., 1997, Maori values for land-use planning. Broadsheet newsletter of New Zealand Association of Resource Management, February, 37–52.

Harmsworth, G.R., 1998, Indigenous values and GIS: a method and framework, *Indigenous Knowledge and Development Monitor*, **6**(3), 3–7.

Harmsworth, G.R. and Page, M.R., 1991, *A Review of Selected Storm Damage Assessments in New Zealand*, DSIR Land Resources Scientific Report 9.

Harmsworth, G., Warmenhoven, T., Pohatu, P. and Page, M., 2002, Waiapu catchment technical report. Maori community goals for enhancing ecosystem health. Landcare Research Contract Report LC0102/100 for Te Whare Wananga o Ngati Porou, Ruatorea (unpublished).

Harp, E. and Savage, W., 1997, *Landslides Triggered by the April 1997 Tropical Storms in Pohnpei, Federated States of Micronesia*, US Geological Survey. Open-File Report 97–696. Denver, Colorado.

Haun, A., 1984, *Prehistoric Subsistence, Population, and Socio-Political Evolution on Ponape, Micronesia*. Ph.D. Dissertation, University of Oregon.

Jessen, M.R., Crippen, T.F., Page, M.J., Rijkse, W.C., Harmsworth, G.R. and McLeod, M., 1999, Land Use Capability Classification of the Gisborne–East Coast region: A report to accompany the second edition New Zeland Land Resource Inventory. Landcare Research Science Series No. 21.

Laird, W., 1982, *Soil Survey of Island of Ponape, Federated States of Micronesia*, US Department of Agriculture, Soil Conservation Service.

Landcare Research New Zealand Ltd GIS tables, 2001, Landcare Research New Zealand Ltd National Environmental GIS Databases, Palmerston North.

MacLean, C., Cole, T., Whitesell, C., Falanruw, M. and Ambacher, A., 1986, The vegetation of Pohnpei, Federated States of Micronesia. Resource Bulletin PSW-18, US Department of Agriculture, Forest Service, Albany, CA.

Marsden, M., 1988, *The Natural World and Natural Resources. Maori Values Systems and Perspectives*, Resource Management Law Reform Working Paper 29. Part A. Ministry for the Environment, Wellington, New Zealand.

McGlone, M.S., 1983, Polynesian deforestation of New Zealand. A preliminary synthesis, *Archaeology in Oceania*, **18**(1), 11–25.

McGlone, M.S., 1989, The Polynesian settlement of New Zealand in relation to environmental and bioitic changes, *New Zealand Journal of Ecology*, **12** (supplement), 115–129.

Ministry for the Environment, 1997, *New Zealand's State of the Environment Report 1997*, Ministry for the Environment, Wellington, New Zealand.

NWASCO (National Water and Soil Conservation Organisation), 1970, *Wise Land Use and Community Development*. Report of Technical Committee of Enquiry into the problems of the Poverty Bay–East Cape District of New Zealand. Wellington, Water & Soil Division, Ministry of Works.

Newsome, P.F.J., 1987, *The Vegetative Cover of New Zealand*, National Water and Soil Conservation Authority.

OTA, 1987, *Integrated Renewable Resource Management for US Insular Areas*, Office of Technology Assessment, Congress of the United States.

Page, M.J. and Trustrum, N.A., 1997, A late Holocene lake sediment record of the erosion response to land use change in a steepland catchment, New Zealand, Zeitschrift für Geomorphologie N.F., **41**(3): 369–392.

Page, M.J., Reid, L.M. and Lynn, I.H., 1999, Sediment production from Cyclone Bola landslides, Waipaoa catchment, *Journal of Hydrology (NZ)*, **38**(2), 289–308.

Page, M.R., Harmsworth, G.R., Trustrum, N., Kasai, M. and Muratani, T. (2001a) Waiapu River (North Island, New Zealand), in T. Marutani, G.J. Brierley, N.A. Trustrum and M. Page (eds), *Source-to-Sink Sedimentary Cascades in Pacific Rim Geo-Systems*, Matsumoto Sabo Work Office, Ministry of Land, Infrastructure and Transport, Japan, 102–111.

Page, M.R., Trustrum, N., Brackley, H., Gomez, B., Kasai, M. and Muratani, T. (2001b) Waipaoa River (North Island, New Zealand), in T. Marutani, G.J. Brierley, N.A. Trustrum and M. Page (eds), *Source-to-Sink Sedimentary Cascades in Pacific Rim Geo-Systems*. Matsumoto Sabo Work Office, Ministry of Land, Infrastructure and Transport, Japan, 86–100.

Petersen, G., 1982, Ponapean matriliny: production, exchange, and the ties that bind, *American Ethnologist*, **9**(1), 129–142.

Petersen, G., 1990, Some overlooked complexities in the study of Pohnpei social complexity, *Micronesica (Supplement)*, **2**, 137–152.

Rau, C., 1993, *100 years of Waiapu*, published by the Gisborne District Council, Gisborne Herald Co Ltd.

Raynor, B., 1996, Developing a Community Approach to Watershed Management Planning on Pohnpei, in *Consultants' Reports. Prepared for the Asian Development Bank*. TA FSM-1925. Watershed Management and Environment, 1–28.

Spengler, S., 1990, Geology and hydrogeology of the Island of Pohnpei, Federated States of Micronesia, unpublished Doctoral Dissertation, University of Hawaii, Honolulu.

Spengler, S., Peterson, F. and Mink, J., 1992, *Geology and Hydrogeology of the Island of Pohnpei, Federated States of Micronesia*, Report prepared for the Water Resources Research Center, University of Hawaii, Honolulu.

Statistics New Zealand, 1996, *1996 Census data*, Wellington, New Zealand.

Te Puni Kokiri, 2000, *Progress Towards Closing Social and Economic Gaps between Maori and Non-Maori: A Report to the Minister of Maori Affairs*, May, Te Puni Kokiri – Ministry of Maori Development.

The Nature Conservancy, 1996, *Pohnpei's Watershed Management Strategy 1996–2000: Building a Sustainable and Prosperous Future*, prepared by the Pohnpei Watershed Project Team. Funding and assistance provided by The Nature Conservancy, Asian Development Bank, South Pacific Regional Environment Hazard Programme.

Trustrum, N.A., 1996, Pohnpei's Watershed Spatial Plan and Management Guidelines, in *Consultants' Reports. Prepared for the Asian Development Bank*. TA FSM–1925. Watershed Management and Environment: i, ii, 1–62.

Trustrum, N.A., Whitehouse, I.E. and Blaschke, P.M., 1989, *Flood and Landslide Hazard, Northern Guadalcanal, Solomon Islands. A Report for United Nations Technical Development, New York.* DSIR Contract Report 89/07. Palmerston North, New Zealand.

Trustrum, N.A., Whitehouse, I.E., Blaschke, P.M. and Stephens, P.R., 1990, Flood and Landscape Hazard Mapping, Solomon Islands. DSIR Land Resources, Palmerston North, New Zealand, in *Proceedings of International Symposium 'Research Needs and Applications to Reduce Erosion Sedimentation in Tropical Steeplands'*, Suva, Fiji, 11–15 June 1990. IAHS Publication 192, 138–146.

Trustrum, N.A., Gomez, B., Reid, L.M., Page, M.J. and Hicks, D.M., 1999, Sediment production, storage and output: the relative role of large magnitude events, *Zeitschrift für geomorphologic. Suppl.* **115**, 71–86.

USDA, 1995, *Pohnpei Island Resource Study*, Natural Resources Conservation Service and Forest Service, US Department of Agriculture.

USDA, 2001, *Kosrae Island Resource Study*, USDA–Natural Resource Conservation Service, Pacific Basin Area.

Wards, I. (ed.), 1976, *New Zealand Atlas*. Government Printer Wellington, New Zealand.

Zeimer, R. and Megahan, W., 1991, Erosion and sedimentation control on roads and construction sites in the Federated States of Micronesia. Environment and Policy Institute, East–West Center. Honolulu, HI. Unpublished manuscript.

# 8

# Reply of Insurance Industry to Landslide Risk

*Hans-Leo Paus*

## 8.1 Introduction

### 8.1.1 Natural Hazards Cause More and More Damage

In the Italian holiday paradise of Meran on 17 July 2001, shortly after 3 p.m., there was a minor earthquake of $M_w = 4.7$ in strength. Although the tremors were perceivable even in Munich, Vienna and Venice, the earthquake did not cause any significant damage to buildings. But a rockfall, which was triggered by it, killed four unsuspecting mountain hikers. Against the background of an ever more complex world, increasing global connectedness and an ongoing concentration of economic values in hazard regions, chain reactions like this will gain in importance.

Globally acting insurance companies are increasingly subject to these developments, since the insurance sector as a financial services provider suffers mostly from the immense increase in the cost of such natural catastrophes. Only a few decades ago, local insurance markets were easy to chart out and were characterized by a small set of specific features (e.g. there was a small risk stemming from natural hazards in Germany, yet a high one in Japan). Today's worldwide connections between insurers and clients through international programmes as well as the global jungle of corporate cross-ownership make risk assessments an increasingly tough exercise. In consequence, insurance losses have risen sharply.

### 8.1.2 Structure of Insurance Business

In order to understand who has to bear which losses at a given moment in time, one has to take into account the basic structure of the insurance business (see Figure 8.1).

*Landslide Hazard and Risk*   Edited by Thomas Glade, Malcolm Anderson and Michael J. Crozier
© 2004 John Wiley & Sons, Ltd   ISBN: 0-471-48663-9

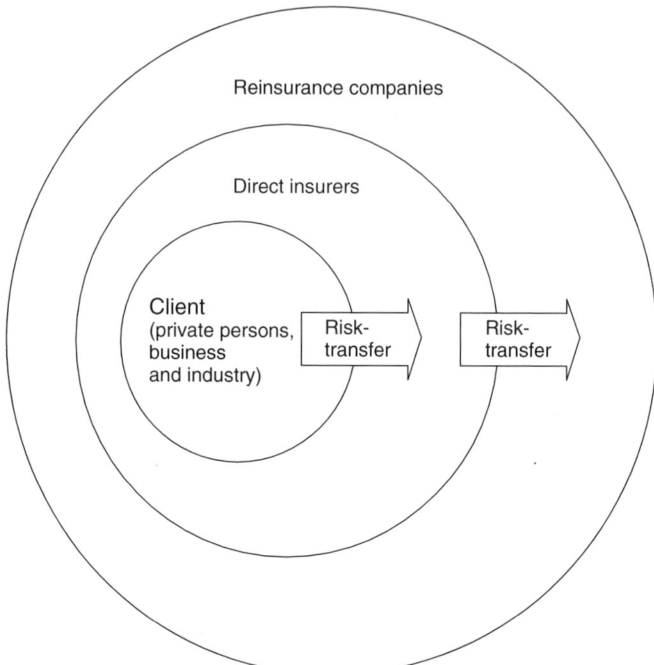



**Figure 8.1**   *Structure of the insurance business*
Source: Munich Re (2000)

For this reason, this section will provide a brief overview of the subject. The causal chain starts with the customer who closes a contract with a direct insurer, which is called an insurance policy. Direct insurers are generally well known, because everybody from private customers to huge industrial corporations deals with them to insure specific risks, such as car insurance polices or credit insurances in international trade. The policy spells out that the customer receives a compensation for its losses, which is called coverage. The size of the coverage is subject to negotiation. In the case of life insurances or private property insurances, coverage is usually defined up to a maximum amount of money, yet in industry insurance policies, coverage may be unlimited in some singular cases. In order to filter out minor claims and to give incentives to the customer to take precautions, sublimits are usually introduced into the policies (sometimes referred to as the 'excess'). In other words, the insurance cover only kicks in when the loss exceeds a specified sublimit. The customer has to pay all losses below this threshold. The difference between the actual loss and the sublimit is covered by the insurer.

In exchange for assuming their risk, the customer pays the direct insurer a premium. The amount of the premium is calculated based on the probability with which a certain loss will occur on the one hand, and on the amount of customers' premiums on the other. The premiums and the total risk will be low if a large group of customers shares the risk for improbable losses, but will be high in the opposite case. There are small margins in the calculation, since the insurance companies have to obey the laws of the market and competition is sharp.

Primary insurers do not automatically make money from this system, but carry a significant risk themselves. The amount of money insurers have to cover may reach existence-threatening levels in cases of large claims or natural catastrophes. Such risks are, in turn, covered by reinsurers. Primary insurers acting as customers of reinsurers, have to pay premiums and to accept sublimits. Reinsurers generally act on a worldwide scale and have the capacity to spread widely the risks they accept from primary insurers. It is this spread of risks that makes large-scale losses insurable in the first place.

However, the margins in the insurance sector are slim across the board, and private customers, primary insurers and reinsurers have to calculate carefully. In consequence, this insurance model is not too stress-resistant. It may collapse if risks are overlooked or underestimated, if premiums are fixed too low or coverage too high, if reinsurance cover is insufficient, or if the premium that is agreed between reinsurers and primary insurers are too small. Indeed, these are the problems the international insurance sector is facing given the rise of insurance claims arising from natural catastrophes. An important means of solving the problem can be applied at the beginning of the insurance chain when primary insurers underwrite such critical risks. The task of geo-scientific risk management is to steer and regulate that underwriting practice.

### 8.1.3    Insurance Companies are Influenced by Various Factors

Experts agree that the pressure on insurance companies will increase in the future. The key factors in this development are as follows.

### 8.1.3.1    Globalization

A significant part of mankind lives in regions close to the boundaries of the Pacific Ocean. At the same time, there is a dynamic increase in population in those regions, and many newly industrializing economies. As a result, the wealth of those countries is growing, the political situation is generally stabilizing, and foreign capital is attracted to the region. One implication of such prosperity is an increase in the demand for insurance services. The insurance density rises, and coverage increases; in consequence the financial risk that the insurance sector has to bear in those regions is rising.

### 8.1.3.2    Geological setting

It is particularly on the Pacific fringe that geological hazards are most concentrated, and natural hazards are therefore highest. From a geological point of view, the region is highly hazardous due to active tectonic processes, combined with recent volcanism, the danger of massive earthquakes, large differences in altitude between sea level and mountain regions at small distance, as well as the massive precipitation in tropical climates and the resulting deep-reaching decay of mountains flanks. A comparable concentration of economic power and unstable natural setting may only be found around the Mediterranean Sea, the Alps, and in some parts of Asia.

### 8.1.3.3    Land use

The stability of the land, that is, the soil used by man for living, agriculture and other economic activity, suffers from human activity: deforestation and settlement on mountainsides are triggering erosion by wind, water and frost. This in turn increases the risk of mountain slides and reduces the amount of land that can be used for agricultural purposes. The reduced capacity to store water in mountain regions also increases the risk of flooding in the valleys and lower regions, which in turn heightens the risk of dams breaking or hillsides becoming unstable. Additional factors such as the adaptation of waterways for economic purposes and river traffic further increase the risk of inundation.

### 8.1.3.4   Climate change

The global rise in temperature will shift vegetation zones and permafrost limits in the medium term. This will contribute to the weakening of the underlying bedrock. The changed climatic patterns, most of all the regionally increased amount of precipitation, will change the hydrogeological system and accelerate the decay of rock in some regions. The expected rise in sea level may cause the backing up of groundwater levels in the hinterland, and could thus change the static conditions in the soil close to the shoreline. This may affect the stability of cliffs as well as sea and river banks.

### 8.1.3.5   Cost explosion

The increase in settlement density of hazardous regions and the concentration of values in those areas have caused the amount of financial damages that are due to natural catastrophes to rise dramatically during the last 50 years. If the development represented in Figure 8.2, which was deduced from claims statistics provided by Munich Re, were to continue unmitigated, we would have to expect losses in the region of US$1000 billion by 2010, and an additional US$2000 billion by 2020 in current terms. This is a rather scary prospect, because this sum of US$3000 billion is about 37 times as much as the total losses connected to the terrorist attacks on the World Trade Center in September 2001. Put differently, if the predictions are correct, it is equivalent to the attack occurring every six months. A slowing or reversion of the trend is neither perceivable nor expected by the experts.

### 8.1.4   Losses Caused by Landslides will Gain Importance

Fortunately, rockfalls, mudflows and landslides have so far caused only few extreme insurance claims. There is, however, little reason to expect that this situation will endure in the future, because there are plenty of potential risks: there is simply no way to control, on a worldwide scale, whether and where pipelines will be built through or under unstable mountainsides, and whether those pipelines will emit toxic substances into the environment if a mountain slide occurs. In the end, who could rule out the risk that, through a chain of unfortunate incidents, a second environmental disaster like the one in Seveso might occur?

The experience of claims managers in the insurance business indicates that the above scenario is not a plot for a horror movie, but a real possibility which needs to be evaluated seriously. Especially critical in this respect is also the rapid expansion of megacities, a process in which housing areas and industrial sites move ever close towards mountainsides. This process occurs mostly without preceding geological evaluations or systematic urban planning in the sense that Europeans would think of it. Judging from the number of casualties (20 000) and the economic damage (US$15 billion), the mudflows in Venezuela of December 1999 (Munich Re, 2000b) were the worst natural catastrophe on global scale of that year. Because the insurance density in this industrializing country

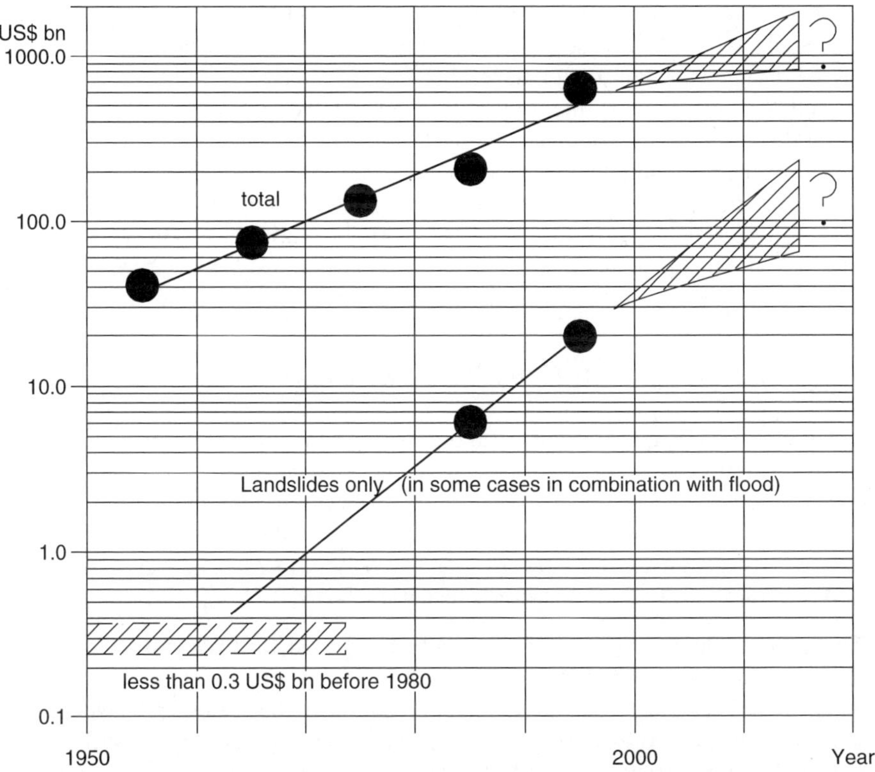


**Figure 8.2**   *Losses caused by natural disasters*
Source: Munich Re 2000b

is relatively low, the insurance claims were not very high, amounting to some US$500 million. But this fortunate outcome from the insurers' point of view is not always certain. It is an insurer's nightmare that a comparable catastrophe might occur in a more developed region with a vulnerable infrastructure, and a correspondingly higher insurance density, as in Central Europe, Japan or the United States. The rise of insurance claims due to landslides represented in Figure 8.2 seems to indicate that such a fear may be well founded. While landslide claims accounted for a marginal share in the total claims before 1980, they already made up some 3% in the total economic losses that were due to natural catastrophes between 1990 and 1999. There are signs that this percentage may rise over the period of the next 10 to 20 years to some 10 percentage points, which would correspond to losses of approximately US$300 billion. One should note that many damages that are due to landslides do not show up explicitly in the statistics of the insurance industry, because they are subsumed to the primary natural phenomenon that causes them. For example, Munich Re publishes historic lists of catastrophes (which are also partly available on CD), in which landslides are generally mentioned in connection with the natural phenomenon that triggered them. The loss figures in this list also only reflect the aggregated losses. This observation holds true, for instance, with regard to the 1970 earthquake in Peru, which also caused devastating mountain slides, as well as to flooding that was connected to landslides in Brazil (1988), Japan (1990) and Indonesia (1999), or to the 1987 torrential rains in Spain, which also caused landslides. Since such 'chain reaction catastrophes' can hardly be disaggregated at a later stage, the share of landslides in the total loss figures could well be above the levels indicated in Figure 8.2, because this only cites landslide loss figures if they were clearly and unambiguously attributable to such incidents. These rough calculations clearly highlight the importance of minimizing future landslide insurance claims by means of a scientifically oriented risk management. Because landslides risks are already a significant part of the total claims figure, and because one cannot rule out that the actual damage is much larger than reflected in the statistics, landslide risks are likely to be one focus of future risk management activities.

## 8.2   Recent Developments Call for New Strategies

The global insurance industry must face the fact that the probability of suffering major losses as a result of natural catastrophes is rising. The danger to the business sector is, however, generally underestimated (Paus, 2002). The current crisis in the capital markets has huge implications for many insurance companies; thus these two developments combined must give cause for concern. For this reason, measures are needed to reduce the risk exposure of the entire insurance sector (Figure 8.3):

- **The peak of the loss probability needs to be shifted towards the sector of lower losses**    This can only be achieved if the underwriters of the primary insurers, that is, the sellers of insurances that are in direct contact with the customers, have the means to identify potential risks as early as possible and to react accordingly. Such a strategy is bound to fail sometimes, but that should occur less often than in the past. Since underwriters are rarely geoscientists, they will need the back-up of scientists in the

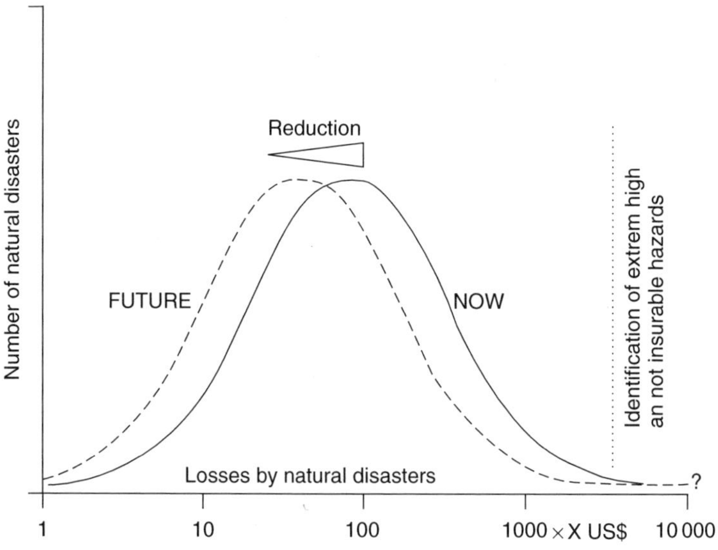


**Figure 8.3**   *Objectives of natural risk management*

form of tools that are easy to use. These tools should aim to transmit the necessary basic information.

- **Extremely large risks need to be excluded**    In the right (upper) end of the probability curve represented in Figure 8.3, there are extremely large and improbable risks. If such a singular incident occurs, it may create claims that are equal or higher than the sum of all other risks. Because such incidents are rare, the corresponding risks are often overlooked or even deliberately ignored. But such practices constitute a real risk to insurance companies, and they and their customers must learn how to deal with extreme risks, even if this involves long-term considerations that exceed the own life span. This is especially important if the frequency–magnitude relationship (understanding the magnitude not in the seismological sense, but as the energy discharge or destructive potential of an incident in general) is so extreme that catastrophes can only be expected in intervals of centuries or millennia.

- **The combined ratio needs to be lowered**    This figure is the quotient of all profits and expenses of a company. It is defined to be equal to or lower than 1 (or 100%) if the profits correspond or exceed the expenses. If insurers are forced to resort to their capital rents in order to satisfy their coverage claims, the combined ratio is clearly above the 100% level. Such practices may be tolerated to a certain point, as was the case during the boom in the stock markets over recent years. However, the stock market bubble has since burst, and the trend towards larger natural catastrophes implies that the future financial burden on insurers is growing (UNEP, 2002). The only way to react under those circumstances is to adjust insurance premiums, introduce higher sublimits, or to deny cover for the highest risks. The last strategy is, however, countered by the fact that the rise in natural catastrophes is bound to foster the demand for insurance cover. Insurers will have to ponder to which point they are willing to satisfy the demand for insurance against natural catastrophes in the future, in order to maintain the functioning of the insurance system, and thus to fulfil their function towards society. They can only continue this service if they succeed in generating profits in an increasingly difficult market environment. If insurers fail to strike this balance, the insurance sector will have to cease covering elementary risks in the middle and long term, which in turn would force society to shoulder the risk. Devastating natural catastrophes without the additional cover provided by the insurance sector might therefore cause wider economic and societal repercussions, such as an increase in companies going out of business, higher unemployment levels, or the economic decline of entire regions.


## 8.3   Insurance Policies

### 8.3.1   Science Replaces Competition

For a long time, insurance obeyed the rules of game theory, because this is exactly what the 'law of large numbers', which constitutes the cornerstone of insurance probability calculation, is about (Dacunha-Castelle, 1997): the insurer is taking a bet in exchange for the premium that nothing is probably going to happen. If something does indeed happen, the money collected from all participants in the game (i.e. insurance customers) will cover the loss of the unlucky participant who lost his personal bet. Of course, the betting pool also needs to cover expenses for the organization of the lottery, that is, the expenses of the insurer. However, it seems that the time for betting is running out, because the insurance business is becoming tougher and risks are becoming ever higher and more concrete. Game theorists could assume in earlier days that risks were incalculable and thus subject to the principle of randomness, but this notion no longer holds in many cases. Human nature strives to select risks through which the probability of suffering damage is growing. If, however, the number of losers in the betting pool increases, and inversely the number of winners decreases, there is little point in continuing to call the arrangement a game. For instance, in a region where large losses are caused by flooding every other year, an insurance solution against flood risks would make no sense because the two partners, customer as well as insurer, could not find a profitable arrangement to set it up. In this case, an unpredictable (random) risk has turned into a calculable risk. Such anti-selection, however, goes against the insurance principle of spreading. The only option insurers have under these circumstances is to introduce exclusion clauses, sublimits, coverage ceilings, and to exclude calculable risks or to limit their coverage to the incalculable residual risk. Obviously, the insurance business is thereby becoming more complicated, because scientifically sound risk analyses have to replace former 'gambling practices'. This is exactly where geoscientists could make a most valuable contribution, because only they have the skills to comprehend the causal chain between latent hazards and potential losses, be it with regard to earthquakes, volcano eruptions, flooding, droughts, storms, and of course gravitational mass movements.

### 8.3.2  Liability Insurers are also at Risk

The fact that natural catastrophes turn out to be calculable risks in principle adds another perspective: if risks can be calculated and potential hazards can be identified more precisely, the chance of counteracting must increase. In other words, the possibility of minimizing the risk, avoiding it, or taking appropriate precautions is growing. However, it would only be consistent to assume negligence if somebody failed to counteract identified risks. This in turn may justify liability claims if third parties suffer injury or losses. For example, an architect or equivalent professional adviser who fails to advise his customer against settling in an area that is affected by landslide risk clearly commits a severe planning error for which he must assume liability. Before the background of the tightening of personal and corporate liability by the so-called Basel II regulations, this problem is bound to affect top managers in the foreseeable future. With regard to the insurance market, these developments imply that damages from natural catastrophes may not necessarily be limited to material values, but could be extended to the liability sector, which has so far only been a minor concern for insurers in this respect.

### 8.3.3  Thinking on the Right Scale

The underwriters of primary insurers need tools to facilitate their decisions, and to help them deal with natural hazards. Those tools should be designed as precisely as necessary and as simply as possible. Neither the detailed assessment of each singular risk that engineers may prefer, nor the very global view on risks of the reinsurers, who tend to accumulate individual risks, is appropriate in this respect. The first approach looks at risks on a scale between 1 and 100 metres; the second does not achieve precision below some 10 kilometres or more. The first strategy is financially viable only in exceptional cases; the latter is ill equipped to assess individual risks. What is needed is a method between these two orders of scale, and which may be called a pragmatic scale risk analysis. The method should reflect what is scientifically and technically possible on the one hand and financially feasible on the other. In the context of the risk management practice such a method would combine publicly available data and information based on application of scientific tools and experience in such a way that the most important natural hazards could be inferred for every point on the globe. The validity of such an assessment will depend on the technical state of the art and on the quality of the available data. At the moment, an accuracy of $1\,\mathrm{km}^2$ appears achievable on a global scale, and thus a worthwhile goal. Naturally, this proposal is no more than a compromise, and many people would wish to achieve a higher resolution. It may well be that in future this could be achieved globally. At the moment, however, the pragmatic approach might lead to the identification of regions that should be, and need to be, investigated in detail.

The form such a comprehensive tool might take in the future will be discussed in the subsequent section using two case studies. They will focus on the relation between earthquakes and the landslides that are caused by them. In addition, I shall discuss whether those tools and principles could also be applied to mountain and hill slides in alpine regions before the background of the current climatic changes.

## 8.4 Better Information for Insurers: Examples, Techniques and Case Studies

### 8.4.1 Methodology

There are basically three risk factors which increase the likelihood of landslides in connection with earthquakes:

- The degree of rock looseness needs to be sufficiently high to allow masses to move at all.
- The relief energy needs to be sufficiently high to allow loosened rock masses to move.
- The ground acceleration caused by the energy of an earthquake needs to be sufficiently high to trigger the landslide process.

In order to quantify the landslide hazard, one therefore needs to identify the temporal probability of soil movement as well as the thresholds above which landslides would be triggered, which in turn depend on the local rock looseness and relief energy.

Local rock looseness may be estimated in qualitative terms based on geological descriptions of the region. In the case study of the island of Taiwan, Liou and Hsiao (1999) and Ceri (2000) formed the basis of a rough categorization regarding rock looseness (consolidation high, medium or low). In connection with the second case study concerning the expected increase in landslide hazard due to the retreat of the permafrost limit in Switzerland, a generally high level of rock looseness was assumed.

In order to quantify the relief, a specific method was developed by the author, which follows the ideas of Dikau (1988, 1994, 1996) and Dikau *et al.* (1995), based on a global digital GTOPO30 elevation model. The model, based on a decimal degree scale, was converted by the author into a metric system in a sinusoidal projection of the investigated regions. The resulting data were used to determine the local relief energy ($_{\mathrm{cal}}\mathrm{RE}$).

In order to determine local ground accelerations, an earthquake model was applied, which the author had developed during the period 2000 to 2003. This contribution introduced the said earthquake model for the first time to a larger public; for this reason, the following sections will give some more detail about the model. The statistical frequency–magnitude relationships that determine the probability of the earthquakes were compiled based on publicly available earthquake catalogues (NOAA, 1996; USGS, 2003).

Based on the combination of the risk factors rock looseness, relief energy and typical ground accelerations, hazard maps regarding the likelihood that landslides may be triggered by earthquakes were computed.

The electronic compilation of the data was carried out by a number of programmes developed by the author, as well as by the geographical system ArcView® by ESRI.

### 8.4.2   Earthquake Modelling, a General Simplified Approach

In order to construct a relationship between a mountain slide and the ground acceleration impulse that triggered it, one should know the value of the latter. To measure soil movements, so-called strong-motion accelerometers are valuable tools. These instruments indicate the ground acceleration in units of $m/s^2$ (PGA = peak ground acceleration) or of $cm/s^2$ (GAL). Unfortunately, such measuring systems are very seldom set up in areas where mountain or hill slides occur. If they are set up there, the danger of the instruments being damaged or covered is high. The lack or scarcity of field data may, however, be compensated by computer simulations of the triggering earthquake. If simulation data are utilized, the ground acceleration is consequently not measured but calculated.

### 8.4.2.1   Model description

The following briefly outlines the functional principles of the earthquake model used here: Super-regional factors:

- The $M_w$ magnitude and the basic parameters (depth, position and spatial orientation) are given.
- The wave energy of the largely devastating surface waves expands elliptically.
- The decrease of the amplitude caused by vertical and horizontal absorption in rocky soils is taken into account.
- The calculation of the decrease of the spectral acceleration is based on the mathematical description according to Abrahamson and Silva (in Sadegh-Azar, 2002).

Regional factors:

- Estimations of the thickness $_{cal}TH$ of the overlying sedimentary rocks were made using geo-morphographic criteria based on a digital elevation model GTOPO30, which was converted into a metric system (resolution $1\,km^2$).
- Estimations of the resonance effects of the sedimentary overlay was done according to the ideas of Bormann *et al.* (2001).
- The calculation of the amplitude amplification was based on resonance effects in the sediment body.
- The calculation of amplitude decreases were based on the absorption in the sediment body according to Budny (1984).

In theory, this method allows prediction of the expected soil movements for each km$^2$ during an earthquake which is characterized by the basic and most important parameters.

### 8.4.2.2   Model calibration

The model has already been verified using data from earthquakes; that is, it was tested for coincidence between predicted and actually measured ground accelerations. Figure 8.4 provides some exemplary figures concerning four typical earthquakes:

- **Verona (Italy), 3 January 1117**   $M_w c$. 7.0; major quake in historical time with significant long-range impact; damages are documented in Northern Italy (collapse of numerous cathedrals), Bamberg (damages at the Michaelskirche), damages in buildings in Rottenburg am Neckar, Konstanz, Meersburg and Fenis (Aostatal) as well as, at a large distance from the epicentre, damages at the abbey in Brauweiler near Cologne (Germany). The descriptions of the damages (Gieszberger, 1924; Sieberg, 1940; Montadon, 1953; Carozzo *et al.*, 1972, Schreiner, 2001) were correlated to macroseismic intensities according to the MMI scale and converted into peak ground acceleration (PGA).
- **Roermond (Netherlands), 13 April 1992**   $M_w = 5.9$; quake of medium intensity in recent history with significant long-range impact; perceivable as far as London

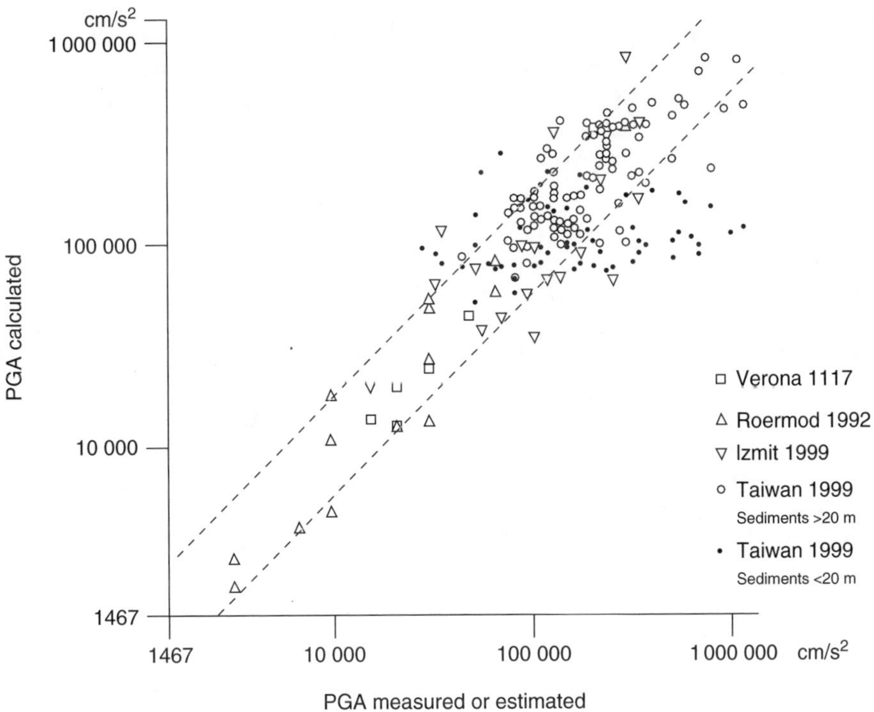


*Figure 8.4*   *Calibration of the earthquake model*

and Zurich (Ahorner, 1994); macroseismic intensities were derived for characteristic locations from intensity maps and converted into ground accelerations.

- **Izmit (Turkey), 17 August 1999**   $M_w = 7.4$; major quake with devastating impact at short ranges; real data about ground accelerations were available from strong-motion accelerometers (Ambraseys *et al.*, 2000).
- **Chi-Chi (Taiwan), 21 September 1999**   $M_w = 7.7$; major quake with devastating impact at short ranges; ground acceleration data were available from strong-motion accelerometers (Taiwan Central Weather Bureau, 2000); the quake triggered numerous mountain and hill slides (Olsen, 1999; Liao and Lee, 2000), which is one reason why it was included in the subsequent case study.

The comparison of computed and measured ground acceleration data in Figure 8.4 shows that the model at the current state of calibration (summer 2003) is fit to predict local soil movement rates with an accuracy of 70% within a range of $\pm 0.75$ intensity levels (MMI), taking into account local soil properties. In the future, this margin of error will be reduced by means of a fine calibration using a large number of documented earthquake incidents on the one hand, and a sensitivity analysis regarding the applied parameters on the other. In addition, I intend to improve my geomorphographic modelling method which estimates the local $_{cal}$TH thicknesses. Possible improvements in the more remote future could be the availability of higher-resolution terrain models at 100 m scales, since variations in the structure of the geological subsoil, especially with regard to the thickness of sedimentary overlays ($_{cal}$TH), are currently not possible below scales of $1 \, km^2$.

The correlation diagram represented in Figure 8.4 also illustrates that, if sedimentary overlays are slim or missing, there may be significant differences between the computed and actual figures. This is especially pointed in the case of the Taiwan earthquake of 1999. The possible reasons for this phenomenon will be discussed in detail below.

### 8.4.3   Case Study 1: Landslides in Taiwan after the Earthquake of Autumn 1999

#### 8.4.3.1   General situation

On 21 September 1999 a massive earthquake took place at the centre of the island of Taiwan. The quake had a magnitude of $M_w = 7.7$. According to the data provided in the earthquake catalogue published by the USGS, the epicentre of the quake was located at 23.77 N and 120.89 E at a depth of approximately 10 km. Other sources state localizations between 23.69 N and 23.87 N, and 120.75 E and 121.31 E respectively. Since the coordinates of the USGS correspond by and large with the mean of the other sources, they were assumed to be accurate and were selected to form the basis for the further analysis.

The quake triggered a number of mountain slides, which were documented during overflights of the area (Olsen, 1999), as well as by satellite pictures. Hung (2000) reports that the quake caused a total of 9272 massive landslides. Liao and Lee (2000) developed a map of the regions that were especially affected by landslides on the basis of satellite picture reconnaissance, which is reproduced in a simplified version in Figure 8.5. As an initial observation, the accumulation of earthquake-induced landslides in the western part of the island should be noted. The underlying reason for this pattern becomes apparent if one considers the individual factors that determine the occurrence of landslides.

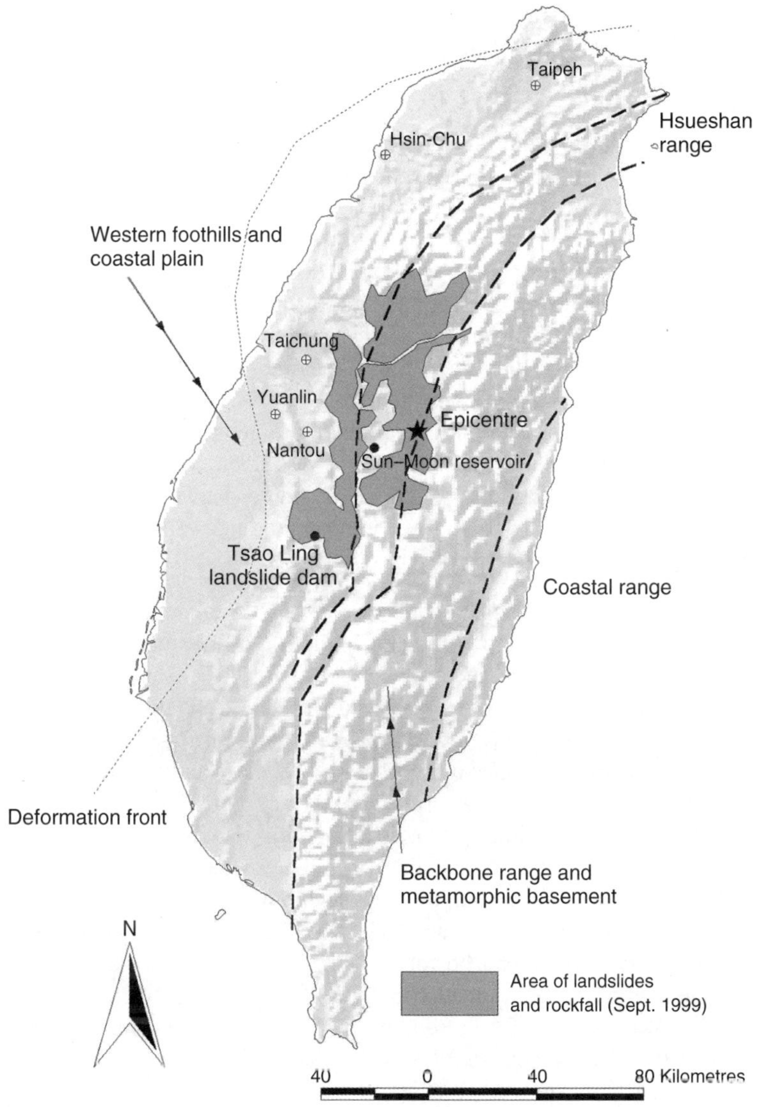


**Figure 8.5**   *Taiwan: geological overview and proliferation of landslides, 1999*

### 8.4.3.2   Geology, tectonic and consolidation

According to Petley (1996), Liou and Hsiao (1999) and Ceri (2000), the geological foundation of Taiwan is almost exclusively made up of Mesozoic and Cainozoic rock, deformed by intensive tectonic processes that continue to the present. Going from west to east, the island may be divided into the following geotectonic units:

- Western foothills and coastal plain: Quaternary to Miocene strata, which are deposited flat in the west and increasingly folded towards the east. A deformation front, which divides sediments that are not yet affected by tectonic activity in the west from young folded structures in the east of the island, cuts through this area following a wide arc from northeast to southwest. The thickness of the Miocene to Quaternary sediments in the area of the coastal plain is approximately 1500 to 3000 m. A low degree of consolidation was attributed to this zone.

- Hsueshan range: folded Palaeogene strata that reach up to heights of more than 3000 m in the central region of Taiwan. This region is considered to have a medium degree of consolidation.

- Backbone range and metamorphic basement: this region consists of heavily folded shales of the pre-Cainozoic basement, and overlying sedimentary Tertiary strata. The degree of metamorphism of the rock increases towards the east. At the border towards the coastal range, there are highly metamorphosed green shales. This zone was attributed a high degree of consolidation.

- Coastal range: along the eastern shoreline there are neogenous volcanic rocks, which should be understood as the continuation of the Luzon arc to the south of Taiwan that reaches up to the Philippines. The tectonic border between the coastal range and the continental part of Taiwan follows the longitudinal valley. This region is also considered to have a high degree of consolidation.

The epicentre of the earthquake of September 1999 lay in the region of the Lishan fault between the Hsueshan range and the Backbone range. This line marks the transition from shales of higher metamorphic degree in the eastern part of the island to relatively young folded or flat sediments of Tertiary or Quaternary origin in the west.

Based on the geological setting, one can assume that east of the Lishan fault, rocks are relatively more resistant to mechanical stress than the younger and less consolidated sedimentary rocks in the western part of the island. As Figure 8.5 illustrates, the distribution pattern of the earthquake-induced landslides confirms this notion: east of the Lishan fault, there are only few traces of landslides in the area of the Backbone range, even in the direct vicinity of the epicentre. An increase accumulation of landslides only manifests itself beyond the Hsueshan range, some 40 km northwest of the epicentre, where rocks of lower mechanical stiffness can be found. The widest spread of landslides occurred, however, in the area of the geologically young folded sediment structures of the western foothills, some 30 to 50 km west or southwest of the epicentre. This is also where the Tsao Ling landslide dam was created through a massive mountain slide, some 40 km to the south of the city of Nantou (Figure 8.6).

The earthquake of 21 September 1999 thus revealed a qualitative relation between the geological setting and the probability of landslides, since the properties of the geological underground, defined as the degree of consolidation or looseness of rock, is known to be a primary risk factor.

### 8.4.3.3  Earthquake-induced ground accelerations

*Comparison of computed and measured figures.*  During the earthquake nearly 200 strong-motion accelerometer stations documented the horizontal and vertical ground movements in the central part of Taiwan. As X-, Y- and Z-components these data are available in the internet (Taiwan Central Weather Bureau, 2000). For this case study only the resulting horizontal figures were used. According to the evaluation of the digital elevation model

***Figure 8.6***   *The massive Tsao Ling landslide dam (photo by R.S. Olsen, ERDC-WES). Reproduced by permission of R.S. Olsen*

following geomorphographical criteria, the positions of the 200 stations were described as being on rocky or sedimentary underground. Concerning the approximate thickness of sedimentary coverage, the estimation of the model gave figures between $_{cal}TH = 1\,m$ and $_{cal}TH = 20\,m$ for mountain regions and smaller valleys. In flat regions such as the coastal range in the western part of Taiwan (Ceri, 2000) the thickness of sedimentary coverage may reach 3000 m.

As shown in Figure 8.7, the acceleration measurements are widespread. The reduction of the earthquake intensity dependent on the distance to the epicentre is just slightly pronounced. Particularly striking is the fact that the registered ground accelerations on rock basement without considerable sedimentary coverage vary from a distance of roughly 50 km to the epicentre between $25\,cm/s^2$ and $700\,cm/s^2$. This corresponds to more than four levels on the MMI scale. The commonly held opinion that on rocky ground the earthquake intensities are generally lower than in areas with sedimentary coverage has not been confirmed by the earthquake in Taiwan. Regardless of the distance from the epicentre, the ground accelerations measured on rock show a scale similar to those measured on sedimentary basement. One of the possible reasons for this similar and unexpected reaction might be the inhomogeneous geological structure of the island, but it cannot be the sole explanation for the phenomenon.

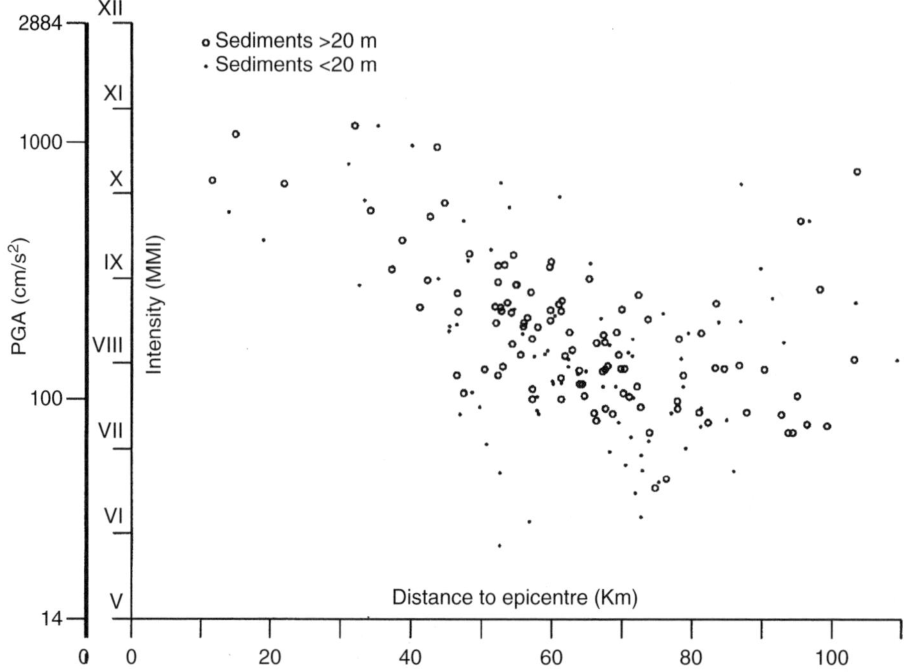



***Figure 8.7***  *Chi-Chi earthquake, 1999, spreading of earthquake intensities in comparison to distance to epicentre*

Source: Taiwan Central Weather Bureau

Contrasting the measured ground accelerations against the computed ones – which have been calculated according to the procedure explained above – in a correlation diagram (Figure 8.8), it becomes clear that ground accelerations generated by the earthquake in areas with considerable sedimentary coverage are more accurately characterized by the model than those in areas with basement rocks outcropping.

Comparing the measured ground accelerations (Taiwan Central Weather Bureau, 2000) with those calculated by the author according to the previously explained earthquake model yields results as shown in Figure 8.9.

The filled circle symbol on the figure means a good correlation between the model and reality – good correlation defined here as calculated ground acceleration that should not differ from the measured one by ±0.75 levels on the MMI scale – has been found throughout all regions, those close to the epicentre as well as those far away. This applies especially to areas with thick sedimentary coverage ($_{cal}$TH > 20 m). Correct results were calculated for 84 out of 108 stations. That is a hit rate of roughly 78%. In rocky areas ($_{cal}$TH <= 20 m) the degree of accordance is only 54%. For 8 out of 78 stations located in rocky areas the ground accelerations calculated according to the model were too high. During the earthquake further 28 stations registered higher ground accelerations than those that can be calculated according to the model. This is a difference of up to three levels on the MMI scale. This shows that ground accelerations calculated according to the model are smaller by a factor of ten than those actually registered at these places.

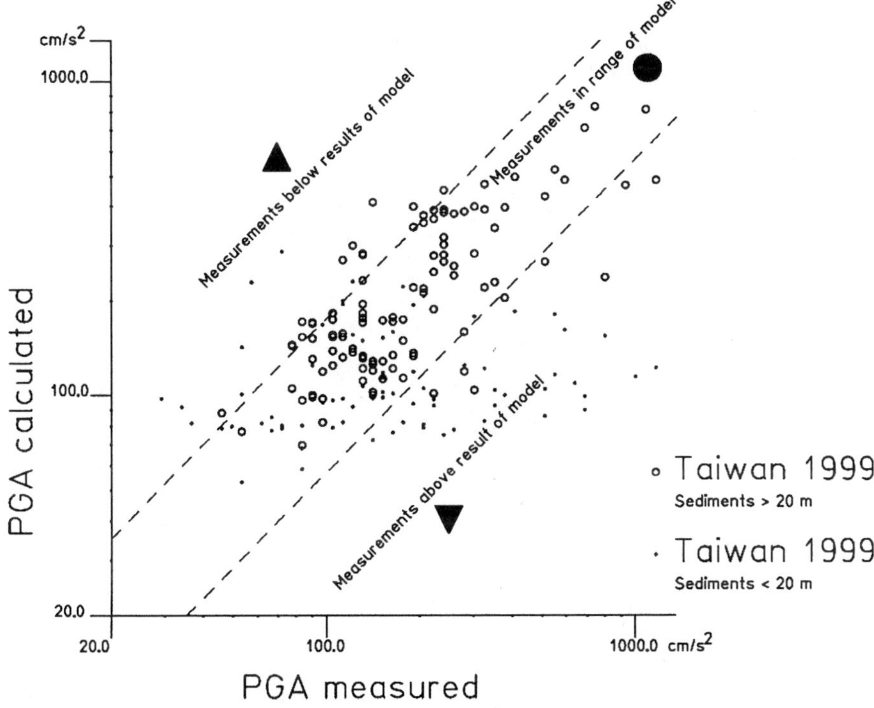



**Figure 8.8**   *Chi-Chi earthquake, 1999, measured and calculated ground accelerations*
Source: Taiwan Central Weather Bureau

The filled triangle symbol means, as Figure 8.9 shows, ground accelerations calculated according to the model fall below the registered ones. This mainly happened in the west of Taiwan and is limited to the area of the western foothills. Accumulations of this phenomenon can be found close to the Tsao Ling landslide dam (15 to 30 km to the west and to the south) as well as between the epicentre and the town of Nantou. A considerable number of underestimated ground accelerations can be found north of the town Taichung, where the western foothills and the coast of Taiwan meet. It is remarkable that there are good correlations between measurements registered at three stations situated close to the epicentre and ground accelerations calculated according to the model. According to Liao and Lee (2000), this region has hardly been affected by rockfalls and landslides. Ground movements affected mainly the area east of one of the three seismological stations, the area around the epicentre, and the belt to the west of the western foothills, the last area with a higher frequency. We have acceleration measurements registered by 8 stations from this area highly affected by landslides – it stretches 80 to 100 km from the Tsao Ling landslide dam in the south to the region in the east, northeast, and north of the town Taichung. Each measurement shows ground accelerations significantly above measurements expected according to the applied earthquake model. This points to a causal relationship between the underestimated ground accelerations according to the earthquake model and the frequent occurrence of rockfalls and landslides.

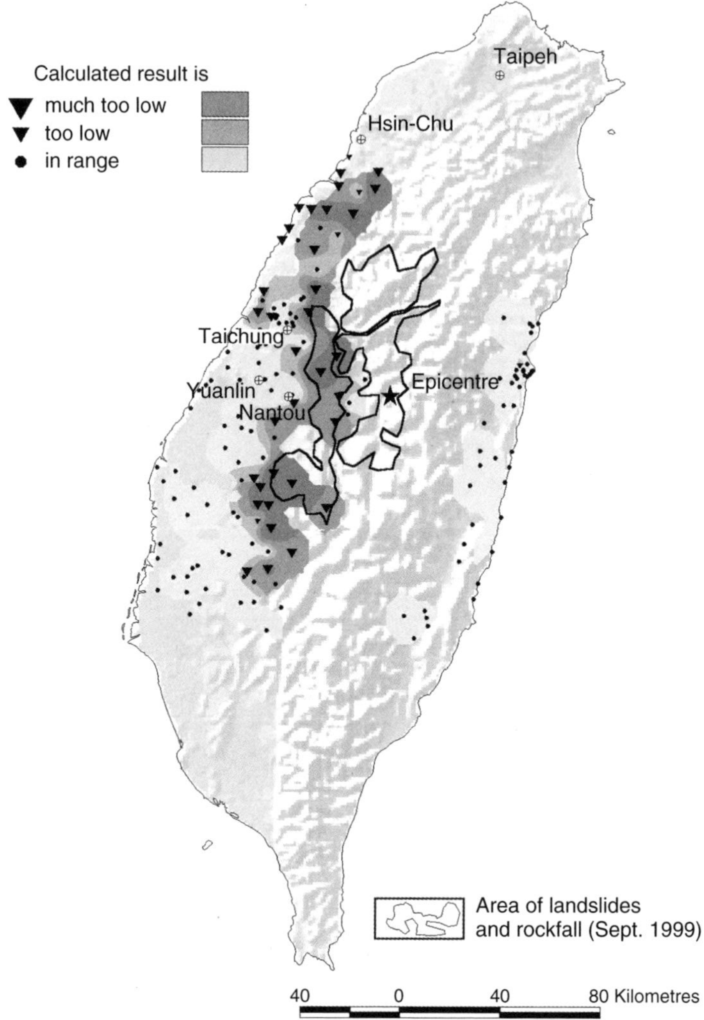


**Figure 8.9** *Chi-Chi earthquake, 1999, distribution and quality of calculated ground accelerations*

In fact, ground accelerations higher than those calculated according to the earthqauke model are rare. These were registered by 23 out of altogether 186 stations (12%). Eight of the stations were situated in rocky areas and 15 in areas with a sedimentary coverage of $_{cal}$TH > 20 m. As Figure 8.10 shows, the stations are partly in the region of the deformation front – that is at the transition of the western foothills into the coastal plain – furthermore at a part of the coastal plain in the southwest of the town Yuanlin, which is close to the coast, and at least along the longitudinal valley in the eastern part of the island. All regions have one thing in common: a very complex geology together with local variations of sedimentary coverage which cannot be resolved by the elevation model – used as data basis – with an accuracy of $1\,km^2$, or derived from geomorphographic

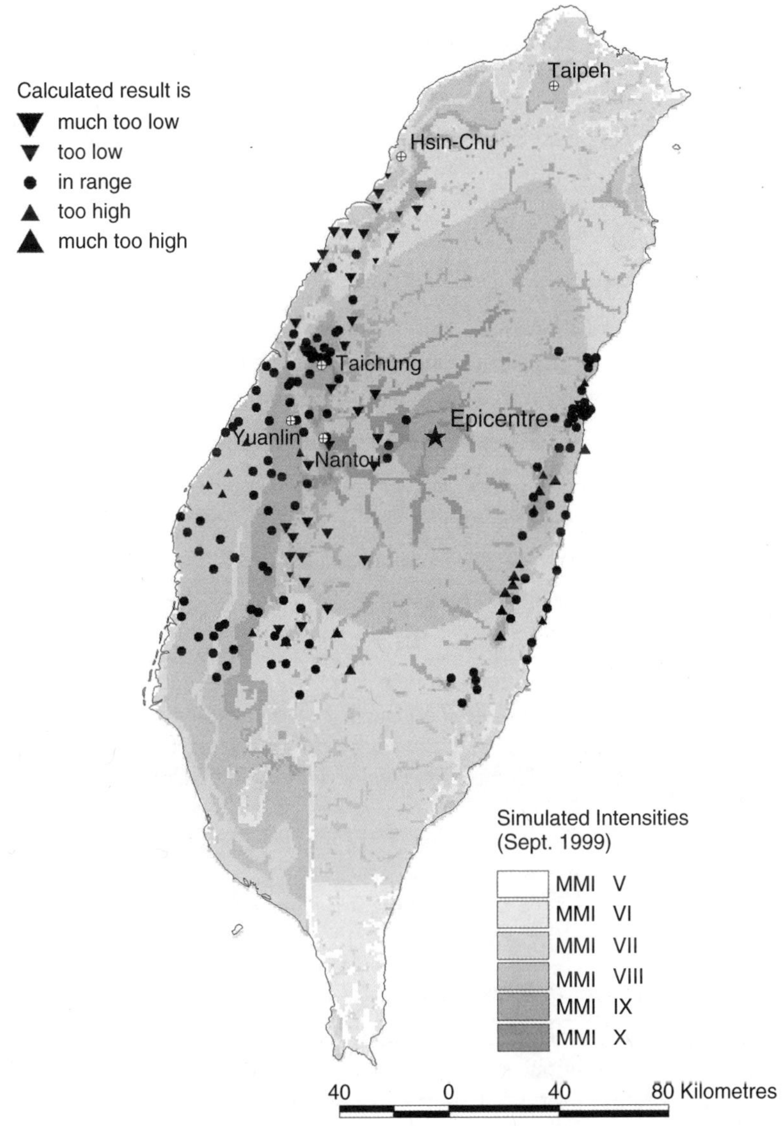


**Figure 8.10**  *Chi-Chi earthquake, 1999, computed intensities (MMI). Reproduced by permission of R.S.Oslen*

criteria. Similarly in the longitudinal valley, the complexity of the geotectonic structures is setting limits to the applied computer models, at least for the moment. In the area of the coastal plain, southeast of Nantou, probably false estimation of the thickness of sedimentary coverage plays a role. It is likely that here the standards for $_{cal}$TH have been set too high; therefore the earthquake model shows an unrealistic reinforcement of amplitude caused by resonance effects.

*Results.* All in all, the earthquake model provides correct results, as can be seen from Figure 8.10. Apart from the differences previously explained and verified, there is a high degree of correspondence between the calculated and the measured ground accelerations even in areas with a complex geological structure: almost throughout the complete area of the coastal plain in the west of Taiwan and the coastal range alongside the east coast, most measured ground accelerations are correctly reported within the tolerable margin of error of ±0.75 levels on the MMI scale. At this point it should be emphasized that the comparatively high ground accelerations in the surroundings of Taichung, Yuanlin and Nantou caused by very special geological conditions were correctly reproduced, as well as the maximum ground accelerations of more than $600\,\text{cm/s}^2$ (intensity > X on the MMI scale) in the area of the valleys at a distance of about 10 km from the epicentre.

The following ground accelerations can be derived as threshold values for mass movements (lowest limiting values and arithmetic mean within the affected areas):

**Consolidated sedimentary rocks and metamorphic shales:**
PGA = 177 to 251 cm/s$^2$ (MMI $\gg$ VIII)

**Moderately consolidated sedimentary rocks:**
PGA = 117 to 225 cm/s$^2$ (MMI $\cong$ VIII)

**Slightly consolidated sedimentary rocks:**
PGA = 116 to 200 cm/s$^2$ (MMI VII to VIII)

This means that ground accelerations of $100\,\text{cm/s}^2$ have to be exceeded to trigger landslides which change the landscape and therefore are recognizable on satellite images. This corresponds to an intensity of VII–VIII and applies to slightly or moderately consolidated rocks. For consolidated or solid rocks this value increases to $150\,\text{cm/s}^2$, exceeding an intensity of VIII.

This way of looking at the problem does not include minor earthslides or rockfalls; obviously these can be triggered by much lower ground accelerations under unfavourable conditions – even ground accelerations of 7 to $15\,\text{cm/s}^2$ (intensity IV–V) might be sufficient.

### 8.4.3.4   Morphology and relief energy

*Method of calculation.* Provided the mountains are sufficiently unconsolidated and the rock loosened through an impulse by an earthquake, the influence of gravity is required to initiate a slide, slump or rockfall. Without relief provided by the morphology of the area the loosened masses of rock would not be able move downslope.

Within the framework of this examination based on the digital elevation model with a resolution of $1\,\text{km}^2$, it is not possible to calculate definite values for the relief energy of single occurrences of slides which are far below this resolution. Therefore a parameter had to be found by which the regional relief energy could be characterized, and which should be derived from the globally available elevation model gridded in $\text{km}^2$. Values were taken in areas of $49\,\text{km}^2\,(7 \times 7\,\text{km})$ and the calculated average difference of the 49 heights from the shared arithmetic mean proved to be a suitable index. The result corresponds with the average topographic gradient with reference to a standard distance of 1000 m. According to this method, in alpine areas one gets values of $>1000\,\text{m}$ (Montblanc massif), in German's low mountain range one gets values between 30 and 300 m, in the hilly area at the foot of the low mountain range one gets values between 5 and 300 m, and in the plains one gets values of $< 5$ m. Following this model the parameter will be called $_{cal}RE$. This stands for a measurement of the typical potential energy of a region which can be turned into kinetic energy during an earthslide under the influence of gravity. This model offers an advantage as it can be applied globally due to the general availability of its data basis.

*Results.* The calculated $_{cal}RE$ values for Taiwan go as high as 700 m for the central high mountain range, they are between 10 and 400 m for the western foothills, and below 10 m for the region of the coastal plain. Taiwan's areas affected by landslides can be characterized as follows (lowest limiting values and arithmetic mean):

**Consolidated sedimentary rocks and metamorhic shales:**
$_{cal}RE = 300$ to 444 m

**Moderately consolidated sedimentary rocks:**
$_{cal}RE = 100$ to 360 m

**Slighly consolidated sedimentary rocks:**
$_{cal}RE = 100$ to 213 m

Areas with a $_{cal}RE$ value of more than 100 m have to be categorized as risk zones for extensive and hazardous landslides and rockfalls. This applies to solid, not loosened, consolidated rock from a value of $_{cal}RE > 300$ m. As mentioned before, it should be pointed out that smaller earthslides or those that cannot be identified by satellite image can already occur at $_{cal}RE \ll 100$ m.

### 8.4.3.5   Liquefaction

Liquefaction should be seen as a special kind of mass movement of rocks triggered by an earthquake. Liquefaction and landslides have a similar origin. For the sake of completeness the phenomenon will be dealt with here: vibrations change the mechanical characteristics of soil material so that under the influence of load a mass movement alongside zones of weakness, faults, or slip circles might occur.

Everybody who has ever tried to dance the twist on a wet sandy beach knows the effect: the initially solid ground starts to liquefy, becomes viscous and eventually one sinks into in up to one's ankles. Something similar happens below the foundations of a building during an earthquake. The load-bearing capacity of the basement is reduced through the liquefaction of the soil structure caused by shaking and the building sinks into the ground or tips over (Figure 8.11). Preconditions for the appearance of liquefaction are a fine sand as well as a small distance between groundwater level and the surface of the earth. Therefore plains in the close vicinity of rivers and lakes, especially at the waterfront, are highly endangered.

Several sources from the towns Taichung, Nantou and Yuanlin report liquefaction as a consequence of the earthquake on 21 September 1999. Numerous websites about this and other earthquakes provide photographs of buildings that have tipped over although their substance has not been completely destroyed. In any case, such buildings are no longer useable and have to be seen as a complete write-off. Applying the earthquake model to this case, ground accelerations of 410 (Taichung), 426 (Nantou) up to 430 cm/s$^2$ (Yuanlin) must have been necessary. The actually measured PGA values in these areas

**Figure 8.11**  *Turkey, 1999: the effect of liquefaction (photo: R.S. Olsen, ERDC-WES). Reproduced by permission of R.S. Olsen*

show a similar scale. During an earthquake in 1999 a slightly lower ground acceleration of roughly $300\,cm/s^2$ caused liquefaction: on 17 August a severe earthquake shook the west of Turkey; meanwhile in the town of Adapazari, situated at the waterfront of a lake, numerous buildings sank into the muddy basement. Information that liquefaction might occur with low ground accelerations can be found in the characterization of the 12-level MMI scale in Bolt (1993). Referring to the characterization which corresponds according to the conversion formula of Gutenberg and Richter (1956) to a ground acceleration of $150\,cm/s^2$, effects such as the eruption of sand and mud or modifications of the groundwater can be observed from level VIII. In the framework of a risk analysis and with a certain security factor one should assume that ground accelerations of roughly $100\,cm/s^2$ are sufficient to trigger liquefaction, if the ground structure shows features such as fine wet sand, silt, clay, or a high groundwater level. Therefore areas endangered by liquefaction are often found in plains or wide valleys in which fine-grain sediments with a reasonable thickness ($>20\,m$) are deposited, and where due to the topgraphical features a small difference between the earth's surface and the groundwater level has to be assumed.

### 8.4.3.6  Result evaluation and hazard mapping

The quantitative and qualitative relation between the degree of consolidation of mountains, relief energy, and local ground acceleration during earthquakes can be used to draw up hazard maps. Initially it is necessary to quantify the seismic hazard. This tells us how often, at certain places according to statistical return periods, earthquake magnitudes occur and what ground accelerations can be reached or exceeded. According to common agreement, magnitudes with an exceedance probability of 10% within 50 years – this corresponds to a return period of 475 years – are used. To quantify the earthquake hazard for Taiwan in this way, data from earthquakes in history and in more recent times in and around Taiwan were entered into a statistical frequency analysis. Using methods well known from specialist literature (Schick and Schneider, 1973), it was possible to determine the expected magnitudes for any return period and to use them for the simulation of 'synthetic' earthquakes in a computer model. These formed the basis for the calculation of the expected maximum ground acceleration for the statistic return period of 475 years all over Taiwan, as used to describe the single earthquake on 21 September 1999. The results are shown in Figure 8.12.

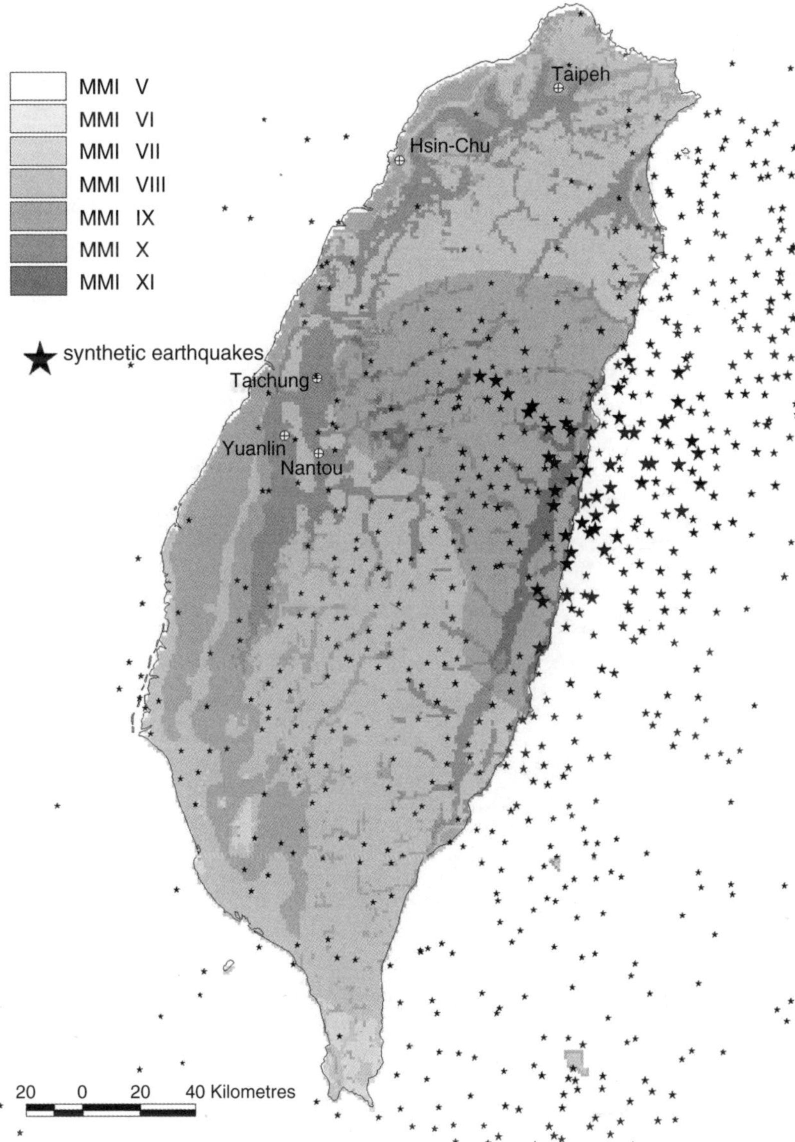


**Figure 8.12**   *Taiwan, computed hazard map of earthquake intensities (return period of 475 years)*

Based on this, and considering the relations derived above, it was possible to draw up a hazard map of gravitational ground movements – landslides and liquefaction – which can be triggered by an earthquake (Figure 8.13). Initially, it was examined whether the threshold value for the local degree of soil disaggregation for ground acceleration was exceeded. In a second step it was checked whether the necessary relief was available. The lowest limiting values were used, as any hazard analysis should have the character of a worst-case scenario.

According to this analysis, the areas that are highly threatened by gravitational earth movement during and after an earthquake are the central and the northern part of the island. The area east of the towns of Taichung and Nantou has to be considered as

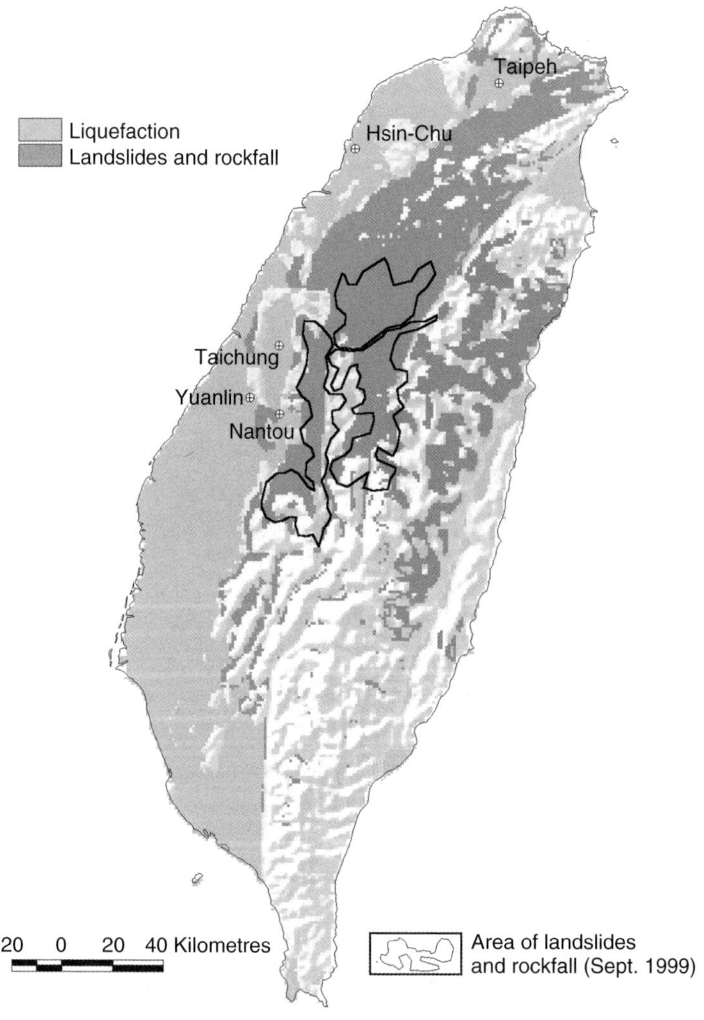


**Figure 8.13** *Taiwan, computed hazard map of earthquake-induced landslide and liquefaction*

especially threatened: there, even in the highly consolidated rocks of the Backbone range, which escaped any damage in 1999, huge mass movement is likely. Mega earthquakes off Taiwan's east coast have to be seen as a possible trigger for rockfalls and landslides in this area.

### 8.4.3.7   Analysis and results

There is good agreement between the threatened zones calculated according to the model and the regions that were actually affected in 1999, although the evaluations shown in Figure 8.13 are based on numerous 'synthetic' earthquakes and do not have any reference to the single earthquake in 1999, in relation to both position of the epicentres and the imputed magnitudes. This points to the fact that the hazard factors due to the local geology and topography contribute far more to the occurrence of landslides than the impulse of an earthquake. To put it simply: in threatened zones destabilized masses of rocks are waiting for a final impulse to go down. At the same time this means that the initiation of earthslides by earthquakes has been underestimated in the past. For example Bolt (1993) explains in his description of the MMI scale that considerable earthslides on embankments and steep slopes only occur from level X; this corresponds to a PGA value $> 590\,\mathrm{cm/s^2}$. Comparable definitions can be found in the reports of several authors. The consequences of the earthquake in Taiwan in 1999, however, show that even much lower ground accelerations, corresponding to intensities between VII and VIII, are sufficient to trigger tremendous mass movements. Paradoxically, there were no significant earthslides in the only region in Taiwan in 1999 where ground accelerations of $600\,\mathrm{cm/s^2}$ were significantly and evidently exceeded, because the necessary geological and morphological preconditions were not present. In Figure 8.13, which shows the general landslide hazard in this area, the region around the Sun–Moon reservoir remains danger-free.

Finally, the earthquake in Taiwan indicated that specific vibration measurements by strong-motion accelerometers in the area might include hints at latent earthslide-threatened areas. This applies above all to the western foothills (Figure 8.13) because the whole of this low mountain range has been marked – beginning in the southeast third of the island up to the region west of Taipeh – as an area of high hazard. At the same time it is the area where during the earthquake in 1999 – at least at places with seismic stations – ground accelerations were registered that were significantly above the values calculated according to the earthquake model (Figure 8.9). A plausible explanation for this phenomenon is that at the affected seismic stations it was not the primary effects of the earthquake that were registered, but the secondary ground accelerations caused by the earthquake. Obviously, these were more violent than the earthquake itself. It might be that latent tectonic faults were mobilized by the seismic shock. Secondary tectonic movement during the 1999 earthquake may have been sufficient to weaken these slopes without actually initiating landslides. It is therefore advisable to consider this area as extremely threatened, as it is to be feared that it might collapse completely with the next violent earthquake.

With regard to liquefaction, the complete western part of Taiwan (Figure 8.13) must be acknowledged as a danger zone. In addition to this, the plain in the surrounding area of Taipeh, the coastal plain in the northeast of the island, and some parts of the longitudinal valley in the east are also danger zones. Throughout these areas one has to reckon with construction failures by liquefaction if the local conditions are as mentioned above (ground accelerations $>100\,\mathrm{cm/s^2}$, sedimentary fine-grain subsoil with a thickness of $>20\,\mathrm{m}$), and if there is a groundwater level near the surface.

Taiwan served as an example of a simple way of hazard mapping from relatively few general data and little information, but with a high degree of detailed information, much more detailed than the maps usually used in the insurance business. This case study shows that it is possible to recognize specific hazards before an event, as well as to delimit zones of danger more precisely than with the procedures used by insurance companies at present.

As every hazard factor which has been taken into account (e.g. mechanical stability of mountains, relief, ground acceleration, and thickness of the sedimentary coverage) applies outside Taiwan as well, it should be possible to apply the same standards to identify landslide hazards in other parts of the world.

### 8.4.4   Case Study: Potential Landslide Hazard in the Swiss Alps

#### 8.4.4.1   General situation

Switzerland, located on the tectonic borderline between the Eurasian Plate and the African Plate, is regarded as an earthquake zone. Devastating catastrophes have not occurred recently, but Basel was reduced to ruins by an earthquake of magnitude $M_w \cong 7$ in the year 1356. Due to the enormous period of time that has passed since, people are not aware of the risk. However, that does not change anything about the existing risk – like the sword of Damocles hanging above one's head. In addition to that, there are earthquake zones in the south and the east of Switzerland, and seismic long-distance effects from the very active northern part of Italy should not be neglected. At this point, one should mention the earthquake in Verona in 1117 which sent its seismic waves far into Germany and probably caused damage as far away as Cologne.

Other risk factors increasing the occurrence of earthslides of larger dimensions are also present, as everybody knows. The geologically complex structured high mountain area shows every degree of density – from completely unconsolidated structures to solid rocks. That the relief energy to be expected here will be match the one in Taiwan does not need any further explanation.

#### 8.4.4.2   Permafrost and climate change

In Switzerland an additional factor has to be taken into account that plays no role in Taiwan. In Switzerland as well as in other high mountain areas throughout Europe, there are regions with continuous or discontinuous permafrost, where pore- and cleftwater are frozen the whole year and give cohesion to the disaggregated rock structure. According to Nutz (1999), permafrost can go as deep as $100\,\mathrm{m}$ into the rock. Permafrost mainly occurs where a protecting and isolating blanket of snow is missing or is of minimal thickness, mainly on unvegetated steep slopes. Nutz (1999) gives $2600\,\mathrm{m}$ above sea level as the limiting border for discontinuous permafrost in the central Alps, and $2400\,\mathrm{m}$ for permafrost in the eastern Alps. From this, one can derive a value of $2500\,\mathrm{m}$ for the south of Switzerland. The limiting border for continuous permafrost is $500\,\mathrm{m}$ up to $1000\,\mathrm{m}$ higher. Against the background of global warming, which according to IPCC (2001) will possibly reach between 1 and $4\,^{\circ}\mathrm{C}$ this century, it is to be expected that with increasing mean annual temperatures the limiting permafrost border will shift to higher regions.

Even taking the uncertainty of climate prognosis into account, and one estimates an increase of only 1–2 °C, there will be a shift of the permafrost borderline between 100 and 200 m upwards. In this way, at heights between 2500 m and 2700 m above sea level newly structured areas will come into being as the melting permafrost leaves a highly unconsolidated rock structure, because the ice that is now operating as a binding agent will have disappeared in the foreseeable future. *A priori*, mountainous regions have all the preconditions that encourage rockfalls and landslides: unstable rocks, high-relief energy, and the complete lack of protecting vegetation with roots to give cohesion to the rocks. In these conditions, the smallest impulse may trigger a disaster.

As an analysis of the digital elevation model carried out by the author (basis: GTOPO30) reveals, the high mountain areas of Switzerland and the bordering countries within the detail of the map shown in Figure 8.14 (about 8470 km$^2$) are above a height of 2500 m above sea level and can therefore be regarded as permafrost regions (compare von der Mühll *et al.*, 2001). With a shift in height of the permafrost zone of only 200 m upwards, the area would be reduced by approx. 3600 km$^2$ to 4870 km$^2$, as summing of the parts shown in Figure 8.14 shows. This means that with a relatively small increase in temperature of 1–2 °C, about 42% of the now existing permafrost area will disappear and leave disaggregated steep slopes and unstable debris waiting to be initiated on the first impulse. The eastern part of Switzerland will be highly affected because the permafrost areas there will be reduced by more than 50% and only relics of the present stands will remain. Of course, this hazard won't become obvious all of a sudden. With a warming of 0.2–0.3 °C each decade (Loster in Munich Rea, 2000b), permafrost degradation will

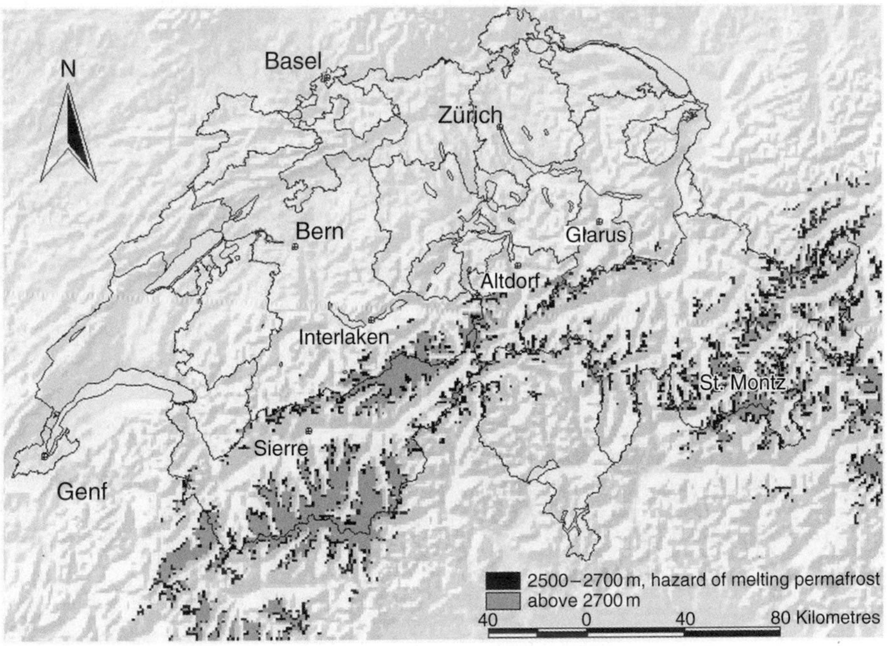


**Figure 8.14**  *Switzerland: distribution of permafrost areas*

take years or decades, therefore the process will be slow, and be manifest by a gradual increase in rockfalls, mudflows and landslides.

### 8.4.4.3   Distribution of earthquake hazards

To determine the ground accelerations expected during an earthquake the method described in the case study of Taiwan was used. Figure 8.15 shows the results. The method indicates that earthquakes of intensity VII on the MMI scale could recur within time intervals of 500 years throughout many areas of Switzerland. A slightly lower hazard can be found in the belt extending from the southwest to the northeast between the cities of Geneva, Bern, Zürich, and Lake Constance, where there are numerous areas in which intensities of V probably won't be exceeded. On the other hand centres of higher hazard are to be found throughout the entire canton of Valais, in the southern part of the canton of Bern, in the region around Glarus, in the surroundings of Lake Neuchatel and Lake Bieler in the northwest of Bern, as well as in Geater Basel, where locally and, under unfavourable underground conditions, earthquake intensities up to level VIII on the MMI scale in a return period of 475 years are possible. An even higher hazard has been acknowledged for numerous valleys of the Alps in the south and in the east of Switzerland, where it can be assumed that under the influence of increasing amplitudes caused by resonance effects within the unconsolidated sediments at the bottom of the valley, intensities up to IX might occur. This applies especially to the canton of Valais (the surroundings of Sierre) and in the region around Interlaken.

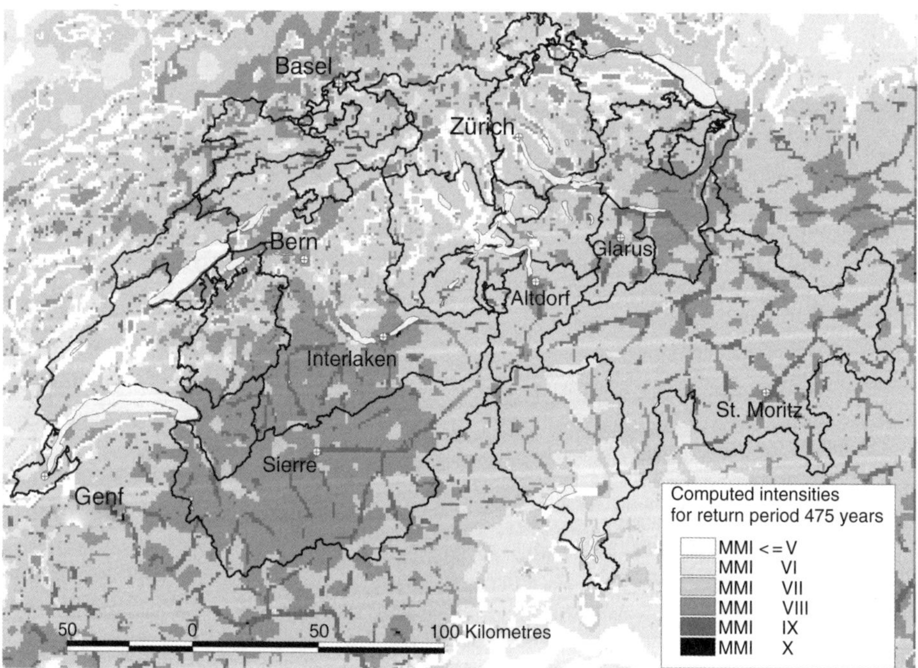



***Figure 8.15***   *Switzerland: computed hazard map of earthquake intensities*

### 8.4.4.4   Identification of earthquake-triggered landslide risks

Applying the criteria necessary to trigger a landslide (limiting values for ground accelerations and relief energy together with a highly loosened rock structure due to a retreat of the permafrost zone) worked out in the the case study of Taiwan to the shown spatial distribution of general seismic hazard (Figure 8.15), the hazard map shown in Figure 8.16 results. According to today's knowledge, shaded areas have to be regarded as threatened by landslides if there are seismic shakes. With the exception of Ticino, the valleys of the Swiss Jura Moutains in the northwest as well as almost every large valley of the Swiss Alps have to be considered as hazard zone. Among them the Matter Valley, where on 18 April 1991, about 10 km from Zermatt, one of the largest rockfalls in the recent past took place, in which 30 million m$^3$ of rocks descended and dammed the river Mattervispa to produce a lake (Glade and Dikau, 2001). On the same day, at 4.37 a.m., in this region, a weak earthquake, magnitude 2.8, was recorded by the international earthquake catalogue of the USGS. It is most likely that the rockfall was not triggered by the earthquake, which was poor in energy. It is more likely that in this case the rockfall caused the vibration which was recorded and included in the international earthquake catalogue.

### 8.4.4.5   Implications for insurance companies

The areas acknowledged as risk zones in Figure 8.16 comprises all in all 2516 km$^2$, essentially along the main valleys. First and foremost, important and frequently used traffic routes are threatened. In addition to that, rivers and streams with a high level of water could turn into lakes after damming by large landslides. On the one hand, this

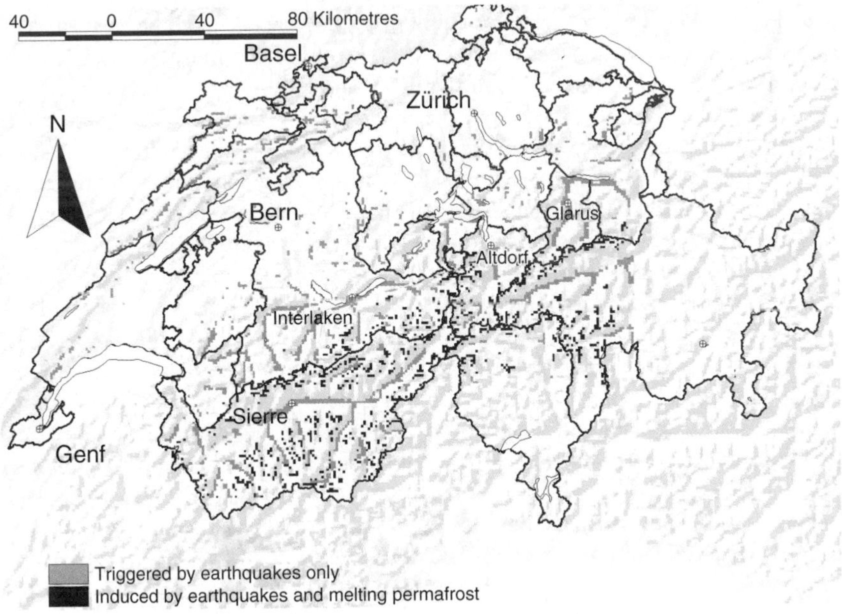


*Figure 8.16*  *Switzerland: computed hazard map of earthquake and melting-permafrost-induced landslides*

could lead to the flooding of settlements and arable land, and on the other, to disastrous torrents in the lower reaches when the dam breaks.

In addition, it should be taken into account that due to melting permafrost huge areas with highly unconsolidated rocks are coming into existence which on steep slopes react extremely sensitively to seismic shakes and will cause landslides and rockfall of greater dimensions as the risk zone expands by $1775\,km^2$, marked black in Figure 8.16. In relation to triggering by earthquakes, the landslide hazard throughout Switzerland will increase by 70% as a result of the expected global warming. The new risk zones are far away from traffic routes and rivers, which, however, does not mean there is no risk. This is because in the areas of high mountain valleys and mountain pastures, landslides are a threat to the major source of income of the country: tourism. In the long run, the winter sports regions in the surroundings of St Moritz and the canton of Valais will be highly affected.

## 8.5    Lessons Learned – Perspectives for Insurance Companies

A road sign pointing out that there is danger of fog, risk of skidding or black ice does not mean that one will inevitably have an accident. It only warns about the danger of reckless driving. The mapping of risk and danger zones has to be understood in exactly the same sense. The knowledge of hazard zones provides only instructions as to level-headed behaviour. Anybody who does not act according to the risks must take the consequences in the case of an emergency. This applies to a driver who despite the road sign drives over the speed limit into a bank of fog, and to anybody who thinks it necessary to settle on the slope of an active volcano, on the waterfront within the flood zone, or below a steep rock face. People who act imprudently can not expect anybody else to bear the costs of the damage. This aspect exactly matches one of the most important principles of the insurance business: anybody who wilfully and deliberately takes such a high risk should either be willing to pay a high premium or to go without insurance coverage. On the other hand, anybody who avoids high risks or takes other kinds of precautions will be rewarded with moderate premiums and fair coverage. On this basis the insurance business can work to the advantage of both parties – client and insurer – only if all persons involved are able to assess realistically the risks taken. If they are not able to do so, financial losses mostly to the disadvantage of the insurance company are inevitable. The benefit of high-quality ranking maps of hazard zones is their ability to offer help with decisions in the field of natural hazards and to minimize the risks.

Information such as 'Risk of skidding between Munich and Nuremberg' is not of great help for a driver who is on this stretch. He needs more specific information. The insurance business is tackling the same problem because the atlases available at present do not give the information needed. But the two case studies outlined above demonstrated that it would be relatively easy to remedy these shortcomings. If it is actually possible to describe earthquake and landslide hazards as well as the dangers caused by melting permafrost accurately to the $km^2$ and to extend the principle to other natural hazards, then this means a great advance on the procedures we had before. It is also clear that the applied database should be extended and that some of the assumptions presented in this chapter certainly must be further refined. However, these are tasks for the future, and everybody who is able to do so is called upon to make a contribution.

## 8.6   Conclusion: Planet Earth is Going its Own Way With or Without Human Beings

The dynamic processes going on the earth's surface are regarded as a matter of course by geoscientists: on the earth forces have effects which can build mountains and other forces which can raze them to the ground. Without volcanism and earthquakes the surface of the earth would be as plain as the desert of central Australia. Each elevation is an obstacle to nature which it wants to eliminate. Rockfalls, mudflows and landslides contribute to that and present therefore only a few facets of a natural process we cannot stop or even reverse. Of necessity, we have to live with it and to come to terms with it. We should do so in a way that protects us from serious damage. Investments in scientific research as well as measures to make research results available for the welfare of the general public are therefore well-invested capital. Against the background of exponentially increasing damages in all fields of natural hazards, insurance companies would be well advised to make such investments and to orientate themselves towards the huge time dimensions to which geoscientists have long since adapted.

## References


Ahorner, L., 1962, Untersuchungen zur quartären Bruchtektonik in der Niederrheinischen Bucht, *Eiszeitalter und Gegenwart*, **13**, 24–105.

Ahorner, L., 1983, Historical seismicity and presentday microearthquake activity in the Rhenish Massif, Central Europe, in K. Fuchs *et al.* (eds), *Plateau Uplift* (Berlin and Heidelberg: Springer-Verlag), 198–221.

Ahorner, L., 1994, Fault-plane solutions and source parameters of the 1992 Roermond, The Netherlands, mainshock and its stronger aftershocks from regional seismic data, *Geologie en Mijnbouw*, **73**, 199–214.

Ahorner, L., 1998, Möglichkeiten und Grenzen paläoseismologischer Forschung in mitteleuropäischen Erdbebengebieten. *DGEB-Publikation* 9, Paläoseismologie, Eurocode 8 und Schwingungsisolierung (Hrsg. S.A. Savidis), 9–42.

Ambraseys, N., Smit, P., Berardi, R., Rinaldis, D., Cotton, F. and Berge, C., 2000, Dissemination of European Strong-Motion Data, *CD-ROM Collection, European Commission, DGXII*, Science, Research and Development, Brussels, Belgium.

Bolt, B.A., 1993, Earthquakes and geological discovery, *The Scientific American Library*, A Division of HPLP, New York (New York: W.H. Freeman and Company).

Bormann, P., Parolai, S. and Milkcrcit, C., 2001, Erdbebenmikrozonierung zur Kartierung standortspezifischer Erschütterungsübertragung. *Deutsches Forschungsnetz Naturkatastrophen (DFNK)*, Jahresbericht 2001.

Budny, M., 1984, Seismische Bestimmung der Bodendynamischen Kennwerte von oberflächennahen Schichten in Erdbebengebieten der Niederrheinischen Bucht und ihre Ingenieurseismologische Anwendung, *Geologisches Institut der Universität Köln*, Sonderveröffentlichung Nr 57.

Carozzo, M.T., de Visentini, G., Giogetti, F. and Iaccarino, E., 1972, General catalogue of Italian earthquakes, *Comitato Nazionale Energia Nucleare*, Rome.

Ceri, 2000, Ceri Team in Taiwan, Selections of Paul's Collection, *CERI WWW Server*, http://www.ceri.memphis.edu/taiwan/initialpics.shtml.

Dacunha-Castelle, D., 1997, Spiele des Zufalls: Instrumente zum Umgang mit Risiken, *Gerling Akademie Verlag*, München.

Dikau, R., 1988, Entwurf einer geomorphographisch-analytischen Systematik von Reliefeinheiten, *Heidelberger Geographische Bausteine*, Heft **5**, Im Selbstverlag des Geographischen Instituts der Universität Heidelberg.

Dikau, R., 1994, Computergestützte Geomorphographie und ihre Anwendung in der Regionalisierung des Reliefs, *Petermanns Geographische Mitteilungen*, **138**(2), Justus Perthes Verlag Gotha GmbH, 99–114.

Dikau, R., 1996, Geomorphologische Reliefklassifikation und analyse, *Heidelberger Geographische Arbeiten*, **104**.

Dikau, R., Brabb, E.E., Mark, R.K. and Pike, R.J., 1995, Morphometric landform analysis of New Mexico, *Z. Geomorph. N.F.* Suppl.-Bd **101** (Berlin and Stuttgart: Gebrüder Borntraeger), 109–126.

Gieszberger, H., 1924, Die Erdbeben Bayerns, *Abh. Bayer. Akad. Wissensch*, Math.-phys. Kl., 29.

Glade, T. and Dikau, R., 2001, Gravitative Massenbewegungen – vom Naturereignis zur Naturkatastrophe, *Petermanns Geographische Mitteilungen*, **145**(6) (Justus Perthes Verlag Gotha GmbH), 42–53.

Gutenberg, B. and Richter, C.F., 1956, Earthquake magnitude, intensity, energy and acceleration, *Bulletin of the Seismic Society of America*, **46**, 105–145.

Hung, J.-J., 2000, Chi-Chi Earthquake Induced Landslides in Taiwan, *Earthquake Engineering and Engineering Seismology*, **2**(2), 25–33.

IPCC, 2001, Third Assessment Report: Climate Change 2001 available at www.ipcc.ch/pub/reports.htm.

Jibson, R.W., Harp, E.L. and Michael, J.A., 1998, A Method for Producing Digital Probabilistic Seismic Landslide Hazard Maps: An Example from the Los Angeles, California, Area, *USGS*, Open-File Report, 98–113.

Liao, H.W. and Lee, C.T., 2000, Landslides triggered by the Chi-Chi Earthquake, National Central University, 32045 Chung-Li, Taiwan, www.gisdevelopment.net/aars/acrs/2000/ts8/hami0007pf.htm.

Liou, J.G. and Hsiao, L.Y., 1999, Report 4 on the Chi-Chi (Taiwan) Earthquake, Tectonic Setting and Regional Geology of Taiwan, Dept. of Geological and Environmental Sciences, Stanford University, Stanford, CA (1 October).

Montadon, F., 1953, Les tremblements de terre destructeurs en Europe, Geneva.

Munich Re, 2000a, *topics 2000, Natural catastrophes – The current position*, special millennium issue.

Munich Re, 2000b, *topics, Jahresrückblick Naturkatastrophen 1999*.

NOAA, 1996, Seismicity Catalogs, vol. 2: Global and Regional. 2150 B.C. – 1996 A.D. (CD), National Geophysical Data Center Boulder, Colorado, National Earthquake Information Center, Golden, Colorado

Nutz, M., 1999, *Permafrost im Hochgebirge*, Institut für Geographie und Raumforschung, Karl Franzens-Universität, Graz.

Olsen, R.S., 1999, Documenting the effects of the Chi-Chi (Taiwan) and Izmit (Turkey) Earthquakes to Earth and Concrete Dams, US Army Corps of Engineer Research and Development Center (ERDC), http://www.liquefaction.com/eq99/eq99/lectures/chichi/landslide_dams.htm.

Paus, H.-L., 2002, The real and perceived significance of serious risks, *International Conference on Probabilistics In Geotechnics*, Graz/Austria 2002 (Essen: Verlag Glückauf GmbH).

Paus, H.-L., 2003, Naturgefahren – ein Zukunftsmarkt mit Hindernissen? *Zeitschrift für Versicherungswesen*, Nr **3**/1, S. 73, Allgemeiner Fachverlag Dr Rolf Mathern, Hamburg.

Petley, D., 1996, *The Geomorphology of Taroko Gorge, First Interim Report and Data Summary*, Department of Geology, Taiwan Research Projects, University of Portsmouth.

Plate, E.J. and Merz, B., 2001, Naturkatastrophen, Ursachen – Auswirkungen – Vorsorge. E. Schweizerbart'sche Verlagsbuchhandlung (Nägele u. Obermiller), Stuttgart.

Reiter, L., 1991, *Earthquake Hazard Analysis, Issues and Insights* (New York: Columbia University Press).

Sadegh-Azar, H., 2002, Schnellbewertung der Erdbebengefährdung von Gebäuden, Dissertation RWTH Aachen, Germany.

Sauer, H.D., 2002, Zwischen Ohnmacht und Risikomanagement, Das Ringen mit der Naturgefahr Bergstürze, *Neue Zürcher Zeitung*, 3 April.

Schick, R. and Schneider, G., 1973, *Physik des Erdkörpers* (Stuttgart: Ferdinand Enke Verlag).

Schreiner, P., 2001, Die Geschichte der Abtei Brauweiler bei Köln, *Pulheimer Beiträge zur Geschichte und Heimatkunde*, **21**. Sonderveröffentlichung, Verein für Geschichte und Heimatkunde e.V., Pulheim.

Sieberg, A., 1940, Beiträge zum Erdbebenkatalog Deutschlands und angrenzender Gebiete für die Jahre 58–1799, *Veröffentl. Reichsanstalt für Erdbebenforschung Jena*, Heft 2.

Taiwan Central Weather Bureau, 2000, PGA data of Chi-Chi earthquake. http://www.cwb.gov.tw/V4e/index.htm.

UNEP, 2002, Climate Change and the Financial Services Industry, available at www.unepfi.net.

USGS, 2000–2003, GTOPO30 – Global Topographic Data, available at http://edcdaac.usgs.gov/gtopo30/gtopo30.html.

USGS, 2003, Earthquake Hazards Program, Past and Historical Earthquakes, National Earthquake Information Center, World Data Center for Seismology, Denver, available at http://neic.usgs.gov/neis/epic/epic.html.

Von der Mühll, D., Delaloye, R., Haeberli, W., Hölzle, M. and Krummenacher, B., 2001, Permafrost Monitoring Switzerland PERMOS, *1. Jahresbericht 1999/2000*, Swiss Academy of Sciences SAS.

Wells, D.L. and Coppersmith, K.J., 1994, Empirical relations among magnitude, rupture length, rupture area, and surface displacement, *Bulletin of the Seismic Society of America*, **84**, 974–1002.

# 9

# The Role of Administrative Bodies in Landslide Risk Assessment

*Kurt Hollenstein*

## 9.1 Introduction

Analysing the role of administrative bodies in landslide risk assessment is as much an organizational and political issue as a technical one. The present chapter therefore focuses mostly on the questions of why, where, by whom and in what context risk assessments are performed rather than on how this is done (i.e. what models are used etc.). Specific techniques for assessing landslide risks and especially landslide hazard are discussed in detail elsewhere in this book.

In this chapter, I make numerous references to legal regulations, guidelines, recommendations and so on that apply to or were developed in Switzerland. I am aware that similar things also exist for many other countries. However, being Swiss, I am most familiar with the situation in that country, and I therefore try to illustrate certain aspects based on examples from Switzerland. This by no means implies that the management of landslide risks in Switzerland is superior to that practised elsewhere.

## 9.2 Administrative Bodies and Risk Assessment

### 9.2.1 The Concept of Risk Assessment

It may be helpful to briefly define and outline the concept of risk and risk assessment to clarify the subsequent use of terms.

---

*Landslide Hazard and Risk*   Edited by Thomas Glade, Malcolm Anderson and Michael J. Crozier
© 2004 John Wiley & Sons, Ltd   ISBN: 0-471-48663-9

- *Risk* is a characterization of the potential negative effects with regard to the frequency of occurrence and the extent of damage. This implies that two components are involved: one or more hazards and elements at risk that can sustain damage.
- *Hazards* are processes or states that result in the generation of impacts or stresses that can have potentially adverse effects. The hazard itself, however, is not yet causing damage. Hazards are generally characterized by a relation between their frequency of occurrence and a spatial and temporal distribution of their intensity.
- *Elements at risk* are subjects or objects that possess a certain (monetary or non-monetary) value, that can coincide temporally or spatially with the hazard and that are vulnerable to the hazard's impacts or stresses.
- *Risk assessment* is the activity of investigating risk on a scientific (i.e. objective) basis. Its major building blocks are illustrated in Figure 9.1.
- *Risk evaluation* is the activity of judging the acceptability of risks.
- *Risk management* comprises all activities to handle risks.

For further discussion of these and other related terms refer to Chapter 1 and the Glossary.

### 9.2.2  Factors Affecting the Current (Non-)Use of Risk Assessments for Natural Hazards by Administrative Bodies

When trying to interpret the potential role that administrative bodies currently play in landslide risk assessment, one should recall some characteristic aspects that distinguish natural hazards in general and landslides in particular from other potentially dangerous events (such as technological or societal risks).

The first and probably most important difference is the cause of the threat. The potential for a certain natural hazard to occur usually exists with no or only little human contribution (although man may play a significant role in modifying the probability and/or the frequency of individual events; see Section 9.3). In contrast to other risks,

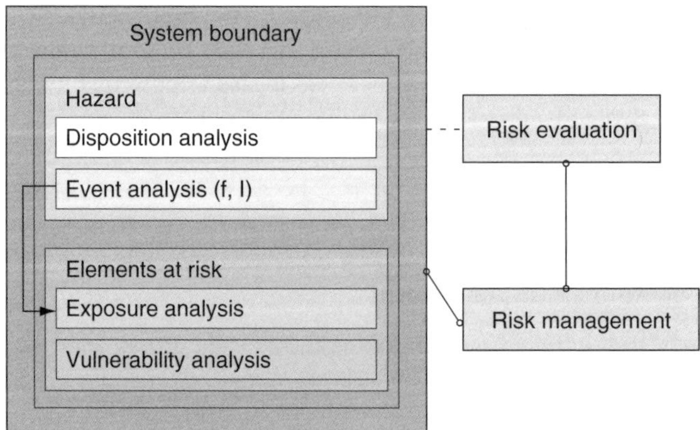


**Figure 9.1**  *The risk assessment concept. The less shaded a box, the more advanced the state of knowledge*

the burden of proof of safety and the responsibility for meeting certain standards cannot be shifted to the designer or operator of a potentially hazardous facility. The concept of liability (which, in the absence of specific risk assessment provisions or safety standards, has largely the same effect as those regulations) is equally not applicable to natural hazards. Thus administrative bodies find themselves confronted with the task of assessing and managing risks that they did not generate and that can have a significant impact on the availability of their resources. In addition, those risks are characterized by their widely varying spatial and temporal distribution and extent, which often make a clear identification of source and impact areas impossible. Since administrative bodies are generally organized as multi-level hierarchies, with each level having its own legal requirements, this often results in unclear territorial and procedural competences.

A second important aspect related to natural hazards is that the means to influence their frequency and extent are usually quite limited (except perhaps for some superficial mass transport phenomena). This means that risk assessments for natural hazards (and particularly for landslides) cannot focus solely on the hazardous process, but must also consider the exposure and vulnerability of the elements at risk. This is in contrast to many technological hazards, whose associated risk can be drastically reduced or even eliminated by altering the hazardous process (e.g. by changing procedures or through the use of containment strategies). However, in spite of the importance of exposure and vulnerability analysis within the risk concept (see Figure 9.1), administrative entities traditionally have a focus that is more hazard-oriented than risk-oriented. This perspective is shifting only slowly, and it will take some time before policy statements such as 'From Hazard Mitigation Towards a Culture of Risk Awareness', as issued by the Swiss 'Platform for Natural Hazards' will be implemented in daily practice. This is probably not primarily because the administrative entities are reluctant to change their practices, but because the concept of hazard (and hazard management) is more easily defined and implemented than the concept of risk (and risk management).

A third aspect is the complexity of the hazardous processes and their interaction with the elements at risk. Traditional risk assessment techniques such as known from technological applications can also be applied to natural risks, as has been shown by Hollenstein (1997), Heinimann *et al.* (1998), Borter (1999) and others. However, compared to technical systems with well-defined components and system states, the system definition as well as the identification of hazardous events becomes more demanding. Not surprisingly, formal assessment methodologies for natural risks were therefore introduced only in the last decades of the twentieth century in most industrialized countries, and their widespread application is only just beginning.

A fourth factor that needs to be remembered is that all actions taken by administrative bodies require a legal basis. Risk assessment and risk management therefore depend on some sort of risk-based regulation. However, legislation dealing with natural risks has so far focused almost exclusively on the hazard, and so was whatever action administrative bodies took. Explicit risk-based regulation was only instrumentalized recently (see, e.g. Seiler, 2000), even though some of its concepts have already been implemented in earlier legal frameworks (see Bundesamt für Forstwesen, 1984; Loat and Petraschek, 1997; Kanton Uri, 1992). The problem with applying risk-based regulation is that in order to be optimal it needs to be cross-sectoral (i.e. applicable to a variety of risk sources). Administrative entities are organized sectorally with their own budgets, procedures and priorities, and risk-based regulation is therefore difficult to implement, since it would require new ways of collaboration and cooperation between different agencies.

Considering all these compromising aspects, it is not surprising that in most countries there is no administrative tradition of performing comprehensive risk assessments for landslides (and other natural hazards). Consequently, the organizational and procedural pattern on and between different hierarchical levels is not yet very well developed.

### 9.2.3    Potential Involvement of Administrative Bodies in Landslide Risk Assessment

There are numerous ways in which administrative bodies may become involved in landslide risk assessments, and what part different administrative entities actually play depends largely on their mission and on the organizational framework within which the assessment is performed. One can also distinguish between a direct involvement (where administrative bodies take some part in the assessment) and an indirect involvement (where the assessment is not immediately related to the bodies' activities).

Among the most important roles that administrative bodies play are:

1. Public agencies with a technical character may perform risk assessments themselves. This is the most active and direct way in which administrative bodies can influence the assessment process. Such special agencies exist in most industrialized countries, and they often work on behalf of or in close collaboration with other branches of the administration. Landslide risk assessments for issues of public importance (e.g. lifeline safety, strategic land use planning) are quite frequently performed by (or under the auspices of) those special agencies to ensure that quality standards are met and that the assessment is not biased by interest groups.

2. Landslide risk assessments that are to be used for public purposes (e.g. for land use planning) generally have to comply with certain procedural guidelines, and those guidelines are usually developed by administrative bodies, most often by the aforementioned special agencies. While the influence on and the control over the assessment may not be as far-reaching and direct as in the first case discussed, this type of direct involvement still provides a very effective tool for public entities to steer the risk assessment practice.

3. The government often subsidizes the risk assessment activities. This can be a valuable tool for fostering the application of risk assessments, and, if the subsidies are made dependent on the compliance with certain guidelines (see above), it also provides a means of controlling not only the number, but also the quality of risk assessments being made. Even though there is no direct involvement of the administrative body in the assessment activity, experience shows that subsidizing it provides a very strong impetus for the process.

4. Subsidies for the realization of other measures or projects may depend on the availability of risk assessments. Such regulations are common in many laws governing land use or infrastructure development. Since the cost of providing a risk assessment is usually small compared to the amount of subsidies provided (and often itself subsidized; see above), this economic condition effectively acts as a obligation to perform risk assessments without formally interfering with ownership rights. Again, the administrative bodies are not directly involved in the assessment activity, but they

have an effective means of controlling the process, in particular when combined with the compliance with guidelines requirement.

5. Government agencies may approve risk assessment results as part of land use regulations. Generally, risk assessments are then considered by delineating zones that are significantly affected by natural hazards and subject to restrictions regarding their use. This is again an indirect, but none the less a very common, way in which administrative bodies become involved in risk assessments, and there are often far-reaching consequences of such an official approval because of its economic implications.

6. Government agencies often maintain and provide databases that are required or useful for performing risk assessments. In this case, the administrative body is directly involved in the assessment process, but acts only as a service provider. However, considering the high cost of data acquisition (especially in the case of geological and pedological data), it is obvious that the availability of this information is often pivotal for the decision whether or not risk assessments are performed. Administrative entities can therefore make an important contribution to the risk assessment practice by making their information easily accessible at no or low cost.

In reality, the involvement of administrative bodies will usually be a mix of these possibilities, with different agencies playing different roles.

### 9.2.4  Administrative Bodies and Risk Assessments: Different Missions, Diverging Interests

Risk assessments are meant to be the product of objective, unbiased scientific reasoning. However, due to the uncertainties inherent in the process, the results of even the most sophisticated assessments may well vary by several orders of magnitude (in particular for so-called 'low-probability–high-consequence events'). In other words, even the 'exact' part of risk assessments is open to interpretation of results. This is even more pronounced when non-scientific subjective or institutional values come into play. Consequently, even though the assessments may give clear indications about the nature and the extent of the risks encountered in a certain area, it is still not clear what the implications for administrative bodies should be. It is also clear that, due to their different missions and goals, the interests of the various agencies affected by the risk assessment may be conflicting. Therefore, there will also be differences in the interpretation of the assessment results. There are at least three aspects that need to be considered when evaluating an agency's interest and stake in risk assessment and its results:

1. Interpretation of results will be done in an optimistic or a pessimistic way, depending on the role of an agency. Implementing the result of a landslide risk assessment usually results in restrictions with regard to land use. Such restrictions may include special provisions for design and construction of buildings and lifelines (e.g. use of flexible water mains), limitations in land exploitation (e.g. conservation of forest cover) or organizational provisions (e.g. installation of monitoring devices). Now it can be assumed that those agencies that are responsible for the protection of people and property (e.g. a civil defence agency) will try to achieve their mission by favouring a conservative approach to risk assessment, that is, one that 'lies on the safe side' when things are uncertain. Such an approach generally results in many restrictions, thereby

lowering the value of the land. It is obvious that this conflicts with the aim of other agencies that focus on fostering the economic development of the same area (e.g. a chamber of commerce). Considering the large and inherent uncertainties in risk assessments, it is clear that two agencies may reach totally different conclusions about what risks are to be taken into account for land use planning.

2. Results will be interpreted selectively depending on the agencies' mission. Even public agencies that are not directly involved in or affected by the risk assessments may still draw widely different conclusions from it. They will tend to interpret risk assessment results in a way that maximizes their institutional influence and benefit. Thus an environmental protection agency may highlight those landslides that can affect hazardous facilities (resulting in release of toxic substances), while a highway administration is primarily concerned about slides that can block roads – despite the fact that these risks could be several orders of magnitude lower when compared to those affecting residential areas.

3. Agencies will try to mimimize the influence of risk assessments on their own business. Risk assessments are relatively new procedures, and they often require both the participation of and the consideration by multiple agencies in their making and their implementation, respectively. However, dealing with risk assessments may not belong to the core activities of most of these agencies, and may primarily be perceived as a disturbance of their 'normal' operation. Consequently, agencies that are not themselves responsible for the assessment may tend to devote as little time and money to it as possible and may also tend to ignore unwanted implications that the risk assessment has for their own business. As an example, they may be reluctant to adapt internal guidelines to incorporate risk assessment findings, because procedures may become more complicated.

One point needs to be highlighted: these diverging interests are not the result of 'institutional malice', but are simply due to the different missions of the agencies involved. However, if widely differing interpretations of the same risks are presented by different agencies of the same administration, this may give an impression of confusion and inaccuracy and undermine the public's confidence in risk assessments. Ideally, such conflicts of interest between different agencies should therefore first be solved by negotiations within the framework of an administrative process. Subsequently, the results of this process can then be communicated to the public as an unequivocal 'administrative point of view' by a single agency (often the one that has the lead in the assessment process).

### 9.2.5    Landslide Risk Assessment Goals and Scope

Focusing now more closely on the agencies that are directly responsible for landslide risk assessments, there are still considerable differences in the purposes for and scope with which those assessments are performed. Due to the wide spectrum of sizes and displacement velocities that they potentially have, landslides are of interest to a wide variety of administrative bodies. Most of these institutions have developed their own special methodological approaches to assessing the risk of the landslides that they are particularly interested in. The following is an attempt to characterize some typical applications, the main focus (including some methodological requirements) and the actors of landslide risk assessments based on a very broad classification of landslides according to their size and displacement velocities. (Note: the remarks made subsequently about the potential risk of certain combinations of slides and objects assume that there is no modification of the quality of the hazardous process; that is, the movement remains a sliding of earth masses and does not turn into a debris flow. Debris flows can result in significantly higher risks because of their generally higher velocities and particularly their extended runout distances; see Sections 9.4 and 9.5.

*Small, slow-moving slides* are usually not critical with regard to their risk, but they may have a significant effect on the geomorphological properties of the surface and on the possible land use options of an area. If risk assessments are performed at all for this category of slides, it is often as a part of comprehensive land evaluations, for example as done for zoning purposes. For this application, it is usually sufficient to have information about the hazard, in particular an accurate delineation of unstable areas and an approximate order of the displacement velocities. This information is then used in the planning process for two purposes: (i) to exclude unstable areas from being used for establishing structures that particularly require stable conditions (e.g. storage tanks or survey marks); (ii) to identify patches of land for land use purposes that are resilient to minor instabilities (e.g. nature conservation areas). Quantitative risk analysis (i.e. including quantified frequency/intensity relations) is usually not performed since damage from these slides only occurs for particularly sensitive objects, and they should simply be located outside of the unstable areas.

*Small, fast-moving slides* ranging from a few $m^3$ up to a few hundred $m^3$ can cause substantial risk to property and, to a lesser extent, also to life. With regard to the scope and techniques of risk assessment, it makes sense to distinguish between two categories of objects at risk: individual buildings and linear infrastructure elements (both categories including the persons using the objects).

The collapse risk to buildings is high if they are directly located on the sliding patches, but mostly low or moderate if they are in the trajectory or the runout zone of a slide, because the forces exerted by the sliding masses is below the capacity of most buildings. Similarly, the fatality risk to persons in or around the buildings is low because of the low probability of the slides directly hitting people (most people will be able to outrun all but extremely fast slides). However, those slides may still result in significant cost for cleanup and repair. Risk assessments for this type of slides should focus on an accurate delineation of the potential starting zones (to take into account the collapse of buildings located there). In the trajectory and runout zone, the analysis should focus on the characteristics of the sliding masses (speed, geometry, composition) and those of the buildings (value, presence of highly vulnerable components, structural peculiarities). This type of assessment is often performed by agencies dealing with landscape, forestry or geological issues as a basis for, but not necessarily as a part of, land use planning and zoning.

Whereas buildings are relatively resilient to slides of this category, infrastructure lines may be more vulnerable. This is due to two factors: (i) these lifelines are made from components that have small mechanical capacities; and (ii) because of the type of traffic that uses the lifelines. The former is the case, for example, with aerial power and telecommunication lines that are suspended by wooden poles with little lateral load-bearing capacity; the latter for high-speed railroad tracks carrying trains that require long distances to come to a stop and that can derail from relatively minor obstacles on the tracks. Risk assessments for these types of objects will primarily be focused on locating the trajectories of probable slides in order to be able to take preventive and/or reactive measures to minimize the risk. Such measures are, for example, the design of protective structures to prevent slides from reaching the lifelines or the allocation of intervention resources for speeding up the recovery of the lines. The majority of the assessments of this type are performed by (or on behalf of) the line operators.

*Large, permanently slow-creeping slides* with displacement velocities of a few cm/yr$^{-1}$ are usually not a risk to life (unless there is a sudden acceleration; see Section 9.3). However, they pose a significant risk to almost all types of rigid structures. In contrast to the small slow-moving slides, it is often not possible to avoid the location of sensitive objects in the unstable area because of its extent. The focus for this type of risk assessment is on the delineation of the unstable area, the overall and particularly the differential displacement velocities and on the vulnerability of different objects to absolute and relative (differential) movements. This information is then utilized for purposes of site selection and structural design, for example by placing sensitive objects outside of areas with significant differential movements or by using reinforced or flexible designs for structures within those areas. However, these assessments are usually performed as hazard rather than risk assessments; that is, the potential damage caused by the slides is not characterized in terms of its frequency of occurrence and its extent, but the sliding is merely taken into account as an additional planning/design requirement specification. Consequently, the focus is usually not on minimizing the associated risk, but on preventing damage from occurring at all or on keeping it below certain performance-related thresholds (e.g. by properly scheduling maintenance and repair).

*Large, fast-moving slides* with sliding volumes ranging anywhere from a few thousand to several billion m$^3$ can destroy virtually all types of structures due to the enormous energy and pulse peaks they develop. This type of slide usually also results in the disintegration of the sliding masses, causing massive damage or destruction to objects (and people) on the sliding surface. In addition, these slides often have deep sliding planes, and consequently technical means of stabilization are very limited. In other words, there is little that can be done to influence either the extent of the hazard or its intensity once a slide of this type occurs. However, as will be seen in Section 9.3, there are cases where the frequency of occurrence can be influenced (in both a positive and a negative way), and there is often a possibility to reduce damage to non-permanent elements at risk, particularly humans.

Risk assessments for these slides are relevant for both planning and emergency management purposes. In planning, the extent of the potential slides as well as the associated trajectories and runout areas are of particular interest. Also of interest is the order of magnitude of the probability of occurrence; however, this value is often difficult to determine because these events are rare (i.e. there is no possibility to derive statistical relations between frequency and magnitude). As long as this can be done, planning will usually focus on avoiding the affected areas. This is particularly important for the siting of hazardous facilities that could result in secondary consequences (e.g. release of toxic or radioactive substances). However, in some instances it will not be possible to avoid the potentially unstable or threatened zones. Under such circumstances, the focus is on the probability of occurrence. If this probability is low enough (i.e. below a critical threshold), an unrestricted development of the area under consideration is usually possible, potentially in combination with a monitoring system. If, however, the probability of occurrence is higher, there will usually be some restriction on how the area may be used, and plans for emergency measures (evacuation, closure of the area) should be considered. In addition, emergency management responsibilities will usually also be interested in the potential outcome of an event to develop provisions for search and rescue as well as for recovery measures.

### 9.2.6   Summarizing the Role of Administrative Bodies

The previous sections present an attempt (i) to characterize potential ways in which administrative bodies may get involved in landslide risk assessment; (ii) to evaluate how the institutions' views about risk assessment differ depending on their mission; and (iii) to characterize typical risk assessment applications and identify potential actors. However, in most cases it is simply not possible to make detailed statements about the exact role of administrative bodies in landslide risk assessment that are generally valid. What role an administrative institution plays in risk assessment needs to be investigated on a case-by-case basis. Aspects to consider include:

- **Legal basis**   The potential role that administrative entities can take in landslide risk assessment depends largely on the legal basis that applies to the given situation.
- **Acceptance of the risk concept**   The degree to which administrative bodies engage in risk assessments depends on the acceptance of the concept and the methodology within an institution. Risk assessments are relatively new and demanding procedures that are still looked upon with scepticism by many people in the administration, especially in cases where experience is lacking.
- **Resource availability**   Risk assessments are probably among the more expensive types of studies that are performed by administrative bodies. For example, the typical cost for assessing superficial mass wasting hazards (snow avalanches, debris flows, superficial slides) in Switzerland ranges from about 1 to more than €100 per hectare with an average of about €10 per hectare (J. Hess, 2002, pers. comm.), and this number increases if deep slides or vulnerability assessments are included. Due to budget constraints, administrative bodies are often limited in the number and extent of projects they can realize, and risk assessments may therefore not have a high priority on their agenda.
- **Type of involvement**   Besides a direct involvement in risk assessment, there is also an indirect role that administrative bodies might play, for example by providing or witholding information or by considering or not considering risk in their decision making. These effects are often difficult to identify and they are usually not based on legal provisions.
- **Organizational setting**   The type and number of agencies and hierarchical levels involved in the process influence the way risk assessments are performed.

As a consequence, there is not a great deal that can be said about the *actual role* of administrative bodies in landslide risk assessments. All that is possible is to consider the *potential role* and illustrate this reasoning using practical examples. The following sections present such example case studies.

## 9.3   Administrative Risk Assessment Blunders: The Vaiont Reservoir Disaster

The Vaiont landslide and the subsequent floodwave caused the death of perhaps more than 2500 people in the town of Longarone and other surrounding villages in 1963. It is an example of a very large, fast-moving slide interacting with a technological system that resulted in catastrophic consequences.

The disaster occurred on the evening of 9 October 1963, when the sliding masses (whose volume is estimated to be around $2.7 \times 10^8 \, \text{m}^3$) started to acelerate from their previous displacement velocity of $<1 \, \text{m} \times \text{d}^{-1}$ to a speed of approximately $30 \, \text{m} \times \text{s}^{-1}$. The sliding masses entered the reservoir and displaced the water there, resulting in a floodwave that overtopped the dam by up to 250 m. Confined in a narrow gorge downstream of the dam, the wave inundated the villages in the Piave Valley. The fact that no evacuation orders or timely warning about the impending disaster had been given and that the event occurred at night meant that a large number of people was exposed to the impact of the hazard.

It is important to note that neither the landslide itself nor the dam itself would have resulted in such dire consequences. The slide hit an area that was only sparsely populated, and the dam did not fail; that is, had the water level in the reservoir been low enough to prevent the overflow, the resulting damage would have been much lower.

The cause and the mechanism of the slide have been the topic of a vast number of research works, and even today there are still some controversial issues (e.g. whether or not it was a reactivation of an old slide). Important geotechnical issues of the slide are presented in, for example, Müller (1964), Hendron and Patten (1985) or Voight and Faust (1992), as well as in many other publications. Although it is clear that a correct assessment of the hazard is necessary as a basis for every risk assessment, these issues will not be evaluated any further in the present chapter. Instead, the focus is on methodological aspects of the way hazard and risk assessment was performed before the disaster occurred. Geotechnical processes also play a role in this context, but (as will be seen) many of the critical errors of commission or omission are only marginally influenced by misinterpretations of the geotechnics of the area. Procedural and organizational inadequacies are primarily responsible for the things that went wrong in the Vaiont case.

The following sections discuss the main deficiencies in handling the Vaiont situation *from a risk assessment and management perspective*. The aim is not to judge people for their mistakes, but to show how an appropriate use of risk assessment techniques can contribute to preventing such tragedies from happening again.

### 9.3.1   Errors in the Assessment System Definition

The first step in performing a risk assessment is the definition of the system under investigation. Such a definition includes spatial and temporal dimensions, but also a decision about what hazards and what probabilities are to be considered. In the Vaiont example, such an explicit system definition had not been made. This omission can perhaps be attributed to the fact that (atleast to the author's knowledge) no formal risk assessment was ever performed for the reservoir *and* its surroundings.

Frequently, an explicit worst-case scenario with extent $E$ is used to define the lower bound of the probabilities $p$ that are considered, because there is generally a relation of the form $E \times f^{-1}$. Applied to the Vaiont situation, such a worst case would be the catastrophic failure of the dam or an event with equal consequences, including a landslide-induced floodwave. Therefore, it has to be assumed that the extent of the worst-case event could have been known. However, its probability of occurrence was obviously estimated to be low enough as not to be included in the assessment. In hindsight, this was a clear misjudgement, but on the basis of the information available at the time of the design and construction of the dam, it may very well have been a reasonable conclusion. The error was not that this worst case was not included in the assessment in the beginning (since it was deemed too unlikely to be considered), but that subsequent assessments and decisions were made on the basis of a system definition whose validity was no longer critically evaluated.

What would be a motivation for revising the system definition and re-evaluating prior assessment conclusions? Besides the most obvious reason, that is, a change in the important system parts (e.g. an increase in dam height), there is also the recognition that the system behaviour does not correspond with what would be expected based on the assessor's understanding of the system. In the case of the Vaiont reservoir, there were numerous pre-failure indications for such discrepancies between what was expected to happen and what actually happened, and there were also good reasons why those discrepancies existed (see Carloni, 1995 for citations of original documents). First, it was known that the area has a complex geology that was not very well understood. Second, there was a previous, although smaller, slide that had not been expected. Third, monitoring showed a pattern of slope movement that was often neither expected nor understood. Fourth, there were even early speculations that a large sliding mass of more than $10^8 \, \text{m}^3$ could enter the reservoir, but it was assumed that this would only happen at velocities low enough that overflowing could be prevented. Together, these should have been motivation enough to redefine the system to be assessed to include the worst case of a catastrophic failure.

### 9.3.2   Errors in Considering the Hazard

Hazard assessments (as part of risk assessments) are performed to identify and, if possible, eliminate potential processes that could affect the safe operation of the system assessed. Regarding the Vaiont reservoir as an engineering system, such an assessment should apply engineering design and management principles that include, among others, the following concepts:

- **Reserve capacity under reasonable load conditions**   An engineering system should be able to operate in a non-dangerous way when it is subject to load conditions that must reasonably be expected under given environmental conditions; that is, it must be designed to withstand 'normal' impacts without damage that affects its safety.
- **Use of conservative estimates for design and operation purposes**   In the design phase, the capacity estimates should tend to underestimate actual values while the demand estimates should tend to overestimate them. In the operation phase, measured capacities should be interpreted as upper bounds of actual values, while measured demands should be understood as lower bounds of actual values.

- **Monitoring of capacity and demand**    Actual values of capacities and demands should be monitored and compared on a regular basis to diagnose developments that could lead to excess demands that affect the safe operation of the system.
- **Gradual degradation, prevention of catastrophic failure**    Since excess demands cannot always be prevented completely, the system should be designed such that it degrades gradually rather than failing in an abrupt and catastrophic way.

With these principles in mind, and considering the large uncertainty with regard to the subsurface geotechnical conditions, it seems that the landslide hazard (i.e. the demand) for the Vaiont case has been assessed in an optimistic rather than in a conservative way. On the other hand, the ability of the reservoir-dam system to withstand certain impacts (i.e. the capacity) was assumed to be very high. This non-compliance with engineering principles was aggravated by the fact that many of the responsible persons failed to take indicators into account that implied that the actual system performance was more problematic than forecast by their assessment. Parameters that were clearly not in line with the optimistic interpretations were overlooked, misinterpreted or sometimes even manipulated (see Carloni, 1995 for examples).

The question is why this could happen. One would think that building one of the world's highest dams required engineering skills and techniques that depend on compliance with the above-mentioned principles. In fact, the principles have been observed, but unfortunately only with regard to the reservoir itself. The dam even withstood the enormous excess demand of the slide and floodwave, and provisions were made before the event to allow for the operation of the reservoir for the case of the lake being completely blocked by sliding masses. *The dam and reservoir system itself performed superbly, but little attention was paid to the effects reaching beyond those components.* In other words, the hazard was considered, but the scope of the consideration was insufficient.

### 9.3.3    Errors in Assessing Exposure and Vulnerability

When looking at the pre-failure history of the events at the Vaiont reservoir site (as described by Carloni, 1995 based on original documents), it is surprising how little (if any) consideration was given to the exposure and vulnerabilty of the area downstream of the dam.

Vulnerability assessments are among the most difficult problems in risk assessments, and very little is known about the quantitative relationship between the parameters characterizing a landslide or flood hazard and the associated damage to structures and people. However, it is reasonable to assume (and confirmed by experience from other hazards, e.g. earthquakes) that vulnerability functions are an S-shaped relation between intensity of impact and degree of damage such as displayed in Figure 9.2. Considering the enormous masses of rock or water associated with the slide and the flood in the worst-case scenario, it is clear that the intensity will always be in the rightmost part of the figure; that is, destruction would be complete in the area affected by either the slide or the resulting floodwave. While this aspect may not be relevant to a pure hazard assessment, it is certainly crucial for risk assessment and management purposes: if neither the hazard (once the sliding got beyond a certain threshold) nor the vulnerability of the elements at risk can be reduced, it is of paramount importance that a spatial or temporal separation of the two is achieved to reduce the risk. In the Vaiont case, permanent spatial

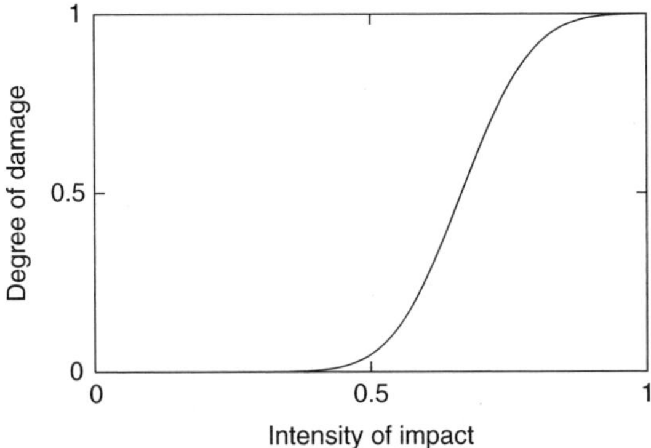


**Figure 9.2**   *Generalized vulnerability function*

separation was not possible at a reasonable cost, so the main focus should have been on reducing the temporal exposure (in case of increasing intensity of the hazard) by removing as many of the elements at risk as possible out of the potentially affected area.

With today's forecasting techniques (see Kilburn, 2002), it would be possible to predict the failure as much as a month in advance based on the displacements. Such a long period would allow the removal of virtually all elements at risk except for the buildings from the potentially affected area. (Naturally, such a pre-event preparation time would also be used to reduce the flood hazard by performing an emergency drawdown of the reservoir – something that could in turn influence the length of the period available.) However, it must be taken into account that most of these modelling techniques were not yet available at the time of the Vaiont disaster. The possibilities of reducing the exposure were thus much more limited, but the fatalities could certainly have been avoided.

### 9.3.4   Errors in Analysing Precursory Events

The fact that no evacuation orders were given for the area downstream of the dam (only an area immediately above the water level of the lake was closed to the public) shows that those responsible for safety misjudged the situation up to the very end. One aspect that contributed to this erroneous assessment was the failure to analyse precursor events thoroughly. Unlike many other situations, there was a series of such events that could have been used to predict the exact extent (although not the time) of the disaster:

1. There was the small ($7 \times 10^5$ m$^3$) slide that happened on 4 November 1960, entered the lake and caused a small floodwave. This slide was relevant with regard to two things: (i) it detatched abruptly, showing that at least a part of the slope had the potential to fail catastrophically; and (ii) it coincided with the opening of an M-shaped perimeter scar that exactly delineated the giant slide that was to happen three years later.
2. There was noticeable seismicity of initially unknown origin at the reservoir site. It was later found that the microquakes were most probably locally induced. This seismicity was an indicator that underground deformations and displacements involving large masses were probably happening.

3. Measurements of the velocity of seismic waves showed a very sharp decrease over a relatively short period. This is a clear indicator that parts of the rock were losing their integrity due to the displacements (and, consequently, also their strength, although this may have been less important since the actual failure most likely occurred in clay layers with even lower initial shear capacity).
4. It was quite obvious that there is a relation between the displacement velocity of the sliding mass and the water level in the reservoir. The reservoir operators were even hoping to use this interaction to bring the slide to a standstill by reaching a new equilibrium as a result of slow movements. However, the character of the interaction between water level, pore pressure and displacements was never sufficiently understood to turn the water level into a reliable control parameter for the sliding.
5. It was known from other events that particularly large sliding masses can fail catastrophically and subsequently reach very high velocities in the order of $30\,\mathrm{m} \times \mathrm{s}^{-1}$. While this is not exactly a precursor for the Vaiont event, it is still an indicator of what the kinetic potential of the slide could be once it failed.

Today, precursor studies are part of most comprehensive risk or safety assessments for technical systems. They include a systematic collection and interpretation of data as well as a history of events that are potentially relevant for the safety of a system. This information is then evaluated in 'what if' scenarios, for example assuming that unfavourable external conditions could aggravate the situation and turn a precursor into an accident. Had such reasoning been applied to the above-mentioned precursors in the Vaiont case, it would have become clear that an appropriate safety level could not be guaranteed with the regime that was used.

### 9.3.5   Failure in Revising the Risk Assessment

One of the main problems of risk assessments is their static nature; that is, they reflect (at best) conditions at the time of the assessment, but they do not take changes into account. This shortcoming has often led to criticism, and newer approaches such as 'living' PSAs for technical applications try to meet the requirement for constant updates. Unfortunately, such revisions of the assessment whenever new relevant information was available did not take place in the case of the Vaiont reservoir. There are numerous examples of inaccurate or outdated data that were kept in the assessment in spite of contradictory or differing new findings.

- **Changes in design**   The Vaiont Valley was considered as a site for a reservoir as early as 1928 (see Paolini and Vacis, 1997), and initial geological and geotechnical assessments that were based on the assumption of a dam height of 200 m were not unfavourable. When the projected dam height was later increased to 260 m, it was recognized by some experts that this would require a thorough revision of the assessment; however, such a revision was never performed, and only minor adaptations were made to the initial assessments to take the design changes into account.
- **Unrealistic model assumptions**   Drilling and piezometer measurements showed that the depth of the sliding plane was most likely at more than 150 m (Carloni, 1995) and the volume of the moving mass more than $2 \times 10^8\,\mathrm{m}^3$. However, for

the hydraulic experiments, the maximum volume was assumed to be no more than $4 \times 10^7 \, \text{m}^3$ (assuming only superficial parts or detached pockets were moving at once). Consequently, the height of the floodwave was drastically underestimated by the model experiments.

- **Inappropriate layout of hydraulic experiments**   Once it was obvious that a slide could enter the reservoir, a series of experiments on a hydraulic model was performed. Accidentally, early runs showed that maximum floodwave was not caused by the sand or gravel used to model the sliding mass, but by the boards used to retain this material (Paolini and Vacis, 1997). In other words, the critical event was the displacement of a large slide as a block. However, subsequent experiments were not adapted to account for these findings. Only after the disaster were such experiments performed in a laboratory in France, and those results reflected the pattern of the real slide well.
- **Inadequate scope of the hydraulic model**   The hydraulic experiments were limited to a range of water levels and slide velocities that were below what was to be expected in a worst case. It was known that a floodwave could top the dam in case of high water levels and sliding velocities, but the experiments were not extended to cover such conditions. Besides, the model only investigated what the effect on the reservoir and the dam would be, but no attention was paid to the villages downstream. Suggestions to include those areas in the modelling were rejected.

These are only a few examples of new findings that should have been considered in the risk assessment and that would normally require a revision of the whole assessment, not just a change in the parameter value (to make sure that no assumptions are retained that are invalidated by the new findings). In the Vaiont case, such a timely revision of the assessment (especially the worst-case scenario) would have given ample time for protective measures.

### 9.3.6   The Core of the Problem: A Lack of Clear Procedures and Responsibilities

The previous sections highlight some methodological and technical aspects of risk assessments that were not appropriately considered during the design and operation of the Vaiont reservoir. To some extent, these errors can be explained (and excused) by the non-availability of certain methods and techniques in 1960. One must also be aware that no formal risk assessment in a modern sense was ever performed at Vaiont; that is, the above observations refer to an idealized hypothetical risk assessment and not to one that actually exists.

However, there were so many things that went wrong at Vaiont that the state of knowledge alone is not sufficient as an explanation. The core problem was another one that can render even the most sophisticated risk assessments methods useless: incompetent organization. A look at the literature that refers primarily to the organizational rather than the geotechnical aspects of the disaster (e.g. Carloni, 1995; Paolini and Vacis, 1997) reveals a lack of clear procedures and responsibilities for many of the crucial issues and decisions. In addition, the separation of powers and competences was inadequate, resulting in a situation where vital steps to prevent the disaster were not taken due to ignorance and neglect or, even to contrary interests.

The organizational inadequacies in the Vaiont case are simply too numerous to list comprehensively. There are, however, a few points that proved to have particularly disastrous consequences:

- **Assignment of assessment task**   Almost all of the hazard assessment work was performed by or on behalf of SADE, the company that was constructing the dam and that sold the facility to the state when power plants were nationalized in 1963. It is clear that SADE had a highly partisan interest in the outcome of the assessment: safety-related restrictions in the operation would lower their revenues. It is not uncommon for an operator to assess the safety of its own facilities, but administrative bodies usually define strict requirements that such an assessment must meet. This was clearly not the case at Vaiont: SADE employees were actively influencing the assessment process, assessors were not independent from SADE, and results were probably manipulated in some cases.

- **Unclear competences**   The involvement of several government agencies (Servizio Dighe, Genio Civile, Commissione di Collaudo etc.) in the control and authorization process resulted in confusion about the exact functions and competences each one of these entities had. There were cases where an agency did not request a proof of safe performance because its officials assumed that another agency had already issued such a request.

- **Communication failures**   With the SADE in charge of the assessment, authorities should have ensured that they were fully and immediately informed about all safety-relevant findings. The lack of a well-established communication pattern allowed the wilful or accidental witholding of information that could have played a major role in preventing or reducing the consequences of the slide. As an example, the results of the hydraulic modelling were never transferred to the authorities, although they could have served as a basis for evacuation plans.

- **Ineffective control instances**   The approving committee (Commissione di Collaudo) that was responsible for supervising the construction and the pre-commissioning tests was working in a very questionable way: after the first visit on site, it had to request technical data about the dam from SADE because its own data were 'somehow lost' (Paolini and Vacis, 1997). Following another on-site visit in late 1961, members of the committee were involved in a fatal car accident. After that, the committee was more or less no longer functional as a controlling body (Carloni, 1995).

- **Lack of emergency management provisions**   This is probably the darkest chapter in the whole Vaiont tragedy. Seeing clearly that things were continuously deteriorating, officials remained inactive and did nothing to prevent thousands from being killed. A simple emergency management plan comprising a set of alert thresholds and associated procedures (whom to alert, what to communicate) would have been sufficient to warn people around the lake and in the Piave Valley and advise them what to do to protect themselves from the impact of a potential floodwave. Such a provision would have been inexpensive and easy to implement – all it had required was the admission by SADE that not everything was perfectly under control.

The Vaiont disaster was not a natural disaster. It was a hazard that was not made, but triggered by man, and its disastrous consequences could have been avoided. What turned the Vaiont into a disaster was the combination of three factors: (i) the geological predisposition of the area for large-scale landslides; (ii) the addition of a technological component whose failure could lead to a much bigger damage than the potential natural slide had; and (iii) human ignorance, neglect, arrogance and irresponsibility that brought geology and technology together under the worst possible circumstances.

## 9.4   Risk Assessment for a Reactivated Large-scale Landslide: The Chlöwena (Falli Hölli) Slide

The Chlöwena slide is located in the Canton of Fribourg in the Swiss Prealps. The slide was a reactivation of a post-glacial landslide complex. It is located in a Flysch formation, and as such it was known to be and declared as an unstable area (Bonnard *et al.*, 1995). From sedimentological, lithological and dendrochronological investigations it is known that local sliding activity occurs with a probability $p > 0.01/yr^{-1}$, and that events with a magnitude similar to the one in 1994 have a return period of approximately $10^3/yr$ (Raetzo and Lateltin, 1996). Between 1981 and 1993, the higher parts of the sliding area experienced displacements of $10–40\,cm/yr^{-1}$. High rainfall, probably in combination with repeated snowmelt events in winter 1993/94, reduced the slope strength below a critical level, and in April the slide started to move at speed increasing to $6–9\,m/d^{-1}$ in August. Several phases could be distinguished spatially and temporally for the sliding, and parts of the sliding material turned into debris flows that posed a secondary (though not necessarily a smaller) risk. The total area of the sliding reached $1.5\,km^2$, and with sliding reaching as deep as $60\,m$, the volume is estimated to approximate $4 \times 10^7\,m^3$ (ECAB, 1995). The sliding mass was displaced by $200\,m$, before its head reached and temporarily dammed the Höllbach River, resulting in a slowdown of the movement (see Figures 9.3 and 9.4).

The slide affected a group of vacation houses that were situated on the sliding mass and were subsequently destroyed as a result of absolute and differential displacements. The total damage exceeded 15 million Swiss francs (€10 million). Nobody was hurt or killed by the slide, although there was some concern that rockfall in the upper area and debris flows with velocities of some $5\,m/s^{-1}$ could pose a risk to people living and working in the area.

### 9.4.1   Pre-Event Risk Assessment Activities by Administrative Bodies

The earliest risk assessment activity related to the landslide was a delineation of slide-prone areas in the Canton of Fribourg. The resulting map of potential and known unstable zones showed Falli Hölli as a characteristic slide area. However, the direct risk from landslides was low before the mid-1970s since there were no residential structures in the slide area. Then, in 1976, a construction permit was requested for a vacation camp. The cantonal building insurance ECAB expressed its concern because of the sliding risk, and requested that the construction permit should not be issued. However, permission for the building was granted in 1977. ECAB then decided that landslide-induced damage would not be eligible for compensation and informed the owner of the structure about this decision. The owner challenged this with an appeal at the cantonal government, and the latter finally decided in 1979 that insurance coverage had to be provided for landslide

**Figure 9.3**   *Chlöwena Landslide. Individual building destroyed by the slide. (photo courtesy of Dr B. Loup, Canton of Fribourg; reproduced by permission)*

damage as well. This regulation also applied to all the other structures (mainly vacation houses) later constructed in the area.

This is an example that shows how an agency's mission affects its assessment of landslide risk. ECAB is responsible for covering economic damage to buildings, and its recommendation regarding the permission as well as its decision regarding coverage of landslide damage show that, from a purely economical point of view, the risk was higher than acceptable to achieve a break-even between premiums and compensations. Other authorities, including the cantonal government, applied a different perspective: they had to evaluate not only direct revenues and cost from insuring the buildings, but also indirect benefits and risks deriving from the operation of the vacation camp, and this evaluation gave a more favourable image. The benefits can be economic, but also sociological or political, including increased spending in the area, prevention of depopulation and avoidance of legal battles (e.g. if no legal basis is available to prevent building or refuse insurance coverage). On this broader level, some factors obviously tipped the balance in favour of the development.

Once it was clear that the development would take place and that ECAB would have to pay landslide-induced damage, no further risk assessment activities were undertaken according to the author's knowledge. This is not surprising: no one at that moment had an interest in performing additional studies. For the developers, it was clear that potential damage was to be paid for by ECAB, so they saw no specific need to investigate the risk in more detail. ECAB on the other hand could have done nothing else than repeat

***Figure 9.4***  *Chlöwena Landslide. The hamlet of Falli-Hölli immediately before its destuction by the slide. (photo courtesy of Dr B. Loup, Canton of Fribourg; reproduced by permission)*

its original assessment, and knowing that this argumentation was not supported by other government agencies, it had also no specific need for additional information.

### 9.4.2   Risk Assessment Activities During and After the Slide

When the sliding started to accelerate and extend, the situation changed. With the occurrence and debris flows, rockfalls in the detachment zone and the possibility of the Höllbach River being dammed by the slide, the characteristics of the hazard became altered in a way that required a reassessment. Now there was a real possibility of persons being at risk both in the sliding area itself and downstream. Consequently, the scope of the risk assessment was extended to include these 'secondary' risks.

The rockfall risk was investigated using field observations to determine maximum block sizes and model calculations as well as simulations to estimate extreme runout distances. It was found that the maximum block size was no reason for concern (Raetzo and Lateltin, 1996).

The risk due to debris flows was more difficult to assess because of their uncertain volume and velocities and their dependence on meteorological conditions. Scenarios were applied to determine critical conditions that could lead to large debris flows, and an emergency warning organization was drawn up to provide safety for structures and particularly for people (Raetzo and Lateltin, 1996).

To assess the risk of a damming of the Höllbach River with a subsequent collapse of the dam and a downstream floodwave, three aspects needed to be studied: dam formation, collapse of the dam and floodwave propagation. The dam formation process was investigated using a numerical non-linear viscous model (Vulliet and Bonnard, 1996). The collapse of the dam was studied based on hydraulic models using different dam geometries and material properties. The temporal and spatial propagation of the floodwave and potential protective measures were again investigated with a numerical hydrodynamic model (Bezzola *et al.*, 1996). The calculations showed that the worst case, a narrow breach in a short dam superposed by a high discharge in the receiving Äergera River, would pose a serious threat to the town of Marly approximately 15 km downstream. However, the probability of this worst case was expected to be very low, particularly in the first six months, and as a result, sufficient time for emergency preparedness measures was available. The stabilization of the slide in late 1994 with subsequent continuous erosion of the dam and drawdown of the lake made such measures unnecessary.

The sliding itself was during this period continuously monitored with a surface GPS survey and clinometer measurements, and modelled using a numerical model (Bonnard *et al.*, 1995); (Vulliet and Bonnard, 1996). Sudden unexpected developments could thus quickly be taken into account had they happened.

### 9.4.3 Organizational Aspects of Risk Assessment and Management

When it became evident that a large slide was imminent at Fälli-Hölli, the authorities quickly organized an emergency management team that comprised administrative as well as technical entities. The duties of the team included a continuous monitoring and assessment of the situation as well as the design and implementation of measures that were required to ensure the safety of people and (as far as possible) property.

The team was headed by a regional governor who reported upwards to the cantonal government. Together with the communal council he was responsible for making and enforcing decisions. To support them in decision making, they could depend on scientific and technical experts for all relevant aspects of the sliding and its secondary effects. The involvement of experts happened on a situational basis; that is, as soon as new issues arose, experts were called upon to provide advice as to what the options were for handling the situation. Most of the scientific experts were not members of the emergency management team, but they usually had their technical counterparts in the team to ensure correct interpretation and execution of their recommendations. Figure 9.5 is an abstract representation of the organization of the emergency management team.

### 9.4.4 Evaluation of the Risk Assessment Activities and Organization

The Chlöwena landslide is an example for the complex and difficult questions that need to be addressed in most risk assessments for natural hazards in general and landslides in particular. Starting with the unknown characteristics of the hazard (probability and extent of the sliding), continuing with its unknown parameters (velocity, direction of displacements) and finishing with the possible secondary hazards, it involved a wide spectrum of issues that were critical for the management of the event.

However, Chlöwena is also an example of successful risk assessment and management. Safety-relevant aspects were considered in a timely and thorough way. Even though the formalism of risk assessments may not have been followed stringently throughout the investigation, it comprises all necessary activities. The definition and consideration of

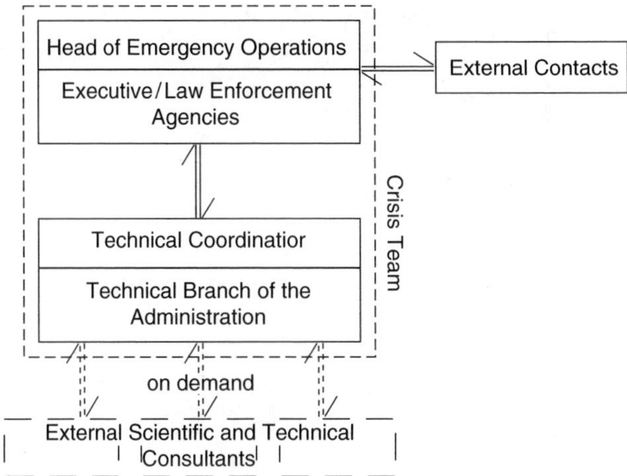


**Figure 9.5**  *Organization of the crisis team. External, non-technical contacts are ideally exclusively handled by the executive agencies (preferably the head of emergency management operations or a specially designated official) to maintain the unity of command. Reproduced by permission of Dr B. Loup*

worst-case scenarios (even if they were not acted upon) and the constant surveillance of the problem helped to ensure that as much safety as reasonably achievable was provided at all times during the event. The fact that previous assessments were revised and additional expertise was actively looked for on demand shows that the safety of people was taken seriously.

But, as is the case with most things, the risk assessment process for the Chlöwena slide was still not perfect: there remains the failure to consider appropriately the actual risk when the building permits were issued. The Falli-Hölli area was, as ECAB correctly argued, not suitable for the construction of buildings that are vulnerable to slides. The damage of the slide could have been prevented almost completely if the houses had not been built. Thus the initial risk assessment by the cantonal authorities that granted the permission to develop the area has to be considered incorrect.

### 9.4.5   Re-evaluating Risk Assessments Following Landslides: The Sachseln Slides

Sachseln is a village in the Canton of Obwalden in central Switzerland. It is located on the shore of Lake Sarnen at the foot of a northwest-facing slope comprising a total of seven small torrent catchment areas, five of which directly threaten the village. On 15 August 1997, a stationary thunderstorm cell discharged up to 140 mm precipitation on the upper parts of the watershed. The thunderstorm triggered a total of 413 mostly small slides with a total volume of $1.1 \times 10^5 \, \text{m}^3$ in the five critical catchments (see Figures 9.6 and 9.7). Many of these superficial slides then turned into debris flows that loaded the torrents with $6.5 \times 10^4 \, \text{m}^3$. Additional channel erosion brought the total

***Figure 9.6**   Sachseln slides. Shallow slides triggered by the thunderstorm. (photo courtesy of Sepp Hess, Agency for Forest and Landscape OW, CH-Sarnen)*

volume of solids to approximately $1.4 \times 10^5 \, \text{m}^3$, of which $5.4 \times 10^4 \, \text{m}^3$ were retained behind sedimentation dams. The remaining volume were either deposited in or beside the channels or transported into the lake. The total cost of the events (including damage and restoration cost) is estimated at about 100 million Swiss francs (€65 million) (Petraschek *et al.*, 1998).

The catchment areas had already (in 1955, 1975 and 1984) been the scene of landslides that turned into debris flows and thus affected the village of Sachseln. The event in 1984 had led the authorities to the conclusion that sediment transport was a core problem, and sedimentation dams were built thereafter to reduce or eliminate the risk from landslides and debris flows. These dams retained large amounts of solids in the 1997 events, and thus at least partially confirmed the 1984 findings. However, new erosion downstream of the dams (due to non-saturated transport capacity) resulted not only in sediment deposition in the village, but also in massive damage to some parts of the channel. As a consequence of this somewhat surprising outcome, a post-event documentation and evaluation (Hess, 1998; Petraschek *et al.*, 1998) as well as several research studies were undertaken (Rickli *et al.*, 2000; Liener and Kienholz, 2000). As a result, there is today a better understanding of the processes that influence both the landslide disposition and the actual landslide events.

The 1984 event was also at the root of a comprehensive hazard assessment and watershed restoration project initiated in 1988. It is now interesting to compare the findings of this project with what happened in 1997 and to see which aspects of the hazard or risk assessments must be revised since the last event.

**Figure 9.7** *Sachseln slides. Shallow slides triggered by the thunderstorm (photo courtesy of Sepp Hess, Agency for Forest and Landscape OW, CH-Sarnen)*

Hess (1998) compares calculated and actual values for water discharge and sediment transport. He shows that the actual sediment transport exceeded the previous estimates significantly, but that this excess was at least partially compensated by the retention of solids behind the retaining dams that was also significantly higher than what was expected. Thus the total effect of the sliding and the channel erosion on the downstream section of the channels was quite accurately predicted, even though the assessment of the upstream process was rather optimistic. What was clearly underestimated is the intensity of channel erosion that occurred *in* the downstream section itself. This happened primarily because of the excessive sediment deposition behind the retaining dams. Before the construction of these dams, the problem had always been too much sediment, and the goal of the protective measures was to reduce this surplus. It is understandable that little attention is paid in general to the possibility that a measure might be too effective, but this process highlights the need to reconsider previous reasoning if there are significant changes in the assessment system. Although accurate in what was covered, the pre-1997 hazard (and therefore also the risk) assessment in the Sachseln case has to be considered incomplete with respect to the effect of varying sediment saturation. Consequently, the sediment balance has come under increased scrutiny in new assessments that have been made since 1997 to account for the risk resulting from channel erosion in the downstream section.

The pre-1997 assessment work in Sachseln was focused almost exclusively on the hazard, and new information about the sliding and the sediment transport process (as e.g. presented by Liener and Kienholz, 2000; Rickli *et al.*, 2000) is still a welcome improvement of current assessments. However, the massive damage caused by spatially confined events and the extent of the potential sliding areas clearly indicate that a significant reduction of the risk is only achievable if the exposure and the vulnerability of the elements at risk are taken into account. Managing the sediment input into and its transport in the channels to prevent damage is a complex and delicate task with often uncertain outcome because of the randomness with which critical events (slope and channel bank failure, channel blockage) occur throughout the catchment area. On the other hand, it is pretty straightforward as to what can be done to reduce the exposure and especially the vulnerability of the elements at risk, and these measures are effective regardless of what exactly happens in the catchment. It is therefore not surprising that in Sachseln, as in many other places, there is a shift away from 'pure' hazard assessments towards more comprehensive risk assessments (Hess, 2002, pers. comm.), and in accordance with this is also an shift in risk management away from measures primarily designed to reduce the hazard towards concepts that include measures to reduce the vulnerability of the objects at risk.

The sliding at Sachseln differs from that at Vaiont and at Chlöwena in one important respect: it is not a very rare or even a one-time event, but one that did and will occur repeatedly, and advanced hazard assessments were therefore already available before the 1997 event. This is certainly a good basis for risk management and planning purposes. However, repeating events are also a test for the quality of risk assessments, and the consequences (both negative and positive) of the 1997 event have shown that even the most sophisticated assessments still leave room for surprises. Such surprises are either due to the uncertainty that is inherent in any assessment or to processes that are not appropriately taken into account. To prevent the latter requires that the validity and the scope of the assessments is checked regularly and thoroughly. No assessment will ever be perfectly accurate, but incomplete or outdated assessments are of little use or perhaps even dangerous.

## 9.5   Summary and Concluding Remarks

What has been said about the three real-world landslides and the corresponding risk assessments amounts to a collection of important individual facts and aspects and not a complete and thorough investigation of the assessment processes. It is also a hindsight evaluation, and some of the decisions and conclusions that are criticized may have seemed absolutely reasonable to the people who made them.

When looking at the consequences of the three landslide events, one must remember that the discrepancies are not only attributable to the quality of the risk assessment and management, but also to the different character of the hazard. The Vaiont slide was significantly larger and it had a much higher displacement velocity than the others. Even if people and a part of the property had been evacuated in time, the damage from the slide would still have exceeded that of the Chlöwena and Sachseln slides. Once it was triggered, the Vaiont slide was bound to become a high-consequence event.

Assessing landslide risks is a difficult and complex task. Neither the characteristics of the hazard nor the behaviour of the objects at risk will ever be known completely, and consequently, risk assessments will never produce exact results. However, the main reason for and benefit of performing risk assessments should not be the generation of numbers, but the systematic and thorough consideration of the risks that are associated with a given situation or system. Only on the basis of this knowledge is an efficient management of risks possible, and this is where administrative bodies enter the scene: landslides, whether truly natural or (as in the Vaiont case) partly man-made, are perceived as negative external impacts on public and private values, and the associated risk must be limited to an acceptable level. This is a public task, because there is usually nobody who has a direct private benefit related to the hazard. Therefore, if administrative bodies want to manage landslide risks efficiently, then they must also take a leading role in the assessment of these risks.

Knowing the risk is one thing, but taking appropriate action is another. The Vaiont tragedy has clearly revealed that risk assessments alone do not make a difference. What is probably more relevant is that people working with administrative bodies understand risk management as their personal responsibility and not as something that is done by someone else. This sometimes includes thinking what may seem unthinkable or going beyond assigned duties and competences to ensure that potential threats don't go unattended until they turn into real disasters.

## Acknowledgements

I would like to express my thanks to Dr B. Loup and Mr P. Ecoffey of the Canton of Fribourg and Mr J. Hess of the Canton of Obwalden for their help in the preparation of this chapter. They have provided a wealth of valuable information without which this work had not been feasible.

## References

Bezzola, G. R., Näf, F., Roth, M. and Zurbrügg, C., 1996, Dammbruchund Flutwellenszenarien als mögliche Konsequenzen der Grossrutschung Chlöwena, in INTERPRAEVENT (Garmisch-Partenkirchen: Tagungspublikation), vol. 3, 141–150.

Bonnard, C., Noverraz, F., Lateltin, O. and Raetzo, H., 1995, Large landslides and possibilities of sudden reactivation, *Felsbau*, **13**(6), 401–407.

Borter, P., 1999, Risikoanalyse bei gravitativen Naturgefahren, Number 107 in *UmweltMaterialien* (Bern: BUWAL).

Bundesamt für Forstwesen, 1984, Richtlinien zur Berücksichtigung der Lawinengefahr bei raumwirksamen Tätigkeiten (Bern: Bundesamt für Forstwesen).

Carloni, G. C., 1995, *Il Vaiont trent'anni dopo. Esperienza di un geologo* (Bologna: CLUEB).

ECAB Fribourg, 1995, Etablissement Cantonal d'Assurance des Bâtiments Fribourg: Rapport Annuel, unpublished.

Heinimann, H. R., Hollenstein, K., Kienholz, H., Krummenacher, B. and Mani, P., 1998, Methoden zur Analyse und Bewertung von Naturgefahren, vol. 85 of *UmweltMaterialien* (Bern: BUWAL).

Hendron, A.J. and Patten, F.-D., 1985, *The Vaiont Slide, Technical Report GL 85–88*, US Army Corps of Engineers.

Hess, J., 1998, Die Unwetterkatastrophe vom 15. August 1997 in Sachseln, Kt. Obwalden, *Schweizerische Zeitschrift für Forstwesen*, **149**(9), 707–714.

Hollenstein, K., 1997, *Analyse, Bewertung und Management von Naturrisiken* (Zürich: vdf).

Kanton Uri, 1992, *Hochwasserschutzrichtlinien für den Kanton Uri*, Kanton Uri.

Kilburn, C., 2002, Forecasting the collapse and runout of giant, catastrophic landslides, http://www.bghrc.com/Geolhaz/Runout/Landslides.pdf.

Liener, S. and Kienholz, H., 2000, Modellierung von flachgründigen Rutschungen mit dem Modell SLIDISP, in INTERPRAEVENT (Villach: Tagungspublikation), vol. 1, 259–269.

Loat, R. and Petraschek, A., 1997, *Empfehlungen zur Berücksichtigung der Hochwassergefahr bei raumwirksamen Tätigkeiten* (Biel: Bundesamt für Wasserwirtschaft).

Müller, L., 1964, The rock slide in the Vaiont valley, *Felsmechanik und Ingenieurgeologie*, **2**(3–4).

Paolini, M. and Vacis, G., 1997, *Il racconto del Vajont* (Milan: Garzati).

Petraschek, A., Lopes, J. B., Mani, P. and Zarn, B., 1998, *Ereignisdokumentation Sachseln*, Studienbericht 8, Bundesamt für Wasserwirtschaft, Tiefbauamt des Kantons Obwalden, Oberforstamt des Kantons Obwalden.

Raetzo, H. and Lateltin, O., 1996, Rutschung Falli Hölli, ein ausserordentliches Ereignis?, in INTERPRAEVENT (GarmischPartenkirchen: Tagungspublikation), vol. 3, 129–140.

Rickli, C., Zimmerli, P. and Zürcher, K., 2000, Waldwirkungen auf oberflächennahe Rutschungen anlässlich der Unwetterereignisse vom August 1997 in Sachseln, Schweiz, in INTERPRAEVENT (Villach: Tagungspublikation), vol. 1, 305–316.

Seiler, H., 2000, Risikobasiertes Recht: Wieviel Sicherheit wollen wir? Abschlussbericht zum Projekt Risk based regulation – ein taugliches Konzept für das Sicherheitsrecht? Technical report, Schweizerischer Nationalfonds.

Voight, B. and Faust, C., 1992, Frictional heat and strength loss in some rapid landslides: error correction and affirmation of mechanism for the Vaiont landslide, *Géotechnique*, **42**(4), 641–643.

Vulliet, L. and Bonnard, C., 1996, The Chloewena landslide: Prediction with a nonlinear viscous model, in *Proceedings of the 7th International Symposium on Landslides*.

# 10

# Addressing Landslide Hazards: Towards a Knowledge Management Perspective

*Susan Michaels*

Although we live in a knowledge society (Machlup, 1962; Drucker, 1989) and information is valued in hazards planning and mitigation (Olshansky and Rogers, 1987; Williamson *et al.*, 2001), utilizing knowledge management to address natural hazards, including landslides, is in its infancy. Knowledge management highlights the institutional arrangements for preparing and responding to landslide risk by actively managing the creation, acquisition, representation, transfer, incorporation and application of knowledge (Bukowitz and Williams, 1999). Applying knowledge management to mitigating natural hazards focuses on how people create and use knowledge to reduce collective vulnerability.

In the context of decision making, knowledge is derived from making information actionable, from using information productively (Niles and Michaels, 2002). Drucker (1989: 209) considers information to be 'data endowed with relevance and purpose'. In turn, Bukowitz and Williams (1999) define data as a set of discrete, objective facts, bits of raw material that have not been put in context.

While knowledge management is rooted in information management, it is distinct. Information or data management is the means through which information is made trackable and accessible by being categorized and systemized (Simard, 2000). In contrast, knowledge management is about utilization, incorporating both individual and institutional learning into the collective knowledge base (Bukowitz and Williams, 1999).

Exploring the potential of knowledge management to address natural hazards builds on a tradition of seeking to reduce loss by improving the dissemination of hazards information (Spiker and Gori, 2000; Fothergill, 2000). In the USA the centrality of

*Landslide Hazard and Risk*   Edited by Thomas Glade, Malcolm Anderson and Michael J. Crozier
© 2004 John Wiley & Sons, Ltd   ISBN: 0-471-48663-9

generating and disseminating earth science research to inform landslide hazard mitigation has long been recognized. This has led to creative and innovative initiatives to integrate earth science research into decision making. These efforts have spanned federal, state and local jurisdictions. While consistent infusion of earth science information into planning has not been achieved, successful examples do suggest what is needed for science to be incorporated into policy.

This chapter begins with a brief section on basic types of landslides and an indication of the consequences of landslide events. This discussion provides the rationale for the search for effective mitigation techniques that fully exploit current geological understanding of landslide phenomena. Those addressing landslide hazards have been frustrated by the question of why information is not more of a factor in mitigating landslide hazards. In this chapter selective explanations of those working in the sphere of landslides and those outside of it are provided as to why science is not more of a consideration in decision making. Examples from California illustrate where earth science has been incorporated into planning and where creative approaches have been employed to disseminate sound science to decision makers in a format that meets their needs. In the USA, the promotion of a national strategy to mitigate landslides goes back over two decades. The strategy most recently put forward envisions a critical role for information generation and dissemination. The final section of the chapter presents three complementary views of information that shed light on the assumptions underpinning the application of information to mitigating landslide hazards.

## 10.1 Landslide Hazard

The term 'landslide hazard' is used as an umbrella term for the wide range of complex landslide phenomena that interact with the environment. Different threats are posed by different types of slope movement (Guzzetti *et al.*, 1999). Each of these different types of slope movement is associated with different degrees of understanding of basic processes, technology transfer and information dissemination needs.

Very large, fast-moving landslides, such as rock avalanches, are probably the most destructive mass movements (Guzzetti *et al.*, 1999). In the United States, the Committee on the Review of National Landslide Hazards Mitigation Strategy (2002) of the National Research Council ('the Committee') considers that rockfall processes are reasonably well understood. The Committee suggests that widespread dissemination of the substantial progress made by the Federal Highway Administration and some state highway departments in technology integration and transfer relating to these processes would promote implementing effective mitigation techniques.

Slow-moving, deep-seated failures can cause significant property damage while rarely resulting in casualties (Guzzetti *et al.*, 1999). The Committee suggests that the mechanics of bedrock slide initiation are well understood. What is required is mapping such landslides in high-risk areas to assist regulation, improve mitigation methods and establish appropriate risk assessment techniques.

Fast-moving soil-slip debris flows initiated by intense rainfalls are very destructive, causing both loss of life and widespread physical damage (Guzzetti *et al.*, 1999). The Committee (2002) recommends improving the understanding of the basic science of debris-flow initiation and movement before proceeding with more work in technology integration and transfer. Advances in basic science would contribute to improving mapping, a priority requirement. Risk assessment and mitigation, including regulation, would benefit from anticipated clarification of magnitude–frequency–runout characteristics.

Landslides can occur as independent events or as part of interrelated multiple natural hazard processes where an initial event causes secondary events or when more than one natural hazards process happens at the same time. Examples are when volcanic eruptions, earthquakes and landslides occur as interrelated processes (Schuster and Kockelman, 1996). Increasingly, mitigating natural hazards emphasizes a multi-hazard approach with sustainability as the overarching objective (Mileti, 1999).

While landslides occur in all of the states of the United States, they are a significant hazard in more than half the states, including Alaska and Hawaii. The most seriously affected regions are the Appalachian Mountains, the Rocky Mountains and the Pacific Coast (Schuster, 1996). In the states of Washington, Utah and Colorado the threat of landslides is increasing significantly because of urbanization. California is the most urbanized of the nation's landslide-prone areas (Olshansky and Rogers, 1987).

Landslide-triggered casualties and economic losses are greater in many countries than is commonly recognized (Guzzetti *et al.*, 1999; Schuster, 1996; Schuster and Highland, 2001). In the United States alone landslide fatalities are estimated to be approximately 25–50 people a year and result in total annual losses of approximately US$2 billion (Schuster and Highland, 2001; Spiker and Gori, 2000). Advances in recognizing, predicting, mitigation measures and warning systems are occurring even while landslide activity is increasing (Schuster, 1996). Greater urbanization and development may result in modifying surface drainage, poorly placed fills and overly steep slope cuts that increase the number of landslides (Olshansky and Rogers, 1987; Schuster, 1996). The upsurge in landslide activity is also a function of ongoing deforestation in landslide-prone areas and increasing regional precipitation resulting from changing climatic patterns (Schuster, 1996). Even though landslide activity is on the upswing, the extent of the problem has not been well appreciated by many earth scientists and public officials (Howell *et al.*, 1999 citing Brabb and Harrod, 1989).

With total costs exceeding US$400 billion, the 1983 Thistle, Utah earthquake illustrates the enormous costs, both direct and indirect, of damaging landslides. It is the single most costly landslide event in US history (Spiker and Gori, 2000). The 21 million m$^3$ debris slide dammed the Spanish Fork River and destroyed both US Highway 6 and the mainline of the Denver and Rio Grande Western Railroad. Floodwaters behind the landslide dam inundated the town of Thistle, including railroad switching yards (Spiker and Gori, 2000; Schuster, 1996). It took an engineered drain system to avert a potential disaster by controlling the release of the floodwaters. Residual sedimentation from the floodwaters partly buried the town and few residents returned to Thistle. The highway was realigned around the landslide and the railway constructed a tunnel around the slide zone (Spiker and Gori, 2000).

More modest slope failures can still have significant adverse economic effects. Slope failures on private property can have public cost repercussions. They can indirectly affect the local economy by affecting neighbouring properties and public infrastructures. When a failure occurs and the original developer and owner cannot be located, a public agency may step in and assume some of the costs. General funds have been used to alleviate problems confined to unstable hillside areas. Frequently, repairs and maintenance of roadways and pipelines are paid for by local governments or utility districts, with the costs carried by the entire population of the jurisdiction, while only hill dwellers benefit. In the USA, where the wealthy tend to live in the hills and the less wealthy in the flatlands, the inequity of who pays and who benefits is further exacerbated (Olshansky and Rogers, 1987).

A widening circle of actors has become engaged in addressing landslide hazards. No longer are landslides primarily the concern of transportation engineers designing and maintaining public roads on unstable slopes. Landslides have become the concern of professionals engaged in urbanizing or developing areas and, increasingly, private property owners contending with landslide-prone sites (Olshansky and Rogers, 1987). Decision makers must consider three fundamental options available for how to address landslide hazards in their communities: (1) they can do nothing; (2) they can provide post-event relief and rehabilitation; (3) they can work to contain or control the hazard before serious damage occurs (Rossi *et al.*, 1982).

## 10.2   Needed: More Science in Mitigation

Howell *et al.* (1999) contend that the geological community has a critical role to play in educating local planners and engineers about the types of hazards facing their communities, the extent, place and economic consequences of these hazards and how to reduce exposure to them. The Committee (2002) recognizes that a major constraint to providing improved mitigation is the lack of information about landslide distribution and degree of hazard. Adequate information about the mechanisms of landslide hazard and mitigation alternatives, the Committee (2002) contends, needs to be available to all sectors of society as a prerequisite for informed decision making.

Likewise, Olshansky and Rogers (1987) note that effectively implementing landslide reduction measures requires increased knowledge of landslide processes. Kockelman (1986, as cited by Schuster and Kockelman, 1996) identifies four approaches to reducing landslide risk.

1. Restricting development in landslide-prone areas;
2. Developing and implementing excavation, grading, landscaping and construction codes;
3. Implementing physical measures to prevent or control landslides, such as drainage, slope geometry modifications and structures; and
4. Developing monitoring and warning systems.

Olshansky and Rogers (1987) explain that designing land use policies and grading codes would be simplified by accurate hazard zone designation and quantified probabilities of landslides. They also note that better engineering designs on unstable slopes could result from improved technical knowledge of landslide processes.

## 10.3   Why Isn't Information More of a Factor in Addressing Vulnerability to Landslide Hazards?

Howell *et al.* (1999), Olshansky and Rogers (1987) and Guzzetti *et al.* (1999) turn their attention to the third of Rossi *et al.*'s (1982) options for addressing natural hazards: what can be done to mitigate landslide hazards. They consider why information is not more of a factor in addressing vulnerability to landslide hazards. As scientists and planners in the landslide hazard field, their explanations revolve around specific attributes of the landslide hazard issue.

Howell *et al.* (1999, citing Brabb, 1996) contend that worldwide, earth science information that could mitigate natural hazards, including landslides, is poorly used. For example, hazard maps based on geological data could be better utilized. Howell *et al.* (1999) attribute these missed opportunities to the highly contextualized perceptions people have of the threats posed by natural hazards.

1. Hazard maps may be seen by policy makers to be anti-business, jeopardizing economic development.
2. Residents don't accept or comprehend the threat because many have long return periods.
3. While recognizing the threat, an individual or community may determine that the benefits of inaction outweigh the costs of the expected loss.
4. Hazard awareness may not exist in the minds of the public or government agencies, because recurring remediation costs are buried in budget line items such as maintenance.
5. Media references to 'mudslides' invoke images not of wholesale destruction of homes or communities, but of localized, almost trivial nuisances.
6. The facts surrounding many landslide disasters are guarded by attorneys and couched in liability terms because it is usually through judicial process, where each case is unique, that redress for landslide damage is settled.
7. Partly out of ignorance, those who suffer losses from landslides are seen as getting what they deserved because of their decision to be in homes on steep slopes.

Olshansky and Rogers (1987: 949) discuss five layers of 'inconsistent availability of hazard information', a phenomenon they consider a key dimension of the landslide problem.

1. Many people don't realize they live in a hazardous location.
2. Few mechanisms, legal or statutory, require that where relevant information is available it is transmitted to prospective home or land buyers.
3. Landlords who have access to hazard information may not necessarily pass this information on to tenants.
4. Land use planning agencies do not always have landslide hazard information or necessarily act upon it when they do.
5. Many citizens probably wouldn't incorporate information made available to them into their actions.

Olshansky and Rogers (1987) also offer five possible explanations why slope stability information is not used more often in local land use planning. These explanations reflect the political reality of local decision making.

1. Local officials face political pressure to approve questionable developments on potentially unstable lands;
2. Apathy about landslide hazards by local officials;

3. Other environmental and social considerations receive higher priority;
4. Landslides are not important considerations in land use decision making because they can be prevented by better engineering; and
5. Local officials fear being subjected to 'taking' claims. The fifth amendment of the US constitution provides that private property shall not be 'taken' for public use without just compensation. 'Taking' refers to any situation where the value of a person's property has been substantially diminished and that in fairness the public should share the burden.

Olshansky and Rogers (1987) point to the low frequency of landslide occurrence in a given area, the geographically limited extent of landslide events and individuals' perception of the landslide risk as reasons why federal and state agencies are challenged in enabling local governments to develop sound hazard mitigation policies. Historically, local governments have not been interested in addressing the federal- and state-scale concerns over the frequency and cost of landslides, and the resulting concerns of public safety, costs and equitable use of state resources. Rather, local officials tended to ignore the risk of landslides because of their low annual probability of occurrence in most local jurisdictions. Local residents, meanwhile, frequently equated low annual risk on their lots with no risk.

Beyond the landslide hazards domain, understandable reasons are offered as to why those who do not generate the science do not invest in acquiring scientific findings. While scientists may assume that scientific knowledge is vital to lay people, lay people may regard it as irrelevant, particularly if it seems unrealistic given existing conditions. Without the necessary opportunity or resources, even a technically literate individual may reject or ignore scientific information (Wynne, 1995). In obtaining information, people place a high premium on ease of access and will settle for minimally adequate information that suits their needs (March and Simon, 1958). As information satisficers (March and Simon, 1958), people weigh the amount of effort required to use a source against its anticipated usefulness. Effort may be physical, such as travelling to a source, intellectual, such as learning a classification system or computer application, or psychological, such as dealing with an unpleasant source (Choo, 2000). What is vital is how non-scientists experience and perceive the relevant institutions that generate and purvey the information (Wynne, 1995). That the US Geological Survey is well regarded is a contributing factor in the acceptance of the earth science research on the mechanisms of landslides, landslide mapping and risk estimates associated with landslides.

The need for planning and policy-making tools for evaluating landslide hazards and the testing of numerous models in a range of physiographic settings is widely acknowledged. Still, this acknowledgement has not resulted in agreement among earth scientists and decision makers on the intent and use of landslide hazard evaluations. This lack of agreement, Guzzetti *et al.* (1999: 210) explain, is why it is exceptional for knowledge about landslide hazards to actually 'become an integral part of building codes, planning policies, or civil protection regulations'. Sarewitz (2000), looking at the capacity of science to help resolve environmental conflicts, explains why agreement among decision makers and earth scientists or even consensus among earth scientists is not the norm: decision makers and scientists have fundamentally different aims. Decision making, grounded in democratic debate, is about achieving operational consensus that enables action. In contrast, the process of scientific investigation at its most robust is designed to reject premature consensus. Science is a process of expanding our understanding of nature by 'questioning, hypothesizing, validation, and refutation' (Sarewitz, 2000: 84).

Earth scientists engaged in analysing landslide inventories and testing and developing either functional, statistically based models or geotechnical or physically based models employ indirect and quantitative methods for ranking slope instability factors and assigning different hazard levels. These approaches epitomize what Sarewitz (2000) describes as the 'physics view', where nature's complexity must be held in abeyance as nature is parsed into its component parts and governing laws. The aim in this view is to develop predictive hypotheses and theories through real or imagined controlled experiments. By generating predictions that dictate action, the physics view when applied to policy-making promises to relieve non-scientists from decision-making responsibility. Science comes across as an authoritative voice, suggesting a way to proceed that does not involve the search for agreement characteristic of the policy-making domain.

The notion that scientific expertise can be 'a neutral, mediating force, contributing to good or even "correct" decisions' and that regulation can 'be a pocket of technocracy, working towards publicly defined ends' appeals to many citizens, scientists and elected officials (Cozzens and Woodhouse, 1995: 542). Yet the reality of scientific uncertainty, conflicting interests and the need to make trade-offs, such as between cost and safety, results in science being one among other factors in policy making (Cozzens and Woodhouse, 1995).

An alternative to the 'physics view' that is equally committed to creating a true picture of nature is the 'geological view'. It recognizes nature as 'the evolving product of innumerable complex and contingent processes and phenomena' (Sarewitz, 2000: 92). The tools of discovery are those employed in field geology, such as historical reconstruction and analogy. From this vantage science is a source of insight, rather than an authoritative voice, in decision making. The resulting emphasis is on developing policies that favour adaptability and resilience, and that incorporate an appreciation of the inevitable constraints to what we know and can know given the reality of diversity, change and surprise (Sarewitz, 2000).

Where political consensus exists or is likely to be realized, science can contribute to beneficial action (Sarewitz, 2000). Where a workable context has been constructed through political structures and processes, science can help guide action (Cozzens and Woodhouse, 1995). The application and utility of science as a factor is more likely where special interests are less, costs of action are lower and consequences from inevitable mistakes are reduced (Sarewitz, 2000).

Science contributes to decision making by providing diagnosis and assessment (Sarewitz, 2000) such as hazard assessment, a cornerstone of hazard management programmes. These can be conducted at three different levels. The most basic, hazard identification, defines magnitudes (intensities) and probabilities of threats. Vulnerability assessment characterizes population and property exposure and the potential extent of injury and damage from the occurrence of an event. Risk analysis is the most sophisticated of the three and provides a more complete picture by incorporating probability estimates of varying levels of injury and damage (Daley, 1998). Results from this work can contribute to fine-tuning or redesigning policies and programmes, evaluating options and informing how decision makers pursue longer-term goals (Sarewitz, 2000).

## 10.4   Dissemination Success

While there is understandable concern about the failure to incorporate earth science information into reducing landslide losses (Howell *et al.*, 1999; Guzzetti *et al.*, 1999), there are success stories. Local governments in California have a longer history of using landslide information than those in other states, partly because California's planning laws explicitly encourage the consideration of landslides when communities make their land use plans. Decisions around future development of each community must be documented in its general plan. The plan must address the potential of slope instability that may lead to mudslides and landslides (Olshansky and Rogers, 1987). It must be noted that these laws are not applied uniformly. No specific regulations relating to debris flow exist in most counties and cities because maps showing this hazard are few and cover only a limited area (Howell *et al.*, 1999).

The California Environmental Quality Act (CEQA) provides another means to incorporate landslide hazard information into land use planning. Under CEQA, all potentially significant projects must have environmental documents prepared for them. As part of the CEQA environmental review process, landslide hazards must be considered. Local agencies must consider potential impacts of landslides and how they might be mitigated if landslide information is readily available for a proposed development site (Olshansky and Rogers, 1987).

In 1983 the California legislature enacted the first landslide statute of any state in the United States. The Landslide Hazard Identification Program was established within the California Division of Mines and Geology (CDMG) to independently develop landslide hazard maps within urban and urbanizing areas of the state and to provide local agencies with other technical assistance in making land use decisions in landslide-prone areas (Olshansky and Rogers, 1987).

In the United States, the federal government is a key player in making relevant, landslide information available for incorporation into planning. It functions primarily as a source of funding for state and local control works, expertise and research support (Schuster and Kockelman, 1996). At the turn of the twentieth century, the US Geological Survey (USGS) published geological maps of the San Francisco Bay region. Not one landslide was depicted on them. Relatively little attention was paid to landslides through the end of World War II, despite the dramatic demonstration from the 1906 San Francisco earthquake that landslide processes are regionally significant. The postwar building boom, beginning in the late 1940s, prompted the USGS and the CDMG to become concerned with engineering geology issues, including landslides (Howell *et al.*, 1999). By 1970, 1200 landslides had been mapped. By documenting landslide-prone areas, these maps became a means to raise awareness of the landslide risks in the San Francisco Bay region. Significantly, they influenced plans for development in some landslide-prone areas (Howell *et al.*, 1999 citing Brabb, 1985).

Portola Valley provides a model of the effective incorporation of geological hazard information in land use planning. Numerous landslides have occurred in the vicinity. Much of the town lies in a valley, in the southern San Francisco Bay region, formed by the San Andreas fault. In the late 1960s, the number of houses planned for the Bovet property in Portola Valley was reduced significantly because of municipal decision making based on landslide mapping (Howell *et al.*, 1999 citing Mader and Crowder, 1969). In 1974 a geological map and slope stability map were incorporated into the town's general plan. The plan requires that in all decision making by staff, commissions and councils the maps and associated policies be used. The plan also specifies for each land stability category permissible uses and residential density (Olshansky and Rogers, 1987).

The Federal Department of Housing and Urban Development (HUD) in the 1970s funded a landslide mapping programme in the San Francisco Bay region carried out by USGS and CDMG. The experimental programme was intended to help get earth science information used in regional planning and decision making. Specifically, to help planners and decision makers, the programme was to

1. identify important earth-sciences-based problems related to growth and development in the region;
2. provide earth science information needed to solve the problems identified;
3. interpret and publish findings in forms understandable and usable by non-scientists;
4. launch new ways of communicating between scientists and users; and
5. consider alternative means of using earth science information in planning and decision making (Olshansky and Rogers, 1987 after Brown in Nilsen and Wright, 1979).

The logic behind the programme was that identifying and mapping more than 70 000 landslides would lead to avoiding construction in risk-prone areas and provide critical information needed to mitigate the hazard. The landslide inventory proved insufficient for these purposes. Planners wanted the level of hazard for all land surfaces characterized on maps, regardless of whether landslides were present. Consequently, maps were prepared for 10 San Francisco Bay region counties that provided a basis for characterizing the landslide susceptibility of areas along a spectrum from most prone to least prone (Howell *et al.*, 1999).

Based on the map produced for San Mateo County in the southern San Francisco Bay region, the zoning in the most landslide-prone areas in the county was changed from five houses per acre to one house per 40 acres (Howell *et al.*, 1999). The county can approve higher densities where the geological report required for structures in these zones concludes that on a particular parcel it is safe to exceed the density requirements (Olshansky and Rogers, 1987). With other earth science data, the landslide susceptibility maps were used to prepare and evaluate environmental impact reports, to design public facilities, to plan open space and to require submission of geotechnical reports before allowing development (Howell *et al.*, 1999 citing Brabb 1984 and Brabb 1995). Up until a high-intensity storm in 1982, new landslides, mostly slips, slumps and slides, that occurred in San Mateo County were in areas mapped as either landslides or highly susceptible to landslides. Consequently, until this storm, the zoning procedure was considered a great success (Schuster and Kockelman, 1996).

The 1982 storm generated 18 000 debris flows in the San Francisco Bay region. Twenty-five people died and there was at least $65 million worth of damages (Howell *et al.*, 1999 citing Ellen and Wieczorek, 1988). Thousands of debris flows were triggered by heavy rainfall where few had been seen before (Schuster and Kockelman, 1996: 96 citing Brabb, 1984). The 1972 landslide susceptibility maps were based on interpreting aerial photographs that showed evidence only of deep-seated landslides, and so new debris flows had not been expected (Schuster and Kockelman, 1996). The unanticipated nature of the 1982 event led to a demand for new maps that would indicate where future debris flows might take place. Consequently, an experimental debris-flow susceptibility map for San Mateo was generated (Howell *et al.*, 1999 citing Mark, 1992).

Landslide research declined in the early 1990s and most of the regional landslide work was abandoned. This was, in part, a function of the lack of human tragedy and major property damage associated with storms that generated debris flows in the late 1980s. Subsequently, there has been renewed interest in providing landslide-related information. It has been fuelled by demand for such information coupled with advances in computer mapping systems (Howell *et al.*, 1999). California continues to be home to innovative, funded approaches to disseminating contextualized geological information that can be used by state and local officials, prospective developers and property owners.

Much of the discussion about the contribution of geological information to landslides focuses on its use in land use planning and mitigation. Geological information effectively conveyed to emergency personnel can contribute to preparedness. An early forecast of a major El Niño phenomenon by the National Oceanic and Atmospheric Administration made during the spring and summer of 1997 led to consideration of how losses of life and property from climatically induced landslides could be reduced through publicizing the potential hazard. The USGS, the National Weather Service (NWS), an agency within the National Oceanic and Atmospheric Administration, and the State of California's Region #2 Office of Emergency Services (OES) linked programmatic mandates and coordinated scientific expertise. The collaborative intent was to provide relevant information about areas of possible landslide and debris-flow activity that might result from major storm activity. The interagency arrangement enhanced USGS work on hazardous slope movement, facilitated USGS and NWS research efforts on hazardous natural processes, and provided a communications process through OES channels to effectively use new USGS information for emergency response (Howell *et al.*, 1999).

Recognizing that neither hazards maps nor an interagency agreement were sufficient to ensure the use of the information generated, early on in the collaborative process geologists, planners and emergency response personnel in the 10 counties of the San Francisco Bay region discussed with each other their particular problems regarding landslide hazard mitigation, response and remediation. During these discussions it became apparent that the limited resources of local government made it difficult for them to be concerned with issues that would not immediately impact their jurisdiction. The fine-scale maps required to meet the expressed needs of local government could not be prepared in the few months immediately preceding the onset of the 1998 El Niño winter. Technically, county staff could download the information from the Internet (http://wrgis.wr.usgs.gov/open-file/of97-745/), and make larger-scale plots. The maps, displaying hazard data in overlay form on shaded relief maps, explicitly carried warnings about not using them at scales larger than warranted by the data (Howell *et al.*, 1999).

After the 1997–98 El Niño rainy season, a small sample of consulting geologists, county planners and emergency planning/response personnel were asked about the utility of the USGS maps from the San Francisco Bay landslide folio and how the information on them could be made more useful. The maps provided a regional perspective helpful to consulting geologists in increasing their credibility in assessing particular landslides. County personnel also noted the utility of a regional overview. The most enthusiastic group was OES personnel, who used the maps to help educate people about preparing for possible storm damages. The debris-flow rainfall threshold maps were useful for reminding OES personnel that large storms can result in life-threatening debris flows. More specifically the maps show rainfall amounts linked to a particular hazard for a defined area. Having the maps delivered to them before the onset of El Niño storms meant they had time to disseminate the information. The intensity of their preparation varied on a county-by-county basis, depending on the extent to which officials perceived landslides as a threat in their counties. The overwhelming need was for more detailed, larger-scale formats, an issue recognized early in the project but unable to be addressed in the time available given the volume of maps required (Howell *et al.*, 1999).

One example of a document that provides property owners with directly relevant information in an accessible form is a digital USGS publication by Brabb *et al.* (2000) entitled *Possible Costs Associated with Investigating and Mitigating Some Geologic Hazards in Rural Areas of Western San Mateo County, California*. The three-part document consists of

1. the distribution of landslides, landslide susceptibility and slope in an ARC/INFO formatted database;
2. interactive landslide hazard maps and tools that enable property owners to estimate the cost of investigating potential landslide hazards available at http://kaibab.wr.usgs.gov/geohazweb/intro.htm; and
3. landslide and landslide susceptibility maps, digital orthophoto quadrangles, digital orthophoto quadrangles, digital raster graphic quadrangles, geological maps and slope maps in plot files.

Property owners can determine which hazards might affect their properties and estimate how much it would cost to investigate the effects of these hazards. They can do so by using either paper plots of the map layers or interactively on the web, superimposing the landslide hazard map layers on property lines. The ability to access the information in both these forms reflects the San Mateo County decision makers' request that the information be readily available to the public. Decision makers in coastal San Mateo County asked the USGS to indicate what the cost might be to investigate and mitigate geological hazards in areas where San Mateo County requires plans for proposed one-dwelling units that include a geological report indicating a safe and suitable site is available. As indicated previously, areas with landslides and other geological hazards are restricted to one dwelling unit per 40 acres in large parts of rural San Mateo County that have been zoned 'Resource Management' (RM) by the County (Brabb *et al.*, 2000).

To determine the cost estimates, the USGS convened a panel of consulting geologists. Their summarized deliberations were circulated to approximately 200 Bay Area region consultants for comments. The summarized deliberations and the consultants' responses to them formed the basis for the USGS publication (Brabb *et al.*, 2000).

Both the product and the process of producing Brabb *et al.* (2000) are noteworthy from a knowledge management perspective. The County called upon the highly regarded federal agency with geological experience. A great deal of attention was given to how and what material would be made available to the public. An implicit recognition was that potential users were likely to have the wherewithal to utilize the results. The process of determining costs took advantage of the knowledge of local practitioners.

Without question the experience of incorporating landslide information into planning in California reflects favourable conditions for dissemination of earth science information: federal provision of expertise, a willing and able state partner, a number of progressive municipalities, high-property values and motivated and well-resourced property owners. Even under such conditions fundamental dilemmas relevant to other settings still exist. For example, one challenge in presenting landslide information is what scale to use. Home owners want to know if their residences fall within or outside the area of a mapped landslide; city and county government agencies are focused on their respective jurisdictional limits. To advance such dynamics as how to calculate susceptibility, understand processes, and determine the role of rainfall and earthquake shaking in triggering landslides requires investigating landslides in a broader context (Howell *et al.*, 1999).

## 10.5   Pursuing the Implementation of a Comprehensive National Strategy for Mitigating Landslide Hazards

The pursuit of a comprehensive national strategy for mitigating landslide hazards builds upon aims in landslide research and education. The longstanding aims of organized landslide research and education programmes have been to understand landslide mechanisms, to synthesize these findings and to make them accessible to practitioners. The general aim of basic research is to investigate where landslides take place, what are the causes, rates, processes and magnitude of past events and to assess future landslide risks. Defining hazard zones and developing cost-effective engineering solutions can be the outputs of synthesis of the basic research. Technical agencies, such as the USGS and the CDMG, publish the information, often in map form, to educate and influence local planning agencies (Olshansky and Rogers, 1987). In 1982 the USGS identified four factors usually shared by successful landslide hazard reduction programs (Schuster and Kockelman, 1996: 92):

1. An adequate base of technical information on the hazards and risks;
2. A technical community able to apply and enlarge upon this database;
3. An able and concerned local government; and
4. A citizenry that realizes the value of and supports a program that promotes the health, safety, and general welfare of the community.

Also in 1982 the USGS developed a comprehensive national programme for landslide hazard reduction that has yet to be activated in a comprehensive fashion. It set forth goals and tasks for conducting landslide studies, evaluating hazard mapping, disseminating information and how to evaluate the use of that information (Schuster and Kockelman, 1996).

The latest iteration of a comprehensive strategy for addressing landslide hazards in the United States was prompted by a directive from the House of Representatives to the USGS. The directive led to a report outlining key elements of a national strategy for reducing losses across the country, including activities in both the public and private sector and at the different scales of government. The direction of the report, promoting a more aggressive approach to landslide mitigation nationally, was endorsed by the Committee that reviewed the strategy (Spiker and Gori, 2000).

The strategy's long-term mission 'is to provide and encourage the use of scientific information, maps, methodology, and guidance for emergency management, land-use planning, and development and implementation of public and private policy to reduce losses from landslides and other ground failure hazards nationwide' (Spiker and Gori, 2000: vii and 9). The strategy has nine major elements ranging from research to mitigation. All of the elements involve managing knowledge. One of the elements is to establish an effective system of information transfer. Systematically collecting and distributing scientific and technical information is still in the early stages of development. At the beginning of the twenty-first century, no nationwide, systematic collection and distribution of landslide hazards information existed in the USA. Objectives of establishing an effective information transfer system include evaluating and using advanced technologies to disseminate technical information, including maps and real-time warnings of potential landslide activity and developing and implementing a national strategy for systematically collecting, interpreting, archiving and distributing this information (Spiker and Gori, 2000).

There is a need to think beyond information management to knowledge management. Information management does not necessarily lead to a new knowledge order that will lead to the society envisioned in the strategy, 'a society that is fully aware of landslide hazards and routinely takes action to reduce both the risks and costs associated with those hazards' (Spiker and Gori, 2000: 9).

The strategy recognizes that only through understanding the nature of the threat, its potential impact, options for reducing risk or impact and how to execute specific mitigation measures can individuals and communities reduce their risk (Spiker and Gori, 2000). Likewise the Committee reviewing the strategy recognized that widespread outreach, education and technology transfer were essential for the success of a national landslide mitigation strategy. The assumption is that the public and private sector as well as individuals will derive tangible economic benefits from having better information about the onset and consequences of natural disasters (Williamson *et al.*, 2001).

## 10.6   Information as Object, Human Construction and Actionable Advice

While collecting, interpreting and disseminating information are regarded as integral elements of mitigating landslide hazards (Spiker and Gori, 2000), less attention has been paid in the landslide mitigation literature to the assumptions underlying how information is conceptualized. Table 10.1 compares the assumptions behind the complementary views of information as object, as constructed by people and as actionable advice.

When we treat information as an object – a thing that resides in documents or information systems – we focus on how to get it and how to represent it so it is easier to use (Choo, 2000). The concern becomes ensuring that repositories remain current, accessible and reliable – essential characteristics if users are to be confident about the information they are accessing (Davenport and Prusak, 1998).

The model of information as object is well entrenched in the realm of hazards. This is exemplified by Williamson *et al.*'s (2001: 8) observation that earth science information is 'quite easily transmitted and copied' after it has been disseminated. Accessibility to a range of different users, applicability to the differing needs of those users and effectiveness in aiding in their decision making, such as for managing risk, become highly sought-after attributes of how to present information (Daley, 1998).

**Table 10.1**  *Information as object, human construction and actionable advice*

| Information | Information as object | Information constructed by people | Information as actionable advice |
| --- | --- | --- | --- |
| Understood as | Thing that resides in document, information system or other artifact; constant, unchanging | Outcome of people constructing meaning out of messages and cues; resides in individuals' minds; individuals actively create meaning of information through their thoughts, actions and feelings | Input into decision making; context dependent |
| Utility derived from | Meaning fixed by representation in artifact | Value determined by social setting in which it is encountered | Relevance to situation |
| Concerns | How to acquire needed information? How to represent information to make it easier to use? | How to understand social and behavioural processes through which information is created and used? | How to assess validity, utility and applicability of information to decision making? |
| Management implications | Clearer and more complete representation of information leads to more accessible information systems | Fuller understanding of information seeking helps to design better information processes and information systems | End-user needs are paramount |
| Generated by | Expert creation | Participatory environmental decision making reflecting different, legitimate realities of stakeholders; no ultimate source of knowledge able to dictate 'correct' action under complex conditions; uncertain characterization of problem and consequences of addressing problem (Sarewitz, 2000: 95) | Solicited and unsolicited contributions from inside and outside decision-making forum |
| Discourse | Monologue | Conversation | Debate |

*Note:*
The contents of the second and third column are derived from Choo (2000: 245) unless otherwise noted.

With more options than ever for disseminating, sharing, accessing and using scientific and technical information, the information as object approach with its emphasis on acquisition and representation becomes even more appealing. For example, advances in Geographical Information Systems make possible highly informative, interactive hazards maps based on geological data.

As applied in the landslide mitigation field, the information as object approach serves to focus concern on the information to be provided to prospective users. Schuster and Kockelman (1996) recognize that awareness and understanding of the landslide problem in the area is key. They develop this point by explaining how the landslide problem can be delineated through classification of mass movements and landslide-hazards-related zoning principles and practices.

Innes (1998) argues that information becomes more influential through the process of creating knowledge than when it is compiled into a report. In the process of creating knowledge it becomes embedded into individual and organizational understanding through debating what data are required, what data collection techniques are most appropriate and through producing the data. Data acquire their value to key players through discussion, debate and agreement to ensure that those produced will be meaningful. By becoming embedded in individual and organizational understanding, information acquires its influence.

Focusing on information as the outcome of people constructing meaning out of messages and cues emphasizes understanding the social and behavioural processes through which it is created and used (Choo, 2000). An interactive approach highlights how science communication, at a minimum, is a two-way process, depending on the interests and concerns of scientists and others in social authority as well as the interests and concerns of the targeted recipients of information (Lewenstein, 1995). Understanding information seeking as social behaviour helps in designing better information processes and information systems (Choo, 2000).

Motivation and effort to seek out information can vary with its intended use (Choo, 2000). Choo (2000: 248–249), drawing on citing the work of Brenda Dervin and Robert Taylor, lists eight general categories of how people use information:

1. to develop a context;
2. to understand a particular situation;
3. to know what to do and how to do it;
4. to get the facts about something;
5. to confirm another item of information;
6. to project future events;
7. to motivate or sustain personal involvement; and
8. to develop relationships and enhance status or personal fulfilment.

Research remains to be done to explore the applicability of cognitive approaches to understanding how people process landslide hazards information.

Knowledge as actionable information, the definition provided at the outset, is the third view of information presented in Table 10.1. This vantage highlights the extent to which diffusing knowledge is about transposing and adapting it to the local setting (Callon, 1995). Borgmann (2000: 103) suggests that because we usually cherish and comprehend where we live, when it comes to providing a foundation for our understanding of the world, 'all geology must be local'. While a global knowledge of geology needs to underpin local knowledge, local knowledge 'can be selective and respond to geology where the contingencies of life suggest it' (Borgmann, 2000: 104). Actionable information presents the pragmatic perspective of decision makers seeking to employ information as one factor in complex policy making. Two aspects of pragmatism in this sphere involve bundling information and addressing imperfect landslide information.

### 10.6.1   Bundling Information

To use landslide-related information productively requires linking it to decision making where there is a strong incentive to factor it in. Linking information to regulatory mechanisms has proven a powerful means to achieve mitigation (Olshansky and Rogers, 1987).

Ideally, market transactions, such as buying and selling real estate, lending for mortgages and property insurance, should internalize landslide information. Risk could be reflected in lower property values. Mortgage lenders made aware of landslide areas may choose to avoid loans in those areas, since abandoning property or discontinuing loan payments is not unheard of among owners of badly damaged premises (Olshansky and Rogers, 1987). Non-subsidized landslide insurance could also signal the relative costs of exposure to landslide risk. Use of land susceptible to landslides could be influenced by advice from insurance organizations (Schuster and Kockelman, 1996). The goal of offering liability coverage is to promote wise and prudent development on hillsides. Embedding earth science information in loss avoidance strategies will enable property owners to better understand their risk exposure (Howell *et al.*, 1999).

### 10.6.2   Addressing Imperfect Landslide Information

Understanding of how and when landslides occur is inevitably incomplete. Likewise engineering solutions cannot be expected to provide definitive fixes. Consequently those who use the best available understanding must accept an element of uncertainty. A local government, in recognizing the incompleteness of available information, may choose to reduce exposure by restricting how many structures can be built in a hazardous area. It can also develop flexible regulations where the onus is on prospective developers to generate site-specific information that the municipality will use in making decisions. Engineering reports could be required for potentially unstable sites. The information generated in these reports could in turn inform how the municipality refines its regulations. Site-specific engineering reports could be accepted as the basis for waiving grading or uniform building regulations (Olshansky and Rogers, 1987).

## 10.7   Conclusion

While this chapter has featured selected experiences initiated at the local, state and federal level in the USA, a knowledge management perspective on addressing landslide hazards may prove valuable within the context of other countries. The process of beginning to consider the potential of such a perspective need not be daunting. An instructive first step is to assess in any jurisdicition whether information is viewed primarily as an object, as constructed by people, as actionable advice or in a manner not discussed in this chapter.

Each of these views has implications for embarking on knowledge management because knowledge management is about utilizing what is known. Utilization is a function of how information is conceptualized. Recognizing the limitations of what can be accomplished with a current conceptualization may lead to reconceptualizing information to expand what might be achieved through knowledge management.

This chapter has presented a highly selective discussion of the intertwining of science and mitigation through information dissemination. While the critical role information might play in lessening the adverse impacts of landslides is well recognized, those who address natural hazards are aware of why information does not feature more prominently in hazard mitigation decision making. Examples from California illustrate success in disseminating information. Success refers to incorporating earth science into local land use planning as well as adapting the substantive presentation of research findings to the needs of planners and decision makers. Still outstanding is the need to implement a national strategy in the USA that comprehensively injects earth science information into decision making. To move beyond the current successes in disseminating landslide hazard information will require adapting relevant knowledge management concepts and practices from outside the sphere of landslide hazards.

# References


Borgmann, A., 2000, The transparency and contingency of earth, in R. Frodeman (ed.), *Earth Matters: The Earth Sciences, Philosophy, and The Claims of Community* (Upper Saddle River, NJ: Prentice-Hall), 99–106.

Brabb, E., Roberts, S., Cotton, W., Kropp, A., Wright, R. and Zinn, E., 2000, *Possible Costs Associated with Investigating and Mitigating Some Geologic Hazards in Rural Areas of Western San Mateo County, California*, Open File Report 00–127 (Menlo Park, CA: United States Geological Survey).

Bukowitz, W.R. and Williams, R.L., 1999, *The Knowledge Management Fieldbook* (Harlow, England: Financial Times Prentice-Hall).

Callon, M., 1995, Four models for the dynamics of science, in S. Jasanoff, G.E. Markle, J.C. Petersen and T. Pinch (eds), *Handbook of Science and Technology Studies*, rev. edn (Thousand Oaks, CA: Sage Publications), 29–63.

Choo, C.W., 2000, Closing the cognitive gaps: how people process information, in D.A. Marchand and T.H. Davenport (eds), *Mastering Information Management* (London: Financial Times Prentice-Hall), 245–253.

Committee on the Review of National Landslide Hazards Mitigation Strategy, Board on Earth Sciences and Resources, Division of Earth and Life Studies, National Research Council, 2002, *Assessment of Proposed Partnerships to Implement a National Landslide Hazards Mitigation Strategy*, Interim Report (Washington, DC: National Academy Press).

Cozzens, S.E. and Woodhouse, E.J., 1995, Science, government and the politics of knowledge, in S. Jasanoff, G.E. Markle, J.C. Petersen and T. Pinch (eds), *Handbook of Science and Technology Studies*, rev. edn (Thousand Oaks, CA: Sage Publications), 533–553.

Daley, M., 1998, Improving the effectiveness of natural hazard information: case studies from the Auckland Region, in D.M. Johnston and P.A. Kingsbury (compilers), *Proceedings of the Natural Hazards Workshop 4–5 November 1998*, Institute of Geological & Nuclear Sciences Information Series 45 (Lower Hutt, New Zealand: Institute of Geological & Nuclear Sciences), 6–10.

Davenport, T. and Prusak, L., 1998, *Working Knowledge: How Organizations Manage What They Know* (Boston: Harvard Business School Press).

Drucker, P.F., 1989, *The New Realities* (New York: Harper & Row).

Fothergill, A., 2000, Knowledge transfer between researchers and practitioners, *Natural Hazards Review*, **1**(2), 91–98.

Guzzetti, F., Carrara, A., Cardinali, M. and Reichenbach, P., 1999, Landslide hazard evaluation: a review of current techniques and their application in a multi-scale study, Central Italy, *Geomorphology*, **31**(1–4), 181–216.

Howell, D.G., Brabb, E.B. and Ramsey, D.W., 1999, How useful is landslide hazard information? Lessons learned in the San Francisco Bay region, *International Geology Review*, **41**(4), 368–381.

Innes, J.E., 1998, Information in communicative planning, *Journal of American Planning Association*, **64**(1), 52–63.

Lewenstein, B.V., 1995, Science and the media, in S. Jasanoff, G.E. Markle, J.C. Petersen and T. Pinch (eds), *Handbook of Science and Technology Studies*, rev. edn (Thousand Oaks, CA: Sage Publications), 343–360.

Machlup, F., 1962, *The Production and Distribution of Knowledge in the United States* (Princeton, NJ: Princeton University Press).

March, J.G. and Simon, H.A., 1958, *Organizations* (New York: Wiley).

Mileti, D., 1999, *Disaster by Design* (Washington, DC: The Joseph Henry Press).

Niles, J.R. and Michaels, S., 2002, Knowledge management, flooding, the watershed approach and the City of Waterloo, Ontario, Canada, in R. Newkirk (ed.), *Facing the Realities of the Third Millennium: 9th Annual Conference Proceedings of The International Emergency Management Society* (Waterloo, Ontario, 14–17 May 2002), 383–388.

Olshansky, R.B. and Rogers, J.D., 1987, Unstable ground: landslide policy in the United States, *Ecology Law Quarterly*, **13**(4), 939–1006.

Rossi, P.H., Wright, J.D. and Weber-Burdin, E., 1982, *Natural Hazards and Public Choice – The State and Local Politics of Hazard Mitigation* (New York: Academic Press).

Sarewitz, D., 2000, Science and environmental policy: an excess of objectivity, in R. Frodeman (ed.), *Earth Matters: The Earth Sciences, Philosophy, and the Claims of Community* (Upper Saddle River, NJ: Prentice-Hall), 79–98.

Schuster, R.L., 1996, Socioeconomic significance of landslides, in A.K. Turner and R.L. Schuster (eds), *Landslides: Investigation and Mitigation*, Transportation Research Board, National Research Council, Special Report 247 (Washington, DC: National Academy Press), 12–35.

Schuster, R.L. and Highland, L.M., 2001, *Socioeconomic and Environmental Impacts of Landslides in the Western Hemisphere*, US Geological Survey Open-File Report 01–0276 (Denver, CO: US Geological Survey).

Schuster, R.L. and Kockelman, W.J., 1996, Principles of landslide hazard reduction, in A.K. Turner and R.L. Schuster (eds), *Landslides: Investigation and Mitigation*, Transportation Research Board, National Research Council, Special Report 247 (Washington, DC: National Academy Press), 91–105.

Simard, A.J., 2000, *Managing Knowledge at the Canadian Forest Service* (Ottawa: Science Branch, Canadian Forest Service, Natural Resources Canada).

Spiker, E.C. and Gori, P.L., 2000, *National Landslide Hazards Mitigation Strategy: A Framework for Loss Reduction*, Open-File Report 00–450 (Reston, VA: US Geological Survey).

Williamson, R.A., Hertzfeld, H.R., Cordes, J. and Logsdon, J.M., 2001, *The Socioeconomic Benefits of Earth Science and Applications Research: Reducing the Risks and Costs of Natural Disasters in the United States* (Washington, DC: Space Policy Institute, Elliott School of International Affairs, The George Washington University).

Wynne, B., 1995, Public understanding of science, in S. Jasanoff, G.E. Markle, J.C. Petersen and T. Pinch (eds), *Handbook of Science and Technology Studies*, rev. edn (Thousand Oaks, CA: Sage Publications), 361–388.

# PART 3
# MANAGEMENT OF LANDSLIDE RISK

# 11

# Management Frameworks for Landslide Hazard and Risk: Issues and Options

*Michael J. Crozier*

## 11.1 The Human Dimension

The juxtaposition of landslides and human habitation exacts a cost. That cost can be attributed variously to damage from actual physical impact, to the loss of opportunity from actions taken as a result of recognizing the threat, or to the expense of sustaining measures to mitigate potential impact. In a sense, there is no escaping the cost; it can be transferred and transformed but not removed; there always remains a price for living within a hazardous environment.

Hazard and risk management is about identifying, calculating and evaluating the risks, assessing and implementing risk reduction options, and balancing the different components of cost in an acceptable way. The realization of risk, the options for reducing risk, the individual and political will, and the resources for reducing risk vary hugely throughout the world. At one end of this global spectrum we are faced with the Malthusian acceptance of disaster and, at the other, the New World moral compulsion to do something about it. The global discrepancy in risk reduction capabilities and the differential exposure to risk of the elite compared to the proletariat are realities in the universal equation of risk and the incentive and capabilities to respond. How often have catastrophic earthquakes been astutely portrayed as 'classquakes' and disasters as 'acts of man', not 'acts of God'?

Biological evolutionary imperatives dictate that human sensory perception is attuned to risk. Human physiological, neural response and behavioural systems have developed

*Landslide Hazard and Risk*   Edited by Thomas Glade, Malcolm Anderson and Michael J. Crozier
© 2004 John Wiley & Sons, Ltd   ISBN: 0-471-48663-9

a heightened involuntary sensitivity to perceived risk. 'Fight or flight' responses have evolved to preserve the individual, the species and its means of survival. Curiosity and awareness of risk is thus embedded in our psyche and involuntary and voluntary individual responses can be expected to act to minimize risk. Social and individual response to risk may be biologically driven (women and children first), economically, ethically, religiously, litigiously or legislatively driven.

Whatever the ultimate response to risk, there needs to be a will to confront risk in the first place, an imperative to reduce it and the means to do so. As societies have evolved, so has the organized response to risk. Risk reduction, however, is a complex task that needs careful management (Alexander, 2002). The success of that management will depend on how it scopes the issues and options, how it balances competing and conflicting interests, and ultimately the degree to which its outcomes are accepted.

## 11.2   Who is Responsible? Who Pays? Internalization/Externalization

A modern well-resourced society has the capability and options of managing risk any-where along the spectrum from total state ownership of the problem to the *laissez-faire* approach, where the individual has complete responsibility. This spectrum is sometimes represented for the individual as the range from maximum externalization to maximum internalization of risk. Politically, these extremes are also equated to the differences between socialism (collective responsibility) and right-wing individualism. Although the spectrum of choice is available to well-resourced societies, in less-organized and poorly resourced societies such a choice may be an unobtainable luxury.

Increasingly, in Western societies, there is a tendency for less government intervention and more internalization of risk by the individual: a shift to the right. This direction is generally presaged by the transfer of responsibility from central government to local government and ultimately divestment to the individual. The argument is not so much about the control of hazard costs but rather the philosophy of cost distribution and cost efficacy.

Many arguments for collective responsibility are generic and closely related to social and religious philosophy. General taxation, government relief measures or insurance can spread the localized risks, event costs, and the cost of risk reduction measures throughout the whole population. There are ethical rationales that mandate the larger more fortunate group to support the less fortunate or the less capable.

Pragmatic arguments for collective responsibility involve the view that 'next time it may be me'; in other words gradual payments to a collective fund are easier to sustain than a one-off full event payment. The rationale of acceptance is that the gradual collective contribution will be recompensed by an eventual event payout to the long-term contributor.

It can be argued that if the risk faced by an individual is voluntary, then they must accept all the fiscal responsibility. Presumably that risk has been accepted because the perceived benefits of exposure to that risk outweigh the consequences. This argument, of course, presupposes that sufficient information exists to allow a realistic evaluation of risk. This in turn raises the question of who is responsible for generating and communicating the information required to make such decisions. The concept of 'voluntary' risk, however, is a very culturally specific term. Most individuals do not have the luxury of accepting or rejecting the degree of exposure to risk: they must live with it, unless government or wider communities intervene.

One other thread in the recognition of responsibility can also be detected in modern legislation and business procedures. A number of Western market-driven democracies, while giving wide licence to private industry, have enacted comprehensive 'Health and Safety' legislation. This in effect is another mechanism of divesting state stewardship of potential hazard victims to other parties. The underlying intentions of this legislation are generously described as making everyone in an organization aware and responsible for safety – a culture of safety where everyone has ownership of the risk and it is hoped that the net effect is a less vulnerable community overall. Thus there is a linear chain of responsibility from top management to the floor worker. However, there is a danger that the chain of responsibility becomes, instead, a top–down divestment of responsibility. Once the devolution boxes have been ticked, the legal responsibilities are seen to have been carried out and the safety system is seen to have been enacted. This clearly only partially addresses the problem because it is usually the owners or managers (not the workers on the shop floor) who know the full spectrum of risks and benefits involved in the entirety of any enterprise. Furthermore, the risk may relate to the structure of the enterprise as much as to the behaviour of those involved in the operation.

Another argument for collective responsibility is that many risks are inherited and may not rest easily within the purview of the individual. For example, the New Zealand government at one time introduced a policy (Land Development Encouragement Loan scheme, 1978) that involved fiscal incentives to encourage forest and scrub clearance from marginal hillslopes in an effort to increase farm productivity (Willis, 1991). The net physical effect has been to destabilize steep headwater catchment slopes and initiate downstream sediment impacts that can last for decades. The question now is whether current inhabitants and particularly those downstream should carry the consequential risk, first, from that generated by a historic government policy and second, from an off-site activity for which many received no benefits and in fact have been adversely affected. This case would seem to stretch the credibility of *caveat emptor* and call for a wider social responsibility.

Advocates of risk internalization are either driven by the contemporary economic imperative of 'user pays' or the philosophy that exposure to market forces can be a powerful tool in reducing risk. User pays philosophy can be an instrument of equity. For example, if general taxation is used to provide disaster relief, it has been argued that those who are prudent and have individually pre-paid to reduce risk should not have to subsidize the ignorance or lack of foresight and investment by others. Even loss-sharing schemes such as insurance can promote internalization of risk by premium loading (IDNDR–UK, 1995). In other words, insurance is designed to be more expensive in higher-risk areas. The fundamental principle of internalization is that individuals or units pay for risk reduction measures in proportion to the exposure to risk. This produces an economic incentive for those affected to take measures to reduce risk.

In one sense, the answer to the question 'who should pay?' may be better addressed by determining who is responsible for generating and enhancing the risk on the one hand and who is benefiting from the elements at risk on the other. If, for example, the level of hazard is constant and we consider that any risk exposure is voluntary, those who own and maintain the elements at risk can be considered responsible for that risk. However, before the cost of that risk is attributed to the owners it is important to determine who actually benefits from the elements at risk. In the simplest case, the owners may be the main beneficiaries, but in other situations the wider community may receive a substantial amount of benefit. The prevailing degree of community benefit therefore may be a useful determinant as to whether the costs are internalized or shared. Generally the various beneficiaries would be expected to assume the proportional costs of risk and associated responsibility for risk management.

The situation is often treated differently when an event actually occurs and risk is realized, particularly if impact exceeds design limits or planning horizons, or overwhelms established coping mechanisms. These situations, by definition, are often considered to be disasters. In such instances the non-affected communities contribute to the costs through relief aid and cross-jurisdictional assistance.


## 11.3   What Went Wrong; Who is to Blame?

When a damaging event occurs, responsibilities are not always automatically assumed by the owners or beneficiaries. In many countries an event inquest may take place. Various levels of government inquiry, peer review, investigation or litigation may be instigated not just to learn lessons and improve systems for the future but also to apportion blame and award damages. Such scrutiny, often referred to as 'post-event/disaster review', is likely to include both the physical factors and the human factors. The inquiry might seek answers to questions such as why a building was constructed in a certain locality, whether earthworks were properly carried out, why the structure failed, or why the community was not adequately prepared. Accordingly attention may be directed at the adequacy of the management system, the appropriateness of procedures invoked to carry out the system, diligence in execution of procedures, and competence of the personnel involved.

The cause of the physical event is also likely to be investigated. The outcome can potentially vary between the landslide being seen as an unpredictable and inexplicable 'act of God' to the other extreme where it is viewed as a clearly understood process where movement can be attributed to an identifiable factor or even a specific human action. The causes of landslides are complex and multivariate and consequently even expert opinions can be sufficiently at variance for the legal system to be faced with 'reasonable doubt'. The variability of natural systems inevitably expresses evidence in terms of probabilities, whereas human constructs such as the legal system demand high degrees of certainty. Investigations into the causes of landslides are consequently often inconclusive. In an increasingly litigious society, because the legal and scientific community often struggles with establishing the cause of landslides, it is worth exploring the different roles and significance of causative factors.

Despite these uncertainties, it is salutary to note that, in Australia, almost half of the landslides causing injury or death have been attributed to human activity such as modification of slopes by construction of roads, railways or buildings on steep slopes (Michael-Leiba *et al.*, 1999).

### 11.3.1   Establishing the Cause of the Landslide

The causes of landslides are mulitvariate and complex. Problems arise in cause assessment when factors are treated in isolation from their context and when there is insufficient distinction between factors that are 'necessary' compared to factors that are 'sufficient'. For example, a certain slope angle may be necessary for movement to take place but not sufficient in itself to initiate movement.

In Chapter 2, it was found useful to characterize a slope in terms of identifiable stability states: 'stable', 'marginally stable' and 'actively unstable' (see Chapter 2, Figure 2.1). These concepts are also helpful in understanding the causes of landslides. For a slope to move from a stable to an actively unstable state, changes must take place that affect the distribution of resistance and shear stress. One way of visualizing this process is to resort to a set of scales as an analogy, with one side representing shear stress and the other resistance. Resting on the 'resistance' pan are several bricks symbolizing equal units of strength (i.e. stress equivalents) that must be overcome to produce movement while on the other pan the bricks represent units of stress tending to promote movement. Each brick can be considered to equate to stress controlled by one particular stability factor. When the balance is loaded so that it is heavily weighted down on the resistance side, the situation is 'stable'. However, when the influence of factors changes so that the scales move to approach the point of balance, the situation can be considered 'marginally unstable' and, finally, 'active instability' (slope movement) is represented by the scales being tipped over in favour of the shear stress side. In the 'marginally stable' state (near the point of balance), as on the real slope, failure can be produced by either removing a 'resistance' brick or adding a 'shear stress' brick.

Returning to the question of responsibility or causes of failure, assume for a moment that there are four bricks on each side: the scales are balanced and subsequently failure is initiated by the action responsible for emplacing a fifth brick on the shear stress side. The question that may be asked in any inquiry is: which of those five bricks on the shear stress side caused the movement – or which of the five factors or actions caused the failure? This is not an idle question considering that the answer might determine culpability and financial responsibility. It can of course be reasonably argued that quantitatively all bricks are equally responsible for the consequential action, as it is the sum of their component weights that is important. However, before the fifth brick was emplaced no movement could occur and therefore, in terms of direct action, the fifth or 'triggering brick' must assume some particular 'functional' significance – yet the fact remains that without the influence of the pre-existing four shear stress bricks, the fifth brick would have no special significance. Furthermore, the fifth brick was only allowed to assume a triggering role because of the existing strength conditions represented by bricks on the strength side of the balance.

In the search for culpability, the quantitative significance of a destabilizing action has less significance than the fact that it was carried out within a pre-existing context of stability that determined the severity of the consequences. The questions can then be legitimately asked whether sufficient attention had been given to the pre-existing conditions, as well as who holds the responsibility for assessing those conditions. Clearly the action of a triggering factor only partially explains the cause of a landslide. Indeed, the triggering factor may represent only a minor destabilizing action compared with action that causes a major lowering of stability without actually causing movement.

For example, as a result of the Abbotsford, New Zealand, landslide disaster of 8 August 1979 (where 69 houses were destroyed and 450 people were affected; Crozier, 1999) a Commission of Inquiry (1980) sat for 58 days in order to determine the cause

(Figure 11.1). It found that the main cause was the 'unstable geology' (despite the fact that the only other landslide in the immediate area had occurred over 10 000 years earlier). The fact that 300 000 m$^3$ of material had been removed from the toe slope (equivalent to a 1% decrease in stability) 10 years before failure was considered only a contributing factor, as was a major water leakage from a water supply pipe that had been going on for 3 years up to the time of failure. The rate of leakage on to the slope of 4–5 litres per minute is equivalent to a 30–40% increase in rainfall over that period. As Hancox (2002, p. 12) concludes in a recent reassessment:

> Whatever these effects on the water table were, however, because of the earlier excavation of sand from the toe of the slope, the water table in the slide area had to rise about 0.3 metres less in order to reach the critical stability in the slope at the time it failed. So in this sense leakage from the water main, together with an inferred long term rise in groundwater levels due to increased rainfall, was probably responsible for timing of the Abbotsford landslide.

Post-event/disaster reviews, such as the Abbotsford inquiry, are an extremely important element of hazard and risk management. They increase the understanding of both the human and physical systems under stress, and should lead ultimately to more effective and efficient ways of reducing risk.

**Figure 11.1a**   *The Abbotsford landslide, 8 August 1979, Dunedin, New Zealand (photo by Allied Press; reproduced by permission of Allied Press Ltd)*

**Figure 11.1b**  The head graben, Abbotsford landslide (photo by W. Brockie)

## 11.4   Hazard and Risk Management Protocols

The fundamental impetus for any form of hazard and risk management is an awareness of the threat, a notion of responsibility and a belief that human action might reduce the risk. The notion of risk is the seminal driver for management action. It will dictate the scope of any such action. The notion of risk can arise from many different circumstances. The most obvious is the history of impact: a damaging event demands action. There are, however, other situations that may instigate action. These include concern arising from the establishment of new developments in a locality where theory or experience might suggest a potential risk. Legislation and its resultant policy, regulations, established protocols or perhaps simply good practice may all, in their own way, demand action. On the other hand, new technological capabilities or newly available databases may be sufficient to promote the start of investigations. Many jurisdictions have the overall planning goal of reducing the risk from natural hazards. However, whatever the initial impetus for concern, before any appropriate management plan can be established, that notion of risk must be investigated, properly estimated and evaluated. The scope of that investigation is influenced by the scale of the area of interest, the notion of risk, and the resources available. In practice, the scope of investigations can range from the broadest indicative national scale (e.g. the 'descriptive atlas' scale) to the detailed scale represented by stability analysis of a building site.

In Chapter 1, various components of the landslide management framework were outlined and illustrated (Chapter 1, Figure 1.1). They include scope definition, hazard and risk identification, consequence analysis, hazard analysis, risk calculation, risk evaluation and the components of risk treatment. The risk treatment aspect of the landslide management framework can be broadened and represented by the more generic hazard management cycle (Figure 11.2).

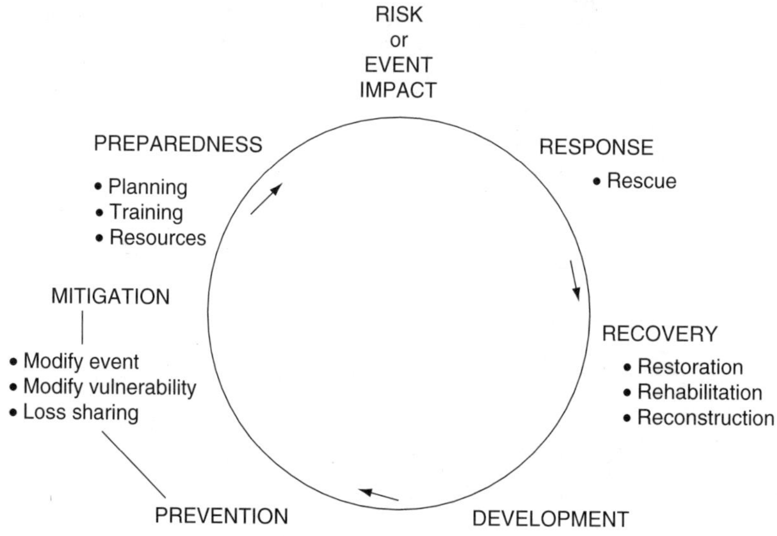


**Figure 11.2**   *The hazard management cycle (based on Carter, 1991, reproduced by permission of Asian Development Bank)*

The hazard management cycle (Carter, 1991) presents a template of the various functions and options that, in an ideal world, might be brought into play to reduce the risk from landslide hazard. This cycle goes beyond the landslide management framework by also incorporating emergency management response and procedures relating to a specific event. The 'mitigation' and 'prevention' phases of the cycle equate to the 'risk treatment' component of the landslide management framework. The phases and components of hazard management schemes will vary in importance depending on the context within which they are applied and should not necessarily be treated separately or in sequence.

'Prevention' is the ultimate form of event modification. It may only be achievable for landslides over a limited period of time. No slope or hazard modification method should be treated as absolutely safe. Most methods developed to address the physical hazard are best described as control measures, reduction measures or mitigating measures – not preventive measures. Mechanical reinforcement can deteriorate, drainage systems can become blocked, and undetected strength deterioration of the rock mass may eventually bring a slope to failure.

Prevention of certain losses may be achieved with respect to vulnerability or behaviour modification by, for example, the employment of land use regulations to prohibit building on susceptible sites. This of course does not preclude other individuals being present on the site and being exposed to the hazard.

## 11.5   Mitigation Options

Essentially, 'mitigation' means lessening the effects of a hazard event. In other words, mitigation is the desired result of risk reduction measures. The approach taken to selecting mitigation options, as previously stated, needs to be informed by such considerations as the risk–benefit ratio of exposure to the hazard, the cost–benefit ratio of any response, as well as cultural factors that might determine the acceptability of the proposed measures.

Nine different approaches to mitigation are listed in Table 11.1. They range from 'hard' engineering measures to 'soft' planning and education measures. The extent to which any particular measure listed is justified requires a careful assessment of the acceptability or tolerance of an existing risk. This in turn depends on the perception of risk versus the benefits accrued while being exposed to that risk, together with a cost–benefit analysis of the risk reduction option (Gough, 1996). The costs involved are often more than the costs of designing and constructing the risk reduction measure. For example, if hazard zoning is invoked, inevitably this will have an effect on the market value and resale value of the properties involved. Similarly, zoning an area to prohibit various land use activities will exact not only an opportunity cost but also lower the rating base for local government. Zoning schemes that have been used to reduce risk include housing density restrictions, reserving areas for activities associated with low population concentration (e.g. recreation or forestry) or low-value commercial activity (e.g. storage facilities, car parks etc.) (Schuster and Kockelman, 1996).

The issue also arises as to where and how these zones should be established in relation to the adequacy of the information base and quality of risk assessment, or predicted trends. An example of where these issues might arise is in the establishment of 'set-back' zones to take into account the potential for effects from future climate change such as increased rainfall or higher water tables.

**Table 11.1**   *Landslide risk mitigation options*

**Physical methods**
Toe buttressing
Slope reinforcement: bolts, anchors, pins, piles
Grouting fissures and joints
Chemical reinforcement of soils
Diverting debris: tunnelling, galleries, net curtains, detainment debris dams, controlled and contained runout zones
Bioengineering

**Hydrological methods**
Diverting surface water away from the site
Impermeable geotextile covers
Drains
De-watering fluid debris
Draining or lowering water bodies that might contribute to impact by adding water or that might allow wave generation

**Site grooming**
Removal of woody and other debris that might aggravate the event
Contouring the land surface to change the form (water dispersion) or to close cracks and fissures
Removal of susceptible material
Bioengineering

**Warning systems**
Periodic survey; continuous monitoring
Alarm systems based on the triggering agent, e.g. accumulated rainfall, seismic shaking. Alarm systems activated by slope movement

**Regulations**
Building codes
Earthwork/foundation and drainage standards
Behaviour safety codes
Specification of 'permitted', 'controlled' or 'discretionary' activities including the ability to place conditions on consents and permits, which may include requirements to mitigate or remedy the effects

**Fiscal incentives**
Tax incentives to leave areas undeveloped
Lending policies to discourage development

**Land use planning schemes**
Activity/building zones, including restrictions on types of activities and/or areas that can be developed (hazard zoning) including the appropriate siting of lifelines

**Education**
Communication, education and advocacy

**Loss-sharing schemes**
Insurance

## 11.6   Responding to Risk Estimates

Section 1.6 of Chapter 1 addresses the fundamental questions of response to risk. The first question is whether the level of estimated risk is *intolerable, tolerable* or *acceptable* (Helm, 1996). The judgement may be made informally by the individual or community or it may be formalized to the extent that the decision rests with an authority. *Intolerable risk* is so high that it cannot be allowed to prevail, despite any benefits accrued from being exposed to the risk. Exposure to this type of risk is unjustifiable and the risk must be avoided or reduced to a level where it can be tolerated. *Tolerable risks* are, by definition, never fully accepted. They represent a level of risk that a society is prepared to live with because there are net benefits in doing so, as long as that risk is monitored and controlled and action is taken to reduce it. *Acceptable risks* represent a level of risk that a given community or authority is prepared to accept without imposing risk reduction measures.

Risk management may aim to reduce all risks to this 'acceptable' level. These different levels of risk are illustrated as components of the 'As low as reasonably practicable' (ALARP) approach in Figure 11.3. The approach recognizes that some risks (the top of the diagram) are so high that they must be avoided or reduced whatever the costs; in other words they are intolerable. At the bottom of the diagram, risks are negligible and acceptable without any specifically designated reduction programme. In between these two extremes, risks are tolerable but only in comparison to the benefits or the costs of reduction. Tolerable risk should be kept as low as reasonably practicable. In some jurisdictions, the onus is placed on those responsible for generating, or introducing, the risk to demonstrate that the ALARP principle has been implemented (Gerrard, 1998).

Management authorities need to have a system in place that will allow the community to participate in the judgement of acceptability. While certain standards of acceptability

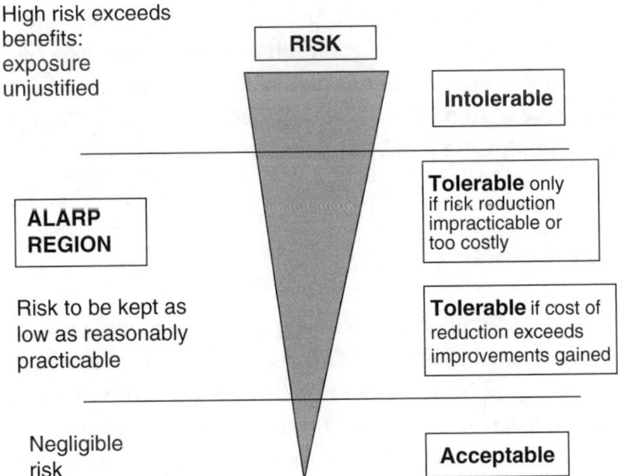


**Figure 11.3**   *Evaluating and responding to risk: the ALARP (as low as reasonably practicable) approach (based on diagram from Helm, 1996, which was sourced from Health and Safety Executive, 1992)*

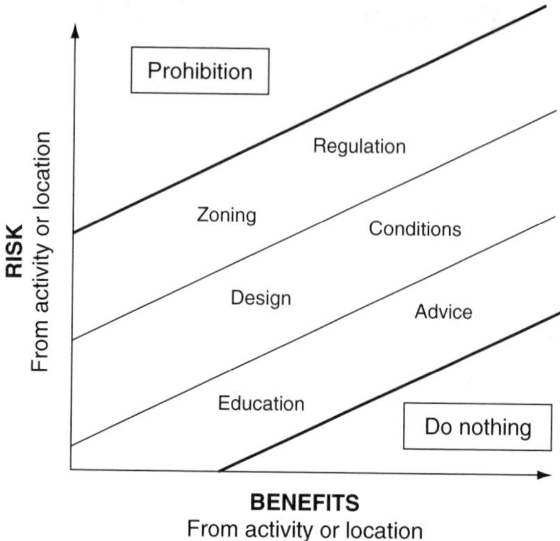


**Figure 11.4**  *The risk–benefit ratio as a guide to adopting risk reduction options (based on Crozier, 1993). Reproduced by permission of New Zealand Geographical Society Incorporated*

have been set by some authorities (see Chapter 1, Section 1.6), they are not universal: perception of risk depends on, among other factors, the source of risk, and the social and cultural context. Acceptability of risk in turn depends on both the explicit and implicit assessment of the risk–benefit ratio. In other words the perceived risk needs to be weighed against the benefits of being exposed to that risk. Figure 11.4 indicates that the choice of mitigating options also depends on the risk–benefit ratio. Essentially if the risk–benefit ratio is extreme, the risk will not be tolerable and stringent measures will be required to avoid, remove or reduce the risk. On the other hand, if the risk–benefit ratio is low, the community may not accept strict regulatory or reduction measures and will opt for moderate or unobtrusive action such as education programmes. Acceptability of risk is an area that is poorly understood and requires further research.

## 11.7  Emergency Response

Emergency response to a landslide should be pre-planned and resourced. Search and rescue of affected persons is paramount. This may require special measures; for example, in the case where people are buried there is usually a need for specific resources such as trained search dogs, thermal sensing techniques and other appropriate measures. Clearly this phase requires provision of emergency food, shelter, transport, evacuation and medical assistance. In some cases, the emergency may necessitate the removal of structures and the allocation of space to receive them. There may also be a need for immediate geotechnical appraisal of the situation to assess the post-event state of slope stability. The common question after a landslide event is whether the slopes are now more stable or less stable than before the event. Even with minor events careful attention needs to be given to those affected.

Commonly those affected have immediate information requirements and communication systems should be set up to meet these. The immediate demand is for those affected to know whom to contact and what assistance is available. They need to know the extent to which they are covered by personal insurance and the degree of assistance they can expect from local authorities. Often the sole information source for affected parties is the media. Clearly authorities need effective systems of communication both with the media and those affected.

## 11.8 Recovery

This phase follows closely and overlaps with the response phase. This phase has both human and physical elements, which need to be treated in tandem. Clearly, recovery requires re-establishment of services and provision for individuals and communities to become self-sustaining once again. Attention also needs to be given to the trauma experienced by individuals and the community as a whole, as well as workers who have been involved in the initial response phase. Rehabilitation measures, however, should not necessarily be aimed at simply re-establishing the *status quo*. For instance, relief measures can be effectively tagged and designed to increase community resilience to future events. For example, financial incentives might be aimed at encouraging movement to less susceptible locations.

## 11.9 Evaluating Emergency Management

As a result of many years of study of disasters throughout the world, Quarantelli (1997) has isolated the main criteria on which emergency management of community disasters can be evaluated. These are commented on as follows.

- *Correctly recognizing differences between agent and response-generated demands* This relates to the ability to distinguish between problems created by the nature of the specific hazard agent itself compared with problems generated by the effort of organizing a response. For example, agent-specific demands in the case of floods might be the supply of sandbags; radiation exposure demands decontamination facilities; landslides might require geotechnical expertise to assess whether further failure is a possibility. Response-generated demands relate, for example, to the capabilities of personnel involved, delegation and coordination.
- *Carrying out generic functions in an adequate way* There are 10 generic functions (usually common to all disasters) that require adequate planning and need to be carried out in the course of any major event or disaster. These are: warnings, evacuation, sheltering, emergency medical care, search and rescue, protection of property, mobilizing emergency personnel and resources, assessing damage, coordination, and restoring essential public services. The success of their implementation can be measured by, for example, time of response and client satisfaction.

- *Mobilizing personnel and resources effectively*   In Quarantelli's experience there usually are plenty of resources and people available: the real challenge is identification and location of the right ones. Efficiency needs to be kept in mind because on occasions more assistance is available than required.

- *Involving proper task delegation and division of labour*   Immediately after an impact, the nature of the required tasks and the scope of organizational involvement is usually confused. For example, at one Canadian fire event 346 organizations appeared on site. In such cases, it is important to distinguish between regular and emergent groups and allocate tasks effectively. A response that tries to involve only established organizations shows poor disaster management.

- *Allowing adequate processing of information*   Means of communication and its content need to be appropriate for the situation. Communication must be effective in the following areas: intra-organization, inter-organization, citizens to organizations, organizations to citizens, and between citizens. One example of where communications systems failed to cope was after the Loma Prieta earthquake, California, 1989. As a result of this event, phone calls jumped from 50 to 80 million per day, causing overload and system failure.

- *Permitting the proper exercise of decision making*   Loss of top echelon personnel because of overwork is a common problem in disaster situations. In addition, any conflict of responsibility needs to be resolved, particularly relating to newly emergent disaster tasks.

- *Focusing on the development of overall coordination*   Unequivocal procedures should be established for decision making: who is responsible for what, as well as establishing a clear hierarchy of control. It is important that these systems are understood and agreed to by all parties before any contingency.

- *Blending emergent and established organization behaviour*   Established organizations must not be so rigid in their functions that they will not recognize the ability of newly emergent problems or new groups. Essentially there are new, unanticipated developments to every disaster.

- *Having a well-functioning emergency operation centre*   This needs clearly defined functions and appropriate human and physical resources. It needs to be located in a physically safe place.

## 11.10   Monitoring, Review and Development

In Section 11.3.1 reference was made to the importance of post-disaster reviews. The lessons from these need to be incorporated beyond the affected locality and recognized by government policy in general. Development in this context means ensuring that lessons are learnt, that these inform policy and that in general the community becomes increasingly resilient. In practice this may mean revising standards and codes of practice by commissioning appropriate research and invoking appropriate training and preparedness measures. Broader questions may also need to be addressed, such as whether an affected land use in a susceptible area is sustainable in the long term.

## 11.11   Preparedness

Preparedness, sometimes referred to as 'readiness', is represented by the ability of a community to put into action established plans and procedures. It means testing the systems and knowing how to put them into operation without delay. Clearly, the occurrence of an actual event is the best test for preparedness. However, in the face of high-magnitude–low-frequency events, preparedness is often best tested by running event scenarios, involving exercises and simulations.

Preparedness needs to be viewed in a broad framework. Quarantelli (1997) offers some advice for good practice in establishing effective preparedness plans:

- View disasters as both quantitatively and qualitatively different from accidents and minor emergencies.
- Highlight a continuing planning process rather than production of an end-product, such as a written plan.
- Adopt a multiple-hazard rather than single-hazard focus, that is generic rather than hazard agent specific.
- Include all four time-phases of the planning process (i.e. mitigation, preparedness, response, and recovery).

As with any planning, preparedness should always incorporate the capacity to monitor and review all established policies and actions.

## 11.12   Planning

All the foregoing management steps need to be planned well in advance of any contingency. Procedures for planning can be established by common sense or alternatively they may be prescribed by legislation.

In New Zealand, for example, the Resource Management Act (1991) not only specifies natural hazards as a subject of concern for local government, but also indicates the management approach that should be adopted. The requirements are to establish an overarching policy statement. This necessitates formulating a goal, setting objectives in accordance with that goal and identifying policies in line with those objectives. Furthermore, each policy should be represented by measurable outcomes. The policies should specify the methods by which they will be attained. When considering those methods, the local government body is required to explore all options, including the 'do nothing option', together with the expected outcomes. The way this procedure might be carried out can be illustrated by an example from the Taranaki Regional Council of New Zealand (Taranaki Regional Council, 1992). After having scoped the range of hazards and risks within their region, the Council formulated their policy towards hazards and risks as follows. The overarching goal is stated as: 'The adverse effects of natural hazards on human life, property and the Taranaki environment will be avoided or mitigated.' The six objectives, together with their associated policies (P) are outlined below (the methods are not listed here).

**Objective 1**: *modify use and behaviour*
**P1**. Make public aware and help them internalize risk and develop and accept risk reduction measures by themselves.
**P2**. Council takes responsibility to reduce vulnerability by managing behaviour with planning tools, avoidance rules, zoning, conditions on permits, plans, preparedness, emergency strategies, and maps.

**Objective 2**: *control the physical process*
**P3**. Invoke engineering works, and establish plantings. Undertake analyses: risk–benefit and cost–benefit of such measures.

**Objective 3**: *response & recovery measures to minimize the losses.*
**P4**. Council supports relief and recovery measures, and sets aside contingency funds to support these.

**Objective 4**: *treaty obligations*
**P5**. Recognize and provide for Maori values when managing hazard and risks: consult, involve and have representation from Maori. (This action is mandatory under the Resource Management Act, which directs local government to take into account the principles of the Treaty of Waitangi. The Treaty represents a long-standing legal agreement between the Crown and indigenous people of New Zealand to share the resources and governance of the country.)

**Objective 5**: *gather the required information*
**P6**. Specifically research or add to other investigations.
**P7**. Monitor physical hazard agents, policy, and works.

**Objective 6**: promote *integrated management*
**P8**. Recognize interconnections, ramifications, overlapping responsibilities and cooperate.

It is essential at the planning stage to involve the affected stakeholders and the public in general in the process. There are sound reasons for this. First, the public may actually be a valuable source of hazard and risk information. Second, compliance with any policies is more likely if the public feel they have some ownership of the process and have had some input into the plan. Third, in the end, it is often the public who have to meet the cost of risk reduction measures.

Many organizations have involved the public by employing Geographical Information Systems (GIS) on the Internet. Hazards, risks and risk reduction measures are spatially constrained and GIS is a particularly appropriate means for displaying this sort of information. Hard copy maps have a sense of permanence that can be counterproductive at the consultative stage. GIS, on the other hand, allows for ongoing and interactive update by readily incorporating new information and presenting alternative scenarios. It also allows material to be readily edited and presented at different levels of generalization. In some cases the spatial differences in hazard and risk may be best conveyed simply in red, orange and green zones; in other cases the detail of underlying factors used for classifying terrain needs to be displayed.

## 11.13   Legislative Framework

There are many aspects of community and individual action that relate to hazards and resulting risk, and they are all subject to potential regulation through legislation. To illustrate the type of legislation that can be invoked to reduce risk, a few examples from the New Zealand (NZ) situation will be briefly reviewed.

The most distinctive aspect of NZ legislation is represented by the Earthquake and War Damages Legislation that set up a government-owned, crown entity now referred to as the Earthquake Commission (EQC) (Earthquake Commission Act, 1993). This body administers an insurance scheme that currently covers, among a number of other hazards, landslide damage. Essentially, all those that take out fire insurance on their properties are compulsorily levied a small amount to support the scheme (currently 5 cents for every $100 of cover).

The scheme was originally set up to cover war damages anticipated as a result of hostile action during World War II and was soon extended to earthquake damage. Other potential hazard damage was subsequently added to the coverage. However, assessment of potential payouts, particularly with respect to earthquake damage, made the EQC realize that they were over-extended (the Government guarantees the Disaster Fund administered by the EQC). In 1998, although the fund has a balance of NZ$3200 million, the EQC was assessed as having an exposure in excess of its current level of assets. In that same year, there were over 800 claims for landslide damage amounting to over NZ$4.7 million. Because of financial over-exposure, the EQC has removed earthquake coverage from commercial buildings. The extent of coverage has been reduced to a maximum of NZ$100 000 for domestic dwellings and NZ$20 000 for personal property; land has no cover limit and there is an excess of NZ$200. The EQC reserves the right to cancel an individual's insurance if it has made a full payout on a claim.

Comprehensive planning and management of hazards is encompassed by the Resource Management Act (1991) (RMA), which sets out the roles and functions of local government with regard to management of natural resources, principally land, air, water and coast. Local government in New Zealand is two-tiered, with 15 overarching regional councils and 74 territorial authorities. Under the RMA regional councils have responsibility for planning for sustainable production, including 'the control of the use of the land for the purpose of the avoidance or mitigation of natural hazards'. Territorial authorities are involved with management of the 'effects of land use' and the protection of the land, including implementation of rules for the avoidance or mitigation of natural hazards. Through plans, policies and resource consent processes local government is able to prohibit, control and regulate activities for the purpose of avoiding or mitigating hazards.

Territorial authorities also have to take account of hazard as part of their land subdivision consent functions specified in the RMA. Under section 106, territorial authorities must refuse subdivision consent where land or structures are likely to be subject to or accelerate damage by erosion, subsidence, slippage, or inundation, unless the territorial authority is satisfied that the hazard will be avoided, remedied or mitigated. The authority can also impose conditions on subdivision consents to protect the land against erosion, subsidence, slippage or inundation (section 220(1) (d)). Esplanade reserves may also be taken on subdivision for the purpose of mitigating natural hazard (section 229).

Under the Building Act (1991) territorial authorities are obliged to keep data on land parcels including hazard information and issue this information to prospective builders in the form of a Project Information Memorandum (PIM). This should include information on: potential erosion, avulsion, falling debris, subsidence, slippage, alluvion or inundation that is likely to be relevant to the design or construction of the building (section 31). A section of another Act governing local government also requires local authorities to issue a Land Information Memorandum (LIM) to any interested party. The LIM contains information similar to a PIM but relates to property as it is and not specifically to a proposed development. These data sources are important in informing owners and users of existing hazard.

Section 36 of the Building Act has some very interesting aspects that contribute to risk mitigation through regulation but at the same time maintain a certain degree of freedom for the landowner. It deals with building on hazard-prone land and contains two main provisions. First, where building work itself is likely to increase the risk for that land or any other property, the territorial authority must refuse building consent, unless it is satisfied that there is adequate provision for protection or restoration of the property. Second, where the territorial authority is satisfied that the building work itself will not add to the risk, it may grant a building consent, but the fact that the land is subject to natural hazard must be entered on to the officially registered title of the property. Thus, by tagging the title, the territorial authority becomes exempt from liability in the event that the building is subsequently damaged by a natural event. This means that an owner can obtain consent to build on hazard-prone land where the building does not add to the risk and the fact is recorded for the benefit of future owners. An interesting consequence of this provision is found in clause 3 (d), Third Schedule of the Earthquake Commission Act, 1993 where it states that the Earthquake Commission may decline (or meet part only of) a claim for damage from natural hazard if the property has been tagged under section 36 of the Building Act.

While the legislative framework for dealing with landslides in particular and natural hazards in general is comprehensive in New Zealand, practice continues to be informed by case law. Policy, practice and legal arguments are often constrained by adequate scientific information, lack of experience with high-magnitude–low-frequency events, and in some cases a regional rather than national approach to the problem.

The lessons from the New Zealand legislative experience with respect to hazards can be summed up as follows. Both national and local government have a role to play in effective risk reduction. Many of the acts under which the authorities operate are effective because they are proactive and enabling. They ensure that the effects of land development and land use are considered before permits are issued. In providing for this evaluation, special reference is made to landslides and related phenomena. In particular, there are requirements to assess not only the impact that landslides might have on human activity but also the effect that human activity might have on destabilizing the land. While the relevant New Zealand legislation tends to be enabling, for the purpose of reducing landslide risk, it also allows for the full range of planning tools from land use zoning to conditions on activity. The New Zealand disaster insurance scheme also has innovative policies for reducing future risk. Above all, the requirement for local government to monitor hazards and provide a database on related land conditions will prove a valuable source of information for future scientific research and management decisions.

## 11.14   Fundamental Requirements to Enable Effective Management

Many of the management procedures and objectives outlined above represent an ideal situation. Not every country or jurisdiction has the capability of achieving these goals. Fundamentally, good hazard and risk management needs to be underpinned by the following capabilities, philosophies and resources:

- a technical and scientific information base
- an informed populace
- an informed and capable local, regional and national government
- a philosophical basis for distribution of costs
- an appropriate statutory and legal infrastructure
- an informed and capable professional and technical community to manage and execute a risk reduction programme
- a philosophical basis for determining the acceptability of risk
- a risk reduction programme with goals, policies, objectives and methods
- practice and experience
- an effective system of communication and education.


## 11.15   Conclusion

This chapter has presented a framework for the ideals and goals, and underlying philosophies, of good hazard and risk management. It encompasses a range of measures that have been distilled from the experience of earth scientists, social scientists, engineers, public policy experts and managers. The success of risk management schemes depends on the continual iterative processes of information input and managerial response. There are many aspects of risk management that are still poorly understood. The most important of these are the frequency–magnitude behaviour of hazards, vulnerability factors, and risk–benefit ratios of exposure to hazard. Ultimately there must be a justifiable and defensible characterization of hazard and risk. This is the ultimate foundation upon which the community, individuals and authorities can base their response. In many cases communities have taken collective responsibility by enacting legislation to guide the risk management process. How the legal principles are manifest in policy and in turn translated into action should be decided by the authorities and stakeholders together.


## References

Alexander, D.E., 2002, *Principles of Emergency Planning and Management* (New York: Oxford University Press).

Carter, W.N., 1991, The disaster management cycle, in W.N. Carter (ed.), *Disaster Management: A Disaster Manager's Handbook* (Manila: Asian Development Bank).

Commission of Inquiry, 1980, *Report of the Commission of Inquiry into the Abbotsford Landslip Disaster* (New Zealand Government Printer).

Crozier, M.J., 1993, Management issues arising from landslides and related activity, *New Zealand Geographer*, **49**(1), 35–37.

Crozier, M.J., 1999, Landslides, in M. Pacione (ed.), *Applied Geography: Principles and Practice* (London: Routledge), 83–94.

Gerrard, S., 1998, Environmental risk, in B. Nath, L. Hens, P. Compton and D. Devuyst (eds), *Environmental Management in Practice*, vol. 1 (London: Routledge), 296–316.

Gough, J., 1996, Natural hazards and risk management, *Tephra*, **15**(1), 18–23.

Hancox, G., 2002, The Abbotsford landslide: its nature and causes, *Tephra*, **19** (June), 9–13.

Health and Safety Executive, 1992, *The Tolerability of Risk from Nuclear Power Stations* (London: HSE).

Helm, P., 1996, Integrated risk management for natural and technological disasters, *Tephra*, **15**(1), 4–14.

IDNDR–UK, 1995, *Preventing Natural Disasters – the Role of Risk Control and Insurance*, Papers from a seminar 6 October 1995 (London: The Royal Society London).

Michael-Leiba, M., Baynes, F. and Scott, G., 1999, *Quantitative Landslide Risk Assessment of Cairns* (Canberra: Cities Project, Australian Geological Survey Organisation).

Quarantelli, E.L., 1997, Ten criteria for evaluating the management of community disasters, *Disasters*, **21**(1), 39–56.

Schuster, R.L. and Kockelman, W.J., 1996, Principles of landslide hazard reduction, in A.K. Turner and R.L. Schuster (eds), *Landslides: Investigation and Mitigation*, Transportation Research Board, National Research Council, Special Report 247 (Washington, DC: National Academy Press), 91–105.

Taranaki Regional Council, 1992, *Regional Policy Statement: Working Paper – Natural Hazards*, Stratford, New Zealand.

Willis, R., 1991, Farming, *Pacific Viewpoint*, **32**(2), 163–170.

# 12

# Reducing Landslide Hazards and Risk in the United States: The Role of the US Geological Survey

*Gerald F. Wieczorek, Paula L. Gori and Lynn M. Highland*

## 12.1  Introduction

Landslides occur in every one of the states of the USA and are widespread in the island territories of American Samoa, Guam, Puerto Rico and the US Virgin Islands, some dramatic examples of which are shown in Figures 12.1–12.5. Landslide deaths in the USA have been estimated at 25 to 50 per year, and total annual economic losses due to landslides estimated to range from $1.6 billion to $3.2 billion (Schuster, 1996; Schuster and Highland, 2001). No single government agency or national programme has responsibility for investigating, mapping, cataloging, or assessing landslide hazards and risk throughout the USA; rather, the responsibility is distributed across many federal, state, and local jurisdictions. Since the Organic Act of 1879 created the USGS, that body has played a key role in reducing geological hazard and risk. Subsequent congressional legislation, mainly the Dam Inspection Act of 1972 and the 1974 Disaster Relief Act (Stafford Act) formalized this role. The USGS derives its leadership role in landslide hazard work from the Stafford Act, which delegated to the director of the USGS the responsibility of issuing disaster warnings for earthquakes, volcanic eruptions, landslides, or other geological catastrophes consistent with the 1974 Disaster relief Act 42 U.S.C. *et seq.* (Spiker and Gori, 2003).

*Landslide Hazard and Risk*   Edited by Thomas Glade, Malcolm Anderson and Michael J. Crozier
© 2004 John Wiley & Sons, Ltd   ISBN: 0-471-48663-9

***Figure 12.1*** *The 1983 Thistle landslide, central Utah. Thistle Lake, which resulted from damming of the Spanish Fork River, was later drained as a precautionary measure. This view, taken 6 months after the slide occurred, shows the realignment of the Rio Grande Western Railroad lines in the lower centre and the large cut for rerouting US Highway 6/50 on the extreme left side of the photograph. Total costs (direct and indirect) incurred by this landslide exceeded $400 million, making this the most costly single landslide event in US history (photo reproduced by permission of R.L. Schuster, US Geological Survey)*

This chapter will examine the role of the USGS in the development of the scientific understanding of landslide processes for devising techniques to assess regional landslide hazard and risk in the USA. Examples of the application of these techniques by local, regional and state governments, and their success and shortcomings on local, regional and national scales, will be presented. Possible improvements in reducing landslide hazards and risk will be discussed in conjunction with future directions of the USGS Landslide Hazards Program. (The majority of references cited in this chapter are from USGS scientists.)

## 12.2   The Role of the USGS in Landslide Hazard Assessment

Although potential hazards posed by individual landslides are commonly evaluated by engineering geological consulting companies, the assessment of landslide hazards on a regional basis is largely undertaken by state or federal organizations, such as the USGS. Through the Landslide Hazards Program (LHP), the USGS is charged by the US Congress to perform scientific research to mitigate the effects of geological hazard, known as landslides. With assistance from and collaboration with several other USGS programmes, the LHP conducts research on landslide processes and methods of landslide hazard and risk assessment. This research includes the development of landslide mapping techniques and landslide management on federal lands. However, regulatory responsibility for the

**Figure 12.2**   *The October 1985 Mameyes landslide, near Ponce, Puerto Rico, caused by a tropical storm, killed at least 129 people, the most fatalities of any single landslide in the United States (photo reproduced by permission of R.W. Jibson, US Geological Survey)*

**Figure 12.3**   *Lahar (volcanic mudflow and debris flow) from 1982 eruption of Mount St Helens, Washington (photo reproduced by permission of T.J. Casadevall, US Geological Survey)*

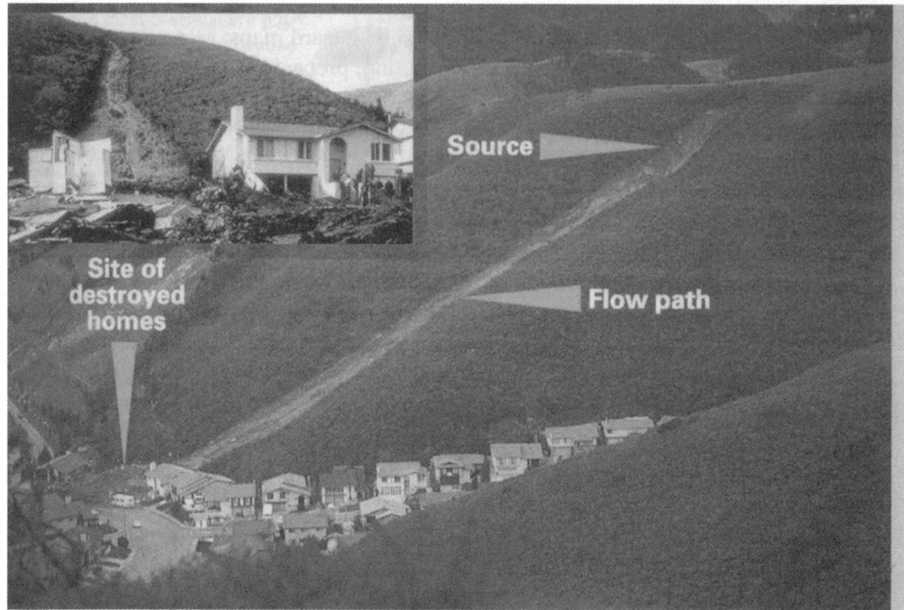


**Figure 12.4**  *Debris flow from Pacifica, California, about 10 miles south of San Francisco, where three children were killed and two homes destroyed on 4 January 1982. Inset, view of destroyed homes from the street. The 3–5 January 1982, storm triggered more than 18 000 landslides in the San Francisco Bay region (photo by G.F. Wieczorek)*

reduction of landslide losses through land use management and the application of building and grading codes is not a function of the USGS, but of other local, state and federal government agencies. Consequently, the results of USGS research to develop methods of identifying and assessing landslide hazards and risks are most effective when implemented by other government agencies.

Scientists in the USGS LHP are located throughout the nation and conduct research, gather and provide outreach information, respond to emergencies and disasters and produce scientific reports and other products for a broad-based user community. Because landslides are often associated with flooding events, there is also landslide expertise in the Water Resources Discipline (WRD) of the USGS, a group of scientists that researches and monitors the nation's surface and groundwater resources. The USGS Volcano Hazards Program also actively studies landslides, primarily those of a volcano-related nature, which most commonly occur as lahars. The Earthquake Hazards Program and the Coastal and Marine Geology Program also direct research towards landslide hazard studies. The National Landslide Information Center (NLIC) of the USGS provides information and literature about landslide hazards to the public, researchers, planners, and local, state and federal agencies through a dedicated website. The NLIC is the outreach and information section for the USGS Landslide Program. The centre is located in Golden, Colorado, along with scientists in the Landslide Program. The NLIC facilitates the distribution of fact sheets and other publications related to landslide research, hazard studies, inventory studies, case studies, emergency management information, and recent landslide event

**Figure 12.5**   *The 27 June 1995, storm in Madison County, Virginia, dropped up to 780 mm of rainfall within 14 hours and triggered an estimated 1000 debris flows within a relatively small area of 130 km². At 11:30 a.m. the debris flow moved the two-storey farmhouse (arrow) more than 10 m from its foundation. The family in the house survived because they took refuge in the second storey (photo copyright by Kevin Lamb, 1995; published with permission)*

information. It also maintains a permanent landslide exhibit and public library of landslide information, accessible during regular business hours. The Internet address for the USGS LHP and the NLIC is: http://landslides.usgs.gov/. The website also links other pertinent agencies such as the Federal Emergency Management Agency, all 50 State Geological Surveys, and other sites of interest. The NLIC is also a repository of landslide publications and maintains several landslide databases. In addition, the NLIC informs the public and media during landslide hazard emergencies.

Although international activities in the LHP are limited, the programme does participate in emergency response to foreign landslide disasters, through cooperative programmes with other agencies such as the Office of Foreign Disaster Assistance (OFDA) of the United States Agency for International Development (USAID). The USGS has provided assistance on foreign landslide disasters in Peru (Plafker *et al.*, 1971), Brazil (Jones, 1973), Guatemala (Harp *et al.*, 1981), and Venezuela (Wieczorek *et al.*, 2002). In October 1998, Hurricane Mitch struck Central America, causing many landslides throughout Honduras (Harp *et al.*, 2002a, 2002b), Nicaragua (Cannon *et al.*, 2001), Guatemala (Bucknam *et al.*, 2001) and El Salvador (Crone *et al.*, 2001). USAID supported the USGS for two years to investigate the distribution of landslides and to assist local communities in dealing with geologic hazards. USAID uses USGS scientific expertise to assist foreign nations with hazard assessment, emergency response, landslide hazard mitigation, and training of local scientists and public officials. Many foreign countries struggle with planning issues related to geological hazards and there is much that can be learned from the mutually shared planning experiences of the United States and other countries.


## 12.3   Investigation of Landslide Processes

Although many different types of landslides have been observed historically throughout the USA, investigative landslide studies did not begin until the end of the nineteenth century. One of the first documented accounts concerns a landslide triggered by a heavy deluge of rain near Carlisle, Pennsylvania, on 19 August 1779, which was described in a letter to Benjamin Franklin (Bell, 1996). Although the observers were not familiar with debris flows, their descriptions were distinctively those of a debris flow – including that of a scar near the source area, a gush of water carrying large-sized rocks and trees in a wide channel down a steep slope, a boulder found several metres high in a tree, and mud and scars extending to a height of about 9 m in trees. Recent (1997) field examination of a steep channel beginning near a source area in this region and boulder levees along the sides of the channel confirmed this much earlier debris flow (Delano and Potter, 1997; Delano *et al.*, 2001).

An investigation of the landslides in the San Juan Mountains in southwestern Colorado was probably the first comprehensive USGS study of specific types of landslides in a local region of the USA (Howe, 1909). Over a period of more than 10 years, Howe and his colleagues examined and classified numerous landslides in this region of volcanic and sedimentary rocks, the majority of which occurred after the retreat of the last glacial ice. Although this study did not directly address the issue of hazards and risks posed by landslides, it did identify a region of significant landslide incidence.

A better understanding of the types of landslide processes is needed to reduce landslide hazards and risk. A variety of techniques has been used to investigate landslides, including: field measurements, instrumentation and monitoring; drilling, sampling, and laboratory testing; examination and detailed mapping from stereo aerial photography; modelling of different modes of failure and movement; and stability analyses. These studies have provided information for better determining the various factors, that is, climatic, hydrological, seismic, geological composition and structure, and so on, that influence the timing, rate of movement, and spatial extent of landslides – all related to landslide hazards and risk.

Through detailed examination and instrumentation of earthflows over several years in the coastal ranges of central California (Keefer and Johnson, 1983), large masses of earthflow materials were rarely found to result from a single episode of movement, but are rather complexes of deposits formed during many episodes of movement. Earthflows were typically mobilized at rates of several centimetres per day by increases in porewater pressure caused by infiltration of water into the soil during and after rainstorms.

The complexity of landslide reactivation and movement was studied by Fleming *et al.* (1988a) by examining the reactivation of the Manti landslide in the Wasatch Plateau of central Utah. In early June 1974, coincident with the melting of a winter snowpack, a rock slump occurred on the south rim of Manti Canyon. Part of the slumped material mixed with meltwater and mobilized into a series of debris flows. A small part of the debris-flow deposit added a load onto the head of the very large, relatively inactive Manti landslide. The upper part of the landslide began moving as cracks propagated downslope. While the upper part of the landslide moved relatively slowly, the lower part was moving rapidly. Consequently, the landslide changed from being in compression, which was caused by loading from the fresh debris-flow deposits, to being in extension, which was caused by the lower part moving faster than the upper part. The rate of downslope movement was generally 4–5 m/h within the first few days of reactivation. Within a year, movement extended through the entire length of the old landslide, involving about 19 million $m^3$ of debris about 3 km long and as much as 800 m wide, threatening to block the canyon. The sequence and rate of reactivation suggested that the movement occurred on a pre-existing failure surface and in part created a new failure surface. After the new failure surface was created, the lower part of the landslide moved more rapidly than the upper part, resulting in extensional cracking and separation of the landslide into two independent parts. During the spring of 1983, when melting of a near-record snowpack was triggering numerous reactivations of large, old landslides on the flanks of the Wasatch Plateau, the Manti landslide remained inactive (Fleming *et al.*, 1988b).

The displacement of earth dams and landslides during seismic shaking was found to be similar. The application of the method developed for analysing displacement of earth dams during seismic events (Newmark, 1965) was evaluated by Wilson and Keefer (1983) for an earthquake-induced landslide on a natural slope and was found to be a valid predictor of the displacement of a landslide during seismic shaking. Analysis of landslides caused by earthquakes worldwide provided data for evaluating the relationship between the maximum distance from faults or epicentres to landslides in earthquakes of different magnitude (Keefer, 1984; Keefer and Manson, 1998).

Based on a detailed examination of joint spacing, roughness, alteration and aperture associated with rock-mass quality near rockfalls triggered by the 1980 Mammoth Lakes earthquake sequence, Harp and Noble (1993) developed a method of assessing regional seismic rockfall susceptibility. Detailed studies of rockfalls in Yosemite National Park,

California revealed that although it is sometimes difficult to determine the triggering event of rockfalls, it is possible to identify the location of areas subject to continuing landslide hazards and risk (Wieczorek *et al.*, 1999). Beginning on 8 March 1987, small rockfalls began from near the top of Middle Brother, on the 900 m high cliff of the northern rim of Yosemite Valley. By 2:20 p.m. on 10 March, the increasing frequency of small rockfalls and audible popping noises forced the National Park Service to close Northside Drive, just below Middle Brother. At 2:47 p.m., 10 March, a large rockfall broke from the face of Middle Brother, dropped 800 m, and spread rapidly across a talus cone, and covered Northside Drive. A second large rockfall occurred from the same site later that day at 5:10 p.m. The combined volume of these two rockfall deposits totalled an estimated 600 000 m$^3$, the largest historical rockfall documented in Yosemite Valley. During and a few days preceding 8–10 March, the weather had been dry and lacked extreme temperature variations that might normally be associated with freeze–thaw cycles or rapid snowmelt that would trigger a rockfall. No earthquakes were detected during this period. Dozens of smaller rockfalls continued during the next several days and weeks following 10 March. Based on the monitoring of the decreasing rate of rockfall activity over the next several months, Northside Drive was reopened in early July 1987. A specific cause for the timing of these rockfalls could not be determined (Wieczorek *et al.*, 1995; Wieczorek, 2002).

## 12.4   Development of Landslide Hazard Maps

Compilation of landslide inventory maps is an initial step in the evaluation of regional landslide hazards. Although a landslide inventory map does not necessarily indicate the frequency of landslide activity or the specific type of landslide relating to velocity or size, it does show where landslides have occurred in the past and consequently where landslides might occur in the future under similar climatic conditions.

One of the first comprehensive regional investigations of landslides resulting in a landslide hazard assessment including inventory maps was made along the Columbia River in the state of Washington (Jones *et al.*, 1961). Geological investigations of more than 300 landslides along 320 km of the Columbia River Valley were conducted intermittently from 1942 to 1948 and continuously from 1948 to 1955, in order to assess the potential instability of land near dams that were being constructed with the consequence of impounding large lakes. Most of the recent landslides in this region had occurred during the slow and intermittent filling of Franklin D. Roosevelt Lake behind Grand Coulee dam (1933–42). The initial studies (1942–48) conducted by Jones for the USGS, in conjunction with the Bureau of Reclamation, classified the stability of lakeshore land. Where privately owned land was found to be potentially dangerous, the US government offered to purchase the property. Subsequently, beginning in 1948, the USGS began technical cooperation on these research studies with the National Park Service and the US Bureau of Reclamation, resulting in a statistical evaluation of slope stability and maps identifying the landslide and potential landslide areas of Franklin Roosevelt Lake. In 1950, in cooperation with the Corps of Engineers, investigations were extended to include a section of the Columbia River between Grand Coulee and Chief Joseph Dams (Jones *et al.*, 1961).

Beginning in the 1970s, more regional landslide inventory maps were prepared, for example Brabb and Pampeyan (1972), Pomeroy and Davies (1975), McGill (1973),

Nilsen (1971) and Pomeroy (1977). Some of these inventory maps identified different landslide-age categories such as 'recent' and 'prehistoric' or 'younger' and 'older' landslides (Pomeroy, 1979). These maps, ranging in scale from 1:4800 to 1:62 500, were largely produced from photo interpretation with field investigations. Consequently, the scale of available aerial photography and the scale of the maps determined the size of individual landslides that were depicted. The smaller-scale geological maps were more general, in that they represented areas with landslide deposits rather than depicting or identifying individual types of landslides. In some cases, surficial geological maps were prepared which identified landslide deposits, as well as other types of surficial deposits, such as glacial deposits (Madole, 1982). In another case, Colton *et al.* (1975) prepared a series of maps of landslide deposits covering the entire state of Colorado at a scale of 1:250 000.

In order to reduce landslide hazards on a local scale, a landslide inventory study in the Pacific Palisades area of the City of Los Angeles was authorized by Congress in the Flood Control Act of 1966. The study was conducted under the direction of the Corps of Engineers in cooperation with the USGS (McGill, 1973). The resulting landslide inventory map (McGill, 1973), prepared at a scale of 1:4800, depicted landslides of different ages, and identified cracks and fresh or relatively unmodified landslide scarps and the direction of landslide movement. A preliminary version of this map, prepared in 1959, was updated because of new landslides that had occurred in the interim, specifically during extremely heavy rains of the winter of 1969, which damaged streets, public utilities and residences.

In order to better understand the relationship between the magnitude of specific landslide-triggering events, for example rainstorms or earthquakes, and the number and distribution of landslides, inventory maps have been prepared following major triggering events (Wilson *et al.*, 1985; Ellen and Wieczorek, 1988; Jacobson, 1993; Morgan *et al.*, 1999; Coe and Godt, 2001; Godt and Coe, 2003). Landslide mapping from a specific triggering event provides useful information for determining what factors, for example geological, hydrologic or topographic, most influence the triggering of landslides and is useful for improving the methodology for assessing landslide hazards. On 28 July 1999, about 480 debris-flows were triggered by an afternoon thunderstorm along the Continental Divide in Clear Creek and Summit counties in the central Front Range of Colorado. Several debris flows triggered by the storm affected Interstate 70 (I-70), US Highway 6, and the Arapahoe Basin ski area. Interstate 70 remained closed for 25 hours. Fortunately no injuries or fatalities resulted from any of the debris flows. An inventory of debris flows in the 240 km$^2$ area was prepared from interpretative mapping of stereo colour aerial photography and by inspecting many of the debris flows in the field (Godt and Coe, 2003).

A more direct assessment of landslide hazard was provided by the development of landslide susceptibility maps. Although landslide susceptibility maps, for example in San Mateo County, California (1:62 500) (Brabb *et al.*, 1972), Oakland, California (1:50 000) (Pike *et al.*, 2001), and in Butler County, Pennsylvania (1:50 000) (Pomeroy, 1977), do not directly assess the frequency or probability of landslides, the representation of areas subject to landslide hazards is very useful for local and regional governments. On a national scale of 1:7 500 000, Radbruch-Hall *et al.* (1983) prepared a landslide overview map of the conterminous states of the USA that summarized geological, hydrogeological and topographical data essential to the assessment of national environmental problems. This map delineates areas where large numbers of landslides exist and areas that are susceptible to landslides. It was prepared by evaluating the geological map of the United States and classifying the geological units according to high, medium or low landslide incidence (number) and high, medium or low susceptibility to landslides. This map has been digitized and is available online (http://landslides.usgs.gov/html_files/landslides/nationalmap/national.html).

Other landslide hazard reduction efforts have been undertaken jointly between the USGS and other state and local governments. In cooperation with the California Geological Survey, the USGS prepared a group of maps showing relative susceptibility of slopes to the initiation sites of rainfall-triggered soil slip–debris flows in southwestern California (Morton *et al.*, 2003). These maps offer a partial answer to one part of the three parts necessary to predict the soil-slip–debris-flow process. These maps empirically show part of the 'where' of prediction (i.e. relative susceptibility to sites of initiation of the soil slips) but do not attempt to show the extent of runout of the resultant debris flows. The susceptibility maps were created through an iterative process from two kinds of information. First, locations of sites of past soil slips were obtained from inventory maps of past events. Aerial photographs, taken during six rainy seasons that produced abundant soil slips, were used as the basis for soil-slip–debris-flow inventory. Second, digital elevation models (DEM) of the areas that were inventoried were used to analyse the spatial characteristics of soil-slip locations.

For improved landslide hazard assessment, details regarding the type of slope movement as well as the recency of movement are important data on landslide inventory maps. A classification system for the type(s) of individual landslides first developed by Heim (1932) was subsequently refined (Varnes, 1958, 1978; Hutchinson, 1968; Cruden and Varnes, 1996). Keaton and DeGraff (1996) combined a landslide classification system based on activity, degree of certainty of identification of the slide boundaries, and the dominant type of slide movement (Wieczorek, 1984) with a landslide age classification system (McCalpin, 1984) into the Unified Landslide Classification System (Table 12.1).

***Table 12.1***  *The Unified Landslide Classification System*

| Age of most recent activity (symbol) | Dominant material* (symbol) | Dominant type of movement (symbol) |
| --- | --- | --- |
| Active (A) | Rock (R) | Fall (L) |
| Reactivated (R) | Soil (S) | Topple (T) |
| Suspended (S) | Earth (E) | Slide (S) |
| Dormant – historic (H) | Debris (D) | Spread (P) |
| Dormant – young (Y) | | Flow (F) |
| Dormant – mature (M) | | Fall and Flow (LF) |
| Dormant – old (O) | | Topple and Flow (TF) |
| Stabilized (T) | | Slide and Flow (SF) |
| Abandoned (B) | | Spread and Flow (PF) |
| Relict (L) | | (Other combinations may be observed) |

* Rock refers to a hard, firm mass in its natural place before initiation of movement. Soil refers to an aggregate of solid particles that was either transported or formed by the weathering of rock in place. Soil is subdivided into earth and debris. Earth refers to soil in which 80% or more of the particles are 2 mm or smaller. Debris refers to soil in which 20 to 80% of the particles are larger than 2 mm.
*Source*: Keaton and Rinne (2002). Reproduced by permission of A.A. Balkema Publishers.

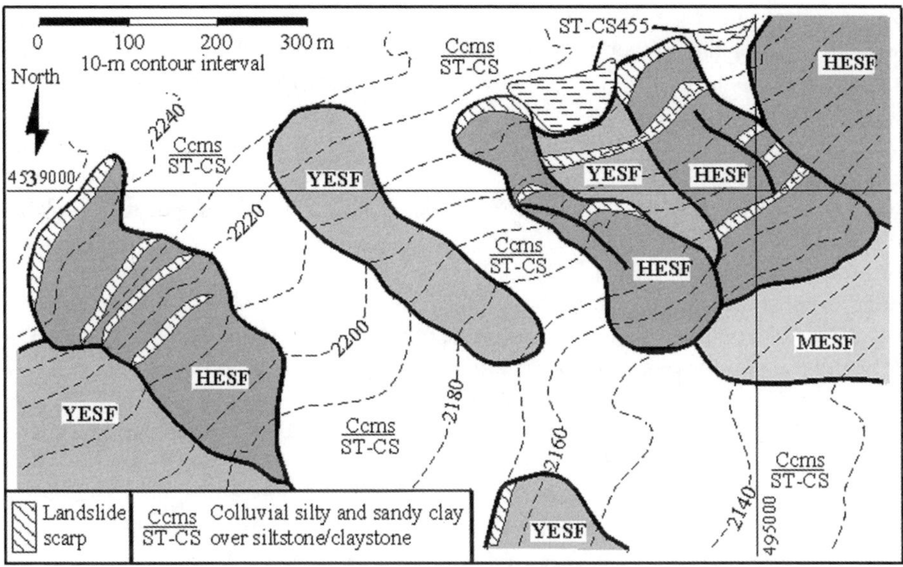


**Figure 12.6**   *Landslide inventory map of the Thousand Peaks area in northern Utah (published with permission from Keaton and Rinne, 2002). Landslides labelled according to the Unified Landslide Classification System (Keaton and DeGraff, 1996; Keaton and Rinne, 2002). Table 12.1 shows the various landslide classifications, with YESF standing for dormant – Young, Earth, Slide and Flow, HESF standing for dormant – Historic, Earth, Slide and Flow, and MESF standing for dormant – Mature, Earth, Slide and Flow). Reproduced by permission of A.A. Balkemer Publishers*

An example from Keaton and Rinne (2002) showing the depiction of landslide classification information on a landslide inventory map for the Thousand Peaks area in northern Utah is shown in Figure 12.6.

Efforts have been made to develop methods for depicting regional landslide hazards incorporating the likely frequency of events or probabilistic depiction of the likelihood of landsliding (Bernknopf *et al.*, 1988; Mark, 1992; Campbell *et al.*, 1998; Jibson *et al.*, 1998; Coe *et al.*, 2000). In December 1991, at the request of the California State Geologist, the USGS began a study to forecast the risk of rainfall-triggered debris-flow damage in the hills northeast of Oakland, California. To develop a method to estimate the spatial distribution of different levels of risk from rainfall-triggered debris flows, Campbell *et al.* (1998) devised a procedure that yields the conditional probability that a soil slip–debris flow will occur in a 100 m map cell at times during a storm in conjunction with rainfall that exceeds identified thresholds. This procedure was applied to an area near Oakland using rain-gauge records for the 3–5 January 1982, storm in the San Francisco Bay region. The results showed the probability of debris flows on a map at 3-hour intervals during the period of the 36-hour storm and showed the relationship between the probability and the post-storm debris-flow inventory.

Some recently developed methods for determining the probability of landslide occurrence have improved landslide hazard assessment (Jibson *et al.*, 1998; Coe *et al.*, 2000; Croveli, 2000). The 1994 Northridge, California earthquake provided data for a regional analysis of seismic slope instability, including (1) a comprehensive inventory of triggered landslides, (2) about 200 strong-motion records of the main shock, (3) 1:24 000-scale geological mapping of the region, (4) engineering properties of geological units, and (5) high-resolution digital elevation models of the topography. These data sets were digitized and rasterized at 10 m grid spacing in an ARC/INFO GIS platform. Combining these data sets in a dynamic model based on a permanent-deformation (sliding-block) analysis (Newmark, 1965) provided estimates of coseismic landslide displacement in each grid cell from the Northridge earthquake. The modelled displacements were then compared with the digital inventory of landslides triggered by the Northridge earthquake to construct a probability curve relating predicted displacement to probability of failure. This probability function can be applied to predict and map the spatial variability in failure probability in any ground-shaking conditions of interest. This mapping procedure can be used to construct seismic landslide hazard maps that will assist in emergency preparedness planning and in making rational decisions regarding development and construction in areas susceptible to seismically induced slope failure (Jibson *et al.*, 1998).

## 12.5   Prediction and Warning of Landslide Hazards

With development and improvements in remote sensing of rainfall, such as Doppler radar, methods have been developed to predict and warn of landslide hazards. Subsequent to the January 1982 storm in the San Francisco Bay region, which triggered more than 18 000 landslides (Ellen and Wieczorek, 1988), a real-time landslide warning system was established and operated by the USGS and the National Weather Service between 1986 and 1995 (Wilson *et al.*, 1993; Wilson, Chapter 17, this volume). Documentation of the rainfall duration and intensity associated with the triggering of landslides identified rainfall thresholds for the triggering of shallow landslides, particularly debris-flows (Wieczorek and Sarmiento, 1983; Cannon and Ellen, 1985). Using a real-time rainfall monitoring system and National Weather Service satellite-based quantitative rainfall forecasts, regional landslide warnings were issued during storms in 1986, 1991, 1993 and 1995. As a verification of the thresholds of Cannon and Ellen (1985), the times of landslide warnings in the storms of February 1986 were found to correspond with documented times of shallow landslides (Keefer *et al.*, 1987). Rainfall thresholds for triggering debris flows have been identified in a number of other regions in the USA, including Puerto Rico (Jibson, 1989; Larsen and Simon, 1993), Hawaii (Wilson *et al.*, 1992), the Blue Ridge of central Virginia (Wieczorek *et al.*, 2000), and Seattle, Washington (Chleborad, 2000, 2003).

The USGS developed an inventory of landslides, debris flows and flooding from the storm of 27 June 1995 in Madison County, Virginia, by using aerial photography, field investigations, rainfall measurements from rain gauges, and National Weather Service Doppler radar observations (Morgan *et al.*, 1999). The inventory data are being used to ascertain the conditions that caused the debris flows and to develop methods of warning of such events in the future (Morrissey *et al.*, 2001). Although Doppler radar can provide detailed spatial depiction of rainfall over large regions during near real time that is useful for prediction of landslide hazards, this technology has yet to be adapted for issuing landslide warnings.

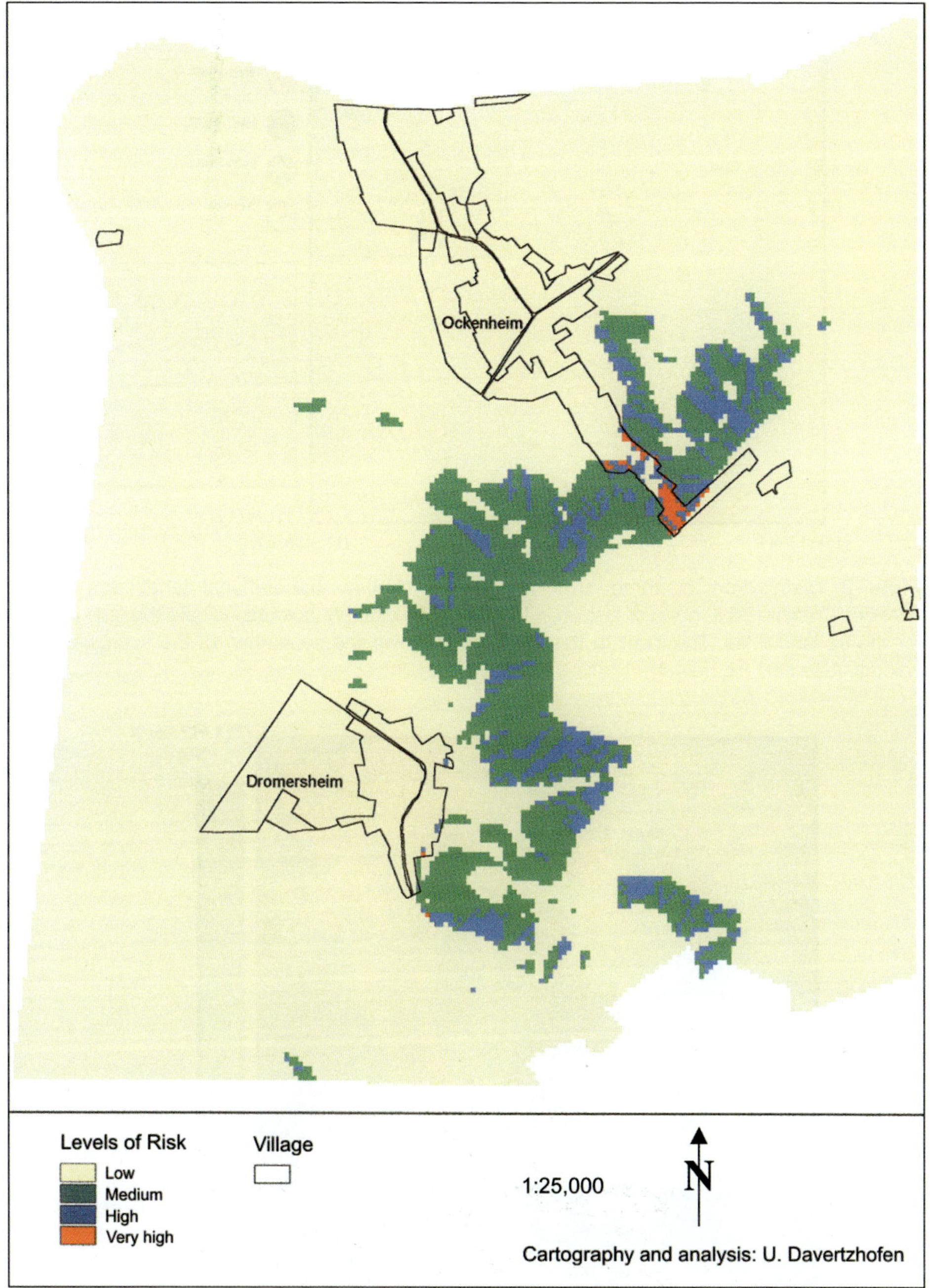



**Plate 1** Regional landslide risk in Rheinhessen, Germany (Glade, *et al.*, in prep.b). Vulnerability to elements at risk is assumed to be 1, referring to total loss if an element is affected by a landslide. (See also Figure 3.9)

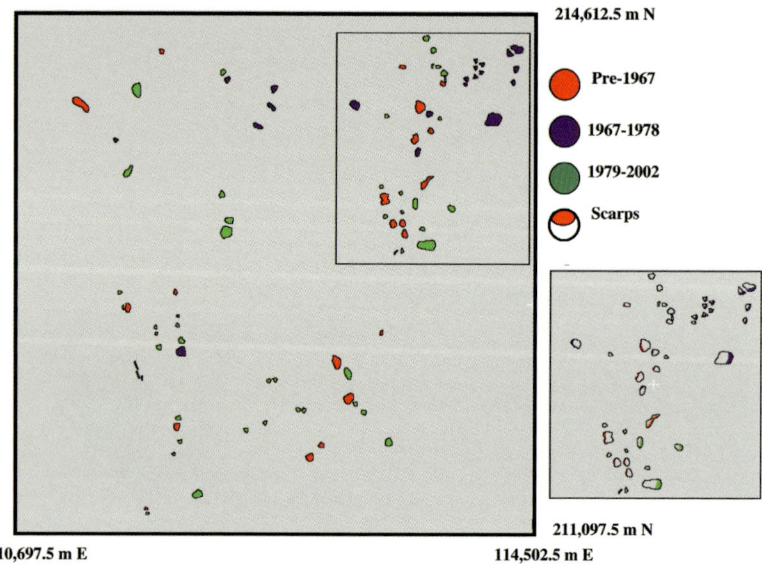



**Plate 2** Distribution of three time periods of shallow translational landslides in the Fanhões–Trancão area, north of Lisbon in Portugal. The different colours indicate the time periods for the 92 landslides. The inset to the lower right shows the separation of the scarps in solid colours. (See also Figure 4.1)

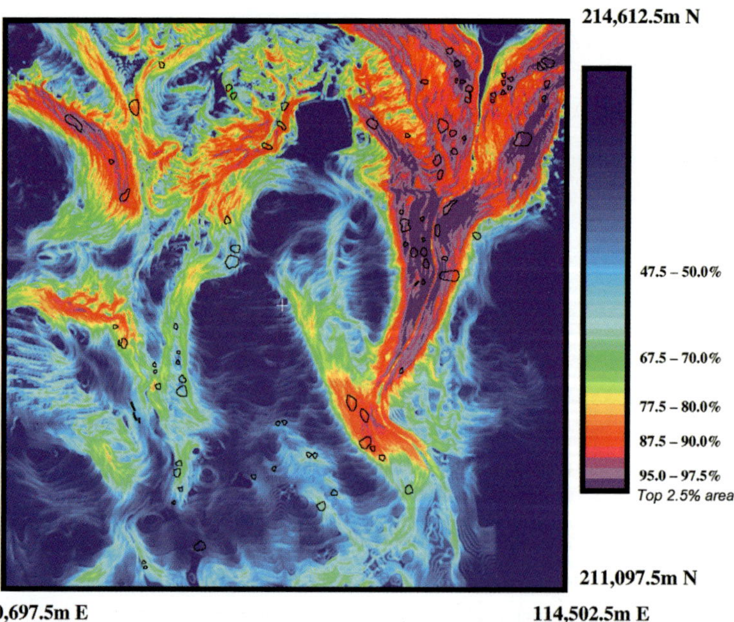



**Plate 3** Landslide hazard prediction map of the Fanhões–Trancão area, Portugal, based on the 92 landslides (shown in Plate 2) and six layers (bedrock geology, land use, surficial material, elevation, aspect angle, and slope angle maps) of geomorphological map information using the linear discriminant analysis model. The 92 black polygons represent the 92 landslides. The pseudo-colours and the associated numbers in the legend refer to the predicted hazard class area percentages. (See also Figure 4.3)

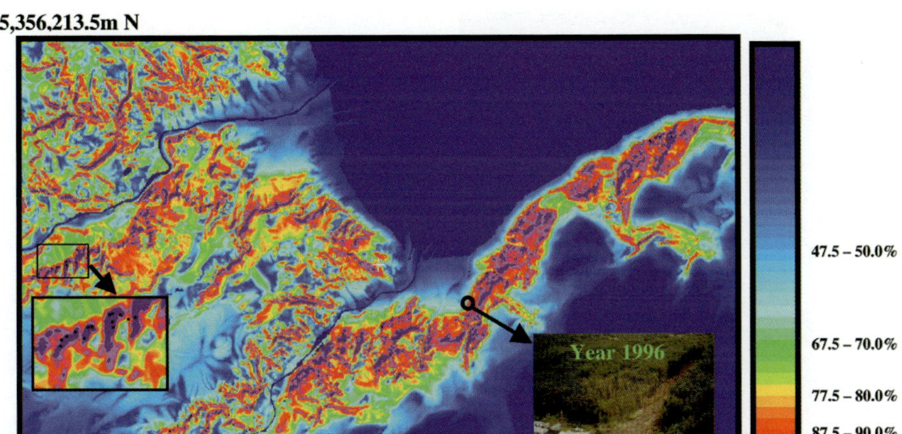



**Plate 4** Landslide hazard prediction map for the La Baie study area, Québec, Canada, based on 22 landslides that occurred in 1967 and five layers (bedrock geology, forest coverage, elevation, aspect angle, and slope angle maps) of geomorphological map information using the likelihood ratio function model. The 51 black dots represent the 51 landslides that occurred in 1976 and 1996. The left-side inset is an enlargement of a small area in the black rectangle in the middle left-side. The right-side inset with 'Year 1996' is a photograph of a landslide that occurred in 1996 at the site of the black circle in the middle of the illustration. The pseudo-colours and the associated numbers in the legend refer to the predicted hazard class area percentages. The classes are regrouped in different percentages from those in Table 4.3. (See also Figure 4.5)

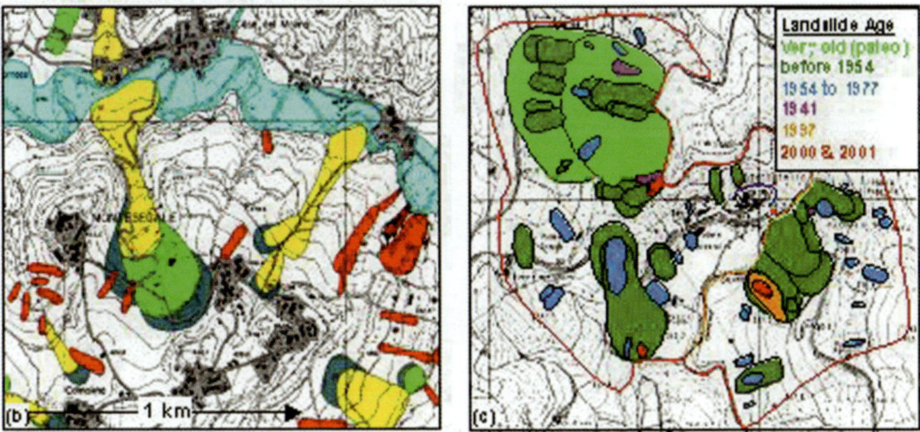



**Plate 5** (b) The various units refer to landslide types and (c) ages. Source: Guzzetti, 1990. (See also Figure 16.1)

**Plate 6** (c) Airborne C-HH SAR Interferogram which facilitates landslide detection, geomorphic mapping and related seismically active structures. (d) Visualization image produced from the interferogram (shown in (c)) provides an accurate representation of the slope geomorphology, and facilitates the identification of landslide features in this difficult high-relief terrain. (See also Figure 16.3)

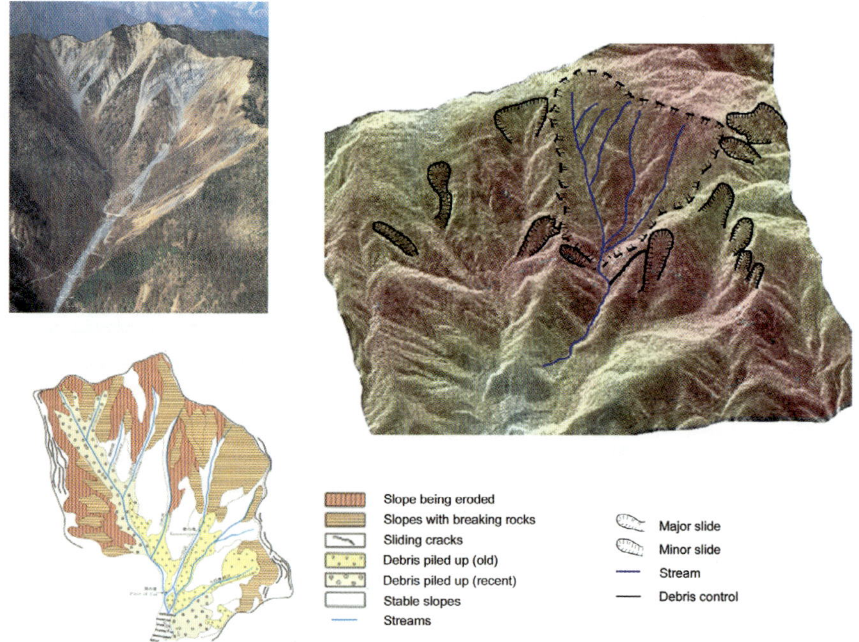



**Plate 7** Interpretation of RADARSAT image draped over DEM for landslide inventory in upper Oya Valley in Japan. (See also Figure 16.5)

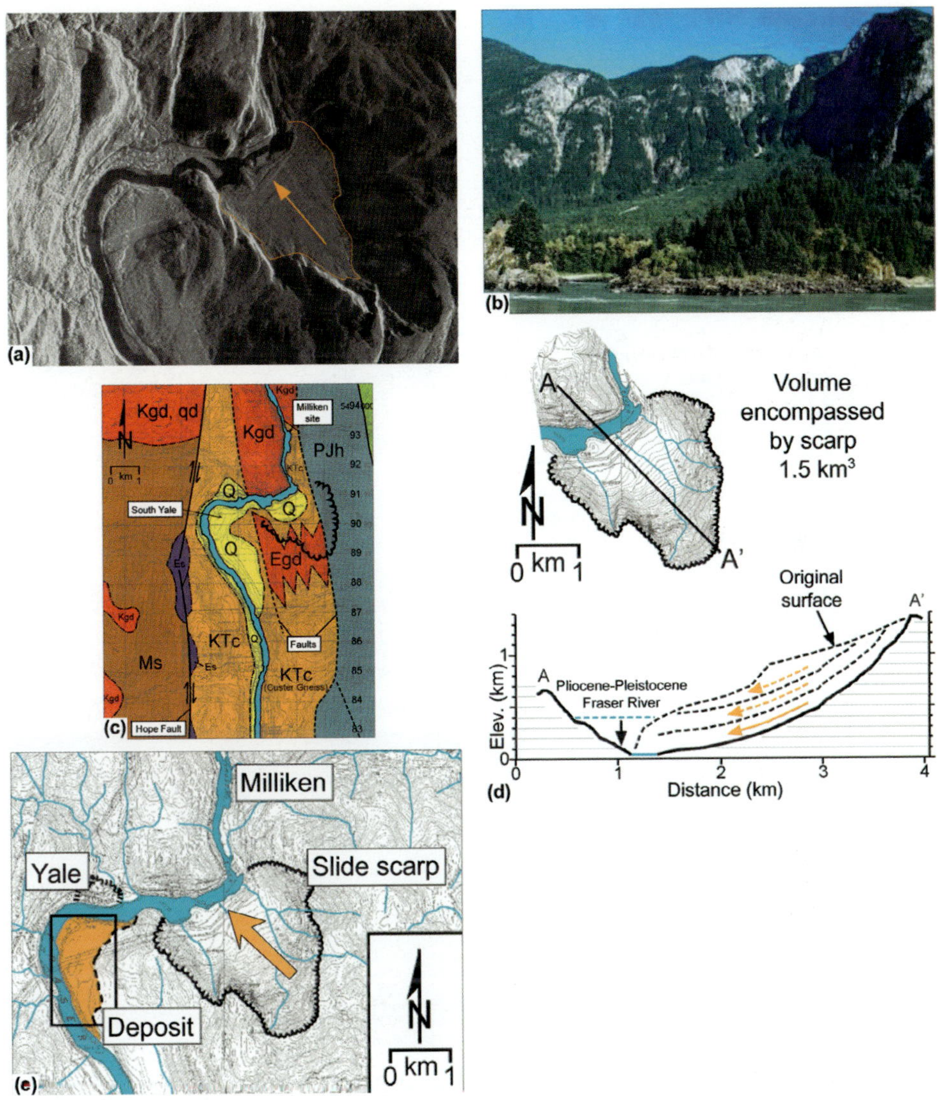



**Plate 8**   (a) South Yale Slide, Hope, B.C., Canada, depicted in a high-resolution airborne radar image and (b) field photograph, (c) geological setting, (d) slide transect, and (e) inventory map. (See also Figure 16.4)

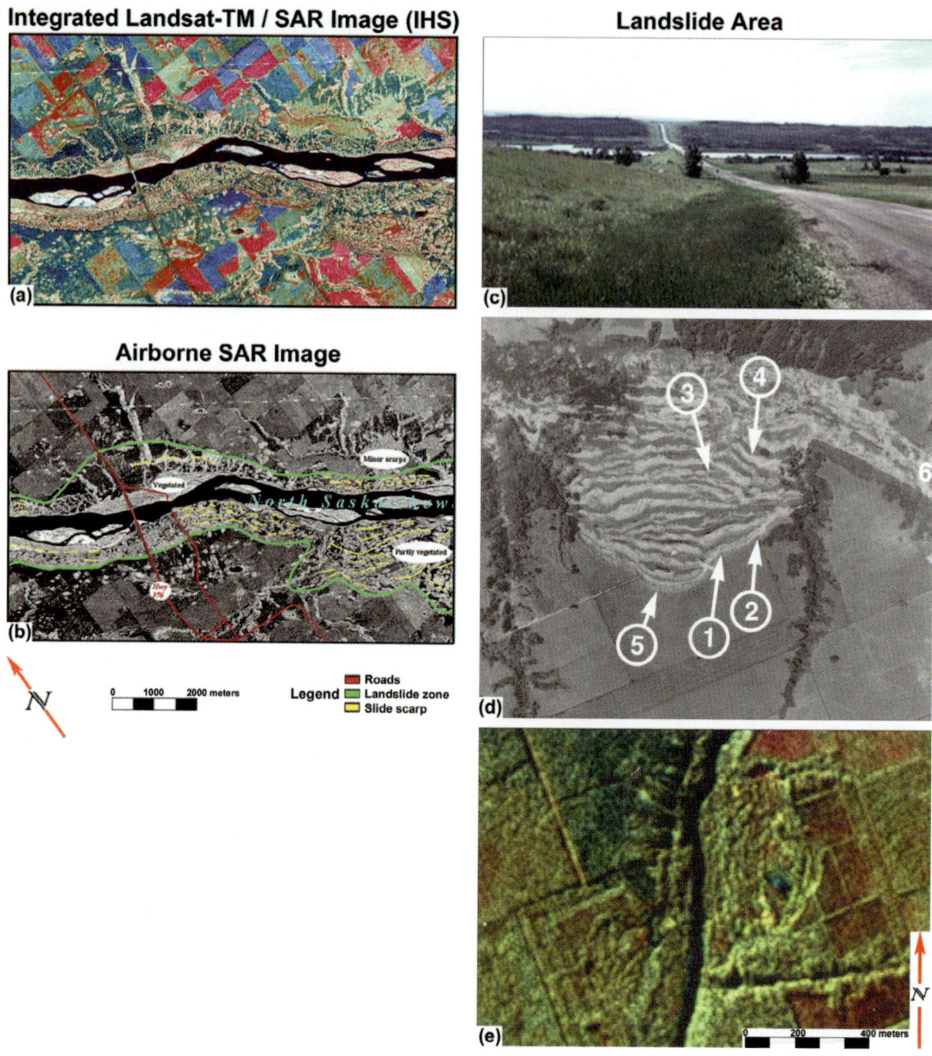


**Plate 9** (a) and (e) Delineation of multiple retrogressive sliding in Cretaceous shales along the Saskatchewan River and in marine clays along the South Nation River using fused SAR and TM, (b) airborne SAR, and (c) field photograph. The airphoto (d) shows backscarp (1 and 2) slump ridges (3 and 4) and headscarp (5) (Mollard & Janes, 1993). Aerial photographs © 1971. Her Majesty the Queen in Right of Canada, reproduced from the collection of the National Air Photo Library with permission of Natural Resources Canada. (See also Figure 16.7)

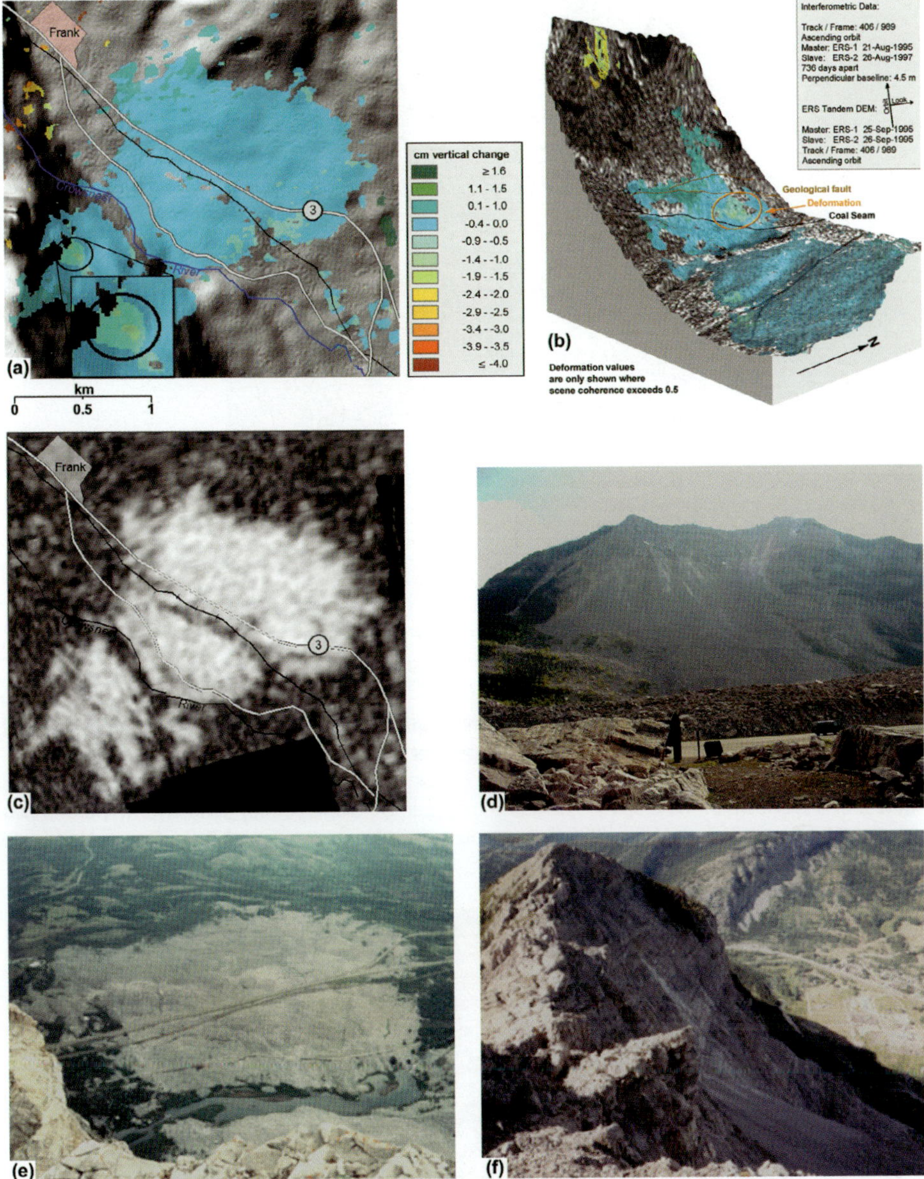

**Plate 10** (a) Vertical surface deformation map, generated from differential SAR interferometry. (b) Result is draped over a high-resolution elevation model. (c) Corresponding coherence image, and (d-f) field photographs. The deformation is located towards the bottom of the slope, just above a coal seam. (See also Figure 16.12)

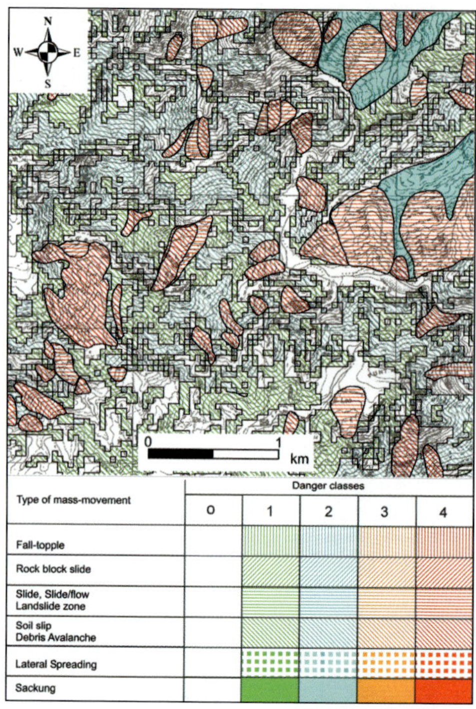

**Plate 11** Part of land classification of Parco Nazionale dell'Aspromonte (Aspromonte National Park, South Calabria) as regards landslide danger, for all the different types of mass movement. Cell size of square-grid DEM is 40 m. In case of superposition of different danger classes, the highest one is shown in the display. The same map can be displayed for each layer (type of mass movement). From Sorriso-Valvo (2001), reproduced with permission of Ente Parco Nazionale dell'Aspromonte. (See also Figure 24.1)

**Plate 12** Example of land classification from a map of the relative hazard (susceptibility) for trigger of and invasion by landslide in the HSP of Autorità di Bacino Nordoccidentale della Campania (2002), South Italy. Key of map colours: blank = no hazard; light green = possible hazard to be defined; dark green = low relative hazard; orange = intermediate relative hazard; red = high relative hazard. Reproduced with permission of Autorità di Bacino Nordoccidentale della Campania. (See also Figure 24.2)

A prediction of abnormally high precipitation that accompanies an El Niño climatic event caused concern that the final months of 1997 and the early months of 1998 might experience exceptional landslide activity in the southern, western and central parts of the USA (Godt, 1999). El Niño events which may occur every few years are characterized by a warming of equatorial waters in the western Pacific Ocean that spreads eastward to the western hemisphere. The El Niño of the winter of 1982–83 was marked by widespread landsliding in different parts of the western hemisphere. According to the prediction by National Oceanic and Atmospheric Administration (NOAA), the 1997–98 El Niño might have been the largest of that century. Although long-term forecasting was generalized, a strong possibility existed for increased precipitation coinciding with the El Niño of 1997–98, leading to increased landsliding. The USGS prepared national maps of the landslide hazard outlook for 1997 and 1998 by combining forecast information for precipitation from NOAA with a USGS map showing landslide incidence and susceptibility for the conterminous states of the USA (Godt *et al.*, 1997). These predictive maps show contours of precipitation anomalies and zones of landslide susceptibility and incidence for large areas of the country.

The strong El Niño of 1997–98 was the wettest season since 1864 and caused more than \$150 million in landslide damage in the ten-county San Francisco Bay region (Godt, 1999; Brabb *et al.*, 2002). Reports of landsliding began in early January 1998 and continued throughout the winter and spring. On 9 February 1998, President Clinton declared all ten counties eligible for Federal Emergency Management Agency (FEMA) disaster assistance. In April and May of 1998, the USGS conducted a field reconnaissance in the area to provide a general overview of landslide damage resulting from the 1997–98 sequence of El Niño-related storms, resulting in landslide damage assessments for ten counties in the Bay area: Alameda, Contra Costa, Marin, Napa, San Francisco, Santa Clara, Santa Cruz, San Mateo, Solano and Sonoma, for which maps were prepared showing the locations of damaging landslides (Godt, 1999).

Real-time monitoring of landslides is used in selected areas by the USGS to reduce risk from active landslides. Continuous monitoring can detect early indications of rapid, catastrophic movement and provide a better understanding of landslide processes, enabling more effective designs for landslide hazards mitigation. During the heavy rains in January 1997, a large landslide occurred in the Sierra Nevada, California, destroying three homes, blocking a major highway (US 50), and briefly damming the adjacent American River. Reopening the highway cost \$4.5 million, and indirect economic losses from the highway closure exceeded \$50 million. To help reduce the risk posed by five large landslides in the same area that continue to threaten US 50, the USGS, in cooperation with the California Department of Transportation, instrumented 58 sites to provide continuous real-time monitoring of landslide activity. Data from a variety of these sensors (precipitation, pore-water pressure, acceleration of slide movement, and ground vibrations associated with movement) are transmitted by radio to USGS computers. Graphs of sensor response are available over the Internet in real time to local officials, geotechnical engineers and emergency managers. The data from one of these landslides are available to the public over the Internet at http://landslides.usgs.gov/hwy50. Near Seattle, Washington, a real-time system monitors a slide threatening a major railway (http://landslides.usgs.gov/woodway), and in Rio Nido, California, another system monitors a large landslide threatening more than 140 homes (Reid *et al.*, 1999).

**Table 12.2**  *Milestones of significant development of methods for evaluation of landslide hazard and risk*

| Signficant developments in landslide hazard and risk | Time period | Reference(s) |
| --- | --- | --- |
| Regional examination of landslide processes and periods of activity | 1890s | Howe (1909) |
| Landslide inventory map and application for local hazards assessment | 1940–1950s | Jones *et al.* (1961) |
| Development and improvement of landslide classification system | 1950–1990s | Varnes (1958, 1978), Cruden and Varnes (1996) |
| Regional landslide susceptibility map | 1970s | Brabb *et al.* (1972) |
| Earthquake-induced landslide susceptibility map | 1980s | Wieczorek *et al.* (1985) |
| Regional rainfall thresholds for triggering landslides | 1980s | Cannon and Ellen (1985) |
| Issuing of regional landslide storm warnings | 1980s | Keefer *et al.* (1987) |
| Regional probabilistic landslide map | 1990s | Campbell *et al.* (1998), Jibson *et al.* (1998) |
| Modeling of debris-flow runout using GIS techniques | 1990s | Iverson *et al.* (1998) |
| Regional rockfall hazard and risk evaluation using rockfall runout model | 2000s | Guzzetti *et al.*(2003) |

Milestones of significant development of methods for evaluation of landslide hazard and risk are listed in Table 12.2. These milestones represent primary or most significant new developments in this field, although many other important contributions generally occurred within the same time period or subsequently.

## 12.6  Utilization of Landslide Hazard Information

The development of landslide inventory maps, susceptibility maps, and regional probabilistic evaluations of landsliding are not effective in reducing landslide hazards unless the information is applied by local or regional government organizations in pursuit of public safety. The following examples demonstrate different applications of USGS landslide information for hazard reduction.

In conjunction with the USGS, a regional governmental organization, the Association of Bay Area Governments (ABAG), involving nine counties within the San Francisco Bay area, established a plan to identify and apply geological hazard information. Within the Bay Area, San Mateo County has used a landslide susceptibility map (Brabb *et al.*, 1972) to reduce the density of development on landslide-prone land to only one dwelling unit per 16 hectares, and to require a geotechnical report indicating that building locations are safe from landsliding even at a lower building density (Brabb, 1995).

Applying a dynamic slope stability method for predicting displacement during earthquakes (Newmark, 1965), a method for developing a seismic landslide susceptibility map was applied to San Mateo County, California (Wieczorek *et al.*, 1985). ABAG used this procedure in their Earthquake Preparedness Program for estimating earthquake damage potential in the San Francisco Bay Area (Perkins *et al.*, 1997; Perkins, 1998). Following the 1989 M6.9 Loma Prieta earthquake, which struck the San Francisco Bay region, the California Division of Mines and Geology tested this USGS method. The Loma Prieta earthquake did not trigger any landslides in San Mateo County that could be used to test the mapping method; therefore, the evaluation of the USGS procedure was performed where many landslides occurred in Santa Cruz County. The predictive results of the procedure were compared to the results of landslides actually triggered by the earthquake in Santa Cruz County. The results indicated that although only a maximum of about 50% of the landslides triggered by the earthquake could be predicted, improvements in the method could be employed utilizing earthquake strong-motion records and improved accuracy of DEM spacing (McCrink, 2001).

To ascertain the rockfall hazard in Yosemite National Park, the USGS and the US National Park Service have compiled an inventory and map of historical landslides (Wieczorek *et al.*, 1992). In addition to damaging roads, trails and other facilities in Yosemite National Park, rockfalls endanger some of the more than 3 million annual visitors. During the period of 1857–2004, 12 people were killed and at least 62 injured by 519 documented rockfalls in the Yosemite Valley (Wieczorek and Snyder, 2004). The minimum shadow angle (Evans and Hungr, 1993) defines the angle extending horizontally from the apex of the talus slope to the farthest outlying boulder. An attempt to assess rockfall hazard based on the shadow angle was completed by Wieczorek *et al.* (1999), and incorporated in the Yosemite Valley Plan prepared by the National Park Service (2000). A physically based three-dimensional rockfall simulation computer program, STONE (Guzzetti *et al.*, 2002), was used for an evaluation of rockfall hazard and risk in the Yosemite Valley (Guzzetti *et al.*, 2003). Visual and statistical comparison of the model results with the mapped rockfall paths confirmed the accuracy of the model. The model results were used to identify the roads and trails most subject to rockfall hazards.

An example of how the USGS, FEMA, and the City of Seattle, Washington, joined forces to mitigate landslide and other natural hazards is FEMA's Disaster-Resistant Communities project (formerly called Project Impact). Beginning in 1997, 200 communities in the USA and more than 1100 businesses were encouraged to mitigate landslide and other natural hazards. Rather than wait for disasters to occur, communities took action to reduce potentially devastating disasters. One of the first communities in the United States to join was Seattle, a city that is exposed to significant landslide hazards. In conjunction with FEMA, the City of Seattle collaborated with the USGS to develop landslide hazard maps that will enable the city to be better prepared for landslide emergencies and to reduce losses resulting from landslide disasters. The City of Seattle made available to USGS scientists hazards data that they had been keeping for almost 100 years to enable the scientists to accurately assess landslide hazards in the area and to produce a computer-based landslide hazard map (coe *et al.*, 2000, 2004). This map includes Seattle's detailed topographic database and related geographical data, detailed precipitation data collected by the National Weather Service, geographical information system support for completing the maps, and a landslide database from city records that date back to the late 1800s. USGS scientists are analysing these records along with other information to determine the degree of landslide hazards throughout the city. The scientists are also conducting studies to determine the probability that landslides will result from storms of different magnitudes. The Disaster-Resistant Communities project has resulted in unprecedented awareness of landslide hazards by the private sector.

Post-wildfire landslide hazards include fast-moving, highly destructive debris flows that can occur in the years immediately after wildfires in response to high-intensity rainfall events. Agencies within the US Department of Agriculture and the US Department of the Interior (US Forest Service, Bureau of Land Management, National Park Service) that are charged with the management of federal lands frequently request technical assistance from the USGS in assessing debris-flow hazards following wildfires in steep terrain, as part of the Burned Area Emergency Rehabilitation (BAER) Team process. In response to these requests, USGS personnel have developed tools and methods for identifying those areas most susceptible to post-fire debris-flow activity and for assessing the potential magnitude of the response. For example, a series of maps that show the probability of debris-flow occurrence and estimated debris-flow peak discharges in response to varying storm conditions have been generated for basins burned by the 2003 wildfires in southern California (Cannon *et al.*, 2003). These maps are used by emergency personnel to aid in the preliminary design of mitigation measures and the planning of evacuation timing and routes.

The assessment of landslides hazards associated with volcanoes was significantly increased following the 1980 eruption of Mount St Helens, Washington by USGS scientists. Eruptions can trigger pyroclastic flows, which are hot avalanches of lava fragments and gas and can rapidly melt snow and ice, resulting in lahars, a term for volcanic mudflows and debris flows. A significant recent development is a model using GIS that predicts the runout path of debris flows that can be used for identifying hazard zones (Iverson *et al.*, 1998). Mount Rainier in Washington State is an active volcano that is currently at rest between eruptions. The USGS has mapped the likely debris-flow pathways on which several communities have been built (Figure 12.7) and has joined with local, county and state agencies to develop a Mount Rainier hazards plan that will address such issues as emergency response operations and strategies for expanded public awareness and mitigation.

## 12.7  Landslide Educational Efforts of the USGS – Current and Future Roles

Currently, educational efforts related to landslides are carried out mainly through the dissemination of USGS publications, exhibits at scientific or local public interest meetings and other public venues, and by providing expertise in areas such as landslide monitoring and research methodology. The LHP supports the NLIC as an advocate for public education to provide education through publication of basic information in the form of short fact sheets, website information and notifications, an interactive bibliographic online database, and by supplying personnel for primary and secondary school presentations and exhibits. The NLIC lists a toll-free phone number for those wanting landslide information and education materials, and assists in wide-application projects such as handbooks on landslide information and mitigation. A handbook for planners is currently being prepared by the USGS and the American Planning Association.

Future efforts to develop information and education programmes for the user community will continue to be led by the Federal Emergency Management Agency (FEMA)

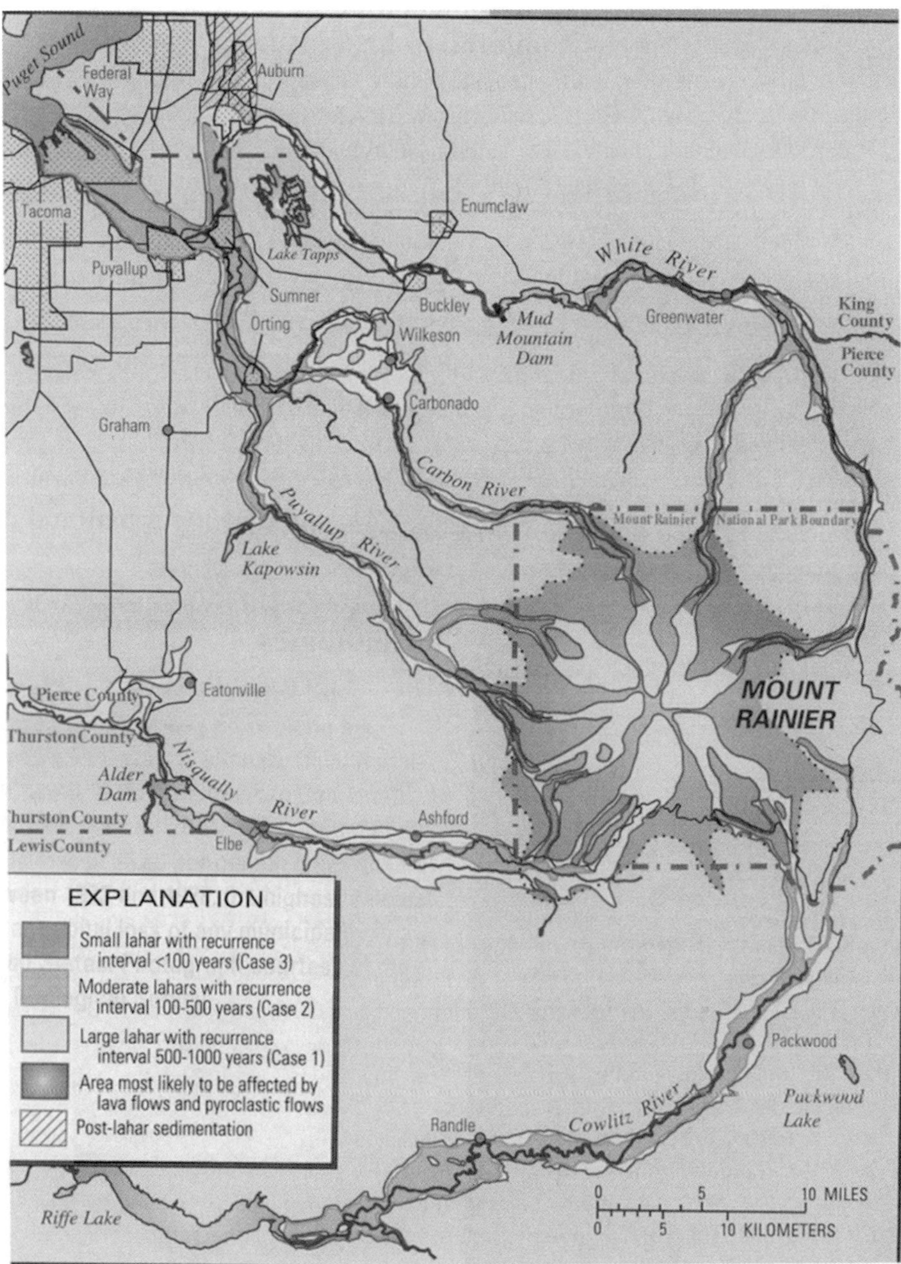


**Figure 12.7**   *Hazard zones from lahars (mudflows and debris flows), lava flows and pyroclastic flows from Mount Rainier, Washington (Scott et al., 1998)*

and the USGS. Before individuals and communities can reduce their risk from landslide hazards, they need to know the nature of the threat, its potential impact on them and their community, their options for reducing the risk or impact and methods for carrying out specific mitigations measures. Achieving widespread public awareness of landslide hazards will enable communities and individuals to make informed decisions on where to live, purchase property, or locate a business. Local decision makers will know where to permit construction of residences, business, and critical public infrastructure to reduce potential damage from landslide hazards. The following actions raise public awareness of landslide hazards and encourage landslide hazard preparedness and mitigation activities nationwide, tailored to local needs:

1. Develop public awareness, training and education programmes involving land use planning, design, landslide hazard curriculums, landslide hazard safety programmes and community risk reduction.
2. Evaluate the effectiveness of different methods, messages and curriculums in the context of local needs.
3. Disseminate landslide-hazard-related curriculums and training modules to community organizations, universities, and professional societies and associations.

## 12.8   Future USGS Efforts for Reduction of Landslide Hazard and Risk

With population increasing and development expanding onto land that is more susceptible to landslides, landslide hazards and resultant losses will increase in the United States unless and until the government adopts a comprehensive strategy to mitigate landslide hazards at the federal, state, local and private level. Today no such strategy exists. States, local governments, transportation departments and numerous federal agencies, including the USGS, handle landslide hazards almost independently of each other. In 1999, the US Congress, concerned over the lack of a comprehensive landslide hazard strategy, directed the USGS to develop a plan to address the widespread landslide hazards facing the nation. The USGS was asked to prepare a strategy that would involve all the parties that have responsibility for dealing with landslides (see Public Law 106–113). As a result the USGS prepared a report titled 'National landslide hazards mitigation strategy: A framework for loss reduction' (Spiker and Gori, 2003). The strategy outlined in the report (hereafter referred to as the 'Strategy') was built on the premise that no single agency, level of government or programme can independently reduce losses from landslide hazards. The Strategy outlines a new public–private partnership that encourages the use of scientific information, maps and monitoring in emergency management, land use planning, and public and private policy decisions to reduce losses from landslides. Drawing on 25 years of experience and suggestions of scientists, public officials and professionals, the Strategy proposes a major, long-term effort and a commitment of all levels of government and the private sector to reduce losses from landslide hazards in the United States. The Strategy calls on the federal government, in partnership with state and local governments, to provide leadership, coordination, research support and incentives in the areas of landslide hazard mitigation. The objective is to encourage communities, businesses and individuals to undertake mitigation measures to minimize potential losses before landslide events and to employ mitigation measures in recovering from landslides.

The primary goal of the Strategy is to reduce the number of deaths, injuries and economic costs caused by landslides over the next 10 years. To accomplish this, the Strategy proposes the following nine major elements (Spiker and Gori, 2003: 2), spanning a continuum that ranges from research to the formulation and implementation of policy and mitigation. These elements are:

1. Research – Developing a predictive understanding of landslide processes and triggering mechanisms,
2. Hazard Mapping and Assessments – Delineating susceptible areas and different types of landslide hazards at a scale useful for planning and decision-making,
3. Real-Time Monitoring – Monitoring active landslides that pose substantial risk,
4. Loss Assessment – Compiling and evaluating information on the economic impacts of landslide hazards,
5. Collection, Interpretation, and Dissemination of Information – Establishing an effective system for information transfer,
6. Guidelines and training – Developing guidelines and training for scientists, engineers, decision-makers and other professionals.
7. Public Awareness and Education – Developing information and education for the user community and the public,
8. Implementation of Loss Reduction Measures – Encouraging mitigation actions, and
9. Emergency Preparedness, Response, and Recovery – Building resilient communities.

The USGS has an important role in each of the above nine elements as a provider of landslide hazard information; however, the lead and key participants in each element differ with the nature of that element.

Implementation of the Strategy will require increased funding, better coordination among levels of government, and new partnerships between government, academia and the private sector. This cooperation will encourage innovative programmes and incentives for hazard mapping and assessment, adoption of loss reduction measures, and implementation of new technology. Specifically, the Strategy proposes:

- expansion of USGS landslide research, mapping, monitoring, and emergency-response activities;
- formation of a new state cooperative landslide programme to map and assess landslides;
- formation by federal agencies of a new cooperative landslide programme to map and assess landslides on federal lands;
- formation of a national grants programme to encourage research and implementation efforts by universities, local governments and the private sector;
- formation of an inter-agency working group to provide guidance and coordination.

The USGS is currently working with state geological surveys and scientific and professional societies to encourage implementation of the Strategy, which to date have not yet been funded by the US Congress. The Landslide Hazards Program has begun to implement certain aspects of the Strategy with available funds. Major thrusts of scientific research and application or research that the LHP is pursuing include:

- Encouraging scientists to produce probabilistic landslide hazard maps that can be used for long-term planning and decision making. These maps can also be used by jurisdictions and private industry as a basis for performing future risk analyses, a long-recognized goal of the USGS (Highland, 1997);
- Encouraging scientists to formulate dynamic landslide hazard models that combine probabilistic maps with rainfall forecasts and geological and geographical information. To accomplish this, scientists are using remote sensing and high-quality weather-related information in the production of these maps and models;
- Refining landslide-warning systems at the regional level (more than one metropolitan area and state);
- Encouraging the compilation of landslide inventories and economic damage at local, state and regional jurisdictions in order to better document the occurrence of landslides and the dollar losses they cause;
- Educating public officials, the general public, and other scientists, engineers and land use planners who use landslide hazard information is an essential component of the LHP;
- Making databases, educational materials, and maps and reports available via the Internet and through the National Landslide Information Center.

The LHP has two projects that represent the initial implementation of the Strategy, albeit on a very small scale. These are joint projects with (1) the American Planning Association, with a membership of more than 30 000 land use planners and planning officials, to produce a guidebook on landslides and land use planning; and (2) the American Association of State Geologists to study the feasibility of collecting information on losses from landslides. Both of these projects should raise the appreciation of landslide hazards by groups and individuals who will work towards mitigating the effects of landslide hazards in the USA. Both of these projects are in progress.

Prospects for improving the scientific understanding of landslides and the ability to reduce the consequences of their effects are promising. Continuing advances in science and technology make it increasingly more feasible to forecast and mitigate landslide hazards and save lives and property. To accomplish this, however, will take a concerted effort by many agencies in the United States and partnerships between scientific agencies and different levels of government and the public and private sector.

## References

Bell, W.J., Jr, 1996, The Carlisle Deluge, 1779, *Cumberland County History*, **13**(1), 30–35.

Bernknopf, R.L., Campbell, R.H., Brookshire, D.S. and Shapiro, C.D., 1988, A probabilistic approach to landslide hazard mapping in Cincinnati, Ohio, with applications for economic evaluation, *Bulletin of the Association of Engineering Geologists*, **25**(1), 39–56.

Brabb, E.E., 1995, The San Mateo County, California GIS Project for predicting the consequences of hazardous geologic processes, in A. Carrara and F. Guzzetti (eds), *Geographical Information Systems in Assessing Natural Hazards* (Dordrecht: Kluwer Academic Publishers), 299–334.

Brabb, E.E. and Pampeyan, E.H., 1972, *Preliminary Map of Landslide Deposits in San Mateo County, California*, US Geological Survey Miscellaneous Field Studies Map MF-344, scale 1:62 500.

Brabb, E.E., Pampeyan, E.H. and Bonilla, M.G., 1972, *Landslide Susceptibility in San Mateo County, California*, US Geological Survey, Miscellaneous Field Studies Map MF-360, scale 1:62 500.

Brabb, E.E., Howell, D.G. and Cotton, W.R., 2002, A thumbnail sketch of what governments around the world are doing to reduce the consequences of landslides, in *Proceedings of the Tenth International Conference and Fieldtrip on Landslides (ICFL)/Polish Lowlands–Carpathians– Baltic Coast*, Poland, 6–16 September 2002, 15–33.

Bucknam, R.C., Coe, J.A., Chavarria, M.M., Godt, J.W., Tarr, A.C., Lee-Ann, B., Sharon, R., Dean, H., Dart, R.L. and Johnson, M.L., 2001, *Landslides Triggered by Hurricane Mitch in Guatemala Inventory and Discussion*, US Geological Survey Open-File Report 01–443.

Campbell, R.H., Bernknopf, R.L. and Soller, D.R., 1998, *Mapping Time-Dependent Changes in Soil-Slip–Debris-Flow Probability*, US Geological Survey, Geologic Investigation Series Map I-2586, scales 1:24 000 and 1:40 000.

Cannon, S.H. and Ellen, S., 1985, Rainfall conditions for abundant debris avalanches in the San Francisco Bay region, California, *California Geology*, **38**(12), 267–272.

Cannon, S.H., Haller, K.M., Ingrid, E., Schweig, E.S., Graziella, D., Moore, D.W., Rafferty, S.A. and Tarr, A.C., 2001, *Landslide Response to Hurricane Mitch Rainfall in Seven Study Areas in Nicaragua*, US Geological Survey Open-File Report 01–412-A.

Cannon, S.H., Gartner, J.E., Rupert, M.G., Djokic, D. and Sreedhar, S., 2003, *Emergency Assessment of Debris-flow Hazards from Basins Burned by the Grand Prix and Old Fires of 2003, Southern California*, US Geological Survey Open-File Report 03–475, http://pubs.usgs.gov/of/2003/ofr-03-475/.

Chleborad, A.F., 2000, *Preliminary Method for Anticipating the Occurrence of Precipitation-Induced Landslides in Seattle, Washington*, US Geological Survey Open-File Report 00–0469, http://pubs.usgs.gov/of/2000/ofr-00-0469/.

Chleborad, A.F., 2003, *A Preliminary Evaluation of a Precipitation Threshold for Anticipating the Occurrence of Landslides in the Seattle, Washington, Area*, US Geological Survey-Open File Report 03–463.

Coe, J.A. and Godt, J.W., 2001, Debris flows triggered by the El Niño rainstorm of February 2–3, 1998, Walpert Ridge and vicinity, Alameda County, California, US Geological Survey Miscellaneous Field Studies Map MF-2384, scale 1:24 000, http://pubs.usgs.gov/mf/2002/mf-2384/.

Coe, J.A., Michael, J.A., Crovelli, R.A. and Savage, W.Z., 2000, *Preliminary Map Showing Landslide Densities, Mean Recurrence Intervals, and Exceedance Probabilities as Determined from Historic Records*, Seattle, Washington, US Geological Survey Open-File Report 00–303, scale 1:25 000.

Coe, J.A., Michael, J.A., Crovelli, R.A., Savage, W.Z., Laprade, W.T. and Nashem, W.D., 2004, Probabilistic assesment of precipitation-triggered landslides using historical records of landslide occurrance, Scattle, Washington, *Environmental & Engineering Geoscience*, **X**(2), 103–122.

Colton, R.B., Holligan, J.A., Anderson, L.W. and Shaver, K.C., 1975, *Preliminary Map of Landslide Deposits, Moab 1° × 2° Quadrangle, Colorado and Utah*, US Geological Survey, Miscellaneous Field Studies Map MF-698, scale 1:250 000.

Crone, A.J., Baum, R.L., Lidke, D.J., Sather, D., Lee-Ann, B. and Tarr, A.C., 2001, *Landslides Induced by Hurricane Mitch in El Salvador – An Inventory and Descriptions of Selected Features*, US Geological Survey Open-File Report 01–144.

Croveli, R.A., 2000, *Probability Models for Estimation of Number and Costs of Landslides*, US Geological Survey Open-File Report 00–249.

Cruden, D.M. and Varnes, D.J., 1996, Landslide types and processes, in A.K. Turner and R.L. Schuster (eds), *Landslides: Investigation and Mitigation*, Transportation Research Board, National Research Council Special Report 247 (Washington, DC: National Academy Press), 36–75.

Delano, H.L. and Potter, N., Jr, 1997, The Carlisle Deluge, 1779 revisited, *Cumberland County History*, **14**(1), 68–70.

Delano, H., Hartman, D. and Potter, N., 2001, Stop 6. The 1779 Carlisle deluge debris flow, in N. Potter Jr (ed.), *The Geomorphic Evolution of the Great Valley Near Carlisle, Pennsylvannia*, Southeast Friends of the Pleistocene 2001 Annual Meeting 26–28 October 2001, 94–101.

Ellen, S.D. and Wieczorek, G.F. (eds), 1988, *Landslides, Floods, and Marine Effects of the Storm of January 3–5, 1982, in the San Francisco Bay region, California*, US Geological Survey Professional Paper 1434.

Evans, S.G. and Hungr, O., 1993, The assessment of rockfall hazard at the base of talus slopes, *Canadian Geotechnical Journal*, **30**, 620–636.

Fleming, R.W., Johnson, R.B. and Schuster, R.L., 1988a, The reactivation of the Manti landslide, Utah, in *The Manti, Utah, Landslide*, US Geological Survey Professional Paper 1311, 1–22.

Fleming, R.W., Schuster, R.L. and Johnson, R.B., 1988b, Physical properties and mode of failure of the Manti landslide, Utah, in *The Manti, Utah, Landslide*, US Geological Survey Professional Paper 1311, 23–42.

Godt, J.W. (ed.), 1999, *Maps Showing Locations of Damaging Landslides Caused by El Niño Rainstorms, Winter Season 1997–98, San Francisco Bay Region, California*, pamphlet to accompany US Geological Survey Miscellaneous Field Studies Maps MF-2325-A-J, scale 1:125 000.

Godt, J.W. and Coe, J.A., 2003, *Map Showing Alpine Debris Flows Triggered by a July 28, 1999 Thunderstorm in the Central Front Range of Colorado*, US Geological Survey Open File Report 03–050, scale 1:24 000, http://pubs.usgs.gov/of/2003/ofr-03-050/.

Godt, J.W., Highland, L.M. and Savage, W.Z., 1997, *El Niño and the National Landslide Hazard Outlook for 1997–1998*, US Geological Survey Fact Sheet 180–197.

Guzzetti, F., Crosta, G., Detti, R. and Agliardi, F., 2002, STONE, a computer program for the three-dimensional simulation of rock-falls, *Computers and Geosciences*, **28**(9), 1079–1093.

Guzzetti, F., Reichenbach, P. and Wieczorek, G.F., 2003, Rockfall hazard and risk assessment in the Yosemite Valley, California, USA, *Natural Hazards and Earth System Sciences*, **3**, 491–503.

Harp, E.L. and Noble, M.A., 1993, An engineering rock classification to evaluate seismic rock-fall susceptibility and its application to the Wasatch Front, *Association of Engineering Geologists Bulletin*, **30**, 293–319.

Harp, E.L., Wilson, R.C. and Wieczorek, G.F., 1981, *Landslides from the February 4, 1976, Guatemala earthquake*, US Geological Survey Professional Paper 1204-A.

Harp, E.L., Hagman, K.W., Held, M.D. and McKenna, J.P., 2002a, *Digital Inventory of Landslides and Related Deposits in Honduras Triggered by Hurricane Mitch*, US Geological Survey Open-File Report 02–61.

Harp, E.L., Mario, C. and Held, M.D., 2002b, *Landslides Triggered by Hurricane Mitch in Tegucigalpa, Honduras*, US Geological Survey Open-File Report 02–33.

Heim, A., 1932, Bertsturz und Menschenleben, *Beiblatt zur Vierteljahrsschrift der Naturforschenden Gesellschaft in Zurich*, **77**, 1–217. Translated by N. Skermer under the title *Landslides and Human Lives* (Vancouver, BC: BiTech Publishers, 1989).

Highland, L.M., 1997, Landslide hazard and risk: current and future directions for the United States Geological Survey's Landslide Program in D.M. Cruden and R. Fell (eds), *Proceedings of the International Workshop on Landslide Risk Assessment, Honolulu, Hawaii, 19–21 February 1997* (Rotterdam: A.A. Balkema), 207–214.

Howe, E., 1909, *Landslides in the San Juan Mountains, Colorado*, US Geological Survey Professional Paper 67.

Hutchinson, J.N., 1968, Mass Movement, in R.W. Fairbridge (ed.), *Encyclopedia of Geomorphology* (New York: Reinhold), 688–695.

Iverson, R.M., Schilling, S.P. and Vallance, J.W., 1998, Objective delineation of lahar-inundation hazard zones, *Geological Society of America Bulletin*, **110**(8), 972–984.

Jacobson, R.B. (ed.), 1993, Geomorphic studies of the storm and flood of November 3–5, 1985, in the upper Potomac and Cheat River Basins in West Virginia and Virginia, *US Geological Survey Bulletin*, 1981.

Jibson, R.W., 1989, Debris flows in southern Puerto Rico, in A.P. Schultz and R.W. Jibson (eds), *Landslide Processes of the Eastern United States and Puerto Rico*, Geological Society of America Special Paper 236, 29–55.

Jibson, R.W., Harp, E.L. and Michael, J.A., 1998, *A Method for Producing Digital Probabilistic Seismic Landslide Hazard Maps: An Example from the Los Angeles, California, Area*, US Geological Survey Open-File Report 98-113, scale 1:24 000.

Jones, F.O., 1973, *Landslides of Rio de Janeiro and the Serra das Araras Escarpment, Brazil*, US Geological Survey Professional Paper 697.

Jones, F.O., Embody, D.R. and Peterson, W.L., 1961, *Landslides Along the Columbia River Valley, Northeastern Washington*, US Geological Survey Professional Paper 367.

Keaton, J.R. and DeGraff, J.V., 1996, Surface observation and geologic mapping, in A.K. Turner and R.L. Schuster (eds), *Landslides: Investigation and Mitigation*, Transportation Research Board, National Research Council, Special Report 247 (Washington, DC: National Academy Press), 178–230.

Keaton, J.R. and Rinne, R., 2002, Engineering-geology mapping of slopes and landslides, in P.T. Bobrowski (ed.), *Geoenvironmental Mapping – Methods, Theory and Practice* (Lisse: Balkema), 9–27.

Keefer, D.K., 1984, Landslides caused by earthquakes, *Geological Society of America Bulletin*, **95**, 406–421.

Keefer, D.K. and Johnson, A.M., 1983, *Earth Flows: Morphology, Mobilization, and Movement*, US Geological Survey Professional Paper 1264.

Keefer, D.K. and Manson, M.W., 1998, Regional distribution and characteristics of landslides generated by the earthquake, in D.K. Keefer (ed.), The October 17, 1989, Loma Prieta, California, earthquake: landslides and stream channel change, in *Strong Ground Motion and Ground Failure*, vol. 2 of *The Loma Prieta Earthquake of October 17, 1989*, US Geological Survey Professional Paper 1551-C, C7–C32.

Keefer, D.K., Wilson, R.C., Mark, R.K., Brabb, E.E., Brown, W.M. III, Ellen, S.D., Harp, E.L., Wieczorek, G.F., Alger, C.S. and Zatkin, R.S., 1987, Real-time landslide warning during heavy rainfall, *Science*, **238**, 921–925.

Larsen, M.C. and Simon, A., 1993, A rainfall-intensity-duration threshold for landslides in a humid–tropical environment, Puerto Rico, *Geografiska Annaler*, **75A**(1–2), 13–23.

Madole, R.F., 1982, *Surficial Geologic Map of the Craig 1/2° × 1° quadrangle, Moffat and Routt counties, Colorado*, US Geological Survey, Miscellaneous Investigation Series Map I-1346, scale 1:100 000.

Mark, R.K., 1992, *Map of Debris-flow Probability, San Mateo County*, California, US Geological Survey Geologic Investigations Series Map I-1257-M, scale 1:62 500.

McCalpin, J., 1984, Preliminary age classification of landslides for inventory mapping, in *Proceedings of the 21st Engineering Geology and Soils Engineering Symposium*, University of Idaho, Moscow, 99–120.

McCrink, T.P., 2001, Regional earthquake-induced landslide mapping using Newmark displacement criteria, San Cruz County, California, in H. Ferriz and R. Anderson (eds), Engineering Geology Practice in northern California, Association of Engineering Geologists Special Publication 12, *California Geological Survey Bulletin*, **210**, 77–93.

McGill, J.T., 1973, *Map Showing Landslides in the Pacific Palisades Area, City of Los Angeles, California*, US Geological Survey Miscellaneous Field Studies Map MF-471, scale 1:4800.

Morgan, B.A., Wieczorek, G.F. and Campbell, R.H., 1999, *Historical and Potential Debris-flow Hazard Map of Area Affected by the June 27, 1995, Storm in Madison County, Virginia*, US Geological Survey Geologic Investigations Series Map I-2623B, scale 1:24 000.

Morton, D.M., Alvarez, R.M. and Campbell, R.H., 2003, *Preliminary Soil-Slip Susceptibility Maps, Southwestern California*, US Geological Survey Open-File Report 03-17, scale 1:100 000.

Morrissey, M.M., Wieczorek, G.F. and Morgan, B.A., 2001, *Regional Application of a Transient Hazard Model for Predicting Initiation of Debris Flows in Madison County, Virginia*, US Geological Survey Open File Report 01-481.

National Park Service, 2000, Yosemite Valley Geologic Hazard Guidelines, in *Final Yosemite Valley Plan Supplemental Environmental Impact Statement*, November 2000, CD.

Newmark, N.M., 1965, Effects of earthquakes on dams and embankments, *Géotechnique*, **15**(2), 139–160.

Nilsen, T.H., 1971, *Preliminary Photointerpretation Map of Landslide and Other Surficial Deposits of Parts of the Mount Diablo Area, Contra Costa and Alameda counties, California*, US Geological Survey Miscellaneous Field Studies Map MF-310, scale 1:62 500.

Perkins, J.B., 1998, *On Shaky Ground – Supplement – A Guide to Assessing Impacts of Future Earthquakes Using Ground Shaking Hazard Maps for the San Francisco Bay Area* (Oakland, CA: Association of Bay Area Governments).

Perkins, J.B., Chuaqui, B. and Wyatt, E., 1997, *Riding Out Future Quakes – Pre-Earthquake Planning for Post-Earthquake Transportation System Recovery in the San Francisco Bay Region* (Oakland, CA: Association of Bay Area Governments).

Pike, R.J., Graymer, R.W., Sebastian, R., Kalman, N.B. and Steven, S., 2001, *Map and Map Database of Susceptibility to Slope Failure by Sliding and Earthflow in the Oakland Area, California*, US Geological Survey Miscellaneous Field Studies Map MF-2385, scale 1:50 000.

Plafker, G., Ericksen, G.E. and Fernández Concha, J., 1971, Geological aspects of the May 31, 1970, Peru earthquake, *Seismological Society of America Bulletin*, **61**(3), 543–578.

Pomeroy, J.S., 1977, *Preliminary Reconnaissance Map Showing Landslides in Butler County, Pennsylvania*, US Geological Survey Open-File Report 77-0246, scale 1:50 000.

Pomeroy, J.S., 1979, *Map Showing Landslides and Areas Most Susceptible to Sliding in Beaver County, Pennsylvania*, US Geological Survey, Miscellaneous Investigations Series Map I-1160, scale 1:50 000.

Pomeroy, J.S. and Davies, W.E., 1975, *Map of Susceptibility to Landsliding, Allegheny County, Pennsylvania*, US Geological Survey, Miscellaneous Field Studies Map MF-685B, scale 1:50 000.

Radbruch-Hall, D.H., Colton, R.B., Davies, W.E., Ivo, L., Skipp, B.A. and Varnes, D.J., 1983, Landslide Overview Map of the Conterminous United States, US Geological Survey Professional Paper 1183, scale 1:7 500 000. Reproduced as J.A. Godt, 1997, USGS Open-File Report 97-289 (http://landslides.usgs.gov/html_files/landslides/nationalmap/national.html).

Reid, M.E., LaHusen, R.G. and Ellis, W.L., 1999, *Real-time Monitoring of Active Landsides*, US Geological Survey Fact Sheet 091-99.

Schuster, R.L., 1996, Socioeconomic significance of landslides, in A.K. Turner and R.L. Schuster (eds), *Landslides: Investigation and Mitigation*, Transportation Research Board, National Research Council, Special Report 247 (Washington, DC: National Academy Press), 12–35.

Schuster, R.L. and Highland, L.M., 2001, *Socioeconomic and Environmental Impacts of Landslides in the Western Hemisphere*, US Geological Survey Open File Report 01-0276, http://pubs.usgs.gov/of/2001/ofr-01-0276/.

Scott, K.M., Wolfe, E.W. and Driedger, C.L., 1998, *Mount Rainier; Living with Perilous Beauty*, US Geological Survey Fact Sheet FS-065-97.

Spiker, E.C. and Gori, P.L., 2003, *National Landslide Hazards Mitigation Strategy: A Framework for Loss Reduction*, US Geological Survey Circular 1244.

Varnes, D.J., 1958, Landslide types and processes. in E.B. Eckel (ed.), *Special Report 29: Landslides and Engineering Practice* (Washington, DC: HRB, National Research Council), 20–47.

Varnes, D.J., 1978, Slope movement types and processes, in R.L. Schuster and R.J. Krizek (eds), *Landslides Analysis and Control*, Transportation Research Board, National Academy of Sciences, Special Report 176 (Washington, DC: National Academy Press), 12–33.

Wieczorek, G.F., 1984, Preparing a detailed landslide-inventory map for hazard evaluation and reduction, *Association of Engineering Geologists Bulletin*, **21**(3), 337–342.

Wieczorek, G.F., 2002, Catastrophic rockfalls and rockslides in the Sierra Nevada, USA, in S.G. Evans and J.V. DeGraff (eds), *Catastrophic Landslides: Effects, Occurrence, and Mechanisms* (Boulder, CO: Geological Society of America Reviews in Engineering Geology, vol. XV), 165–190.

Wieczorek, G.F. and Sarmiento, J., 1983, Significance of storm intensity-duration for triggering debris flows near La Honda, California, *Geological Society of America Abstracts with Programs*, **15**(5), 289.

Wieczorek, G.F. and Snyder, J.B., 2004, *Historical Rock falls in Yosemite National Park*, US Geological Survey Open-File Report 03-491, http://pubs.usgs.gov/of/2003/of03-491/.

Wieczorek, G.F., Wilson, R.C. and Harp, E.L., 1985, *Map Showing Slope Stability During Earthquakes in San Mateo County, California*, US Geological Survey Miscellaneous Investigation Map I-1257E, scale 1:62 500.

Wieczorek, G.F., Snyder, J.B., Alger, C.S. and Isaacson, K.A., 1992, *Rock Falls in Yosemite Valley, California*, US Geological Survey Open-File Report 92-387.

Wieczorek, G.F., Nishenko, S.P. and Varnes, D.J., 1995, Analysis of rock falls in the Yosemite Valley, California, in J.J. Daemen and R.A. Schultz (eds), *Proceedings of 35th U.S. Symposium on Rock Mechanics* (Rotterdam: A.A. Balkema), 85–89.

Wieczorek, G.F., Morrissey, M.M., Iovine, G. and Godt, J., 1999, *Rock-fall Potential in the Yosemite Valley*, California, US Geological Survey Open-File Report 99-578, scale 1:12 000, http://greenwood.cr.usgs.gov/pub/open-file-reports/ofr-99-0578/.

Wieczorek, G.F., Morgan, B.A. and Campbell, R.H., 2000, Debris-flow hazards in the Blue Ridge of central Virginia, *Environmental & Engineering Geoscience*, **VI**(1), 3–23.

Wieczorek, G.F., Larsen, M.C., Eaton, L.S., Morgan, B.A. and Blair, J.L., 2002, Debris-flow and flooding deposits in coastal Venezuela associated with the storm of December 14–16, 1999, US Geological Survey, Geologic Investigation Series Map I-2772, scales 1:6000 and 1:25 000. http://pubs.usgs.gov/imap/i-2772/.

Wilson, R.C. and Keefer, D.K., 1983, Dynamic analysis of a slope failure from the 1979 Coyote Lake, California, earthquake, *Bulletin of the Seismological Society of America*, **73**, 863–877.

Wilson, R.C., Wieczorek, G.F., Keefer, D.K., Harp, E.L. and Tannaci, N.E., 1985, *Ground Failures from the Greenville/Mt. Diablo Earthquake Sequence of January 1980, Northern California*, US Geological Survey Miscellaneous Field Studies Map MF-1711, scale 1:125 000.

Wilson, R.C., Torikai, J.D. and Ellen, S.D., 1992, *Development of Rainfall Warning Thresholds for Debris Flows in the Honolulu District, Oahu*, US Geological Survey Open-File Report 92-521.

Wilson, R.C., Mark, R.K. and Barbato, G.E., 1993, Operation of a real-time warning system for debris flows in the San Francisco Bay area, California, in H.W. Shen, S.T. Su and F. Wen (eds), *Hydraulic Engineering '93, Proceedings of the 1993 Conference*, Hydraulics Division, American Society of Civil Engineers (San Francisco, CA: ASCE), 25–30 July 1993, vol. 2, 1908–1913.

# 13

# Basic Data and Decision Support for Landslide Management: A Conceptual Framework

*Walter Pflügner*

## 13.1 Introduction

This contribution endeavours to present and discuss a conceptual framework – in terms of damage potentials – for preparing decision support for the efficient management of major landslide hazards. Major landslide hazards are distinguished from small-scale hazards. The latter may damage part of a road, a railway line, a single house or production plant (Figures 13.1a–d). Preventive measures and damage repair involve mostly structural engineering work; problem solving frequently does not include strategic planning (as there are not too many feasible options) or comprehensive decision making (as resources needed can be estimated and checked against damage).

The problem setting differs depending on the type of land use activity under threat: housing area, business area, recreation park, agricultural land of significant size and value are under threat; it becomes complex if a land use mix is involved. The more different types of land use affected and the larger the damage potential, the more time-consuming and costly the analysis of the threat. In order to reduce complexity, solve data problems, implement efficient data processing and so on, specific tools and methods are needed to manage the task. Experience with natural hazard studies has shown that these needs become obvious even if only some 50 or 100 objects are involved, or as little as $100\,000\,\text{m}^2$ of land. Problem settings of this size or greater are defined as 'major' landslide hazards throughout this chapter, but also involve large-scale regions

*Landslide Hazard and Risk*   Edited by Thomas Glade, Malcolm Anderson and Michael J. Crozier
© 2004 John Wiley & Sons, Ltd   ISBN: 0-471-48663-9

**Figure 13.1**  Examples of small-scale landslide damage ([a, b] Hamilton County; [c] Holt/ stockpix, 2001; [d] BWG, 2002)

(c)

(d)

**Figure 13.1** (Continued)

where 'mega-events' occur frequently and are documented in insurance industry statistics (Munich Re, 2002). Methods and tools depend on the size of the problem, but should be selected and implemented according to an underlying conceptual framework built on the following reasoning.

Major landslide hazards need some sort of landslide hazard policy (LHP) and management (LHM). A management process involves not only detailed planning of mitigation (prevention and/or reduction) measures, but also planning for their implementation, control of effectiveness, as well as control of residual risks. This has to be achieved by continuous revolving planning, using feedbacks from effectiveness measurement, landslide damage documentation, as well as residual risks assessment. Preconditions for efficiently managing landslide hazards are, in principle, the two stages of pre-planning and planning (Figure 13.2).

The conceptual framework presented below belongs primarily to the pre-planning stage of a landslide hazard policy. Pre-planning means that, once landslide hazards have been identified, they need to be assessed thoroughly. Assessment should focus on landslide damage potentials. Definition should correspond to the insurance industry's term of 'maximum probable loss'. However, assessment should go far beyond monetary terms only, and consider all the socio-economic values under threat.

LANDSLIDE HAZARD POLICY

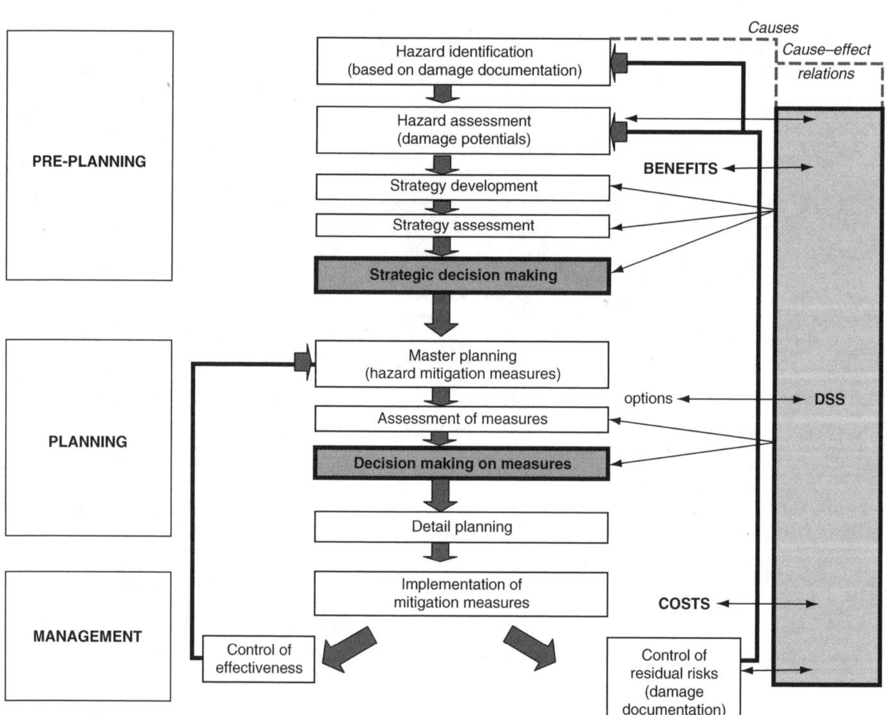


***Figure 13.2***   *A scheme for landslide hazard policy*

Control of damage potentials in time, especially their development due to (i) socio-economic growth and (ii) implementation of mitigation measures, may then not only serve pre-planning, but also support decision making and planning activities at all stages.

The conceptual framework for defining and analysing landslide damage potentials presented below is derived from another field of managing natural hazards. It was developed for flood control as recently as the last decade (Klaus and Schmidtke, 1990). In that time it has been thoroughly tested, optimized, and successfully applied to a dozen flood basins in Germany (Pflügner, 2001). During the last 2 years, the approach has been applied to a landslide hazard area for the first time (Glade and von Davertzhofen, 2004). Experience gathered so far suggests that transfer to this and other natural hazards may prove a useful tool for coping with the challenges at hand.

## 13.2   Problem Setting

Each year major events around the globe as well as the financial balances of the insurance industry demonstrate that landslides are among the most important natural risks to population and economic assets. Year by year many lives are lost and billions of euro worth of damage occur (Munich Re, 2002). They show that decision making concerning the protection and/or development of endangered areas calls for truly comprehensive decision support. This involves taking into account all the economic and other impacts as well as their distribution patterns in terms of benefits and costs, and other implications for social well-being (Eurostat, 1996). Clearly, decision support must explain:

- the cause;
- cause–effect relationships;
- options, with their technical, administrative etc. conditions, and effectiveness; as well as their
- benefits and costs.

Decision making in the field of landslide hazard policy consequently has to deal with all the technical, economic as well as social efficiency aspects of policy options. Dealing with this information basically demands expert evaluation in order to prepare decision support.

To evaluate adequately different policy options in terms of these aspects, a great deal of information is needed, based on different types of basic data. The conversion of basic data into information for decision making is one of the most important aspects of integrated decision support systems (DSS). The value of decision support information content mostly depends on:

(1) data availability (i.e. input);
(2) model availability and features (i.e. processing); and
(3) modelling of DSS features (i.e. output).

The model output has to show the area at risk in sufficient spatial and sector disaggregation so that at least the following questions can be answered:

- Which impacts will result from the cause (= landslide)?
- What is affected?

- Where?
- What is the intensity of the impacts?
- What are the primary effects?
- Which secondary effects will follow?
- To whom do the impacts accrue?

Output user groups, their differing information needs, and usage of the DSS in real-world information systems and decision environments will be discussed in more detail later.

The more land use affected and the larger the damage potential, the more time-consuming and costly the analysis of what is at stake!

In the past, analysis of damage potentials has been limited by application of only a very few methods. Literature research worldwide (Klaus *et al.*, 1994b) has shown that up to the 1980s only (i) 'forecasting' based on analysing earlier events, that is, trying to use empirical averages/extrapolation and/or (ii) 'property-by-property' approaches, that is, evaluating the single objects under threat, have been applied. With major areas under study this approach frequently proved not to be feasible for three main reasons. Principally, these 'micro' approaches such as property-by-property analysis cannot be applied efficiently to larger-scale problems, because:

- Very large numbers of data sets would have to be collected; field surveys are beyond real-world capacities, too time-consuming and too costly.
- Many individuals and businesses nowadays are unwilling to reveal personal and individual data. The same is true for enterprises because of competition risks and so on. For this reason questionnaires on a voluntary basis can only help to gather (small) parts of the data needed. The alternative way of collecting and using the needed individual and site-specific data from public authorities administering these households and business units mostly also fails due to data protection laws, distribution of data among different authorities, incompatibility and incomparability of available data sets and so on.
- Resulting from forward and backward linkages, specifically for businesses, not all, and not necessarily the most important, of the adverse effects can be raised for the single property itself and thus be adequately ascribed to the intra- or extra-regional individuals or economic units affected.

These findings forced the adoption of other approaches. In contrast to micro analysis, with property-by-property approaches the idea was to explore the next level above the single-object layer, namely the regional scale (Table 13.1). The concept therefore was named Regional Scale Analysis (RSA). The research question then became: 'Is it possible to find and process sufficient data sources and data, especially from official statistical systems, to generate the information needed in a feasible and cost-efficient way?'

The core information needed was defined as the damage potential in contrast to single-event damage:

- Damage potential includes, in a narrow sense, (i) all the economic values of all assets (stocks) as well as production outputs (flows) within the impact area. In a broader sense it must also include (ii) all life and limb, all other non-economic assets and so-called 'intangible' assets which cannot be assessed in monetary terms (e.g. loss of

**Table 13.1**   *Scaling levels in natural hazard policy and management*

| Policy/management levels | Planning level | | |
| --- | --- | --- | --- |
| | International/ national | Regional | Local |
| Comprehensive mitigation policies | Macro-analysis (aggregates) | | |
| Large-scale mitigation strategies | | Meso-analysis – RSA – (land use types) | |
| Small-scale planning: single protection measures | | | Micro-analysis property-by-property (objects) |

memorabilia) or should not for moral reasons (e.g. loss of a life). Moreover, (iii) all the other costs of emergency operations, damage repair, damage compensation and so on to society should be included in order not to underestimate the damage potential, as often they constitute a significant multiplier (2 or more) to the values previously mentioned. (Sometimes this damage potential is called 'vulnerability'; however, IPCC has generated a different definition of 'vulnerability'; therefore these terms should not be confused (IPCC, 1992)). 'Damage potential' as described here equates to the total threatened value of the elements at risk. In conventional use 'damage ratio' for a given event equates to 'vulnerability'.

- Damage, in contrast to 'damage potential', is related to a specific, single landslide event. Each single event is characterized by specific event parameters such as debris masses, impact route, velocity, time of occurrence and so on. The probability that two natural events will produce completely the same damage seems, in general, to be very low because of different parameter settings.

Forecasting of a specific event is always difficult; the same is true for event-specific damage. Most adequate handling therefore appears to work with a bandwidth or corridor, and describes minimum probable loss, maximum probable loss, and the difference between them, to be a better reference. Considering only parts of the damage potential may lead to suboptimal planning and decisions.

A further point is that statistically based asset values can, of course, only describe average conditions within the statistical collection unit they stem from. The latter may be defined by political or administrative borders. The impact area under study may be situated within one such border, so that statistical data can describe the assets completely. Otherwise, the impact area may stretch over two, three or more such collection units with differing values, making it necessary to (i) handle the sub-areas differently; or (ii) find out the most relevant averages in order to analyse them comparatively and comparably.

Seen in this way, damage potential includes all the assets under threat in total, regardless of cause–effect differences. For example, two housing objects lying on the impact trail may be completely comparable in terms of present value, household property, inhabitants and so on, but may have different protection: one has a bulwark, the other not. Damage from the specific event may therefore vary significantly, but this need not be specified in the according object data set at the earliest possible time of analysis in order to avoid misinterpretation later on. The damage potential reference ensures that both objects are dealt with; neither has to be sorted out at an early stage of analysis, as data sets may be refined from planning stage to planning stage, so that analysis of completely different options can always start from the same damage potential reference.

## 13.3   History

This concept of damage potential together with the regional scale approach was developed and applied for the first time during the second half of the 1980s in Germany for solving a coastal flood protection project (Klaus and Schmidtke, 1990). It was a long-term research and development (R&D) activity with much trial and error, and carried out without sufficient tools. None the less, in the end it delivered valuable support to the decision processes and received very good marks from all the parties involved.

This was why the approach was selected for further R&D during EUROflood, a Europe-wide cooperation project with partners from eight member states co-sponsored by CEC. Cross-country comparison of planning and decision approaches, application of tools, analysis of DSS in use, and worldwide literature retrieval verified the hypothesis that neither comparable systems nor needed tools existed elsewhere (Klaus *et al.*, 1994b).

This led to the RSA methodology being developed, optimized and implemented within a GIS environment in the context of the EUROflood project in the years 1992–96. Starting from the experience that no two planning environments are directly comparable, a generic tool was the objective – embedded in the GIS environment – which facilitates the analysis and geographical presentation of damage potential, with specific emphasis on the 'interactivity' features of the RSA–DSS model (Klaus *et al.*, 1994a). This resulted in a DSS prototype being built with only modest requirements as to hard and software, and tested with project data from the UK, the Netherlands and Germany (Gewalt *et al.*, 1996).

Since that time the conceptual framework has been applied to approximately a dozen projects. It has been used to optimize one or the other feature, to develop new approaches to data retrieval or to implement different GIS systems according to whatever data were available. Apart from the above-mentioned RSA–DSS prototype, experience has been gained with ArcInfo, ArcView, MapInfo, Intergraph, to name just the most prominent; many other systems exist. Nowadays there is a tool box of specific approaches suited for the analysis of fairly large impact areas as well as quite small ones with case-specific and adequate tools; to deal with all the heterogeneous data formats available; to bring into use the relevant interfaces; to suit the output system to fit to end-user needs (which is now at quite a sophisticated development stage); and so on.

The next two sections deal with basics, background, and RSA methodology in detail first, and subsequently discuss the most important specifications to suit the generic system to a specific application. Section 13.5 will serve to explain its most valuable contributions to integrated policy and management from our point of view today, concluding with an indication of important possible extensions for future application in landslide hazard management.

## 13.4   Conceptual Framework

### 13.4.1   Generic Model and DSS Design Requirements: Approach

Based on the aforementioned fundamentals, RSA–DSS was prototyped, and all the later model developments were organized so to ensure this principle as a generic model, that is, not case- or site-specific but readily applicable to any project region under study. In general terms such a model needs inputs of different categories (see Figure 13.3):

- topographical and physical data (i.e. area size, terrain, elevations etc.);
- land use data (objects/object types, polygons, etc.);
- socio-economic data, especially values of land use (statistical averages or site-specific/type-specific);
- specifics of landslide hazards (scenarios);
- susceptibility of socio-economic values to damage (cause–effect relationships).

Note that specifics of the landslide hazard are: the input needed; type (debris flow, earthslide, rockfall), impact trail, speed (slow, rapid), dimension (mass of soil in m$^3$ expected) and so on. These should be specified in scenarios to show forecast limitations and allow bandwidth analysis.

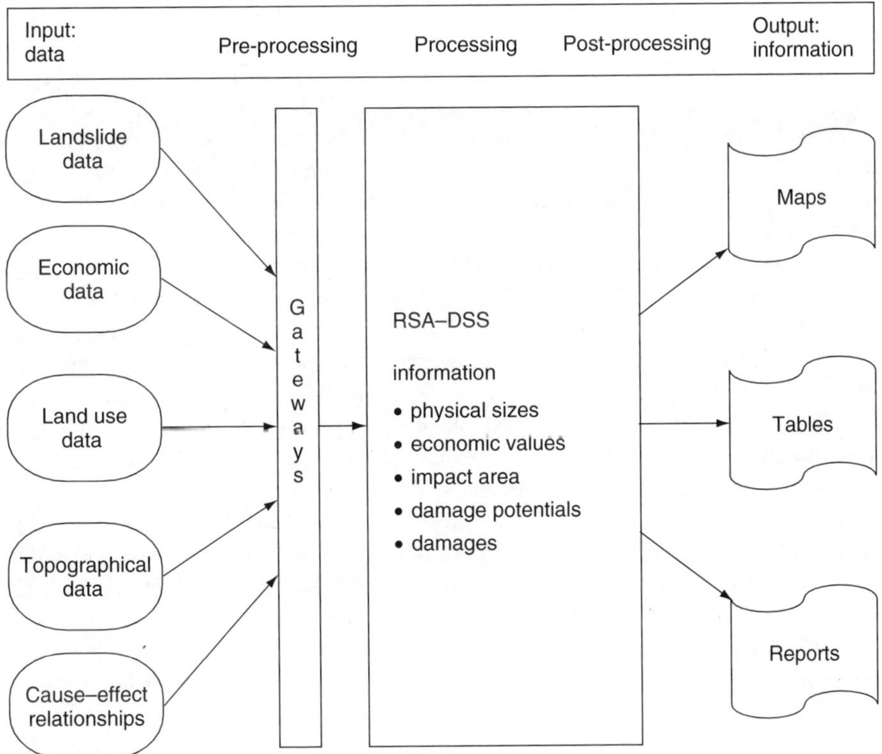



**Figure 13.3**   *A generic model of data flows and data linking*

When input data have been made available, the process is quite simple: a landslide event causes damages if it affects vulnerable land use, for example residential buildings, business facilities, agricultural land, or maybe public gardens with rare species, representing the damage potential.

The main problem in modelling land use on the regional scale consists of finding and processing suitable data sources for (i) the different land uses and (ii) their socio-economic values (where socio-economic means more than pure economic market value). Both should not only be as complete and up to date as possible, but also detailed enough to ensure the estimation of socio-economic damage potentials with a high resolution as to space and land use activities.

The task simply consists of combining (i) land register or other land use data with; (ii) data from socio-economic statistical systems, using such sources as far as possible in order to derive good data and high information content. The better the data in terms of completeness, timeliness and so on, the less work is necessary for quality assurance, for data transformation, for digitizing, and for implementing additional field surveys to fill gaps and so on. To minimize effort spent on data collection, the analysis should make use of existing, official data sources whenever possible, and of digitized data as much as possible in building the application model for the real-world case out of the generic model.

Design of the DSS application model for the real-world case depends on the problem setting under study, that is, the landslide hazard specifics, the endangered area, and so on. At the design stage, budget considerations should not dominate goal setting and specifying of requirements for establishing the real-world DSS, as this may lead to severe shortcomings in terms of limiting its benefits from the outset. This argument can be made with respect to the importance of scaling levels (see Table 13.1).

Top-level policy formulation on how to deal with major societal problems often appears as a medium- or long-term activity, dealing with the problem of assessing the consequences of a do-nothing option in contrast to benefit–cost ratios of action (with and without comparisons). As the problem at hand has been observed for a certain time period, available aggregate information is often regarded as a proper information base for decision making in terms of (i) undertaking problem solving; (ii) setting overall goals; (iii) defining milestones and timeline for goal achievement; and (iv) allocating budgets to deal with this problem. Such decisions in real-world cases are not necessarily linked to or based on adequate insight as to whether the goals are achievable under these constraints or not, and to what extent the resources allocated will suffice to solve the problem. The motto is often 'time will tell', based on hoping that future growth will allow the allocation of additional resources in case the measures prove inadequate.

Lower-level action, dealing with planning and management (see again Figure 13.2), however, should allocate highest importance to integrating all the activities in order to gain from synergy wherever possible and avoid double work. Seen from this viewpoint, DSS architecture cannot aim at a one-time application. Most of the data will be used again and again during the evolving process, and serve different purposes. It would be short-sighted to specify DSS requirements for only strategic-level application, as some months or years later the same data and information generated during this stage will be needed for the master planning and detailed planning levels. Experience gathered so far shows that this is especially true with handling of land use data and with defining the study area adequately.

With regard to the latter, socio-economic analysis often leads to the realization that restricting attention to the impact area itself has constrained the study unduly, because links to neighbouring regions (e.g. economic interdependencies) and impacts on them (e.g. burden of emergency assistance) have not been considered adequately from the beginning.

Regarding land use, it is particularly important to decide from the start what the adequate approach to decision support is. The proper resolution must be promptly defined. This may be best explained by means of a few examples:

First, dealing with project regions throughout Europe, nowadays we can use, readily available for GIS implementation, CORINE satellite data (European Environment Agency, 1985). But as the classification system has been primarily developed for environmental policy purposes, there is only one category describing urban land use: 'built-up areas', that is all dwellings and residential areas, as well as all businesses. Only plots of over $250\,000\,m^2$ are registered. Using these data, as was done for analysing a transnational problem affecting five European member states (IKSR, 2001), sets the condition that additional data are available for describing at least the percentage distribution of residential areas, industrial facilities, commercial centres and so on within such a 'built-up area'. Of course, all these data cannot be transferred to small-scale planning, as this will need far more detailed information at a resolution below the 25 ha scale.

Second, dealing with project regions throughout Germany, land use is defined in a five-level land register system. The area of a single municipality is built up from the three lowest levels of land register classification. Level 1 land use applies to units from 50 to $5000\,m^2$, depending on their location (small in residential areas; large in commercial zones, agricultural areas). Level 2 land use units are aggregates of level 1 units; on average they apply to areas of $50\,000$ to $200\,000\,m^2$. Level 3 land use units then apply to areas of $5-10\,km^2$. The importance for an application is easy to explain: land use within a single rural municipality may be described based on very few, for example three to ten level 3 data sets. If, however, you decide to take level 1, then you have to handle, for example, $25\,000-35\,000$ data sets. (Fortunately, the technical problems and costs following from this decision are far less important today than 5 or 10 years ago. This means that in most cases today best practice is to use the most detailed data available if there is sufficient probability that detailed planning will follow and need 'property-by-property' object data, which is included in level 1.)

Third, as these data sets are available with three different specifications according to (i) land survey purposes; (ii) ownership registration; or (iii) regional planning information content, DSS design considerations should focus more on whether one such specification will suffice all future needs, or if two or all should be input from the very beginning.

Therefore, design must correspond to the problem under study: the bigger the area affected, the more important a comprehensive design allowing for multiple use of data in the course of the study work.

### 13.4.2   Data Linking and Application Data Needed: Methodology

The above remarks also show that handling of land use data is not a fundamental problem. The most important achievement reached with RSA was to develop an adequate methodology for allocating socio-economic present values of assets to land use, the results of data linking representing the core information for calculating damage potentials (see Table 13.2).

**Table 13.2**   *Allocation of statistical land use values to land use types*

| Land use types (LUT) | Agricultural building values | Production industry assets | Commercial + services sector assets | Transport business assets | Value of stocks of businesses | Value of residential buildings | Household inventory values | Motor vehicle values | Livestock values (cattle etc.) |
|---|---|---|---|---|---|---|---|---|---|
| Agricultural buildings | ■ |  |  |  | ■ |  |  |  |  |
| Production facilities |  | ■ |  |  | ■ |  |  |  |  |
| Commercial facilities |  |  | ■ |  | ■ |  |  |  |  |
| Service sector businesses |  |  | ■ |  | ■ |  |  |  |  |
| Transport sector companies |  |  |  | ■ |  |  |  |  |  |
| Communication facilities |  |  |  | ■ |  |  |  |  |  |
| Transport infrastructure |  |  |  | ■ |  |  |  |  |  |
| Water, heat, wastewater lines |  |  |  | ■ |  |  |  |  |  |
| Residential housing objects |  |  |  |  |  | ■ | ■ | ■ |  |
| Farmland |  |  |  |  |  |  |  |  | ■ |

The solution simply developed from finding out that there is an adequate level in land use register information on the one hand, and aggregate data from official statistical systems providing socio-economic present values on the other, which allows allocation and results in data links.

It is fair to state at this point that it is not by chance that this methodology first became applicable in Germany. This stems from the fact that relevant national and federal laws and data-raising procedures have since the early 1970s provided for (i) harmonized data mining and statistical calculations applying a bottom–up approach; and have been prepared (ii) with good correspondence to state-of-the-art socio-economic theory. This led not only to the gathering of real-world business figures (gross values), but, based on them, calculations and procedures to calculate net values and report them regularly for socio-economic analysis purposes, namely separating time data load from taxes, transfers, depreciation and so on. During the last years Eurostat Europe-wide harmonization efforts have adopted some of these very important concepts. Cross-country comparison in 2001 has shown that some of the necessary information systems are already developing, and some of the data used and mentioned below are already available in other member states today (Pflügner, 2001).

In Table 13.2 land use types are represented in rows, and value components in columns. The black cells indicate which combinations need to be considered in a particular case study. The table explains only the most fundamental specifications as they apply to German cases (terms for land use types and land use capital stocks may be different in other countries, but the appropriate items can be found and will be statistically harmonized over the next few years). Having processed all the land use within the study area, mostly based on level 1 data, all the single plots and objects are available. Each single plot of land (e.g. agricultural use for corn-growing) and each single object (e.g. a personal computer assembly factory) carries a certain present value. The agricultural plot has (i) a market value in case its owner sells it; (ii) a rent value in case he lends it to a tenant farmer; and (iii) an annual revenue from corn-growing. The industrial facility carries (i) a market value of the plot of land; (ii) a present value of the production and administration buildings established on it; (iii) a present value for all the production facilities inside; (iv) a present value for all the input materials and products on stock; and (v) annual revenues and value added from all these operations.

All these data are held, of course, at least by their owners. All or parts of them are also transferred to and known by tax authorities, chambers of commerce, chambers of agriculture, land owners' associations, municipal administrations and so on. But it is most important that owners as well as all these institutions have, on the one hand, the right to reveal their data and the duty to practise data protection. On the other hand, due to relevant laws, they have to report certain, specified data to statistical authorities. This enables the latter to analyse them, and to aggregate and report them anonymously. In Germany, the socio-economic present value data on land, buildings, production facilities, materials stocks and so on are aggregated bottom up and processed by the federal statistical offices of the 16 *Länder* (most of them representing an NUTS 'region' according to the Europe-wide Nomenclature on Units for Territorial Statistics – Eurostat NUTS); however, a central quality assurance system was set up long ago to cross-check these data before reporting. This ensures that they are completely comparable between different regions; for example if you compare average present values of housing of a specified type between Bayern, Nordrhein-Westfalen, and Hessen, then you can be sure that the statistical differences between them are checked, and do reflect real-world regional differences.

As concerns observing land use and land use changes, and bottom–up aggregation to the *Länder* level, the same applies. All the regional statistical offices prepare annual land use reports according to a common, unified classification system. The data are completely comparable (with the exception of some unsolved reclassification problems at the five eastern *Länder* due to land survey updating).

This means that in Germany on the regional level one can find, for example, the total residential area given in $m^2$, as well as the present value of residential housing. This allows calculation of the present value of housing per $m^2$. And this is the basic principle for all the data links shown in Table 13.2 (case studies using the systems of the Netherlands and the UK have shown that comparable links can be established in these different statistical environments).

Of course, the real world is much more complicated: there are small houses and big ones, old and new structures, residences of wealthy and poor quarters. But, fortunately, statistical systems are detailed enough to allocate the needed data adequately, for example using distances from the average.

This is why a DSS real-world application needs far more data than can be shown in Table 13.2, which only aims to show the generics, namely that for a full damage potential allocation two or more sources must be used to build up the total present value. To describe one single cash and carry centre, for example, building, shop inventory, market goods, remote stocks and so on must be added. For residential housing the value of the building, data on how many households are living in it, household inventory values, and average present values of the inhabitants' cars are needed.

Many other basic data, only the most prominent of which are mentioned in Table 13.3, are necessary to link the data mentioned above in order to derive further information.

*Table 13.3   Examples of basic data used for damage potentials analysis*

| A | | Socio-economic base | |
|---|---|---|---|
| | I | Population | (1) status and long-term development<br>(2) age pattern, growth rate etc. |
| | II | Housing | (1) residential houses, flats, residents<br>(2) housing capital stock<br>(3) income status, purchasing power etc. |
| | III | Infrastructure | (1) education, recreation, social etc.<br>(2) power supply and environmental services<br>(3) traffic and communication |
| B | | Business sites and production factors | |
| | I | Production plants and workplaces | (1) employers and employees in economic sectors<br>(2) size of production plants |
| | II | Land as a factor of production | (1) land use categories and property shares<br>(2) values of built-up and agricultural land |
| | III | Productive capital stock and other property values of economic sectors | (stocks)<br>(1) gross capital invested<br>(2) stocks and other values |
| C | | Results of economic activities | (flows) |
| | | | (1) gross value added<br>(2) tax revenues |
| D | | Environmental assets | |
| | | | (1) environmental structures + conditions<br>(2) conservation areas and other protected areas/assets |

In addition to basic input data, it is necessary to feed 'background' data into RSA–DSS, for example information on constraints for certain strategies, or locations of neighbouring towns (e.g. for analysis of evacuation distances), roads (e.g. for analysis of evacuation routes) (vulnerable) supply lines, emergency aid bases and so on.

For example, in order to characterize a specified residential area, community or county statistics can be used to find out how many inhabitants on average are allocated to one single residential house. Combined with private car register data, average figures of cars per household, cars per residential building, employees per total inhabitants, within a specified quarter and so on can be obtained. Some parts of this information content may also qualify directly as important output information: for example, totalling the inhabitants within a quarter yields an estimate of affected population.

Adding some other statistical data, for example on age pattern, yields estimates of those parts of the population most vulnerable to the event under study.

Postal codes and address data files yields information on all the small businesses in a specified residential area, and their value to society.

Adding the number of workforce in a single business object, for example the cash and carry centre mentioned above, and then adding the statistical value added per employee per day for this specific business allows estimation of the damage potential per day due to closedown of operations, and so on (the importance of information production of this kind will be discussed later on in more detail).

### 13.4.3   RSA-DSS Modular Construction: Architecture

In order to apply RSA (as presented in Sections 13.4.1 and 13.4.2) efficiently we have developed a related computer-based tool called Regional Scale Analysis–Decision Support System (RSA–DSS) which allows adding and linking modules specifically for the application at hand (Penning–Rowsell *et al.*, 1996). Figure 13.3 shows that this system supports pre-processing, processing and post-processing of data and information.

Basically, the DSS–GIS has to link physical, technical, event-specific and economic data, which may have a geographical dimension and/or connection, but not necessarily unambiguous coordinates. Unified geo-references, with topographical and land use data, normally do have standardized formats.

This is why there is a section for pre-processing called the 'gateways' section. Its task is to pre-process all the input data, whose formats may vary significantly (alphanumeric, text information, point data, polygons) and need to be prepared for further use. Details need not be discussed here. The only important point to note is that pre-processing not only re-formats, but the larger parts should provide testing, and – most important – quality assurance.

The main activities in the three stages of analysis may be summarized as follows:

1. Pre-processing:

   - collecting, judging, quality control and preparing of

     - terrain model
     - topographical, land use, economic data (geo-referencing)
     - event scenarios
     - data on vulnerability/damage susceptibility of objects, land use plots, etc.

- updating of economic data with differing time reference to a common reference year;
- definition of spatial resolution of the model in terms of smallest regional units to study;
- definition of areas of interest (collections of the above units);
- preparation of gateway files for GIS system input;
- organizing of themes, views, layers, etc.
- testing.

2. Processing:

- blending of the smallest land use unit data with terrain altitudes and elevations;
- updating and refinement of land use and elevation data (e.g. add newly digitized data);
- feeding the system with scenarios and information about damage susceptibility of objects, land use plots etc.
- computing of landslide events and impacts (physical and economic dimensions).

3. Post-processing: generating outputs to support planning, decision making, management: here we differentiate between primary outputs, directly processed, and secondary outputs, produced by linking and comparing direct outputs and other (basic) information available.

The output side of an RSA–DSS has to supply information in both physical and economic terms, namely:

- area affected, impact borders (event-specific);
- physical size: quantities and sizes of objects/activities, with additional parameters;
- economic values: of objects, land use types, activities in the study area;
- damage potential: the economic value within the area identified as being under threat;
- damage: the economic damage induced by a certain landslide scenario, with and without a specified protection strategy, taking into account specific cause–effect relationships.

The last term will be discussed in more detail later. The impacts are presented on different spatial and economic aggregation levels, basically in maps, tables, and reports [but also using other media such as photos, slide shows, motion pictures (e.g. to show landslide event steps according to a specific landslide scenario studied or, e.g. for analysing a specified development scenario for the study area in time steps related to (i) growth of economic activities; (ii) retreat from high-risk areas, and so on)]. The results can be combined with geographical and other information, for example, in order to obtain a ranking of land use units according to their specific value per $m^2$ in terms of damage potential and/or damage. Such figures help a great deal to identify economic hot spots and to find the most adequate solutions.

The above outline of the model architecture should suffice to show those familiar with any GIS application how to organize their own system for a real-world application; this is work, not art. Far more important are the explanations of the 'philosophy' of DSS application and experience gathered so far.

## 13.5   Contributions to Integrated Landslide Policy and Management

### 13.5.1   Information to Decision Makers, Stakeholders and the Public

Reflecting on all the specific applications in the field of natural hazards, we have to realize that it is very rare that only one single institution or administration or only a few of them are involved. On the contrary, within the single steps of planning and decision making in most cases there are many essential linkages, interrelations, communication processes and so on with the 'outside world'. This includes, for example, local and regional politicians and administrations (environment, transport, economic, etc.) needed for giving inputs or potentially affected by the planning activities or measures under study. It also encompasses the affected population, especially inhabitants of the study area, stakeholders and pressure groups of all kinds (economic, environmental, recreation facilities, hunting associations, etc.). If not already occurring at the policy stage, at least planning and management are multi-institutional in most cases.

This requires that a truly 'integrated' landslide mitigation policy and management within such a complex array of relations, participation and involvement is implemented. It means not only producing isolated partial solutions for the 'customer' and 'contractor', but also establishing a 'dialogue environment' enabling involvement of the different parties in the following stages:

- setting objectives;
- defining criteria for assessment and decision making;
- generating and assessing options and solutions;
- formulating policies, strategies and measures;
- selecting and implementing policies, strategies and measures.

The different stakeholders have widely different information needs as well as diverse understanding of the topics at hand. Landslide damage potentials and landslide damage are essential information for all of them; however, the required details are interest-defined and heterogeneous. This is why the DSS should be built up as a basic information system serving all the different needs in a systematic, comprehensive and unbiased manner.

In particular, the post-processing module and its features for generating output information must serve these needs. The top-level decision makers' only question may be: 'What is the maximum probable damage with the maximum probable event?', and the answer expected may consist of a single figure or bandwidth.

But one single authority in one of several communities affected may say: 'Please show me exactly the border of landslide impacts within my community, and list how many people, buildings, etc. will be affected. Tell me about the mitigation margin of these impacts in case we implement measure x, not y.' Of course, this needs many more features to answer quickly and correctly. In order to show the socio-economic impacts of landslides if the event were to happen tomorrow, as well as damage reduction resulting from different strategies implemented in the future, a feature for with–without comparison is needed. This may be somewhat tricky to implement as it needs a great deal of technical input in terms of changes in impact trail, durability/withstanding forces of structures, efficiency of planned measures and so on, but with tools that now implement such processing or post-processing features this is not a fundamental problem.

### 13.5.2   Conflict Management, Mediation, Participation and Support
###                for Action

Apart from answering such questions, we have to note that, within the decision environments developing today, the importance of feeding the right information 'online' and 'just in time' into such processes is growing rapidly. The parties involved become used to this work style and demand it. Willingness to wait for potentially decisive information is vanishing rapidly. In other words, the probability of biased bargaining and decision results will grow without sufficient decision support.

In many cases, there are different stakeholders, whose interests conflict. They need to be addressed, for example, in formulating appropriate strategies. 'Online' DSS output comparison of excellent quality is a means of rationalizing such strategies and 'integrating' stakeholders in decision processes. Transparency and interactivity must be considered the most essential features supporting this. The 'users' must be helped to understand the different steps of analysis and to interfere where they want to. Answers to their questions must be generated quickly in order to be efficient.

Experience shows that another feature of increasing importance is the possibility of taking a closer look at the questions of who and what is affected. Evidence shows that residential housing and business capital stocks are the value components dominating the economic damage potentials and damages (Pflügner, 2001; Glade and von Davertzhofen, 2004). The subjects and objects affected have been called 'land user groups', 'sectors' or 'land use types' above. In the real world, however, this means inhabitants of endangered sites, individual businesses and so on. Their economic assets have been called 'value components'; for most of them, it is their economic base. Most of them do not know about the hazards at all, or if they know, cannot assess their individual effects.

This is why DSS outputs, for example in the field of flood hazard policy in Germany and Switzerland, are not only used within the policy and management environments. Instead, they are published widely as 'action plans', press releases and Internet presentations (especially 'fact sheets' for the individual communities) in order to further awareness, to motivate individual action, and to generate a new willingness to take part. Many reactions observed so far promise significant quality improvement for hazard management. Using comparable information policies in the field of landslide hazards should elicit the same benefits.

But despite these promising developments we should not neglect the fact that we may need some more features and tools to answer questions in a real-world case. The next chapter will discuss four such loose ends.

## 13.6   Future Research and Development

### 13.6.1   Cause–Effect Relationships

Throughout the sections above we have dealt with the perhaps most important DSS feature rather superficially, that is 'cause–effect relationships'. This reflects the fact that we do not have available the necessary tools for assessing event-specific landslide damage to the damage potentials identified. The first German application study (Glade and von Davertzhofen, 2004) has scrutinized this question, and as result of an in-depth literature study concluded that available data do not suffice to forecast quantitatively the degrees of damage to a specific land use from a defined landslide event based on a given set of independent variables.

This means that, at present, risks, event parameters and land use vulnerability can only be assessed qualitatively. As long as the factors, parameters, variables and formulae cannot be given in mathematical terms, analysis will be limited. It seems somewhat unrealistic to assume 100% damage, at least for smaller landslide events, for the total impact area. In other words, event-specific damage may be overestimated. The only means of avoiding this at present would be to work with assumptions, some sort of 'simulation', and/or risk classifications (e.g. Glade and von Davertzhofen, 2004).

Therefore, the state of the art in the flood hazard field may serve as an example of what is needed. Within some European member states, for example, the UK, France, the Netherlands and Germany, activities started during the 1980s to build up empirical databases, to gather data sets from observed events, and to establish sophisticated statistics in order to allocate appropriate data sets to specific events. This has been done after realizing that other sources, especially data from the insurance industry, do not fit the purpose. Insurance companies mostly cover a certain impact area only partly (the most prominent exceptions in the flood hazard field are France and the USA, with full coverage approaches/state-owned systems, etc.). In documentation, the insurance industry limits efforts to raising only such data absolutely necessary for their business. Systematic data gathering is restricted to major events, and nowadays done primarily by international reinsurance companies. Not all the companies and business associations have associated research and development departments with capacities to conduct statistical analyses. During the last few years there has been a great deal of cooperation and discussion with these experts on 'data mining', but the overall result is that support from these data is rather limited. These remarks may suffice to show that only an independent institution, capable of doing in-depth scientifically sound work, may provide the data needed.

In Germany, the LAWA, the National–Federal Working Group of *Länder* Authorities responsible for water management, has therefore established guidelines and fact sheets for flood damage field surveys. People sent to an impact area for damage assessment have an unambiguous task description at hand. *Länder* authorities are responsible for these surveys when major events take place. The data are then fed into a central database, HOWAS. At present this contains more than 3000 data sets from different events, with damage to all kinds of land uses and objects. This is not the right place to discuss its features and problems in more detail. Instead, the message simply is: for most cases we can infer adequate damage functions, carrying information on the susceptibility of specified assets to damage from a specific event. Comparable activities, preferably cross-country and supranational, seem to be absolutely necessary for supporting landslide hazard analysis in the future.

But apart from empirical research some other approaches may help, based on the following hypotheses:

1. Average figures for 100% damage of residential houses, inventory, cars, small businesses and so on can be derived from official statistics, the insurance industry, and the like. This is also true for other countries, at least the ones mentioned above.

2. Average figures for other 100% damages, especially small and medium-sized businesses, have become available during recent years, as questionnaires have been exploited in the field of flood hazard management.
3. Assessment of 100% damages in landslide studies is not a fundamental problem. Some of the average figures derived for flood hazards may be transferred directly to landslide hazard studies in other regions, at least within Europe; others may be derived empirically using the approaches given above.
4. Damage functions used in natural hazard studies need not necessarily be as sophisticated as the ones available for flood hazard management in Germany and the UK.
5. There is at least a sectional plane between flood and landslide damage functions: some flood events in alpine regions, especially events occurring in combination with debris flows, and so-called flash floods in hillside terrain, seem in their effects to be quite comparable to the specific class of landslides consisting of debris flow or earthslides, with slow or medium velocity, and comparatively small impact area. Comparison between these two specific types of flood damage functions and other flood damage functions from low-lying areas indicates that the first are significantly higher, especially in damage to the construction section (Vischer, 1996; Jäger, 1997; BUWAL, 1999; BWG, 2002). This may be a good starting point for future landslide damage function research and development.


### 13.6.2   Prosperity Damage

A second question concerns the assumption of stability and persistence of the socio-economic system of a study area. As RSA methodology will primarily be used to support finding out the overall best mitigation options for tackling larger-scale events, we should also consider what prosperity damage means. This term stems from disaster impact analysis and indicates that there may be far-reaching effects beyond the immediate, direct, short-term ones. The point is that the assumption of stability and persistence of a socio-economic system may not be valid for larger events, but that they may result in shocks which the system cannot recover from completely.

Take, for example, the case of a country heavily devastated by long-time war activity, as in present-day Afghanistan. Of course, it is easy to state that such an area has, in comparison to neighbours or other countries that were in a comparable development state 20 or 30 years ago, obviously comparative disadvantages. Not only has the economic system been destroyed (which could be described in terms of monetary damage), but, more important, also the currency system, the majority of cultural assets, the education system and so on.

The questions are (i) could any natural disaster lead to comparable devastation? (ii) how will people react in that specific case? (iii) is it adequate to assess possible long-term effects only looking at the affected region? and so on.

The scientific community has proposed defining prosperity damages as impacts that will result in a structural breaking point in long-term future economic growth (IPCC, 1992). The practitioners' question is simply: 'What percentage of damage of assets represents such an important shock that it exceeds the absorptive capacity of the regional economy/society?'

Currently we cannot apply the prosperity damage approach to practice because of at least three major problems:

1. As of now there are no widely accepted statistical indicators to analyse economic time series data according to 'sound' or 'sustainable' development. Any existing definitions may vary from region to region.
2. In many cases it will not suffice to analyse the region under study 'internally' in order to discover the percentage of damage which can be seen as representing the point of breakdown. Instead one would have to include at least the most important backward and forward linkages between the regional and the outside economy.
3. The two problems described above lead to the third application problem, namely that it will be extremely difficult to provide decision makers with sufficient information on landslide impacts leading to such an irreversible system deterioration.

Degree of dependence on only a few businesses may be a criterion if the latter are endangered. But who can forecast if a destroyed business will be built up again or not (which means that it will move to another facility, leaving the affected area without its annual contributions in terms of workplaces, taxes, etc. in the future)?

Considering potential prosperity damage may add some important arguments for explaining the 'without' case, but today there is no readily available analytical approach for answering the question. Aggregation of gross product losses and their secondary impacts on the economic system may be the right starting point for further R&D.

### 13.6.3 Cost–Benefit Analysis Integration

Up to now, RSA methodology and DSS system design have been presented as being primarily 'output production facilities'. But Figure 13.2 also contains the terms 'benefits' and 'costs', and shows some important data flows between these tools.

Considering that cost–benefit analysis (CBA) is essentially based on with and without comparison of the options under study, it becomes apparent that there is not too much difference. In fact, DSS output, as discussed above, can serve to fill up at least the benefit side of the balance. Implementing it within the evolving processes, and adding some close links to engineering dealing with planning and cost calculation, will generate a new synergy.

Of course, the (mix of) measures necessary under various scenarios will incur different costs. Calculating them early and step by step allows balancing them with the benefits. This will provide additional decision support for (i) optimization of mitigation and (ii) cost minimization. Preparing CBA as a comparative analysis in terms of damage avoided/cost ratios will produce additional planning and decision support.

### 13.6.4 Policy 'Toolbox'

In comparison to flood hazard applications, the landslide hazard field seems far less 'standardized' today. This means that, from the economic point of view, a set of different 'policies' may be defined and used for shaping landslide mitigation policies. Experience has shown that starting the analysis with such different goal sets, as for example,

- total safety policy;
- hot spots' safety policy only;
- economic optimization policy (e.g. protection only, if benefit–cost ratio > 1).

may help to identify what will be feasible at all, to analyse differences in performance, to receive responses as to acceptability, compatibility with societal goals, and so on.

Elaborating these policies step by step leads to differentiation according to 'who pays' (central or local governments, communities, individuals or groups of individuals affected), 'what does fair burden sharing look like' and similarly 'who is to decide about the degree of safety'.

Experience shows that the latter questions are not only quite interesting in real-world discussions, as safety levels are not fixed precisely by law, but also of high importance for bargaining processes between the parties involved. Therefore it is useful to be prepared to answer such 'standard questions' in order to support not only with compulsory, but also with optional, strategies. The differences can be explained in some more detail as follows:

**A total safety policy**, clearly, represents the standard approach. However, if it shows that the required budget is not available, or full protection is not feasible for technical reasons, then some parts of the impact area will receive it, and other parts will not. The latter will not be prepared to accept a proportional cost share, for example based on an area protected criterion. Bargaining then becomes difficult. The same is true if there is a central budget, though without determined cost sharing, and in the case this budget is limited and cannot finance 100% of costs. This means that other, somewhat more difficult, but perhaps fairer models for cost sharing must be presented.

**The hot spots' safety only policy** would represent another solution, leading, for example to the question whether it is fair to assess all land use equally. The present value content of the single plots (namely damage potential) and, if possible, comparison with damage expectancy per plot might then help to find differentiated safety levels by, with for example applying a classification system (highly, very, moderately, less vulnerable etc.).

**The economic optimization policy** would determine the optimal level of protection for each 'protection unit' separately, according to marginal cost equalling marginal productivity in the sense of damages avoided. This may end up with different safety levels and/or decisions as to for which protection units the according measures will be financed, and which not (or later). This leads the latter into a situation where they must themselves decide about their individual willingness to pay.

Willingness to pay should play an important role in all such bargaining processes and may help in cases with limited budgets, differing damage expectancy, problems with technical feasibility, finding new options, or mixed policies (e.g. basic protection in combination with 'insurance' or 'solidarity fund' solutions, future land use restrictions or land use change as the cheapest means, and so on.

## 13.7   Conclusions

Geographical Information Systems have become standard tools for natural hazard studies because they support handling of terrain models, land use data, and other geo-referenced basic data efficiently, that is, because of their benefits in data processing, especially in the pre-planning and planning stages.

Pre-planning and planning procedures in the natural sciences and engineering are, of course, core activities. But real-world cases demand more: the decision environment demands decision support; the project management environment expects support in terms of a management system. They have a common sectional plane, but also several different requirements.

This chapter has focused on decision support, and presented a conceptual framework in terms of damage potentials and socio-economic values. The sections on methodology, data handling and model architecture tried to explain the most important features in terms of software modules, functions, and DSS output for the decision environment, not only or primarily consisting of institutional and administrative decision makers, but also of other stakeholders and especially the affected population, businesses and the like.

This was done to emphasize that a real-world DSS needs to be more sophisticated than a simple GIS; the latter is a core tool only, the DSS must be tailor-made according to the problem under study, but most important is that decision support is integrative and dialogue-based: if it becomes apparent during the processes of analysis and decision making that, for example, a more detailed or different approach with respect to the scenarios investigated is required, then the DSS has to ensure an appropriate step-by-step implementation. Pairwise comparisons of different solutions help to narrow down the options. The options can be evaluated with different decision criteria corresponding to different attitudes as to who should decide and who should pay for landslide mitigation. Discussion of results may lead to new solutions; then the DSS again has to ensure an appropriate step-by-step implementation.

But the most important effect of all such considerations and discussions during the evolving processes may be to change

- inadequate perception of risks by the owners/land users and/or;
- inadequate land use or land use planning.

This may be the most important precondition for discussing resource allocation questions on a rational basis.

Arguments in support of the suggested framework and design, especially insight into which DSS features are needed most, have been derived mainly from real-world DSS applications on flood hazard problems up to now, and from a pilot study that attempted to transfer them to a landslide hazard problem (Glade and von Davertzhofen, 2004). Based on the latter, throughout the text comments are added on transferability, especially the need for specific developments. This led to the conclusion that most parts are transferable; some crucial questions have been identified in Section 13.6 together with proposals for future R&D.

## References

BUWAL, 1999, *Risikoanalyse bei gravitativen Naturgefahren* (Bern: BUWAL).

BWG (Bundesamt für Wasser und Geologie), 2002, *Hochwasser 2000 – Ereignisanalyse/ Fallbeispiele* (Bern: BWG).

European Environment Agency, 1985, Corine Land Cover Classification, available from EU – RAMON server.

Eurostat, 1996, *European System of Accounts – ESA 1995* (Brussels/Luxemburg: Eurostat).

Eurostat, NUTS – Systematik der Gebietseinheiten für die Regionalstatistik, available from Eurostat server 1993 ff.

Gewalt, M., Peerbolte, E.B., Pflügner, W. and Verhage, L., 1996, *Regional Scale Analysis – Decision Support System (RSA-DSS), User Manual* (München: Emmeloord).

Glade, T. and von Davertzhofen, U., 2004, GIS-based landslide risk analysis in Rheinhessen, Germany, *Natural Hazards* (submitted).

IKSR, 2001, Atlas der Überschwemmungsgefährdung und möglichen Schäden bei Extremhochwasser am Rhein, Koblenz.

IPCC, 1992, *Preliminary Guidelines for Assessing Impacts of Climate Change* (Geneva: IPCC).

Jäger, S., 1997, *Fallstudien zur Bewertung von Massenbewegungen als geomorphologische Naturgefahr* (Heidelberg).

Klaus, J. and Schmidtke, R.F., 1990, *Bewertungsgutachten für Deichbauvorhaben an der Festlandküste – Modellgebiet Wesermarsch* (Bonn).

Klaus, J., Pflügner, W. and Schmidtke, R.F., 1994a, Technical Annex 3 – Regional Scale Analysis (RSA) Final Report EUROflood I, München.

Klaus, J., Pflügner, W., Schmidtke, R.F., Wind, H. and Green, C., 1994b, Models for Flood Hazard Assessment and Management, in E.C. Penning-Rowsell and M. Fordham (eds), *Floods Across Europe* (London), 69–106.

Munich Re, 2002, *Topics – 50 bedeutende Naturkatastrophen 2002* (annual publ.).

Penning-Rowsell, E.C., Pflügner, W., Peerbolte, B., 1996, RSA-DSS application studies, unpublished.

Pflügner, W., 2001, Hochwasserschadensdaten und Wasserstand-Schaden-Funktionen, unpublished.

Vischer, D., 1996, Hochwassergefahr im Gebirge, in Institut für Wasserwesen der Uni-Bw München, *Klimaänderung und Wasserwirtschaft* (München), 293–306.

# 14

# Instability Management from Policy to Practice

*Robin McInnes*

## 14.1 Introduction

Instability management is being recognized increasingly by those in regional and local government and by other practitioners as one of the most effective means of addressing the impacts of ground movements. This involves the interpretation of field and desktop studies and ground investigations and use of the information gained to address the health, safety, economic and social issues in the area affected by instability. Rather than facing the problem of how to provide an 'emergency response' to a particular instability event, the focus of instability management is concentrated on pre-planning and preparation, allowing appropriate longer-term planning decisions to provide the policy framework and a 'landslide management strategy' to address the different mitigation opportunities and to put 'policy into practice'.

In recent years scientific research has provided a wealth of additional technology and techniques that have assisted the understanding of ground instability problems. Lessons learnt from a number of major landslide events have also highlighted the role that human activity can play in instigating ground movements. As pressures have increased the occupation of more stable and commercially attractive locations, there is increasing demand to extend new development into adjacent areas where there may be greater physical constraints because of problems such as instability (see Figure 14.1 for a breakdown of factors promoting urban landslides). Such developments, particularly in the absence of effective planning policies, may lead to costly mistakes posing risks to life and property. These can usually be avoided if planning and management systems are in place which have taken full account of ground conditions in their formulation.

*Landslide Hazard and Risk*   Edited by Thomas Glade, Malcolm Anderson and Michael J. Crozier
© 2004 John Wiley & Sons, Ltd   ISBN: 0-471-48663-9

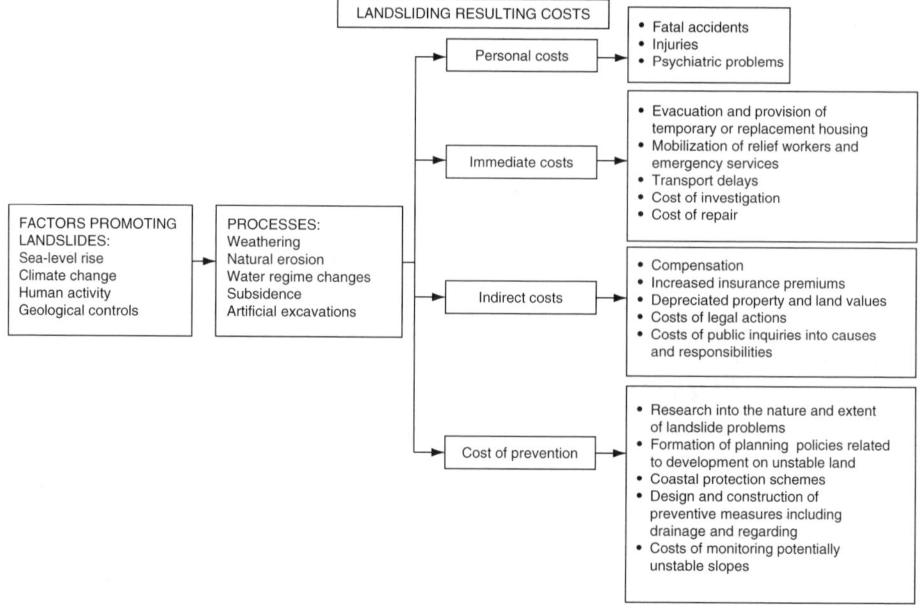


**Figure 14.1**  *Factors promoting urban landslides – causes and effects. Adapted from Jones and Lee (1994). Reproduced by permission of Mark Lee*

## 14.2  Developing the Planning Policy Framework

It is clear that the important relationship between the nature of ground conditions and the land's capacity to accommodate development requires effective legislation, guidance and planning policies to be put in place. Recognizing the wide variety of instability problems, ranging from rapid, high-energy events to slow, ongoing movements, it is necessary to have in place a system that can take account of the varying geological situations and which is supported by a coordinated planning framework.

In Europe, for example, European Union and national legislation and guidance for the most part provide a strategic framework for land use and development. Increasingly a strategic perspective is being applied, especially with respect to the concept of sustainable development, which is becoming the major influence on all aspects of planning policy.

The World Commission on Environment and Development (Bruntland Commission, 1987) described sustainable development as 'development that meets the needs of the future without compromising the ability of future generations to meet their own needs'. This commitment is being reflected increasingly in planning policies and development control decisions as awareness of ground instability and other environmental issues continues to grow. Many states are now publishing strategies for sustainable development which will provide a framework for setting strategic objectives and the implementation of policies to achieve sustainable development.

Within this framework planning systems are designed to regulate the development and use of land in the public interest. Most planning systems are intended to:

- provide guidance which will assist in planning the use of land in a sensible way and enable planning authorities (regional and local government) to interpret the public interest wisely and consistently;
- provide an incentive, with local government stimulating development by the allocation of land in statutory plans;
- implement development control to ensure that development does not take place against the public interest and to allow people affected by development to have their views considered.

Planning systems do of course play a vital role in promoting the principle of sustainable development and are often supported by guidance and advice to assist local government with the implementation of legislation (Thompson *et al.*, 1998).

In contrast to some other parts of the world, it is only comparatively recently that planning authorities across Europe have started to consider ground instability issues within the planning framework. More recently there has been a change in perception about the way in which instability and planning are being addressed. These changes reflect a growing appreciation that the past approach was not always in the public interest. In coastal areas, for example, there is recognition that:

- development in vulnerable locations can lead to demands for expensive public civil engineering works (e.g. slope stabilization and coast protection);
- coastal defence works can have significant adverse effects on the interests of other aspects of the coastal zone (e.g. environmental and aesthetic impact);
- coastal defence works can encourage future development in vulnerable areas, increasing the potential for greater losses when extreme events occur.

In the UK, for example, the planning system is designed to regulate the development and use of land in the public interest. It is an important instrument for protecting and enhancing the environment and reconciling the interests of conservation with development. Generally the two most important functions of planning authorities are the preparation of 'development plans' and the control of development through the determination of 'planning applications'. All planning decisions should be made in accordance with the policies and proposals presented within the development plan unless material considerations indicate otherwise. The former Department of the Environment (DoE) published planning guidance which stressed the need to take instability issues into account at all stages of the planning process (DoE, 1991). Planning Policy Guidance Note 14, 'Development on Unstable Land', states that 'the stability of the ground, insofar as it affects land use, is a material consideration which should be taken into account when deciding a planning application'. In the UK, therefore, the aim of this advice is not to prevent development on unstable land, though in some cases this may be the appropriate response. Rather it is to ensure that development is suitable and that both natural and man-induced physical constraints on land use are considered at all stages of the planning process. Where development is proposed on unstable or potentially unstable land, the Planning Authority should ensure that a number of issues are adequately addressed by the development proposal; these should include:

- the physical capability of the land to be developed;
- any possible adverse effects of instability on the proposed development;
- the possible adverse effects of the development on the instability of adjoining land;
- possible effects on local amenities and conservation interests and the impacts of any remedial or precautionary measures proposed.

The assessment of ground conditions and the preparation of geomorphological, ground behaviour (Clark *et al.*, 1994; Lee *et al.*, 1991) and planning guidance maps will assist in the implementation of policies of this kind (e.g. Clark *et al.*, 1996).

In Switzerland the government has published advice on how hazards caused by ground movement should be taken into account in the land use planning framework. There is clear evidence from the Swiss experience that hazards caused by ground movements can often not be managed by building and safety measures alone. The risk of potential damage should first be reduced by planning measures. Construction-related protection measures should only be implemented, in terms of reducing a potential hazard, in cases where there is already land use worthy of protection or in cases where, after complete consideration of all the interests, a change in allocation of land use proves to be absolutely essential (Lateltin, 1997, 2002).

The Swiss believe that there needs to be a conscious perception of natural hazards before it is possible to take on the responsibilities that they imply. Long-term results can only be expected once all the stakeholders have readily taken on board the nature of the present hazard. In Switzerland the drawing up of hazard maps is regarded as an essential precondition to the implementation of some of its key legislation. The threat created by a natural hazard is part of the particular characteristic of a site; it is an element comparable to soil fertility or gradient because it dictates a certain land use or indeed may render it impossible.

Within the planning framework in Switzerland, the local authority assigns different types of allocation to the zones considered, according to their particular properties. In order to protect human life and prevent risk to property and the environment, certain types of land use can be forbidden in high or medium hazard zones, or only allowed under certain precise conditions. Owners can further reduce potential risks arising from their own actions. The Swiss government is encouraging hazard maps to be drawn up as far as possible for all types of natural hazards. The division into degrees of hazard is worked out independently of the current allocation of land. The government has recommended that hazards caused by ground movements should be taken into account in all activities affecting development, and this process has been aided by the publication of national guidance (Lateltin, 2002).

In France responsibility for mapping natural hazards rests with the state. France benefits from one of the most comprehensive administrative and legislative frameworks for addressing the prevention of natural risks, with respect to both provision of preventive information and insurance compensation. Risks Prevention Plans are legal documents which have to be taken into account in urban planning management when associated with the Urban Planning Code through 'Plans d'Occupation des Sols' (POS) or the new 'Plan Local d'Urbanisme' (PLU). Communes share the responsibility for the prevention of natural risks with the state (Leroi, 1996; McInnes and Jakeways, 2000).

## 14.3    From Policy to Practice: an example from the Isle of Wight, UK

The identification and assessment of instability problems usually begins with a desk study of existing information; this general assessment allows an overview to be prepared of ground conditions in the area concerned. With this information in place, it is possible to start developing plans and policies to assist in addressing instability, including the identification of those areas where risks may affect existing and proposed developments.

The most commonly used techniques available to assist a preliminary landslide investigation include a review of existing sources of literature together with geological and other hazard maps, ground investigation and civil engineering reports, aerial photographs and satellite images. The detail of a particular assessment will depend on resources available. However, it is usual to undertake a geomorphological interpretation which may be based on an overview of existing information or field mapping or a combination of both approaches.

A comprehensive examination of coastal landslide potential within part of the Ventnor Undercliff on the Isle of Wight was commissioned by the former UK Department of the Environment (DoE) in 1987 (Doornkamp *et al.*, 1991). This three-year study began with a review of available records, reports and documents, followed by preparation of geomorphological maps, a survey of damage caused by ground movement, a land use survey and a review of local building practices. This information, together with a survey of the results of past site investigations, assisted in identifying the nature and extent of landslide systems, the types of contemporary movement taking place, and the magnitude and frequency of events and their impact on development. Furthermore, the information gathered assisted in assessing the nature of land use at risk and the vulnerability of structures to ground movements of different intensities (Lee and Moore, 1991).

All this information was incorporated within a geographical information system which allowed the factors influencing the distribution and frequency of contemporary movements to be summarized on a ground behaviour map. A simplified version of this map was prepared to offer guidance to the planning process and to assist in the development of a landslide management strategy. A flow chart for the Ventnor Study is illustrated in Figure 14.2.

One of the challenges is to present instability-related information in such a way that it can provide support to the strategic planning process. Experience gained from a range of studies in the UK (Lee, 1995a, b, 1996) suggested that information of this kind could most helpfully be provided in three forms:

- as a series of *element maps* illustrating, for example, rock types, geomorphology, soils, slope steepness, etc.
- as *derivative maps* which make use of the base data to define the character of particular conditions of interest, e.g. landslide potential, etc.
- as *summary maps* which draw information from the element and derivative maps and provide a general view of the area in terms of availability of land for development and constraints on development.

An assessment of ground conditions, which may comprise the preparation of geomorphological, ground behaviour and planning guidance maps as previously described, does,

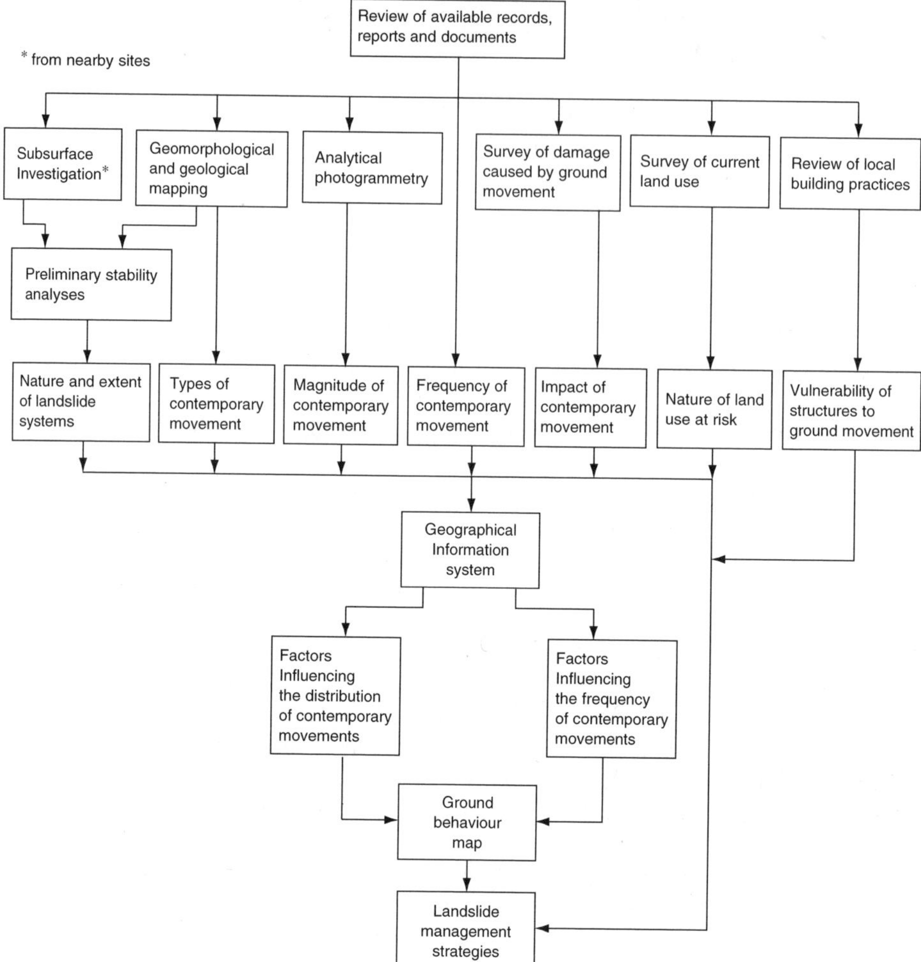



**Figure 14.2**   *The work programme for the study of landsliding at Ventnor, Isle of Wight, UK. After Geomorphological Services Ltd (1991). Reproduced by permission of Dr Alan Clark*

therefore, provide a foundation for assessing risk together with the development of management policies allowing decisions to be made on areas suitable for development or otherwise.

## 14.4   Developing a Landslide Management Strategy

In areas where urban development has taken place on degraded landslides, the problems tend to be related to very slow ground movement and progressive damage to property, services and infrastructure. In such circumstances many problems can be reduced if there is a programme of active landslide management where the local community is able to come to terms with the situation and learn to 'live with landslides' (Lee and Moore, 1991; Lee and McInnes, 2000).

An example of this approach is the case of the Ventnor Undercliff on the south coast of the Isle of Wight, where a landslide management strategy has been in place since 1993. Before considering this it is important to stress that landslide management needs to be understood within the context of the legislative and administrative framework where landslide issues are set against other factors (e.g. conservation, the environment and socio-economic issues) in the decision-making process. This framework can be very complex and will be unique to individual countries (Lee, 1993, 1997).

The coastal town of Ventnor and the villages of Bonchurch, St Lawrence, Niton and Blackgang are built on a large landslide complex known as the Undercliff, but fortunately the geological setting and the style of landsliding are such that movements are often concentrated in a few locations, and the intervening areas show negligible movement (Lee and Moore, 1991; Moore *et al.*, 1995). An Undercliff 'Landslide Management Strategy' has been developed which aims to reduce the likelihood of future movements by controlling the factors that cause ground movement and limiting the impact of future movements through the adoption of appropriate planning and building controls. A variety of approaches have been adopted to address the ground movement problems, including:

- preventing unsuitable development through planning controls and building control;
- improving ground conditions through the control of water in the ground and coast protection measures;
- monitoring ground movement and weather conditions at automatic and manual recording stations;
- raising professional and public awareness through displays, meetings and dissemination of information.

To assist in ensuring that a coordinated approach is adopted by professionals working in related areas of activity, a Landslide Management Committee meets up to twice a year to enhance professional and public awareness of how the strategy is being implemented and to monitor its effectiveness. This Committee comprises representatives from the local authority including the coastal manager, planners, building control officers and highway engineers, the water company and other service industries, the Association of British Insurers, the Builders Employers Confederation and local estate agents as well as technical observers (geotechnical consulting engineers who have been involved in previous or current studies in the area).

An important element of the landslide management strategy for the Undercliff is to ensure that future development is compatible with ground conditions and is discouraged where the likelihood of movement is high. New property within the Undercliff must be capable of withstanding movement and must not lead to a worsening of slope stability at the site or on adjoining land. These requirements are overseen by the Council through the planning system and application of the Building Regulations.

Developments in the UK require planning consent. Local planning authorities are required and empowered under the Town and Country Planning Act 1990 to control most forms of development and are responsible under the Building Regulations and the Housing Acts for ensuring standards of construction and development. When considering an application for planning consent, the local authorities, in England and Wales, have a duty to take into account a range of material considerations, which include potential land instability problems (e.g. ground movement and landsliding). However, the responsibility for determining whether land is physically suitable for a proposed development, and the appropriate technical measures to protect that development, lies with the developer and the landowner (Thompson *et al.*, 1998).

In addressing all instability matters throughout the Isle of Wight, the local authority approach is set out in their Unitary Development Plan, which states 'development of areas known to suffer from instability will not normally be permitted, unless the Local Planning Authority can be satisfied that the land can be developed and used safely and will not add to the instability of the site or adjoining land'. Detailed areas have been identified on 1:2500 scale maps of ground behaviour and planning guidance (based primarily upon geomorphological field studies but supported by ground investigation data where available) that have been prepared for the whole of the Undercliff (Lee and Moore, 1991; Moore *et al.*, 1995). Areas are identified which are likely to be physically capable of development along with areas that are either subject to significant constraints or are likely to be unsuitable.

In recent years landslides have become the focus of increasing attention by the planning system in the UK. This has led to a major shift in consideration of landslide issues from site-specific problems to addressing the physical constraints over broad areas and increased emphasis on managing landslides rather than simply relying on engineering solutions (Clark *et al.*, 1996). This allows a greater range of responses to be practised that attempt to minimize landslide risk in an area. In this way, the traditional engineering solutions can be viewed as just one of a range of landslide management strategies that can be adopted. The range of responses available for managing potential landslide problems includes:

- accepting the risk;
- avoiding vulnerable areas;
- reducing the likelihood of potentially damaging events;
- providing advance warning of potentially damaging events;
- protecting against potentially damaging events through modifications to building designs and the use of slope stabilization measures.

In most cases the response would be complex, involving a variety of measures adopted by residents, landowners and other interested parties. However, it should be recognized that the complexity of landslide problems dictates that conventional engineering solutions cannot be guaranteed to prevent the occurrence of potentially damaging events, especially on many coastlines. More realistic aims will involve reducing the frequency of damaging events and minimizing their impact.

It should also be appreciated that landslide management can involve reconciling a number of conflicting demands, including reducing risk to vulnerable properties, important economic resources and facilities, and protecting areas of scenic, geological or ecological importance. It is important, therefore, that management decisions are based on the best possible understanding of landslide systems and the wider environment, and how to manage and protect it. Landslide management will generally involve a partnership between a wide range of interests, including planners, developers, insurers, environmental managers and the public, together with engineers and geoscientists.

## 14.5   Informing the General Public – A Case Study: Ventnor, Isle of Wight, UK

On the Isle of Wight a number of the residents of the town of Ventnor were aware that they resided within an ancient landslide complex known as the Undercliff (see Figure 14.3). The landslide complex was thought to be initiated as a result of sea-level rise during the Flandrian Transgression, a period of sea-level rise about 7000–8000 years ago, although recent evidence suggests the landslide may be significantly older (Moore *et al.*, 2003). Although considerable areas of the Undercliff are relatively stable, some areas are close to a threshold of instability as through various processes and mechanisms the landslide slowly degraded. Historically, ground movement had been slow and episodic with more severe movements concentrated in a limited number of locations.

An analysis of historical information suggested that ground movement problems had increased over the last century or so. A key factor was no doubt the rapid urban development that took place as Ventnor established itself as a popular seaside resort and spa; this development continued on through the Edwardian period and up until the middle of the last century. The increased urbanization of the area has had an impact on the landslide stability through human activity such as excavations for building sites, inappropriate 'cut and fill' operations affecting slope geometry and additional drainage into the ground from sewerage and leaking pipes, all of which contributed to an increase in instability.

***Figure 14.3*** *Ventnor looking west. The town of Ventnor, population 7000, is situated within the Isle of Wight Undercliff landslide complex. The town developed from the 1830s and was built on the steep southern face of St Boniface Down. Reproduced by permission of Dr Alan Clark*

In 1986 the former UK Department of the Environment (DoE) was investigating suitable sites for research that could lead to the publication of advice and guidance to assist the planning process for development on unstable land. The DoE decided that the town of Ventnor would form an excellent case study for a coastal landslip potential assessment and this was started in 1987 (see Figure 14.4).

This study (Lee and Moore, 1991) proved to be of enormous benefit for Ventnor as in particular it assisted in developing cost-effective approaches for mitigating landslide problems through the development of planning control procedures which take instability into full account. This broad-based study involved:

- determining the nature and extent of the landslide problems;
- understanding the past behaviour of separate parts of the Undercliff;
- formulating a range of management strategies to reduce the impact of future movements.

The results of the Department of the Environment study were presented both as a technical report (Lee and Moore, 1991) which included a series of 1:2500 scale maps (land use, geomorphology, ground behaviour and planning guidance) and as a summary report (Doornkamp *et al.*, 1991) which was aimed at non-specialist professionals and the 'educated layman'. In addition a four-page colour leaflet was prepared to aid dissemination of results to the public (McInnes *et al.*, 1996).

The technical report emphasized that 'a positive approach to coordinating the community's response to the landslide problems is considered essential. Central to this should be a programme of public awareness. The aim should be to ensure that the purpose and potential benefits of the proposed management strategies is fully understood by and acceptable to the general public, developers, builders, estate agents, solicitors and financial institutions.'

The local authority was keen to work with the Department of the Environment over dissemination of the results of this study and ensure that the findings were acceptable to the local community. A potential problem was that the results could give rise to fears that might have an adverse impact on property values in some parts of the landslide system. This could well be acceptable in areas where landslide damage had long been recognized. However, in other developed areas where instability problems were not so readily apparent the logic of the designation would need to be clearly explained to those landowners who might be affected. These problems were likely to be exacerbated as a result of the difficulties in obtaining insurance cover in Ventnor. As the results of the study clearly indicated that serious problems were confined to a small number of locations, it was felt that the study could have a major influence on the availability of insurance cover, the ability to raise mortgages and ultimately house values. It was also recognized, therefore, that it was of great importance to provide the local residents with the opportunity to see the results of the study and judge for themselves how they might be affected.

The final report concluded that much could be done to minimize the impact of ground movement problems if appropriate management procedures were put in place. It was fully recognized by the local authority that the community should be made aware of the ways in which the instability problems could be managed, both as a means of emphasizing the local authority's commitment to tackling the problems and of highlighting what individual residents could do themselves.

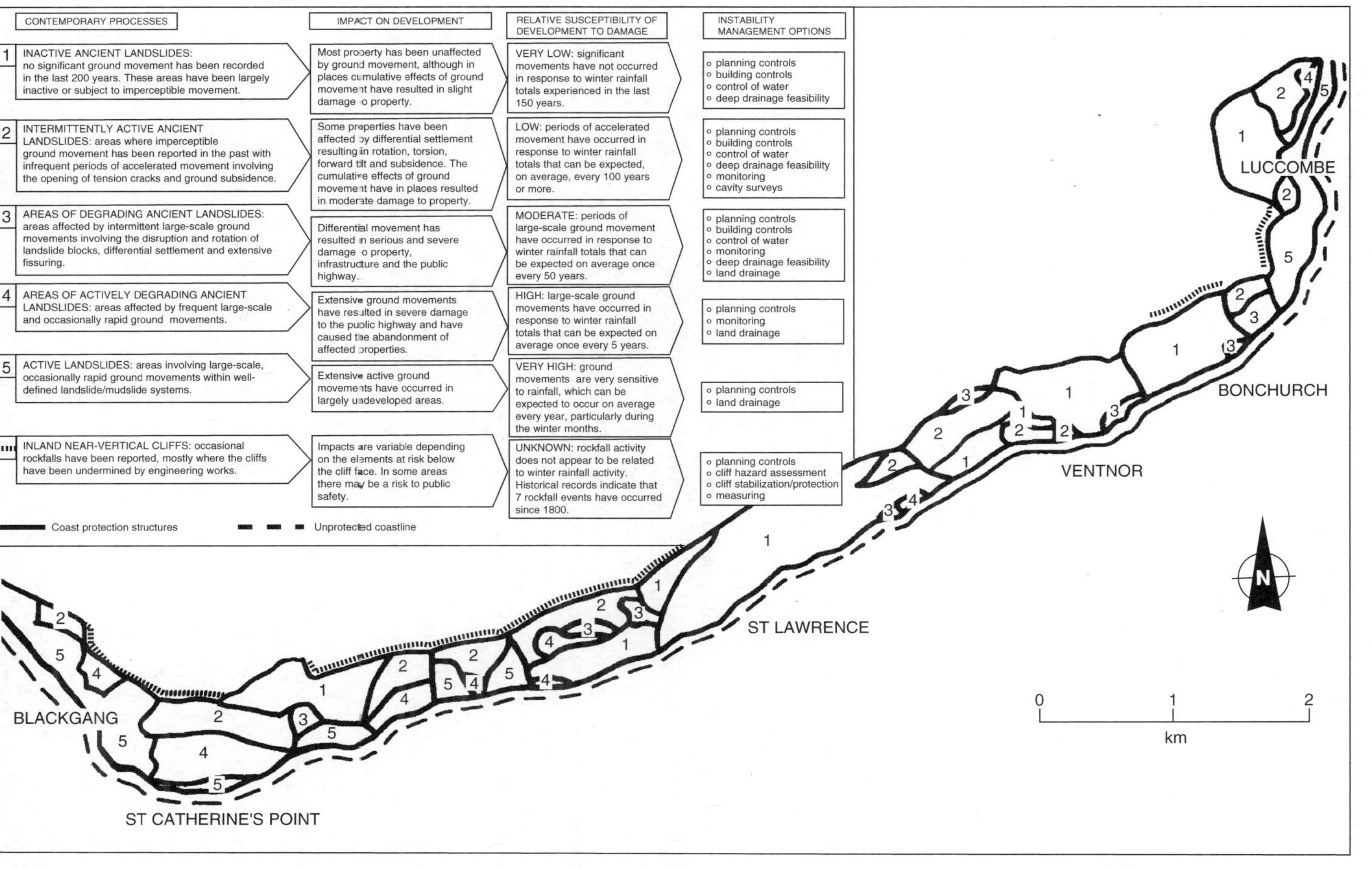


**Figure 14.4** *From policy to practice: geomorphology – ground behaviour and management options. Adapted from High-Point Rendel (1995). Reproduced by permission of Dr Alan Clark*

One way of disseminating advice and information is to establish an information centre which can provide a range of advice and information for residents and which is presented in an interesting, non-technical format (Doornkamp *et al.*, 1991). Examples of topics to be discussed could include:

- What is the history of ground movement in the area?
- What is the scale of the problem?
- Why is there a ground movement problem?
- What causes ground movement?
- How can we define landslide hazard?
- How can the landslide problems be managed most effectively?
- What is the local authority doing to help?
- What can developers do?
- What can property owners do?
- What can estate agents, solicitors and insurers do?
- What does the future hold if the community works together with the local authority?

In the case of Ventnor this kind of display was accompanied by a four-page explanatory leaflet entitled 'Land Stability in Ventnor and You'. An information centre of this kind provides the opportunity for interested residents not only to read the display boards, but also to ask questions and discuss any problems or concerns that they had with the local authority technical staff or its consultants, confidentially if required.

The information being disseminated by those staffing the Geological Information Centre was not entirely straightforward. However, the final paragraph from the summary report (Doornkamp *et al.*, 1991) provided a basis for explanation:

> There is no reason why there should not be confidence in Ventnor from a building, insurance or financial development point of view. This is true so long as sensible use is made of the technical information presented in the report and obtained from future monitoring exercises, and that the proposed landslide management strategies are practised. Of course, unstable areas must be avoided where possible. More stable areas may be successfully developed, as long as necessary stabilisation measures are adopted and the developer is willing to accept, in some locations, a higher level of risk than would be expected in normal circumstances.

As a result of the success of the temporary information centre as a dissemination point, a permanent display has now been provided since 1998 within the Isle of Wight Coastal Visitors' Centre based in Ventnor (Figure 14.5). The town has been able to turn the geological situation to its advantage and capitalize on the interest in geological, coastal and environmental tourism and education.

Following the completion of the public consultation on the first phase of the Ventnor ground investigations, the Council commissioned its consultants to formulate a 'landslide management strategy'. In due course a strategy was developed which included the following main elements:

- civil engineering measures (coastal protection works and drainage improvements);
- avoidance of seriously affected areas through planning and development control policies;
- public education and information;
- further research and monitoring.

**Figure 14.5**  *The Isle of Wight Coastal Visitors' Centre in Ventnor was opened in 1998. It provides a focus for information on coastal and geotechnical issues and is an important educational resource. Reproduced by permission of Brian Bradbury*

Furthermore, the Council, with the assistance of the Association of British Insurers, extended the original Department of the Environment study by including the remainder of Bonchurch (to the east) and working westwards along the Undercliff from central Ventnor to include the villages of Steephill and St Lawrence and the eastern part of Niton (Moore *et al.*, 1995). On completion of this further study an additional six maps were produced, together with a further information leaflet entitled 'Land Stability in the Undercliff and You'. This extension study was launched in a similar way but in addition presentations were given to groups such as estate agents, insurers, surveyors and builders. In the following year, 1996, the final phase of the Undercliff geomorphological mapping was undertaken and a further leaflet was launched, 'Advice to Homeowners in the Undercliff' (McInnes *et al.*, 1996). This particular leaflet aimed to provide specific information to assist the public in problems that they might need to address, including maintenance of slopes and vegetation, retaining walls, drainage systems and also advice with respect to insurance. The leaflet provided information on what the Council was doing on major infrastructure issues such as drainage and coastal protection.

This leaflet was distributed to each of the 2600 property owners in the Undercliff and was accompanied by a programme of publicity and information. Over the intervening period ongoing efforts were made to develop the Landslide Management Strategy through, in particular, addressing important infrastructure requirements such as protecting the toe of the landslide system with coast protection and encouraging the water company to replace public drainage systems to prevent leakage.

In September 2000 a slope stability study in Cowes and Gurnard on the northern coast of the Isle of Wight was completed (Moore *et al.*, 2000). This study successfully followed the model developed through the Ventnor studies and produced geomorphology, ground behaviour and planning guidance maps in addition to the technical report. The Cowes study was important because it illustrated the transferability of the approach developed in the Undercliff. This point was highlighted in a major study led by the Isle of Wight called 'Coastal Change, Climate and Instability', which received financial support from the EU LIFE Environment Programme (McInnes and Jakeways, 2000). The approach adopted in the Undercliff was outlined in this study to assist other regions facing similar instability management problems.

In order to establish the success of the Landslide Management Strategy in terms of the local community response, a questionnaire was sent to all 2600 households in the Ventnor Undercliff in February 2000. Its purpose was to obtain feedback on a range of instability, planning and management initiatives undertaken by the local authority (McInnes and Jakeways, 2000, 2002).

The results confirmed a high degree of satisfaction with the approach being adopted by the Isle of Wight Council. A high percentage of responding local residents (over 60%) had lived in the area over 10 years and the majority were aware of ground instability issues at the time of moving to the area (82%). Approximately half of those intending to move to the Undercliff sought professional advice on ground instability, the majority of which was obtained from surveyors, consulting engineers or local estate agents. It is encouraging to note that of those who sought advice from the local authority over 90% found the advice very helpful or helpful and two-thirds of those who responded had read the report published by the Council in 1995 entitled 'The Undercliff of the Isle of Wight – a review of ground behaviour'. Again, it was pleasing to note that 100% of those who read the report found it to be either very informative (55%) or informative (45%) (McInnes and Jakeways, 2000, 2002).

In 1996 the Council distributed a leaflet to all homeowners relating to instability management (McInnes *et al.*, 1996). Twenty per cent of those who responded had actually moved to the Undercliff area since the leaflet had been published but of those who had received a copy over 70% found the leaflet to be very helpful or helpful. The questionnaire also sought advice on whether further instability guidance would be of value. Eighty-one per cent of residents felt that further advice would be very useful and a further 15% felt it would be useful. Finally, residents were asked whether they considered ground instability in the Undercliff to be an issue of concern and 96% of those questioned regarded this as of great importance. The Council is actively responding to these results, recognizing that much is still to be achieved in a potentially worsening scenario as a result of predicted climate change impacts. A sustained effort will be required to maintain the status quo in terms of ground instability in the Undercliff.


## 14.6  Landslide Management – Drainage Measures

Many ground movement problems can be linked to high groundwater levels, which in combination with other factors such as coastal erosion or human activity can promote landsliding. Measures which control these factors will assist in reducing the likelihood of future movements but they will not, however, eliminate the risk. Rainfall and groundwater can act in a number of ways in promoting slope failure, first as *preparatory factors* which work to make the slope increasingly susceptible to failure without actually initiating it (i.e. causing the slope to move from a stable state to a marginally stable state, eventually resulting in a low factor of safety), and second, as *triggering factors* which actually initiate movement, that is, shift the slope from a marginally stable state to an actively unstable state (Crozier, 1986; Lee and Moore, 1991; Jones and Lee, 1994).

A common scenario is for a transient event (e.g. an intense storm) to trigger movement after there has been a gradual decline in stability (e.g. a prolonged wet period), so rainfall and groundwater can have both a long-term and a short-term influence on slope stability.

Management of groundwater and drainage can be approached in a number of different ways, including the provision of cut-off drains to intercept groundwater before it enters a landslide system with the aim of achieving a reduction in porewater pressures, as well as the drainage and disposal of surface water runoff and groundwater from within the landslide area itself. The ability to have an impact on groundwater will depend on the extent, depth and processes of landsliding taking place, as well as the underlying geology, the nature of the soils and their degree of permeability. In areas where the principal form of drainage is surface water runoff, shallower drainage measures are likely to be most suitable while deep drainage systems which may include cut-off drains, wells and adits may be more appropriate for more extensive and more deeply seated landslides.

Experience has shown that it is rare for drainage works alone to provide a total solution to an instability problem, and in most cases a combination of measures is necessary to achieve a reduction in the extent of the instability problem. Fundamental to the design of a drainage scheme is a knowledge of groundwater conditions which will only be achieved through hydrogeological studies, the obtaining of meteorological data, geomorphological mapping, and/or other investigations.

In the case of the Sirolo landslide on the Adriatic coast of Italy (Figure 14.6), a network of monitoring equipment was installed in order to provide, together with climatic data,

**Figure 14.6**   *The Sirolo landslide on the east coast of Italy near Ancona has been the subject of detailed investigations. Civil engineering measures and instability management strategies are also in place. Reproduced by permission of Dr Alan Clark*
*Source: Angeli et al., (1996)*

the piezometric response deep inside the landslide and to measure the rate of deformation at several points in the interests of public safety (Angeli *et al.*, 1996). This research has led to a preliminary hypothesis on the landslide mechanism and to the introduction of new measures to mitigate the displacement occurring in the sea cliff and coastal slope. The slope was strengthened to mitigate a continuous subsidence of the portion of the village directly facing the sea. Engineering measures comprising coastal protection works, the installation of high-strength steel anchors and three systems of long, sub-horizontal tubular drains have all assisted in reducing ground instability, including the lowering groundwater levels. A strategy of repairing leaking water supply and drainage pipes has also been put in place. A range of instrumentation provides continuous meteorological data, and information on both groundwater levels and stress within the body of the landslide has assisted in the design of the engineering works and an ongoing assessment of their effectiveness (Angeli *et al.*, 1996).

Many towns and villages throughout the world have inadequate or ageing drainage systems (both surface water and sewerage), and leakage can aggravate instability problems (McInnes and Jakeways, 2000). Where key service pipes cross known areas of instability, flexible materials including joints can be provided that can accommodate some ground movements.

The water company will wish to ensure that the new pipelines are watertight, flexibly jointed and routed around areas of greatest ground movement. A relative risk assessment can be undertaken for main supply pipes and a preferred route option can be identified. Such an assessment should take account of information gained from geomorphological and ground behaviour mapping (Clark *et al.*, 1996). This can help in assessing hazard, leading to a simple subjective method for determining the relative risk or 'hazard rating' for each pipeline route under consideration. This enables the various route options to be compared and ranked from best to worst with respect to ground conditions and the magnitude and frequency of recorded landslide and ground movement events.


## 14.7   Landslide Management – Slopes, Cliffs and Walls

The problem of landslide hazard on sloping ground is most commonly addressed through a range of structural solutions which attempt to address problems associated with ground-water levels, loading or excavation of slopes and past human activity. Generally speaking works of this kind reduce risks to development from slope movements or failure but do not prevent risk entirely. For this reason preventive measures are often accompanied by programmes of inspection or monitoring.

In many cases cliff and slope stability engineering works will require planning consent and in some locations it will be necessary to reconcile the demands for improved levels of protection with landscape, nature and earth science conservation interests (e.g. Lee, 2000a, b). Issues of maintaining biodiversity, geological exposures and habitats will have to be weighed against the socio-economic and sustainability arguments for undertaking civil engineering works at each site.

A range of slope stabilization options exist, including 'cut and fill' operations which may assist in unloading a slope or providing toe support or slope fill (e.g. Hutchinson, 1977; Bromhead, 1986). Drainage works which will divert surface and groundwater or ensure more effective drainage through the slope may be done by means of either drainage blankets or relatively shallow land drains or deeper cut-off drains which intercept groundwater at the top of the slope landward of the area of instability. In some locations horizontal drains have been drilled into the slope to assist in removal of groundwater in accordance with an engineering design. Finally, it is possible to remove water through pumping mechanisms, again by means of drainage or siphons.

Weathering and erosion of slopes can be controlled by means of a number of solutions, including provision of toe support measures or protection of the surface, for example with turfing and netting. Where slopes consist of rocky outcrops and exposures, public safety may be improved through the provision of netting or rock trap fencing (Figure 14.7).

In the case of walls, these may comprise structural retaining walls on suitably designed foundations, sheet or bore-piled walls or soil strengthening measures, for example ground and rock anchors or soil nails, which may be pre-stressed.

Slope treatment comprising reprofiling or drainage is a favoured method for stabilization. However, space may not always be available for techniques of this kind. Where space is confined, particularly in developed urban areas, the use of retaining walls, reinforced earth, soil and rock anchors and other engineered structures may be appropriate. These measures may also accompany slope reprofiling and drainage where they are deemed to be inadequate to provide the necessary level of slope stability on their own. Typically, retaining structures are composed of concrete or stone and may take the form of masonry or concrete walls, criblock structures or gabions (wire baskets filled with stones or rock). In some locations reinforced earth has been used, which involved the incorporation of

**Figure 14.7** *At Wheeler's Bay, to the east of Ventnor, it has been necessary to carry out extensive slope stabilization and coast protection works in order to reduce the impacts of coastal erosion on the landslide system. Reproduced by permission of David Bowie*

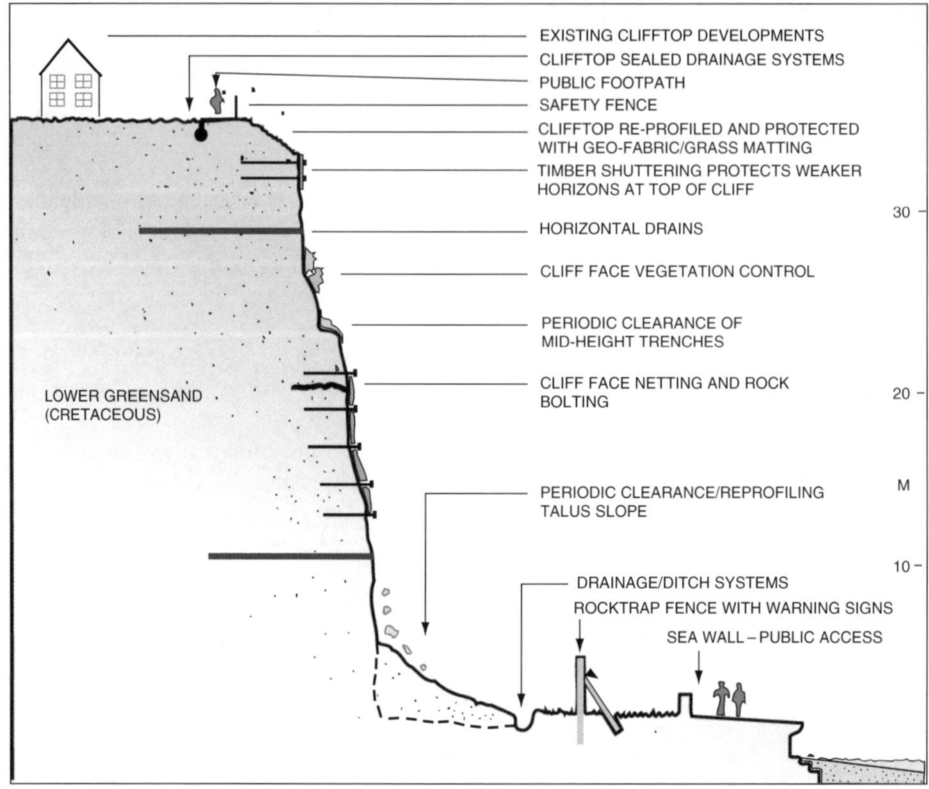


**Figure 14.8**  *Cliff stabilization measures, Sandown Bay, Isle of Wight, UK*
*Source: McInnes (2000)*

geotextile material within the supporting soil mass. Some of these structures may be accompanied by the provision of vegetation for both stabilization and aesthetic reasons (Figure 14.8).

## 14.8  Landslide Management – Construction and Maintenance of Buildings

Damage to buildings, roads and other infrastructure from instability events is a common problem but much can be done to try to mitigate damage caused by ground movements. Some of these measures have been described earlier, for example, with respect to management of slopes, walls and drainage systems. Having established the planning framework, which will provide a basis for decision making in terms of the general suitability of a site for development, it is usually the developer who will undertake appropriate investigations and studies to demonstrate that the particular site concerned is suitable to accommodate the proposed construction activity. In some locations local authorities hold more detailed maps and reports, which can assist in this process first by guiding development to the most suitable areas in terms of ground stability and second by giving advice on the type of further survey or investigation that may be required to support the development proposal.

It is clear, therefore, that the local authority, the developer, the architect and the builder and in due course the occupier of the building all have a role to play in terms of ensuring that the development is constructed in the most suitable manner to take account of ground conditions and is maintained adequately in the future to try to reduce the impact of any ongoing ground movements.

The developer has, of course, a key interest in ensuring that the site can support his development and that any ongoing or subsequent ground movement will not adversely affect the site and, therefore, reduce its value or marketability. Unfortunately, in many cases proper site investigation does not begin until problems have arisen, often during the construction stage. The undertaking of investigation and remedial works during construction as opposed to before the start can be extremely costly in terms of delays, contractual claims and adverse publicity which may affect the future marketability and insurance of the development. The developer must take account not only of the ground conditions on the development site itself but also of the possible impact of the development in terms of adjacent sites which may be adversely affected by the proposal. Again, this is a matter that must be addressed at the planning stage, and the development proposal must be supported by sufficient information to allow the local authority to undertake a proper review of the proposal. Studies of this kind should be undertaken by a competent geotechnical specialist who is likely to be registered with an appropriate professional institution in the country concerned. The planning authority may wish to receive proof of the competence of the person undertaking the survey and complete a checklist of important issues that must be addressed in the stability report as an aid to assessing the development application.

In terms of construction techniques, there are a number of opportunities for minimizing the impacts of ground movement through appropriate design. The foundations of buildings are particularly important. Traditional strip foundations can easily fracture, causing significant structural damage. Rafts can accommodate slight movements and span minor fissures and voids that may form beneath the raft over the lifetime of the building.

In general, simple designs are preferable to more complex structures, with some degree of interlocking or articulation to accommodate slight movements. Light-weight framed buildings, which may be of timber with brick or concrete infill or sheet construction materials, are likely to be least problematic in the future. Tile- or slate-hung features can cover minor damage whereas more rigid construction methods or rendering tend to show cracking quite quickly. If movements do occur that affect the raft foundation, this can be accommodated in some situations through the provision of adjustable jacking points as part of the ring beam or foundation design. This method creates the possibility of re-levelling the raft at some time in the future.

In summary, therefore, a residential property on unstable ground could be constructed on a raft foundation with jacking points and should be light-weight, low-rise and composed of materials that will not be prone to visible cracking and damage.

Particular attention should be paid to the use of concrete. This should be constructed in small bays with frequent use of expansion joints, and should contain reinforcing mesh. For paths and hardstanding areas, block paving may be appropriate. Externally, particular attention should be paid to the design of guttering and rainwater downpipes to ensure that these are of sufficient capacity to accommodate more intense rainfall events and to take account of possible changes in climate which may result in more prolonged winter rainfall. Ideally roof water and surface water runoff from hard areas should be connected to sealed drainage systems or existing ditches. Some of these issues may well be set out by the planning authority as 'conditions for approval' for the development.

During the course of construction, the developer needs to pay particular attention to on-site management, proper control of earth works in sequence to avoid inappropriate excavations or leaving slopes inadequately supported; unsupported trenches, for example for service supply pipes, should be avoided. Many instability problems arise from poor site management and lack of proper earth works control generally. During the course of construction, groundwater should not be allowed to pond on site as this may cause problems during construction or at a later date.

## 14.9    The Role of the Homeowner

Whilst individual property owners may be able to exert only a minimal influence on ground instability problems, the cumulative effect of efforts by many homeowners on the landslide system may be significant. Activities such as vegetation clearance, slope regrading, cut and fill operations, lack of maintenance or inattention to leaking pipes can all adversely affect ground stability. Residents, both individually or in groups (by area or by road), can work together to ensure that issues such as the maintenance of drainage systems are addressed. A good time for inspections is in the early autumn before the onset of the autumn/winter rainfall period, when guttering and downpipes should be checked for blockages and leakage and road drainage systems and ditches should be cleared.

For properties in shared ownership there is an individual but also a collective responsibility to contribute towards management of the building and its grounds. It is sometimes more difficult for tenants, particularly in large buildings that have been divided into apartments, to ensure that a coordinated approach is taken to addressing structural maintenance and drainage problems; this may be resolved through a residents' management committee. Lack of maintenance will make the building all the more susceptible to slight ground movements, and so regular maintenance is particularly important. Figure 14.9 illustrates some of the practical problems associated with maintenance of properties in an area of instability and provides guidance with respect to key issues; further information may be obtained from the references at the end of this chapter.

## 14.10    The Suitability of Land for Development

Some areas of unstable land are quite clearly less suitable for development and should be avoided (see Figure 14.10). A range of organizations and individuals may be involved in this decision-making process, including the planning authority, mortgage lenders and

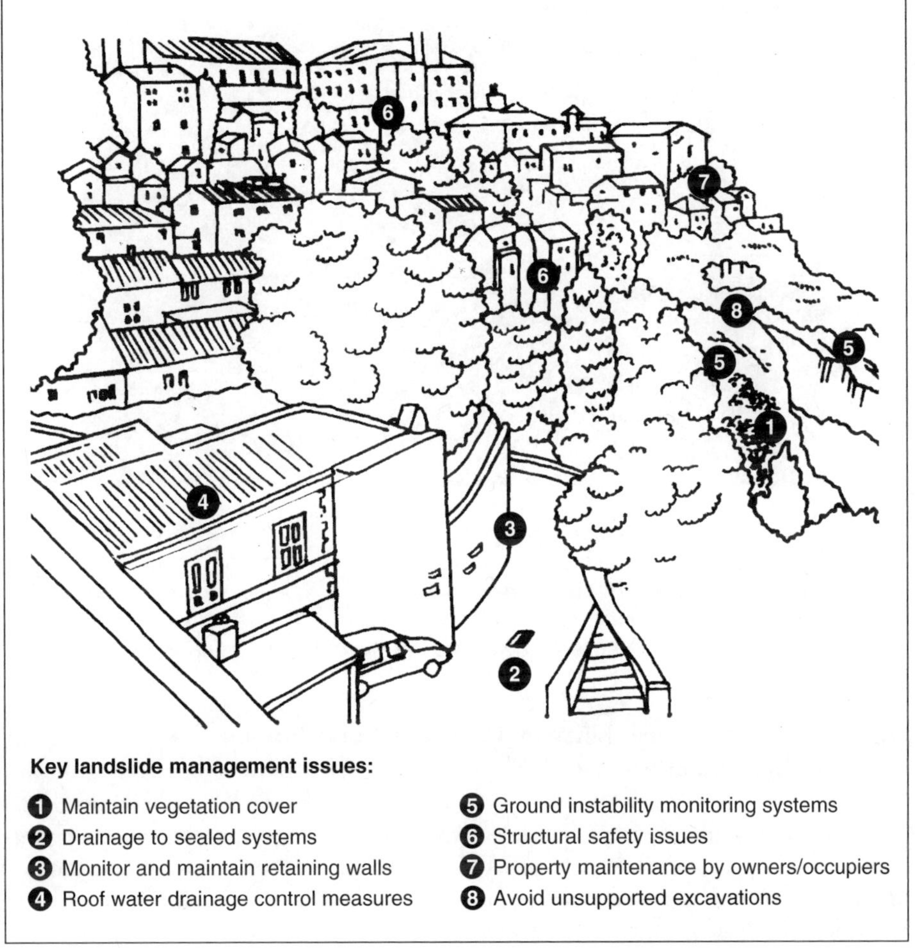


**Figure 14.9** *Instability management in an Italian hill town*
Source: McInnes (2000)

insurance companies, as well as the purchaser. In most cases the responsibility lies with the owner to ensure that the property or site is safe to develop. However, increasingly, information is being made available by local authorities to aid planning, and in addition vendors are required to provide any information they may have on ground conditions relating to the land or property that they are selling.

A balance needs to be struck between avoiding development in certain areas and being too restrictive. A decision-making process has been outlined whereby, following a geomorphological appraisal, it is possible to assess the impact of ground conditions on development and thereby allocate land in terms of suitability or unsuitability for development; some decisions cannot be finalized until further site investigations or studies have been undertaken.

***Figure 14.10*** *The Blackgang landslide at the western end of the Isle of Wight Undercliff. The combined effects of coastal erosion and instability have led to the loss of properties and infrastructure during the last 150 years; over that period the coastline has retreated by about 300 m. For economic, environmental and technical reasons coast protection and stabilization is not a viable option. Reproduced by permission of Isle of Wight County Press Ltd*

## 14.11   Monitoring Instability as Part of a Landslide Management Strategy

Monitoring is an integral part of landslide investigation and ongoing management because it provides a means of accurately and objectively gauging the stability conditions of unstable or potentially unstable slopes; it can also fulfil an important role in assessing risk. The objectives of monitoring include:

- providing information to assist landslide investigation;
- determining the rate and scale of ground movements, particularly in vulnerable locations;
- identifying links between ground movement, rainfall and groundwater levels that can be used to develop a methodology for landslide forecasting;
- providing early warning in areas where ground movements could affect life and property;
- monitoring the effectiveness of landslide management strategies. In addition, instrumentation can help deduce the mechanisms of failure as well as assisting in design verification, construction control, quality control, performance of structures and can provide legal protection against claims from owners of properties adjacent to construction sites.

Ground conditions often determine the choice of a specific instrument, but in addition instrument performance must be considered. It is necessary to consider minimum performance requirements and the cost of the instrumentation, which may increase with its resolution, accuracy and precision.

In a number of unstable locations, where monitoring has taken place over several years, a monitoring strategy has been established which includes the gathering of data on both meteorology and ground movements (Fort *et al.*, 2000). Increasingly taking into account the cost of manual data gathering, electronic systems are being used, allowing easier data acquisition, interpretation and storage. A comprehensive approach is that adopted at Lyme Regis in Dorset, UK (Fort *et al.*, 2000). A detailed study of the town of Lyme Regis was undertaken because of concerns about ground instability and the need for upgrading of coastal defences along the frontage. Monitoring systems of this kind can, therefore, support manual inspections, assist landowners by increasing their awareness of any hazards that may exist and which may cause nuisance to adjacent owners, as well as allowing the provision of baseline data against which climatic change can be assessed. For all monitoring programmes it is essential that accurate records are kept of inspections and due attention is given to trends or variations in the rate of change of readings. Not only will monitoring systems allow the implementation of an emergency response if required; they can also provide baseline information and increase scientific knowledge in the area concerned.

## 14.12   Coordinating the Community Response to Instability Problems

It has been established that an understanding of earth science information, including the geological and geomorphological setting, can assist the planning and development process by safeguarding assets as well as reducing costs during the development stage. It is equally clear that both national and local government and all those living and working in areas of instability have a role to play in assisting hazard and risk reduction. In many cases it may not be possible to undertake engineering measures that will eliminate landslide hazard but it may be possible to modify or reduce future movement through a more coordinated approach.

A strategy for landslide management is likely to comprise a range of initiatives, including engineering measures (such as coastal defence works, drainage improvements and repair of leaking pipes) alongside construction activity, undertaken in an appropriate manner, together with planning measures and ongoing monitoring and further research. In the Isle of Wight the Undercliff landslide complex has been managed in this way for the last 9 years. A strategy comprising these key elements is overseen by the technical management committee which has the aim of coordinating the overall response to the instability problem in order to try to reduce the impact of landsliding, thereby minimizing its effects on the local economy and population. The structure of the Management Strategy is illustrated in Figure 14.11, together with the responsibilities of key agencies involved in this coordinated approach.

## 14.13   The Role of the Insurance Industry in Landslide Mitigation and Insurance

The insurance industry can play an important role in assisting mitigation of the costs associated with landslides. Most insurance cover is based on a broad-brush assessment of instability risk by the particular company concerned; this usually takes account of past

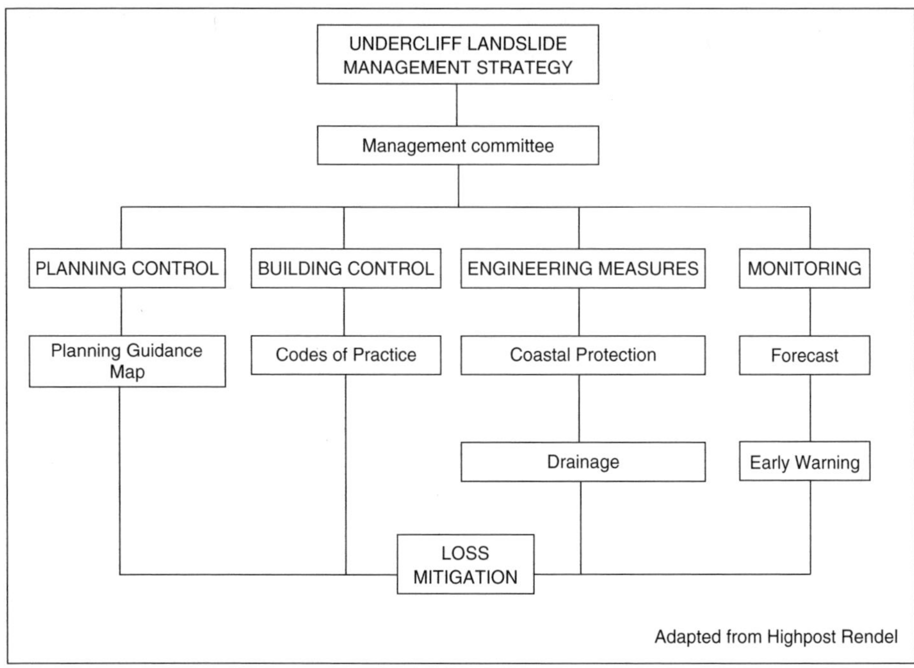



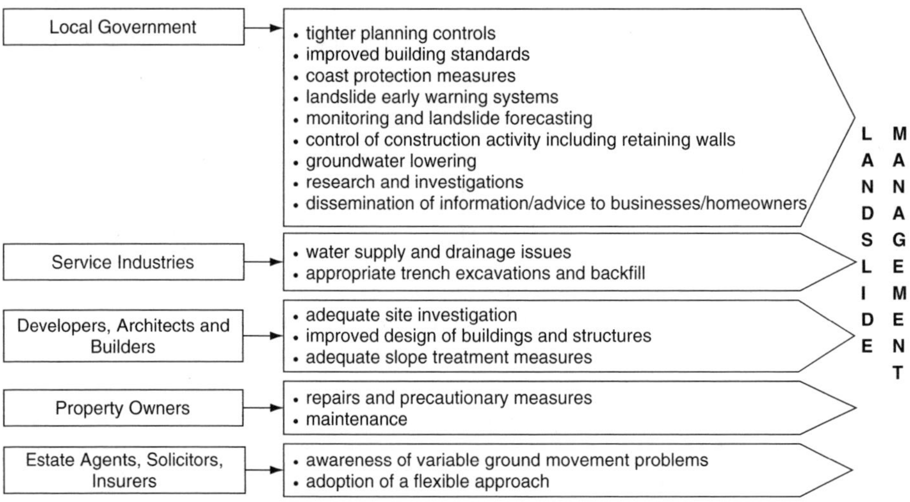



**Figure 14.11**   *Isle of Wight Undercliff landslide management strategy. Adapted from High-Point Rendel (1995). Reproduced by permission of Dr Alan Clark*

claims. Enhanced premiums may be charged where the insurance company believes that a higher level of risk exists. In some areas, taking account of the risk, insurance cover may not be available at all.

It is important that those involved with instability management open a dialogue with the insurance industry in order to provide advice and information on initiatives that are being considered and implemented to try to reduce the impact of landsliding on local economies and the community in general. Likewise the insurance industry should play a greater role by becoming involved in the strategic planning process as a consultee, and by obtaining more reliable information on ground conditions, thereby allowing an improved assessment to be made of insurance risk.

Much damage that occurs in landslide-prone areas is of a relatively minor nature. Insurance companies can lend support to landslide management strategies by requiring homeowners, as a condition of insurance cover, to maintain their properties in a satisfactory condition. Mortgage lenders can also assist in this respect by requiring surveys to be undertaken of particular risk elements such as drainage systems when properties are changing hands.

The provision of information regarding ground instability without doubt forms a very valuable contribution towards improving practice and management in the areas concerned. The availability of this information, however, imposes responsibilities on all those who may be using it, including property developers, builders, planners, estate agents and insurers. With the information provided in reports and maps, all those involved in land management are expected to take account of this information during the decision-making process. In the same way homeowners would be expected to disclose to a prospective purchaser any information they are aware of regarding ground instability affecting their property.

A major objective of geotechnical studies is to try to restore confidence in the area concerned through an improved understanding of ground conditions and instability potential. In areas where it is difficult to obtain insurance, insurance companies may reconsider the question of providing cover if information provided by their customers is available which can demonstrate a reduced level of risk. Insurance is, therefore, a further additional, important aspect that should be considered as part of the development of a landslide management strategy.


## 14.14    Transferability of the 'Landslide Management' Approach

In this chapter a detailed description has been provided of some aspects of landslide management implemented in the UK and elsewhere. How transferable are these initiatives to other situations and locations? From consultations with engineering geologists and local authorities within Europe and elsewhere as part of a major study supported by European Union funding, it is clear that there is widespread interest in this kind of approach (McInnes and Jakeways, 2000). The key elements of a landslide management strategy which aims to reduce landslide hazard should, therefore, involve:

- identifying and understanding the nature and extent of the instability hazard;
- modifying the hazard to the community by means of engineering works (including coastal defence, drainage measures or slope stabilization) and improved building practice;

- effective planning controls to guide development to suitable areas and to control the nature of new development;
- improving the understanding of landslide behaviour;
- mitigating the losses resulting from ground movement through insurance and other measures;
- coordinating the community response to the problems.

Sustainable development requires wise decision making taking full account of past and present ground conditions. This can be achieved most effectively by means of a coordinated approach to instability management, thereby minimizing risks to vulnerable communities by:

- guiding development towards suitable locations;
- ensuring that existing and future developments are not exposed to unacceptable risks;
- ensuring that development does not increase the risk for the rest of the community.

The implications of climate change and sea-level rise also present a significant challenge to future instability management. On the coast an understanding of long-term evolution will allow the identification of areas where management problems are likely to arise in the future. The existing arrangements in place for strategic examination of coastal issues within the framework of guidance provided in the UK form a good model which could be replicated elsewhere. This would assist in avoiding poor planning and siting of developments that have, in the past, made people and property more vulnerable to natural hazards.

The relevance of natural and man-made risks to the planning system is already recognized in guidance published by a number of countries. It is in the interest of insurance companies and the other stakeholders to ensure that guidance is realistic and appropriate in the way that it considers risk.

New technology provides excellent opportunities to ensure that better exchange and dissemination of information takes place both within the scientific community and between engineers and practitioners. There is a need for closer integration between the actions of engineers, planners and the construction industry. Local authorities can play a pivotal role by coordinating landslide management strategies 'on the ground', and they are best placed to maintain the momentum following the development of a strategy.

It is hoped that the advice and information provided in this chapter will prove of practical assistance in reducing the impact of instability on local communities and economies in areas affected by instability problems.

## References

Angeli, M.-G., Gasparetto, P. and Pontoni, F., 1996, Longterm monitoring and remedial measures in a coastal landslide (Italy), *Proceedings of the VIIth ISL*, Trondheim, Norway (Rotterdam: Balkema).

Bromhead, E.N., 1986, *The Stability of Slopes* (Guildford: Surrey University Press).

Brundtland Commission, 1987, *Proceedings of the World Conference on Environment and Development.*

Clark, A.R., Lee, E.M. and Moore, R., 1994, The development of a ground behaviour model for the assessment of landslide hazard in the Isle of Wight Undercliff and its role in supporting major development and infrastructure projects, in *Proceedings of the 7th International IAEG Congress*, vol. VI, 4901–4913.

Clark, A.R., Lee, E.M. and Moore, R., 1996, *Landslide Investigation and Management in Great Britain: A Guide for Planners and Developers*, Department of the Environment. (London: HMSO).

Clark, A.R., Palmer, J.S., Firth, T.P. and McIntyre, G., 1996, The management and stabilisation of weak sandstone cliffs at Shanklin, IW, in Cripps *et al.* (eds), *Engineering Geology of Weak Rocks Conference* (Rotterdam: Balkema), pp. 375–384.

Crozier, M.J., 1986, *Landslides: Causes, Consequences and Environment* (London: Croom Helm).

DoE (Department of the Environment), 1991, Planning Policy Guidance Note 14 – Development on Unstable Land (London: HMSO).

Doornkamp, J., Lee, E.M. and Moore, R., 1991, *Ground Movement in Ventnor, Isle of Wight*, Non-technical Summary Report by Geomorphological Services Ltd for the Department of the Environment, UK (London: HMSO).

Fort, D.S., Clark, A.R., Savage, D.T. and Davis, G.M., 2000, Instrumentation and monitoring of the coastal landslides at Lyme Regis, Dorset, UK, in *Landslides in Research, Theory and Practice* (London: Thomas Telford).

Hutchinson, J.N., 1977, The assessment of the effectiveness of corrective measures in relation to geological conditions and types of slope movement, *Bulletin of the International Association of Engineering Geology*, **16**, 131–155.

Jones, D.K.C. and Lee, E.M., 1994, *Landsliding in Great Britain* (London: HMSO).

Lateltin, O.J., 1997, *Natural Hazards: Recommendations Taking Account of the Hazards Caused by Ground Movements in the Land Use Planning Framework* (Berne). Translated from German by Isle of Wight Centre for the Coastal Environment, 2000.

Lateltin, O.J., 2002, Landslides, land use planning and risk management – Switzerland as a case study, *Proceedings of the International Conference 'Instability – Planning and Management'*, Ventnor, IW, UK (London: Thomas Telford).

Lee, E.M., 1993, Coastal planning and management: Planning and Management. Policy responses to the implications of sea level rise, *The Geographical Journal*, **159**(2), 169–178.

Lee, E.M., 1995a, *The Investigation and Management of Erosion, Deposition and Flooding in Great Britain* (London: HMSO).

Lee, E.M., 1995b, Coastal Planning and Management : A review of Earth Science Information Needs (London: HMSO).

Lee, E.M., 1996, Earth science information in support of coastal planning: the role of the Shoreline Management Plans, in C.A. Fleming (ed.), *Coastal Management 1995: Putting Policy into Practice* (London: Thomas Telford), 54–65.

Lee, E.M., 1997, Landslide risk management: key issues from a British perspective, in D. Cruden and R. Fell (eds), *Landslide Risk Assessment* (Rotterdam: Balkema), 227–237.

Lee, E.M., 2000a, The management of coastal landslide risks in England: the implications of conservation legislation and commitments, in E.N. Bromhead, N. Dixon and M.-L. Ibsen (eds), *Landslides: In Research, Theory and Practice* (London: Thomas Terford), 893–898.

Lee, E.M., 2000b, Landslide hazard in Great Britain, *Geoscientist*, **5**(4).

Lee, E.M. and McInnes, R.G., 2000, Landslide hazard mapping for planning and management: the Isle of Wight Undercliff, UK, in *Living with Natural Hazards, Proceedings of the CALAR Conference, Vienna*.

Lee, E.M. and Moore, R., 1991, *Coastal Landslip Potential Assessment – Isle of Wight Undercliff*, Technical report by Geomorphological Services Ltd for DoE (London: HMSO).

Lee, E.M., Brunsden, D., Moore, R. and Siddle, H.J., 1991, The assessment of ground behaviour at Ventnor, Isle of Wight, in R.J. Chandler (ed.). In *Slope Stability Engineering, Developments and Applications*. In *Proceedings of the Shanklin, Isle of Wight Conference*. (London: Thomas Telford), 207–212.

Leroi, E., 1996, Risk maps at different scales: objectives, tools and developments, *Proceedings of the VIIth ISL*, Trondheim, Norway. (Rotterdam: Balkema).

McInnes, R.G. and Jakeways, J., 2000, *Coastal Change, Climate and Instability*, EU LIFE Environment project for the European Commission, Ventnor.

McInnes, R.G. and Lee, E.M., 2000, Living with landslides: the management of the Isle of Wight Undercliff, UK, in *Living with Natural Disasters, Proceedings of the CALAR Conference, Vienna.*

McInnes, R.G. and Jakeways, J., 2002, Managing ground instability in the Ventnor Undercliff, Isle of Wight, UK, in *Proceedings of the International Conference 'Instability – Planning and Management', Ventnor, IW* (London: Thomas Telford).

McInnes, R.G., Clark, A.R. and Moore, R., 1996, Advice to Homeowners in the Undercliff, summary leaflet for South Wight Borough Council.

Moore, R. and Lee, E.M., 1991, *Ventnor Information Centre – Getting the Message Across, Ground Movement and Public Perception*, Report for South Wight Borough Council, Ventnor, UK.

Moore, R., Lee, E.M. and Noton, N., 1991, The distribution, frequency and magnitude of landslide movements at Ventnor, Isle of Wight, in R.J. Chandler (ed.). In *Slope Stability Engineering, Developments and Applications, Proceedings of the Shanklin, Isle of Wight Conference,* 213–218.

Moore, R., Lee, E.M. and Clark, A.R., 1995, *The Undercliff of the Isle of Wight – A Review of Ground Behaviour*, Report for South Wight Borough Council, Ventnor, UK.

Moore, R., Lee, E.M. and Brundsden, D., 2002, *Cowes to Gurnard Coastal Slope Stability Study – Ground Behaviour Assessment*, Report by Halcrow for Isle of Wight Council.

Moore, R., Siddle, H. *et al.*, 2003, Ventnor landslide investigation and risk assessment, draft report for Isle of Wight Council (unpublished), Isle of Wight Council.

Thompson, A., Hine, P.D., Poole, J.S. and Griegg, J.R., 1998, *Environmental Geology in Land Use Planning*, Report by Symonds Travers Morgan for the Department of the Environment, Transport and the Regions, UK.

# 15

# Geomorphological Mapping to Assess Landslide Risk: Concepts, Methods and Applications in the Umbria Region of Central Italy

*Paola Reichenbach, Mirco Galli, Mauro Cardinali, Fausto Guzzetti and Francesca Ardizzone*

## 15.1  Introduction

Following a catastrophic landslide disaster in the Campania region on 5 May 1997, when tens of people were killed by debris flows, on August 1998 the Italian government passed new legislation on landslide and flood risk assessment and mitigation (*Gazzetta Ufficiale della Repubblica Italiana*, 1998). The new legislation requires that the regional governments and the National River Basin Authorities identify and map areas where landslide risk is most severe, and take action to reduce societal risk and economic damage. The government of the Umbria region and the Tiber River Authority commissioned the Italian National Research Council (CNR), Institute for Geo-Hydrological Protection (IRPI) in Perugia to assess landslide hazard and risk in Umbria. To respond to this request, we devised a geomorphological methodology to evaluate landslide hazards and risk at site-specific scale.

In this chapter we briefly review landslide risk assessment methods, illustrate the geomorphological methodology we have devised to determine landslide hazards and risk in Umbria, and demonstrate applications of the method in three study areas representative of the main geological and physiographical provinces in the region. We conclude by discussing advantages and limitations of the proposed approach.

*Landslide Hazard and Risk*   Edited by Thomas Glade, Malcolm Anderson and Michael J. Crozier
© 2004 John Wiley & Sons, Ltd   ISBN: 0-471-48663-9

## 15.2   Concepts and Terminology

Many different triggers cause landslides, including intense or prolonged rainfall, earthquakes and rapid snowmelt. On earth the area of landslides spans nine orders of magnitude and the volume of mass movements spans 15 orders of magnitude; landslide velocity extends over 14 orders of magnitude, from millimetres per year to hundreds of kilometres per hour. Mass movements can occur singularly or in groups of up to several thousands. Multiple landslides occur almost simultaneously when slopes are shaken by an earthquake, or over a period of hours or days when failures are triggered by rainfall or by snowmelt. Landslides can involve flowing, sliding, toppling or falling movements, and many landslides exhibit a combination of these types of movements (Varnes, 1978). The extraordinary breadth of the spectrum of landslide phenomena makes it difficult to define a single methodology to ascertain landslide hazards and to evaluate the associated risk.

Assessing landslide hazards and risk is a complex and uncertain operation that requires the combination of different techniques and methodologies, and the interplay of various types of expertise, not all of which pertain to the realm of the earth sciences. A review of the vast literature on landslide hazard assessment (see Varnes and IAEG Commission on Landslides and other Mass-Movements, 1984; Hutchinson, 1995; Soeters and van Westen, 1996; Guzzetti *et al.*, 1999b; and references therein), and of the literature on landslide risk evaluation (Einstein, 1988; Fell, 2000; Cruden and Fell, 1997; and references therein) is beyond the scope of this chapter. In this section we briefly review the concepts and terminology related to landslide hazard and risk assessment.

Landslide hazard refers to the natural conditions of an area potentially subject to slope movements. In a well-known report, Varnes and the IAEG Commission on Landslides and other Mass-Movements (1984) (henceforth Varnes – IAEG) proposed that the definition adopted by UNDRO for all natural hazards be applied to landslide hazard. Landslide hazard is therefore 'the probability of occurrence within a specified period of time and within a given area of a potentially damaging phenomenon'. Guzzetti *et al.* (1999b) amended this definition to include the magnitude of the event, that is, the area, volume, velocity or momentum of the expected landslide. The amended definition incorporates the concepts of location, time and magnitude. Location refers to the ability to forecast where a landslide will occur; magnitude refers to the prediction of the size and velocity of the landslide; and frequency refers to the ability of forecasting the temporal recurrence of the landslide event (Guzzetti *et al.*, 1999b). Quantitative (probabilistic or deterministic) and qualitative approaches to ascertain landslide hazards are possible. Reviews of quantitative approaches are given by Soeters and van Westen (1996) and Guzzetti *et al.* (1999b). In this work we utilize a qualitative approach.

Landslide risk evaluation aims to determine the 'expected degree of loss due to a landslide (*specific risk*) and the expected number of lives lost, people injured, damage to property and disruption of economic activity (*total risk*)' (Varnes and the IAEG Commission on Landslides and other Mass-Movements, 1984). Quantitative (probabilistic) and qualitative (heuristic) approaches are possible (Einstein, 1988, 1997; Michael-Leiba *et al.*, 1999; Fell, 1994, 2000; Fell and Hartford, 1997).

Quantitative risk assessment aims to establish the probability of occurrence of a catastrophic event, for example, the probability of loss of life, or the probability of a landslide causing a given number of casualties or fatalities. The method requires a catalogue of landslides and their consequences. A few such lists have been prepared for landslides with human consequences, that is, deaths, missing people and injuries (Evans, 1997; Guzzetti, 2000; Kong, 2002; Guzzetti *et al.*, 2003). To compile accurate and complete lists of landslides that have caused other types of damage is more difficult, due to the lack of relevant information. When this information is available, levels of individual and societal risk can be determined. Individual risk is the risk posed by a hazard to any identified individual, and is expressed using mortality rates, which are given by the number of deaths per 100 000 of any given population over a predefined period. Societal (collective) risk is the risk imposed by a hazard (i.e. a landslide) on society, and is established by investigating the relationship between the frequency of the damaging events and their intensity, as measured by the number of fatalities. Acceptable risk levels are determined by comparison with other natural, technological, social and medical hazards for which acceptable risk levels have already been established (Fell and Hartford, 1997; Salvati *et al.*, 2003). The completeness and time span of the landslide catalogue greatly affect the reliability of such quantitative risk assessments.

When attempting to evaluate landslide risk for a site or region where slope movements are likely to take various forms or pose various types of threat, the quantitative approach often becomes impracticable. As an example, the quantitative assessment of the risk posed by mass movements to structures and the infrastructure is defined as the 'product' of landslide hazard and the vulnerability of the structure or infrastructure (Varnes – IAEG, 1984). The latter ranges from 0 to 1, 0 meaning no damage and 1 representing complete destruction. The definition of landslide risk requires that both hazard and vulnerability be defined as independent probabilities (of occurrence, for hazard; and of damage, for vulnerability). In practice, it is rarely possible to define hazard and vulnerability as probabilities, limiting the rigorous application of the definition of landslide risk.

Considering that it may not be easy to ascertain the magnitude, frequency and forms of evolution of landslides in an area, that detailed information on the vulnerability of the elements at risk is often lacking, and that accurate and reasonably complete catalogues of historical events with consequences may not be readily available, in some areas a qualitative approached can be pursued so as to establish qualitative levels of landslide risk. This can be accomplished by investigating the impact of mass movements in a given area, and by designing landslide scenarios.

The impact that slope failures have had, or may have, in a given area can be established in two ways. First, where a historical catalogue of landslides and their consequences is available, the sites repeatedly affected by catastrophic events can be determined and the vulnerability of the elements at risk ascertained (Budetta, 2002; Kong, 2002). Alternatively, where a detailed landslide inventory map and a map of structures (houses, buildings, etc.) and infrastructure (roads, railways, lifelines, etc.) at risk are available in GIS form, simple geographical operations allow one to determine where landslides may interfere with the elements at risk. For the Umbria region, Guzzetti *et al.* (2003) intersected in a GIS a detailed geomorphological landslide inventory map showing more than 45 000 landslides with maps of the built-up areas and of the transportation network. The operation revealed 6119 sites where known landslides intersect (i.e. may interfere) with built-up areas, and 4115 sites where landslides intersect roads or railways. At these localities damage due to landslides can be expected, particularly during major landslide-triggering events (e.g. prolonged rainfall, snowmelt events, etc.).

The design of a landslide scenario is a complex, largely empirical procedure that involves: (i) identification of the types of slope processes present in the study area (ii) accurate mapping of the existing landslides; and (iii) assessment of the possible (or probable) evolution of the slope movements. The last can be ascertained qualitatively by experts, or determined quantitatively using mathematical or physically based models which simulate the expected evolution of a landslide. Multiple landslide scenarios can be prepared, one scenario for each landslide or landslide type present in the study area. By combining multiple landslide scenarios with information on the location and the type of structures and infrastructure and their vulnerability, one can determine qualitative levels of landslide risk. We accomplished this in 79 villages in Umbria, where we have adopted a qualitative, heuristic approach to ascertain landslide risk based on the definition of multiple landslide scenarios.

## 15.3　Settings and Previous Landslide Studies

The Umbria region covers $8456\,km^2$ in central Italy (Figure 15.1), with elevation ranging from 50 to 2436 m a.s.l. The landscape in the region is hilly or mountainous, with open valleys and large intra-mountain basins, drained by the Tiber River, which flows to the Tyrrhenian Sea. Rainfall occurs mainly from October to December and from March to May, with cumulative annual values ranging between 500 and 2100 mm. Snowfall occurs every year in the mountains and about every five years at lower elevations.

Due to the lithological, morphological and climatic setting, landslides are abundant in Umbria (Servizio Geologico d'Italia, 1980; Guzzetti *et al.*, 1996). Mass movement occurs almost every year in the region in response to prolonged or intense rainfall, rapid snowmelt and earthquake shaking. Landslides in Umbria can be very destructive and have caused damage at many sites. In the twentieth century a total of 29 people died or were missing and 31 people were injured by slope movements in Umbria in a total of 13 harmful events (Salvati *et al.*, 2003).

Research on slope movements is abundant in Umbria. Landslide inventory maps were compiled by Guzzetti and Cardinali (1989), Antonini *et al.* (1993) and Cardinali *et al.* (2001). Such studies revealed that landslides cover about 8% of the territory. Locally, landslide density is much higher, exceeding 20%. Geomorphological relationships between landslide types and pattern, and the morphological, lithological and structural settings were investigated among others by Guzzetti and Cardinali (1992), Barchi *et al.* (1993) and Cardinali *et al.* (1994), and were summarized by Guzzetti *et al.* (1996). Site-specific, geotechnical investigations on single landslides or landslide sites were conducted at several localities, mostly in urbanized areas (Crescenti, 1973; Tonnetti, 1978; Diamanti and Soccodato, 1981; Calabresi and Scarpelli, 1984; Lembo-Fazio *et al.*, 1984; Canuti *et al.*, 1986; Cecere and Lembo-Fazio, 1986; Righi *et al.*, 1986; Tommasi *et al.*, 1986; Ribacchi *et al.*, 1988; Capocecere *et al.*, 1993; Felicioni *et al.*, 1994). Landslide hazard assessments have been completed in test areas and for different landslide types by Carrara *et al.* (1991, 1995) and Guzzetti *et al.* (1999b, 2004). A regional landslide hazard map for the upper Tiber River basin, most of which lies in Umbria, was completed by Cardinali *et al.* (2002a). Historical information on the frequency and recurrence of failures in Umbria was compiled by a nation-wide project that archived data on landslides and

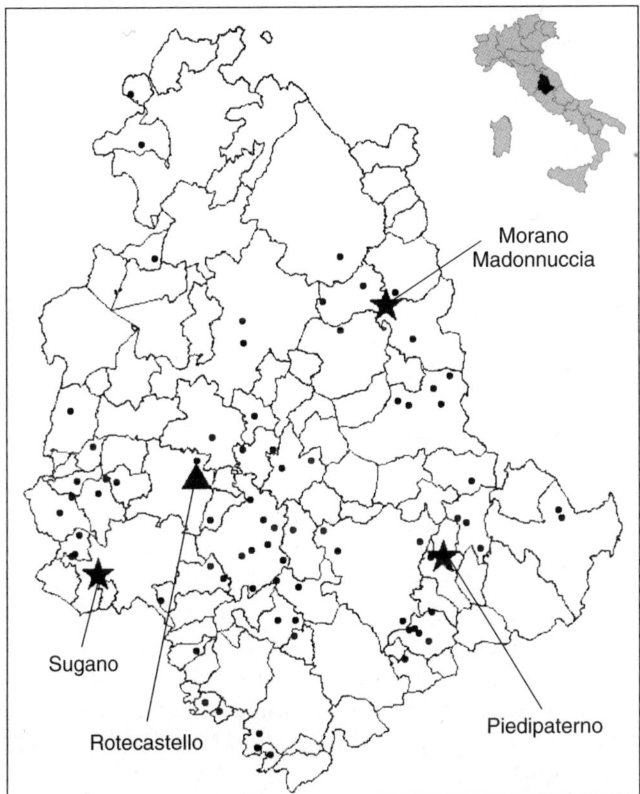



**Figure 15.1**   *Umbria region with municipality boundaries. Black dots show the location of 79 sites where landslide risk was ascertained. Stars show the location of the three study areas described in the text. Triangle shows the location of the village of Rotecastello where landslide risk was first determined by Cardinali* et al.*(2002b)*

floods for the period 1918–2000 (Guzzetti *et al.*, 1994). This information was recently summarized by Guzzetti *et al.* (2003). A reconnaissance estimate of the impact of landslides on the population, the transportation network, and the built-up areas in Umbria was attempted by Guzzetti *et al.* (2003). Despite these efforts, definition of landslide risk at site scale remains an open problem.

## 15.4   Methodology

We devised a methodology to assess landslide hazard and risk at site scale using a geomorphological approach based on the interpretation of multiple sets of aerial photographs, combined with the analysis of site-specific and historical information (Antonini *et al.*, 2002a; Cardinali *et al.*, 2002b). We start with the definition of the study area and the careful scrutiny of the 'state of nature', that is, of all the existing and past landslides that can be identified in the study area. Based on the observed changes in the distribution and pattern of landslides, we infer the possible short-term evolution of the slopes, the probable type of failures, and their expected frequency of occurrence. We use this information to estimate the landslide hazard, and to evaluate the landslide risk. The methodology involves the following seven steps:

(a) definition of the extent of the study area,
(b) compilation of a multitemporal landslide inventory map, including landslide classification,
(c) definition of landslide hazard zones,
(d) assessment of landslide hazard,
(e) identification and mapping of the elements at risk, and assessment of their vulnerability to different landslide types,
(f) evaluation of specific landslide risk, and
(g) evaluation of total landslide risk.

### 15.4.1   Definition of Study Area

Preliminary to any landslide hazard or risk assessment is the definition of the area to be investigated. This apparently trivial problem is essential to the analysis and the application of the results. In our study, we define a 'site' as an area bounded by drainage and dividing lines around the place selected for the landslide risk assessment. A site is an ensemble of one or more adjacent 'elementary slopes' or watersheds (Carrara *et al.*, 1991). To outline the site we select major dividing and drainage lines, wherever possible. Where this is not feasible, we select minor dividing or drainage lines. Mapping of elementary slopes or watersheds is accomplished at 1:10 000 scale, using large-scale topographic base maps, locally aided by the analysis of large- and medium-scale aerial photographs. At each site, the number and the extent of the elementary slopes depend on the local geological and morphological setting, and on the type, number and extent of landslides.

### 15.4.2   Multitemporal Landslide Map

In Umbria region landslides show a remarkable spatial recurrence (Cardinali *et al.*, 2000; Guzzetti *et al.*, 2003). Mass movements tend to repeat where they have occurred in the past, within or in the vicinity of other landslides, or in the same slope or watershed. Guzzetti *et al.* (2003) pointed out that in Umbria region detailed knowledge of the location of past failures is the key to forecast future landslide occurrence. This is not unique to Umbria or central Italy. Nilsen and Turner (1975) determined that most landslides that caused damage to man-made structures in the urbanized parts of Contra Costa County, California, from 1950 to 1971, occurred on pre-existing ancient landslide deposits.

Within a study area we ascertain the spatial distribution of landslides through the interpretation of multiple sets of stereoscopic aerial photographs and detailed field surveys. In Umbria, for a period of about 60 years (from 1941 to 2001) seven sets of aerial photographs taken in different years are available. The oldest photographs were taken in 1941, and the newest in 1997. Nominal scale of aerial photographs ranges from 1:13 000 to 1:73 000. Only three sets of photographs cover the entire territory, whereas the other flights are limited to specific areas. Field surveys used to test the methodology were carried out mostly in the years 2000 and 2001 (Cardinali *et al.*, 2002b).

To compile the multitemporal landslide inventory map, we began by identifying landslides on the 1954–55 aerial photographs. We selected this set because it is the oldest flight covering the entire region, because the aerial images were taken in a period when intense cultivation of the land by machine had not started, and the forms of old and recent landslides were clearly visible on the photographs (Guzzetti and Cardinali, 1989). We then analysed the other sets of aerial photographs, separately and in conjunction with the 1954–55 photographs and with the other flights. In this way we prepared separate landslide inventory maps, one for each set of aerial photographs and for the field surveys.

Landslide information collected through the interpretation of aerial photographs or mapped in the field is transferred to large-scale topographic base maps (at 1:10 000 scale). The different landslide maps are then overlaid and merged to obtain a single, multitemporal landslide inventory map. The process is not straightforward and requires adjustments to eliminate positional and drafting errors. The multitemporal landslide inventory map is then digitized and stored in a GIS database.

In the separate inventory maps and in multitemporal inventory maps, landslides are classified according to the type of movement, and the estimated age, activity, depth and velocity. A degree of certainty in the recognition of the landslide is also attributed. Landslide type is defined according to Varnes (1978) and the WP/WLI (1990, 1993, 1995). Landslide age, activity, depth and velocity are decided based on the type of movement, the morphological characteristics and appearance of the landslide on the aerial photographs and in the field, the local lithological and structural settings, the date of the aerial photographs, and the results of site-specific investigations carried out to solve local instability problems (Cardinali *et al.*, 2002b). Landslide relative age is defined as recent, old or very old, despite some ambiguity in the definition of the age of a mass movement based on its appearance. Landslides are classified as active where they appear fresh on the aerial photographs, or where movement is known from monitoring systems. Mass movements are classified as deep-seated or shallow, depending on the type of movement and the landslide volume. The latter is based on the type of failure, and the morphology and geometry of the detachment area and the deposition zone. Landslide velocity is considered as a proxy of landslide type, and classified accordingly. Rotational or translative slides, slide earthflows, flows and complex or compound slides are classified as slow-moving failures. Debris flows are classified as rapid movements. Rockfalls and topples are classified as fast-moving landslides (WP/WLI, 1995).

The adopted classification scheme, and in particular the evaluation of landslide age, activity, velocity and depth, includes uncertainty and suffers from simplifications. The classification requires geomorphological inference, but fits the available information on landslide types and process in Umbria (Felicioni *et al.*, 1994; Guzzetti *et al.*, 1996, 2003).

### 15.4.3   Landslide Frequency

Information on landslide frequency is essential to assess landslide hazard. The frequency of landslides can be obtained through the analysis of historical records of landslide occurrence (Guzzetti *et al.*, 1999b). In general, a complete record of past landslides from which to derive the frequency of occurrence of landslide events is difficult to obtain for a single landslide or a slope (Ibsen and Brunsden, 1996; Glade, 1998; Guzzetti *et al.*, 1999b). In this work we ascertain landslide frequency based on the analysis of the multitemporal inventory map. We define four classes of landslide frequency, based on the number of events recognized in the 47-year observation period from 1954 to 2001

**Table 15.1**  *Frequency of landslide events*

| Landslide frequency | | | Events in the observation period |
| --- | --- | --- | --- |
| Category | Index | Ratio | |
| Low | 1 | 1/47 (0.02) | 1 |
| Medium | 2 | 2/47 (0.04) | 2 |
| High | 3 | 3/47 (0.06) | 3 |
| Very high | 4 | >3/47 (>0.06) | >3 |

(Table 15.1). We ascertain the frequency of landslides, $F_L$, during the observation period based on: (a) the number of events inferred from the analysis of the aerial photographs; (b) the landslide events observed in the field; and (c) the information on landslide events obtained from technical reports, historical accounts and chronicles. We do not make a distinction between events inferred through the interpretation of aerial photographs and events identified in the field or described in technical or historical reports.

### 15.4.4  Landslide Intensity

Definition of landslide hazard requires information on landslide intensity (or magnitude; Guzzetti *et al.*, 1999b). In contrast to earthquakes, volcanic eruptions or hurricanes, no unique measure of landslide intensity is available (Hungr, 1997). Since our goal is to estimate landslide risk, we assume landslide intensity ($I_L$) as a measure of the destructiveness of the landslide (Hungr, 1997), and we define it as a function of the landslide volume ($v_L$) and of the landslide expected velocity ($s_L$), that is, $I_L = f(v_L, s_L)$. Table 15.2 shows how we assign the intensity to each landslide based on the estimated volume and the expected velocity. We estimate landslide volume based on the landslide type defined in the inventory map. For slow-moving landslides volume depends on the estimated depth of movements; for rapid-moving debris flows on the size of the contributing catchment and on the estimated volume of debris in the source areas and along the channels; for fast-moving rockfalls on the maximum size of a single block, obtained from field observations, or from the estimated volume of rockslide deposits. The expected

**Table 15.2**  *Landslide intensity, in four classes, based on the estimated landslide volume and the expected landslide velocity*

| Estimated volume (m³) | Expected landslide velocity | | |
| --- | --- | --- | --- |
| | Fast-moving rockfall | Rapid-moving debris flow | Slow-moving slide |
| <0.001 | Slight (1) | | |
| 0.001–0.5 | Medium (2) | | |
| 0.5–500 | High (3) | Slight (1) | |
| 500–10 000 | High (3) | Medium (2) | Slight (1) |
| 10 000–500 000 | Very high (4) | High (3) | Medium (2) |
| >500 000 | | Very high (4) | High(3) |
| ≫500 000 | | | Very high (4) |

landslide velocity depends on the type of failure, its volume and the estimated depth of movement. For any given landslide volume, rockfalls have the highest landslide intensity, debris flows exhibit intermediate intensity, and slow-moving landslides have the lowest intensity.

### 15.4.5   Landslide Hazard Zones

We evaluate landslide hazard in the areas of evolution of existing (i.e. mapped) landslides (Cardinali *et al.*, 2002b). For this purpose we define a 'landslide hazard zone' (LHZ) as the area of possible (or probable) short-term evolution of an existing landslide, or a group of landslides, of similar characteristics (i.e. type, volume, depth, velocity), identified from the aerial photographs or observed in the field. In an LHZ an existing landslide can grow upslope, develop downslope, or expand laterally. An LHZ is therefore a form of landslide scenario designed using geomorphological inference.

To map an LHZ we use again the multitemporal landslide inventory map. Within each elementary slope, we map the area of possible evolution of each landslide, or group of landslides, based on the observed location, distribution and pattern of landslides, their style of movement and activity, and the local lithological and morphological setting. To design an LHZ we consider the observed partial or total reactivation of the existing landslides, the lateral, head (retrogressive) or toe (progressive) expansion of the existing landslides, and the possible occurrence of new landslides of similar type and intensity.

We identify separate landslide scenarios for the different type of failures observed in the elementary slope, for example fast-moving rockfalls and topples, rapid-moving debris flows, and slow-moving slump earthflows, block slides or compound failures. LHZ includes the crown area and the deposit of the existing landslides, and the area of possible direct or indirect influence of the landslide. LHZs were identified based on the local topographic, morphological and geological settings, and the type and extent of landslides. For slow-moving failures (e.g. slide, slump earthflow, block slides and compound failure) LHZ is limited to the surroundings of the existing landslide, or group of landslides. This limitation is justified in Umbria where the evolution of these landslides is predictable in space (Cardinali *et al.*, 2000). For relict (i.e. very old) landslides the LHZ overlaps in places with the entire elementary slope. For debris flows the LHZ includes the source areas, the river channels and the depositional areas on alluvial or debris fans. For rockfalls, topples and minor rockslides LHZ includes the rock cliffs from where landslides detach, and the talus, debris cones, or debris slopes along which rockfalls travel, and the places where they are deposited.

### 15.4.6   Landslide Hazard Assessment

Landslide hazard depends on the frequency of landslide movements $(F_L)$ and on the landslide intensity $(I_L)$, $H_L = f(F_L, I_L)$. We obtain an estimate of landslide hazard by combining the value of landslide frequency, ascertained based on the number of landslide events of the same type observed within each LHZ (Table 15.1), and landslide intensity, in four classes, based on the estimated landslide volume and the expected landslide velocity (Table 15.2).

Levels of landslide hazard are shown using a two-digit, positional index (Table 15.3, Cardinali *et al.*, 2002b). The right digit shows the landslide intensity $(I_L)$ and the left

**Table 15.3**  *Landslide hazard classes based on estimated landslide frequency, $F_L$ (Table 15.1) and landslide intensity, $I_L$ (Table 15.2)*

| Estimated landslide frequency | Landslide intensity | | | |
|---|---|---|---|---|
| | Slight (1) | Medium (2) | High (3) | Very high (4) |
| Low (1) | 1 1 | 1 2 | 1 3 | 1 4 |
| Medium (2) | 2 1 | 2 2 | 2 3 | 2 4 |
| High (3) | 3 1 | 3 2 | 3 3 | 3 4 |
| Very high (4) | 4 1 | 4 2 | 4 3 | 4 4 |

digit shows the estimated landslide frequency ($F_L$). The index expresses landslide hazard, keeping distinct the two components of the hazard. This facilitates landslide hazard zoning, allowing us to understand if hazard is due to a high frequency of landslides (i.e. high recurrence), to a large intensity (i.e. large volume and high velocity), or to both. It is worth noticing that values of the landslide hazard index in Table 15.3 do not provide an absolute rank of hazard levels. Although extreme values are easily defined, intermediate conditions of landslide hazard are more difficult to rank. A landslide that exhibits a low frequency and a slight intensity ($H_L = 1\ 1$) will certainly have a hazard much lower than a landslide exhibiting very high frequency and intensity ($H_L = 4\ 4$). Deciding if the hazard of a landslide with a very high frequency and a slight intensity ($H_L = 4\ 1$) is higher (or lower) than that of a landslide with a low frequency and a very high intensity ($H_L = 1\ 4$) is not straightforward, and may be a matter of local judgement.

### 15.4.7  Vulnerability of Elements at Risk

To ascertain risk, one needs to know the type and location of the vulnerable elements. We prepare a map of the elements at risk, including built-up areas, structures and the infrastructure, at 1:10 000 scale by analysing large-scale topographic base maps, and recent aerial photographs. Care is taken in locating precisely the elements at risk within or in the vicinity of the landslides and the LHZ. The map is then digitized, registered to the multitemporal landslide map, and stored in a GIS database.

To classify the elements at risk we adopt a legend with 11 classes (Table 15.4), of which six refer to built-up areas and structures (houses, buildings, industries and farms, sports centres, and cemeteries), four to the transportation network (roads and railways), and one to mining activities (quarries). For the risk to the population, we assume that houses, buildings and roads in the study area are a proxy for population density, and we consider the population to be vulnerable because of the presence of structures and infrastructure. As an example, in a densely populated zone vulnerability of the population is considered higher than for sparse, farming structures. Along a secondary road vulnerability of the population is lower than along a high-transit road.

Evaluating the vulnerability of the elements at risk to different landslide types is a difficult and uncertain operation. To estimate vulnerability we adopt a simple approach, based on the inferred relationship between the intensity and type of the expected landslide, and the likely damage the landslide can cause. Table 15.5 illustrates the expected damage to buildings and roads, and to the population, if affected by landslides of different type

**Table 15.4**   *Types of elements at risk (for structures and infrastructure)*

| Code | Elements at risk |
|---|---|
| HD | Built-up areas with a high population density |
| LD | Built-up areas with a low population density and scattered houses |
| IN | Industries |
| FA | Livestock farms |
| SP | Sports facilities |
| Q | Quarries |
| MR | Main roads, motorways, highways |
| SR | Secondary roads |
| FR | Farm and minor roads |
| RW | Railway lines |
| C | Cemeteries |

**Table 15.5**   *Vulnerability, the expected damage to elements at risk (i.e. buildings, structures and infrastructure) and population*

| Landslide intensity | Elements at Risk | | | | | | | | | | | Population |
|---|---|---|---|---|---|---|---|---|---|---|---|---|
| | Structures and infrastructure | | | | | | | | | | | |
| | Buildings | | | | | | Roads | | | | Others | |
| | HD | LD | IN | FA | SP | C | MR | SR | FR | RW | Q | |
| **Slight** | | | | | | | | | | | | |
| Rockfall | A | A | A | A | A | A | A | A | A | A | A | N |
| Debris flow | A | A | A | A | A | A | A | F | F | A | A | N |
| Slide | A | A | A | A | A | A | A | F | S | A | A | N |
| **Medium** | | | | | | | | | | | | |
| Rockfall | F | F | F | F | F | F | F | F | F | F | F | D, I, H |
| Debris flow | F | F | F | F | F | F | F | F | F | F | F | D, I, H |
| Slide | F | F | F | F | F | F | F | S | S | F | F | I |
| **High** | | | | | | | | | | | | |
| Rockfall | S | S | S | S | S | S | S | S | S | S | S | D, I, H |
| Debris flow | S | S | S | S | S | S | S | S | S | S | S | D, I, H |
| Slide | S | S | S | S | S | S | S | S | S | S | S | I, H |
| **Very high** | | | | | | | | | | | | |
| Rock fall | S | S | S | S | S | S | S | S | S | S | S | D, I, H |
| Debris flow | S | S | S | S | S | S | S | S | S | S | S | D, I, H |
| Slide | S | S | S | S | S | S | S | S | S | S | S | I, H |

*Note:*
For elements at risk: A-aesthetic (or minor) damage; F-functional (or medium) damage; S-structural (or total) damage. For population: N-no damage; D-direct damage (fatalities); I-indirect damage; H-homeless people. For classes of elements at risk see Table 15.4. For landslide intensity see Table 15.2.

and intensity. The table is based on the information of the damage caused by slope failures in Umbria (Felicioni *et al.*, 1994; Alexander, 2000; Cardinali *et al.*, 2000; Antonini *et al.*, 2002b), our field experience and judgement, and on the review of the scant literature (Alexander, 1989; Michael-Leiba *et al.*, 1999; Fell, 2000). A crude estimate (i.e. few, many and very many) of the number of people potentially subject to landslide risk is considered, based on the extent and type of the built-up areas.

Damage to structures and infrastructure is classified as:

- Aesthetic (or minor) damage, where the functionality of buildings and roads is not compromised, and the damage can be repaired, rapidly and at low cost.
- Functional (moderate or medium) damage, where the functionality of structures or infrastructure is compromised, and the damage takes time and large resources to be fixed.
- Structural (severe or total) damage, where buildings or transportation routes are severely or completely damaged, and require extensive work to be fixed, and demolition and reconstruction may be required.

Damage to the population is classified as:

- Direct damage, where casualties (deaths, missing persons and injured people) are expected.
- Indirect damage, where only socio-economic damage is expected.
- Temporary or permanent loss of private houses (i.e. evacuees and homeless people).

In Umbria, direct damage to the population is foreseen for fast-moving landslides, or for high-intensity, slow-moving slides. Indirect damage to the population is expected where landslides can cause functional or structural damage to infrastructure, with negative socio-economic effects on public interests. Homeless are expected where functional or structural damage to buildings is foreseen.

### 15.4.8   Specific Landslide Risk

Landslide risk is the result of the complex interaction between the 'state of nature' (i.e. landslide hazard, $H_L$) and the expected vulnerability to the elements at risk ($V_L$), or $R_S = f(H_L, V_L)$. We use this general relationship to ascertain the specific landslide risk, $R_S$, that is, the risk at which a set of elements (e.g. building, roads, etc.) is subject when a landslide occurs (Einstein, 1988). We define the specific landslide risk separately for each class of elements at risk and for each landslide type present in each LHZ. If more than a single type of elements at risk is present in an LHZ, a different value of specific risk is computed for each class.

To determine specific landslide risk we use Table 15.6, which correlates the expected damage to the landslide hazard, loosely ranked from low (1 1) to high (4 4) values. Construction of Table 15.6 required extensive discussion, and it is largely based on the analysis of damage caused by two recent regional landslide events in Umbria: a rapid snowmelt that triggered thousands of failures in January 1997 (Cardinali *et al.*, 2000), and the Umbria–Marche earthquake sequence of September–October 1997 that caused mostly rockfalls (Antonini *et al.*, 2002b; Guzzetti *et al.*, 2003). Information on past landslide damage in Umbria region was also considered (Felicioni *et al.*, 1994; Alexander, 2000).

**Table 15.6** *Levels of specific landslide risk, based on landslide hazard (Table 15.3) and vulnerability (Table 15.5)*

| Landslide hazard | Vulnerability (expected damage) | | |
|---|---|---|---|
| | Aesthetic (minor) damage | Functional (major) damage | Structural (total) damage |
| Low  1 1 | A 1 1 | F 1 1 | S 1 1 |
| 1 2 | A 1 2 | F 1 2 | S 1 2 |
| 1 3 | A 1 3 | F 1 3 | S 1 3 |
| 2 1 | A 2 1 | F 2 1 | S 2 1 |
| 1 4 | A 1 4 | F 1 4 | S 1 4 |
| 2 2 | A 2 2 | F 2 2 | S 2 2 |
| 2 3 | A 2 3 | F 2 3 | S 2 3 |
| 3 1 | A 3 1 | F 3 1 | S 3 1 |
| 3 2 | A 3 2 | F 3 2 | S 3 2 |
| 2 4 | A 2 4 | F 2 4 | S 2 4 |
| 3 3 | A 3 3 | F 3 3 | S 3 3 |
| 4 1 | A 4 1 | F 4 1 | S 4 1 |
| 4 2 | A 4 2 | F 4 2 | S 4 2 |
| 3 4 | A 3 4 | F 3 4 | S 3 4 |
| 4 3 | A 4 3 | F 4 3 | S 4 3 |
| High  4 4 | A 4 4 | F 4 4 | S 4 4 |

*Note:*
Only damage to elements at risk (structures and infrastructure) is considered. Landslide hazard is loosely ranked from low (1 1) to high (4 4) values.

To show the level of specific risk we add to the left of the two-digit landslide hazard index a third digit describing the expected damage (i.e. aesthetic, functional or structural, see Table 15.5). Thus, the specific risk index shows, from right to left, the landslide intensity, the landslide frequency, and the expected damage caused by the specific type of landslide. As for the hazard index, the landslide specific risk index $(R_S)$ does not provide an absolute ranking of risk levels. The extreme conditions are easily ranked: a house having an $R_S = $ A 1 1 (i.e. aesthetic damage due to a low-frequency and slight-intensity landslide) poses a lower risk than a dwelling with $R_S = $ S 4 4 (i.e. expected structural damage caused by a very high-frequency and very high-intensity landslide). Deciding for the intermediate conditions may not be straightforward. A decision should be made on a case-by-case basis, considering the type of elements at risk, their vulnerability, the possible defensive measures, and the economic and social implications of landslide risk.

### 15.4.9 Total Landslide Risk

Where an absolute ranking of landslide risk is required, total risk has to be determined (Varnes – IAEG, 1984; Einstein, 1988). Total landslide risk is the ensemble of all the specific landslide risk levels. Different strategies can be used to lump the detailed information given by the specific landslide risk index into a limited number of classes of total landslide risk. In this chapter we use a system that attributes to each LHZ a value of total landslide risk, in five classes, based on the type and severity of the largest specific landslide risk attributed in the LHZ (Table 15.7). This is different from what was

**Table 15.7**  *Relationships between classes of total landslide risk, type of landslides, and expected damage to structures, infrastructure and the population in Umbria*

| Total risk | Type of landslides | Damage to structures and infrastructure | Damage to the population |
|---|---|---|---|
| Very high | Rapid and fast-moving landslides | Structural and functional damage | Casualties and homeless people expected, indirect damage expected |
| High | Slow-moving landslides | Structural and functional damage | Casualties not expected. Homeless people and indirect damage expected |
| Medium | Slow-moving landslides, fast and rapid-moving landslides of slight intensity | Aesthetic damage | Homeless people and indirect damage expected |
| Low | Relict, large, slow-moving landslides of very low frequency | Structural and functional damage | Homeless people and indirect damage expected |
| Very low | Landslides are present | Nil (elements at risk are not present) | Nil (population is not present) |

proposed by Cardinali *et al.* (2002b), which attributed a value of total landslide risk to the entire study area, based on the largest specific landslide risk attributed in the study area.

Very high total landslide risk is assigned where rapid and fast-moving landslides can cause direct damage to the population. These are the areas where debris flows and rockfalls can result in casualties or homeless people. High total landslide risk is assigned to the areas where slow-moving landslides can cause structural and functional damage to structures and infrastructure. In these areas casualties are not expected. Moderate total landslide risk is attributed where aesthetic damage to vulnerable elements is expected, as a consequence of slow-moving slope failures and fast or rapid-moving landslides of slight intensity. Large or very large, relict deep-seated landslides can cause structural and functional damage to structures and infrastructure, homeless people and indirect damage to the population. However, such areas are assigned low total landslide risk, because they are not expected to move entirely under the present climatic and seismic conditions. Lastly, a very low value of total landslide risk is assigned where landslides were identified and landslide hazard was ascertained, but elements at risk or the population are not present in the LHZ.

## 15.5   Examples of Landslide Risk Assessment

We utilized the described methodology to ascertain landslide risk in 79 towns in Umbria region (Antonini *et al.*, 2002a). Cardinali *et al.* (2002b) previously illustrated the application of the geomorphologically based methodology to the village of Rotecastello, in central Umbria. In the following, we describe specific and total landslide risk assessments for three sites: Morano Madonnuccia, in northeastern Umbria, Sugano, in southwestern Umbria, and Piedipaterno, in southeastern Umbria (Figure 15.1). The three sites were selected to illustrate the application of the risk assessment methodology in different physiographical and geological environments, and for different types of slope failures. At Morano Madonnuccia, landslide risk is due to slow-moving slides, slide earthflows and flows, of slight to medium intensity that involve marl and shale, and their associated soils. At Sugano, fast-moving rockfalls and rockslides coexist with slow-moving, shallow and deep-seated slides and slide earthflows, involving volcanic rocks and marine clay. At Piedipaterno, landslide risk is due chiefly to fast-moving rockfalls, and rapid-moving debris flows, involving limestone, talus and debris deposits.

### 15.5.1   Morano Madonnuccia

The landscape in central Umbria is hilly, with elevation ranging from 300 to about 1200 metres. Drainage density and pattern are controlled by lithology and the structural setting. Valleys are asymmetrical and controlled by the attitude of bedding planes. Slopes are long and rectilinear along dip-slopes, and short and steep where bedding planes dip into the slope. In the area, there are outcrops of flysch deposits, heterogeneous sequences of well-stratified, graded deposits composed of soft and weak layers of marl, sandy shale and clay, orderly interbedded with coarse and fine sandstone, graywackes and calcarenites, in various percentages.

Due to the lithological and structural setting, landslides of various types are abundant in the area. Large, deep-seated slides, block-slides, and complex or compound mass movements predominate where bedding is nearly parallel to the slope, or less steep than the slope. Along reverse slopes, rotational slides, rockslides, and slide earthflows are most abundant. On the surface of large landslides, on the soils mantling the bedrocks, and on the locally thick colluvial deposits, shallow landslides predominate. The latter comprise soil slips, minor slumps, mudflows and minor earthflows. Badlands and areas of intense surface erosion are locally present, chiefly where clay and marl predominate and bedding is sub-horizontal (Guzzetti *et al.*, 1996).

Morano Madonnuccia is a small village in Gualdo Tadino Municipality constituted by sparse houses built along road SS 444, connecting Assisi to Gualdo Tadino (Figure 15.2). The village is located on a SW–NE trending divide, at an elevation of about 650 metres. In the area there are outcrops of thinly bedded marl pertaining to the Schlier Formation [Fm.], Burdigalian in age, and well-bedded marl, sandstone and calcarenite pertaining to the Marnoso Arenacea Fm., lower to upper Miocene in age. Locally, a chaotic mixture of clay and exotic rocks (i.e. an olistostrome) is embedded within the Marnoso Arenacea Fm.

In the Morano Madonnuccia study area we identify three 'elementary slopes', for a total area of about 2 km$^2$ (Figure 15.3). For the study area, we compiled a multitemporal landslide inventory map by analysing three sets of aerial photographs, and through field surveys (Figure 15.3). Aerial photographs were flown in May 1956 at 1:33 000 nominal scale, in June 1977 at 1:13 000 nominal scale, and in October 1997 at 1:13 000 nominal scale. Field surveys were carried out in March 1997, shortly after a major landslide-triggering event (Cardinali *et al.*, 2000), and in February 2001.

Superficial and deep-seated landslides were identified in the three sets of aerial photographs, and were classified based on relative age, inferred from the date of the aerial photographs, and on the prevalent landslide type (Figure 15.3). A total of 70 landslides

***Figure 15.2*** *Panoramic view of the Morano Madonnuccia study area. Photograph taken from south*

was mapped, for a total landslide area of $1.10\,\mathrm{km}^2$. The territory affected by slope movements extends for $0.58\,\mathrm{km}^2$, equivalent to 26.7% of the study area. In the study area, landslides originate from the upper part of the slopes, and where topography is concave. In the latter areas, soils and weathered deposits are thick, allowing for shallow failures to develop. Small, shallow failures take place also on pre-existing landslide deposits. Flows and slide earthflows predominate in the areas where chaotic rocks crop out. In these areas, transitional and rotational slides are present. Relict landslides are uncommon in the area. Historical information and reports on landslide events indicate that slope movements in the area are triggered chiefly by prolonged rainfall and by rapid snowmelt, which are relatively frequent in this part of Umbria.

In the Morano Madonnuccia study area we identify 22 LHZs (grey areas in Figures 15.3 and 15.4), of which 16 are for slow-moving, shallow landslides of slight intensity (2–5, 7–10, 12–13, 15–18, 20–21 on Figure 15.3C), six for slow-moving, deep-seated, rotational slides of medium (1, 6, 11, 13, 14, 19) and high (14) intensity (Figure 15.3B), and one for very old (relict), deep-seated rotational landslides of medium intensity (22 in Figure 15.3A). We obtain landslide frequency for each LHZ through the interpretation of the available sets of aerial photographs, and from field surveys in 1997 and 2001 (Table 15.8). In the 46-year observation period, landslide frequency ranges from low (one event) to very high (more than three events). The highest frequency was observed in three LHZs where shallow soil slips and earthflows occurred repeatedly (8, 13 and 15 in Figure 15.3C). Several landslides were mapped as active at the date of the photographs or the field surveys. Active landslides are shown by stars in Figure 15.3.

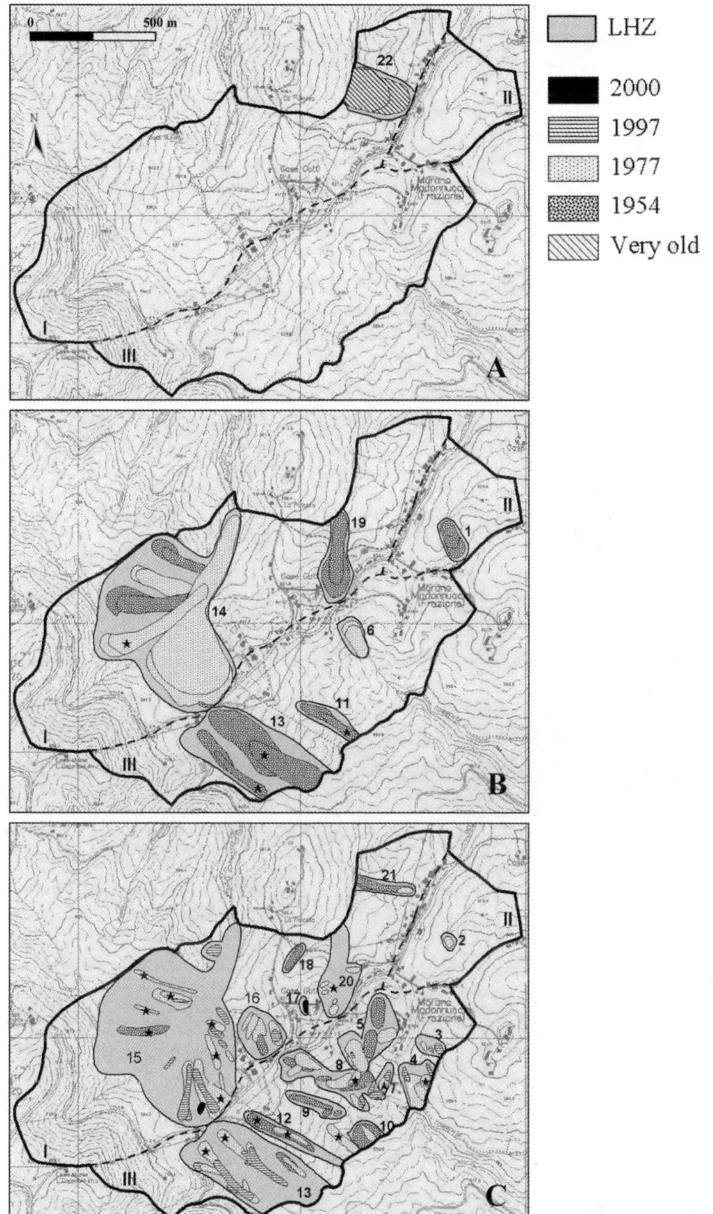


**Figure 15.3** *Morano Madonnuccia. Multitemporal landslide inventory map. Landslides classified by prevalent type of movement: A, relict, deep-seated slide; B, deep-seated slides and earthflows; C, shallow slides and flows. Patterns indicate relative landslide age, inferred from the date of the aerial photographs and from field surveys. Stars indicate active landslides. Grey areas with Arabic numbers are LHZs (see Table 15.8). Roman numbers indicate elementary slopes*

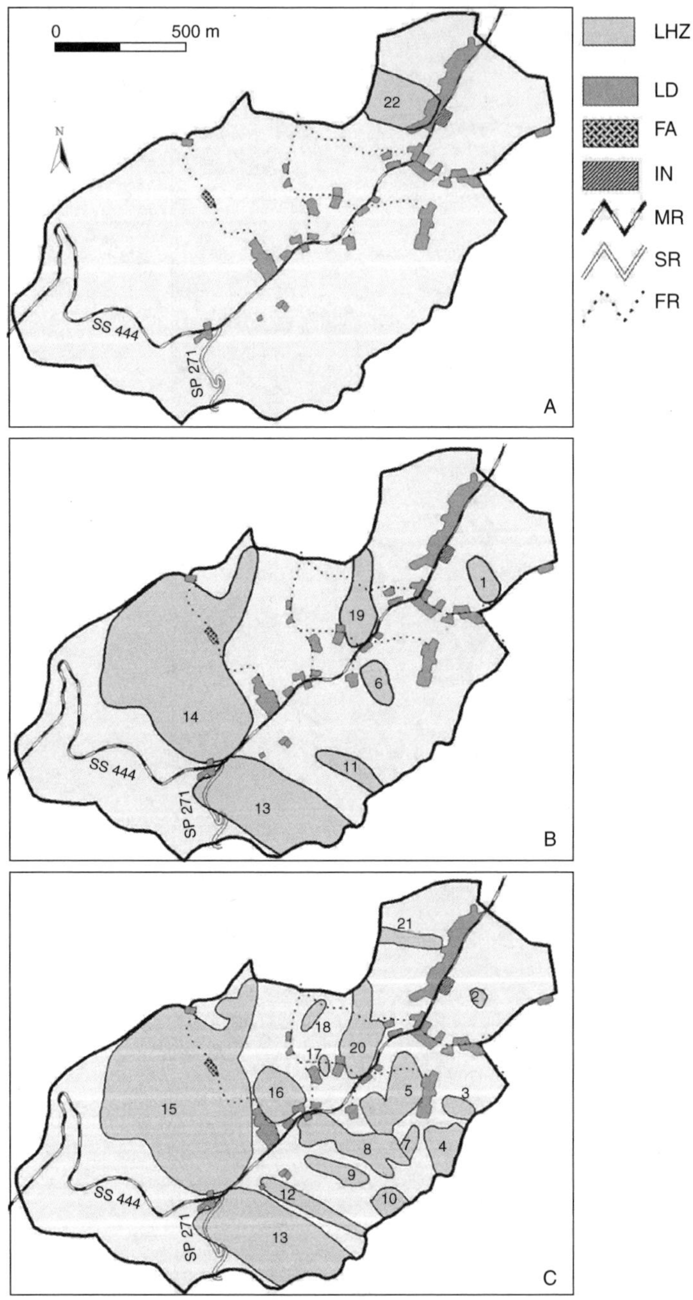



**Figure 15.4** *Morano Madonnuccia. Specific landslide risk assessment (see Table 15.8). A, B and C show prevalent landslide types, as in Figure 15.3. Grey areas with Arabic numbers are LHZs. Patterns indicate types of elements at risk (see Table 15.4). Topography shown in Figure 15.3 is not shown here for clarity*

**Table 15.8** *Morano Madonnuccia study area: classification of specific, $R_S$, and total, $R_T$, landslide risk*

| LHZ # | Landslide type | $F_L$ | $I_L$ | $H_L$ | E | V | P | $R_S$ | $R_T$ |
|---|---|---|---|---|---|---|---|---|---|
| 1 | Deep-seated slide | 1 | 2 | 1 2 | N | – | – | – | Very low |
| 2 | Shallow slide | 1 | 1 | 1 1 | N | – | – | – | Very low |
| 3 | Shallow slide | 2 | 1 | 2 1 | N | – | – | – | Very low |
| 4 | Shallow slide | 3 | 1 | 3 1 | N | – | – | – | Very low |
| 5 | Shallow slide | 3 | 1 | 3 1 | LD<br>FR | A<br>S | N<br>N | A 3 1<br>S 3 1 | Medium |
| 6 | Deep-seated slide | 1 | 2 | 1 2 | N | – | – | – | Very low |
| 7 | Shallow slide | 2 | 1 | 2 1 | N | – | – | – | Very low |
| 8 | Shallow slide | 4 | 1 | 4 1 | LD | A | N | A 4 1 | Medium |
| 9 | Shallow slide | 1 | 1 | 1 1 | N | – | – | – | Very low |
| 10 | Shallow slide | 2 | 1 | 2 1 | N | – | – | – | Very low |
| 11 | Deep-seated slide | 1 | 2 | 1 2 | N | – | – | – | Very low |
| 12 | Shallow slide | 2 | 1 | 2 1 | LD | A | N | A 2 1 | Medium |
| 13 | Deep-seated slide | 1 | 2 | 1 2 | LD<br>MR<br>SR | F<br>F<br>S | H<br>I<br>N | F 1 2<br>F 1 2<br>S 1 2 | High |
| 13 | Shallow slide | 4 | 1 | 4 1 | LD<br>MR<br>SR | A<br>A<br>F | N<br>N<br>N | A 4 1<br>A 4 1<br>F 4 1 | Medium |
| 14 | Deep-seated slide | 3 | 2 | 3 2 | LD<br>FA<br>MR<br>FR | F<br>F<br>F<br>S | H<br>I<br>I<br>N | F 3 2<br>F 3 2<br>F 3 2<br>S 3 2 | High |
| 14 | Deep-seated slide | 3 | 3 | 3 3 | LD<br>FA<br>MR<br>FR | S<br>S<br>S<br>S | H<br>I<br>I<br>N | S 3 3<br>S 3 3<br>S 3 3<br>S 3 3 | High |
| 15 | Shallow slide | 4 | 1 | 4 1 | LD<br>FA<br>MR<br>FR | A<br>A<br>A<br>S | N<br>N<br>N<br>N | A 4 1<br>A 4 1<br>A 4 1<br>S 4 1 | Medium |

**Table 15.8**   *(Continued)*

| LHZ # | Landslide type | $F_L$ | $I_L$ | $H_L$ | $E$ | $V$ | $P$ | $R_S$ | $R_T$ |
|---|---|---|---|---|---|---|---|---|---|
| 16 | Shallow slide | 3 | 1 | 3 1 | LD<br>FR | A<br>S | N<br>N | A 3 1<br>S 3 1 | Medium |
| 17 | Shallow slide | 1 | 1 | 1 1 | FR | S | N | S 1 1 | Medium |
| 18 | Shallow slide | 1 | 1 | 1 1 | FR | S | N | S 1 1 | Medium |
| 19 | Deep-seated slide | 1 | 2 | 1 2 | LD<br>FR | F<br>S | H<br>N | F 1 2<br>S 1 2 | High |
| 20 | Shallow slide | 2 | 1 | 2 1 | LD<br>FR | A<br>S | N<br>N | A 2 1<br>S 2 1 | Medium |
| 21 | Shallow slide | 2 | 1 | 2 1 | N | – | – | – | Very low |
| 22 | Very old, deep-seated slide | 1 | 2 | 1 2 | LD | F | H | F 1 2 | Low |

*Note:*
LHZ = landslide hazard zone; $F_L$ = landslide frequency (Table 15.1); $I_L$ = landslide intensity (Table 15.2); $H_L$ = landslide hazard (Table 15.3); $E$ = type of element at risk (Table 15.4); $V$ = vulnerability of elements at risk (Table 15.5); $P$ = vulnerability of population (Table 15.5).

We obtained information on the location and type of the vulnerable elements in the study area from large-scale topographic base maps at 1:10 000 scale, prepared in 1993 from aerial photographs taken in 1997, and from aerial photographs flown in October 1997. By combining this information with the landslide hazard assessment, we ascertained specific landslide risk for each vulnerable element or group of vulnerable elements. For vulnerable elements that are subject to hazards posed by different landslide types, we attributed separate levels of specific landslide risk. Table 15.8 lists the results of the risk assessment.

Our analysis indicates that deep-seated slides of medium to high intensity and high frequency (14 in Figure 15.4B) pose the highest threat to the vulnerable elements in the study area (Table 15.8). In LHZ 14 structural and functional damages to built-up areas (LD, FA) and to the transportation network (MR, FR) are expected. Indirect damages to the population and homeless people are possible, chiefly as a consequence of the expected damage to the transportation network on mobility. Significant levels of specific landslide risk are also expected where shallow landslides with high (5, 16 in Figure 15.4C) or very high (8, 13, 15 in Figure 15.4C) frequency of occurrence are present. In these LHZs, damage ranges from structural to aesthetic, and affects mostly low-density built-up areas (LD) and roads of various categories (MR, SR, FR). Based on the available information, in the Morano Madonnuccia study area shallow landslides are not expected to threaten the population. The only very old (relict), deep-seated slide identified in the area (22 in Figure 15.4A) exhibits very low frequency and medium intensity, and is expected to produce functional damage to low-density build-up areas (LD).

It is worth noticing that in the Morano Madonnuccia study area, for 10 LHZs (i.e. 1–4, 6–7, 9–11, 21 in Figure 15.3) levels of landslide hazard were ascertained, but vulnerable elements are not present (Figure 15.4). Hence landslide risk does not presently exist in these LHZs (N in Table 15.8). If houses, roads, or other elements at risk are constructed in the LHZs, landslide risk will materialize. It will then be straightforward to determine levels of specific landslide risk, based on the type of vulnerable elements, and of values of landslide hazard.

Table 15.8 also illustrates levels of total landslide risk for the 22 LHZs identified in the Morano Madonnuccia study area. Total landslide risk is estimated to be high for low-density built-up areas (LD) and for livestock farms (FA), where deep-seated landslides are present (13–14, 19 in Figure 15.4). This is chiefly because indirect damages to the population and homeless people are expected. Where low-density settlements (LD) and roads (MR, SR, FR) are affected by shallow slides of high and very high frequency, total landslide risk is medium (5, 8, 12–13, 15, 15–18, 20 in Figure 15.4). We attribute very low levels of total landslide risk to LHZs where landslide hazard was determined but that are currently free of vulnerable elements (1–4, 6–7, 9–11, 21 in Figure 15.4). In these LHZs the estimate of total landslide risk will change significantly if building, roads, and other structures are constructed.

### 15.5.2 Sugano

The landscape in southwestern Umbria has the tabular morphology of a mesa rimmed by an articulated escarpment formed by resistant pyroclastic rocks. Towns and villages are built on buttes and isolated mesa remnants, and at the edge of the escarpment that bounds the volcanic plateau. In the area, there are outcrops of lava flows, ignimbrites and pyroclastic deposits pertaining to the Mt Vulsini volcanic complex, Quaternary in age. The volcanic rocks overlie marine sediments, chiefly clay and subordinately sand and gravel, upper to middle Pliocene in age. The contact between the volcanic cap and the underlying marine deposits is low angle, and almost everywhere covered by thick talus deposit.

In the area, mass movements affect the edge of the volcanic cap and extend laterally for several kilometres, to form a continuous belt of landslide deposits (Guzzetti *et al.*, 1996). Landslides occur within the volcanic complex, in the jointed pyroclastic cap, and in the underlying marine clay. The relative position of stiff and deformable rocks accounts for the widespread landsliding. Failure of the volcanic cap has two main causes: the increase of tensile stresses in the volcanic rock due to the different deformability of the pyroclastic sediments in contrast to the more plastic marine sediments; and the reduction of the resisting stresses at the base of the cliff, due to landsliding in the clay (Lembo-Fazio *et al.*, 1984).

Sugano, a small town in Orvieto Municipality, extends along the edge of a tabular promontory bounded by a sub-vertical escarpment, 50–100 meters in height, where volcanic rocks crop out (Figure 15.5). The study area is bounded by a single 'elementary slope', which extends for about $1\,km^2$ (Figure 15.6). For the study area, we made a multitemporal landslide inventory map by studying three sets of aerial photographs, and by studying field surveys. Aerial photographs were taken in September 1954 at $1:33\,000$ nominal scale, in June 1977 at $1:13\,000$ nominal scale, and in March 1994 at $1:36\,000$ nominal scale. Field reconnaissance of the area was completed in May 2000.

The multitemporal inventory map shows 39 landslides, for a total landslide area of $0.41\,km^2$. The territory affected by slope movements extends for $0.29\,km^2$ (including $0.035\,km^2$ of rockfall source areas), 35.2% of the study area. Different types of mass

**Figure 15.5**   *Panoramic view of the Sugano study area. Photograph taken from southeast*

movements coexist in the area. Marine clays underlying the volcanic rocks are affected by large, very old (relict), deep-seated block slides (Figure 15.6A), by deep-seated slides and slide earthflows (Figure 15.6B), and by shallow rotational and translational slides, and slide earthflows (Figure 15.6C). Rockfalls, topples and rockslides (Figure 15.6D) erode the volcanic cliff, and endanger the houses located at the edge of the escarpment, some of which had to be abandoned.

Rockfall, topple and minor rockslide deposits and talus deposits form a continuous belt of coarse landslide debris at the toe of the volcanic escarpment and in the crown area of the relict, deep-seated block slides. At the base of the volcanic escarpment, deposits of disrupted rockslides, of high to very high intensity (Table 15.2), were identified under the thick forest cover. Shallow and deep-seated landslides are abundant at the toe of the relict block slides, where the Leone Creek undercuts the toe of the landslides. Shallow landslides locally affect the softened soils on the pre-existing landslide deposits.

In the Sugano study area we identify eight LHZs (grey areas in Figures 15.6 and 15.7), of which two are for very old (relict), deep-seated block slides of high intensity (7, 8 in Figure 15.6A), two for slow-moving, deep-seated, rotational slides of medium intensity (4, 5 in Figure 15.6B), three for slow-moving shallow landslides of slight intensity (2, 3, 6 in Figure 15.6C), and one for fast-moving rockfalls, topples and rockslides, of low to very high intensity (1 in Figure 15.6D).

For the eight LHZs, we obtain landslide frequency through the interpretation of the three sets of aerial photographs and the field surveys. In the 47-year observation period, landslide frequency ranges from low (one event) to high (three events). Frequency of rockfalls is assigned high, based on historical information and field observations carried out along the volcanic escarpment and on the talus slope. In the most recent set of aerial

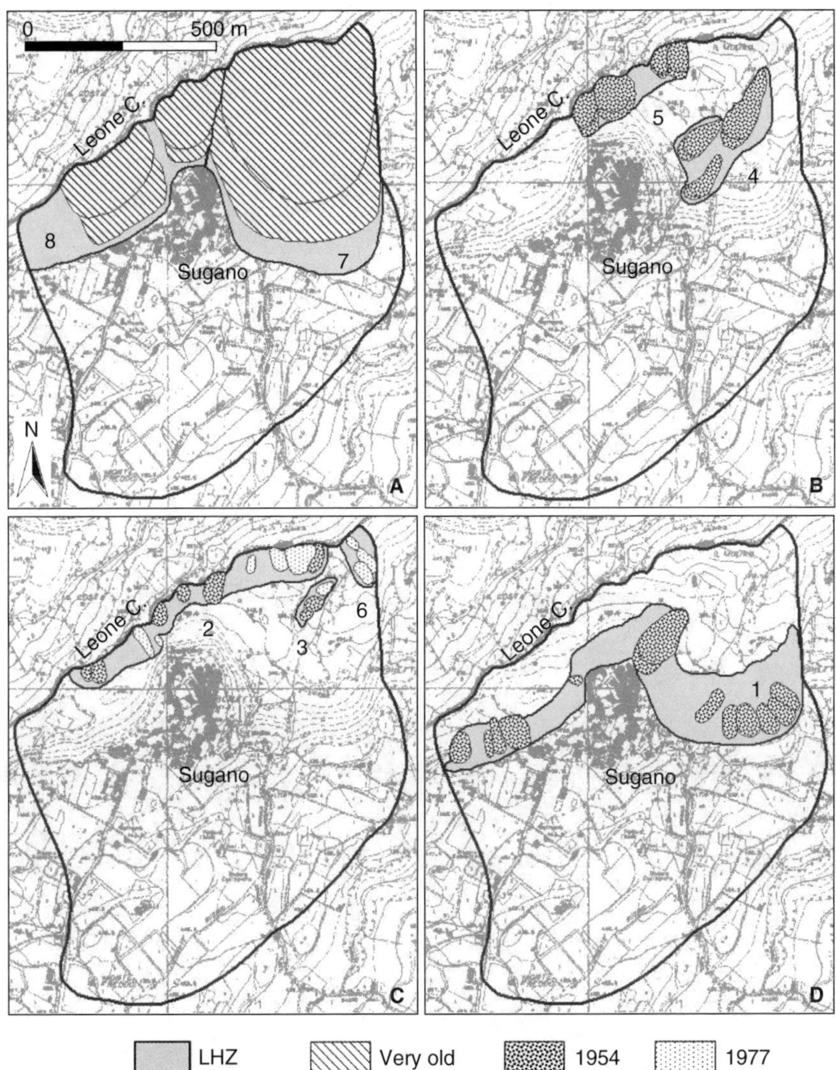



**Figure 15.6** *Sugano. Multitemporal landslide inventory map. Landslides classified by prevalent type of movement: A, relict, deep-seated block slides; B, deep-seated slides and earthflows; C, shallow slides and flows; D, fast-moving rockfalls, topples and rockslides. Patterns indicate relative landslide age, inferred from the date of the aerial photographs and from field surveys. Grey areas with Arabic numbers are LHZs (see Table 15.9)*

photographs (March 1994) no new landslides were identified. Small, mostly shallow slope movements have almost certainly occurred in the period 1977–94, but they may have been rapidly hidden by intense mechanical ploughing. In addition, rockfalls and topples occurring from the volcanic escarpment may not have been visible on the aerial photographs, or their deposits may have been concealed by the forest cover (Figure 15.5).

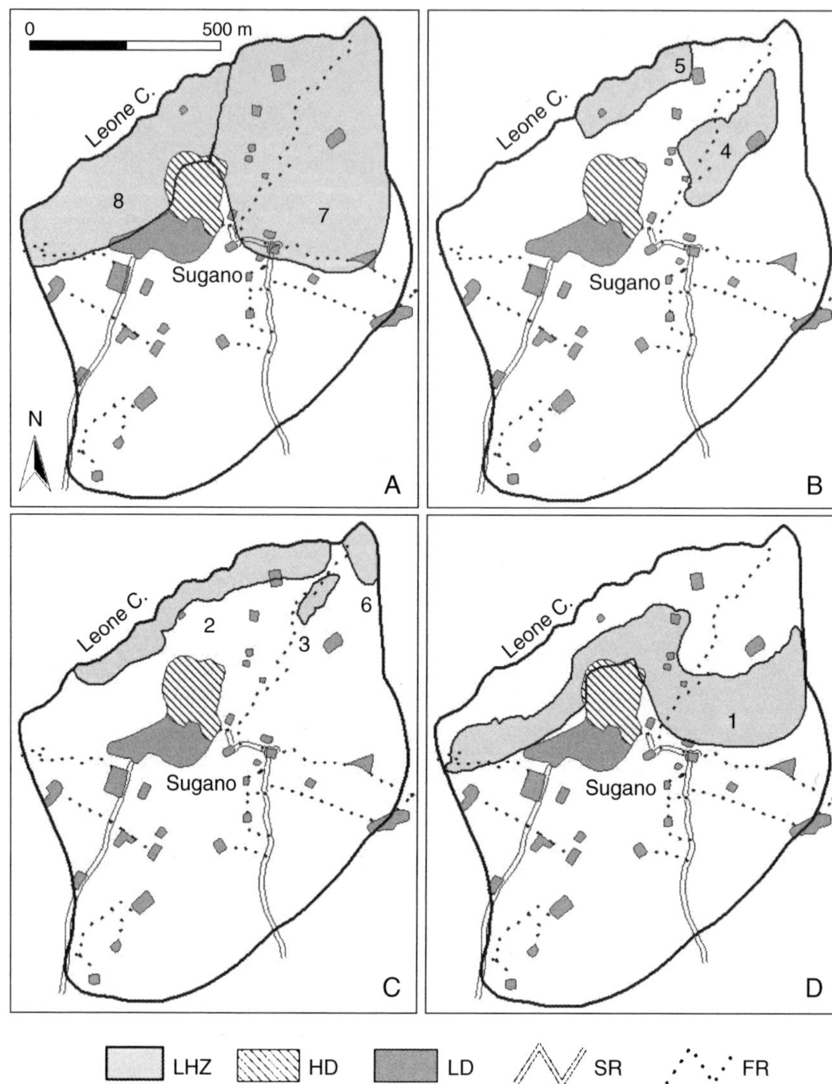



**Figure 15.7** *Sugano. Specific landslide risk assessment (see Table 15.9). A, B, C and D show prevalent landslide type, as in Figure 15.6. Grey areas with Arabic numbers are LHZs. Patterns indicate types of elements at risk (see Table 15.4). Topography shown in Figure 15.6 is not shown here for clarity*

We obtained information on the location and type of the vulnerable elements in the study area from the available large-scale topographic maps at 1:10 000 scale, prepared in 1993 from aerial photographs taken in 1987, and from aerial photographs flown in March 1994. By studying together the landslide hazard and the spatial distribution of the vulnerable elements (Figure 15.8), we ascertained specific landslide risk, $R_S$, for each vulnerable element in the Sugano study area. For vulnerable elements that are subject to hazards posed by different landslide types, we attributed separate levels of specific landslide risk. Table 15.9 summarizes the result of the risk assessment.

Fast-moving landslides pose the highest landslide treat in the Sugano study area. Where rockfalls, topples and rockslides affect the edge of the volcanic cap, structural damage is expected, including total destruction of buildings (HD and LD) and roads (FR) (1 in Figure 15.7D). In this area, fatalities and homeless people are expected if fast-moving landsides should occur. Deep-seated slides and slide earthflows of medium intensity may produce functional damage to low-density settlements (LD) and structural damage to farm and other minor roads (FR) (4, 5 in Figure 15.7B). Shallow mass movements of slight intensity and high frequency are expected to cause aesthetic damage to low-density built-up areas (LD) and structural damage to minor roads (FR). Relict block slides, of high and very high intensity (7–8, Figure 15.7, Table 15.9), are expected to cause structural damage to high-density (HD) and low-density (LD) built-up areas, and to secondary and farm roads (SR, FR). In these areas, indirect damage to the population and

**Table 15.9**   *Sugano study area, classification of specific, $R_S$, and total, $R_T$, landslide risk*

| LHZ # | Landslide type | $F_L$ | $I_L$ | $H_L$ | E | V | P | $R_S$ | $R_T$ |
|---|---|---|---|---|---|---|---|---|---|
| 1 | Rock falls | 3 | 3 | 3 3 | HD<br>LD<br>FR | S<br>S<br>S | D, H<br>D, H<br>D | S 3 3<br>S 3 3<br>S 3 3 | Very high |
| | | 1 | 4 | 1 4 | HD<br>LD<br>FR | S<br>S<br>S | D, H<br>D, H<br>D | S 1 4<br>S 1 4<br>S 1 4 | Very high |
| 2 | Shallow slide | 3 | 1 | 3 1 | LD | A | N | A 3 1 | Medium |
| 3 | Shallow slide | 2 | 1 | 2 1 | FR | S | N | S 2 1 | Medium |
| 4 | Deep-seated slide | 1 | 2 | 12 | LD<br>FR | F<br>S | H<br>N | F 1 2<br>S 1 2 | High |
| 5 | Deep-seated slide | 1 | 2 | 1 2 | LD | F | H | F 1 2 | High |
| 6 | Shallow slide | 1 | 1 | 1 1 | FR | S | I, H | S 1 1 | Medium |
| 7 | Very old, deep-seated slide | 1 | 4 | 1 4 | HD<br>LD<br>SR<br>FR | S<br>S<br>S<br>S | I, H<br>I, H<br>I<br>N | S 1 4<br>S 1 4<br>S 1 4<br>S 1 4 | Low |
| 8 | Very old, deep-seated slide | 1 | 3 | 1 3 | HD<br>LD<br>FR | S<br>S<br>S | I, H<br>I, H<br>N | S 1 3<br>S 1 3<br>S 1 3 | Low |

*Note:*
*LHZ* = landslide hazard zone; $F_L$ = landslide frequency (Table 15.1); $I_L$ = landslide intensity (Table 15.2); $H_L$ = landslide hazard (Table 15.3); E = type of element at risk (Table 15.4); V = vulnerability of elements at risk (Table 15.5); P = vulnerability of population (Table 15.5).

homeless people is expected, mostly as a consequence of the damage to the transportation network.

Table 15.9 illustrates total landslide risk, $R_T$, in the Sugano study area. Total landslide risk is estimated to be particularly severe where fast-moving landslides, including rock-falls, topples and rockslides, are expected. In these areas casualties and structural damage to high-density settlements (HD) and minor and farm roads (FR) are possible (1 in Figure 15.7 and Table 15.8). A high level of total landslide risk exists where low-density settlements (LD) are affected by deep-seated slides of medium intensity, which can cause functional damage (5 in Figure 15.7 and Table 15.8). Where low-density settlements (LD) and farm and other minor roads (FR) are affected by shallow landslides, total landslide risk is medium (2–3, 6 in Figure 15.7). Where very old (relict), deep-seated slides have been identified (7–8 in Figure 15.7), total landslide risk is ascertained as low, despite the fact that structural damage to structures and infrastructure, and indirect damage to the population is possible. This is done because reactivation of the entire, very old (relict) landslides is not considered likely under the present climatic and seismic conditions.

### 15.5.3  Piedipaterno

Massive and layered limestone, marl and clay crop out in southeastern Umbria. These rocks pertain to the Umbria–Marche stratigraphic sequence, Lias to Eocene in age. Recent travertine sediments, talus deposits, debris cones and alluvial sediments are locally present. The structural setting of the area is determined by the superposition of two tectonic phases. A compressive phase, of late Miocene to early Pliocene age, was followed by an extensional tectonic phase of Pliocene to recent age. The compressive deformation

**Figure 15.8** *Panoramic view of the Piedipaterno study area. Photograph taken from southeast*

produced large, east-verging folds associated with thrusts and transcurrent faults. The extensional tectonic phase produced normal faults with large vertical displacements. Morphology of the area is controlled by the lithological and structural settings. Major divides coincide with the trend of the largest anticlines, and valleys parallel to the major tectonic elements were formed along synclines, grabens or semi-grabens (Guzzetti *et al.*, 1996).

Regional landslide inventory maps for the Umbria region (Antonini *et al.*, 2002a) reveal that where layered limestone and marl crop out landslides affect about 7.43% of the area. Landslides are both deep-seated, complex or compound mass movements, and shallow failures, chiefly fast-moving rockslides, topples, rockfalls and channelled debris flows. Slow-moving failures are transitional or rotational slides and slide flows, and are most abundant where marly limestone crops out. Rockfalls, topples and rockslides are abundant where cliffs in hard rocks are present. Fast-moving landslides are triggered by different causes, including rainfall and freeze–thaw cycles, but they are most abundant during earthquakes (Antonini *et al.*, 2002b). Debris flows originate where loose debris is abundant, and in particular along the shear zone of regional faults, from talus and scree slopes, or from landslide deposits (Guzzetti *et al.*, 1996; Guzzetti and Cardinali, 1991, 1992).

Piedipaterno is a village in Vallo di Nera Municipality, at the confluence of the Lagarelle Torrent with the Nera River (Figure 15.8). The Lagarelle Torrent drains a small and steep catchment that extends for $2.4\,km^2$ on the eastern slope of Mount Galenne. Elevation in the area ranges from 320 m, at the confluence with the Nera River, to 1217 at the top of Mount Galenne. Slopes are very steep, locally sub-vertical to the north and northwest of the village, and along the Nera River Valley, where rock cliffs and pinnacles are present. In the area, layered limestone and marl crops out that is Cretaceous to upper Eocene in age, pertaining to the middle to upper section of the Umbria–Marche stratigraphic sequence. In the middle and upper part of the Lagarelle catchment, bedding dips into the slope (i.e. towards the west) with an angle of about 30°. In the lower part of the catchment tectonic folds are present.

In the Piedipaterno area we identify two 'elementary slopes', for a total area of about $2\,km^2$ (Figure 15.9). The first elementary slope comprises the catchment of the Lagarelle Torrent (I in Figure 15.9). The second elementary slope is constituted by steep rock slopes along the Nera River Valley (II in Figure 15.9).

For the study area, we obtained a multitemporal landslide inventory map by studying three sets of aerial photographs, and by looking at field surveys. Aerial photographs were taken in August 1954 at 1:33 000 nominal scale, in June 1977 at 1:13 000 nominal scale, and in October 1997 at 1:13 000 nominal scale. Field reconnaissance of the area was completed in October 1997, following the September–October earthquake sequence (Antonini *et al.*, 2002b) and in June 2000.

A total of 17 landslides was mapped in the area, for a total landslide area of $0.87\,km^2$. Landslides cover $0.61\,km^2$ (including six debris-flow areas and three rockfall source areas), 23.1% of the study area. Figure 15.9 illustrates the various types of landslides identified in the study area. Landslides are classified based on the prevalent type of movement and estimated age, inferred from the date of the aerial photographs. Inspection of Figure 15.9 reveals that the lower part of the Lagarelle catchment hosts a large, deep-seated and complex slide (Figure 15.9A). Based on the morphological appearance, we classify the very old landslide as relict. The slope movement affects the limestone and marl bedrock and the thick talus deposit. Numerous deep-seated (Figure 15.9B) and shallow (Figure 15.9C), rotational and translational landslides are present inside or at the edge of the relict landslide. Some of the slope failures are partial reactivations of the relict landslide, whereas others are new landslides that have affected the softened cover of the very old landslide deposit. A small rotational slide identified in the 1977 aerial photographs (Figure 15.9C) produced functional damage to Provincial Road 465. Fast-moving rockfalls (Figure 15.9D) occur from the rock cliffs above the northern side of the village and along the Nera River Valley. These fast-moving landslides threaten

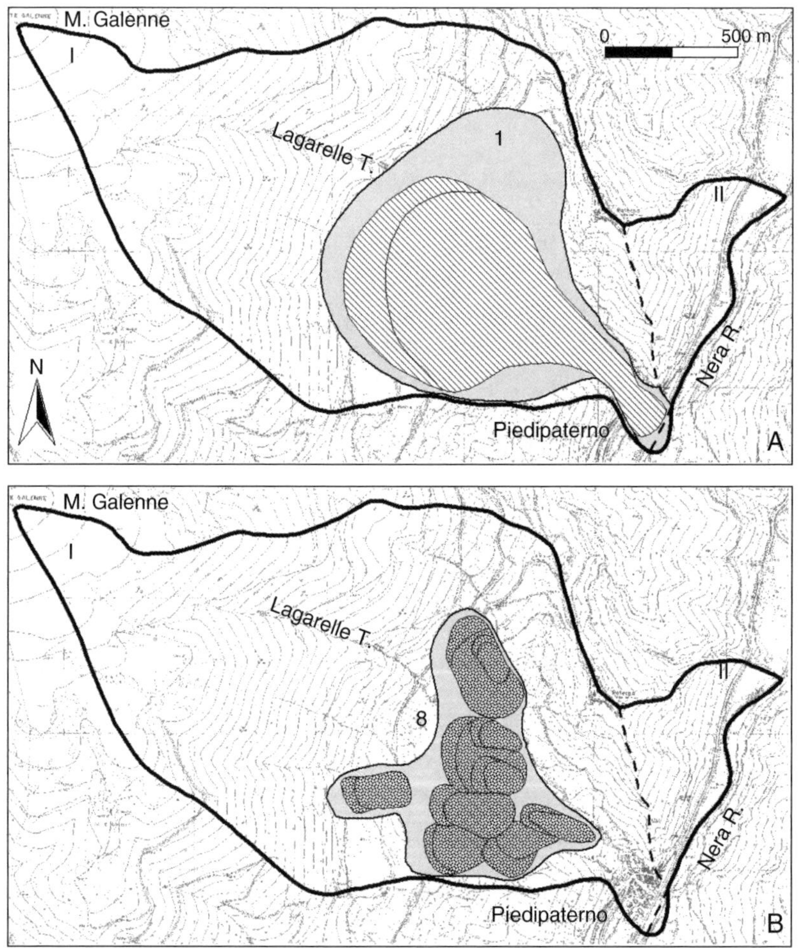


**Figure 15.9**  *Piedipaterno. Multitemporal landslide inventory map. Landslides classified by prevalent type of movement: A, deep-seated slide; B, deep-seated slides and earthflows; C, shallow slides and flows; D, fast-moving rockfalls; E, rapid-moving debris flows. Patterns indicate relative landslide age, inferred from the date of the aerial photographs and from field surveys. Grey areas with Arabic numbers are LHZs (see Table 15.10). Roman numerals indicate elementary slopes*

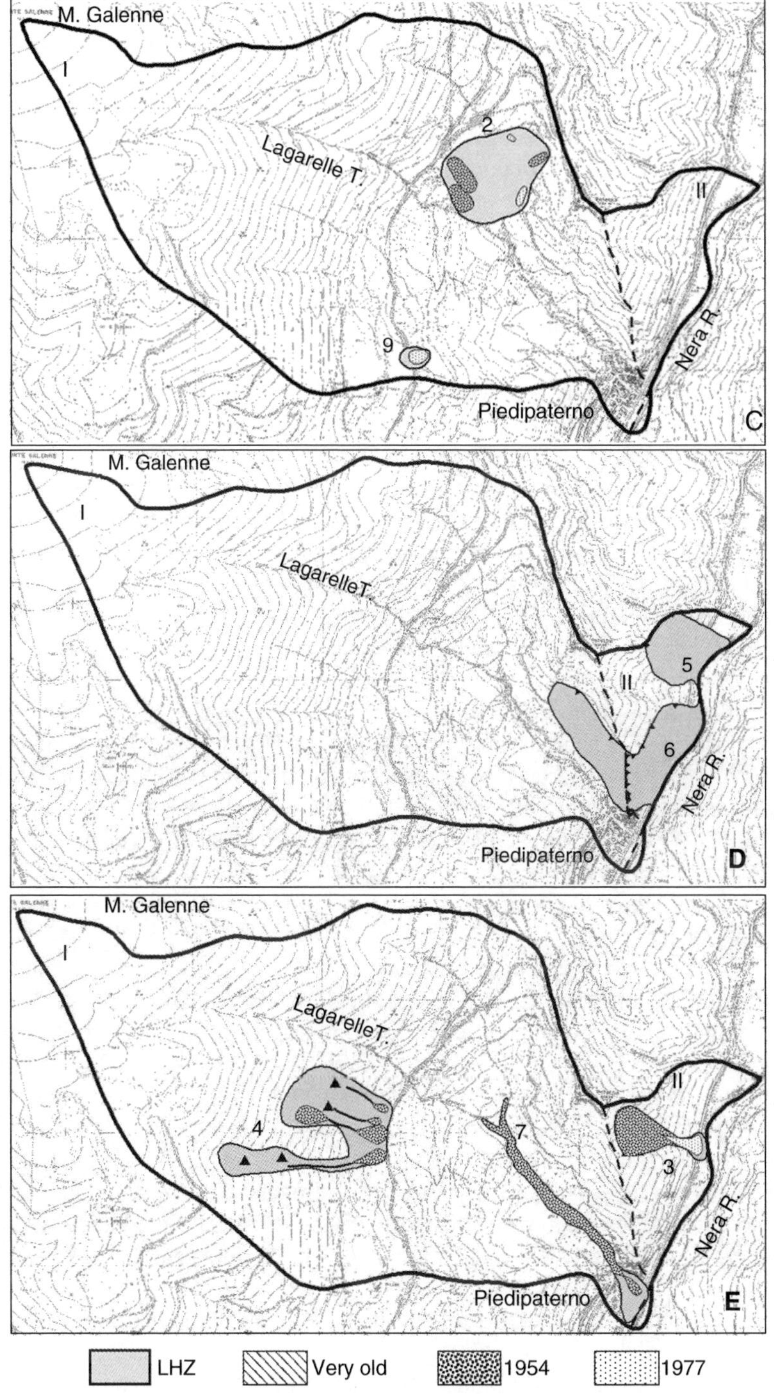



**Figure 15.9**    *(Continued)*

the northern part of the village and a 900 m long section of State Road 209. In April 1992, boulders detached from the rock cliff above the oldest part of the village, severely damaging at least a house. Following this event, elastic nets were installed to protect the houses from rockfalls. The September–October 1997 earthquake sequence apparently did not trigger significant rockfalls in the Piedipatero area (Antonini *et al.*, 2002b). Morphological evidence for debris flows was observed along the lower part of the Lagarelle Torrent on the 1954 aerial photographs. In 1965, a flash flood with associated debris flow occurred along the Lagarelle Torrent, inundating the village and producing extensive damage. Following the event, check dams were built upstream of the village to prevent further damage. Minor debris-flow sources and channels were mapped on the 1954 aerial photographs on the steep slopes of Mount Galenne, and along the Nera River Valley (Figure 15.9E).

In the Piedipaterno study area we recognize nine LHZs (grey areas in Figures 15.9 and 15.10), of which one is for a relict, deep-seated landslide (1 in Figure 15.9A), one for deep-seated, rotational slides of medium intensity (8 in Figure 15.9B), two for slow-moving shallow landslides of slight intensity (2, 9 in Figure 15.9C), two for fast-moving rockfalls (5, 6 in Figure 15.9D), and three for rapid-moving debris flows (3, 4, 7 in Figure 15.9E). For each LHZ we obtain landslide frequency through the interpretation of the available aerial photographs, the historical information, and the field surveys. In the 47-year observation period, landslide frequency ranges from low (one event) to high (three events). For rockfalls and debris flows, landslide frequency is assigned as high, based on the available historical information and on field observations. We obtained information on the location and type of the vulnerable elements in the study area from large-scale topographic maps at 1:10 000 scale, prepared in 1992 from aerial photographs taken in 1986, and from aerial photographs flown in October 1997 (Figure 15.10). By combining this information with the landslide hazard assessment, we estimated levels of specific landslide risk for each vulnerable element or group of vulnerable elements. Where vulnerable elements are subject to hazards posed by different landslide types, we attribute separate levels of specific landslide risk. Table 15.10 summarizes the result of the risk assessment.

In the Piedipaterno study area, fast-moving rockfalls and, subordinately, rapid-moving debris flows pose the highest threat. Field evidence, information on past landslide events, and the location of the vulnerable elements with respect to the rockfall source, travel and deposition areas, together contribute to very high levels of specific landslide risk to vulnerable elements and the population (Figure 15.10E). In the areas where rockfalls have occurred in the past, total or partial destruction of buildings (HD) and functional damage of roads (MR) is expected in the future. Our assessment does not consider the mitigating effect of the existing rockfall defensive structures (i.e. retaining nets and elastic fences) because the existing structures may not be adequate in stopping all rockfalls (Guzzetti *et al.*, 2004). Where rockfalls are expected, fatalities and homeless people are possible.

The section of the Piedipaterno village nearest to the Lagarelle Torrent is subject to flash floods and debris flows (7 in Figure 15.10D). In this area, debris flows can produce functional damage to buildings (HD) and roads (MR), and can cause direct and indirect damage to the population, including homeless people. Historical information indicates that this occurred in 1965. Since then the situation has worsened. Check dams upstream of the village are filled by debris and vegetation, and the Lagarelle Torrent was channelled and covered where it goes through the village (Figure 15.11), reducing significantly its water and sediment discharge capability. In addition, debris flows can cause functional damage to State Road 209 (3 in Figure 15.10D) and to Provincial Road 465 (4 in Figure 15.10D). In both areas, direct and indirect damage to people is possible.

Where deep-seated slides of medium intensity (8 in Figure 15.10) and shallow landslides of slight intensity (9 in Figure 15.10) were identified, structural damage to secondary roads (SR) is expected. The relict, deep-seated landslide (1 in Figure 15.10) can cause severe structural damage to both high-density (HD) and low-density (LD) built-up

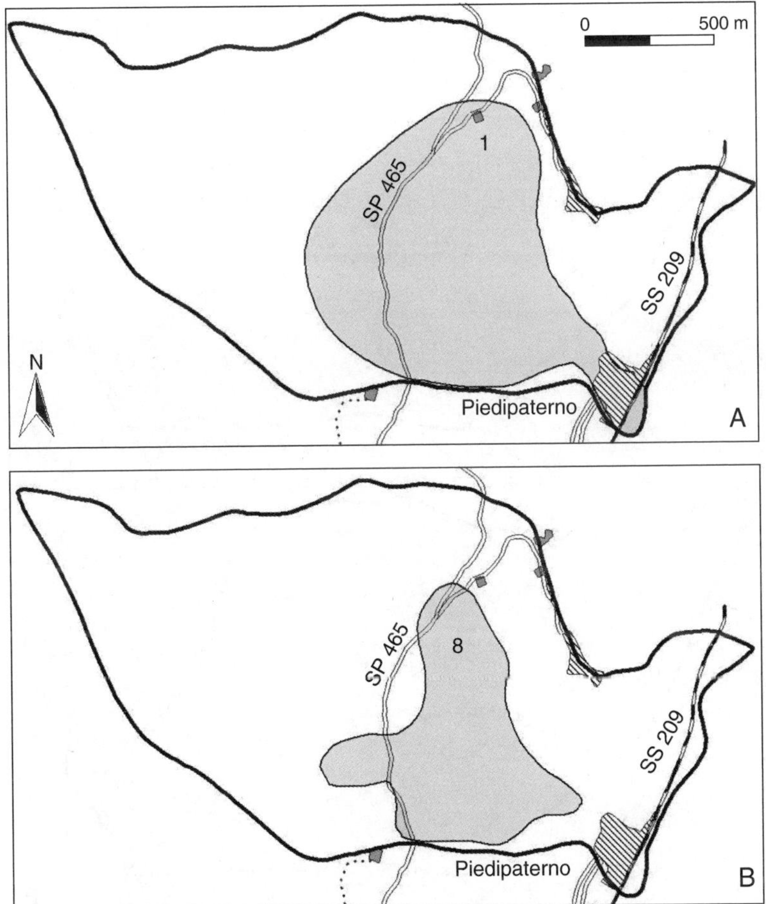


**Figure 15.10** *Piedipaterno. Specific landslide risk (see Table 15.9). A, B, C, D and E show prevalent landslide type, as in Figure 15.9. Grey areas with Arabic numbers are LHZs (see Table 15.10) Patterns indicate types of element at risk (see Table 15.4). Topography shown in Figure 15.9 is not shown here for clarity*

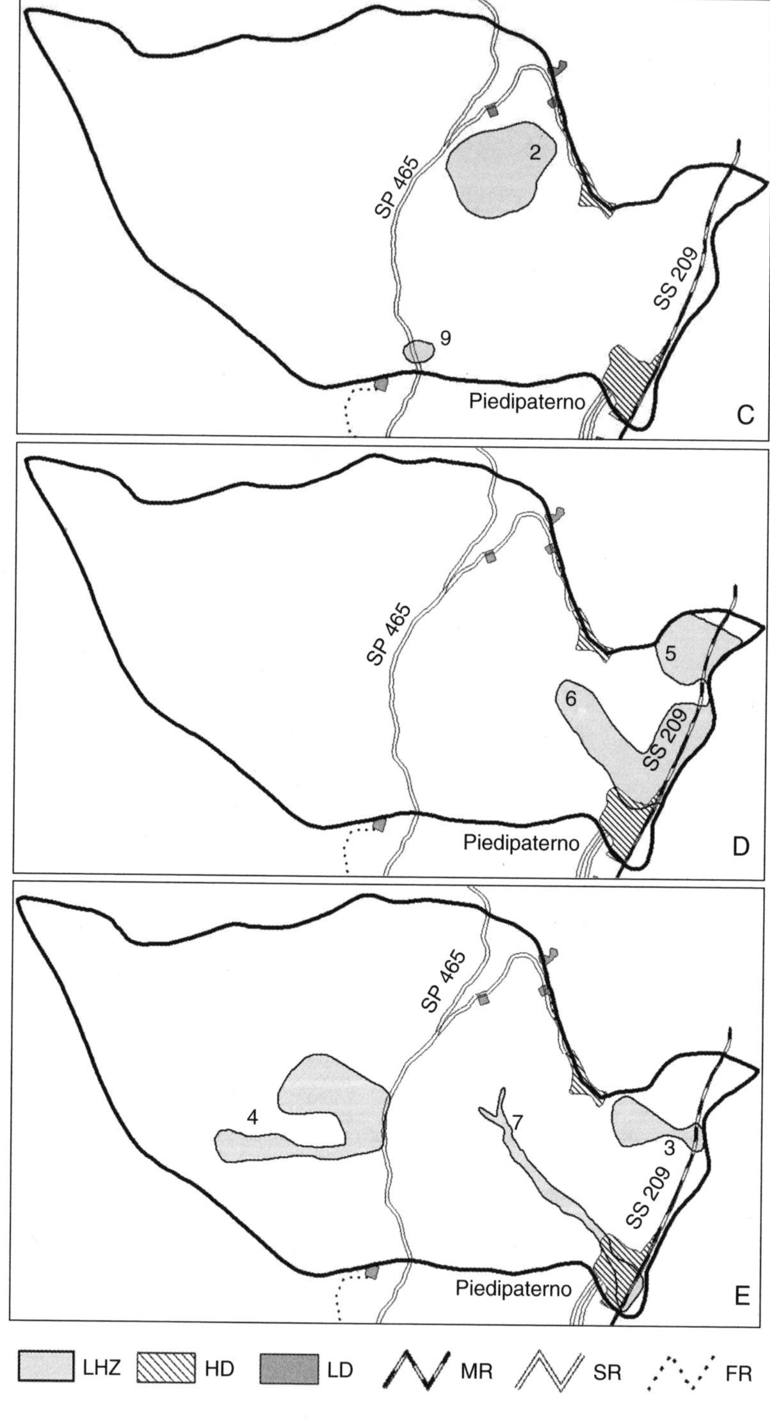


**Figure 15.10**   (Continued)

**Table 15.10**   *Piedipaterno study area, Classification of specific, $R_S$, and total, $R_T$, landslide risk*

| LHZ # | Landslide type | $F_L$ | $I_L$ | $H_L$ | E | V | P | $R_S$ | $R_T$ |
|---|---|---|---|---|---|---|---|---|---|
| 1 | Very old, deep-seated slide | 1 | 4 | 1 4 | HD<br>LD<br>MR<br>SR | S<br>S<br>S<br>S | H<br>H<br>I<br>N | S 1 4<br>S 1 4<br>S 1 4<br>S 1 4 | Low |
| 2 | Shallow slide | 2 | 1 | 2 1 | – | – | – | – | Very low |
| 3 | Debris flow | 2 | 2 | 2 2 | MR | F | D, I | F 2 2 | Very high |
| 4 | Debris flow | 2 | 2 | 2 2 | SR | F | D | F 2 2 | Very high |
| 5 | Rock falls | 3 | 2 | 3 2 | MR | F | D, I | F 3 2 | Very high |
| 6 | Rock falls | 3 | 2 | 3 2 | HD<br>MR | F<br>F | D, I, H<br>D, I | F 3 2<br>F 3 2 | Very high |
| 7 | Debris flow | 2 | 2 | 2 2 | HD<br>MR | F<br>F | D, I, H<br>D, I | F 2 2<br>F 2 2 | Very high |
|  |  | 1 | 3 | 1 3 | HD<br>MR | F<br>F | D, I, H<br>D, I | F 1 3<br>F 1 3 | Very high |
| 8 | Deep-seated slide | 1 | 2 | 1 2 | SR | S | N | S 1 2 | Medium |
| 9 | Shallow slide | 1 | 1 | 1 1 | SR | F | N | F 1 1 | Medium |

*Note:*
*LHZ* = landslide hazard zone; $F_L$ = landslide frequency (Table 15.1); $I_L$ = landslide intensity (Table 15.2); $H_L$ = landslide hazard (Table 15.3); $E$ = type of element at risk (Table 15.4); $V$ = vulnerability of element at risk (Table 15.5); $P$ = vulnerability of population (Table 15.5).

areas, and to the transportation network (MR, SR). This is mostly because of the large volume of the landslide (i.e. intensity). If the landslide should move, indirect damage to the population and homeless people is possible.

Table 15.10 illustrates total landslide risk for the Piedipaterno study area. Total landslide risk is estimated to be very high where fast-moving rockfalls are expected, and in the areas that can be affected by rapid-moving debris flows. Where rockfalls and debris flows are expected, casualties and structural damage to high-density settlements and major roads are possible (3, 5–7 in Figure 15.10). The very high level of total landslide risk is assigned mostly because casualties are expected. In the case of rockfalls, the observed frequency of events results in a very high total landslide risk. Medium levels of total landslide risk arise in the areas where deep-seated and shallow slope failures can cause structural damage to the transportation network (8, 9 in Figure 15.10). Where the relict, deep-seated slide was identified (1 in Figure 15.10), total landslide risk is ascertained as low, despite possible damage to structures and infrastructure, and indirect damage to the population. As in the case of Morano Madonnuccia, this is done because reactivation of the relict landslide in considered unlikely under the present climatic and seismic conditions. For one LHZ (2 in Figure 15.10), total landslide risk is assessed as very low, despite landslide hazard being not negligible ($H_L = 21$). This is because elements at risk are not present in this LHZ.

**Figure 15.11**  *The Lagarelle Torrent canalized and partly covered where it crosses the Piedipaterno Village. The canalized cross section is too small to contain flash floods and debris flows*

## 15.6   Discussion

The proposed method complies with the existing and widely accepted definitions of landslide hazard (Varnes – IAEG, 1984; Guzzetti *et al.*, 1999a, 1999b) and of landslide risk (Varnes – IAEG, 1984; Einstein, 1988; Cruden and Fell, 1997; Guzzetti, 2002). The method is empirical and subject to various levels of uncertainty, but has proved to be reliable and cost-effective, allowing for detailed and comparable assessments of landslide hazard and specific and total landslide risk levels in urban and rural areas in Umbria. The method allows the comparison of landslide hazard and risk in distinct and distant areas, and where different landslide types are present.

The method was successfully applied in 210 landslide hazard zones located around or in the vicinity of 79 towns and villages in Umbria (Antonini *et al.*, 2002a; Cardinali *et al.*, 2002b). In these areas landslide hazard was determined, vulnerable elements were identified, and specific and total risk levels were evaluated. The time and human resources required for completing the risk assessment procedure at each site varied, depending on the extent of the study area, the number, type and scale of the aerial photographs, the available thematic and historical information, the extent and type of landslides present in the study area, and the local geological and morphological setting. On average, completion of the risk assessment procedure at each site required five days for a team of three to four geomorphologists, including bibliographical investigation, interpretation of the aerial photographs, field surveys, storage of landslide and thematic information in the GIS database, and the production of the final hazard, vulnerability and risk maps.

The method requires extensive geomorphological judgement. For this reason it should only be used by skilled geomorphologists. If the extent, type, distribution and pattern of past and present landslides are not correctly and fully identified, errors can occur and thus affect the estimate of landslide hazards and risk. With this in mind, the definition of the temporal frequency of landslides from the analysis of the multitemporal inventory map is particularly important. The map covers a period of about 47 years (from 1954 to 2001), which is long enough to evaluate the short-term behaviour of slopes in the areas investigated. Should information on landslide frequency be available only for a shorter period of time (e.g. 10–15 years or less), the reliability of the hazard forecast will be reduced. If a landslide event fails to be recognized, the frequency of occurrence is underestimated, and hazard and risk estimates are negatively affected. It should be noted that the method estimates the expected landslide frequency based on what has happened (and was observed) in the recent past. If low-frequency, high-magnitude events did not occur or were not recognized in an LHZ, the hazard assessment in the area may be biased, and the actual landslide risk is underestimated. This is a limitation of the method.

The method allows for detailed and articulated hazard and risk assessments. Landslide hazard is determined separately for all the different landslide types that may be present in the study area. Specific landslide risk is determined independently for each type of vulnerable element in the study area. Thus vulnerable elements may be subject to multiple levels of specific landslide risk, and landslide hazard may be ascertained where vulnerable elements are not present. The proposed method assesses landslide hazard in the areas of probable evolution of the existing landslides (i.e. in a landslide hazard zone). The method says nothing about the hazard outside an LHZ, even within the same elementary slope. In these areas minor landslides, mostly superficial failures, can occur with a low frequency. For a regional, spatially distributed landslide hazard and risk assessment, other methods should be used (see for example Soeters and van Westen, 1996; Guzzetti *et al.*, 1999b; and references therein), possibly in combination with the method proposed here.

Uncertainty varies with the different steps of the method. The production of the separate landslide inventory maps and of the multitemporal landslide map is less uncertain than the identification of the landslide hazard zones, or the possible spatial evolution of the existing landslides, which is obtained mostly through geomorphological inference. Landslides mapped through the interpretation of aerial photographs were carefully checked in the field, whereas the identification and mapping of LHZs was based on the observation of other landslides and on the inferred geomorphological behaviour of slopes. Evaluations of landslide frequency, which determine landslide hazards, are conditioned by the availability of aerial photographs and historical information, and by our ability to recognize past and present landslide events. Estimates of landslide volume and velocity, which are essential for the evaluation of landslide intensity, also exhibit uncertainty.

Uncertainty also arises because a difference exists between geomorphological and historical information on landslides in Umbria. Geomorphological information obtained through field mapping and the analysis of aerial photographs provides the basis for determining the prevalent landslide types and location, but provides poor constrain on the date (or the age) of the slope failures. Historical and archived information provide precisely the date (or even the time) of occurrence of some of the landslide events, and the type and extent of damage caused by mass movements, but provide little information on the type of failures and the precise location and extent of the landslides. As pointed out by Guzzetti *et al.* (2003), combining of geomorphological and historical information on mass movements is not straightforward, and may be a matter of local interpretation and local judgement.

The proposed method relies on a set of correlation tables which are used to define landslide frequency (Table 15.1) and intensity (Table 15.2), to ascertain landslide hazard (Table 15.3), to evaluate the expected damage to the vulnerable elements (Table 15.5), and to attribute levels of specific landslide risk (Tables 15.6). The tables used in this work are based on empirical observations and on our own experience, but are also the result of a heuristic approach. Whenever possible we tried each of several possibilities and evaluated the difference after each attempt. We believe that the tables fit the present understanding of landslide processes and match landslide damage in Umbria satisfactorily. However, the tables are not definitive, and should not be used unconditionally in all settings. If applied to other sites, or in other study areas, they should be carefully checked with the local information on landslide types and damage. If one, or more, of the tables is changed significantly, the hazard and risk assessment will vary, and may not be comparable with the one we have prepared. This is particularly important for Tables 15.5 and 15.6.

Landslide hazard and risk are expressed using a multiple-digit, 'positional' index that shows, in a compact and convenient format, all the variables used to ascertain landslide hazard and risk (i.e. intensity, frequency and vulnerability). The index allows for the ranking of risk conditions at the end of the risk assessment process, when all the necessary information is available, and not *a priori*, based on predefined (often ill-formalized) categories. We consider this a major advantage of the method, giving risk managers and decision makers great flexibility in deciding which area exhibits the highest risk, and providing geologists and engineers with a clue about why any given vulnerable element is at risk. In addition, the use of a simple index to express levels of specific landslide risk makes it possible to adopt different schemes to determine levels of total landslide risk, depending on the priorities or the specific interests of the investigators or the end-users.

When determining landslide risk, in places we decided not to consider the mitigating effects of the existing defensive measures. Structural measures may be at least partially ineffective in mitigating landslide risk. This is often the case for rockfall-retaining nets or elastic fences that can be jumped by high-flying boulders or that can be destroyed by large blocks (Guzzetti *et al.*, 2004). Check dams and channels built to prevent or to mitigate debris flows may be filled by debris and vegetation, due to lack of maintenance. The effects of structural measures on deep-seated landslides (e.g. retaining walls, piles, earthworks, drainage systems, etc.) are difficult to evaluate without accurate, long-term monitoring.

Lastly, the proposed method is not simple or straightforward. Dependable and consistent prediction requires multiple sets of aerial photographs and a team of experienced geomorphologists to interpret them. This cannot be considered a limitation: landslide hazard and risk assessments are difficult tasks, and require proper expertise and skills.

## 15.7  Conclusion

We have presented a geomorphological method to ascertain landslide hazard and to evaluate the associated risk, at site scale, and we have shown three applications of the method in Umbria. The method is based on careful recognition of present and past landslides, scrutiny of the local geological and morphological settings, and analysis of site-specific and historical information on past landslide events. Most of the information used to ascertain landslide hazard is obtained from the analysis of a multitemporal landslide inventory map that portrays information on the distribution, type and pattern of landslides and on their changes in time. The multitemporal map is obtained by merging landslide inventory maps prepared through the analysis of stereoscopic aerial photographs of different ages, and by use of field surveys.

The method was applied extensively in the Umbria region of central Italy, and allowed determining and comparing specific landslide risk in 210 landslide hazard zones around or in the vicinity of 79 towns and villages in Umbria. Results of the hazard and risk assessment procedure were recently incorporated in the river basin geo-hydrological hazard reduction plan, prepared by the Tiber River Basin Authority. Results were also distributed by the Umbria regional government to municipalities in Umbria. At present it is not possible to judge quantitatively how good the proposed method is, and how reliable our hazards and risk assessments are. Municipalities in Umbria are checking the landslide hazard and risk assessments, and they are comparing them to prior information and to existing planning schemes and building codes. We look forward to knowing what problems they will discover, and to investigating how they will use the new information on landslide hazards and risk. While we wait for the results of this independent analysis, we have started applying the methodology to an area extending to about $70\,km^2$ in central Umbria, that comprises several tens of elementary slopes, hundreds of landslide hazard zones, and thousands of shallow and deep-seated landslides. We hope to bridge the existing gap between hazards and risk assessments at the site scale, as presented in this study, and regional assessments of the impact of mass movements on the population, the built-up areas and the infrastructure in Umbria (Guzzetti *et al.*, 2003).

## Acknowledgements

Research for this chapter was supported by CNR GNDCI and CNR IRPI funds, and by a specific grant of the Regione dell'Umbria and of the Tiber River Basin Authority. We are grateful to Earl E. Brabb for reviewing the manuscript. The paper is CNR GNDCI publication number 2806.

## References

Alexander, D., 1989, Urban landslides, *Progress in Physical Geography*, **13**, 157–191.
Alexander, D., 2000, Landslide risk estimation in Umbria region, unpublished technical report for CNR IRPI, Perugia.

Antonini, G., Cardinali, M., Guzzetti, F., Reichenbach, P. and Sorrentino, A., 1993, Carta Inventario dei Fenomeni Franosi della Regione Marche ed aree limitrofe, CNR GNDCI Publication number 580, map at 1:100 000 scale (in Italian).

Antonini, G., Ardizzone, F., Cacciano, M., Cardinali, M., Castellani, M., Galli, M., Guzzetti, F., Reichenbach, P. and Salvati, P., 2002a, *Rapporto Conclusivo*. Protocollo d'Intesa fra la Regione dell'Umbria, Direzione Politiche Territoriali Ambiente e Infrastrutture, ed il CNR IRPI di Perugia per l'acquisizione di nuove informazioni sui fenomeni franosi nella regione dell'Umbria, la realizzazione di una nuova carta inventario dei movimenti franosi e dei siti colpiti da dissesto, l'individuazione e la perimetrazione delle aree a rischio da frana di particolare rilevanza, e l'aggiornamento delle stime sull'incidenza dei fenomeni di dissesto sul tessuto insediativo, infrastrutturale e produttivo regionale (in Italian).

Antonini, G., Ardizzone, F., Cardinali, M., Galli, M., Guzzetti, F. and Reichenbach, P., 2002b, Surface deposits and landslide inventory map of the area affected by the 1997 Umbria–Marche earthquakes, *Bollettino Società Geologica Italiana*, **121**, 843–853.

Barchi, M., Cardinali, M., Guzzetti, F. and Lemmi, M., 1993, Relazioni fra movimenti di versante e fenomeni tettonici nell'area del M. Coscerno – M. di Civitella, Val Nerina (Umbria), *Bollettino Società Geologica Italiana*, **112**, 83–111 (in Italian).

Budetta, P., 2002, Risk assessment from debris flow in pyroclastic deposits along a motorway, Italy, *Bulletin of Engineering Geology and the Environment*, **61**, 293–301.

Calabresi, G. and Scarpelli, G., 1984, A typical earthflow in a weathered clay at Todi, *Proceedings IV International Symposium on Landslides*, Toronto, vol. 2, 175–180.

Canuti, P., Marcucci, E., Trastulli, S., Ventura, P. and Vincenti, G., 1986, Studi per la stabilizzazione della frana di Assisi, *Atti XVI Convegno Nazionale Geotecnica*, Bologna, 14–16 May 1986, vol. 1, 165–174 (in Italian).

Capocecere, P., Martini, E. and Peronacci, M., 1993, Sistemi di monitoraggio del colle di Todi, *Studio Monitoraggio e Bonifica dei Centri Abitati Instabili, ENEA*, Rome, 67–71 (in Italian).

Cardinali, M., Galli, M., Guzzetti, F., Reichenbach, P. and Borri, G., 1994, Relazioni fra movimenti di versante e fenomeni tettonici nel bacino del Torrente Carpina (Umbria settentrionale), *Geografia Fisica e Dinamica Quaternaria*, **17**, 3–17 (in Italian).

Cardinali, M., Ardizzone, F., Galli, M., Guzzetti, F. and Reichenbach, P., 2000, Landslides triggered by rapid snow melting: the December 1996–January 1997 event in central Italy, in P. Claps and F. Siccardi (eds), *Proceedings Plinius Conference '99*, Maratea, 14–16 October 1999, CNR GNDCI publication number 2012, 439–448.

Cardinali, M., Antonini, G., Reichenbach, P. and Guzzetti, F., 2001, Photo-geological and landslide inventory map of the Upper Tiber River basin, CNR GNDCI publication number 2116, map at 1:100 000 scale.

Cardinali, M., Carrara, A., Guzzetti, F. and Reichenbach, P., 2002a, Landslide hazard map for the Upper Tiber River basin, CNR GNDCI publication number 2634, map at 1:100 000 scale.

Cardinali, M., Reichenbach, P., Guzzetti, F., Ardizzone, F., Antonini, G., Galli, M., Cacciano, M., Castellani, M. and Salvati, P., 2002b, A geomorphological approach to estimate landslide hazard and risk in urban and rural areas in Umbria, central Italy, *Natural Hazards and Earth Systems Science*, **2**(1–2), 57–72.

Carrara, A., Cardinali, M., Detti, R., Guzzetti, F., Pasqui, V. and Reichenbach, P., 1991, GIS techniques and statistical models in evaluating landslide hazard, *Earth Surface Processes and Landform*, **16**(5), 427–445.

Carrara, A., Cardinali, M., Guzzetti, F. and Reichenbach, P., 1995, GIS technology in mapping landslide hazard, in A. Carrara and F. Guzzetti (eds), *Geographical Information Systems in Assessing Natural Hazards* (Dordrecht: Kluwer Academic Publishers), 135–175.

Carrara, A., Guzzetti, F., Cardinali, M. and Reichenbach, P., 1999, Use of GIS Technology in the Prediction and Monitoring of Landslide Hazard, *Natural Hazards*, **20**(2–3), 117–135.

Cecere, V., and Lembo-Fazio, A., 1986, Condizioni di sollecitazioni indotte dalla presenza di una placca lapidea su un substrato deformabile, *Atti XVI Convegno Nazionale Geotecnica*, Bologna, 14–16 May 1986, vol. 1, 191–202 (in Italian).

Crescenti, U., 1973, Studi di conservazione territoriale. I movimenti franosi in Comune di Montone, Perugia, *Geologia Applicata & Idrogeologia*, **12** (in Italian).

Cruden, D.M. and Fell, R., 1997, *Landslide Risk Assessment, Proceedings of the International Workshop on Landslide Risk Assessment*, Honolulu, 19–21 February 1997 (Rotterdam: Balkema).

Diamanti, L. and Soccodato, C., 1981, Consolidation of the historical cities of San Leo and Orvieto, *Proceedings of the X International Conference on Soil Mechanics and Foundation Engineering*, Stockolm, June 1981, vol. 3, 75–82.

Einstein, H.H., 1988, Special lecture: landslide risk assessment procedure, *Proceedings of the V International Symposium on Landslides*, Lausanne, vol. 2, 1075–1090.

Einstein, H.H., 1997, Landslide risk – systematic approaches to assessment and management, in D.M. Cruden and R. Fell (eds), *Landslide Risk Assessment, Proceedings of the International Workshop on Landslide Risk Assessment*, Honolulu, 19–21 February 1997 (Rotterdam: Balkema), 25–50.

Evans, S.G., 1997, Fatal landslide and landslide risk in Canada, in D.M. Cruden and R. Fell (eds), *Landslide Risk Assessment, Proceedings of the International Workshop on Landslide Risk Assessment*, Honolulu, 19–21 February 1997 (Rotterdam: Balkema), 185–196.

Felicioni, G., Martini, E. and Ribaldi, C., 1994, Studio dei Centri Abitati Instabili in Umbria. Atlante regionale, CNR GNDCI publication number 979 (in Italian).

Fell, R., 1994, Landslide risk assessment and acceptable risk, *Canadian Geotechnical Journal*, **32**(2), 261–272.

Fell, R., 2000, Landslide risk management concepts and guidelines, *Australian Geomechanics Society*, Sub-committee on landslide risk management.

Fell, R. and Hartford, D., 1997, Landslide risk management, in D.M. Cruden and R. Fell (eds), *Landslide Risk Assessment, Proceedings of the International Workshop on Landslide Risk Assessment*, Honolulu, 19–21 February 1997 (Rotterdam: Balkema), 51–109.

*Gazzetta Ufficiale della Repubblica Italiana.* Misure urgenti per la prevenzione del rischio idrogeologico ed a favore delle zone colpite da disastri franosi nella regione Campania. Serie Generale, Anno 139, n. 208, 7 September 1998, 53–74 (in Italian).

Glade, T., 1998, Establishing the frequency and magnitude of landslide-triggering rainstorm events in New Zealand, *Environmental Geology*, **35**, 160–174.

Guzzetti, F., 2000, Landslide fatalities and evaluation of landslide risk in Italy, *Engineering Geology*, **58**, 89–107.

Guzzetti, F., 2002, Landslide hazard assessment and risk evaluation: overview, limits and prospective, in *Proceedings of the 3rd MITCH Workshop Floods, Droughts and Landslides – Who plans, who pays*, 24–26 November 2002, Potsdam, available at http://www.mitch-ec.net/workshop3/Papers/paper_guzzetti.pdf

Guzzetti, F. and Cardinali, M., 1989, Carta Inventario dei Fenomeni Franosi della Regione dell'Umbria ed aree limitrofe, CNR GNDCI publication number 204, map at 1:100 000 scale (in Italian).

Guzzetti, F. and Cardinali, M., 1991, Debris-flow phenomena in the Central Apennines of Italy, *TerraNova*, **3**, 619–627.

Guzzetti, F. and Cardinali, M., 1992, Debris-flow phenomena in the Umbria–Marche Apennines (central Italy), *Proceedings Interpraevent 1992*, Bern, vol. 2, 181–192.

Guzzetti, F., Cardinali, M. and Reichenbach, P., 1994, The AVI Project: a bibliographical and archive inventory of landslides and floods in Italy, *Environmental Management*, **18**, 623–633.

Guzzetti, F., Cardinali, M. and Reichenbach, P., 1996, The influence of structural setting and lithology on landslide type and pattern, *Environmental and Engineering Geoscience*, **2**(4), 531–555.

Guzzetti, F., Cardinali, M., Reichenbach, P. and Carrara, A., 1999a, Comparing landslide maps: a case study in the upper Tiber River Basin, central Italy, *Environmental Management*, **25**(3), 247–363.

Guzzetti, F., Carrara, A., Cardinali, M. and Reichenbach, P., 1999b, Landslide hazard evaluation: an aid to a sustainable development, *Geomorphology*, **31**, 181–216.

Guzzetti, F., Reichenbach, P., Cardinali, M., Ardizzone, F. and Galli, M., 2003, Impact of landslides in the Umbria region, central Italy, *Natural Hazards and Earth System Sciences*, **5**, 1–17.

Guzzetti, F., Reichenbach, P. and Ghigi, S., 2004, Rockfall hazard and risk assessment in the Nera River valley, Umbria region, central Italy, *Environmental Management*, **34**(2), 191–208.

Hungr, O., 1997, Some methods of landslide hazard intensity mapping, in D.M. Cruden and R. Fell (eds), *Landslide Risk Assessment, Proceedings International Workshop on Landslide Risk Assessment*, Honolulu, 19–21 February 1997 (Rotterdam: Balkema), 215–226.

Hutchinson, J.N., 1995, Keynote paper: landslide hazard assessment, in D.H. Bell (ed.), *Landslides* (Rotterdam: Balkema), 1805–1841.

Ibsen, M.L. and Brunsden, D., 1996, The nature, use and problems of historical archives for the temporal occurrence of landslides, with specific reference to the south coast of Britain, Ventnor, Isle of Wight, *Geomorphology*, **15**, 241–258.

Kong, W.K., 2002, Risk assessment of slopes, *Quarterly Journal of Engineering Geology Hydrogeology*, **35**, 213–222.

Lembo-Fazio, A., Manfredini, G., Ribacchi, R. and Sciotti, M., 1984, Slope failure and cliff instability in the Orvieto tuff, *Proceedings of the IV International Symposium on Landslides*, Toronto, vol. 2, 115–120.

Michael-Leiba, M., Baynes, F. and Scott, G., 1999, Quantitative landslides of Cairns, Australian Geological Survey Organisation, Department of Industry, Science & Resources, *AGSO Records* 1999/36.

Nilsen, T.H. and Turner, B.L., 1975, Influence of rainfall and ancient landslide deposits on recent landslides (1950–71) in urban areas of Contra Costa County, California, *US Geological Survey Bulletin*, **1388**.

Ribacchi, R., Sciotti, M. and Tommasi, P., 1988, Stability problems of some towns in central Italy: geotechnical situations and remedial measurements, in *Proceedings of the International Symposium IAEG on Engineering Geology of Ancient Works, Monuments and Historical Sites*, Athens, vol. 1, 27–36.

Righi, P.V., Marchi, G. and Dondi, G., 1986, Stabilizzazione mediante pozzi drenanti di un movimento franoso nella città di Perugia, *Atti XVI Convegno Nazionale Geotecnica*, Bologna, 14–16 May 1986, vol. 2, 167–179 (in Italian).

Salvati, P., Guzzetti, F., Reichenbach, P., Cardinali, M. and Stark, C.P., 2003, Map of landslides and floods with human consequences in Italy, CNR GNDCI publication number 2822, map at scale 1:1 200 000.

Servizio Geologico d'Italia: Carta Geologica dell'Umbria. Map at 1:250 000 scale, 1980 (in Italian).

Soeters, R. and van Westen, C.J., 1996, Slope instability recognition, analysis and zonation, in A.K. Turner and Schuster, R.L. (eds), *Landslide Investigation and Mitigation*, National Research Council, Transportation Research Board Special Report 247 (Washington, DC: National Academy Press), 129–177.

Tommasi, P., Ribacchi, R. and Sciotti, M., 1986, Analisi storica dei dissesti e degli interventi sulla Rupe di Orvieto, *Geologia Applicata & Idrogeologia*, **21**, 99–153 (in Italian).

Tonnetti, G., 1978, Osservazioni geomorfologiche sulla frana del Fosso delle Lucrezie presso Todi, Perugia, *Memorie Società Geologica Italiana*, **19**, 205–213 (in Italian).

Varnes, D.J., 1978, Slope movements, type and processes, in R.L. Schuster and R.J. Krizek (eds), *Landslide Analysis and Control*, Transportation Research Board, National Academy of Sciences, Washington, DC, Special Report 176, 11–33.

Varnes, D.J. and IAEG Commission on Landslides and other Mass Movements, 1984, *Landslide Hazard Zonation: A Review of Principles and Practice* (Paris: UNESCO Press).

WP/WLI – International Geotechnical societies' UNESCO Working Party on World Landslide Inventory, 1990, A suggested method for reporting a landslide, *International Association Engineering Geology Bulletin*, **41**, 5–12.

WP/WLI – International Geotechnical societies' UNESCO Working Party on World Landslide Inventory, 1993, A suggested method for describing the activity of a landslide, *International Association Engineering Geology Bulletin*, **47**, 53–57.

WP/WLI – International Geotechnical societies' UNESCO Working Party on World Landslide Inventory, 1995, A suggested method for describing the rate of movement of a landslide, *International Association Engineering Geology Bulletin*, **52**, 75–78.

# 16

# Remote Sensing of Landslides

*Vern Singhroy*

## 16.1 Introduction

Globally, landslides cause approximately 1000 deaths per year with property damage of about US$4 billion (Alexander, 1995). Landslides pose serious threats to settlements and to structures that support transportation, natural resource management and tourism. They cause considerable damage to highways, railways, waterways and pipelines. They commonly occur with other major natural disasters such as earthquakes (Keefer, 1984), volcanic activity (Kimura and Yamaguchi, 2000), and floods caused by heavy rainfall. For instance, earthquake-induced landslides occur in various geological environments, ranging from steep rock slopes to gentle slopes with unconsolidated sediments. The area affected by landslides in an earthquake correlates with the magnitude, geological conditions, earthquake focal depth, and specific ground motion characteristics (Keefer, 1984, 1994). Damage from landslides and other ground failures has sometimes exceeded damage directly related to earthquakes. In many cases, expanded development and human activity, such as modified slopes in urban areas and deforestation, can increase the incidence of landslide disasters.

## 16.2 Landslide Mapping

Photogeologic techniques (Varnes, 1974) provide a framework for developing mapping strategies which will assist in the interpretation of remote sensing data. Currently, airphotos are used extensively to produce landslide inventory maps, because they allow for features demonstrating slope movement that range from small terracettes, indicating soil creep, to large landslides to be resolved, as demonstrated by Mollard and Janes

*Landslide Hazard and Risk*   Edited by Thomas Glade, Malcolm Anderson and Michael J. Crozier
© 2004 John Wiley & Sons, Ltd   ISBN: 0-471-48663-9

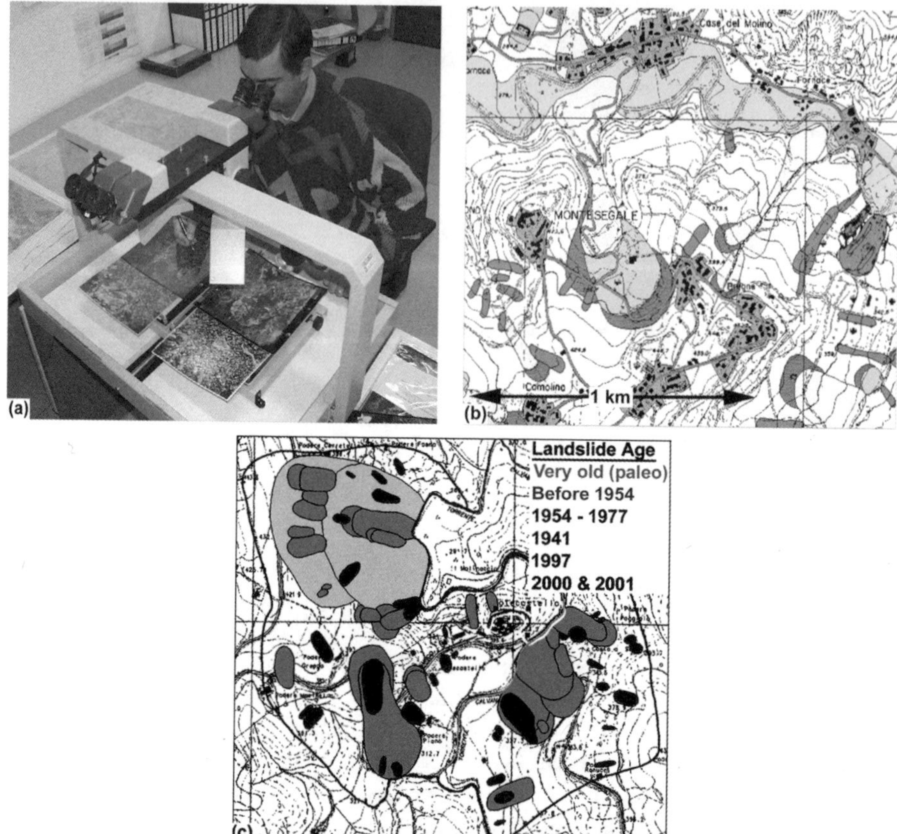



**Figure 16.1**  *(a) Interpretation of stereo airphotos to produce 1:100 000 scale landslide inventory maps of Italy. (b) The various shaded units refer to landslide types and (c) ages Source: Guzzetti, 1990. (See also Plate 5)*

(1993), Guzzetti (1990) (Figure 16.1; see also Colour Plate section, Plate 5), and others. In addition, geological and geomorphological units can be interpreted on the basis of morphological, textural and structural characteristics using stereo aerial photos and remote sensing images.

The majority of landslide research carried out by remote sensing to date falls into the category of inventory mapping. In general, the working scale for inventory mapping and slope instability analysis is determined by the requirements of the user. Planners and engineers use the following examples of scales:

- National scale (<1:1 000 000) provides a general inventory of problem areas for an entire country, which can be used to inform national policy makers and the general public.
- Regional scale (1:100 000–1:500 000) is used in the early phases of regional development projects to evaluate possible constraints, due to instability, in the development of large engineering projects and regional development plans.

- Medium scale (1:25 000–1:50 000) is used for the determination of hazard zones in areas affected by large engineering structures, roads and urbanization plans.
- Large scale (1:5000–1:15 000) is used at the level of site investigations prior to the design phase of engineering works.

Remote sensing images generally provide information on regional inventory (1:100 000). However, there are significant limitations on the uses of EO data for large-scale landslide studies. The principal problem is that current remote sensing systems do not have the very high spatial resolution stereo capability (less than 3 m stereo) required for landslide studies, and therefore only large landslides and regional inventory mapping can be done using remote sensing techniques (Table 16.1).

For the evaluation of the suitability of remote sensing images for landslide inventory mapping the size of individual slope failures in relation to the ground resolution cell is of crucial importance. Although sizes of landslides vary enormously according to the type of slope failure, some useful information can be found in the literature. The total map area for a failure of $42\,000\,\mathrm{m}^2$ corresponds to $20 \times 20$ pixels on a SPOT Pan image and $10 \times 10$ pixels on SPOT multispectral images. This would be sufficient to identify a landslide displaying a high contrast, but it is insufficient for a proper analysis of the elements pertaining to the failure to establish characteristics and type of landslide. It is believed that if 1:15 000 is the most appropriate scale, then 1:25 000 should be considered as the smallest scale to analyse slope instability phenomena on aerial photographs. Using smaller scales, a slope failure may be recognized as such if size and contrast are sufficiently large. However, the amount of analytical information enabling the interpreter to make conclusions on type and causes of the landslide will be very limited at scales smaller than 1:25 000. For this reason, 3 m stereo images will be most useful for detailed interpretation. The practical or operational use of the currently available EO data in engineering geology site-specific landslide investigations is considerably limited (Wasowski and Gostelow, 1999). Future satellite systems providing less than 3 m stereo images such as RADARSAT-2 (2005) will permit detailed geomorphological mapping.

Bulmer (2002) used two distinct approaches to determine the characteristics of different landslides from remotely sensed data. With these approaches it is possible to derive qualitative and quantitative parameters on landslides that are necessary for improved understanding of landslide processes. The first approach is to determine the number, distribution, type and geomorphology of landslides using remotely sensed data. The second approach measures the dimensions (length, width, thicknesses and local slope) along and across the landslides using imagery and topographic profiles. However, when selecting and using remotely sensed data, Bulmer (2002) noted that the goal should be to determine: (1) the local lithology, (2) aerial extent of landslide deposits at each site, (3) local age relationships, (4) evidence for the cause and frequency of emplacement, (5) differences in landslide morphologies as keys to the magnitude and types of mass movement events, and (6) dimensions, slopes (local and regional), volumes, and material sizes.

Current research has shown that high-resolution stereo SAR and optical images, combined with topographic and geological information, have assisted in the production of landslide inventory maps (Singhroy *et al.*, 1998; Singhroy and Mattar, 2000; Bulmer and Wilson, 1999). The multi-incidence angle, stereo and high-resolution capabilities of RADARSAT are particularly useful for landslide inventory maps. High-resolution

**Table 16.1** *Overview of current and past medium- and high-resolution satellites equipped with active (a) over passive (b) sensors*

(a)

| Satellite/sensor | Country | Bands | Resolution | Revisit time | Image swath | Weblink |
| --- | --- | --- | --- | --- | --- | --- |
| Terra Aster | USA<br>Japan | 4 × VNIR[1]<br>6 × SWIR<br>5 × TIR | 15 m<br>30 m<br>90 m | 16 days | 60 km | http://asterweb.jpl.nasa.gov/default.htm<br>http://imsweb.aster.ersdac.or.jp/ims/cgi-bin/dprSearchMapByMenu.pl |
| Landsat 5 TM[1] | USA | B, G, R, NIR, 2 × MIR, TIR | 30 m<br>60 m | 16 days | 185 km | http://landsat7.usgs.gov/wrsconvert/index.html<br>http://edcdaac.usgs.gov/dataproducts.html |
| Landsat 7 ETM+[1] | USA | pan (0.5–0.9 μm)<br>B, G, R, NIR, 2 × MIR<br>TIR | 15 m<br>30 m<br>60 m | 16 days | 185 km | http://edcdaac.usgs.gov/dataproducts.html |
| Spot 5 | France | pan (0.51–0.73 μm)<br>G, R, NIR, SWIR | 10 m<br>20 m | 26 days | 60 km | http://www.spotimage.fr/html/_.php |
| IRS1-D | India | pan (0.5–0.75 μm)<br>G, R, NIR | 5.8 m<br>23 m | 24 days | 70 km | http://www.nrsa.gov.in/engnrsa/satellites/irs1d.html |
| Ikonos | USA | pan (0.45–0.9 μm)<br>B, G, R, NIR | 1 m<br>4 m | 3 days (40°)[3] | 11 km | http://www.spaceimaging.com/products/ikonos/ |
| Quickbird | USA | pan (0.45–0.9 μm)<br>B, G, R, NIR | 0.61 m<br>2.44 m | <11 days[3] | 16.5 km | http://www.digitalglobe.com/products/quickbird.shtml |

(b)

| Satellite/sensor | Country | Band | Polarization | Resolution | Revisit time | Image swath | Weblink |
| --- | --- | --- | --- | --- | --- | --- | --- |
| ERS-1[2]/ERS-2 | Europe | C-band (5.6 cm) | VV | 30 m | 35 days | 100 km | http://earth.esa.int/ers/ |
| ENVISAT | Europe | C-band (5.6 cm) | HH/VV/HV/VH | 30 m | 35 days | 57–100 km | http://envisat.esa.int/ |
| RADARSAT | Canada | C-band (5.6 cm) | HH | Fine mode: 7.6 m × 8.3 m | 24 days | 50 km | http://www.space.gc.ca/ asc/eng/csa_sectors/earth/ earth.asp |
| | | | | Standard/wide mode: 23 m × 27 m | | 100 km/ 150 km | |
| JERS[2] | Japan | L-band (23.5 cm) | HH | 18 m | 44 days | 75 km | http://www.eorc.nasda.go.jp/ JERS-1/ |

Notes:
[1] **B** Blue, **G** Green, **R** Red, **NIR** Near infrared, **MIR** Mid infrared, **TIR** Thermal Infrared, **Pan** Panchromatic, **VNIR** Visible and near infrared, **SWIR** Short wave infrared
[2] Only archival imagery.
[3] Satellite's off-nadir capability allows for higher revisit frequencies, i.e. satellite can be rotated and pointed to image a target area.

optical systems such as IKONOS, IRS and the stereo capability of SPOT 4 are useful for landslide recognition and related land use mapping. Other planned high-resolution stereo systems such as RADARSAT-2 and ALOS will be useful to map landslide features. Where possible, the highest-resolution data that are available should be obtained and used to identify a range of geomorphic features and dimensional data on landslides of interest.

Singhroy *et al.* (1998) have provided some simple guidelines for the selection of multi-incidence SAR to facilitate the mapping of geomorphic features of large deep-seated slides in the upper Fraser Valley in British Columbia, Canada. Figures 16.2 and 16.3 (see also Colour Plate section, Plate 6) show various RADARSAT acquisition and processing techniques which facilitate geomorphic mapping of landslides in high-relief terrains.

Figure 16.2 shows a comparison of RADARSAT standard beam mode, S1 (20–27°), and an extended high-beam mode, EH6 (57–59°) of the same landslide areas in the Frazer Valley. A model showing of which SAR incidence angles are the most useful for

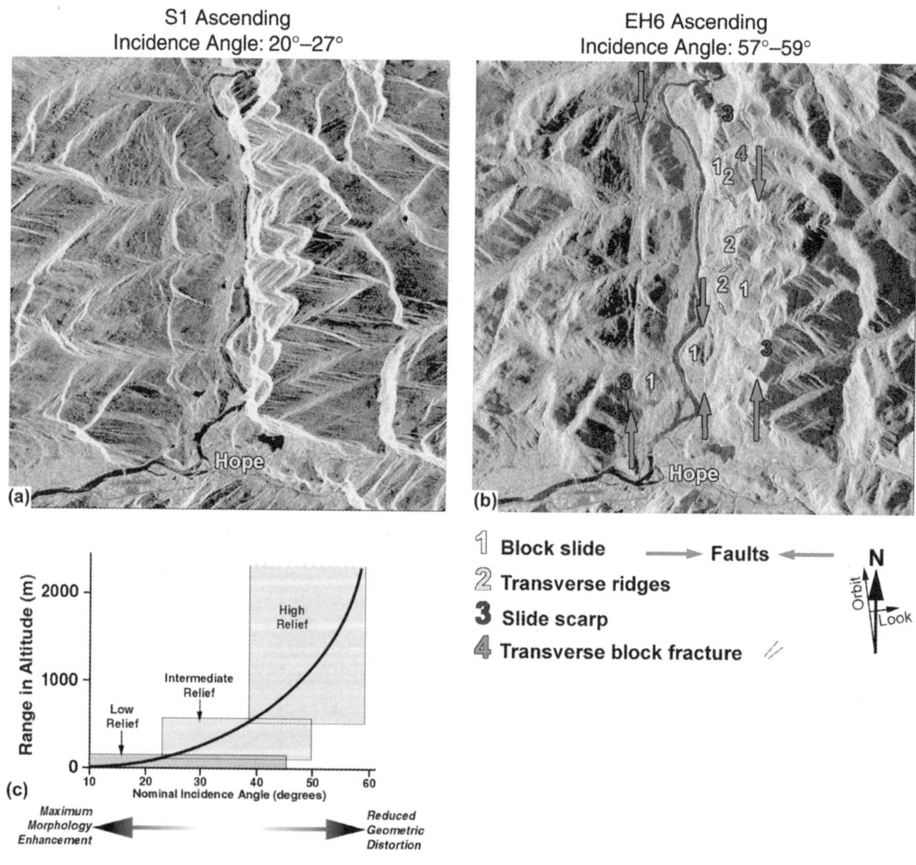


*Figure  16.2   Comparison of RADARSAT (a) standard beam mode, S1 (20–27°), and (b) an extended high beam mode, EH6 (57–59°) of the same landslide areas in the Fraser Valley (British Columbia, Canada). The graph (c) shows the most useful SAR incidence angles for geomorphic mapping*

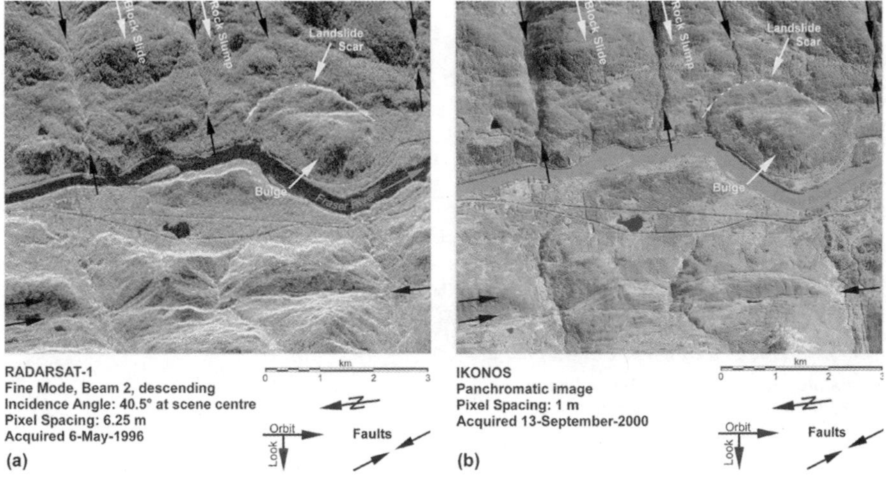


**Figure 16.3** *(a) RADARSAT fine beam mode 2 image of the Fraser Valley landslide area at 8 m spatial resolution, with incidence angles of 39–42°. More detailed geomorphic features, such as block slide, rock slumps, scars, bulge and the fault are identified. (b) The Fraser Valley in a 1 m IKONOS panchromatic image. Although the IKONOS image has a higher resolution than the RADARSAT one, the lower resolution 8 m pseudo-stereo RADARSAT SAR image with a 40° incidence angle reveals more details of the landslide geomorphology*

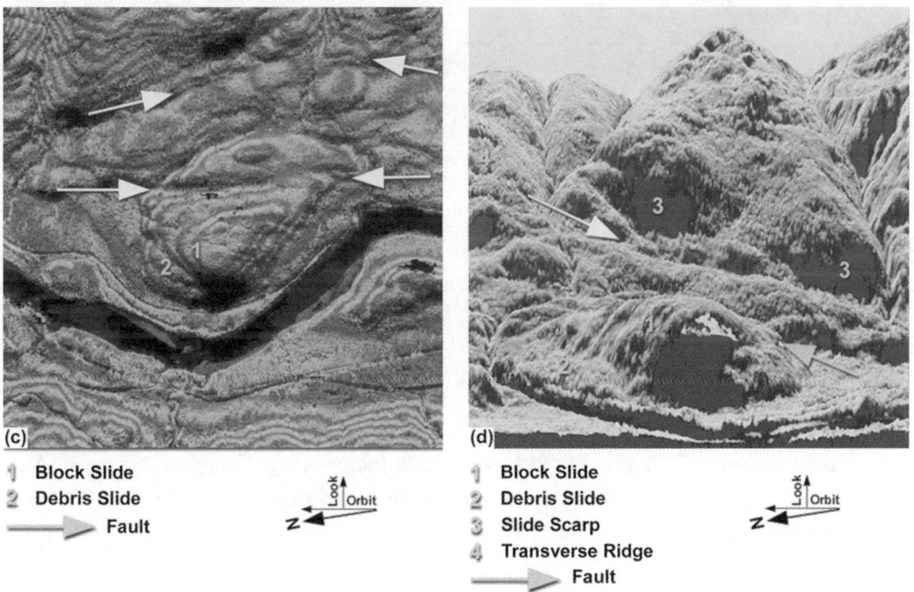


**Figure 16.3** *(c) Airborne C-HH SAR interferogram which facilitates landslide detection, geomorphic mapping and related seismically active structures. (d) Visualization image produced from interferogram (Figure 16.3(c)) provides an accurate representation of the slope geomorphology, and facilitates the identification of landslide features in the difficult high-relief terrain. (See also Plate 6)*

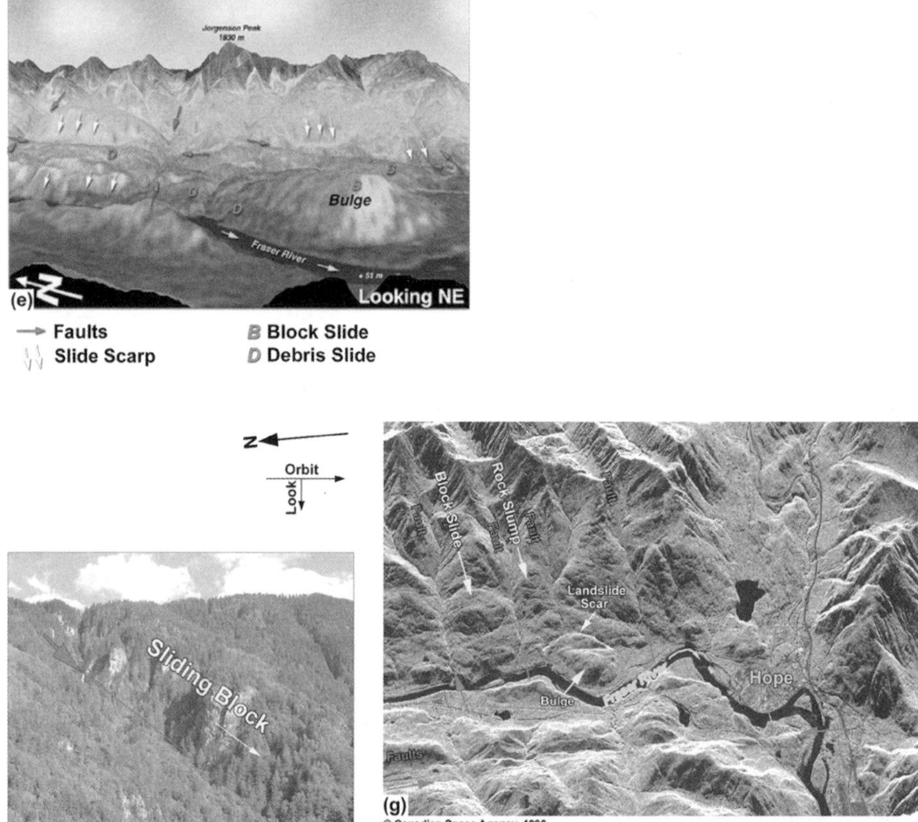


*Figure 16.3*  *(e) RADARSAT-1 image draped over a DEM of the same Frazer Valley area shown in Figure 16.7. This image shows the capability of RADARSAT visualization techniques to map regional geomorphology in high-relief landslide-prone areas. (f) and (g) Field photo of sliding block and geomorphic interpretation of a descending fine mode (8 m) RADARSAT image*

geomorphic mapping. It is clear that the landslide features and faults on the steep slopes of the Fraser Valley are more easily recognized on the EH6 images than from the S1 image. Both images have a resolution of 30 m. The eastern slopes, where most of the slides are occurring, have a local slope, which varies from 40° to 60°, and as such are suitable for RADARSAT viewing at the higher angles between 40° and 60°. From the EH6 image, block slide scarps and transverse ridges associated with rock slumps and faults were identified. On the image, block slides (1) are characterized by a number of parallel fault blocks and steps. The rock slumps are characterized by numerous closely spaced arcuate and transverse ridges (2) with considerable rock exposure and sparse vegetation. Slide scarps (3) appear as light tone areas on steep convex slopes. Faults associated with slides appear as arrows which show linear depressions on the image. The S1 image with its steep viewing geometry resulted in considerable layover, which restricts interpretation of landslide features. It is important to select the most suitable SAR incidence angle, particularly with the availability of current and future multi-incidence angle SAR satellites.

A series of visualization, fusion and enhancement techniques is shown to demonstrate the uses of both high- and medium-resolution SAR, and optical images can be used to map landslide and landslide features. In high-relief areas, it is necessary to take advantage of appropriate SAR geometry and look direction. Figure 16.3(a) shows a RADARSAT fine mode beam 2, 8 m spatial resolution, with incidence angles of 39–42° of the same Frazer Valley landslide area. More detailed geomorphic features, such as block slide, rock slumps scars bulge and the fault are identified. Figure 16.3(b) shows the same area using a 1 m IKONOS panchromatic image. It is clear that although the IKONOS image has a higher resolution (2.5 m), the lower resolution 8 m pseudo-stereo RADARSAT SAR image with an appropriate incidence angle of 40° is more suitable for landslide geomorphology. Figure 16.3(c) shows an airborne C-HH SAR interferogram which was used to facilitate the detection of landslide geomorphology and related seismically active structures. Figure 16.3(d) shows the three-dimensional visualization image produced from interferogram (Figure 16.3(c)). This provides an accurate representation of the slope geomorphology, and facilitates the identification of landslide features in the difficult high-relief terrain. Sometimes, when SAR images are draped on a DEM; the image product can be very useful for mapping regional landslide geomorphology (Figure 16.3(e)) or detailed landslide features as shown in Figure 16.5 (see also Colour Plate section, Plate 7). Figures 16.3(f) and 16.3(g) show a field photograph of a sliding block and geomorphic interpretation of a descending fine mode (8 m) RADARSAT image. It is important to note that in this case the descending orbit provides the line of sight and grazing angle which can facilitate interpretation. High-resolution (3 m) airborne SAR (Figure 16.4; see also Colour Plate section, Plate 8) is particularly useful to interpret landslide features and related geological units and structural features.

Image fusion techniques, combining optical and SAR images, are particularly useful for regional inventory mapping. Large landslides are easily recognized on the TM images, and SAR provides the geomorphic context (Figure 16.6). In areas of low relief with characteristic retrogressive slides, optical and SAR fusion techniques are especially useful for inventory mapping (Figure 16.7; see also Colour Plate section, Plate 9).

Disaster response comprises rapid damage assessment and relief operations once the disaster has occurred. Currently, damage assessment related to landslides and other disasters is done using aerial photography, videos and ground checks. In order to be able to use EO data for landslide damage assessment, two criteria should be met: high temporal and high spatial resolution (*c.* 3–10 m stereo) is essential for landslide damage assessment and relief efforts. Images taken at the time of disaster or days after the event similar to other geohazards – earthquakes and volcanoes – are a requirement to support relief efforts. Figure 16.8(a–f) shows the uses of high-resolution IKONOS (2.5 m) and fine mode RADARSAT (8 m) images to assess damage and produce revised inventory maps related to high-velocity mudflows resulting from high-intensity rainfall in coastal Venezuela. The landslide, which occurred on 30 December 1999, killed 50 000 people, and caused and US$1.5 billion in property damage.

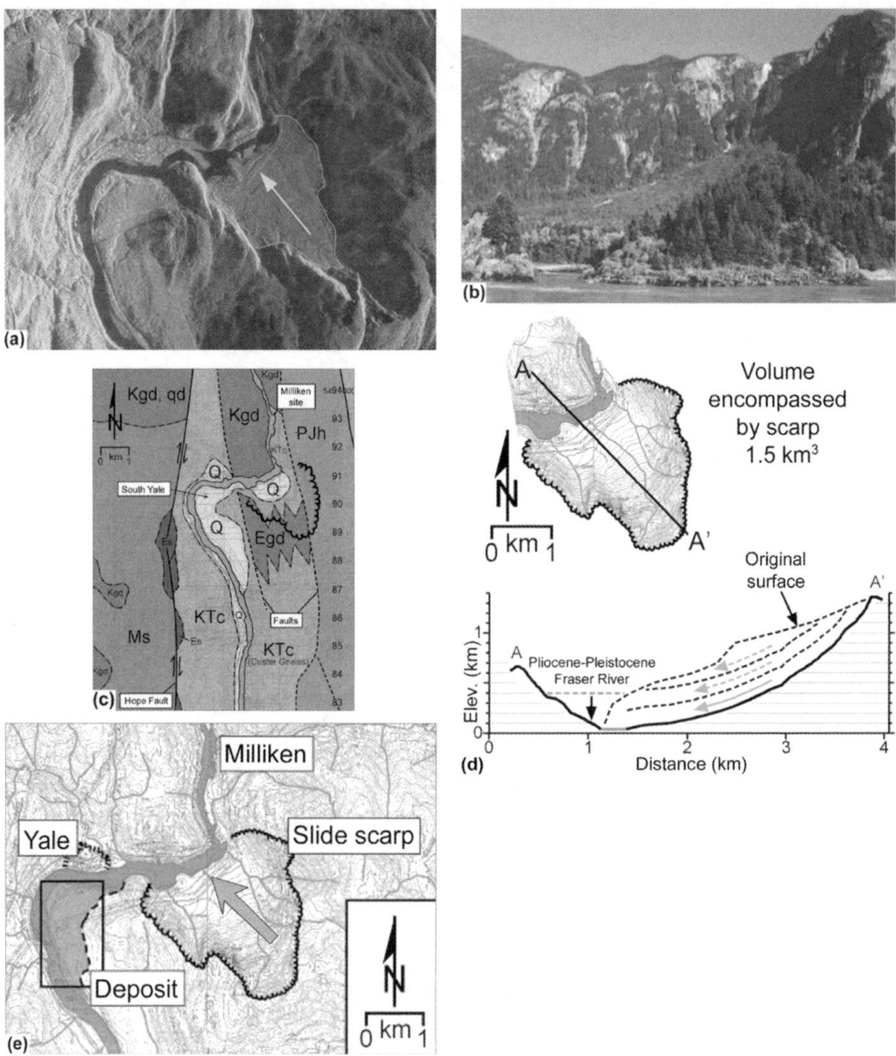



***Figure 16.4*** *(a) South Yale Slide, Hope, B.C., Canada, depicted in a high-resolution airborne radar image and (b) field photograph, (c) geological setting, (d) slide transect, and (e) inventory map. (See also Plate 8)*

## 16.3  Landslide Monitoring using InSAR

Interferometric synthetic aperture radar (InSAR) can be applied for measuring displacements at the Earth's surface with very high accuracy, and for topographic mapping. Both capabilities are of high relevance for landslide hazard assessment. Possibilities and constraints of spaceborne SAR for these applications are briefly reviewed.

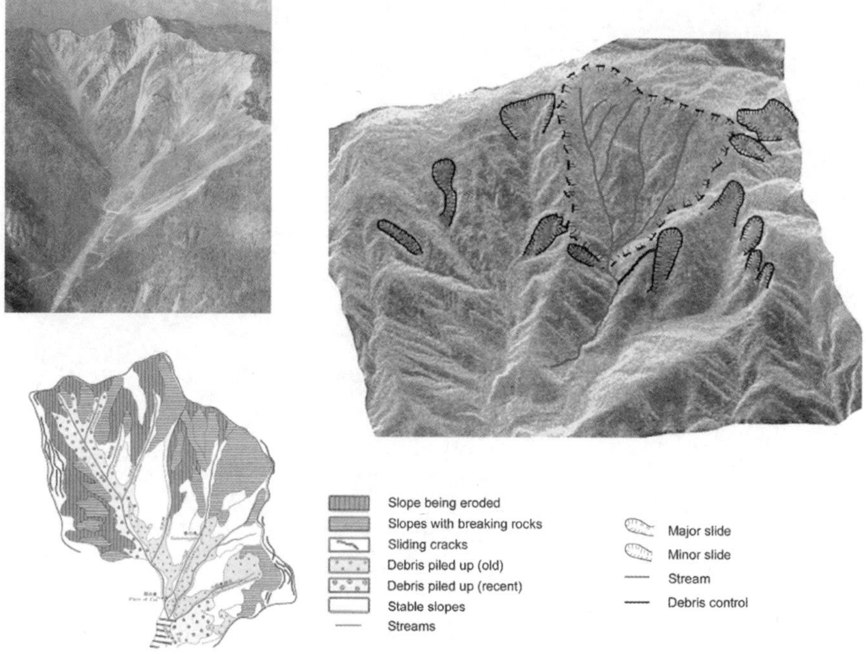



**Figure 16.5** *Interpretation of RADARSAT image draped over DEM for landslide inventory in upper Oya Valley in Japan. (See also Plate 7)*

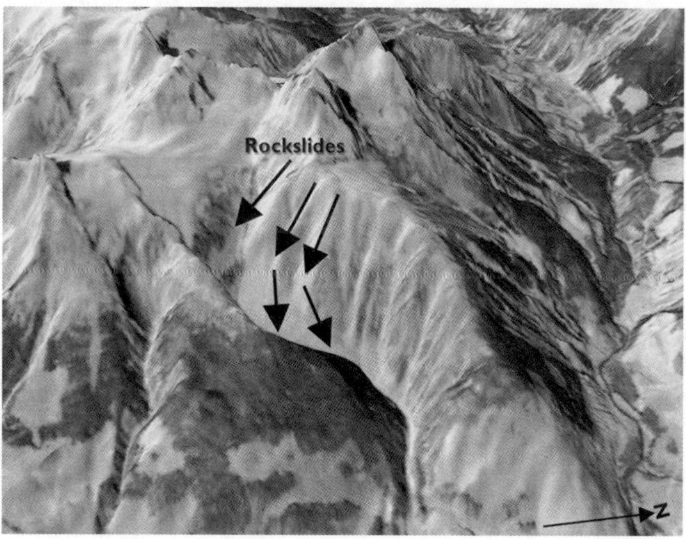



**Figure 16.6** *(a) Mt Meager, B.C. Three-dimensional visualization from satellite imagery. Landsat 7 ETM+, acquired 23 September 2000, bands 5 (R), 4 (G), and 3 (B), sharpened through IHS integration with 15 m pan band; DEM: 25 m pixel spacing*

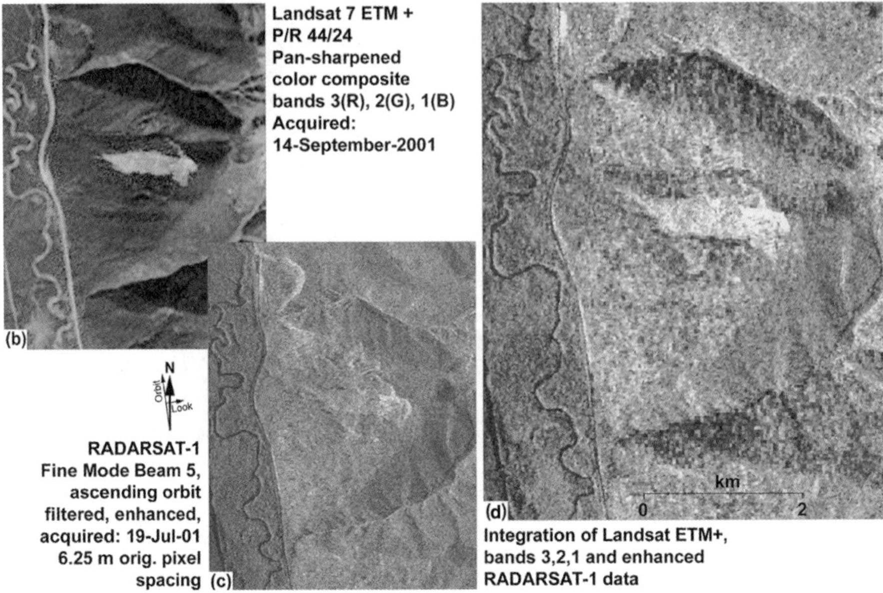



**Figure 16.6**   *(b–d) Large debris flow at Beaver River, B.C., identified from Landsat TM (a), RADARSAT (b) and fused TM and SAR (d). The fused image reveals more information on the dimension and geomorphology of the landslide*

In an SAR image the location of a target is represented in a two-dimensional coordinate system, with one axis in flight direction (along-track) and the other axis across-track (slant range), in which the target position (distance) is measured by the round-trip travel time from the SAR antenna to the target and back. Because the across-track position represents a range measurement, the SAR image is distorted in this direction. Steep slopes facing in the direction of the antenna appear shortened or are affected by layover which often inhibits the interferometric analysis on these slopes. To minimize the distortion produced by steep slopes in high-relief areas it is useful to choose the most appropriate SAR incidence angles. Figure 16.9(a–c) provides some useful guidelines regarding data acquisition geometry with respect to the orientation of the slope to be monitored (Ferretti *et al.*, 2000; Fruneau *et al.*, 1996; Vachon *et al.*, 1995).

An interferometric image represents the phase difference between the reflected signal in two SAR images obtained from similar positions in space (Hanssen, 2001; Massonnet and Feigl, 1998; Rosen *et al.*, 2000). In the case of spaceborne SAR, the images are acquired from repeat-pass orbits. For the European ERS, for example, the standard orbital repeat interval is 35 days; for the Canadian RADARSAT it is 24 days. The phase differences between two repeat-pass images result from topography and from changes in the line-of-sight distance (range) to the radar due to displacement of the surface or change in the atmospheric propagation path length. For a non-moving target the phase differences can be converted into a digital elevation map if very precise satellite orbit data are available (Ferretti *et al.*, 1999a). Effects of noise due to changes of atmospheric propagation between various images can be strongly reduced by combined

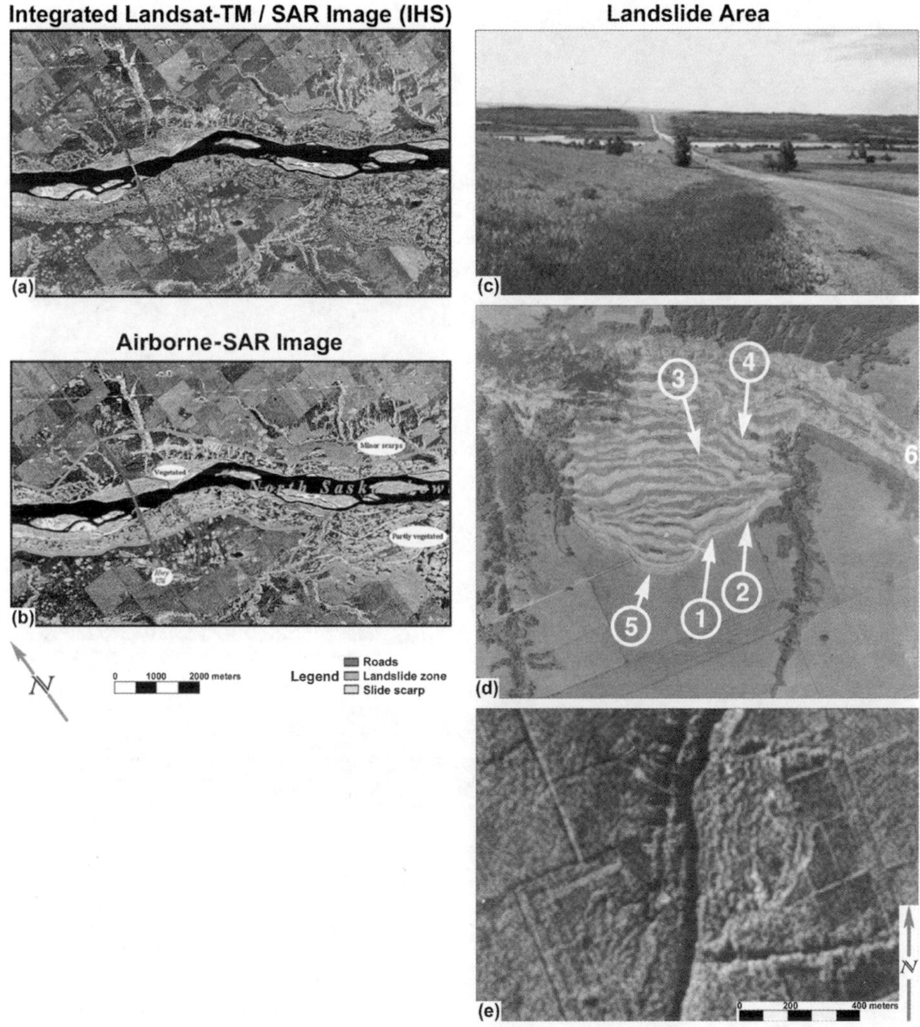



**Figure 16.7** *(a) and (e) Delineation of multiple retrogressive sliding in Cretaceous shales along the Saskatchewan River and in marine clays along the South Nation River using fused SAR and TM, (b) airborne SAR, and (c) field photograph. The airphoto (d) shows backscarp (1 and 2) slump ridges (3 and 4) and headscarp (5) (Mollard and Janes, 1993. Aerial photographs ©1971. Her Majesty the Queen in Right of Canada, reproduced from the collection of the National Air Photo Library with permission of Natural Resources Canada). (See also Plate 9)*

processing of several interferometric image pairs with different baselines (multi-baseline interferometry) (Ferretti *et al.*, 2001).

For motion mapping by means of InSAR it is necessary to separate the motion-related and the topographic phase contributions (Ferretti *et al.*, 1999c). This can be done by differential processing using two interferograms of different time periods calculated from two or three images if the motion was constant in time. If the motion is slow, the topographic phase can be taken directly from an interferogram of a short time span

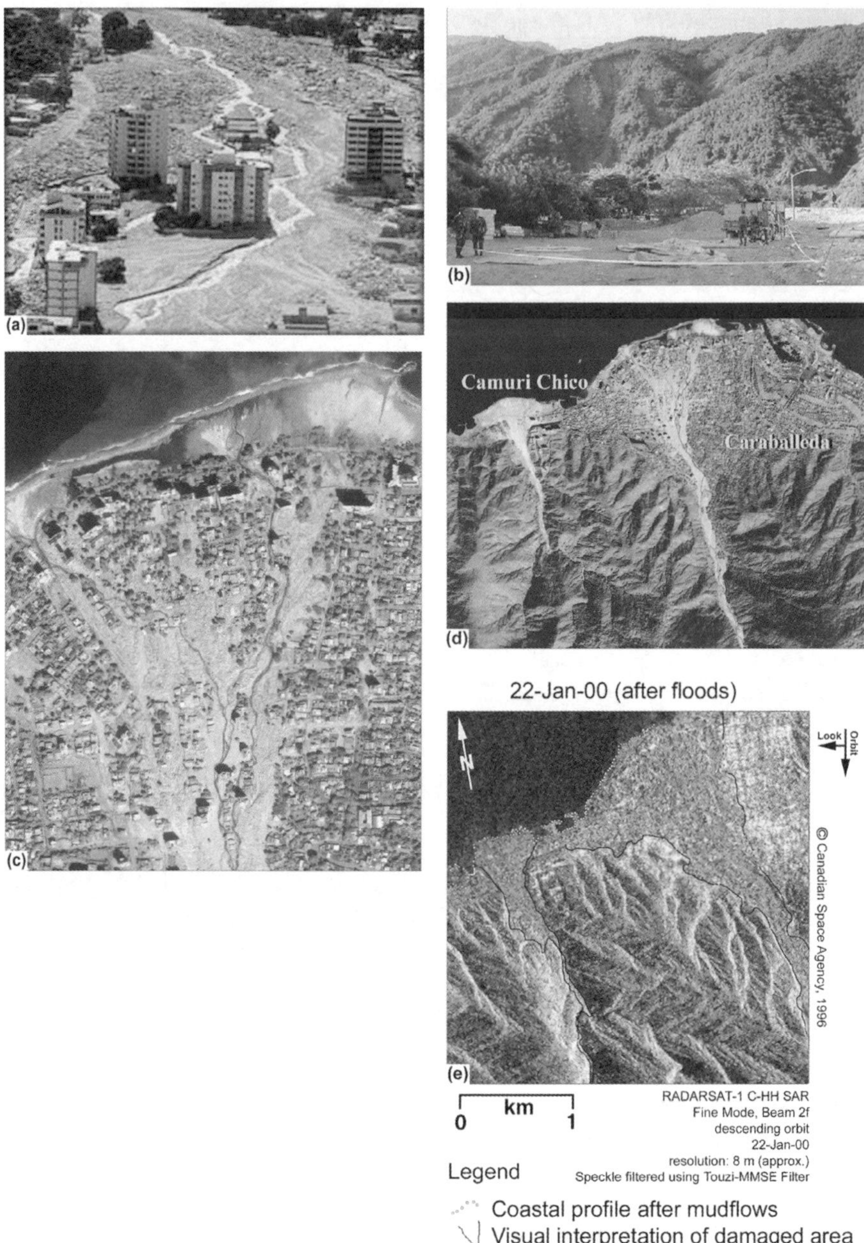



***Figure 16.8(a–e)***   *(a) Landslides in Venzuela resulting from high-intensity rainfall and high-velocity water and debris flow which buried residential areas resulting in 50 000 deaths and $1.5 billion in damage on the north slope of Avila Mountain range. The photograph (b) shows the resulting landslide scars on the steep slopes. (c) and (d) show high-resolution Ikonos 2.5 m imagery. (e) shows an 8 m RADARSAT fine mode image used for damage assessment*

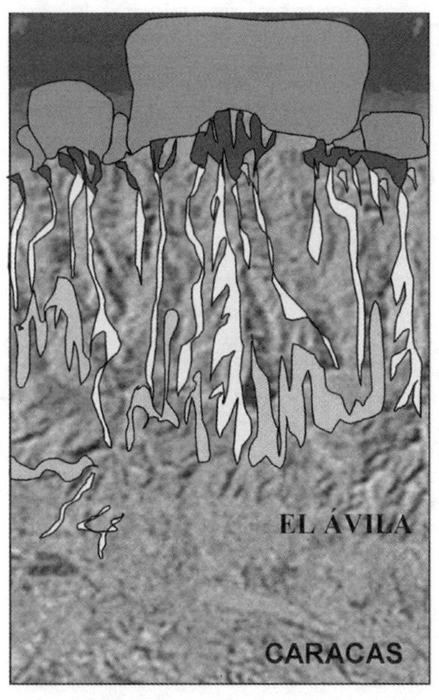
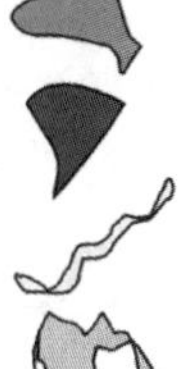


**Figure 16.8f**   *Interpretation of various types of debris flow and deposits and areas of landslide scars*

(e.g. the one-day time span of the Tandem Phase, when ERS-1 and ERS-2 operated simultaneously).

With the advent of repeat-pass interferometry, it has become possible to detect subtle changes (at mm scales) in the landscape such as seismic displacement (e.g. Massonnet *et al.*, 1996). However, landslides are difficult to study using radar interferometry (e.g. Fruneau *et al.*, 1996) because they can experience ground deformations in excess of the phase gradient limit (Carnec *et al.*, 1996) which eliminate interferometric correlation (Massonnet and Feigl, 1998). Attempts are being made to better integrate radar interferograms, field measurements, and ancillary remote sensing of landslides to obtain 'calibrated' interferograms which will provide useful geological and geophysical information to the landslide monitoring community (e.g. Bulmer *et al.*, 2001). However, even such improved technologies are rarely utilized to their full potential in hazard assessment.

There are two important constraints on the application of InSAR to slope motion monitoring: (1) InSAR measures only displacements in slant range; the component of the velocity vector in flight direction cannot be measured; (2) InSAR can only map the motion at characteristic temporal and spatial scales (Massonet and Feigl, 1998), related to the spatial resolution of the sensor and the repeat interval of imaging. Typical scales for ERS interferometry application to landslide movements are mm to cm per month (with 35 day repeat-pass images) down to mm to cm per year (with approximately annual time spans). Faster landslides could only be studied during special orbital repeat configurations of ERS in previous years (Fruneau *et al.*, 1996), such as the Tandem Phase or the three-day repeat cycle during the Commissioning Phase and the Ice Phase of ERS-1 during a few months of 1992, 1993 and 1994. With the resolution of ERS

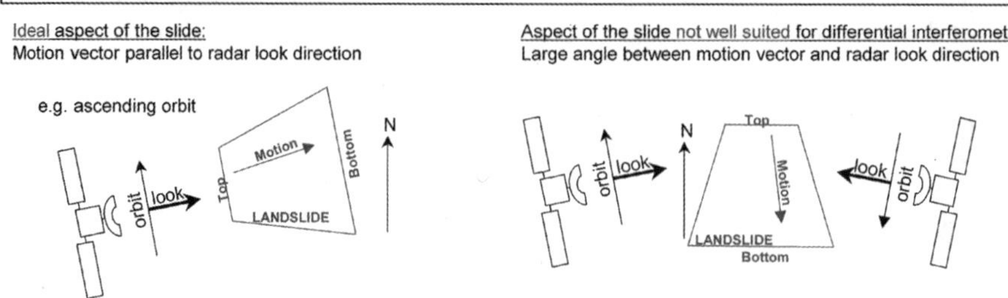



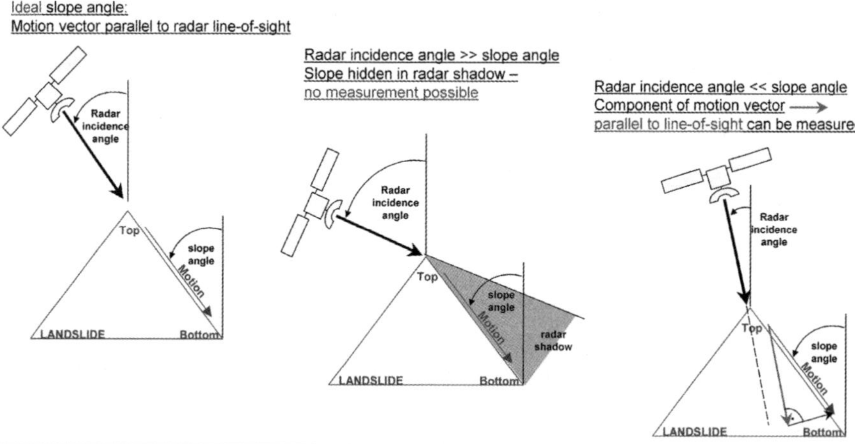





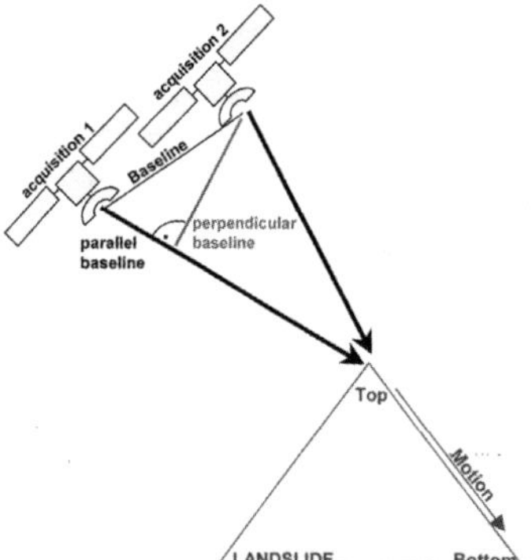



**Figure 16.9** *Some guidelines regarding data acquisition geometry with respect to the orientation of the slope to be monitored. Of a three-dimensional displacement vector only the component parallel to the line of sight of the satellite can be measured (Ferretti et al., 2000; Fruneau et al., 1996; Vachon et al., 1995)*

(23 m in ground range, 4.9 m along track, 5.6 cm wavelength) the minimum horizontal dimension of a landslide for area-extended interferometric analysis, which can be applied with a single image pair, is about 200 m across- and along-track. Future SARs with higher resolution (RADARSAT-2) will enable the mapping of smaller slides. With the Permanent Scatterer Technique the movement of small objects (down to about 1 m$^2$) can be monitored (Ferretti *et al.*, 1999b).

A precondition for the generation of an interferogram is coherence, which means that the phase of the reflected wave at the surface remains the same in the two SAR images. The loss of coherence (decorrelation) is the main problem for interferometric analysis over long time spans, as required for mapping of very slow movements. Whereas the signal of densely vegetated areas decorrelates rapidly, the phase of the radar beam reflected from surfaces that are sparsely vegetated or unvegetated often remain stable over years. This has been utilized for mapping very slow slope movements in high alpine terrain (Rott *et al.*, 1999, 2000).

Motion analysis in vegetated areas is only possible if a few stable objects (usually man-made constructions such as houses, roads etc.) are located within these areas. Using long temporal series of interferometric SAR images (typically about 30 or more repeat-pass images over several years), objects with stable backscattering phase are determined by statistical analysis. Only some of the man-made objects reveal long-term phase stability. The analysis of the SAR time series with the Permanent Scatterer Technique (Ferretti *et al.*, 2000, 2001) enables the detection of very small movements of individual objects (e.g. single houses (Figure 16.10)). A certain number density of stable objects (at least about 5 per km$^2$) is needed to enable accurate correction of atmospheric phase contributions. This method has been applied to map subsidence in urban und rural areas in various countries.

The future availability of spaceborne InSAR data for slope motion monitoring is not yet clear. The European ERS SAR is a useful system for repeat-pass SAR interferometry

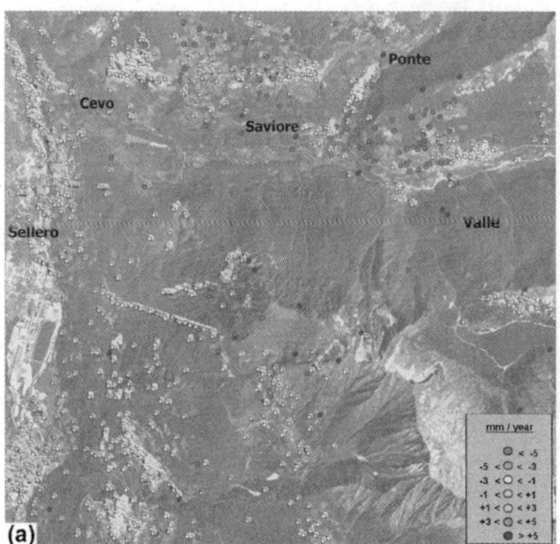

**Figure 16.10**  *(a) Permanent scatterer analysis. Surface displacement values detected at locations of permanent scatterers (single houses). (b) shows a field photograph (Ferretti* et al., *2001). Reproduced by permission of IEEE*

because of the high stability of the sensor, good orbit maintenance and the fixed operation mode. However, a system failure that occurred on ERS-2 on 17 January 2001 has resulted in the orbit deadband being relaxed from $+/-1$ km to $+/-5$ km. As a result interferometry can only be performed on a few random occasions. The European follow-on sensor ASAR on board the ENVISAT, as well as other planned SARs, provide many different operation modes, which will reduce the availability of repeat-pass interferometric data. On the other hand, the higher spatial resolution of some of these sensors would also be of interest for mapping small slides. The important contributions of InSAR to hazard management and to a range of other environmental monitoring tasks would justify a long-term SAR mission optimized for InSAR applications.

Due to the typical SAR repeat orbits of the order of 25 to 35 days, InSAR is mainly suitable for monitoring very slow movements of slopes and individual objects, and for mapping of subsidence. Thus it is able to fulfil specific information needs for landslide monitoring, complementary to other information sources. The main advantage over conventional techniques is the possibility of very precise displacement measurements over large areas at reasonable costs, thus providing an excellent tool for reconnaissance.


## 16.4   Case Study: Frank Slide

As previously noted, remote sensing techniques are increasingly being used in slope stability assessment (Murphy and Inkpen, 1996; Singhroy *et al.*, 1998; Singhroy and Mattar, 2000). Recent research has shown that differential interferometric SAR techniques can be used to monitor landslide motion under specific conditions (Vietmeier *et al.*, 1999; Rott *et al.*, 1999). Provided coherence is maintained over longer periods, it is possible, for example in non-vegetated areas, to observe surface displacement of a few cm per year. Using data pairs with short perpendicular baselines, short time intervals between acquisitions, and correcting the effect of topography on the differential interferogram, reliable measurements of surface displacement can be achieved.

A case study is provided to demonstrate the capability of InSAR to monitor gradual motion on a large rock avalanche in the Canadian Rockies. Our study focused on the Frank slide, a $30 \times 10^6 \, \text{m}^3$ rockslide avalanche of Paleozoic limestone, which occurred in April 1903 from the east face of Turtle mountain in the Crowsnest Pass region of southern Alberta, Canada (Figure 16.12(d), (e) and (f); see also Colour Plate section for (e) and (f), Plate 10). Seventy fatalities were recorded.

ERS data were used for the InSAR analysis. In order to select a set of suitable scenes a thorough baseline analysis of all ERS-1 and ERS-2 ascending scenes acquired over the location (track 406, frame 989) during summer between 1992 and 2001 was performed. It was of interest to find as many data pairs as possible during that time period, yet keep the perpendicular baselines below 100 m, thus reducing contributions of topography on differential phase values. Ascending orbit was chosen so that the look direction (right) would correspond to the aspect of the slope. Seven scenes were finally selected for this initial reconnaissance study, which yielded five data pairs with perpendicular baselines below 100 m. The scenes span the time period between 1993 and 1997. The  interferometric DEM used was generated from an ERS tandem pair of 25/26 September 1995 (B$\perp$ 191 m). Geocoding and elevation values were refined

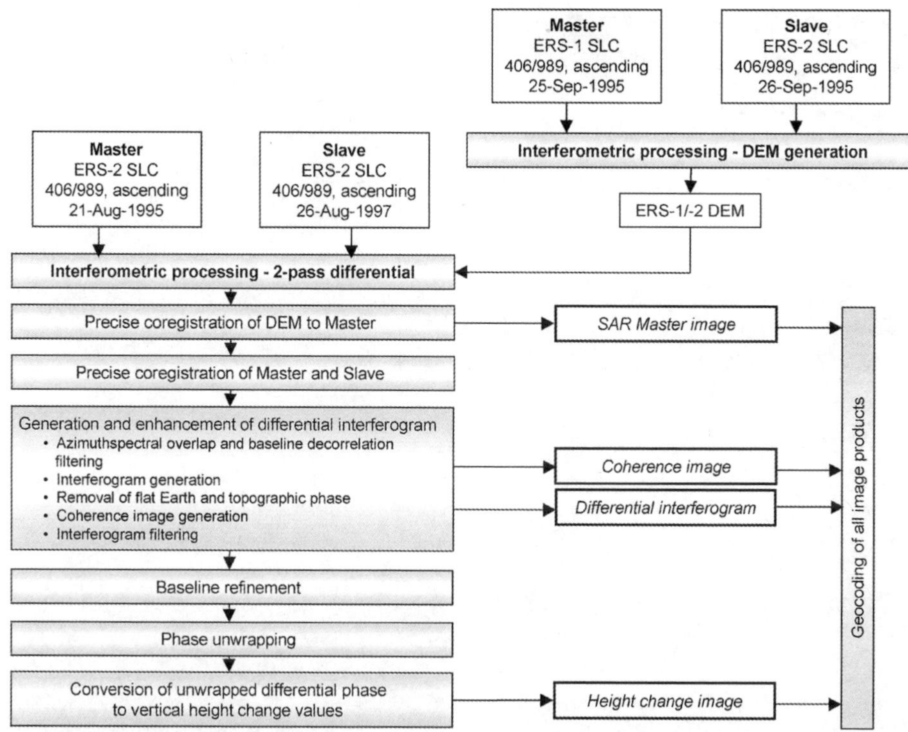


**Figure 16.11**   *Sample processing flowchart, differential interferometry*

using ground control points taken from 1:10 000 scale topographic maps. All data pairs with baselines below 100 m were processed to geocoded vertical elevation change maps using the software package EarthView-InSAR. Figure 16.11 provides an overview of the processing steps involved, as they are implemented in the software. Due to the small perpendicular baseline of only 4.5 m for this InSAR pair (Aug. 95/Aug. 97), the topographic contribution to the differential phase was minimal and was removed during the processing.

The InSAR investigation revealed the presence of a near-circular fringe in the refined differential interferogram (Figure 16.12(a); see also Plate 10), the maximum displacement values of −1.3 cm indicate gradual motion of the rock face before the 2001 rockfall. Minor deformation along parts of the geological structure (Figure 16.12(b), see also Plate 10) suggests that the slide is still active.

These findings assist in targeting current *in situ* motion detectors at the Frank slide. The textural analysis of the high-resolution RADARSAT image provides an accurate characterization of the debris size and distribution, suggesting that SAR image texture is a useful parameter to map the distribution of slide debris. The InSAR techniques will assist in the understanding of landslide processes, post-failure mechanism and mobility.

At the Frank slide we also use the SAR backscatter of the rough debris surface to assist in characterizing debris texture. Contrary to motion on the detachment zone, roughness and distribution of landslide debris and their post-slide stability have not been

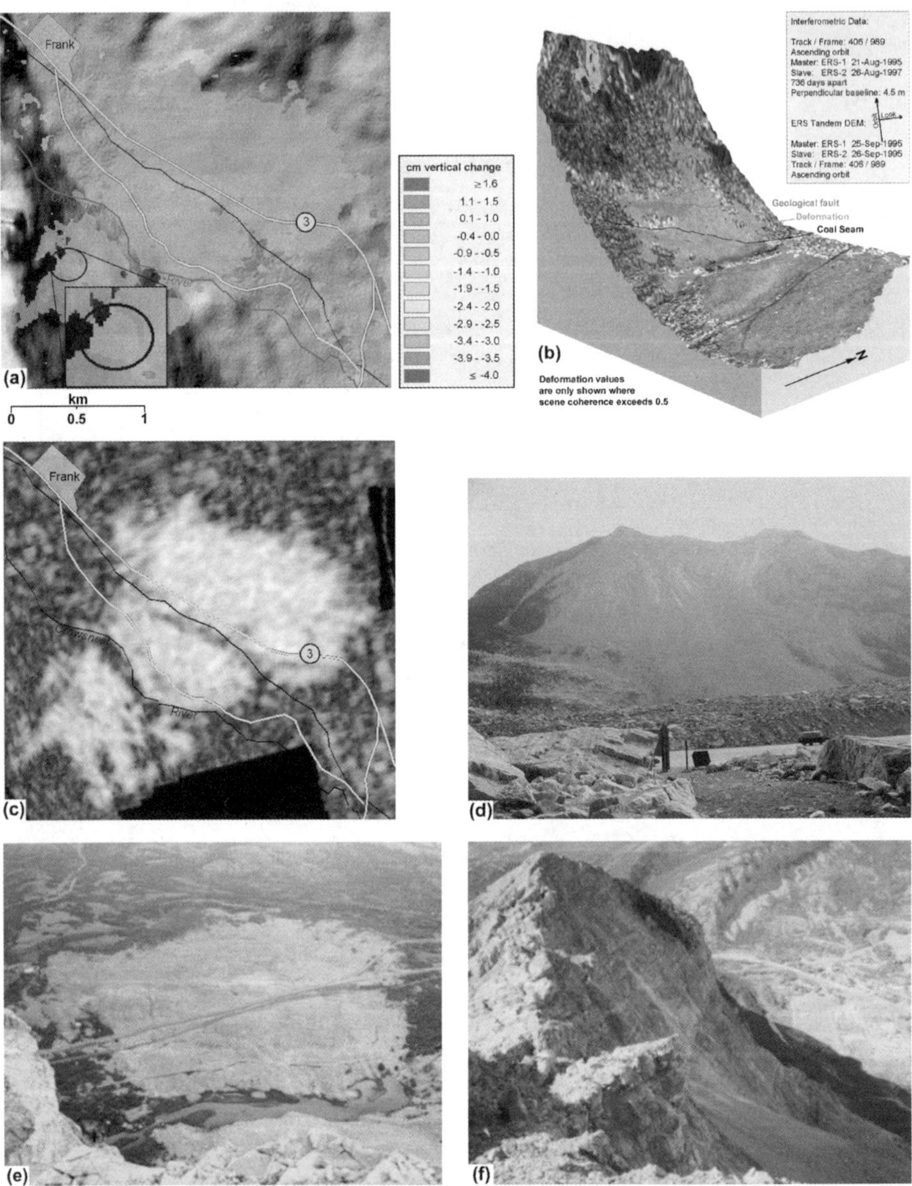



**Figure 16.12**  (a) Vertical surface deformation map, generated from differential SAR interferometry. (b) Result is draped over a high-resolution elevation model. (c) Corresponding coherence image, and (d–f) field photographs. The deformation is located towards the bottom of the slope, just above a coal seam. (See also Plate 10)

studied in detail using remote sensing. This is due in part to the lack of topographic data for blocky landslides and therefore the link between debris roughness and radar backscatter ($\sigma_0$) has remained elusive. Roughness is defined as the topographic expression of surfaces at horizontal scales of centimetres to a few hundred metres. Landslide surface structures and roughness provide information on flow emplacement parameters (such as emplacement rate, velocity and rheology). Several investigations have focused on characterizing grain size and distribution of this rock avalanche, in order to understand post-failure mechanism and mobility (Couture *et al.*, 1998; Cruden and Hungr, 1986). For the slide debris distribution study, a RADARSAT fine mode, beam 4, (incidence angle 43°–46°) ascending, acquired on 1 September 2001, was used. It was verified that there had been no precipitation on the acquisition date as well as several days before in order to eliminate ground-moisture-induced effects on the radar backscatter. The data were processed to a 16-bit path-oriented, ground-range single-look product with 6.25 m pixel spacing (SGF). The image data were not filtered or rectified in order to avoid any disturbance of pixel neighbourhood relationships introduced through the resampling procedures involved.

Several methods are available for evaluating SAR texture parameters and for subsequent classification; well established are texture measures retrieved from co-occurrence matrices and the analysis of local histograms (Keil *et al.*, 1997; Lohmann, 1994). In this study the latter method was used.

From field and aerial photographs of the accumulation zone, three areas of predominantly coarse, medium-textured, and fine debris were selected for local histogram analysis. After linearly scaling the RADARSAT data to 8 bit, the pixel values for three windows of approximately 1400 pixels each, falling within the selected areas, were extracted from the SAR image (Figure 16.13a). Histograms were generated for each sample, depicting grey value frequencies. Various statistical parameters can be used to describe the local histograms' distribution (Figure 16.13a–c).

The results from our debris distribution investigation at the Frank slide have shown that a SAR textural map of a large rock avalanche can be a useful first step in the understanding of post-failure mechanism and mobility, since there is a close relationship between the SAR textural measurements from local histograms and the debris size distribution and ridge morphology.

## 16.5    Research, Challenges and Associated with EO Data

The difficulties associated with interpretation of EO data can require a high level of user knowledge in remote sensing systems. Characterizing form, size, causative and triggering factors, pre-monitory signs, mechanisms and post-failure evolution will require both ground-truth knowledge and advanced technical skills in remote sensing processing. Although any InSAR-sensed deformation is very useful for unstable slopes, a change detection in both vertical and horizontal distances is needed to evaluate landslide mechanisms. Furthermore, some other phenomena such as settlement or subsidence of engineering structures, or shrink and swell of some geological materials, needs to be taken into account to correctly interpret the significance of the ground deformation from EO data (Wasowski and Gostelow, 1999).

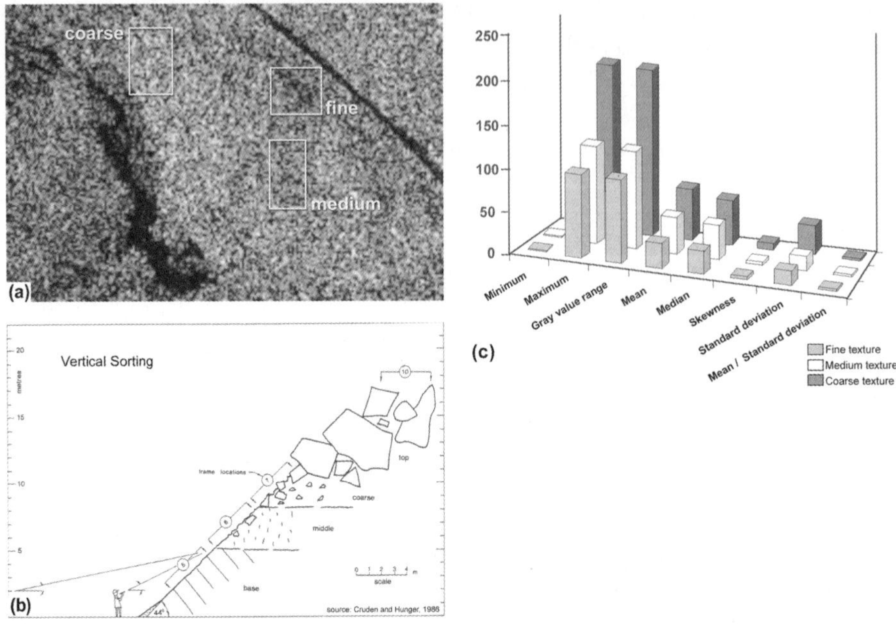



***Figure 16.13*** *(a) Colour-coded RADARSAT-1 fine mode image of the accumulation zone, Frank Slide, Alberta, Canada. (b) Textural differences in the radar image correspond to differently textured surface material, in this case to varying grain sizes within the landslide debris. (c) Local histograms were generated, the distribution of which can be described using various statistical parameters*

It follows that, in general, the information obtained from InSAR (or other EO) methods will need to be correlated with ground data and detailed survey controls in order to be correctly evaluated and to provide reliable relevant information to a disaster management community or to engineering geologists and geotechnical engineers.

There are two aspects of EO data that are important for landslide mitigation. First, it has been shown that multitemporal EO data can be used to determine the changes in landslide distribution, and as such are useful to produce landslide inventory maps. Second, EO data can be used to map factors that are related to the occurrence of landslides, such as lithology, faults, slope, vegetation and land use, and the temporal changes in these factors, which can be used within a GIS in combination with a landslide inventory map for landslide hazard assessment. To facilitate the use of EO data for landslide inventory maps more research needs to be done.

High-resolution (<8 m) remote sensing data needs to be carefully integrated with existing information. This task is particularly challenging in high-relief slopes where most landslides occur.

Current landslide interpretation, data fusion and InSAR techniques need to be tested in different topographic and geological environments, and correlated with *in situ* information.

Standardized landslide inventory mapping procedures using high-resolution RS data as an image base needs to be developed. This is possible at a scale of 1:50 000 using current techniques.


## Acknowledgements

I would like to thank Katrin Molch for her valuable contributions in compiling all the figures and assisting in editing and contributing to the section on InSAR applications. As chairman of the CEOS Landslide Hazard Working Group (1999–2002) I had access to the ideas resulting from the reports of team members, especially Drs Wasowski and Bulmer, which were used in this contribution.


## References

Alexander, D., 1995, *Natural Disasters* (New York: Chapman and Hall).

Bulmer, M.H., 2002, Studies of Landslides using Remote Sensing Data, CEOS Landslide Hazard Team Report.

Bulmer, M.H. and Campbell, B.A., 1999, Topographic data for a silicic lava flow – A planetary analog, *Lunar and Planetary Science Conference*, **XXX**, 1446–1447.

Bulmer, M.H. and Wilson, J.B., 1999, Comparison of Stellate Volcanoes on Earth's Seafloor with Stellate Domes on Venus using Side Scan Sonar and Magellan Synthetic Aperture Radar, *Earth and Planetary Science Letters*, **171**, 277–287.

Bulmer, M.H., Campbell, B.A. and Byrnes, J., 2001, Field studies and radar remote sensing of silicic lava flows, *Lunar and Planetary Science Conference*, **XXXI**, 1850.

Carnec, C., Massonet, D. and King, C., 1996, Two examples of the use of SAR interferometry on displacement fields of small extent, *Geophysics Research Letters*, **23**(24), 3579–3582.

Couture, R., Locat, J., Drapeau, D., Evans, S. and Hadjigeorgiou, J., 1998, Evaluation de la granulometrie à la surface des debris d'avalanche rocheuse par l'analyse d'images, *Proceedings 8th International IAEG Congress*, Vancouver, 1383–1390.

Cruden, D.M. and Hungr, O., 1986, The debris of Frank Slide and theories of rockslide-avalanche mobility, *Can. J. Earth Science*, **23**, 425–432.

Ferretti, A., Prati, C. and Rocca, F., 2000, Nonlinear subsidence rate estimation using permanent scatterers in differential SAR interferometry. *IEEE Transactions of Geoscience of Remote Sensing*, **38**, 2202–2212.

Ferretti, A., Prati, C. and Rocca, F., 2001, Permanent scatterers in SAR interferometry, *IEEE Transactions of Geoscience of Remote Sensing*, **39**, 8–20.

Ferretti, A., Prati, C. and Rocca, F., 1999a, Multibaseline InSAR DEM reconstruction: the wavelet approach, *IEEE Transactions of Geoscience of Remote Sensing*, **37**, 705–715.

Ferretti, A., Prati, C. and Rocca, F., 1999b, Permanent scatteres in SAR interferometry, *Proceedings IGARSS '99 Conference*, Hamburg.

Ferretti, A., Prati, C. and Rocca, F., 1999c, Monitoring terrain deformations using multi-temporal SAR images, *Proceedings of FRINGE '99*, Liège, Belgium.

Fruneau, B., Achache, J. and Delacourt, C., 1996, Observations and modelling of the Saint-Etienne-de-Tinée landslide using SAR interferometry, *Tectonophys.*, **265**, 181–190.

Fruneau, B., Delacourt, C. and Achache, J., 1996, Observation and Modelling of the Saint-Etienne-de-Tinée Landslide Using SAR Interferometry, *Proceedings of FRINGE '96*, Zurich, Switzerland.

Guzzetti, F., 1990, Carta Inventario Dei Ovimenti Franosi Della Region Marche Ed Aree Limitarofe: Scala 1:100 000. CNR-IRPI, Purugia, Italy.

Hanssen, R.F., 2001, *Radar Interferometry* (Dordrecht: Kluwer Academic Publishers).

Keefer, D., 1984, Landslides caused by earthquakes, *Geological Society of America Bulletin*, **95**, 406–421.

Keefer, D., 1994, The importance of earthquake-induced landslides to long-term slope erosion and slope failure hazards in seismically active regions, *Geomorphology*, **10**, 265–284.

Keil, M., Scales, D., Schwäbish, M., Eigemeier, E., Siegmund, R., Kanellopoulos, I., Winter, R., 1997, Investigation of SIR-C/X-SAR data for vegetation mapping in the Harz mountains and in the region of Oberpfaffenhofen, Germany, *SIR-C/X-SAR Final Report*, JPL, Pasadena.

Kimura, H. and Yamaguchi, Y., 2000, Detection of landslide areas using Radar Interferometry, *Photogrammetric Engineering and Remote Sensing*, **66**(3), 337–344.

Lohmann, G., 1994, Co-occurence based analysis and synthesis of textures, *Intern. Conf. on Pattern Recognition*, Oct. 1994, Jerusalem, Israel.

Massonnet, D. and Feigl, K.L., 1998, Radar interferometry and its application to changes in the Earth's surface, *Review of Geophysics*, **36**(4), 441–500.

Massonnet, D., Vadon, H. and Rossi, M., 1996, Reduction in the need for phase unwrapping in the radar interferometry, *IEEE Transactions of Geoscience of Remote Sensing*, **34**(2), 489–497.

McKean, J., Buechel, S. and Gaydos, L., 1991, Remote sensing and landslide hazard assessment, *Photogrammetric Engineering and Remote Sensing*, **57**, 1185–1193.

Mollard, J.D. and Janes, J.R., 1993, Airphoto Interpretation of the Canadian Landscape. Energy, Mines and Resources, Canada, Ottawa.

Murphy, W. and Inkpen, R.J., 1996, Identifying landslide activity using airborne remote sensing data, *GSA Abstracts with Programs*, **A-408**, 28–31, Denver.

Rosen, P.A., Hensley, S., Joughin, I.R., Li, F.K., Madsen, S.N., Rodriguez, E. and Goldstein, R.M., 2000, Synthetic Aperture Radar Interferometry, *Proceedings of the IEEE*, **88**(3), 333–385.

Rott, H. and Siegel, A., 1999, Analysis of mass movements in Alpine Terrain by means of SAR interferometry, *Proceedings of IGARSS '99*, Hamburg.

Rott, H., Scheuchl, B., Siegel, A. and Grasemann, B., 1999, Monitoring very slow slope movements by means of SAR interferometry: a case study from a mass waste above a reservoir in the Ötztal Alps, Austria, *Geophysical Research Letters*, **26**, 1629–1632.

Rott, H., Mayer, C. and Siegel, A., 2000, On the operational potential of SAR interferometry for monitoring mass movements in Alpine areas, *Proceedings of the 3rd European Conference on Synthetic Aperture Radar* (EUSAR 2000), Munich, 23–25 May 2000, 43–46.

Singhroy, V. and Mattar, K., 2000, SAR image techniques for mapping areas of landslides, *ISPRS 2000. Proceedings*, Amsterdam, 1395–1402.

Singhroy, V., Mattar, K.E. and Gray, A.L., 1998, Landslide characteristics in Canada using interferometric SAR and combined SAR and TM images, *Advanced Space Research*, **3**, 465–476.

Vachon, P., Geudtner, D., Gray, A.L. and Touzi, R., 1995, ERS-1 Synthetic Aperture Radar Repeat-Pass interferometry studies: implications for RADARSAT, *Canadian Journal of Remote Sensing*, **21**(4), 441–454.

Varnes, D.J., 1974, *The Logic of Geologic Maps, with Reference to their Interpretation for Engineering Purposes*, US Geological Survey Professional Paper 837.

Vietmeier, J., Wagner, W. and Dikau, R., 1999, Monitoring moderate slope movements (landslides) in the southern French Alps using differential SAR interferometry, *Proceedings of Fringe '99 Workshop: Advancing ERS SAR Interferometry from Applications towards Operations*, Liège, Belgium.

Wasowski, J. and Gostelow, P., 1999, Engineering geology landslide investigations and SAR interferometry, *Proceedings of FRINGE '99*, Liège, Belgium.

# 17

# The Rise and Fall of a Debris-flow Warning System for the San Francisco Bay Region, California

*Raymond C. Wilson*

## 17.1 Introduction

### 17.1.1 Why a Debris-flow Warning System is Needed

As a generic term, 'landslide' simply means 'any gravitational movement of earth materials' (Cruden and Varnes, 1996). Landslides come in a wide variety of types and sizes, depending on the slope steepness, the parent material, the degree of wetness and the triggering mechanism (earthquake, rainstorm, undercutting and so forth). Selection of the most appropriate methods for preventing or mitigating landslides depends on the type and size of landslide involved, the types of structures and number of people to be protected, and the frequency of recurrence of the landslide. In particular, the efficacy of a warning system depends on the potential for life-threatening movements or side effects of the landslide (e.g. flooding).

Rapidly moving debris flows, triggered by severe rainstorms, are among the most numerous and dangerous types of landslides. Debris flows can begin suddenly, accelerate quickly, reach high velocities (up to 60 km/hr) and flow down streams or other channels for distances of several kilometres. They can smash homes and other structures, wash away roads and bridges, sweep away cars, knock down trees and, finally, lay down thick deposits of mud, rock and other debris where they come to rest, obstructing drainages and roadways. Because debris flows occur abruptly and move swiftly, public warnings,

*Landslide Hazard and Risk*   Edited by Thomas Glade, Malcolm Anderson and Michael J. Crozier
© 2004 John Wiley & Sons, Ltd   ISBN: 0-471-48663-9

to be effective, must be provided promptly (hours to minutes) as a major storm develops or approaches.

The state of Oregon, for example, in the chapter of its state emergency plan dealing with landslides and debris flows (Oregon, 2000), denotes as a special sub-class 'rapidly moving landslides . . . those which are difficult or impossible for people to outrun or escape'. Debris flows are placed in this category. The Oregon emergency plan notes, in contrast, that 'slumps/earthflows are relatively intact landslides . . . which move downslope at slow to moderate velocities (a person can normally walk away from these landslides)'. The emergency plan also notes that while both rapid and slow-moving landslides cause property damage, the rapidly moving slides and debris flows have caused most of the injuries and deaths related to landslides in Oregon in recent years. In addition to recommending landslide inventory mapping and timber harvest restrictions, therefore, the Oregon emergency plan establishes a debris-flow warning system, described in a later section of this chapter.

While debris-flows occur during or immediately after a period of excessive rainfall, other types of landslides, such as deep-seated slumps or slides, may exhibit a delayed reaction, where significant movement occurs days, weeks or even months later. These delayed reactions may hamper or confuse a warning system focused on synchronous rainfall forecasts and observations. Fortunately, these delayed landslides usually move more slowly than debris-flows, and are therefore less hazardous to personal safety. More traditional forms of mitigation may be more appropriate for the delayed types of landslides.

## 17.1.2   An Early Attempt at a Warning System for Los Angeles

Campbell (1975), studying the 1969 debris flows in Los Angeles, had already suggested a debris-flow warning system for Los Angeles, based on National Weather Service (NWS) forecasts and (pre-Doppler) radar imagery. Campbell noted (p. 29) that 'The hillside sites where soil slips may generate future debris flows are so small, numerous, and widely scattered, that the safety of all downslope residents cannot be ensured merely by the construction of defensive works such as check dams, debris basins, and levees.' Further, the specific locations of the soil-slip sites would be very difficult to predict. Also, once mobilized, debris flows move very rapidly, creating a risk of death or injury. On the other hand, Campbell noted that the 'general time of greatest debris-flow hazard' could be recognized by monitoring rainfall intensity and duration, and comparing them to (pre-established) threshold levels so that residents of areas with steep terrain could be advised when the rainfall reaches critical levels for debris-flow generation.

Campbell listed three major elements for an effective debris-flow warning system: (1) a system of rain gauges, recording rainfall on an hourly basis; (2) a weather-mapping system capable of recognizing centres of high-intensity rainfall in the storm areas and, at frequent intervals, plotting the locations of these centres with respect to locations of gauges with adequate registry for accurate transfer to slope maps or topographic maps; and (3) an administrative and communications network to collate the data, recognize when critical factors have been exceeded in a particular area, and inform the residents there. Campbell (1975: 32) noted that 'Such a system is probably well within the capability of existing technology.'

### 17.1.3  A Tragedy in Pacifica

In the early January of 1982, a disastrous rainstorm struck the San Francisco Bay region, triggering thousands of debris flows and other shallow landslides across the region, causing many millions of dollars in property damage and 25 deaths. Out of the many stories of grief and hardship from this storm, I was particularly touched by the death of three children, crushed by a debris flow which struck the back of their home at the base of a steep hillslope in Pacifica, a suburb of San Francisco (Figure 17.1). When the debris flow struck, shortly after 11:00 p.m., the children were asleep in rear bedrooms, but the parents were still awake, watching the evening news on television in the front living room. The parents survived. When interviewed later, one of the parents noted that the lead news story that night had been about flooding from the storm, but that nothing had been said about mudslides. When I heard this, I remember thinking that here was a modern family, connected to a real-time news source, yet there is no warning of a mortal danger in their own (literal) backyard. Although Campbell's proposal had not been pursued in Los Angeles, the 1982 storm in the San Francisco Bay region made a debris-flow advisory system seem like an urgent necessity.

At the Landslide Working Group at the USGS, we decided to begin with the concept of a 'threshold' – that a critical amount of rainfall is required to trigger debris flows on susceptible slopes. Thresholds are an essential element of any debris-flow warning system. In this context, a 'threshold' is a defined set of values of rainfall intensity and duration that predicts debris-flow initiation within a specified region. The next section describes the development of empirical rainfall/debris-flow thresholds for the San Francisco Bay region.

**Figure 17.1**  *Debris flow in Pacifica, California, 4 January 1982. At the base of the slope (lower left) two houses were destroyed and three children killed. Reproduced by the permission of USGS*

## 17.2    The Evolution of Rainfall/Debris-flow Thresholds

### 17.2.1    The Campbell Model of Debris-flows as Soil Slips

Campbell (1975) proposed that many rainfall-triggered debris flows in southern California are actually shallow landslides that remobilize into debris flows. These landslides are initiated by a loss of shear strength resulting from an increase in pore pressure, which is, in turn, created by intense rainfall. Campbell's postulated scenario begins with a permeability discontinuity at some depth, roughly parallel to the slope (Figure 17.2). Under low or moderate rainfall conditions, the hillslope soil moisture remains below saturation, and drainage is dominated by deep percolation into the underlying bedrock, with little or no surface runoff. Under intense rainfall conditions, however, when the rate of rainfall infiltration from the surface exceeds the rate of percolation across the discontinuity, a temporary zone of saturation forms, with excess infiltration drained as downslope seepage and surface runoff. As this saturated zone grows, the piezometric surface rises, increasing the pore pressure on the potential slide plane. If the slope is sufficiently steep, and the shear strength of the slope materials sufficiently weak, then intense, prolonged rainfall may ultimately trigger slope failure, producing debris flows and other shallow landslides. For the Santa Monica Mountains in southern California, Campbell (1975: 20) proposed a value of 0.25 in/hr (6.35 mm/hr) as a preliminary estimate of the critical rainfall intensity, sustained for a duration of at least several hours.

### 17.2.2    Development of Empirical Thresholds

After Campbell's work, several other investigators published empirical estimates of the intensity and duration of rainfall required to trigger debris flows. Caine (1980) collected a worldwide set of rainfall data recorded near reported occurrences of debris flows

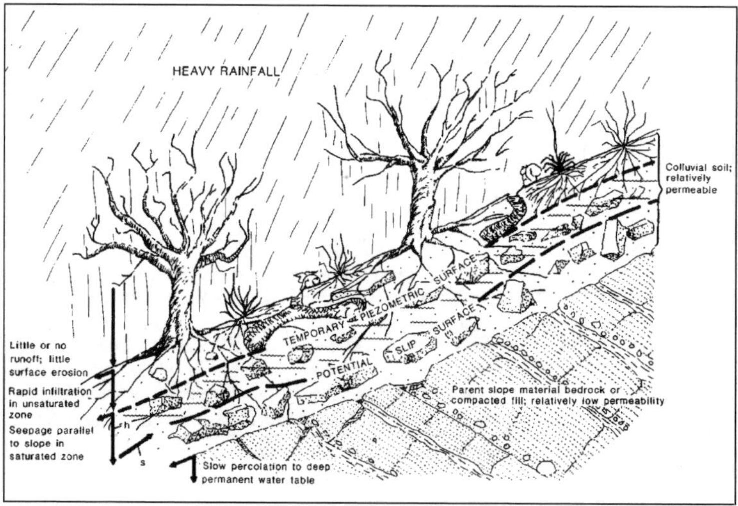


**Figure  17.2**  *Cross section of a typical hillslope as described by Campbell (1975)*

and estimated a threshold from the lower bound of a log–log plot of rainfall intensity, $I_r$ (mm/hr), versus the duration, $D$, in hours, such that:

$$I_r = 14.82\, D^{-0.39}. \tag{1}$$

During the late 1970s and early 1980s, Wieczorek (1987) studied a small ($12\,\mathrm{km}^2$) area near La Honda, in San Mateo County, that appears to be especially susceptible to debris flows. He estimated the threshold in terms of a minimum duration,

$$D = (9.0\,\mathrm{mm})/(I_r - 1.7\,\mathrm{mm/hr}), \tag{2}$$

of continuous rainfall equal to or exceeding the specified intensity, $I_r$ (e.g. continuous rainfall intensity at or above 3.2 mm/hr for a duration of at least 6 hours). Wieczorek's La Honda threshold, based on individual debris-flows in a highly susceptible area, is quite low compared with the Caine threshold, which is based on significant debris-flow disasters on a global scale.

Cannon and Ellen (1985) also developed thresholds for the San Francisco Bay region, using data from the January 1982 storm and several other major storms. Initially, Cannon and Ellen (1985) separated the historical rainfall data into two groups on the basis of whether the mean annual precipitation (MAP) in the area of the rain gauge was above or below 660 mm (26 in). They found that abundant debris-flow activity in the more humid upland areas required storm rainfall with a minimum duration and average intensity of 4 hours at 15 mm/hr (0.6 in/hr), 12 hours at 10 mm/hr (0.4 in/hr), or 20 hours at 8 mm/hr (0.3 in/hr).

In developing thresholds that could be used on a regional scale, however, a complication arose. Because it is a coastal region with high topographic relief, the San Francisco Bay region exhibits significant local climatic variations due to orographic effects, distance from coastline, slope aspect, prevailing winds and so forth. To correct for these effects, Cannon (1988) subsequently refined her empirical thresholds by dividing the storm rainfall intensity recorded at a given gauge by the MAP of that gauge, then using these data to prepare a normalized debris-flow threshold. (As discussed in a later section, this normalization by MAP, while a great advance at the time, has only limited application on broader scales.)

The Cannon and Ellen (1985) threshold formed the basis for the debris-flow warning system in the San Francisco Bay region when it was initiated formally in February 1986. Two public warnings were issued, through the NWS radio broadcast system, during a sequence of intense storms that triggered hundreds of debris flows in the San Francisco Bay region, causing one fatality and approximately $10 million in property damages (Keefer *et al.*, 1987). These were the first public warnings of debris-flow hazards issued for any region in the United States and they accurately predicted the times of major debris-flow events (ibid.: 925).

### 17.2.3   Refining the San Francisco Thresholds

In a further development (Wilson, 1986), I described the slope failure criterion in terms of a critical amount of rainfall, here denoted as $Q_c$, that must be retained within the slope. It is conceivable that a brief, but very intense, rainstorm could deposit an amount of rainfall just equal to $Q_c$ on the hillslope and trigger a debris flow. This would, however, require the slope to have an unusual combination of very high infiltration capacity, so that all of the rainfall would be absorbed, and very slow drainage, so that all of the rainfall would be retained. Hillslopes susceptible to debris flows generally do have a high infiltration capacity, but they also tend to drain rapidly, so that a significant amount of rainfall is likely to be lost during a storm. Thus retention of rainfall within the hillslope depends on a dynamic balance between the rate of infiltration and the rate of drainage.

Generally, an amount of storm rainfall well in excess of $Q_c$ will be required to produce a debris flow on such slopes. Therefore the rainfall threshold is not simply a constant rainfall total equal to $Q_c$, but is composed of a set of inversely related combinations of rainfall intensity and duration necessary to overcome the drainage while retaining a rainfall amount equal to or greater than $Q_c$. Thus, a rainfall/debris-flow threshold may be expressed in the generic form,

$$R_t = Q_c + I_o D \tag{3}$$

where $R_t$ = threshold rainfall total (mm) for the period $D$ (hr), $Q_c$ is the critical quantity (mm) of rainfall stored within the slope when failure is imminent and $I_o$ = the average drainage rate of the hillslope (mm/hr). For a given location, therefore, the calibration of the threshold becomes a search for those values of $I_o$ and $Q_c$ that best correlate historical data on rainfall and the occurrence of debris flows.

This relationship was used a few years later (Wilson *et al.*, 1993) to reconcile and consolidate the two local San Francisco Bay region empirical thresholds into a pair of relationships between the duration and cumulative amount of peak rainfall bursts that, together, outlined a spectrum of debris-flow activity (Figure 17.3). The lower, 'safety' threshold was adapted from Wieczorek's (1987) threshold for the initiation of individual debris flows in the La Honda study area to represent a rainfall level below which significant debris flow hazards were considered unlikely. The upper, 'danger' threshold was adapted from the threshold of Cannon and Ellen (1988), and was intended to represent a rainfall level above which abundant debris flows are likely to occur across broad areas in the San Francisco Bay region.

### 17.2.4    The Antecedent Rainfall Requirement

The relationship between rainfall, soil moisture and slope failure in climates with a strongly asymmetric distribution of rainfall through the year, such as the Pacific coast of California, creates an additional complication, the so-called 'antecedent condition', that has important implications for the operation of a landslide warning system. At the beginning of the winter rainfall season, the hillslope materials have been dehydrated by evaporation and transpiration during the long summer dry season. Any remaining soil moisture is held under strong negative pore pressures (soil suctions). Until this moisture deficit is restored by early seasonal rainfall, conductivity will be slow and high soil suctions will prevent the formation of the positive hydrostatic pore pressures necessary for slope movement. This moisture deficit may far exceed the rainfall likely from any single storm. Thus debris flows are very unlikely early in the rainfall season, even if early storms are severe.

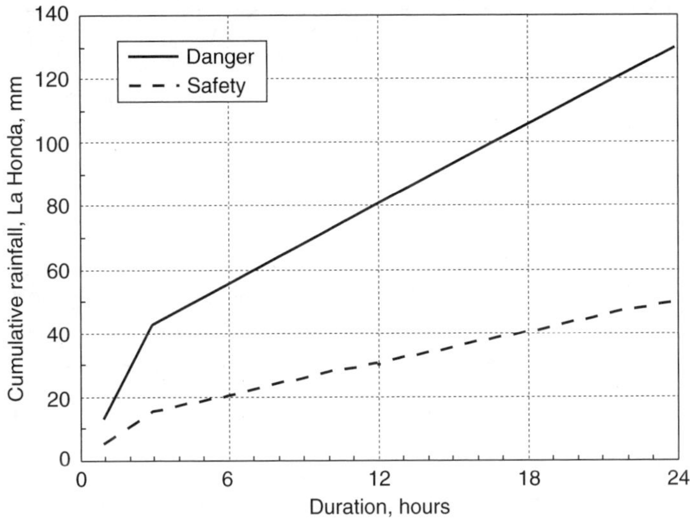


**Figure 17.3**  *Rainfall/debris-flow thresholds determined for La Honda, California. There is slight chance of significant debris-flow activity below the Safety threshold, a likelihood of damaging debris flows above the Danger threshold*

As the winter rains rehydrate the soil, at some point the hydrological behaviour of the hillslope is transformed from a moisture deficiency where additional water is absorbed by capillarity and clay rehydration to a moisture surplus where added water creates positive pore pressures and rapid drainage. The total amount of antecedent seasonal rainfall required for this transition depends on the initial degree of dehydration, the thickness of material that must be rehydrated, and the subsequent interplay between rainfall and evapotranspiration.

In the San Francisco Bay region, the rainfall and evapotranspiration cycles are about six months out of phase, leading to significant seasonal variations in soil moisture (Figure 17.4). The rainfall begins in late autumn, peaks in mid-winter, then declines, ending in late spring. Low temperatures, short periods of daylight and relatively dormant vegetation minimize evapotranspiration losses during the early and middle part of the winter rainfall season. Evapotranspiration increases rapidly during the spring months as temperatures and daylight hours increase and the grasses and other annuals reach their peak growth periods. In winter seasons with rainfall that is normal or higher, there will be a period of several weeks when the hillslope materials reach and maintain an optimum moisture content where soil suctions are minimized and any surplus moisture is removed by gravitational drainage. This is the time interval in which positive pore pressures may be formed and intense rainfall can trigger debris flows. In a 'typical' rainfall year, this period begins in late December and extends through late March.

In other parts of the world, where the seasonal distribution of rainfall is more uniform through the year, this antecedent moisture requirement may not be as obvious or may take a different form. In tropical areas with high evapotranspiration rates, soil moisture

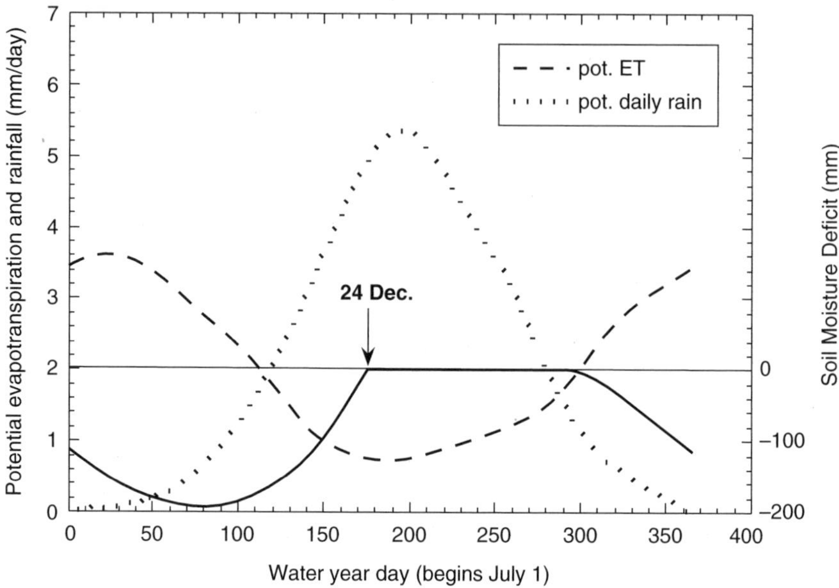


**Figure 17.4**  *Variations in rainfall, evapotranspiration and soil-moisture content in a typical year on a hillslope in the Santa Cruz Mountains. Reproduced by the permission of ASCE*

may be more strongly influenced by variations in rainfall over much briefer periods of time – a few days or weeks, rather than several months.

## 17.3  Operation of the Landslide Warning System in the San Francisco Bay Region

During the 1980s, meanwhile, the NWS had made two key advances that furthered the development of the debris-flow warning system: (1) It refined its procedures for preparing quantitative precipitation forecasts throughout northern and central California and (2) coordinated the development of the Automated Local Evaluation in Real Time (ALERT) system, a network of radio-telemetered rain gauges across the San Francisco Bay region.

When a storm approached the San Francisco Bay region, the local NWS Forecast Office (moved from Redwood City to Monterey in 1994) would attempt to make a quantitative forecast of the storm rainfall. Then the USGS Landslide Initiation and Warning Project in Menlo Park would perform a manual comparison of the observed and forecast rainfall to the estimated thresholds for debris flow initiation. Finally, both groups work together to assess the probable hazard from debris-flows, so that appropriate public statements could be issued.

### 17.3.1  Forecasting the Rainfall

Most of the rainfall in the San Francisco Bay region is produced by North Pacific weather systems that originate either in the Gulf of Alaska or in subtropical latitudes near Hawaii.

In the January 1982 storm, extremely heavy rainfall resulted from a collision of air masses from both regions. The principal tool for tracking storms over the North Pacific has been the imagery from weather satellites, principally the GOES-7 (Geostationary Operational Environmental Satellite), launched in February 1987 (replaced by GOES-10 in 1997). Every 30 minutes, the GOES satellite transmits an image of cloud cover across the northeastern Pacific from the Gulf of Alaska to Hawaii to the west coast of North America. The spatial patterns of the clouds and their movements, revealed by time-lapse sequencing of several images, allow the estimation of speed, direction and intensity of large storm systems. Imagery in the infrared spectrum also indicates the temperatures of the cloud tops and provides important inferences about the expected intensity of rainfall.

Surface and upper-air weather observations, including barometric pressure, wind velocity, temperature and precipitation, from a network of land-based weather stations, combined with reports from aircraft and ships, also furnish important data on approaching storm systems. Additional forecast guidance is provided by computer simulations of long-term weather trends from the NWS National Meteorological Center in Camp Springs, Maryland. These computer simulations, based on models of global atmospheric circulation, are updated frequently with surface and upper-air observations from throughout the northern hemisphere.

At the local NWS Forecast Office, the lead forecaster compiles this information and prepares the Quantitative Precipitation Forecast (QPF). The QPF, issued twice daily, estimates the amount of rainfall expected in each of four 6-hour periods, for the following 24 hours throughout northern and central California.

### 17.3.2   Comparing Forecast Storm Rainfall to Debris-flow Thresholds

In evaluating the chance for an approaching storm to trigger hazardous debris-flow activity, two thresholds had to be considered: (1) the accumulation of antecedent seasonal rainfall and (2) the combinations of rainfall intensity and duration forecast for the approaching storm. In order to evaluate the seasonal progress of soil moisture towards the antecedent rainfall threshold, the USGS Landslide Working Group installed and monitored shallow piezometers at the La Honda study area, which served as a benchmark site for the region. We assumed that the soil moisture had reached the antecedent threshold when the piezometers first responded strongly to storm rainfall, generally within a few weeks after the winter solstice.

Once the antecedent threshold for soil moisture was exceeded, subsequent storms were evaluated as they approached to see if the intensity and duration of the expected rainfall would be sufficient to trigger debris flows. The 1986 debris-flow warnings were based on the empirical rainfall thresholds determined by Cannon and Ellen (1985). By 1989, we had developed the pair of cumulative rainfall/duration relationships for a spectrum of size and frequency of debris flows described above (Wilson *et al.*, 1993).

### 17.3.3   Observing the Rainfall

The ALERT system was designed to collect automatic measurements of high-intensity rainfall at remote locations and transmit these data to central receiving stations for observation and analysis in a near real-time environment. By 1995, there were more than 60 rain gauges in the ALERT network in the San Francisco Bay region (Figure 17.5).

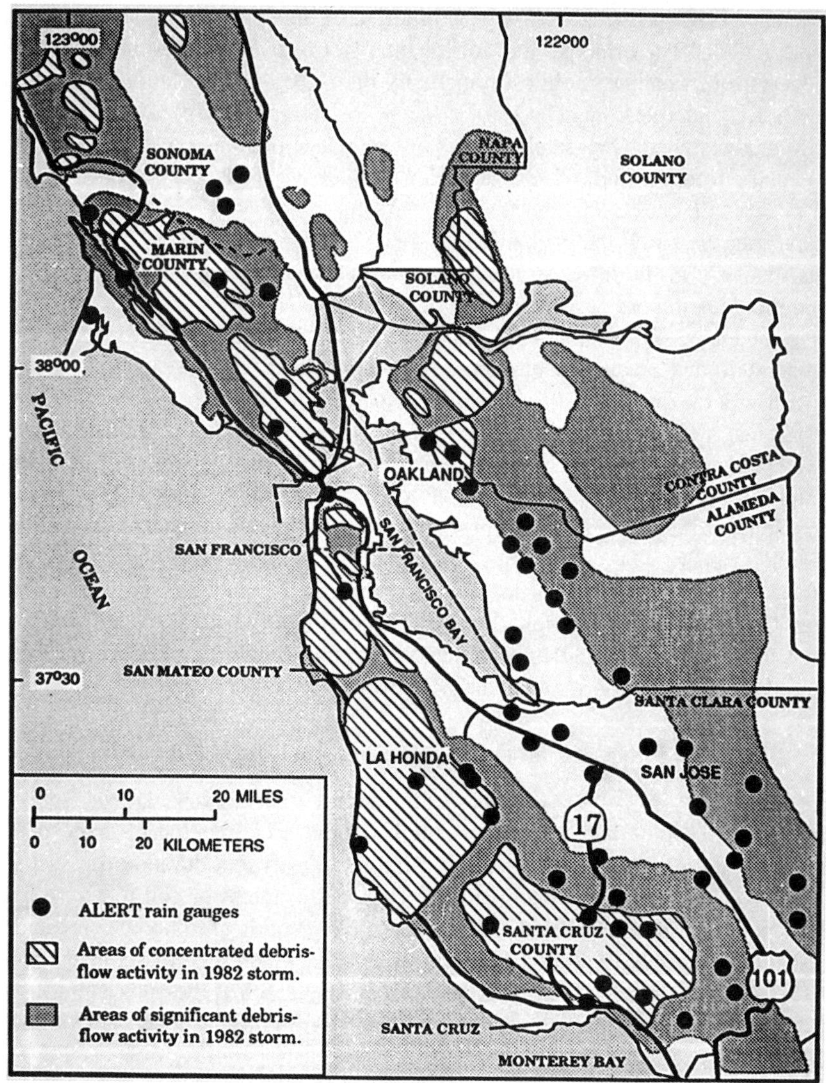



**Figure 17.5**   *Map of the ALERT network in the San Francisco Bay region in 1992*

Although the network was sponsored by the NWS, the individual ALERT stations were purchased, installed and maintained by a number of other federal, state and local government agencies.

Each ALERT station is a self-contained unit consisting of a tipping-bucket rain gauge, a power supply, an electronic data processor and a radio transmitter (Figure 17.6). When an amount of rain equivalent to 1 mm (0.04 in) depth is collected by the rain gauge, the tipping-bucket mechanism engages, tips out the accumulated water, and closes a circuit that increments a data register and prompts the radiotelemetry system to transmit a

***Figure 17.6*** *The ALERT rain gauge operated by the USGS near La Honda, California from 1985 through 1995*

binary-coded sequence, consisting of the station identification code and the total number of 1 mm (0.04 in) rainfall increments accumulated during the season. This signal may be received and processed by anyone with the requisite radio receiver, microcomputer, and software for data collection, analysis and display. The USGS maintained an ALERT receiver and dedicated microcomputer data-processing system in Menlo Park, California, from 1985 through 1995.

### 17.3.4 Issuing Debris-flow Hazard Statements

As a storm began to make landfall, the ALERT network of radio-telemetered rain gauges was used to monitor the rainfall intensities and estimate the speed of advance of the storm front. Observed rainfall amounts, combined with QPF estimates, were compared to the warning thresholds (Figure 17.3) to determine the level of hazard and the type of public statement to be issued. Both the NWS and the USGS participated in this phase of operation.

Storms with peak rainfall periods that fell below the lower threshold ('safety') were considered unlikely to trigger hazardous debris flows and generally required no statements. For storms with rainfall levels just above the lower threshold, brief statements were sometimes added to an NWS 'Urban and Small Streams Flood Advisory', warning motorists that roadways may be obstructed by rockfalls or debris flows. If rainfall was forecast to approach the upper threshold, a Flash-Flood/Debris-Flow Watch was issued, advising people living on or below steep hillsides, or near creeks, to stay alert and be prepared to evacuate, as debris flows were a strong possibility during the watch period. Storms that exceeded the upper threshold could trigger numerous, massive debris flows leading to loss of life and substantial property damage. Therefore, when rainfall was observed to exceed the upper threshold, or if reports of significant debris-flow activity were received, the strongest statement – a Flash-Flood/Debris-Flow Warning – was issued. Sample texts for these debris-flow statements were prepared, with wording agreed upon by both the USGS and the NWS, so that timely, informative advisories with complete, relevant information could be issued with a minimum of preparation time.

### 17.3.5   Summary of Debris-flow Statements Issued During Operating Period (1986–95)

During its period of operation, the debris-flow warning system issued advisory statements in response to several unusual events.

(1)  A Flash-Flood/Debris-Flow Warning was issued on 24 March 1991 after a brief, but intense, rainstorm triggered a large rockfall that closed Highway 17 in the Santa Cruz Mountains.

(2)  A catastrophic wild fire in October 1991 burned a large residential area in the Oakland Hills. Because of the increased risk of debris flows being triggered by subsequent rainfall on the burned area, a special warning threshold was devised. Because the burned area had been completely stripped of vegetation, leaving a loose, easily eroded, layer of ash and soil, the intensity–duration thresholds were reduced and the antecedent threshold was eliminated (USGS Press Release, 28 October 1991). Fortunately, a re-seeding effort was successful, and the following winter brought relatively mild rainstorms, so there were no significant debris-flow problems in the burn area.

(3)  Based on a forecast for heavy rainfall, a Flash-Flood/Debris-Flow Watch was issued on 11 February 1992 for the Oakland burn area. While subsequent rainfall was only moderate in Oakland, it reached levels of 80–120% of the danger threshold in southwestern San Mateo County, triggering numerous, small debris flows.

(4)  Flash-Flood/Debris-Flow Watches were issued on 13 January and 15 January 1993, during a closely spaced sequence of intense rainstorms. A post-storm field reconnaissance located a number of small, widely scattered debris flows on roadways and natural slopes in Marin, San Mateo, Alameda, Santa Clara and Santa Cruz Counties. On our post-storm field reconnaissance, these debris flows appeared to be very fresh, but we were unable to determine the exact timing relative to our warning statements.

(5)  A series of Flash-Flood/Debris-Flow Watches was also issued during the heavy storm period from 6–10 January 1995. A number of small debris flows occurred in the East Bay Hills (Figure 17.7) and the Santa Cruz Mountains, closing several roads for hours to days, and causing some damage to outbuildings. Fortunately, there were no deaths reported. Most of these events followed the earliest of our advisories.

***Figure 17.7***  *Small debris flows triggered by the storm of 13 January 1995 in the hills above Fremont, California*

(6) In the second week of March 1995, an even more powerful storm brought debris flows and flash floods in the Santa Cruz Mountains and East Bay Hills, as well as a major flood on the Salinas River. Debris-flow advisories were issued on 10 March 1995 for Monterey County. The March 1995 storm system also produced many debris flows along the Central California Coast, south of the warning area, and caused significant landslide damage in Santa Barbara and Ventura Counties.

## 17.4  Later Developments (1992–95) in the San Francisco Landslide Warning System (LWS)

In the early years of LWS operation, weather forecasting was based principally on interpretation of GOES weather satellite imagery. The principal tool for monitoring rainfall intensity concentrations across the San Francisco Bay region was the ALERT system of radio-telemetered automatic rain gauges, described above. These were then state of the art, representing a great advance over hand-read rain gauges. Spatial resolution was poor in some areas, however, because of the uneven distribution of ALERT gauges, and the data processing was tedious and complex enough to be difficult in a real-time environment, especially in critical situations. The advent of the nationwide network of high-power Doppler radar stations that the NWS established in the early 1990s promised great improvements in both the spatial resolution and the ease of interpretation of real-time rainfall intensity data in the SFBR.

### 17.4.1   NEXRAD – Hope and Disappointment

In 1994, the NEXRAD weather radar station for the San Francisco region was constructed on Mt Umunhum, south of San Jose, bringing both remarkable new possibilities and serious complications. The NEXRAD (NEXt generation RADar) system was designed to map the velocity structure of the atmosphere through Doppler-effect phase shifts of reflections from dust particles in the atmosphere. Of greater interest for debris-flow forecasting, NEXRAD also uses radar waves reflected off water droplets in rain clouds to estimate precipitation. The radar reflectivity is measured in terms of the logarithm of the energy of the reflected microwave energy received back from a transmitted pulse, expressed in terms of decibels, dBz. The reflectivity is proportional to the volumetric concentration of condensed water droplets within the cloud. If there is precipitation, the intensity of precipitation is also proportional to the concentration of water droplets, so there should be a close relationship between reflectivity and rainfall intensity.

The hope was, therefore, that a NEXRAD based reflectivity image would provide a real-time snapshot of the rainfall intensity distribution within the range of the radar (approximately 150 km). If so, the spatial and temporal resolution of the NEXRAD imagery (6 minutes in time, 1 km$^2$ in space) would be a vast improvement over the rainfall data from the sparse, irregular ALERT network. Unfortunately, while the NEXRAD on Mt Umunhum does generate detailed spatial images of basal radar reflection, the correlation between radar reflectivity and actual rain-gauge data on the ground has been disappointing.

There are several factors that complicate the relationship between radar reflectivity and rainfall. For one thing, ice particles reflect radar energy far more efficiently than water droplets, so that air temperature may become a critical factor. Also, the size distribution of water droplets within the cloud may vary with temperature and/or altitude. Airborne objects other than water droplets may also reflect radar energy – for example dust, insects, birds and so forth. Another common problem is 'ground scatter', the return of radar energy by the ground surface below the radar beam, especially in areas of rugged topography.

The fact that the radar reflectivity samples moisture droplets in high-altitude clouds, not raindrops hitting the ground, introduces further complications. Because horizontal wind velocities may equal or exceed the falling velocity of the raindrops, rainfall may be subject to significant wind drift before it reaches the ground. If the raindrops fall through drier air at lower altitudes, significant evaporation may also occur (leading, in the extreme, to the phenomenon of 'virga', where little or none of the rain actually reaches the ground). These problems are aggravated at the Mt Umunhum NEXRAD station because the high elevation of the site (about 1060 m) further elevates the radar beam. Past a range of a few kilometres, the radar beam samples only the upper layers of the rain-bearing clouds coming off the ocean.

Adding to the complications of radar physics and cloud dynamics is a significant meteorological problem – the complexity of large storm systems. NEXRAD reflectivity images of large storms reveal an intricate pattern of convection cells, squalls and other 'micro-structure', with rainfall intensities varying over orders of magnitude over distances of only a few kilometres.

All of these complicating factors should make one very cautious about using any generic formula for converting radar reflectivity into rainfall intensity. Conversion formulas should be developed only for specific areas, calibrated for local atmospheric conditions and re-calibrated frequently when these conditions change.

## 17.5   The Fall of the San Francisco Regional Landslide Warning System

### 17.5.1   The Burden

When the LWS began operation, the lines of responsibility were fairly diffuse and informal. In the USGS Landslide Research Group, we regarded the LWS as an experiment; the highest priority was to see if we could actually predict debris-flow activity. Public advisories were regarded as a by-product, potentially useful to some (unspecified) clientele, but not the central focus. The NWS forecasters, on the other hand, pressed us to consider seriously the criteria for issuing advisories and exactly how to word them. Over time, a detailed protocol was evolved, with pre-established templates for texts for the various contingencies and expected levels of debris-flow activity (Wilson *et al.*, 1993).

We also realized that the warning system could not be completely automated. A trained, responsible person must assimilate the data (NWS forecasts, ALERT data, news reports), then make an informed, yet subjective, decision about the potential hazard, and finally, choose the appropriate advisory to broadcast to the general public. These are non-trivial judgements and must be made not once but many times in the course of a storm sequence that could last several days. False alarms create nuisances and erode credibility. On the other hand, the absence of an advisory when debris-flows do cause death or destruction becomes a dereliction of duty.

Thus the LWS had to be staffed as a round-the-clock activity, at least during periods of heavy rainfall. While the NWS was already staffed for 24-hour operations, the USGS side of the LWS was staffed on a collateral duty basis. In addition to our regular research duties, we had to provide at least four trained observers – one person per 6-hour 'watch', 24 hours per day – who could not only monitor the data, but also make correct interpretations and take appropriate actions. As the permanent staff of the USGS Landslide Research Group shrank from ten people in 1986 to five in 1994, this four-person staffing requirement became a heavy burden.

### 17.5.2   Termination

In late 1994, the NWS moved its forecast office from Redwood City, about 8 km from our USGS offices in Menlo Park, to Monterey, almost 150 km away, complicating interactions during crisis periods. This move was accompanied by a net reduction in staffing, including the loss of several senior forecasters, who chose retirement over relocation. Gray Barbato, the NWS hydrologist who had helped create the LWS, accepted a new assignment in Reno. Finally in 1995, the USGS also underwent a reduction in force and two of the four people staffing the LWS were let go. Negotiations to have the NWS take over the monitoring duties for the hydrological aspects of the LWS were unsuccessful, and the LWS was officially terminated in December 1995.


## 17.6    Response to Issued Warnings

One of the most important outcomes of the LWS was that it became a focal point for media attention and thereby served to raise public awareness of debris-flow hazards in the San Francisco Bay region. For example, we issued a press release when seasonal rainfall totals reached the 'antecedent condition' – when early seasonal rainfall has replenished the soil-moisture deficit incurred during the long summer dry season (Campbell, 1975). This annual press release, which always received wide coverage in the local news media, served not only to inform concerned local agencies, but also provided a 'wake-up call' to the media and general public that a heavy winter rainstorm could bring a return of debris-flow activity.

### 17.6.1    Getting People to Respond to Warnings

When we began our debris-flow warning system in 1986, we believed that our principal tasks were: (1) to determine when rainfall conditions had reached critical thresholds for significant debris-flow activity in a susceptible area, and then (2) to broadcast, through the NWS Weather Radio network, public advisories warning of the risk of debris flows in the relevant areas, so that (3) appropriate actions could be taken by other public agencies (e.g. fire departments, sheriff's deputies) and private parties (e.g. voluntary evacuations by residents of threatened areas). We thought that the key to the system was the technical question of how to determine the threshold levels of rainfall intensity and duration needed to trigger abundant debris flows. Once that was done, the burden of forecasting the weather would be borne by the NWS and the problem of getting the people out of the way of the debris-flows would fall to emergency response agencies and the residents themselves. Underlying everything was our assumption that, if someone were told that their life or personal safety was threatened, they would then take appropriate action to minimize that threat – such as to leave the threatened area and go somewhere safer. I now realize that these ideas about what a landslide warning system was and how it might operate, and particularly the underlying assumption, were very naïve.

A few problems became obvious right away. Most people, both laymen and emergency workers, have no real understanding of what a debris flow is or what it can do to a house, road or other structures. It is hard for a threat to be credible when it involves something you have never heard of. Fortunately, the news media had taken an interest in our debris-flow warnings and, through many interviews with TV, radio and print reporters, we were able to explain the threat and educate the public, at least to a certain extent, so that by the end of the first season of operation (1986) the level of public awareness of debris-flow hazards was firmly established and grew over the next several seasons.

Once the advisory is broadcast, however, people in the target audience must first hear the warning, realize that they are at risk and determine that they must take some action, before any constructive action can be taken. Although relatively few residents of the California coastal uplands listen to Weather Radio directly, our advisories would be picked up and re-broadcast by commercial radio and TV stations, often within a few minutes of the original NWS broadcast. All-news radio stations are especially useful in this respect, since they respond quickly and reach a wide audience in the region. Further, many local fire and law enforcement agencies monitor the NWS Weather Radio directly. By these means, our debris-flow advisories received fairly wide distribution.

### 17.6.2   What Actions Should be Taken in Response to a Warning

Even if a resident or emergency services worker hears the advisory and realizes that the threat of debris-flow activity applies to their situation, there still remains the need to take appropriate action, but it may not be obvious what that action should be. Temporary evacuation is appropriate – if it can be accomplished without placing the resident and his family at an even greater risk. If the advisory is received well in advance of the heavy rainfall and high winds of an approaching rainstorm, evacuation may be accomplished easily and safely. Once the storm has progressed to its full intensity, however, the heavy rainfall, poor visibility, slippery roadways, and the likelihood that some roads may already be obstructed by debris flows and rockfalls increases the risk considerably.

Finally, there is the danger that the place selected for refuge may be in danger itself. In a storm near Los Angeles in 1969, for example, flash flooding caused 60 residents to seek shelter in a fire station in Silverado Canyon; later that morning, a debris flow swept through the fire station, killing 5 of the sheltering residents and injuring 20 others (Campbell, 1975: 42). Similar incidents have been reported in other times and places.

Once intense rainfall is in progress, the best course of action may sometimes be to 'shelter in place' within the residence. Sheltering residents should select a room that opens to the outside, preferably in a downhill direction, and should remain awake and monitor the situation outside. Campbell (1975: 30) reports, for example, that several fatalities from debris flows were avoided (albeit narrowly) when alert residents were able to recognize approaching debris flows and move quickly out of the way. Therefore, it is necessary for the sheltering residents to remain awake and to listen for unusual sounds (e.g. crashing rocks, falling trees, rapid streamflow), and to be prepared to take prompt action to escape an approaching debris flow.

### 17.6.3   What *not* to do During Warning Periods

Driving through susceptible areas during heavy rainstorms is also hazardous and should be avoided, except to evacuate from the area. Drivers should drive slowly and watch carefully both the roadway and the slopes above it. Embankments and road cuts are especially susceptible to debris flows or rockfalls. The roadbed may be undermined by heavy runoff. Watch the road for collapsed pavement, mud, fallen rocks, fallen trees and other debris (USGS Fact Sheet 112–95).

One of the disturbing things I have learned from talking with residents of areas impacted by debris flows during the El Niño storms of February 1998 in the San Francisco Bay region is how many people had attempted to drive *into* the area during the worst of the storms. These residents had been at work in offices and shops in the lower, urbanized cities around the Bay and were trying to get back home. Some of these people made heroic efforts – fording streams, removing fallen trees, dodging boulders – to get home to feed a pet or to grab a shovel and try to drain mud and water away from their houses, blithely unaware that they were placing themselves at risk from potentially fatal dangers.

## 17.7    Post-1995 Developments in Forecasting Rain-induced Debris-flow Thresholds

### 17.7.1    Climatic Adjustments to Rain/Debris-flow Thresholds

Several of my recent papers (Wilson, 1997, 2000, 2002) have attempted to show that rainfall/debris-flow thresholds are strongly influenced by the local precipitation climate. This exploration began in 1995 with an attempt to unravel one of the perplexing puzzles of the old Landslide Warning System – the curious behaviour of some of the drier areas in the San Francisco Bay region. To begin the story, I will digress briefly to review the problem of adjusting the rainfall/debris-flow thresholds for orographic variations.

In the San Francisco Bay region, winter rainstorms coming off the North Pacific Ocean interact with the local topography in complex patterns, producing significant orographic variations in rainfall. Moist air forced to rise over the local ground surface undergoes adiabatic cooling, increasing precipitation at higher elevations on the windward sides of topographic highs. As the storm clouds descend the leeward flank of the range, however, they undergo adiabatic warming, which significantly decreases precipitation – the so-called 'rain-shadow effect'.

As an illustrative example, we may postulate a typical storm that produces 50 mm of rain as it makes landfall over the San Mateo County coastline. As it continues eastward over the Santa Cruz Range, the same storm might produce over 110 mm of rain near the crest, but then fall back to less than 40 mm in the lee of the range, along the shore of San Francisco Bay. A similar increase and decrease of storm rainfall occurs as the storm passes further eastward over the East Bay Hills (back up to 65 mm), then drops back down into the Central Valley (about 30 mm).

### 17.7.2    MAP- Normalized Thresholds

When Cannon (1988) developed her regional threshold rainfall values for debris-flow initiation, she attempted to correct for orographic effects by first dividing the rainfall intensity data for each rain gauge by its long-term MAP. MAP normalization would allow severe rainstorms that produce significant debris-flow activity to be recognized as 'extraordinary events, when rainfall at a particular site exceeds the commonly occurring conditions' (Cannon and Ellen, 1988: 30). MAP is the parameter most commonly used to describe the long-term precipitation climate and may be obtained from standard climate maps (e.g. Rantz, 1971).

Cannon's MAP-normalized threshold was certainly a step in the right direction and provided, in fact, the primary technical basis for the landslide warning system (Keefer *et al.*, 1987). However, as Cannon (1988) herself noted (p. 38), 'normalization introduces inconsistencies in areas of low MAP'. Indeed, during the period of operation of the LWS, data from ALERT gauges in low-rainfall areas appeared to yield 'false alarms', with an inordinate frequency (R. Mark, unpub. data, 1995). Later, I found that attempting to apply

MAP-normalized thresholds from the San Francisco Bay region to southern California or to the Pacific Northwest produced significant under- or over-estimated thresholds, respectively (Wilson, 1997).

The problem with the mean annual precipitation as a normalizing parameter is that MAP actually reflects a combination of two processes: (1) the distribution of sizes of individual rain storms, and (2) the frequency of rain storms during the year. Rain gauges in the US Pacific Northwest, for example, have a high MAP because of a high frequency of relatively small storms. Many gauges in southern California, where the frequency of rainfall is much lower, record a much lower MAP, but have a higher proportion of large individual storms when it does rain. A climatic normalization scheme for rainfall/debris-flow thresholds should clearly identify those storms that are 'extraordinary events', as opposed to the more frequent, but much smaller, storms that dominate the MAP value for most stations. For estimating rainfall/debris-flow thresholds, therefore, the size distribution of individual storms is more important than the storm frequency.

Size–frequency analyses of daily rainfall were used to re-examine the inconsistencies in MAP-normalized thresholds noted by Cannon (1988: 38) in low MAP areas in the San Francisco Bay region. These low-MAP gauges all showed a strong 'rain-shadow' pattern where the reduced MAP is produced by a reduction in rainfall frequency, but the size distribution of individual storms is altered only slightly. The result is that the larger individual storms occur more frequently than would have been predicted from MAP alone, and so the rainfall/debris-flow thresholds are higher than would be predicted from MAP normalization.

### 17.7.3   Equilibration of the Hillslope to Long-term Climate

In order to better understand the interaction of the hillslope with the precipitation that triggers debris flows, we must go beyond the MAP and make a more fundamental examination of the long-term precipitation climate and its effects on rainfall/debris-flow thresholds. As noted earlier, these thresholds reflect the relative rates of rainfall infiltration and hillslope drainage. Because the steep hillslope soils that are susceptible to debris flows tend to drain rapidly, this process becomes a contest between rapid infiltration of intense rainfall and rapid drainage through a combination of percolation, surface runoff and downslope throughflow (Wilson and Wieczorek, 1995). My working assumption has been that, over a period of time (centuries), the landscape equilibrates itself to the long-term precipitation climate, such that, under 'normal' conditions, the hillslope can balance infiltration with evapotranspiration and surface runoff while maintaining gravitational stability.

The process of long-term equilibration may encompass a number of mechanisms, known and unknown. While some of the geotechnical characteristics of hillslope soils appear to have no obvious correlation with the climate, the drainage rate may be influenced by climatically driven adjustments in hillslope vegetation and the spatial density of surface channels. The native vegetation of a hillslope is very sensitive to annual rainfall and the seasonal range of temperatures, and can adjust relatively rapidly (decades) to changes in the local microclimate. Changes in the variety and abundance of hillslope vegetation may change the rate of evapotranspiration and alter the effective hydrological storage through changes in leaf litter. Also, the contribution of root fibres to the cohesive strength of hillslope soils may greatly influence the rainfall thresholds. The density of surface channels reflects the balance between infiltration during intense rainfall and the drainage delivered by shallow throughflow to a network of ephemeral surface channels, which are, themselves, carved out by erosion from surface runoff.

### 17.7.4   Reference Rainfall

We may use the assumption of climatic equilibration to estimate how hillslope drainage rate scales with rainfall climate. We might imagine that there exists a certain intensity of rainfall, termed the 'reference rainfall', which if continued at a constant rate, would result in a surface discharge equal to the rate of recharge (infiltration). The reference rainfall corresponds to a fairly significant rainfall event that can mobilize sediment and erode or otherwise carve out efficient paths for drainage – whether these paths are visible surface channels or invisible subterranean networks of macropores. Yet the reference rainfall must also have a fairly high frequency of occurrence, so that whatever hydraulic work is done is not completely dissipated between events by diffusive hillslope processes – downslope creep, burrowing, re-deposition into channels. Although somewhat smaller than the 'extraordinary events' that trigger abundant debris flows, an estimation of the reference rainfall could provide a useful reference point for the interaction of rainfall and surficial drainage.

As an empirical approach, I suggested (Wilson, 2000) that a climate-normalized rainfall/debris-flow threshold might be developed from the estimated rainfall for a 'reference storm' with a return period of approximately 5 years (20% annual exceedance probability). Initially, the 5-year return period was chosen simply for convenience – infrequent enough that the small, but numerous, rainfall events that dominate MAP values may be filtered out, yet frequent enough that accurate values may be estimated from a few decades of data from a rain gauge. Later work (Wilson, 2002) has suggested that the 5-year storm represents an optimum combination of rainfall frequency versus erosion rate, such that much of the geomorphic work of creating and maintaining the surficial drainage network would be done by storms in this size range.

The next step was to test the practical value of reference rainfall for normalizing rainfall/debris-flow thresholds. Rain gauge data were collected from several historical storms that triggered destructive debris-flow activity in three different regions with significantly different rainfall climatic patterns – southern California (Los Angeles), the San Francisco Bay area and the Pacific Northwest (Oregon and Washington). Figure 17.8 is a graph of peak 24-hour rainfall amounts, from storms that triggered debris flows, plotted against the maximum 24-hour rainfall expected in a 5-year return period, for the same gauges (Wilson, 2000).

The graph in Figure 17.8 shows a fairly well-constrained distribution of data points with a distinct lower-bound threshold of approximately 4/3 of the reference rainfall value. A line corresponding to an apparent upper bound at twice the reference rainfall value is also plotted for comparison. In areas that lack a detailed history of debris-flow activity, but have a reliable, long-term record of daily rainfall, the lower-bound threshold in Figure 17.8 might be useful as an approximate threshold for a level of debris-flow activity likely to pose a significant threat to the safety of lives and property.

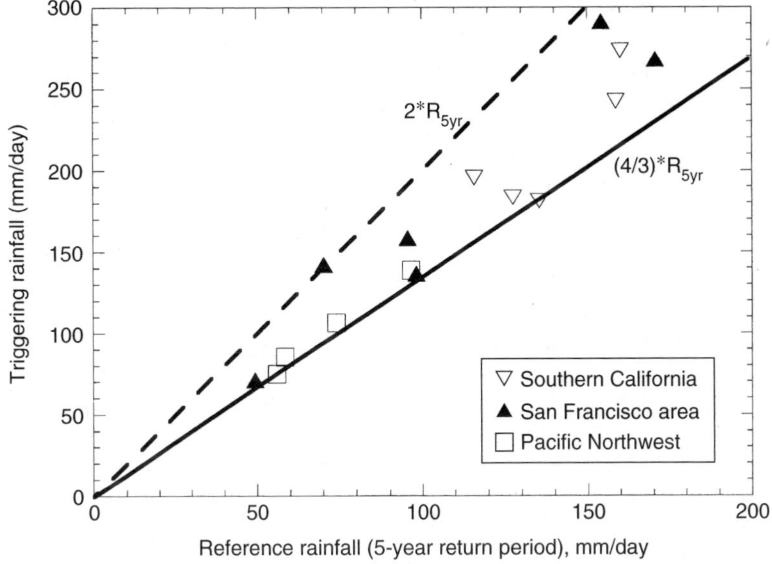


**Figure 17.8**  *Plot of peak 24-hour rainfall from historical storms triggering abundant debris flows versus the reference storm values (R$_{5yr}$) for the corresponding rain gauges (Wilson, 1997, 2000)*

In addition to extrapolation into new areas with undocumented debris-flow histories, the climate-normalized rainfall/debris-flow thresholds in Figure 17.8 also allow more reliable interpolation within a region where thresholds have been developed at specific points (e.g. the La Honda study area in the San Francisco Bay region). This is especially important in mountainous areas where strong orographic effects produce significant variations in storm rainfall over short distances.

## 17.8   A Second Look at the NEXRAD Radar System

Also since 1995, I have gained a greater appreciation of the potential advantages of the NEXRAD weather radar system, to balance my pessimistic discussion of its limitations earlier. After several years viewing NEXRAD radar imagery of large storms in the San Francisco Bay area, I now perceive that, despite the lack of quantitative correlation on a pixel-by-pixel basis, NEXRAD reflectivity images portray the overall shape and structure of the rainfall intensity field in most significant storm systems with reasonable accuracy. Cloud masses with a high radar reflectivity, for example, generally deliver high rainfall intensities as they pass over the site of an ALERT rain gauge, even if the actual amount may vary from the calibration curve. This qualitative agreement between regional patterns of radar reflectivity and the gross structure of rainfall intensities within a major storm system should still be useful in forecasting debris-flow activity, although a significant amount of 'operator judgement' may be required.

## 17.9    Conclusion

A good deal of progress has been made since 1995 in refining rainfall/debris-flow thresholds and gaining a deeper understanding of interactions between rainfall, hillslope hydrology, long-term rainfall climate and slope stability. With these advances, it should be possible to develop a new, advanced debris-flow warning system for the San Francisco region with extended spatial coverage and longer forecast lead times, well beyond those of the 1986–95 Landslide Warning System described here.

## 17.10    Some Other Debris-flow Warning Systems

### 17.10.1    Hong Kong

Probably the most extensive, and successful, debris-flow warning system in the world is that in Hong Kong, jointly operated by the Civil Engineering Department's Geotechnical Engineering Office (GEO), which exercises geotechnical control on new slopes and developments, both public and private, and the Hong Kong Observatory (HKO), which acts as Hong Kong's weather service. Based on HKO forecasts and rainfall data from an extensive network of automatic rain gauges, begun in 1984, Landslip Warnings are issued when it is predicted that numerous (>10) debris flows are likely to incur within the city. The system is described in a series of Internet pages from GEO (http://hkss.ced.gov.hk/hkss/eng/safety/warning/landslip.html) and HKO (http://www.weather.gov.hk/wservice/tsheet/tsflwarn.html).

The rain-gauge network consists of 86 automatic gauges operated by GEO and 24 HKO gauges. The two agencies share data and decisions to issue or cancel Landslip Warnings that are made jointly by the Director of the HKO and the Head of the GEO. The original rainfall thresholds for numerous debris flows were 24-hour rainfall exceeding 175 mm or 60-minute rainfall exceeding 70 mm. These thresholds were revised and tested in the 1999 wet season, with significant improvements found. Unfortunately, the websites contain no details on the new rainfall criteria.

### 17.10.2    Taiwan

Taiwan appears to have had a much less successful experience with debris-flow warnings. According to a newspaper account (Chiu, 2001), the Taiwan government began working with the Disaster Prevention Research Center at National Cheng Kung University in the early 1990s to establish a warning system for mountain areas vulnerable to debris flows, but 'the results of the collaboration were disappointing'. Chiu noted that 'The mudslide warning system, comprising 18 monitoring stations, each costing more than NT\$ 1 million, was criticized for its inaccuracy . . . It often issued warnings but no mudslides materialized. When mudslides occurred, no warnings were issued. Residents living in mudslide-prone areas . . . had no confidence in the system.' The warning system was terminated in 1998. In August 2001, Typhoon Toraji swept across Taiwan, triggering many more debris flows, causing additional loss of life and property damage.

### 17.10.3   Brazil

In 1996, after decades of debris-flow disasters, the city of Rio de Janeiro installed an alarm system for landslides triggered by heavy rainfall. The Institute of Geotechnology in Rio de Janeiro (Insitutek) designed and deployed a network of 30 automatic rain gauges to send data to a computer in the central office of the Fundacão Instituto de Geotechnica do Municìpio do Rio de Janeiro (Geo-Rio). This central computer performs a series of automated evaluations and also provides access to the rainfall data on the Internet. Alarms would be relayed from Geo-Rio to emergency response agencies (police, fire, hospitals) and the general public through news media. Insitutek maintains a Web page describing the rain-gauge network and alarm system at www.insitutek.com.br/pluvi.htm (in Portuguese).

Torrential rainfall in the region in late December 2001 triggered severe flooding and many landslides with more than 60 fatalities across the state of Rio de Janeiro (Environmental News Service story, January 2002). Landslide alarms were issued for parts of the city of Rio de Janeiro, but the heaviest damage occurred in and around the city of Petropolis, about 60 km to the north.

Unfortunately, the Geo-Rio alarm system was shut down in December 2002 after a fire in its office building damaged its computers and other equipment. As of this writing (August 2003), I do not know the current operational status of the Geo-Rio alarm system.

### 17.10.4   Oregon (USA)

In addition to recommending landslide inventory mapping and timber-harvest restrictions in its emergency plan, the state of Oregon (USA) established a debris-flow warning system in 1997. The Oregon system is staffed by meteorologists from the Oregon Department of Forestry, geologists from the state geological survey (DOGAMI) and engineers from the Oregon Department of Transportation (ODOT). Advisories and warnings are broadcast over NOAA weather radio, and over the state Law Enforcement Data System. DOGAMI is charged with providing additional information on debris flows to the media and the interested public. ODOT is also responsible for warnings to motorists during high-risk periods, including installing warning signs along the three roadways judged to be at highest risk from debris flows. This is the only official, operational debris-flow warning system in the United States at the present time (spring 2002).

## References

Caine, N., 1980, The rainfall intensity-duration control of shallow landslides and debris flows, *Geografiska Annaler*, **62A**, 23–27.

Campbell, R.H., 1975, *Soil Slips, Debris Flows, and Rainstorms in the Santa Monica Mountains and Vicinity, Southern California*, US Geological Survey Professional Paper 851.

Cannon, S.H., 1988, Regional rainfall-threshold conditions for abundant debris-flow activity, in S.D. Ellen and G.F. Wieczorek (eds), *Landslides, Floods, and Marine Effects of the Storm of January 3–5, 1982, in the San Francisco Bay region, California*, US Geological Survey Professional Paper 1434, 35–42.

Cannon, S.H. and Ellen, S.D., 1985, Rainfall conditions for abundant debris avalanches, San Francisco Bay region, California, *California Geology*, **38**(12), 267–272.

Cannon, S.H. and Ellen, S.D., 1988, Rainfall that resulted in abundant debris-flow activity during the storm, in S.D. Ellen and G.F. Wieczorek (eds), *Landslides, Floods, and Marine Effects of the Storm of January 3–5, 1982, in the San Francisco Bay region*, California: US Geological Survey Professional Paper 1434, 27–34.

Chiu, Yu-Tzu, 2001, Japanese offer lessons in prevention of mudslides, Taipei Times Online, 3 August 2001, http://www.taipeitimes.com.

Cruden, D.M. and Varnes, D.J., 1996, Landslide types and processes, in A.R. Turner and R.L. Schuster (eds), *Landslides: Investigation and Mitigation*, National Academy of Sciences, Transportation Research Board, Special Report 247, 36–72.

Keefer, D.K., Wilson, R.C., Mark, R.K., Brabb, E.E., Brown, W.M., Ellen, S.D., Harp, E.L., Wieczorek, G.F., Alger, C.S. and Zatkin, R.S., 1987, Real-time landslide warning during heavy rainfall, *Science*, **238**, 921–925.

Oregon, 2000, Landslide and debris flow chapter, in State of Oregon, *Emergency Management Plan*, vol. 1, Natural Hazards Mitigation Plan: Salem, Oregon, USA, June 2000 (URL: www.osp.state.or.us/oem/oem/library/plans/nhmp%20 edited.pdf).

Rantz, S.E., 1971, *Precipitation Depth–Duration–Frequency Relations for the San Francisco Bay Region, California*, US Geological Survey, San Francisco Bay Region Environment and Resources Planning Study Basic Data Contribution 25.

Wieczorek, G.F., 1987, Effect of rainfall intensity and duration on debris flows on central Santa Cruz Mountains, California, in J.E. Costa and G.F. Wieczorek (eds), *Debris-Flows/Avalanches: Processes, Recognition, and Mitigation*, Geological Society of America, Reviews in Engineering Geology, vol. 7, 23–104.

Wilson, R.C., 1986, Estimating rainfall required to initiate debris flows, in Association of Engineering Geologists, 29th Annual Meeting, San Francisco, *Abstracts and Programs*, 69.

Wilson, R.C., 1997, Normalizing rainfall/debris-flow thresholds along the U.S. Pacific coast for long-term variations in precipitation climate, in C.-L. Chen (ed.), *Proceedings of the First International Conference on Debris-flow Hazards Mitigation*, Hydraulics Division, American Society of Civil Engineers, San Francisco, 7–9 August 1997, 32–43.

Wilson, R.C., 2000, Climatic variations in rainfall thresholds for debris-flow activity, in P. Claps and G.W. Wieczorek (eds), *Proceedings of the First Plinius Conference on Mediterranean Storms*, Maratea, Italy, European Geophysical Union, 14–16 October 1999, 415–424.

Wilson, R.C., 2002, Climatic influences on rainfall thresholds for debris flows – a search for mechanisms, in A. Mugnai, F. Guzzetti and G. Roth (eds), *Proceedings of the Second EGS Plinius Conference on Mediterranean Storms*, Siena, Italy, European Geophysical Union, 16–18 October 2000, 449–461.

Wilson, R.C., Mark, R.K. and Barbato, G., 1993, Operation of a real-time warning system for debris flows in the San Francisco Bay area, California, in H.W. Shen, S.T. Su, and F. Wen (eds), *Hydraulic Engineering '93: Proceedings of the 1993 Conference*, Hydraulics Division, American Society of Civil Engineers, San Francisco, CA, 25–30 July 1993, vol. 2, 1908–1913.

Wilson, R.C. and Wieczorek, G.F., 1995, Rainfall thresholds for the initiation of debris flows at La Honda, California, *Environmental & Engineering Geoscience*, **1**, 11–27.

# 18

# Reforestation Schemes to Manage Regional Landslide Risk

*Chris Phillips and Michael Marden*

## 18.1  Introduction

It is widely recognized that vegetation, particularly forest, contributes to improved slope stability and reduces the risk of shallow landslides. It seems a logical extension, therefore, that where a landslide risk exists, forest vegetation might be employed in some manner to reduce that risk. This chapter briefly reviews the role of vegetation in modifying those factors that contribute to landslide risk, using examples largely from New Zealand, and then chronicles efforts by the New Zealand government to manage landslide risk in the East Coast region of New Zealand using a large-scale regional forestry scheme. We discuss the benefits and disadvantages of this scheme, and conclude with a commentary on the role of such schemes in a wider international context.

## 18.2  Role of Vegetation in Modifying Landslide Processes

The role of vegetation in improving slope stability is well recognized, and comprehensive reviews may be found in several publications (e.g. Greenway, 1987; Phillips and Watson, 1994; Gray and Sotir, 1996). We consider the effectiveness of different vegetation covers for influencing landslide risk first by examining the functions that vegetation provides in either promoting or mitigating erosion processes. In general terms, vegetation influences slope stability through either hydrological or mechanical mechanisms. The hydrological mechanisms are those elements of the hydrological cycle that exist when vegetation is present. The mechanical factors arise from the physical interactions of either

---

*Landslide Hazard and Risk*   Edited by Thomas Glade, Malcolm Anderson and Michael J. Crozier
© 2004 John Wiley & Sons, Ltd   ISBN: 0-471-48663-9

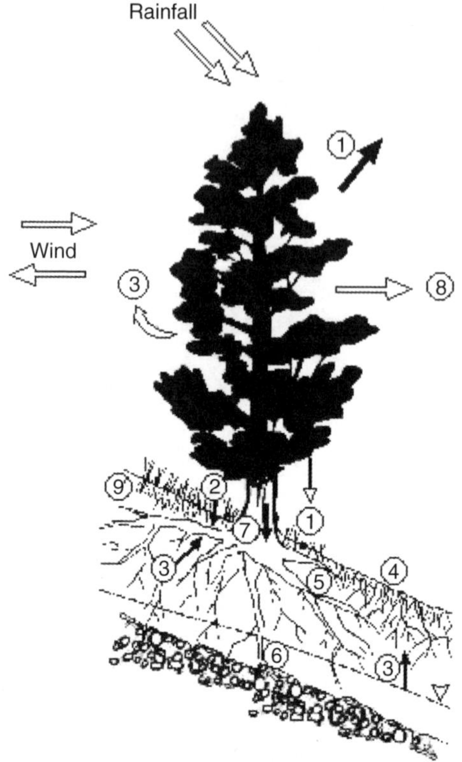



**Figure 18.1**    *Slope–vegetation interactions affecting slope stability (modified after Greenway, 1987). Refer to Table 18.1 for key to numbers. Reproduced with permission of John Wiley & Sons Ltd*

the canopy (e.g. wind effects) or root system of the plant with the slope (Figure 18.1; Greenway, 1987).

Vegetation influences can be classified either as beneficial or adverse to stability (Table 18.1, after Greenway, 1987). Hydrological mechanisms that lead to lower porewater pressures and soil moisture are beneficial, while those that yield higher soil water are adverse. Mechanical mechanisms that increase shear resistance in the slope are beneficial, while those that increase shear stress are adverse. In simple terms, the above-ground components of vegetation (canopy) reduce the ability of rainfall to cause slope failure through the processes of interception and evaporation while the below-ground components (roots) bind the soil and add strength.

## 18.2.1    Hydrological Factors

Rainfall on a vegetated slope is partly intercepted by the foliage, which leads to absorptive and evaporative losses of moisture that ultimately reduce the amount of rainfall available for infiltration. Water is considered 'lost' if it is evaporated back to the atmosphere or is otherwise held unavailable for infiltration. Interception losses are controlled by many factors, not the least of which is the amount of slope covered in vegetation, its type and

**Table 18.1**   *Effects of vegetation on slope stability (after Greenway, 1987)*

| Hydrological mechanisms | Influence |
|---|---|
| 1  Foliage intercepts rainfall, causing absorptive and evaporative losses that reduce rainfall available for infiltration. | B |
| 2  Roots and stems increase the roughness of the ground surface and the permeability of the soil, leading to increased infiltration capacity. | A |
| 3  Roots extract soil moisture from the soil, which is lost to the atmosphere via transpiration, leading to lower porewater pressures. | B |
| 4  Depletion of soil moisture may accentuate desiccation cracking in the soil, resulting in higher infiltration capacity. | A |

| Mechanical mechanisms | Influence |
|---|---|
| 5  Roots reinforce the soil, increasing soil shear strength. | B |
| 6  Tree roots may anchor into firm strata, providing support to the upslope soil mantle through buttressing and arching. | B |
| 7  Weight of trees surcharges the slope, increasing the normal and downhill force components. | A/B |
| 8  Vegetation exposed to the wind transmits dynamic forces into the slope. | A |
| 9  Roots bind soil particles at the ground surface, reducing their susceptibility to erosion. | B |

*Note*:
A – adverse to stability, B – beneficial to stability.
Reproduced with permission of John Wiley & Sons Ltd.

species. Generally, interception losses may involve a large proportion of gross rainfall in dense forests under certain circumstances. Most studies show a diminishing interception loss with increasing storm magnitude and intensity.

Increased permeability and infiltration capacity of the surface soil layers of vegetated slopes may be attributed to the presence of roots, vacant root channels and increased macroscopic surface roughness. The degree to which water can easily move into and within the soil is important in the development and dissipation of porewater pressures, which are a major contributor to landsliding.

The rate at which a plant consumes soil moisture depends on several factors, including the type, size and species of the vegetation; weather; climatic and seasonal factors; and site conditions such as rock type, soil depth and texture, slope aspect and steepness. Roots extract soil moisture within the root zone with the result that porewater pressures may be reduced. Capillary action induced by vegetation, particularly in porous and permeable soils, may reduce porewater pressures well below the root zone. This reduction in porewater pressure improves soil strength and slope stability. Soil moisture extracted by roots is ultimately lost to the atmosphere via transpiration from the foliage. In certain situations, prolonged extraction of moisture can lead to desiccation of the soil and to the formation of shrinkage cracks. Once formed, such cracks may permanently increase the permeability and infiltration capacity of the soil and correspondingly have an adverse effect on stability.

All these changes in water balance components result in a soil profile under forest being substantially drier for much of the year than it would be under grass cover or low-stature vegetation. Further, the period of high soil-water content under forest is often substantially shorter compared with that under grass. Conversely the period of soil moisture deficit is longer under forest.

## 18.2.2   Mechanical Factors

Roots, due to their tensile strength and frictional or adhesive properties, reinforce the soil. Large roots, particularly of trees, may penetrate deeply and become anchored in firm strata, thereby forming a support (buttress) to the soil mantle upslope of the tree. By binding soil particles at the ground surface, roots also reduce the rate of soil erosion that may otherwise lead to undercutting and instability of slopes. Slope surcharge by large trees increases both normal and downhill force components on potential slip surfaces. Vegetation exposed to wind transmits dynamic forces into the slope, and if uprooting or overturning occurs, both increased erosion and infiltration may result.

The contribution of roots to a site's stability increases as a function of the speed and ease at which the roots colonize the soil. The extent to which roots improve slope stability depends not only on the root content, but also the roots' material properties and morphology or architecture. A wide variety of complex root systems has been observed, and from a stability viewpoint these variations are highly significant (Greenway, 1987).

Vegetation roots are defined in several ways (Böhm, 1979; Phillips and Watson, 1994; Gray and Sotir, 1996). Taproots are the main vertical roots below the bole of the tree; lateral roots radiate from the central bole but more or less in the horizontal plane; sinker roots are more or less vertical roots arising from lateral roots. Three morphological types of tree root systems exist: taproot, heartroot and plateroot (Wilde, 1958). Inherent differences in patterns of root development are especially noticeable during early seedling growth and can significantly affect plant establishment and survival. Root morphology may be genetically controlled or modified by environmental and edaphic factors. Generally, the morphologies of root systems and individual roots are strongly determined by the physical soil conditions, particularly stoniness, drainage conditions, depth of water table, bedrock conditions or the strength and permeability of strata. Shallow plate-like root systems are typically found on steep landslide-prone slopes where shallow soils (<1 m deep) directly overlie bedrock (Marden *et al.*, 1991). The size of individual tree-root systems in dense stands can also be adversely affected by competition. Growth rates for an individual species are dependent on a number of variables. These include competition for nutrients, water, light and space, and various adverse environmental site conditions that might affect the growth of the tree. The latter may include persistently high water-table levels, and impenetrable barriers to roots such as bedrock at shallow depth or frozen soils. It is beyond the scope of this chapter to discuss differences in growth rates between individual species, although there is an extensive literature available on the growth performance of commercial tree species and the various factors that limit or promote productivity, both at the individual plant and stand scale (e.g. Van Kraayenoord and Hathaway, 1986).

Furthermore, the success of a particular vegetation strategy for managing landslide risk will be dependent on a number of factors. These include plant spacing, the time it takes until a closed canopy is formed, local site conditions such as the nature of slope materials, and the depth of the rooting zone relative to the potential landslide failure surface.

## 18.2.3   Soil Reinforcement and Slope Stability

A variety of studies of how roots reinforce soil have been made. These include laboratory shear tests of soils with roots (Waldron, 1977) or soils reinforced by fibres that simulate roots (Gray and Ohashi, 1983; Jewel and Wroth, 1987; Wu *et al.*, 1988b; Shewbridge and Sitar, 1989); *in situ* shear tests on soil blocks with roots (Endo and Tsuruta, 1969; O'Loughlin, 1974b; Ziemer, 1981; Abe and Iwamoto, 1988; Wu *et al.*, 1988a; Nilaweera, 1994; Ekanayake *et al.*, 1997; Wu and Watson 1998); and evaluation of root forces from slope failures (O'Loughlin, 1974a; Wu *et al.*, 1979; Reistenberg and Sovonick-Dunford, 1983; Reistenberg, 1994; Preston and Crozier, 1999). All these studies produced data that show increases in shear strength due to soil–root interaction.

Slope stability analyses of vegetated hillslopes show that the stress–strain behaviour of soils with roots is quite different to that of soils without roots (fallow soil). Results of *in situ*, direct, shear box tests (O'Loughlin and Ziemer, 1982; Ekanayake *et al.*, 1997) have shown that the increased peak shear stress of soils with roots produces a broader and flatter-shaped strength vs displacement curve compared to fallow soil (Figure 18.2). Such soils have the ability to undergo larger shear displacements before reaching failure conditions than soil without roots. It is therefore not only the *increased peak shear resistance* that contributes to improved hillslope stability, but also the *increased shear displacement* due to elasticity of the soil–root system (Ekanayake *et al.*, 1999). Shear box testing under saturated conditions of colluvial soils containing roots required three times more shear stress to cause failure than did the same soils without roots (Ekanayake *et al.*, 1999). Thus tree roots are considered to be a major contributor to soil strength and slope stability in the zone where roots are present (O'Loughlin, 1974a, b; O'Loughlin and Ziemer, 1982; Phillips and Watson, 1994).

Various studies have also used the tensile strength of tree roots as a measure of the contribution of vegetation to increasing a slope's stability and thereby reducing the incidence of landsliding (Gray and Sotir, 1996; Montgomery *et al.*, 2000). Most studies of root strength have measured tensile strength of individual tree roots, usually in a laboratory. Wide variations in tensile strength (tensile stress at failure) have been noted

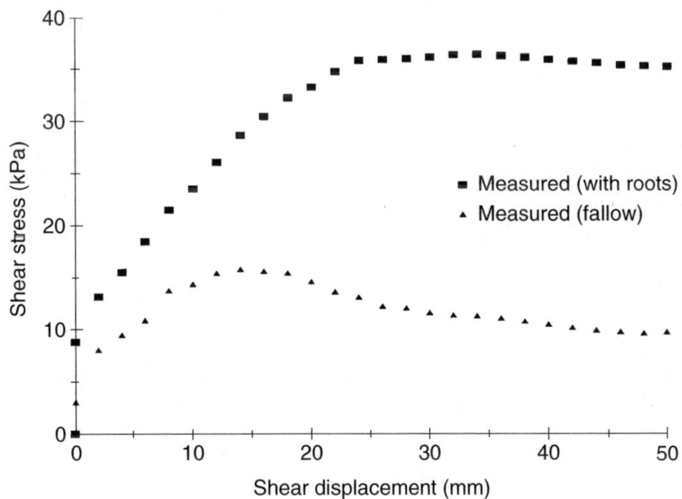


**Figure 18.2**  *Shear stress–shear displacement for soil with roots of 2- to 3-year-old radiata pine and soil without roots (Ekanayake and Phillips, 1999). Reproduced by the permission of NRC Research Press*

among species. Some roots have a high tensile strength (up to 70 MPa), about one-quarter the tensile strength of mild steel. Root tensile strength also varies with growing environment. The grain-size distribution of the growing medium has been shown to be important (Phillips and Watson, 1994). Roots growing in coarse material are weakened by kinks that occur along their length compared with roots growing in fine-grained soils, which are generally straight with a circular cross-section. In some trees growing on slopes, the uphill roots are generally stronger than the downhill roots. Several researchers have noted a decrease in tensile strength with increasing root diameter (Phillips and Watson, 1994).

Our observations of both fresh and old shallow landslides triggered under assumed saturation conditions also suggest that most roots are pulled out during failure rather than breaking in tension (Zhang *et al.*, 1993). This indicates that the actual material strength of the roots does not play a significant part in contributing to the peak shear strength of the soil–root system during shear failure (Coutts, 1983; Ekanayake and Phillips, 1999, 2002). The interactions at the soil–root interface are far more important in terms of the roots' contribution to soil shear strength. The transfer of root strength to enhance soil shear strength takes the form of a cohesive force. Several studies on the effect of vegetation on slope stabilization consider the root strength transfer as such a form of cohesive force (Sidle, 1991; Montgomery *et al.*, 2000). Therefore, the total cross-sectional area of roots at a potential failure plane in the soil, commonly referred to as the root area ratio (Abernethy and Rutherfurd, 1999), becomes important for determining their contribution to slope stability.

It may also be important to look at the decline in tensile strength of roots with time once the above-ground part of the tree or plant has been removed. Many field investigations in steep forested terrain worldwide have noted a two- to ten-fold increase in rates of mass erosion 3–15 years after timber harvesting (Bishop and Stevens, 1964; Endo and Tsuruta, 1969; O'Loughlin and Pearce, 1976; Megahan, 1978; Wu and Sidle, 1995; Jakob, 2000). This increase in landslide frequency and volume is related to the period of minimum root strength after clearcut harvesting and before substantial regeneration. The length of time between the death of the parent tree and the onset of root decay is species dependent. For example, for radiata pine (*Pinus radiata*), a softwood species, the time will be short, possibly a few weeks (O'Loughlin and Watson, 1979), but in more durable timbers it will be much greater. The mean rate of loss of tensile strength of a New Zealand indigenous shrubland forest species, kanuka (*Kunzea ericoides*), over a 4-year post-felling period was 4.2 MPa/year, about a third lower than that of radiata pine at 5.9 MPa/year (Watson *et al.*, 1995, 1999).

### 18.2.4   Landslide–Climate–Vegetation Relationships

Mass movements tend to occur after a triggering event, usually heavy or prolonged rainfall (Crozier *et al.*, 1980; Glade, 1998). Rainfall intensities known to have triggered shallow landslides in many parts of the world exceed about 25 mm/hour (Caine, 1980). Similarly, rainfall totals during triggering events vary widely but are generally in excess of 200 mm over a few days, at least for significant regional landsliding events in New Zealand soft-rock hill country (Page *et al.*, 1999; Reid and Page, 2002). However, whereas regionally specific thresholds have been identified, no universal minimum rainfall total can be identified as a 'threshold' above which damage occurs. Once a threshold has been surpassed, storm severity measured as the amount of area damaged is proportional to rainfall magnitude, and the same extent of damage requires an increase in storm magnitude – a process referred to as event resistance (Crozier, 1986). Thresholds will exist, but these will vary, and are dependent on such factors as the type of terrain (geology, soils and slope aspect and steepness), vegetation cover and type, antecedent moisture conditions, and rainfall duration and intensity during storms (Glade, 1998). While rainfall intensity and duration are the main triggering factors for shallow landslides and debris flows, prolonged wet periods may reactivate other mass movement features such as deep-seated slumps and earthslides and earthflows (Zhang *et al.*, 1993).

The probability of landsliding arises, then, from two main causes: the infrequent occurrence of intense rainstorms and the likelihood that such events occur when other predisposing factors are present, for example times when vegetative cover is absent or low, such as following forest harvesting. It is relatively easy to calculate probability of occurrence related to the first parameter if there are reliable and detailed rainfall records (e.g. Hicks, 1988; Page *et al.*, 1999). Empirical relationships are formulated between total storm rainfall and the density or number of landslides that can be attributed to that event (Page *et al.*, 1999); in that study the authors concluded that storm rainfall alone accounted for 60–90% of the variance in landsliding rates between storms. For the East Coast region of New Zealand, these authors indicated that when the results of all storms are combined, 50% of landslides are found to occur during storms with recurrence intervals less than 7 years, and 75% at recurrence intervals of less than 25 years.

In the same region of New Zealand, another approach used the frequency of flooding in a major river as a surrogate for landslide triggering, from which a probability relationship was derived (Kelliher *et al.*, 1995). For this region, they suggested that there is a 29% chance of at least one extreme landslide-triggering event every year, and a greater than 99% chance that one will occur every 10 years.

In terms of different vegetation types, rainfall-initiated shallow landslides are more likely for grassed hillslopes than for those with intact forest (Marden and Rowan, 1993). In general terms, grassed hillslopes, at least in New Zealand, have been shown to be between 4 and 16 times more susceptible to shallow landsliding than slopes with a mature standing-forest cover (Marden and Rowan, 1993). Studies in New Zealand have also shown that for forested areas there is little difference in the protective value between different forest types, for example indigenous forest or exotic plantation forests, but that forest age has a significant effect on the number of landslides (Marden and Rowan, 1993; Bergin *et al.*, 1995).

The connection between forest cover and initiation or reactivation of deep-seated mass movement is less clear than with shallow landslides; however, any land use or management practice that alters hydrological pathways and timing could potentially influence deep-seated mass movement activity (Sidle *et al.*, 1985; Phillips *et al.*, 1990). In general, though, the duration of rainfall events or the occurrence of long wet periods are contributory factors to the reactivation of deep-seated mass movements (Zhang *et al.*, 1993). With many creeping landslides such as earthflows or earthslides, the presence of roots can act as a reinforcing layer that possesses relatively high lateral (tensional) strength as well as enhanced compressional shear strength (O'Loughlin and Zhang, 1986). The lateral extent of this reinforced layer is highly dependent on tree density. It has been suggested that the roots of individual trees bind topsoil to form a small semi-rigid raft or block that 'floats' on the more plastic material beneath (Zhang *et al.*, 1993). Where tree density is high enough these individual rafts may be connected by long, interlocking, flexible lateral roots to form larger raft blocks constituting a reinforced surface layer.

### 18.2.5   Summary

Vegetation affects the stability of slopes in two major ways: first, by modifying components of the hydrological cycle largely through the processes of interception, evapotranspiration and infiltration; and second, by contributing strength to the soil via roots. The degree to which vegetation mitigates landslide risk is a function of many factors. These include the tree density, age, spatial cover of vegetation and the intensity and duration of rainfall events that exceed the local threshold conditions for landsliding at a regional scale.

## 18.3   Reforestation Schemes to Manage Landslide Risk: Case Study

### 18.3.1   Background to the Erosion Problem

Soil erosion in the East Coast region of New Zealand often related to shallow landsliding is unique in that its magnitude is greater than for any other part of the country. This region (Figure 18.3) comprises 7.8% of the North Island in area yet contains 26% of the land considered to be severely eroding (Bayfield and Meister, 1998). Approximately 57% of the East Coast region is severely erodible (Table 18.2) and the impacts from flooding, landsliding and sedimentation are considered greater than for other areas of New Zealand.

The geology of the East Coast region has been described as 'predisposed towards rapid or accelerated erosion' (Parliamentary Commissioner for the Environment, 1994). It comprises mudstones, sandstones and argillites, in a region that is characterized also by moderately steep terrain. In detail the geology is complex (Mazengarb and Speden, 2000). Broadly, older rocks of Cretaceous age (fractured argillites, greywackes and basalts) are dominant in the northern and western areas, while younger rocks of Tertiary age (soft mudstones, sandy mudstones, sandstones and limestones) are dominant to the south and east. In geological recent times (2000–60 000 years BP), the area has been covered with volcanic ash. While much has subsequently been eroded from unstable slopes, ash cover remains intact on more gentle stable slopes and alluvial terraces. On steep land where the underlying rocks are firm and stable, erosion is largely confined to the soil mantle itself and is manifest as sheet and rill erosion or as shallow landslides. However, where the underlying rocks are unstable, the accelerated erosion involves the basal rock itself, as evidenced by the numerous deeply eroding gullies and accompanying earthslides and rotational slumps.

The East Coast also has a high incidence of large-magnitude earthquakes. Ongoing rapid uplift and associated river downcutting, in conjunction with seismic shaking as a consequence of earthquakes, are mechanisms known to trigger slope instability, including shallow landslides, reactivation of earthslide movement, liquefaction and ground cracking.

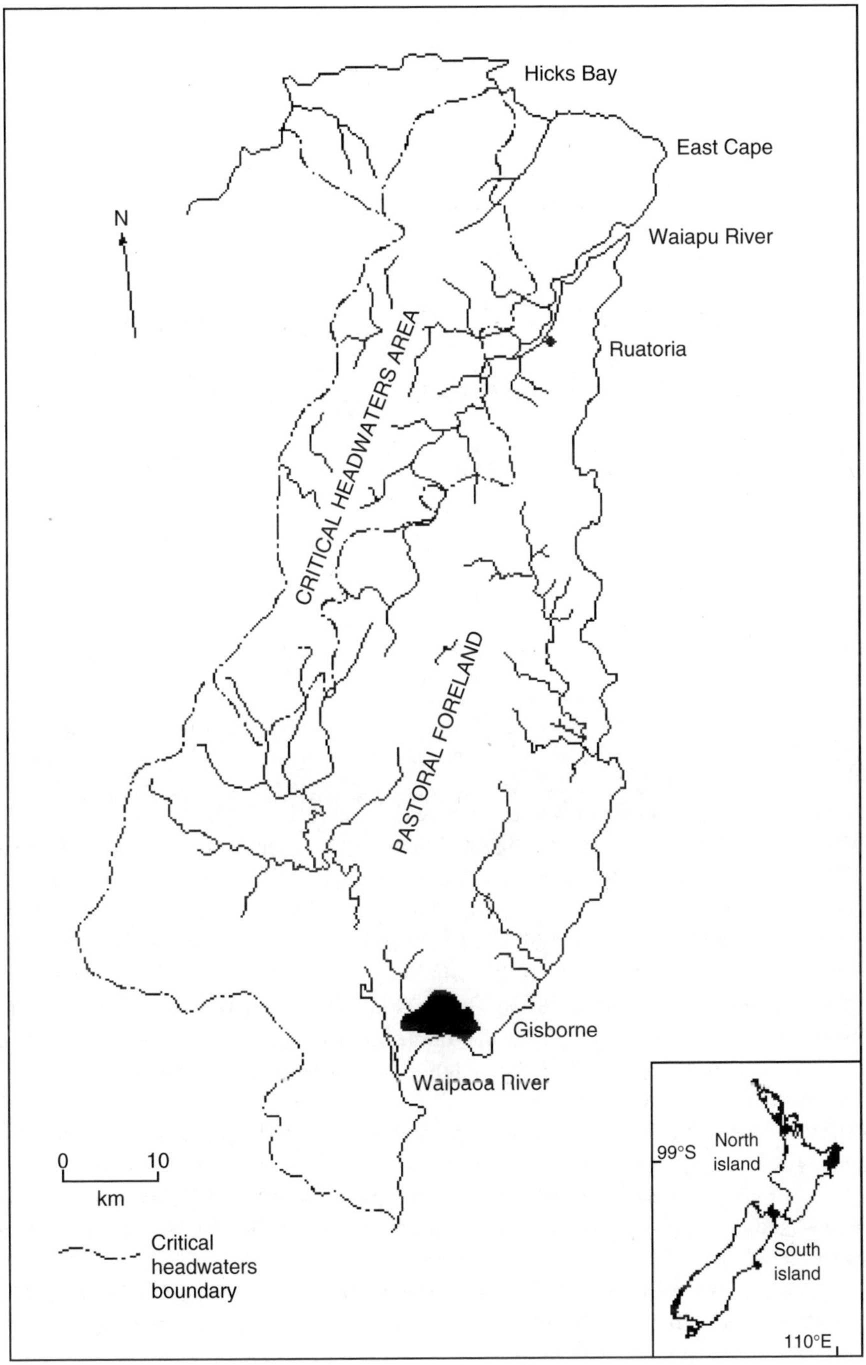

**Figure 18.3**  *Map of East Coast region, North Island, New Zealand*

**Table 18.2**   *Land area in different land use capability classes in the East Coast region 1:50 000 scale (after Jessen et al., 1999)*

| LUC class | Area of Gisborne East Coast region (%) | Comments |
|---|---|---|
| I | 56 km$^2$ (<1%) | Elite soils |
| II | 173 km$^2$ (2%) | Suitable for intensive cash cropping and a wide variety of horticultural crops |
| III | 500 km$^2$ (6%) | Suitable for cash cropping and horticulture, but versatility is reduced by imperfect to poor soil drainage (on floodplains) or erosion hazard (where sloping). Widely used for cash cropping and horticulture on the Poverty Bay flats |
| IV | 265 km$^2$ (3%) | Poor-quality arable land. Rarely cropped |
| V | 0 km$^2$ (0%) | The best non-arable land, without an erosion hazard. Not present in large enough areas to record at the NZLRI scale of mapping |
| VI | 2838 km$^2$ (32%) | Good non-arable land, and is recorded on the better hill country |
| VII | 4088 km$^2$ (46%) | Can be used for pastoral farming or forestry, but 98% has severe or worse erosion potential if used without suitable trees |
| VIII | 907 km$^2$ (10%) | No potential for primary production. Has special conservation/environmental values |

Reproduced by permission of Manaaki Whenua Press.

The region is susceptible to severe storm events and occasionally to cyclones, which contribute to the incidence of erosion. Although such storms are infrequent, they have resulted in extreme flooding on 29 separate occasions since 1900 (Kelliher *et al.*, 1995; Jessen *et al.*, 1999). Many of these flood events were associated with regional or widespread landsliding. For the 6000 km$^2$ of erosion-prone terrain in the region it is probable that a major flood-producing storm and associated extensive hillslope landsliding will occur once every decade (Kelliher *et al.*, 1995). The last major storm to have widespread regional landsliding was in March 1988 (Cyclone Bola). Since then, there have been several events of a similar magnitude to Cyclone Bola but with landsliding restricted to smaller local areas.

Concerns about the region's erosion problem have been long-standing. Various documents, commentaries, reviews and reports have discussed the nature of the problem and solutions developed to tackle it (Table 18.3). Some evidence suggests that the problems began during the early occupation by the indigenous people of New Zealand (Maori) (Wellman, 1962; Wilmshurst, 1997). The rate of erosion was, however, multiplied many times with the arrival of European settlers and their farming practices (1860s onwards) (Hill, 1895; Henderson and Ongley, 1920). Clearance of the indigenous forest by European settlers for pastoral production began in the 1880s and by the 1920s all but the highest and most rugged hill country had been cleared. Though some forested areas were logged, most were burnt, then oversown with introduced grass species. Such a drastic change in vegetation cover from forest to grazed pasture proved to be a powerful

**Table 18.3**  *Chronology of events leading to current reforestation scheme (East Coast Forestry Project 1993, 1994, 1997)*

| Date | Action | Comment |
|---|---|---|
| Pre-1850 | Vegetation clearance by Maori | |
| 1880s–1920 | Vegetation clearance by early Europeans | First signs of erosion becoming apparent |
| 1920s | Early geologists' warnings of erosion | |
| 1938 | Big floods and regional landsliding | Early documentation of problem |
| 1941 | Soil Conservation and Rivers Control Act was passed | Catchment boards were given the task to manage and prevent soil erosion on farmland |
| 1944 | Poverty Bay Catchment Board formed | First attempts to deal with the problem on farmland at a local scale |
| 1947 | Big floods and regional landsliding | |
| 1960 | First plantings by New Zealand Forest Service | First attempt at large-scale afforestation in this region |
| 1963 | Soil Conservation and Rivers Control Council appointed a technical committee of inquiry. The committee reported back in 1967 | First attempts to deal with the problem on farmland at a regional scale |
| 1967 and 1970 | East Coast Project – report back 1967 Formal publication 1970 | Taylor Report |
| Late 1970s | Reviews of East Coast Project | 'Red Report' plus an Economic report (Ministry of Works and Development) |
| 1979–85 | Small-scale farm afforestation known as 'Protection–Production Forestry Encouragement Grants' | NZFS grants for erosion control on small areas of marginal farmland |
| 1987 | Demise of New Zealand Forest Service | Major restructuring led to foreign ownership of ex-state forests |
| 1988 | Cyclone Bola | Largest landslide/flood event in living memory. Return period > 100 years. National disaster declared |
| 1988 | Parliamentary Commissioner for the Environment report | Inquiry into flood mitigation measures following Cyclone Bola |
| 1988–92 | East Coast Conservation Forestry Scheme | Forerunner to the East Coast Forestry Project but restricted to two southern catchments |
| 1993 | East Coast Forestry Project | Region-wide scheme |
| 1993 | Parliamentary Commissioner for the Environment review | Water and soil resource management on the East Coast |
| 1994 | Parliamentary Commissioner for the Environment review | Sustainable land management and the East Coast Forestry Project |

**Table 18.3**    *(Continued)*

| Date | Action | Comment |
|---|---|---|
| 1998 | East Coast Forestry Project review | Significant review by Bayfield and Meister (1998) resulted in changed objectives |
| 2000–2001 | East Coast Forestry Project 2000 | New singular objective |
| 2002 | East Coast Forestry Project 2002 | Revised guidelines and widening of treatment options |

predisposing cause of accelerated erosion. Slope stability declined as the root systems of the indigenous forest began to decay. With the removal of the forest, interception by the forest canopy ceased. Soil moisture levels subsequently increased and soils remained wetter for longer and were thus more prone to mass wasting.

In summary, uplift, relief, rainfall and vegetation clearance all contribute to high rates of total erosion (denudation), giving this region the distinction of being one of the most erosion-prone regions of New Zealand and in the world (Williams, 1980). Estimated annual rates of denudation for hill country typical of the East Coast region are from 150 to 500 $m^3/km^2$ (Crozier *et al.*, 1992).

### 18.3.2    Chronology of Attempts to Control the Problem

By the early 1940s it became apparent that the rate of soil erosion and river aggradation on land in pasture in the East Coast region and also in other parts of New Zealand was far greater than on areas still under native forest. In 1941 the Soil Conservation and Rivers Control Act was passed and catchment boards were given the task to manage and prevent soil erosion (Table 18.3). In 1944, largely in response to growing public concern at the effects that aggradation from the Waipaoa River, one of the major rivers on east coast New Zealand, was having on the productive and fertile Poverty Bay flats and on Gisborne City, the Poverty Bay Catchment Board was formed.

By 1948, the Board had established a large-scale reforestation trial using a variety of tree species to mitigate the effects of erosion. The trial demonstrated that once a canopy had formed, runoff was considerably reduced and the advance of eroding gullies was slowed. In contrast, attempts at stabilizing riverbeds and gullies by check-dams or tree planting failed as these became rapidly overwhelmed with sediment.

In 1953, the Poverty Bay Catchment Board undertook a major flood control scheme in the Waipaoa River to protect the Poverty Bay floodplain, but it was soon realized that this would not succeed unless land degradation in the catchment was arrested and erosion controlled. At the request of the Board, the Soil Conservation and Rivers Control Council in 1955 set up a special committee to report on remedial measures.

The Council, largely on the basis of the success of the early planting trials, particularly of conifers, then urged the New Zealand government to purchase eroding farmland to establish dual-purpose exotic forests, for protection against erosion and for production of timber. In 1959, Cabinet approved an erosion control scheme in which the New Zealand Forest Service (NZFS) was to acquire and reforest 7000 ha of the most severely eroding country in the upper Waipaoa River catchment. Planting began in 1960 in what was to become known as Mangatu Forest, which ultimately became the nucleus of a much larger forest stretching from Mangatu north to Hicks Bay (Figure 18.3). The estimated cost of the scheme at the time was NZ£1 240 000. However, the radical new policy advocated by the committee for the conversion of a large block of farmland back to forest did not meet with immediate acceptance. Reluctance to make the change by many in the community, particularly farmers in the worst eroding areas, was understandable because, although the land was in general unstable, it was very fertile where not badly affected by erosion (Allsop, 1973).

### 18.3.3 East Coast Project (1968–87)

In 1963 the Government established a Committee of Inquiry under the Soil Conservation and Rivers Control Act (1941) 'to inquire into the conservation problems of Poverty Bay–East Cape (an area covering some 600 000 hectares) and to make recommendations on a comprehensive control programme' (Committee of Inquiry, 1970, p. 1). In 1967, the Committee prepared a report entitled *Wise Land Use and Community Development*, known widely as the Taylor Report, published in 1970 (Table 18.3). This report became the baseline document for an understanding of the relationship between land use and erosion in the region. The report acknowledged that the immense erosion problems were beyond the resources of the farming community and required substantial taxpayer help. The Committee stated: 'Where such erosion has developed on a grand scale the cost of erosion control would far exceed the value of pastoral production from the land concerned. There is, therefore, no known way of economically controlling erosion of this kind other than by complete afforestation of the catchment concerned' (Committee of Inquiry, 1970, p. 5). In an effort to identify those areas that would need to be retired from pastoral production and that were in most danger of eroding (the 'critical headwaters'), the Committee introduced a notional 'Blue Line' on the maps accompanying the report separating this area from land where the erosion problem was not as great (the 'pastoral foreland'). The report prescribed conservation farming for the pastoral foreland and complete afforestation for the critical headwaters area.

Following receipt of the Taylor Report, the New Zealand government approved, in May 1968, a programme for the NZFS to progressively plant the unforested parts of the critical headwaters with dual-purpose protection/production forests. This became known as the 'East Coast Project'. Plantings by the NZFS continued up to its corporatization in March 1987. At no stage was there any review to determine if the costs of the work were consistent with those estimated by the Taylor Report. In addition, money for on-farm soil conservation works (non-blanket forestry) was made available via government subsidy through soil conservation programmes approved by the Soil Conservation and Rivers Control Council (SCRCC), or its successor, the National Water and Soil Conservation Authority (NWASCA). The individual property owner paid the balance not covered by the subsidy.

Following widespread dissatisfaction with the 'Blue Line' concept, the Poverty Bay Catchment Board (renamed then to East Cape Catchment Board) in 1977–78 investigated whether modifications were required for implementing the recommendations of the Taylor Report. Their report, which became known as the 'Red Report', found 'the simple decisive zoning between the Pastoral Foreland and the Critical Headwaters Area ... had served its purpose and is now recognized as being too general' (Poverty Bay Catchment Board, 1978). The Red Report categorized the land for uses according to its physical characteristics and susceptibility to erosion. The four categories proposed by the Catchment Board (with various subcategories) were seen as a better approach to formulating erosion priorities than the two categories of the Taylor Report. Nevertheless, this report reaffirmed the major principles presented in the Taylor Report, including the continuation of the afforestation programmes. The NZFS, however, continued its planting programme according to the Blue Line concept.

In December 1986, the then Minister of Works and Development presented a Cabinet paper pointing out that the planned disestablishment of the NZFS would have a profound effect on the East Coast Project. The establishment of the new Forestry Corporation, with its primarily commercial objectives, would mean that the East Coast Project would not continue unless the Corporation were subsidized by government to do so. The paper proposed that interim funding be provided for continued conservation forest planting. Cabinet, on 15 December 1986, directed officials to undertake a review of the East Coast Project; approved NZ\$511 000 in 1986/87 to NZFS for land preparation of 1725 ha of land; and approved the expenditure of up to NZ\$100 000 as a subsidy payment to the Forestry Corporation for the continuation of the East Coast Project in 1987/88 only. The report was released in December 1987 (Ministry of Works and Development, 1987).

The NZFS was corporatized on 31 March 1987. Some 36 100 ha of land, of which 25 800 ha was severely eroded or erodible, had been acquired and planted by the NZFS as part of the East Coast Project. In addition the NZFS established a further 1809 ha of protection/production forest on farmland by way of forest encouragement grants. The total cost of afforestation programmes from 1968 to 1987 was estimated (in 1987 NZ dollars) at \$229 million, of which \$92 million were costs spent directly on planting, thinning and pruning. If the project had continued as envisaged, a further 75 000 ha of erosion-prone land would have been planted in conservation forests at an estimated subsidized cost of \$300–550 per ha (total cost \$22.5–41.25 million) (Ministry of Works and Development, 1987).

In addition, by March 1987, on-farm soil conservation works implemented by the East Cape Catchment Board, with financial support from SCRCC or NWASCA, were undertaken on parts of farms covering a total area of 28 200 ha at a cost (in 1987 NZ dollars) of \$14.3 million, with an overall government subsidy of \$9.2 million (64%). Full implementation of this programme as envisaged on all affected lands would have involved treatment of a further 110 000 ha at an estimated total cost of \$97 million (Ministry of Works and Development, 1987).

### 18.3.4   Cyclone Bola and the East Coast Conservation Forestry Scheme

The review of the East Coast Project (Ministry of Works and Development, 1987) was presented to the Cabinet Committee on State Owned Enterprises on 3 February 1988. The Committee invited the Minister of Regional Development to establish an Officials' Committee, which would report by 31 July 1988 on the review, submissions made on the review, and the future of the East Coast Project.

In spite of extensive afforestation, the erosion/landslide hazard was graphically illustrated in 1988 when Cyclone Bola struck the region. This subtropical cyclone moved across northern New Zealand between 6 and 9 March 1988, bringing torrential rainfall to the East Coast region and causing widespread severe landsliding, erosion, flooding and siltation. During Cyclone Bola, 900 mm of rain fell in only 72 hours (Trotter, 1988). Flood peaks were estimated to have a return period greater than 100 years and 24-hour rainfalls across the region were estimated in the 45–80-year return period. Pastoral land was particularly damaged. Some pasture slopes lost 70% or more of their grass cover to shallow landslides (Figure 18.4). However, on hillslopes protected by mature native forest and older pine forest, landslides were less frequent. In general terms, pastoral hillslopes were shown to be between 4 and 16 times more susceptible to landsliding than slopes with a mature standing-forest cover affected by the same storm event (Marden and Rowan, 1993).

Forest age had a significant effect on the amount of damage to first-rotation exotic plantation forests (i.e. planted on pasture land) (Phillips *et al.*, 1990; Hicks, 1991; Marden *et al.*, 1991; Marden and Rowan, 1993; Marden *et al.*, 1995) and to regenerating indigenous scrub (Bergin *et al.*, 1993, 1995) – the impact declining with increasing age. For example, landslide densities were generally highest in stands over 6 years old (0.49–0.62/ha), intermediate in 6- to 8-year-old stands (0.16–0.21/ha) and least in stands older than 8 years (0.04–0.06/ha). Thus forest stands older than 8 years sustained ten

**Figure 18.4**  *Pastoral land in the East Coast region of New Zealand showing widespread landslide damage as a result of Cyclone Bola in March 1985*

times less damage than did stands younger than 6 years old (Phillips *et al.*, 1990; Marden and Rowan, 1993).

Damage caused by Bola was so great that the government provided a short-term disaster-relief package to help speed recovery and tried to facilitate major land use change. Some 19 months and $111 million later, a review of the Farm Assistance Fund (which accounted for $50 million of the total relief package) was to conclude that the Fund had achieved only limited success and the work carried out would do nothing to prevent similar disasters recurring (McLean, 1994).

In 1988, the government agreed to provide $8 million to directly subsidize a new East Coast Conservation Forestry Scheme. The funding was to be spread over 5 years (1989–93), and was aimed at establishing about 3000 ha of protection forest per year. The government also agreed that the funding would be provided as a subsidy covering two-thirds of the cost of establishment and that the remaining one-third should be met by the region through the East Cape Catchment Board. Oriented purely towards protection forestry, subsidies did not cover ongoing pruning or thinning costs. The government also insisted that all protection forests would come under covenants that precluded logging for at least 25 years after planting, and then only with the permission of the local catchment authority.

The main factors influencing the government to invest in such a scheme included: (1) the real extent of severe erosion being much greater than in other parts of New Zealand and its substantial negative impact on the region's social and economic development; (2) the need to carry out erosion control quickly and comprehensively to reduce future costs of erosion and flood damage, both to the region and to government; and (3) the lack of money and resources within the region for carrying out a comprehensive erosion control scheme (O'Loughlin, 1991).

The East Cape Catchment Board and later the Gisborne District Council administered the East Coast Conservation Forestry Scheme. This scheme was phased out in 1993 but by that time it had established 13 578 ha of forest over the 5 years. The scheme cost nearly $11 million, comprising costs to the government of $7.2 million, to Gisborne District Council of $3 million, and to the landowner $0.5 million.

### 18.3.5  East Coast Forestry Project (ECFP)

The East Coast Forestry Project (ECFP) was announced in the 1992 budget. At that time, 195 500 ha of Class VII land (see Table 18.2) across the whole region was assessed as needing protection (Parliamentary Commissioner for the Environment, 1993). This was the first attempt to tackle the erosion problem across the whole region, unlike earlier schemes that were targeted at specific catchments. The first plantings took place in 1993.

The ECFP and all the preceding schemes were based on a desire to bring about a long-term solution to a severe erosion problem. The overriding aim was to bring about sustainable land use, through either a change in land use (afforestation), soil conservation measures, or retirement and/or soil conservation planting (Bayfield and Meister, 1998).

The ECFP is administered under the Forestry (East Coast) Grant Regulations established in 1992. A major policy change resulted in a new set of regulations in 2000. These stated that the region-wide ECFP aimed to plant 200 000 ha over 28 years (1992–2020). The targets set were a planting rate of 7000 ha per year and a maximum annual budget of $6.5 million.

The region has long had one of the highest unemployment rates in New Zealand, and the terms of reference during the formative years of the ECFP were extended to include the creation of employment and other social benefits through commercial forestry (Cocklin and Wall, 1997). However, no explicit targets for employment or regional development were set. In addition, an explicit interest in promoting opportunities for Maori on the East Coast was stated by the government. This region is also distinguished by having the highest proportional representation of people claiming Máori descent of any region in New Zealand.

Amendments and modifications to the project have occurred since its inception as a result of various reviews, and these have been included in updated guidelines (Ministry of Agriculture and Forestry, 2000, 2001, 2002). These include changes to the nature and description of target land, the range of species that could be planted, the clearance (or not) of indigenous scrub forest (manuka (*Leptospermum scoparium*) and kanuka (*Kunzea ericoides*), inclusion of farm gully planting, and the objective of nature conservation on individual properties was added (Bayfield and Meister, 1998).

In the most recent review, Bayfield and Meister (1998) concluded that if the ECFP was to continue either for the full-intended term or as a strategic wind-down, some modifications needed to be made. They recommended that a new goal and objective be adopted for clarity of purpose. In redefining the goal they concentrated on the major purpose for government intervention – soil erosion control. They acknowledged that although this primary focus denied the importance of employment, regional development, nature conservation and Maori development, these would follow from achieving the overriding goal. They suggested a revised goal, which has continued through to the present: 'To achieve sustainable land management on 60,000 hectares in the East Coast Region on severely erodible land by changing land use' (Bayfield and Meister, 1998, executive summary). They recognized that commercial forestry was only one of the instruments that could be used to achieve the objective, but considered it was likely to be the main one.

The minimum area eligible for tender is 5 ha for all erosion control treatments in the current ECFP. The approved grant areas may also consist of one block or a number of different blocks. For reversion treatments, the minimum areas considered are 2 ha within a forestry treatment and 5 ha if the surrounding land is farmed. A minimum of 2 ha is required for a single gully under the farm gully option. Therefore several gullies may be required to be submitted to achieve the overall 5 ha minimum.

Submitting an application does not automatically mean a grant will be approved. The ECFP operates through a tender process and applicants are required to submit a competitive tender. All applications are weighted and ranked to determine which give the best value for money.

In the 11 years since the ECFP started, 29 965 ha of severely eroding land have been funded and planted with a further 2937 ha of unfunded land planted. In addition, 7181 ha was set aside as reserves (Table 18.4). Thus 52% of the total of 77 000 ha (i.e. 11 years times 7000 ha/y) of land has been treated (includes reserves).

### 18.3.6   Benefits of the ECFP and Earlier Schemes

International estimates of soil erosion damage in recent decades have indicated that off-site damage may be greater than that on site (Fox *et al.*, 1995; Krausse *et al.*, 2001). For New Zealand, Krausse *et al.* (2001) conservatively estimated the damage caused by

**Table 18.4**  *Planting achievements since the East Coast Forestry Project began (R. Hambling ECFP Manager, pers. comm.)*

| Year | Planting year | Grant (ha) | Non-qualifying area | Total planted (ha) | Informal reserves (ha) |
|---|---|---|---|---|---|
| 1 | 1993 | 1 910.7 | 1 021.8 | 2 932.5 | 753.7 |
| 2 | 1994 | 2 967.5 | 696.1 | 3 663.6 | 387.9 |
| 3 | 1995 | 2 497.0 | 55.1 | 2 552.1 | 224.7 |
| 4 | 1996 | 4 763.8 | 335.1 | 5 098.9 | 1 009.9 |
| 5 | 1997 | 4 222.7 | 456.2 | 4 678.9 | 903.7 |
| 6 | 1998 | 3 406.4 | 168.9 | 3 575.3 | 557.1 |
| 7 | 1999 | 3 724.9 | 117.0 | 3 841.9 | 1 004.2 |
| 8 | 2000 | 2 442.7 | 63.3 | 2 506.0 | 797.7 |
| 9 | 2001 | 931.9 | 23.7 | 955.6 | 204.4 |
| 10 | 2002 | 2 129.1 | 0.0 | 2 129.1 | 1 337.7 |
| 11 | 2003 | 968.2 | 0.0 | 668.2 | Not assessed |
| Total | | 29 964.9 | 2 937.2 | 32 902.1 | 7 181.0 |

*Notes:*
Grant = area established under project.
Non-qualifying area = additional areas not paid for by the ECFP but established by the tenderer at their cost (additional benefit).
Informal reserves = parcels of indigenous bush or standing scrub excluded by ECFP, i.e. not allowed to be clearfelled, are not covenanted or fenced and do not receive grants.
Reproduced by permission of Ministry of Agriculture and Forestry.

erosion, both on site and off site, to be $126 million annually. In contrast, expenditure on preventing additional erosion is approximately $26 million. The authors highlight the implications of the demise of centralized funding for soil and water conservation through the National Water and Soil Conservation Organisation and the marked decline in direct government expenditure in this area. They cite the East Coast Forestry Project as one of only two areas of expenditure directly related to soil erosion and conservation. The other was research, but it too has declined in the last 10 years.

Measures used to control erosion are likely to have greater off-site than on-site benefits. Many of the benefits of the ECFP come in the form of intangible values (environmental and social), which are difficult to isolate and measure (Bayfield and Meister, 1998). It is unclear, at least as far as the ECFP is concerned, whether the benefits outweigh the costs, as it has always been understood that the benefits from soil erosion control take a long time to eventuate and are hard to quantify (Bayfield and Meister, 1998). Benefits arising from the ECFP are expanded below but may be divided into:

- On-site benefits:

    - on-site soil retention
    - reduced need for replacing fencing and other infrastructure affected by continued erosion
    - income from wood production.

- Off-site benefits:

    - saved damages of reduced flooding
    - saved maintenance costs of river and marine environment

– increased environmental quality of river water and marine environment
– multiplier effect of forestry estate development activities and value-adding processing (regional development)
– employment creation in forestry and associated industries
– social benefits
– greenhouse gas absorption
– clean and green image for promotion of products from New Zealand in the global marketplace.

These benefits may accrue to the individual landowner, the region or the nation. In terms of overall impact, the ECFP has generally benefited the region and the nation (Bayfield and Meister, 1998).

### 18.3.6.1   Expansion of the exotic forest estate

The establishment of the ECFP coincided with the significant increase in afforestation throughout New Zealand that followed the restructuring of the forest sector in 1987 and which is now commonly referred to as the 'third planting boom' (the first two periods of extensive afforestation were in the 1920s–30s and the 1970s–80s (Roche, 1996)).

Since reforestation started in the early 1960s, progress towards dealing with the problem has been incremental as the various schemes outlined above have come and gone. In the East Coast region, there is now about 25% in reforested land (153 000 ha). However, only about one-half of this area was planted with government assistance and is on land that had been identified, at least at a regional scale, in need of blanket planting because it was severely eroding (Table 18.5).

### 18.3.6.2   Erosion activity

The on-site benefits of afforestation include a marked reduction in shallow landsliding (Pearce *et al.*, 1987; Phillips *et al.*, 1990; Marden *et al.*, 1995), substantially reduced rates of earthflow movement (Phillips *et al.*, 1990; Zhang *et al.*, 1993), and cessation of gullying processes (Bayfield and Meister, 1998; Phillips *et al.*, 2000). One reason for the success of reforestation on the East Coast relates to the higher than normal density of planting. Over the various schemes the planting density has varied between 2500 and 1250 stems/ha. Stocking density affects the timing of canopy closure. As vegetated

**Table 18.5**  *Approximate reforested areas from all sources for East Coast region as at end of 2000 planting season*

| Agent/project | Area (ha) |
| --- | --- |
| New Zealand Forest Service (NZFS) | 36 100 |
| East Cape Catchment Board | 2 017 |
| NZFS protection/production grants | 1 809 |
| East Coast Conservation Forestry Scheme | 13 578 |
| East Coast Forestry Project | 32 902 |
| Private (non-government-assisted) | 70 851 |
| Total | 157 257 |

slopes, before canopy closure, are 'at risk' of damage from landsliding, particularly during heavy rainfall in storms such as Cyclone Bola, the timing of significant reductions in landslide vulnerability coincides with the approximate timing of canopy closure. Hence the importance of stocking densities that produce earlier canopy closure, and thus an earlier reduction in landslide risk.

### 18.3.6.3   Sediment delivery to marine and freshwater ecosystems

Off-site benefits are often more difficult to quantify, particularly those that relate to sediment transport and water quality. Much of the material delivered to valley floors by active mass-wasting processes is often reworked by fluvial action for many years even though the supply of new material is cut off with afforestation. Thus there can be an indeterminate lag of decades to centuries between the change in on-site benefits and the accrual of downstream benefits.

In a recent study, Reid and Page (2002) illustrated that a 50% reduction in landslide-derived sediment to the Waipaoa River could be achieved through reforestation of 12% of the Waipaoa catchment if land were prioritized on the basis of landslide susceptibility, whereas random selection would require 30% of the catchment to be converted for the same result. Given the increase in reforestation throughout the region since 1988, Page *et al.* (1999) suggest that if Cyclone Bola struck today there would be a 15% decrease in landslide-derived sediment than in that event.

Other off-site benefits of reforestation include the impacts of reduced sediment on the quality of the marine and freshwater environments. No detailed data are available to describe or measure sedimentation impacts, but following Cyclone Bola in March 1988 reef communities of the inner shelf were temporarily inundated by a fluid mud layer (Battershill, 1993; Foster and Carter, 1997). The long-term effects on marine benthic communities and especially spawning grounds are largely unknown, although there is anecdotal evidence about the effect of siltation on kai moana (Maori food source). It would be logical, therefore, to expect that a reduction in total sediment load brought about by widespread reforestation would have some direct benefits to both freshwater and marine habitats.

### Flood protection

One of the key off-site benefits of the reforestation scheme is its contribution to reducing the effects of future floods on the Poverty Bay floodplain (and of flooding in the other rivers of the region). The Waipaoa River Flood Control Scheme at present provides a '1 in 100 year' level of protection. Some works were lost as a result of Cyclone Bola in 1988 and it is proposed to raise stopbanks and construct floodways between 2002 and 2008 to increase the level of protection. The scheme will gradually lose capacity through aggradation of the channel and deposition within the stopbanks, and may need further upgrading around the year 2040. Once all structural works are completed on the floodway there will be no more hydraulic efficiencies to be achieved and hence it will not be possible to slow the loss of scheme capacity by this method (Peacock, 1998). By then, the only way to achieve this will be through afforestation in the headwaters, the most cost-efficient method being the targeting of riparian areas and the most landslide-susceptible parts of the landscape (Page *et al.*, 2000).

### 18.3.6.4   Other environmental benefits

An additional benefit of the ECFP relates to the Kyoto Protocol, whereby countries are required to deal with their greenhouse gas emissions. In 2000, it was estimated that East Coast forests would absorb 3% of the national (1990 level) $CO_2$ emissions (Bayfield and Meister, 1998). Further, New Zealand has used its clean and green image (a clean environment and a sustainable agriculture) in promoting itself in the world trading environment. The direct intervention of government in the ECFP shows that it is serious about dealing with a difficult erosion problem and achieving sustainable land management.

The ECFP has also benefited nature conservation. Through the preparation of tender documents and implementation of approved grants, areas of indigenous forest or significant scrublands become protected. In 1998 it was estimated that 3280 ha of such land had been set aside for nature conservation (Bayfield and Meister, 1998). However, this came at a cost of 1360 ha of closed canopy scrub being cleared for forest planting between 1993 and 1997 in the early phases of the ECFP. Such clearance was declared ineligible for funding from around 2000. Instead, in the latest guidelines for the ECFP, reversion or regeneration of indigenous scrub/forest on pastoral land is eligible as one of the treatment options. The key difference is that a covenant must be placed on such land to prevent any future vegetation removal. As a consequence, the area set aside as conservation reserves had grown to 5320 ha in February 2002 (Table 18.4).

### 18.3.6.5   Economic and social effects

Restructuring of the forest industry in New Zealand led to a massive downsizing of government involvement in forestry (Aldwell and Roche, 1992; Roche, 1996). Through the late 1980s and early 1990s the harvesting rights to a large proportion of the forest assets developed by the state were sold to private investors. Because government had at the same time increased the opportunities for foreign investment, much of the forest resource was sold to offshore interests. Investment in the ECFP, however, was/is in keeping with the contemporary national trend involving smaller-scale private investors (including farmers), rather than the large-scale corporate investment in planting that was more common in the earlier planting boom periods.

Even so, the ECFP stands out in New Zealand's contemporary political economy (Cocklin and Wall, 1997). The government subsidization of forest establishment under this scheme sat uneasily amidst other reforms of the forest sector and land-based production at that time. Grants under the ECFP were/are justified in terms of 'levelling the playing field' (Parliamentary Commissioner for the Environment, 1994). This document stated:

> The ECFP grant is intended to 'top-up' the difference between the private rate of return for a commercial forestry investment on eroding and erodible Class VII land in the East Coast region in comparison to the private rate of return for a typical commercial forestry investment in other regions. The commercial rate of return for forestry on this at-risk land is low compared to other regions because of the physical and economic constraints of the East Coast region. (Parliamentary Commissioner for the Environment, 1994: 27)

In the entrepreneurial and market-led economy of New Zealand, this amounted to a rare admission that the market could not adequately internalize the social and environmental objectives required for the East Coast region (Cocklin and Wall, 1997). The reasons for this exception relate to the persistent concern expressed to government over successive decades of the acute susceptibility of this land to erosion, and the relative economic disadvantage of the region.

As a natural follow-on, the region has also experienced investment in the processing of some of the production being harvested from the earliest plantings in the 1960s and 1970s. However, a large proportion of this production is exported as logs for subsequent processing either outside the region or overseas. Development of local processing not only gives regional development benefits, but also provides a stable ongoing market for logs from local growers.

Forestry has contributed to the overall regional development of the Gisborne District. Surveys of employment in forestry were carried out in 1992 and again in 1993 (Wall and Cocklin, 1994) but it is difficult to separate the contribution of the ECFP from that of other 'corporate' forestry. Gisborne District Council in 1997 stated that forestry in the Gisborne District was the fastest growing industry, with employment rising from 251 FTE in 1993 to 460 in 1994 and 540 in 1996. Fairweather *et al.* (2000), however, indicated that the number of FTE employed in forestry in the East Coast region in 1996 was 747, having dropped from 807 in 1986–87 when the NZFS was restructured.

A 1996 survey of the East Coast silviculture workforce concluded: 'the objective of the ECFP in creating employment in the region appears to have been successful, with a larger percentage of locals than non-locals being employed' (Cummins and Byers, 1997, p. 27). Although no current data exist, there now appears to be a higher percentage of non-locals than locals employed in forestry in the region, particularly as a result of the more specialized skills required for harvesting (Tomlinson *et al.*, 2000).

The economic impacts of forestry development in the East Coast region as a result of the ECFP were assessed by Butcher and Partners (Ministry of Forestry, 1996), who took a conversion of 112 000 ha from farmland to forestry as their base scenario. Their analysis revealed a decline in farming but an increase in both employment and economic returns from the forestry development (Table 18.6). The greatest benefits would accrue if processing were to be undertaken within the region.

While the larger forestry companies, particularly those with international ownership, have been perceived to lack interest in the community's welfare, in practice this has

**Table 18.6**   *Predicted economic impacts of the East Coast Forestry Project (from Bayfield and Meister, 1998, Table 3.2, p. 14)*

|  | Gross household income ($ million) | Employment (FTEs) | Value added ($ million) |
|---|---|---|---|
| Farming | −8 | −230 | −11 |
| Forestry | 97 | 2 460 | 129 |
| Forestry + Processing | 135 | 3 980 | 215 |
| Total ECFP impact at Year 30 |  |  |  |
| Without processing | 90 | 2 200 | 120 |
| With processing | 130 | 3 800 | 200 |

Reproduced by permission of Ministry of Agriculture and Forestry.

not proved to be the case. For example, forestry companies have offered scholarships to schools for secondary education (Tomlinson *et al.*, 2000).

When the ECFP was initially developed, Maori development was considered an explicit objective, though this was dropped after the review by Bayfield and Meister (1998). Cummins and Byers (1997) noted indications that Maori were benefiting from the ECFP and were encouraged to find that 58% of those employed were local Maori. In addition, 33% of ECFP grant approvals, as assessed in 1998, were on Maori-owned land. Joint venture arrangements between Maori landowners and foreign investors seem to be having some significant benefits for the local Maori. Also, scholarships for tertiary education have benefited younger Maori.

The impact of forestry investment on land prices in the early part of the ECFP caused increases in land value. However, land was generally regarded as having become considerably undervalued in the region (Wall and Cocklin, 1994), partly as a reaction to the devastation caused by Cyclone Bola. The upward movement at that time might therefore be seen in part as a natural market correction.

### 18.3.7   Downsides of the ECFP and Earlier Schemes

#### 18.3.7.1   Maori interests

A range of external factors unrelated to the ECFP are impediments to Maori participation in forestry development. For example, because much Maori land is covered in closed canopy scrub, it has been difficult for Maori to find a joint venture partner that is not a signatory to the New Zealand Forest Accord 1991 (unpublished – an arrangement between representatives of the forest industry and influential conservation groups in New Zealand, its key principle is to protect defined areas of native vegetation from clearance for planting or disturbance during forestry operations). These external factors have prevented a greater involvement of Maori in forestry development, including the ECFP, and have contributed to their view that they are disadvantaged from developing what they see as sustainable use of their land and development for their people.

#### 18.3.7.2   Decline in agricultural processing industries

Declines in agricultural processing industries have also been attributed to the increasing level of forestry investment in the East Coast region. However, the contribution of forestry, and in particular the ECFP, to the decline of agriculture and associated processing industries is far from clear (Wall and Cocklin, 1994). Other factors may also contribute, such as the decline in international prices for agricultural products.

#### 18.3.7.3   Depopulation of the rural community

Closely related to the economic impact of the ECFP are concerns about the social consequence of land use change. Since the 'second planting boom' of the early 1980s, there have been several analyses of the effects of forestry on rural society in New Zealand, including a number of studies focused on the East Coast region (Aldwell, 1984; Fairweather *et al.*, 2000).

The main concerns highlighted by these studies included shifts in the quality of life, employment and income effects, declines in rural services, and effects upon local decision making. The belief that forestry would encourage rural depopulation was at the heart of much of the concern (Le Heron and Roche, 1985). Depopulation, it was believed, would undermine the viability of rural communities, services would close down and hasten the decline, and power and authority would become vested in private capital (Smith, 1981; Wall and Cocklin, 1994). Many of these concerns have been realized in the East Coast region, but more as a result of restructuring of the NZFS and the expansion of the private forest estate. Little can be attributed directly to the ECFP, though it is hoped that in time the project may assist in turning things around.

### 18.3.7.4 Biodiversity

While alternative species can be planted under the ECFP, the costs (of plant material, labour for planting, and often tree mortality) are usually far greater than for radiata pine. In addition, there is better knowledge of the management and potential market for radiata pine than for some of the alternative species. Because of this, ECFP tenders have mostly been restricted to radiata pine, except for a small amount of Douglas fir (*Pseudotsuga menziesii*) planted at higher altitudes where radiata pine is less vigorous. Small amounts of poplar (*Populus* spp.) have also been planted.

While there have been gains in nature conservation from reforestation, there are also downsides. The extensive 'monoculture' has, in many people's eyes, reduced bio-diversity. Many lay people consider that the preponderance of one species creates an unacceptable risk from an introduced pest or disease, and that there should be some controls to diversify the species used in plantation forestry. However, there are counter-arguments and some data exist on the diversity found in radiata pine plantations (Maclaren, 1996; Brockerhoff *et al.*, 2001). For example, in second-rotation compartments in central North Island forests, 35 species of vascular plants have been recorded (Allen *et al.*, 1995).

It has been well established that afforestation can lead to a reduction in available water downstream of the forest, thereby limiting water for irrigation and other abstractive uses (Rowe *et al.*, 1997). No data exist to predict the likely effects of the ECFP in terms of water availability or reduction in flow for the major river systems of the East Coast. However, at least 20% of a catchment needs to be planted to change streamflow, with the greatest reductions in streamflow occurring in high rainfall areas, and little difference in low rainfall areas (Rowe *et al.*, 1997).

Costs also accrue to the roading infrastructure and the port facilities, particularly when the forests are harvested. Significant damage to roads as a consequence of logging the forests planted in the 1960s and 1970s is now occurring. This is placing an undue burden on both local and national roading funds. In addition, significant investment in log handling facilities at the Port of Gisborne is required to keep pace with the increasing level of cut expected from both the old and the newly planted forests.

### 18.3.8 Summary of ECFP

The ECFP has been operating since 1993 and has planted in excess of 33 000 ha of severely eroding farmland. While the project has 18 more years to run, it is clear that there have been a number of both on-site and off-site benefits, though the success of the ECFP in meeting its primary goal is difficult to assess. This is largely because erosion control benefits are long-term in nature and there is a significant lag period between tree planting and the accrual of off-site benefits. In addition, although the spatial distribution of plantings will create local benefits, the cumulative effects will not be in evidence at a regional level for generations.

The benefits include such things as sustainable land use, income from wood production, reduced flood damage, increased environmental quality of river water and marine environment, enhanced regional development, increased employment, social benefits, greenhouse gas absorption, and a clean and green image for promotion of products from New Zealand in the global marketplace.

There have been several downsides to the project, many of which are perceptions and, if realized, may not have been a direct result of the ECFP alone. These include issues around the participation of Maori in forestry development, biodiversity loss, rural depopulation, loss of rural service infrastructure, and the potential impact on regional water resources.


## 18.4   Discussion

The link between forest vegetation and reduced soil erosion has been well demonstrated for nearly all parts of the globe (e.g. O'Loughlin and Ziemer, 1982). Forest removal generally causes an increase in soil erosion, either from surface or mass erosion processes. Developed nations recognized this, and since the 1950s (and even earlier for some) most have implemented a range of reforestation projects to both restore forest function and minimize erosion risk. There was a boom in reforestation expenditure in many countries during 1960–90, with a slowing in the last decade or so as at-risk areas became treated. However, for many developing countries, reforestation to mitigate erosion has been much later in coming, as social and economic pressures, particularly in rural areas, has limited efforts.

In countries such as Japan, Taiwan, South Korea and China there has been considerable effort at both national and regional government levels to initiate forestry schemes to manage erosion and landslide risk. However, in most cases it is difficult to separate expenditure on reforestation/vegetation management from that on more traditional 'hard engineering' mitigation measures. Policies aimed at stopping forest removal, particularly for fuelwood, in these and neighbouring countries have also contributed to 'management' of erosion.

In South Korea, the first erosion control project was initiated in 1907 around the outskirts of Seoul. Efforts in the mid-1950s and later to rehabilitate denuded land through government programmes were initially aimed at erosion control measures other than reforestation, with planting taking place once some level of control had been achieved. Erosion control projects spread all over the country and as a result the area of denuded forest land had declined from 680 000 ha in the 1950s to 120 000 ha in 1972. The forest area in Korea now (2000) occupies 6.4 million hectares, about 65% of the entire land area. Almost 76%, however, consists of young trees that are less than 30 years old.

In Taiwan, the Taiwan Forest Bureau began a long-term management lent project in 1964 to deal with watershed management to control forest destruction, stabilize riverbeds, prevent landslides and lengthen the life of reservoirs. Successive projects have continued to be initiated up to the present.

In Japan, national and prefectural governments are responsible for forest conservation projects, which include landslide prevention. Projects that take place in privately owned forests that are large in scale, require specialist technologies, and/or involve interests of more than one prefecture are carried out by the national government in addition to projects within the nationally owned forest lands. As in Korea and Taiwan, Japan integrates both 'hard' and 'soft' measures to reduce erosion and spends approximately US$4 billion per year on landslide control works.

Other countries also have programmes to reduce landslide hazard although, as indicated above, it is difficult to assess what proportion of expenditure relates to reforestation or forest management as specific activities aimed at mitigating erosion, particularly arising from landslides.

In countries where the risk of landslide activity is high, forests can play an important role in reducing the hazard. Many countries have adopted a mixture of both hard and soft measures to deal with their erosion problems, but New Zealand remains unique in that it continues to rely on reforestation (the soft measure) to manage regional landslide risk in its erosion-prone areas such as the East Coast region of the North Island.

## Acknowledgements

Randolph Hambling, Manager of the East Coast Forestry Project, is thanked for supplying details relating to the ECFP and its predecessors, as well as providing useful comments on a draft of this chapter. Christine Bezar is thanked for editing the chapter.

## References

Abe, K. and Iwamoto, M., 1988, Preliminary experiment on shear in soil layers with a large direct-shear apparatus, *Journal of Japanese Forest Science*, **68**(2), 61–65.

Abernethy, B. and Rutherfurd, I., 1999, Riverbank reinforcement by riparian roots, in *Proceedings of 2nd Australian Stream Management Conference*, Adelaide, 8–11 February 1999, 1–7.

Aldwell, P.H.B., 1984, *Direct and Processing Effects of Expanding Forestry in Three New Zealand Counties: An Input–Ouput Based Approach* (The Australian and New Zealand Section Regional Science Association), 272–294.

Aldwell, P. and Roche, M., 1992, Dismantling the New Zealand forest service, in S. Britton, R. Le Heron and E. Pawson (eds), *Changing Places in New Zealand: A Geography of Restructuring* (Christchurch: New Zealand Geographical Society), 118–119.

Allen, R., Platt, K. and Wiser, S., 1995, Biodiversity in New Zealand plantations, *New Zealand Forestry*, **39**(4), 26–29.

Allsop, F., 1973, The story of Mangatu: the forest which healed the land, *New Zealand Forest Service Information Series 62* (Wellington: Government Printer).

Battershill, C.N., 1993, What we do to the land we do to the sea: effects of sediment on coastal marine ecosystems, Abstract in *Marine Conservation and Wildlife Protection Conference*, Wellington, 1992.

Bayfield, M.A. and Meister, A.D., 1998, *East Coast Forestry Project review*, Report to Ministry of Agriculture and Forestry.

Bergin, D.O., Kimberley, M.O. and Marden, M., 1993, How soon does regenerating scrub control erosion? *New Zealand Forestry*, **38** (2), 38–40.

Bergin, D.O., Kimberley, M.O. and Marden, M., 1995, Protective value of regenerating tea tree stands on erosion-prone hill country, East Coast, North Island, New Zealand, *New Zealand Journal of Forestry Science*, **25**, 3–19.

Bishop, D.M. and Stevens, M.E., 1964, *Landslides on Logged Areas, Southeast Alaska*, USDA Forest Service Research Report NOR-I, Juneau, Alaska.

Böhm, W., 1979, *Methods of studying root systems, Ecological Studies 33*. (Berlin: Springer-Verlag).

Brockerhoff, E.G., Eckroyd, C.E. and Langer, E.R., 2001, Biodiversity in New Zealand plantation forests: policy trends, incentives, and the state of our knowledge, *New Zealand Journal of Forestry*, May, 31–37.

Caine, N., 1980, Rainfall intensity–duration control of shallow landslides and debris flows, *Geografiska Annaler*, **62A** (1–2), 23–29.

Cocklin, C. and Wall, M., 1997, Contested rural futures: New Zealand's East Coast Forestry Project, *Journal of Rural Studies*, **13**, 149–162.

Committee of Inquiry, 1970, *Wise Land Use and Community Development*, Report of the Technical Committee of Inquiry into the Problems of the Poverty Bay–East Cape District of New Zealand (Wellington: Ministry of Works and Development).

Coutts, M.P., 1983, Root architecture and tree stability, *Plant Soil*, **71**, 171–188.

Crozier, M.J., Eyles, R.J., Marx, S.L., McConchie, J.A. and Owen, R.C., 1980, Distribution of landslips in the Wairarapa hill country, *New Zealand Journal of Geology & Geophysics*, **23**, 575–586.

Crozier, M.J., 1986, *Landslides: Causes, Consequences, & Environment* (London: Croom Helm).

Crozier, M.J., Gage, M., Pettinga, J.R., Selby, M.J. and Wasson, R.J., 1992, The stability of hillslopes, in M.J. Soons and M.J. Selby (eds), *Landforms of New Zealand* (Auckland: Longman Paul), 64–90.

Cummins, T. and Byers, J., 1997, *East Coast Silviculture Workforce 1996*, Project report, PR-64 Liro, Rotorua.

Ekanayake, J.C. and Phillips, C.J., 1999, A method for stability analysis of vegetated hillslopes: an energy approach, *Canadian Geotechnical Journal*, **36**, 1172–1184.

Ekanayake, J.C. and Phillips, C.J., 2002, Slope stability thresholds for vegetated hillslopes: a composite model, *Canadian Geotechnical Journal*, **39**, 849–862.

Ekanayake, J.C., Marden, M., Watson, A.J. and Rowan, D., 1997, Tree roots and slope stability: a comparison between *Pinus radiata* and kanuka, *New Zealand Journal of Forestry Science*, **27**(2), 216–233.

Ekanayake, J.C.,Phillips, C.J. and Marden, M., 1999, A comparison of methods for stability analysis of vegetated slopes, in *Proceedings of the First Asia–Pacific conference on Ground and water Bioengineering for Erosion Control and Slope Stabilization*, 19–21 April 1999, Manila, The Philippines. Published by International Erosion Control Association, 411–418.

Endo, T. and Tsuruta, T., 1969, The effect of the tree's roots on the strength of soil, in *1968 Annual Report*, Hokkoio Branch, Forestry Experimental Station, Sapporo, Japan, 167–182.

Fairweather, J.R., Mayell, P.J. and Swaffield, S.R., 2000, *Forestry and agriculture on the New Zealand East Coast: socio-economic characeristics associated with land* use change, Research Report 247, Agribusiness Economics Research Unit, Lincoln University, Canterbury, NZ.

Foster, G. and Carter, L., 1997, Mud sedimentation on the continental shelf at an accretionary margin – Poverty Bay, New Zealand, *New Zealand Journal of Geology & Geophysics*, **40**, 157–173.

Fox, G., Umali, G. and Dickinson, T., 1995, An economic analysis of targeting soil conservation measures with respect to off-site water quality, *Canadian Journal of Agricultural Economics*, **43**, 105–118.

Glade, T., 1998, Establishing the frequency and magnitude of landslide-triggering rainstorm events in New Zealand, *Environmental Geology*, **35**(2–3), 160–174.

Gray, D.H. and Ohashi, H., 1983, Mechanics of fiber reinforcement in sand, *Journal of Geotechnical Engineering*, **109**, 335–353.

Gray, D.H. and Sotir, R.B., 1996, *Biotechnical and Soil Bioengineering Slope Stabilisation* (New York: John Wiley & Sons).

Greenway, D.R., 1987, Vegetation and slope stability, in M.G. Anderson and K.S. Richards (eds), *Slope Stability* (Chichester: John Wiley & Sons), 187–230.

Henderson, J. and Ongley, M., 1920, The geology of the Gisborne and Whatatutu subdivisions, Raukumara Division, *Bulletin 21* (new ser.), Geological Survey Branch, Department of Mines, NZ.

Hicks, D.L., 1988, *An Assessment of Soil Conservation in the Waihora Catchment, East Coast*, Internal Report 226, Soil Conservation Centre, Aokautere, Palmerston North, NZ.

Hicks, D.L., 1991, Erosion under pasture, pine plantations, scrub and indigenous forest: a comparison from Cyclone Bola, *New Zealand Forestry*, **36**(3), 21–22.

Hill, H., 1895, Denudation as a factor of geological time, *Transactions Proceedings of the New Zealand Institute*, **28**, 666–680.

Jakob, M., 2000, The impacts of logging on landslide activity at Clayoquot Sound, British Columbia, *Catena*, **38**, 279–300.

Jessen, M.R., Crippen, T.F., Page, M.J., Rijkse, M.C., Harmsworth, G.R. and McLeod, M., 1999, *Land Use Capability Classification of the Gisborne–East Coast region: A Report to Accompany the Second-Edition New Zealand Land Resource Inventory*, Landcare Research Science Series 21 (Lincoln: Manaaki Whenua Press).

Jewel, R.A. and Wroth, C.P., 1987, Direct shear tests on reinforced sand, *Géotechnique*, **37**(1), 53–68.

Kelliher, F.M., Marden, M., Watson, A.J. and Arulchelvam, I.M., 1995, Estimating the risk of landsliding using historical extreme river flood data, Note, *Journal of Hydrology* (NZ), **33**, 123–129.

Krausse, M., Eastwood, C. and Alexander, R.R., 2001, *Muddied waters: Estimating the national economic cost of soil erosion and sedimentation in New Zealand.* (Lincoln: Landcare Research).

Le Heron, R. and Roche, M., 1985, Expanding exotic forestry and the extension of a competing use for rural land in New Zealand, *Journal of Rural Studies*, **1**, 211–229.

Maclaren, J.P., 1996, Environmental effects of planted forests in New Zealand; the implications of continued afforestation of pasture, FRI Bulletin 198, New Zealand Forestry Research Institute.

Marden, M. and Rowan, D., 1993, Protective value of vegetation on tertiary terrain before and during Cyclone Bola, East Coast, North Island, New Zealand, *New Zealand Journal of Forest Science*, **23**, 255–263.

Marden, M., Phillips, C.J. and Rowan, D., 1991, Declining soil loss with increasing age of plantation forest in the Uawa catchment, East Coast Region, North Island, New Zealand, in *Proceedings of International Conference on Sustainable Land Management*, Napier, NZ, 358–361.

Marden, M., Rowan, D. and Phillips, C.J., 1995, Impact of cyclone-induced landsliding on plantation forest and farmland in the East Coast of New Zealand: a lesson in risk management, in K. Sassa (ed.), *Proceedings of 20th IUFRO World Congress, Tampere, Finland, 1995.* Technical Session on Natural Disasters in Mountainous Areas, 133–145.

Mazengarb, C. and Speden, I.G. (Comp), 2000, Geology of the Raukumara area. Institute of Geological and Nuclear Sciences 1:250 000 geological map 6. Institute of Geological and Nuclear Sciences, Lower Hutt, NZ.

McLean, V., 1994, Save our soils: East Coast Forestry, *New Zealand Forest Industry* **25**(6), 10–13.

Megahan, W.F., 1978, Erosion processes on steep granitic road fills in central Idaho, *Soil Science Society America Journal*, **42**, 350–357.

Ministry of Agriculture and Forestry, 2000, *The East Coast Forestry Project* (Wellington).

Ministry of Agriculture and Forestry, 2001, *Guidelines for Applicants to the East Coast Forestry Project* (Wellington).

Ministry of Agriculture and Forestry, 2002, *Guidelines for Applicants to the East Coast Forestry Project* (Wellington).

Ministry of Forestry, 1993, *A Guide to the East Coast Forestry Project 1993* (Wellington).

Ministry of Forestry, 1994, *A Guide to the East Coast Forestry Project 1994* (Wellington).

Ministry of Forestry, 1996, The impact of the East Coast Forestry Scheme on employment and economic structure in the Gisborne Region. Unpublished report prepared for the Ministry of Forestry by Butcher and Partners, Christchurch.

Ministry of Forestry, 1997, East Coast Forestry Project. Unpublished report to the Ministry by Owen Cox, Forme Consulting Group, Wellington.

Ministry of Works and Development, 1987, *The East Coast Project Review.* Water and Soil Directorate, Ministry of Works and Development, Wellington.

Montgomery, D.R., Schmidt, K.M., Greenberg, H.M. and Dietrich, W.E., 2000, Forest clearing and regional landsliding, *Geology*, **28**, 311–314.

Nilaweera, N.S., 1994, Effects of tree roots on slope stability: the case of Khao Luang Mountain area, southern Thailand. Doctor of Technical Science dissertation, Asian Institute of Technology, Bangkok, Thailand.

Officials' Committee Report, 1988, *East Coast Project Review*. Ministry for the Environment.

O'Loughlin, C.L., 1974a, The effect of timber removal on the stability of forest soils, *Journal of Hydrology (NZ)*, **13**, 121–134.

O'Loughlin, C.L., 1974b, A study of tree root strength deterioration following clearfelling, *Canadian Journal of Forest Research*, **4**, 107–113.

O'Loughlin, C.L., 1991, Priority setting for Government investment in forestry conservation schemes – an example from New Zealand, in *Proceedings of 19th IUFRO World Congress, 5–11 August 1990, Montreal, Canada*. USDA Forest Service General Technical Report PSW-GTR-130, 6–10.

O'Loughlin, C.L. and Pearce, A.J., 1976, Influence of Cenozoic geology on mass movement and sediment yield response to forest removal, North Westland, New Zealand, *Bulletin International Association Engineering Geology*, **14**, 41–46.

O'Loughlin, C.L. and Watson, A.J., 1979, Root-wood strength deterioration in radiata pine after clearfelling, *New Zealand Journal of Forestry Science*, **9**, 284–293.

O'Loughlin, C.L. and Zhang, X., 1986, The influence of fast-growing conifer plantations on shallow landsliding and earthflow movement in New Zealand steepland, in *Proceedings 18th IUFRO World Congress, Ljublijana, Yugoslavia, September 1986*. Division 1, Volume 1, 217–226.

O'Loughlin, C.L. and Ziemer, R.R., 1982, The importance of root strength and deterioration rates upon edaphic stability in steepland forests, in R.H. Waring (ed.), *Carbon Update and Allocation in Subalpine Ecosystems as a Key to Management*. Proceedings of an IUFRO Workshop P.I. 107-00 Ecology of Subalpine Zones, 2–3 August, Oregon State University, Corvallis, Oregon, USA, 70–78.

Page, M.J., Reid, L.M. and Lynn, I.H., 1999, Sediment production from Cyclone Bola landslides, Waipaoa catchment, *Journal of Hydrology (NZ)*, **38**, 289–308.

Page, M.J., Trustrum, N.A. and Gomez, B., 2000, Implications of a century of anthropogenic erosion for future land use in the Gisborne–East Coast region of New Zealand, *New Zealand Geographer*, **56**(2), 13–24.

Parliamentary Commissioner for the Environment, 1993, *Water and Soil Resource Management on the East Coast* (Wellington: Office of the Parliamentary Commissioner for the Environment).

Parliamentary Commissioner for the Environment, 1994, *Sustainable Land Management and the East Coast Forestry Project* (Wellington: Office of the Parliamentary Commissioner for the Environment).

Peacock, D.H., 1998, Integrated management of an aggrading floodplain, Poverty Bay, East Coast, North Island, New Zealand: a case study, in Cathay Pacific Risk Management (eds), *Proceedings*, Seminar on project risk management, August 1998, Beijing, China.

Pearce, A.J., O'Loughlin, C.L., Jackson, R.J. and Zhang, X.B., 1987, Reforestation: on-site effects on hydrology and erosion, eastern Raukumara Range, New Zealand, in *Forest Hydrology and Watershed Management, Proceedings*, Vancouver Symposium, August 1987, Publ. 167, *International Association of Hydrological Sciences* 489–497.

Phillips, C.J. and Watson, A.J., 1994, *Structural Tree Root Research in New Zealand: A Review*. Landcare Research Science Series 7 (Lincoln: Manaaki Whenua Press).

Phillips, C.J., Marden, M. and Pearce, A.J., 1990, Effectiveness of reforestation in prevention and control of landsliding during large cyclonic storms, in *Proceedings, 19th IUFRO Conference*, Montreal, 358–361.

Phillips, C.J., Marden, M. and Miller, D., 2000, Review of plant performance for erosion control in the East Coast Region. Unpublished Landcare Research Contract Report LC9900/111 prepared for Ministry of Agriculture and Forestry.

Poverty Bay Catchment Board, 1978, Report of land use planning and development study for erosion-prone land of the East Cape region, Poverty Bay Catchment Board, Gisborne ('Red Report'). Unpublished.

PresTon, N.J. and Crozier, M.J., 1999, Resistance to shallow land slide through root–derived cohesion in East coast hill country soils, North Island, New Zealand, *Earth surface Processes & landforms*, **24**(8), 605–676.

Reid, L.M. and Page, M.J., 2002, Magnitude and frequency of landsliding in a large New Zealand catchment, *Geomorphology*, **49**, 71–88.

Reistenberg, M.M., 1994, Anchoring of thin colluvium on hillslopes by roots of sugar maple and white ash, in *Landslides in the Cincinnati area*, US Geological Survey Bulletin 2059E.

Reistenberg, M.M. and Sovonick-Dunford, S., 1983, The role of woody vegetation on stabilising slopes in the Cincinnati area, *Geological Society of America Bulletin*, **94**, 506–518.

Roche, M., 1996, Forestry, in R. Le Heron and E. Pawson (eds), *Changing places: New Zealand in the nineties* (Auckland: Longman Paul), 160–168.

Rowe, L., Fahey, B., Jackson, R. and Duncan, M., 1997, Effects of land use on floods and low flows, in M.P. Mosley and C.P. Pearson (eds), *Floods and Droughts: the New Zealand Experience* (Wellington: New Zealand Hydrological Society), 85–102.

Shewbridge, S.E. and Sitar, N., 1989, Deformation characteristics of reinforced sand in direct shear, *Journal of Geotechnical Engineering*, **115**, 1134–1147.

Sidle, R.C., 1991, A conceptual model of changes in root cohesion in response to vegetation management, *Journal of Environmental Quality*, **20**, 43–52.

Sidle, R.C., Pearce, A.J. and O'Loughlin, C.L., 1985, *Hillslope Stability and Land Use*. (Washington, DC: American Geophysical Union, Water Resource Monograph 11).

Smith, B., 1981, Rural change and forestry, *People and Planning*, **20**, 10–13.

Tomlinson, C.J., Fairweather, J.R. and Swaffield, S.R., 2000, *Gisborne/East Coast Field Research on Attitudes to Land Use Change: An Analysis of Impediments to Forest Sector Development*, Research Report 249, Agribusiness Economics Research Unit, Lincoln University, Canterbury, NZ.

Trotter, C., 1988, Cyclone Bola: the inevitable disaster, *New Zealand Engineering*, **43**(6), 13–16.

Van Kraayenoord, C.W.S. and Hathaway, R.L., 1986, *Plant Materials Handbook for Soil Conservation Volume 2: Introduced Plants*, Ministry of Works and Development, Water & Soil Miscellaneous Publication 94. Soil Conservation Centre, Aokautere.

Waldron, L.J., 1977, The shear resistance of root-permeated homogeneous and stratified soil, *Soil Science Society of America Journal*, **41**, 843–849.

Wall, M. and Cocklin, C., 1994, *Attitudes Towards Large-Scale Afforestation in the East Coast region*, Department of Geography, The University of Auckland, report prepared for the New Zealand Forest Research Institute.

Watson, A., Marden, M. and Rowan, D.R., 1995, Tree species performance and slope stability, in D.H. Barker (ed.), *Vegetation and Slopes – Stabilisation, Protection and Ecology* (London: Thomas Telford), 161–171.

Watson, A., Phillips, C. and Marden, M., 1999, Root strength, growth, and rates of decay: root reinforcement changes of two tree species and their contribution to slope stability. *Plant & Soil*, **217**, 39–47.

Wellman, H.W., 1962, Holocene of the North Island of New Zealand: a coastal reconnaissance, *Transactions of the Royal Society New Zealand of Geology*, **1**, 29–99.

Wilde, S.A., 1958, *Forest Soils, their Protection and Relation to Silviculture* (New York: Ronald Press).

Williams, P., 1980, From forest to suburb: the hydrological impact of man in New Zealand, in A.G. Anderson (ed.), *The Land Our Future* (Auckland: Longman Paul), 103–124.

Wilmshurst, J.M., 1997, The impact of human settlement on vegetation and soil stability in Hawke's Bay, New Zealand, *New Zealand Journal of Botany*, **35**, 97–111.

Wu, T.H. and Watson, A., 1998, In situ shear tests of soil blocks with roots, *Canadian Geotechnical Journal*, **35**, 579–590.

Wu, T.H., McKinnel, W.P. and Swanston, D.N., 1979, Strength of tree roots and landslides on Prince of Wales Island, Alaska, *Canadian Geotechnical Journal*, **16**, 19–33.

Wu, T.H., Beal, P.E. and Chinchun, L., 1988a, In-situ shear test of soil-root systems, *Journal of Geotechnical Engineering*, **114**, 1376–1394.

Wu, T.H., McOmber, R.M., Erb, R.T. and Beal, P.E., 1988b, Study of soil-root interaction, *Journal of Geotechnical Engineering*, **114**, 1351–1375.

Wu, W. and Sidle, R.C., 1995, A distributed slope stability model for steep forested basins, *Water Resources Research*, **31**, 2097–2110.

Zhang, X., Phillips, C. and Marden, M., 1993, A comparison of earthflow movement mechanisms on forested and grassed slopes, Raukumara Peninsula, North Island, New Zealand, *Geomorphology*, **6**, 175–187.

Ziemer, R.R., 1981, *Roots and the Stability of Forested Slopes*, International Association of Hydrological Sciences Publications **132**, 343–361.

# 19

# Geotechnical Structures for Landslide Risk Reduction

*Edward Nicholas Bromhead*

## 19.1  Introduction

Landslides in all their diversity of types and sizes are just one category of potentially damaging natural phenomena that are loosely termed 'geohazards'. The available range of responses to clearly identified and quantified geohazards can be classified using a simple, risk-based, framework. This is explored below, and put into a landslide context.

Risk from a geohazard may be expressed in terms of the following equation:

$$R = PNV,$$

where:
- $P$   is the probability that the damaging event occurs within a particular timescale, expressed as a number ranging from 0 to 1;
- $N$   is the number and value combined of the elements at risk, or inventory of assets, which are threatened;
- $V$   is the vulnerability, or proportion of the total value of the assets liable to be lost should the threat materialize, also expressed as a number ranging from 0 to 1; and
- $R$   is the risk.

Usually, $R$ is expressed in the same units as $N$, and for convenience, this may be either in terms of money or of human lives.

Several factors complicate the risk equation. For example, a given element at risk may be threatened by more than one damaging phenomenon. In the case of landslides, there

*Landslide Hazard and Risk*   Edited by Thomas Glade, Malcolm Anderson and Michael J. Crozier
© 2004 John Wiley & Sons, Ltd   ISBN: 0-471-48663-9

may be a range of severities, or a range of types of landslides, to consider. Accordingly, the risk equation is conventionally expressed as a summation of all the individual risks:

$$R = \Sigma PNV.$$

Clearly, the summation sign is only a shorthand representation of the true complexities of combinations of different probability-based assessments, some of which may be mutually exclusive. Furthermore, the risk may be considerably different for different classes of elements at risk, not least because of different temporal occupancy of the zone where the event may occur. As an example of this, consider a rockfall from a steep face. The assets are a rural road, some agricultural land with fencing, and some animals. To these must be added the occasional road user in a car, and the value of the road as access to a farm. The annual probability of a single rockfall of $1\,\text{m}^3$ size is, from experience, 0.5. The fencing will be destroyed along a short length, but as a proportion of its total value, vulnerability is small. Furthermore, the animals might escape, but if the site is sufficiently remote, they may be recaptured. The chances of a car in motion being hit by the rockfall are small: with few vehicle movements by day, and even fewer at night, the temporal occupancy of the risk zone is small. There is, however, the risk of hitting a fallen block after it has come to rest, although on a low-speed road the risk is lessened (it is increased where the road is unlit). The blocks are not large enough to significantly damage the road, or to block it, and they are readily cleared. In such a situation, the risks of monetary loss are small, to life even smaller, and the remediation after an event is well within the resources of those threatened by it.

Now contrast the equivalent case where the road is a main urban highway leading to an industrial zone. At all times of the day, traffic is heavy – and sometimes at a standstill. The equation contains much bigger risks to life. It also contains losses associated with interruptions to the flow of goods and services during an event, while it is being remediated and, subsequently, if business is lost. Clearly, the relative risks justify different levels of treatment in the two cases.

Adopting a risk-based approach to landslide hazard provides much of the framework needed to classify and assess the possible responses.


## 19.2   A Fourfold Scheme for Classifying Responses to Geohazards such as Landslides

The community faced with any identified and quantified geohazard such as a landslide has a range of strategies for dealing with it. These strategies are as follows:

1. Avoidance
2. Correction
3. Desensitization
4. Acceptance.

The *avoidance* strategy involves the relocation of elements at risk to a less hazardous location. This strategy may be politically difficult to implement for a wide variety of reasons. For example, if the perceived risks are small, the political will to address the problem may be absent. Where the hazardous area has significant attractions for the community, such as an attractive landscape setting, proximity to a harbour or river, agricultural fertility, cultural tradition and so on, these may outweigh even quite appreciable risks. In a densely populated and economically developed country, it is also potentially a very expensive operation both to abandon a favoured location and to acquire and develop a new one. Even in a poor country, the relocation costs may exceed the threatened population's capacity to pay.

*Correction* as a strategy for dealing with a geohazard is the treatment of the underlying source of danger. Most practitioners would recognize it as appropriate in some situations (e.g. in stabilizing a landslide) but impossible or inappropriate in others (the source of earthquakes, for instance). Of course, where the underlying problem cannot be addressed, and here the earthquake example is a good one, it may be more practical to reduce the susceptibility of the elements at risk to damage or loss by appropriate construction technologies or patterns of use. Building to codes which cater for seismic effects (encompassing design and construction) is a perfect example of the third item in the list – the *desensitization* strategy.

Finally, a community may need to *accept* the risks from a given geohazard. This acceptance might be rational acceptance, in which case the risks are perfectly understood, but are offset against the benefits that the community obtains in the particular locality, or it might be poverty-led acceptance. It is not always the case, for instance, that a community with a valuable set of elements at risk can contemplate the necessary further expenditure to fully protect them. This is true of both rich and poor societies.

An acceptance strategy is made more palatable by the availability of insurance or governmental aid in the case of disaster. Risk sharing across a threatened community is useful where the geohazard poses threats at a low level, for example in helping Insurance to form risk-sharing across a wide community: governmental aid (and indeed, private charitable donations) are a form of risk and cost sharing. Care must be taken that the use of compensation schemes does not encourage high-risk activity, thus justifying high-expenditure protection at some time in the future.

All four of these strategies have been adopted in the treatment of landslide hazards. This chapter describes a number of examples of each strategy, and discusses the issues arising from each choice. However, it is essential to stress that an informed debate about the correct strategy to adopt can only be made when the particular geohazard (a) has been fully identified; and (b) is properly quantified. In the landslide context, it follows that both surface and subsurface investigations are required and must be made before an informed decision can be taken on the correct strategy to adopt. The final choice on what to do is often a political one as well as an economic one, but the politics should follow on from the understanding, not control the process of investigation.

Returning to the risk equation, the four strategies may be seen as fitting logically in the same framework. The *avoidance* strategy may be seen as reducing $N$, the *correction* strategy as reducing $P$, and the *desensitization* strategy as reducing $V$. The remaining *acceptance* approach considers $R$ as acceptable, or unavoidable.

Figure 19.1 shows this subdivision diagrammatically.


## 19.3   Identification and Quantification of Landslide Hazard

It is implicit in the above that the specific landslide hazards are recognized, and are quantified, before undertaking the risk assessment. This cannot be done solely by reference to the historic record, since we are operating in an era of, *inter alia*, changing climate and

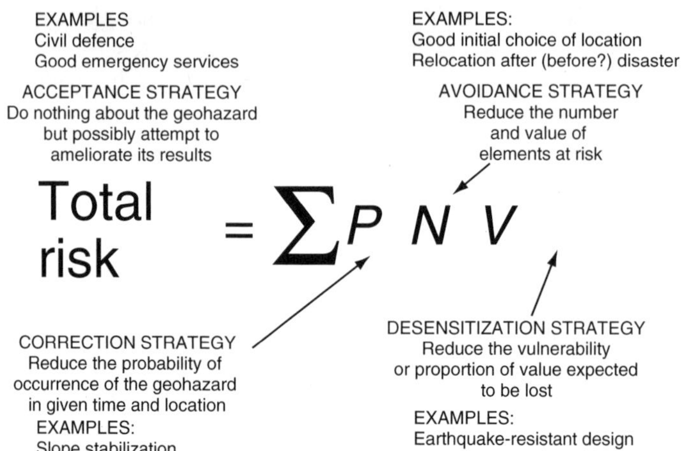



$$\text{Total risk} = \sum P \; N \; V$$

**Figure 19.1**    *The four main responses to natural hazards such as landslides can be classified by reference to the factors that create risk*

sea levels, and at a time when human impacts on the landscape are becoming increasingly felt. Furthermore, a purely geomorphological approach records the record of past events in the landscape, and does not always lend itself to the prediction of future behaviour. Of course, relocation or avoidance strategies, together with acceptance strategies, can be adopted solely on geomorphological criteria, with a high degree of success.

In particular, the economic choice of remedial schemes cannot be made without an integrated geomorphological, geological and geotechnical investigation, where the consequences of all options – including to do nothing – can be assessed on a systematic and consistent basis.

Mostly, landslide damage occurs within the footprint of a landslide, and only during the time that the landslide is in motion. There are exceptions. One critical exception concerns landslide dams, where a breach can have consequences over a significantly larger area than the landslide footprint. Even an unbreached landslide dam causes flooding upstream. A further consequence of landslide activity is blight, which causes loss of value to assets in the area that is at risk, and sometimes outside it. In the UK, for instance, insurance companies assess premiums based on experience, but on a geographical system related to postal codes. Hence properties sharing the same postal code as an active landslide may suffer increased premiums, or simply be uninsurable, while being at no genuine risk. People can also be blighted, suffering nervous ailments or poor health as a result of the fear of landslide activity.

The geotechnical professional dealing with landslide hazard needs to adopt a holistic approach to the investigation of the problem. This is not the place to discuss the steps in a landslide investigation, but in general terms the problem will be studied within a framework initially based on a topographical survey. Engineering geological and geomorphological mapping will have been conducted. The primary mechanisms on the site will have been identified to form a ground model. Archive material will inform the investigators about past activity and its relationship with causal factors. Subsurface investigations will confirm the details of the ground model, and laboratory and *in situ* testing, analyses and instrumentation observations will complete the picture. A quantitative risk analysis will then inform the users of the investigation of the future risks from landslide activity on the site.

In addition to the infrastructure and human population of the unstable area, a wide variety of human activities is considered valuable. Should these activities be terminated or suspended, this would constitute a further loss to add to the losses of more tangible assets. Some losses may not easily be costed purely in financial terms, for example in the fields of leisure and conservation. The broad spectrum of assets and facilities potentially subject to loss constitutes the 'elements at risk' from slope instability (Varnes *et al.*, 1984).

It might be thought that a strategy based on acceptance of the risks could avoid subsurface and other geotechnical investigations, but if this is the approach adopted, then risk must be based on records of past activity. Historical records are often inaccurate, with major gaps in the data. There may be a bias in the interpretation of data – sometimes cynically. An example of this was the post-1945 attribution of clay shrinkage and heave cracking in houses in southern Britain to the effects of enemy air raids! This led to investigations into the true cause. Geotechnical investigations are critical in informing decision makers on the possibility of future changes in pattern of activity. Notwithstanding this caveat, preliminary assessments of future risk, especially in the immediate future, are often based on past behaviour. Related activities such as monitoring, and the provision of warnings, together with evacuation procedures and medical and other assistance, are also often planned largely on the basis of past behaviour.

Similarly, the avoidance (or relocation) strategy is usually adopted before there is a large investment in geotechnical investigations. It is difficult to relocate large communities, or those occupying historic sites, regardless of the risks (actual or perceived), but it is not unknown. There are similarly few instances of structures consciously designed and built to accommodate landslide movements – perhaps the exception is flexible pipes used for services. Instead, repair or replacement of damage is undertaken.

## 19.4   Avoidance Strategies

A full and proper use of the avoidance strategy, in the sense of choosing *ab initio* a location free of hazard, or less subject to it, is not often available. More frequently, it has to be *relocation*, when the particular location initially selected is shown by experience to be unsuitable. Examples abound of inappropriate initial choice of location.

Perhaps the best example of a town being completely relocated in a less hazardous location comes from the town of Valdez, in Alaska, shown in Figure 19.2 (after van Rose, 1983). In its original location, on an outwash delta, the town was ideally located to service the end of the trans-Alaska oil pipeline at Jackson Point. During the earthquake of 1964, the seaward edge of the delta was involved in submarine landsliding, causing extensive damage and loss of life in the town. As part of the reconstruction, the town was relocated to a site at Mineral Creek where buildings could be founded on bedrock or on gravel terraces, and thus they would be significantly less susceptible to coastal landsliding, whether or not earthquakes provoked this.

In southern Italy, the town of Campo Maggiore was simply deserted by its inhabitants in the late nineteenth century after landslide movements. A mid-twentieth-century en masse

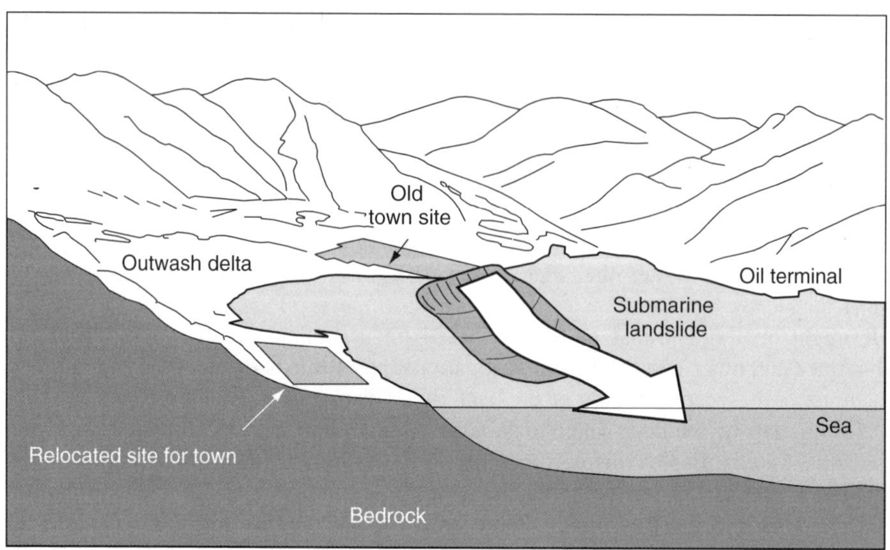



**Figure 19.2**  *The town of Valdez was relocated to a completely new site following the catastrophic landslide during the earthquake of 1964. The old town site has clearly susceptible to future landslides (after Van Rose, 1983). Reproduced by permission of the British Geological Survey. © NERC. All rights reserved. IPR/53–24c*

desertion of the town of Craco, also in southern Italy, took place following landslide activity there. There are numerous examples in the historical and archaeological record.

Classic examples of inappropriate choice of location include the town of Frank, Alberta, destroyed in a landslide from Turtle Mountain in 1903 (McConnell and Brock, 1904). Inappropriate choice of location was also a principal cause of problems with the Sevenoaks Bypass road scheme (Weeks, 1969), which crossed an area of ancient landslides, unrecognized at the time. Both fills and cuts interacted with the pre-existing landslides on the site, requiring extensive additional works. For a treatment of the role of engineering geomorphology in recognizing problem sites, the reader is referred to Hutchinson's Glossop Lecture (Hutchinson, 2001). This also contains other case records of inappropriate initial site location. Related problems arise from inappropriate choice of location for earthworks, particularly those associated with mine waste disposal, such as the disastrous Aberfan coal mine waste tip failure (Bishop, 1973). There, the mine waste was disposed of by uncontrolled tipping on steep slopes above the town. One of a series of tip slides reached the town, causing numerous fatalities. In these cases, there is a considerable effect from anthropogenic activities. A review of case records in this region showed that this was the most severe of a not-uncommon type of failure.

When the choice of location is free from constraint, the avoidance strategy is highly cost-effective, and the location with least hazard can often be found on the basis of a preliminary geomorphological mapping. To provide convincing data on the degree of hazard, however, it may well be necessary to undertake extensive surface and subsurface investigations – an investment that is abandoned along with the site. In those cases where a location has already been exploited for development, the costs of implementing a relocation strategy may rule it out of contention. This is particularly the case where no disaster has yet happened, and even those at most risk may view the risk analysis as scaremongering! The press is often unhelpful in this.

Some forms of construction, for example, passing a landslide area in tunnel, will avoid the problem (Figure 19.3), although some tunnels in a landslide area will simply meet the problem head on. This is demonstrated in the figure for tunnels that cross the slip surface of a pre-existing landslide. Tunnelling is also a method of avoiding rockfall hazard. However, there are different problems at the entrance an exit of a tunnel. At the exit, for instance, drivers have no forewarning of slope failures that could affect them, whereas drivers entering a tunnel have many opportunities to see the problem coming.

Viaducts can avoid landslides by bridging over them. An interesting variant of this approach may be seen at the Lago di Guardialiera, in southern Italy, where the unstable side slopes of a reservoir are avoided by the road that runs for most of the length of the reservoir on a viaduct. Tunnelling to bypass a landslide may be used to reduce subsidiary hazard, for example to drain a landslide-dammed lake (e.g. Byce, 1984).

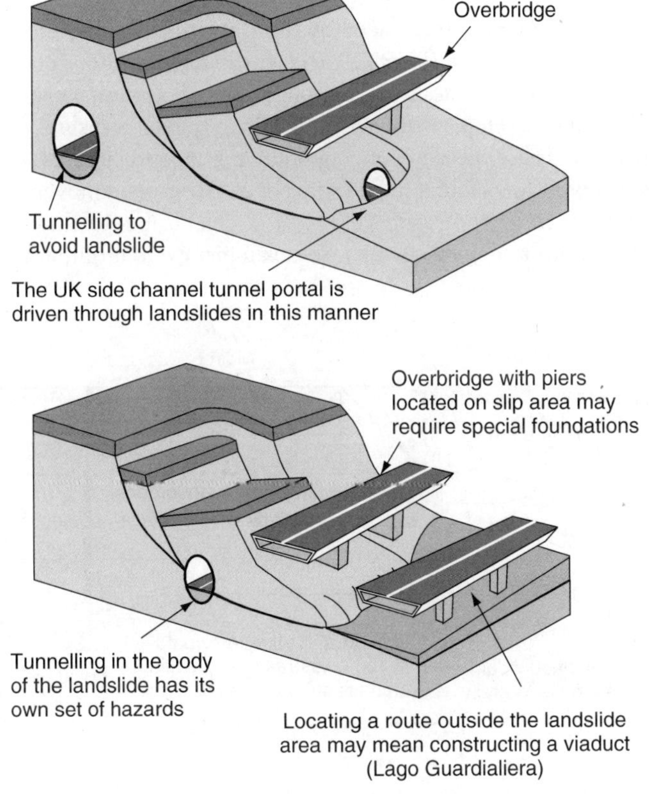


**Figure 19.3** *Tunnelling and bridging solutions help to avoid landslide hazards. However, injudicious design increases risk*

## 19.5    Desensitization

The vulnerability of a particular development to landslide activity must depend both on the nature of the development itself, and on the nature of the landslide hazard. The vulnerability term in the risk equation for most fixed installations (houses, roads etc.) affected by large-scale landsliding is effectively 1 (100%), and for rockfalls is large.

A structure in a landslide area may be designed as an isolated stable 'island', anchored to stable subsoil or bedrock, and capable of resisting the forces installed as landslides move past. All the piers of a bridge, for example, may be similarly designed (Hutchinson, 1984a; Popescu, 1991).

Further examples of the approach include the provision of roller bearings at a bridge abutment to permit the abutment to move without damaging the bridge superstructure (Thomson and Hayley, 1975, describe one case where the rollers were added as a repair measure). For this procedure to be satisfactory, the movements must be in a predictable direction, and at a predictable or at least manageable rate. The bearing is better able to accommodate regular small movements than episodic large ones. Facilities for jacking to restore levels may also be required. Figure 19.4 shows this scheme diagrammatically. It worked in this case because the direction of the movement vector was consistent, as a result of the nature of sliding, and small, as a result of the erosion mechanism operating.

In the UK, Planning Guidance is issued by the Department of the Environment. Planning Policy Guidance Note 14 (DoE, 1990), which has recently been supplemented, counsels against development on unstable ground. In general, this is wise advice. However, it is possible to redevelop existing sites by taking special measures to accommodate landslide movement. These measures include building the structure as a rigid box, supported in such a way as to enable it to be re-levelled after ground movement. In addition, it is highly desirable that other measures are taken which ensure that at worst the development is neutral in respect of slope stability, for example, by controlling water discharges.

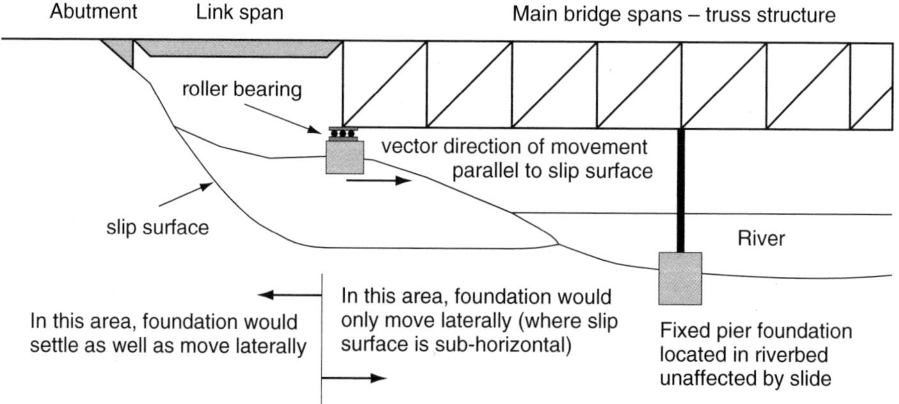


**Figure 19.4** *Thomson and Hayley (1975) describe how a bridge across the Little Smoky River is constructed to accommodate landslide displacements, which are predictable in magnitude and direction. Reproduced by permission of NRC Research Press*

A variety of means is used to protect against the secondary effects of landsliding. For example, the lowering of a reservoir, to improve the resistance to overtopping by landslide-generated waves in the event of landslides from the valley sides is sometimes undertaken, although in such cases it may well be necessary to control the reservoir lowering lest the very incidents one seeks to resist are provoked by rapid drawdown (Jaeger, 1969).

## 19.6   Acceptance Strategies

### 19.6.1   Controlling the Movement of Slide Masses

A selection of landslide countermeasures used to control the movement of debris slide masses without attempting to stabilize them is shown by Baldwin *et al.* (1987) and Hungr *et al.* (1987). They may be classified either as *containment* or *deflection* systems. The check dam is a classic containment system. It is widely used in the Alps, and throughout Italy, both to control runoff erosion, in which case the dam is usually solid, or to trap rockslide debris, in which case it is permeable. Solid check dams are designed as retaining walls. They may reduce landsliding by preventing downcutting of valleys. A typical solid check dam at Barcellonnette in the French Alps is shown in Figure 19.5. When the space upstream of the check dam is filled with material, it offers little obstacle to debris slides, which simply cascade over it. Worse still, accumulations of debris behind check dams provide a readily mobilized sediment source. Periodic maintenance is necessary, and this includes removal of deposited material. Variants of the check dam have been used in Hong Kong to trap boulders released from slopes by deep soil erosion. They may be made from concrete, piling, gabions and geogrid-reinforced earth or rockfill.

Catch fences and trap ditches used on rock slopes (Fookes and Sweeney, 1976; Figure 19.6) also constitute containment systems. Periodic maintenance is necessary, since the usefulness of any barrier is reduced when the storage capacity of the system is full and impact damage must be repaired in order for the system to continue to perform as designed. The provision of a ditch reduces the effective height of rebound of a boulder, and if the ditch is floored with energy-absorbing material, for example gravel or loose crushed rock, the subsequent rebound can be much lessened (Figure 19.6, c and d). For non-vertical slopes, or those with an irregular face, boulders may leap the ditch and fence, the optimum position for which may be some distance from the foot of the slope in this case.

Rock slopes subject to small failures can be covered in mesh. Figure 19.6 (a and b) shows this in diagrammatic form. There seem to be two major approaches: either a loose mesh (Figure 19.6a), where boulders are allowed to fall, but the mesh keeps them from becoming air-launched, or a tightly bolted mesh (Figure 19.6b), which keeps all the rock debris close to the face. Catch fences on the slope, or at its foot, have to contend with high-energy impacts. Figure 19.7 shows a strong fence in Hong Kong. It replaced an earlier system of concrete-filled oil drums (seen at the right of the picture) which were ineffective at protecting a major highway at a lower level to the right of the picture.

Small-area hazardous locations that need to be crossed can be bridged, allowing in this instance landslides to pass harmlessly beneath, or the route can be protected by a

**Figure 19.5**  *Check dams are widely used to prevent erosion, which would increase landslide activity on the valley sides. They may also be employed to contain debris from landslides in the valley. This example from the Super Sauze Valley, Barcellonnette, French Alps*

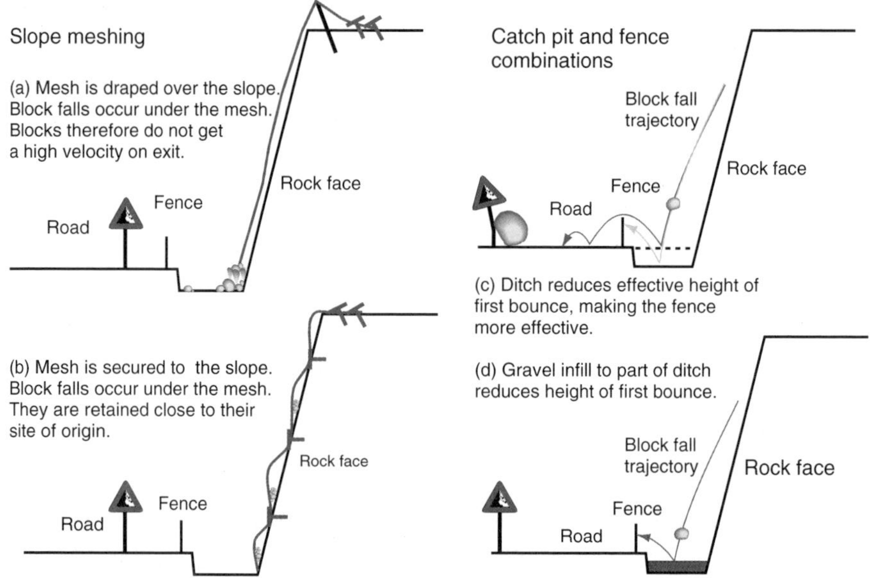


**Figure 19.6**  *Slope meshing may be used to control block falls. Catch pits (with or without 'soft landing' zones) and catch fences are alternative control measures. See also Fookes and Sweeney (1976)*

**Figure 19.7** *A catch fence may need to be a robust structure. This example from Hong Kong replaces the concrete-filled oil drum barrier, which had already proved ineffective in protecting the road to the right of the picture*

reinforced concrete shed to allow the safe passage of landslide debris above. Deflection systems such as the shed are also widely used in the mountainous areas of the world where not only is it inappropriate to try to stabilize the slope, but the volumes of moving material would too rapidly saturate a containment system. Two such sheds are shown in Figures 19.8 and 19.9. Figure 19.8 shows a long shed that deflects two debris streams across a road. This shed is situated on a set of hairpin bends, and the debris it deflects needs to be dealt with further downslope. Where the shed is to deflect a debris slide, the roof is left smooth, but where it is susceptible to impact, it must be protected from impact loading by a bed of gravel, or old tyres as illustrated in Figure 19.9. The engineer designing such a structure must be prepared for the debris slide or rockfall source to move, necessitating extension of the structure. Figure 19.10 shows the devastation following the 1999 earthquake in Taiwan. Virtually the only recognizable part of the cross-island highway is the shed, seen here unaffected by landsliding, but with the road on either side of it destroyed by fresh, unanticipated, debris slides.

A novel deflection system is in use at Lamosano in the Italian Dolomites east of Belluno. A large mudslide has developed. It is deflected past Funes by an earth-fill and concrete-block dam, although Funes is largely situated on a ridge above the general elevation of the mudslide, and is therefore at lesser risk unless the landslide develops to overtop the ridge or to breach it. The path of the landslide takes it through the town of Lamosano, where a concrete-lined channel has been built, complete with a water injection system, so that the encroaching toe of the landslide may be additionally

**Figure 19.8** *In mountainous areas, shed structures deflect rockslides and other mass movements, usually of rock debris. An armoured roof can be an effective deflector. Example from central Taiwan, Cross-Island Highway*

**Figure 19.9**  *Impact loadings on the roof of a shed structure can be reduced with tyres, or a crushed-rock soft landing zone. Example from central Taiwan, Cross-Island Highway*

mobilized, and caused to pass harmlessly through the town (see Figure 19.11, and Angelli *et al.*, 1994). Perhaps fortunately, the landslide came to rest before the system was tested. Water injection is turned on in advance of strict necessity when hazardous conditions, for example heavy rainfall, are experienced. On a recent visit, it seemed that the water injection system was no longer functional: systems must remain serviceable over long periods of time, or the initial investment is wasted.

### 19.6.2  Early Warning Systems

Early warning systems reassure the public in threatened locations. There are four domains in which such systems can be installed (Figure 19.12). They are:

1. The ambient environment
2. The ground surface
3. The landslide body
4. The landslide track.

In the first of these we place weather stations. Any historical correlations of weather with landslide activity are useful in giving advance warning of landslide activity. Weather forecasts extend this warning period. False alarms are a major problem.

The ground surface domain is monitored by survey methods. The problems here include the number of survey stations, the frequency with which they are measured, the sensitivity of the system used, and the need for intervisibility: easy across a valley, but

**Figure 19.10** *A shed structure is an expensive item to construct, and if the slope instability changes its route, the structure may become an expensive, useless, liability. During the September 1999 Chi-Chi earthquake in Taiwan, this rock shed on the Cross-Island Highway was left unscathed as new rockslides swept down the mountainside to destroy the road. Reproduced by permission of Professors J.J. Hung and M.L. Lin*

not always easy on a coastal location which is poorly overlooked. GPS or satellite-based systems have the advantage here, as they merely need access to the sky. At the simplest end of the scale are 'line and level' survey methods, which are low-cost, quick and cheap. Geodetic survey using triangulation, triangulateration and traversing (the last less accurate) have been used to monitor slow-moving landslides with 1- to 10-year intervals. Concrete pillars provide a method of reducing set-up and centring errors. Modern servo-controlled total station instruments can be automatically or remotely controlled, and can search for 'lost' targets.

Satellite systems are increasingly being used. At St Catherine's Point Lighthouse (Hutchinson *et al.*, 2002) the UK's Ordnance Survey (national survey authority) has installed a base station. Information on landslide movement is obtained at a very small time interval (Figure 19.13). The records showed a movement episode starting and stopping, and coinciding with the rainy season. Otherwise, differential GPS systems provide positional information more rapidly than geodetic survey.

Early warning systems in the landslide body include the range of classical geotechnical instruments, and some more novel approaches, including monitoring acoustic emissions, or the sound generated as a landslide moves (Dixon *et al.*, 1997). When remotely interrogated on a small timescale, the data stream from instrumentation is overwhelming, but must be examined for trends if the results are to be used for something other than a forensic investigation. This is a difficult, costly and error-prone exercise (Davis *et al.*, 2002).

**Figure 19.11** *In Lamosano, N. Italy, a flume has been constructed to permit a mudslide safe passage through the town. Water injection in the floor of the flume is designed to assist in maintaining the mobility of the debris*

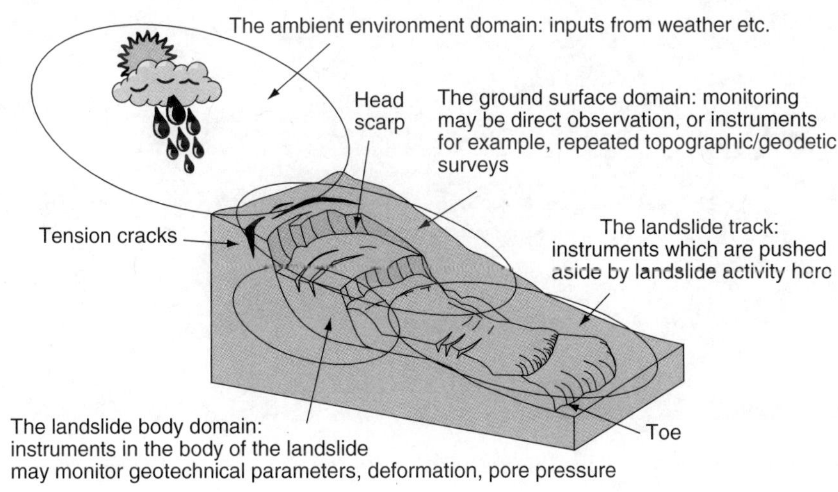


**Figure 19.12** *Early warning systems may operate in one of four domains of a landslide. Illustrated here are the four domains applied to a rotational or translational slide. The landslide track is encroached on in a slow, predictable, manner, and monitoring the track domain is of little significance. The track domain is of much greater significance for rapid movement landslides, e.g. falls and flows*

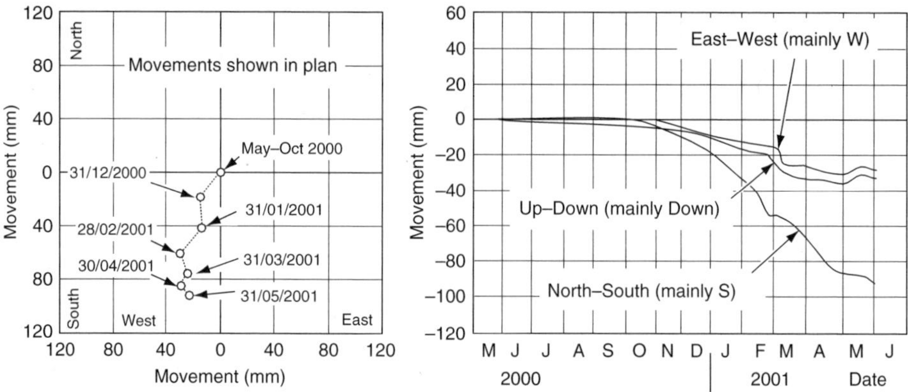



**Figure 19.13**  *Most monitoring merely gives a historical record of what occurred. This record of movements of the landslide at St Catherine's Point, Isle of Wight (Hutchinson et al., 2002) follows the landslide movements through a winter (rainy) season, 2000–2001. To estimate the eventual movement from an early part of this record is a difficult task. Reproduced by permission of Thomas Telford Ltd*

**Figure 19.14**  *In the particular context of this slide at Nothe Point, Weymouth, UK, the warning is superfluous*

Instruments in the landslide track are more novel. At Lamosano, a system has been installed with rod sensors suspended above the landslide, and out of reach of vandals. A rapid increase in landslide activity would reach and deflect these rods to the critical 20 degrees from the vertical for long enough to sound an alarm. The system is supplemented by suspended sounding devices to check on the corresponding levels of landslide debris.

Warning notices such as the familiar sign warning of falling debris are sometimes used on their own, to put the onus of self-protection on *bona fide* users of the site. Opinions vary as to whether or not this constitutes abdication of the engineer's responsibility. It is, however, an essential element of response in relatively remote areas, where people are only at risk during short and infrequent visits, or indeed in the short term while other options are under consideration (Figure 19.14).

### 19.6.3   Civil Defence and Disaster Management

Good access to rescue facilties, arrangements for evacuation and rehousing, and hospital facilities all form part of a disaster management system. Such systems may ameliorate the effects of landslide activity.

## 19.7   Correction of the Underlying Unstable Slope

This strategy is the engineer's favourite recourse. Reviews of available methods are presented, for example, by Hutchinson (1977, 1984a) and by Bromhead (1986, 1991).

The methods of slope stabilization fall naturally into the categories of:

1. Alteration of the slope geometry, either its profile by cut and/or fill, or its internal details, for example by digging out and replacing the landslip mass;
2. Improvement of the soil or rock strength properties, most often by decreasing pore-water pressures;
3. Providing force systems to counteract the tendency to slide.

### 19.7.1   Slope Geometry

The factors within the general area of slope geometry that affect the tendency to slide are the slope height, its mean angle, its shape, and its internal make-up. Most geotechnical engineers will have an intuitive understanding of these factors, especially after a period spent analysing slope stability.

Slope height is a factor that is most significant where the soil or rock strength has a high cohesive component, and is less significant when the strength is dominated by what may loosely be termed 'friction'. It is often not under the control of the geotechnical engineer, although in the design of, for example, spoil heaps, a limiting height criterion may be used to maintain stability. There are also many situations where the slope angle is not readily changed, especially where natural slopes of significant size are involved, or even with embankments where further land cannot be acquired for extension of the toe of the slope, nor may it be lost at the crest. Reducing the slope height is more effective for deep-seated modes of failure of the slope, and reducing the slope angle is more effective for shallow modes of failure. Alterations in the slope profile are possible only where landslides are at a reasonable scale. Earthworks-based schemes operate best where the slide surface position is controlled by geological structure, the landslide is only slow moving, and movements take place only along a small range of surfaces which can be identified *a priori*. Landslides that follow the terrain slope, for example debris slides, are

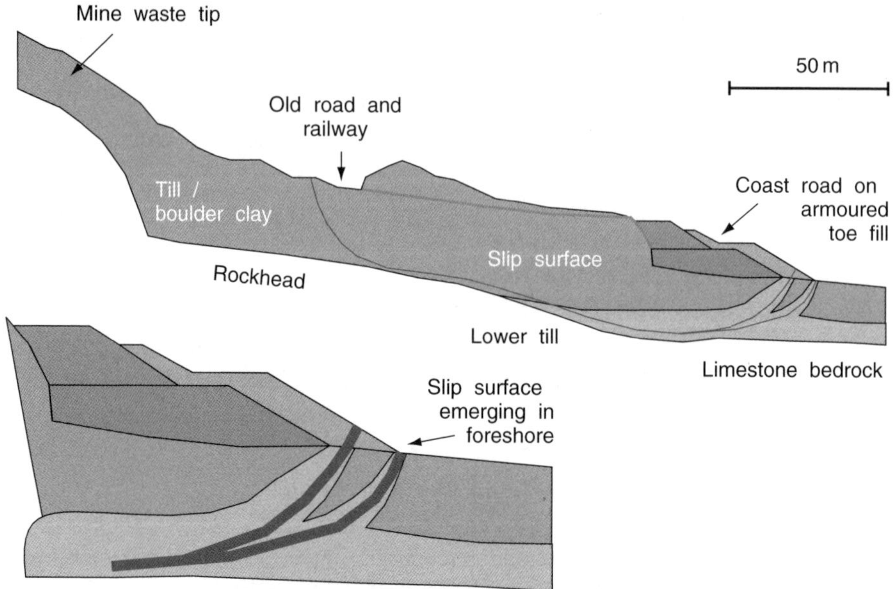



**Figure 19.15**  *Toe loading embankments are a favoured stabilization measure (e.g. Skempton and Hutchinson, 1969). This example from Llandulais, N. Wales, illustrates an armoured embankment used to support a highway as well as providing coast protection and landslide stabilization (Wilson and Smith, 1983, reproduced by permission of Thomas Telford Ltd)*

most difficult to control with earthworks, since it is clearly impractical to contemplate regrading a mountainside.

The shape of a slope in section may influence its stability. For example, undercut toes are obviously adverse. The provision of toe weighting or a berm (Figure 19.15) is often highly effective in improving the stability of a slope against deep-seated failures (e.g. Wilson and Smith, 1983; Viner-Brady, 1955; Kelly and Martin, 1986), and often serves the dual purpose of reducing toe erosion. Although head unloading is less often used as a permanent solution, it has been used as a temporary expedient to give stability to a landslip while other more permanent works are carried out (e.g. Watson, 1956). The reason why head unloading is less common than toe weighting is that it often increases the risk of the retrogression of instability upslope.

Toe fills may be classed as being on top of the toe of a slide, or in front of it (Figure 19.16). In the former case, the weight of the toe fill is the principal stabilizing factor, and comes into play as it is placed, although it may be subject to the dissipation of load-induced porewater pressure. Where a toe fill is placed on a rising section of slip surface (Figure 19.16b), it is more effective than where the slip surface is at a low angle. The stability improvement may be different for slip surfaces which cut through the toe fill relative to those that override it (or underride it). See Figure 19.17 for examples. A toe fill may be 'in front' of a slide in two senses. In the first, it is built in contact with the slide and it acts in a passive earth pressure mode (Bromhead and Harris, 1986) rather than in a deadweight mode, and needs continuing movements of the slide relative to the

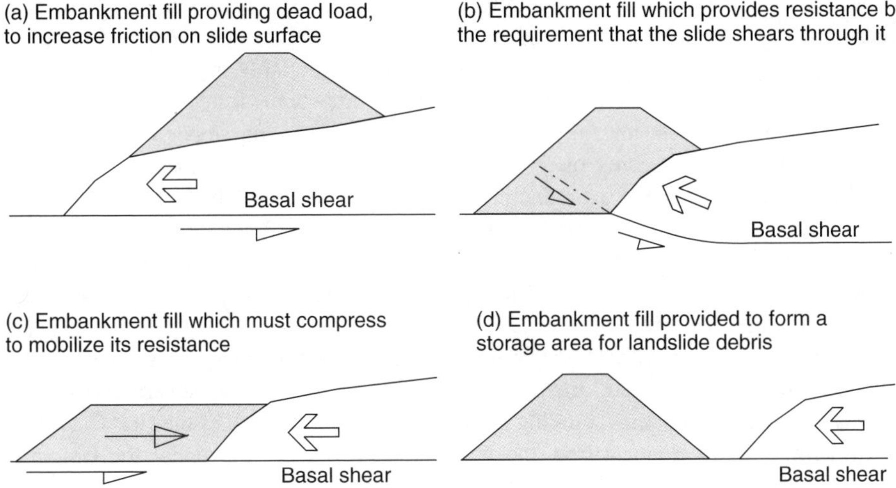



**Figure 19.16**   *Toe loads operate in a variety of ways*

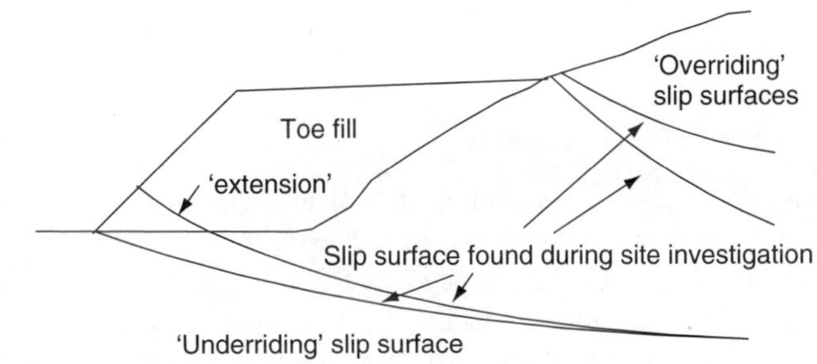



**Figure 19.17**   *Overriding slip surfaces are not stabilized by a toe fill. Underriding slip surfaces, especially those with a slip surface extension through a toe fill, may result in misleading calculated factors of improvement*

toe fill to mobilize this passive resistance. It is therefore less effective, in both the short and long term. A finite element analysis of the Carsington Dam failure (Vaughan *et al.*, 1990) shows how the late addition of a berm could not prevent the development of slip surfaces in the main dam body for essentially the above reason. A larger berm, or one built as the dam was raised, would have been more effective. Natural landslides fall into a less critical category than the brittle failure of an embankment, in the sense that they have already lost the slip surface brittleness, and if still in movement, will eventually respond to the resistance of a toe berm of the latter sort. Finally, an embankment fill may be provided simply to provide a storage area for landslide debris.

Toe fills may accidentally be founded on different subsoils than the slope they are intended to stabilize, and may therefore have their own special instability problems. Similarly, head unloading may expose weaker materials to the risk of local failure.

The shape of the slope in plan may also have a bearing on its stability. All three-dimensional analyses of essentially plane strain sections of a slope show higher factors of safety than the basic two-dimensional case. This is inevitably the result of what may loosely be termed 'side shear', and is found even in quite sophisticated analyses. It ought, therefore, to be the case that failures occur most frequently on slopes that are convex in plan. However, this seems rarely to be the case, with many slides on slopes that are concave instead. Careful examination of a number of these show that the essential three-dimensional features are predicated on the orientation of geological features in the slope, or three-dimensional variations in porewater pressure distribution, together with a subsequent spread of instability laterally from a site where it is initiated to more stable parts of the slope (Vaughan, 1991).

Internally, a slope may have some geometric features that predispose it to instability, as well as controlling the shape of the resulting slide surface. There are several examples in man-made embankment dams, ranging from the 'boot' on the core of the first Carsington Dam (Skempton and Coats, 1985), the layer of 'yellow clay' left under the Chingford embankment (Skempton, 1990) (a similar layer was also left in place under parts of the Carsington embankment) and the Acu Dam (Penman, 1985). Natural slopes may contain bedding, tectonic shearing, joints and other smaller-scale discontinuities, solution cavities, and folding or faulting. The designer of an embankment will have the opportunity to adjust the internal make-up of his slopes, whereas when faced with a natural slope, the engineer has fewer options.

## 19.7.2   Improvement of Soil Strength

The embankment slope designer operating on a clean sheet of paper has the choice of what materials to use where, within the overall constraints of what is available in total, and at what cost. Treating a natural slope or pre-existing earthwork is often largely a matter of accepting what is there and making the best of it. However, the effective stress shear strength properties can be improved by, for example, digging out slip surfaces and replacing the soil with the slip surfaces broken up and the soil recompacted. Slip surfaces in rockslides have on occasion been broken up *in situ* by blasting. The dig-out operation may, for small landslides, be accompanied by replacement with better-quality material, with or without soil reinforcement.

Digging out the complete slide mass, and replacing it with a fill with some improved properties, is normally only possible with small earthslides of the sort that affect highway and railway embankments. It may be too dangerous to attempt in the case of water-retaining structures such as canal banks unless the water levels are drawn down. For natural slopes, the costs of digging out may be prohibitive or may lead to further, retrogressive failures.

Some forms of grouting, with cement (Ayres, 1985a) or chemicals, may improve the fundamental properties of the soil or rock. The effect of natural changes in the soil chemistry is usually a worsening of strength, for example desalinization in quick clays or dissolution of halite or gypsiferous rocks, and these must be protected from the damaging effects of water. Grouting or other techniques for void filling may be used to counteract mining subsidence, which has the possible deleterious effects of fracturing the ground mass and tilting critical discontinuities.

Usually, however, the simplest and most effective way of improving the shear strength of soil is to reduce the porewater pressures in it (joint water pressures in a rock mass) by some form of drainage. Water pressures may arise from ephemeral infiltration, steady or unsteady seepage, or they may be induced by stress changes. Many slopes will have porewater pressure regimes that are in the process of equalizing to a new equilibrium from disturbances experienced in the past (e.g. Bromhead and Dixon, 1984). Often, especially in the case of unloaded slopes, or those formed from compacted clay fills, the pore pressure state immediately post excavation or construction is a state of porewater suction. Interference in the equilibration process may be good or bad, but without some understanding of the processes involved is more likely to be the latter. Shallow trench drains may increase the rate at which suctions are lost (bad) in a cut slope in a stiff clay, but control the final equilibrated level of groundwater (good).

Drainage systems may act to prevent the ingress of water into a landslide area, or may attempt to remove it once it is there. These two alternatives generally correspond to shallow and deep drainage. Some systems are needed for only a limited time span (e.g. those that dissipate stress-induced or construction-induced porewater pressures) or may need to function more or less permanently. Drains in the latter category are susceptible to blocking by silting up, or by chemical precipitation, often assisted by bacteriological action. Vegetation can also cause some forms of blockage, but a good vegetation cover on a slope is most usually beneficial, since it moderates the infiltration characteristics, controls runoff erosion, and contributes through transpiration to reducing the soil moisture.

Where drains are thought to be susceptible to blocking, access is needed for maintenance. Pipe systems are available where a filter geofabric liner can be removed for replacement or cleaning. Since the effectiveness of a drain is largely invisible (except if the discharge is carefully monitored), it may be necessary to instrument and monitor the drainage system.

Figure 19.18 shows a selection of drainage measures applied to a landslide. The figure has been compiled from a number of case records, and all of the drainage measures adopted in the figure have been used successfully, either singly or in combination, to stabilize landslides. Drainage may be used to prevent surface or subsurface water reaching the slide area (e.g. Siddle, 1984), or to remove it from the slide area (e.g. Bianco and Bruce, 1991). Water in the slide area may be the result of steady seepage flows, or it may be released from the soil or rock mass by stress changes or chemical activity, for example loading or the decomposition of refuse (Bromhead *et al.*, 1996) respectively. Contaminated leachate may need to be treated before disposal, and if corrosive, it may attack the drain systems. Drains to remove ephemeral sources of water may not need the same attention to filtration as drains with a longer life expectancy. Electro-osmotic systems appear to have been used successfully to reduce soil moisture at landslide sites.

Care should be taken in the construction of drainage measures to ensure that they do not increase porewater pressures in the ground in such a way as to adversely affect stability. This situation can arise in the temporary case as they are being installed. Drains are very susceptible to blocking, following which they continue to act as a collection network, but instead of channelling water safely away, concentrate it at the blockage.

It is sometimes difficult to predict accurately the effect of drainage systems, and hence their impact on slope stability. Shallow trench drainage systems, used widely in the UK,

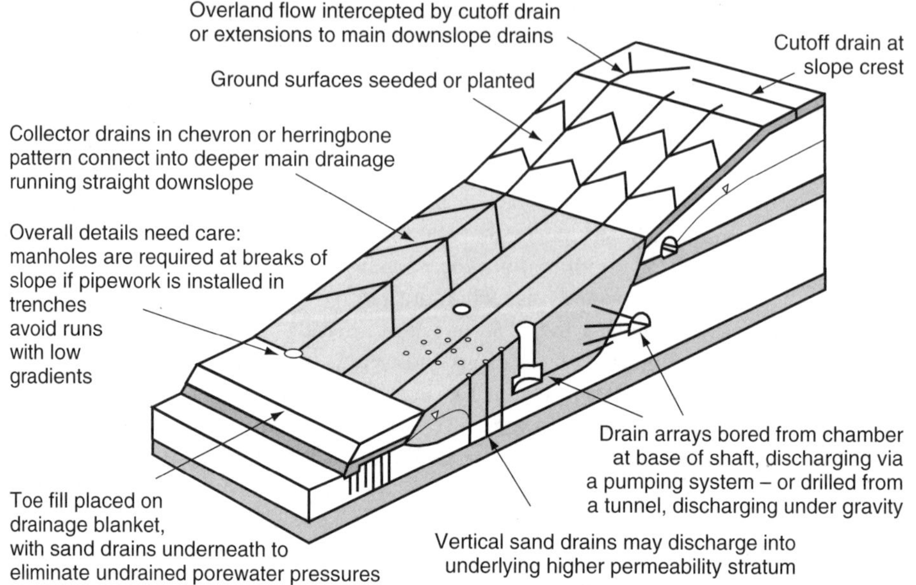



**Figure 19.18**  *Schematic diagram of a selection of drainage measures applicable to rotational and translational landslides. Clearly, not all these systems would be used simultaneously. Drainage measures may be shallow, or deep. They may be expected to at for the short term, or permanently into the future*

are an exception (Bromhead, 1984; Hutchinson, 1977). They are, however, particularly susceptible to blockage, and whereas in some cases it may merely be inconvenient and expensive to dig them out and refill them, where access is difficult or the drains have been built over, it may not be possible. It is important to consider a complete range of water inputs, and how these may change during the lifetime of a project (e.g. including postulated climate change effects). Figure 19.19 shows both natural and anthropogenic water inputs and outputs to a landslide. The latter can be a highly significant factor (Watson and Bromhead, 2000), although the effects are usually localized.

### 19.7.3  Force Systems to Resist Slope Instability

Passive systems to stabilize slopes rely on the construction of piles, piers, buttresses or walls through the slide into the underlying bedrock. Further movements of the slide increase pressures on the obstacle, and the reaction forces put into the landslide lead to stabilization.

Active systems, or ground anchors, utilize pre-loading to put the stabilizing forces into the landslide system *ab initio* (e.g. Barley, 1991). Anchor loads are spread into the slide mass by pads (Figure 19.20) so that bearing capacity failures of the ground are avoided. Pads with a smaller number of anchors are preferred. They may be cast *in situ* or pre-cast: the latter will often deflect more than the former when the anchors are stressed.

Piling systems for slope stabilization are described by, *inter alia*, Allison *et al.* (1991) and Leadbeater (1985). The stiffness of a pile system can be substantially increased by

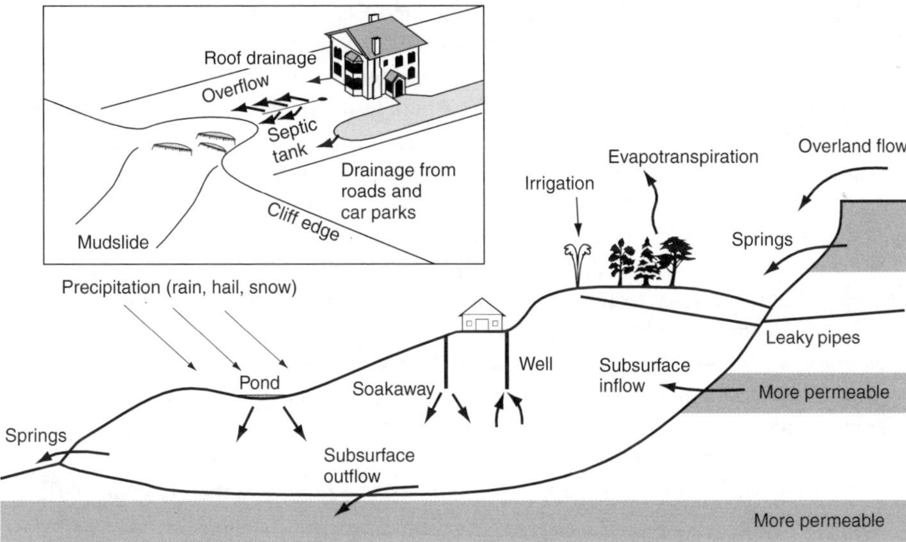


**Figure 19.19**   *The water balance in a landslide is the result of numerous input and output factors. These may change, seasonally. Anthropogenic sources of water are liable to change dramatically if an unstable slope is urbanized. Inset: impact of wastewater disposal on local slope instability (Watson and Bromhead, 2000, reproduced by permission of Thomas Telford Ltd)*

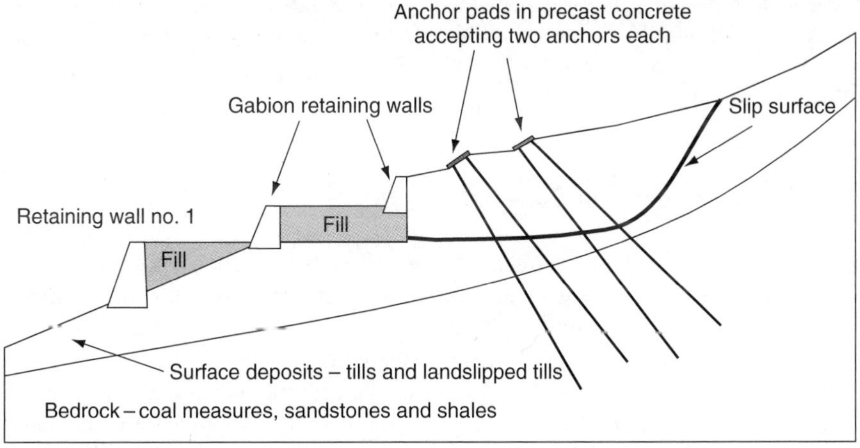


**Figure 19.20**   *Anchors may be used when other systems are ineffective: but at significant cost. Anchored schemes require maintenance (e.g. restressing) and have a comparatively short life (example: Risca-Rogerstone Bypass, S. Wales)*

connecting the pile heads together in an upslope direction. Connecting the pile heads of a row of piles along the contours of a slope, and thus normal to the slide direction, does not have this beneficial effect. Ellis (1985) describes a reticulated pali radice wall where small-diameter piles are used in a three-dimensional fan like the roots of a tree. This method is more widely used in Japan and mainland Europe than in the UK; for example, Ginzburg's authoritative treatment (Ginzburg, 1979) pre-dates most UK usage.

Where the slide mass is not monolithic, then piling will only control deformations in its immediate vicinity, especially in the short term. In the system used by Allison *et al.*, the pile heads were connected by small walls to control soil erosion and shallow mudsliding which took place between the original piles, although as pointed out above, this does not increase the stiffness of the system. A stiffening effect can be obtained by the construction of a shed-like structure (similar in appearance to those shown in Figures 19.8 and 19.9, but without the deflection function), in effect bracing a retaining wall and lending rigidity to the piles which are invariably used in the foundation. This composite structure is widely used in Italy in clay soils as an alternative to anchoring a retaining wall, especially where no suitable stratum can be readily found in which to position the fixed lengths of the ground anchors. In the majority of cases anchoring a wall principally improves the stability of the wall against overturning, so that the wall can passively resist the slide movement. To gain the full benefit of an anchor it must react on a pad that can spread the force into the slide mass, rather than merely counter-flexing a rigid structure.

A variety of materials can be used to form the pile. Beles and Stanculescu (1958) describe piles formed by burning clay *in situ*. Railway practice in Britain in the nineteenth century included elements of this, although the hazard from fires in combustible embankment fills (Ayres, 1985b) was a considerable disincentive to the practice. Ground freezing may operate in a similar way in the short term.

As well as piles, shear keys or counterforts are sometimes constructed. Figure 19.21 shows counterforts and piling schemes to stabilize a landslide. This sketch shows how a variety of ground reinforcement techniques might be employed. Soil nails and rock bolts are small-scale versions of the above, and they are appropriate only in the treatment of small-scale instability.

## 19.8  Choice of Factor of Safety

Where remedial works are undertaken, the engineer has to choose a factor of safety. Guidance in Codes of Practice intended, one suspects, for embankments of moderate height (e.g. 'the factor of safety shall be 1.3') is often of little use in natural slopes, where such a factor of safety may be simply unattainable.

A more suitable approach may be to ensure that the remedial works to a landslide are capable of surviving any 'design'-destabilizing event, or appropriate combination of events in such a way that collapse does not occur, and the situation can be retrieved. For example, a coastal landslide stabilized with a toe weight, armoured against wave attack, must be able to survive the (say) 1 in 100-year storm erosion without being so much disrupted that the slide moves (e.g. Chandler and Hutchinson, 1984). This then gives the opportunity to rebuild the defences. It may, in this case, be necessary to design the toe load so that this event is survivable also with (say) a 1 in 50-year return period groundwater level. A drainage system, not relied upon for the main stabilization works, may be employed to prevent the effects of rainfall or snowmelt adversely interacting with other destabilizing events. On this approach, the actual factors of safety achieved would

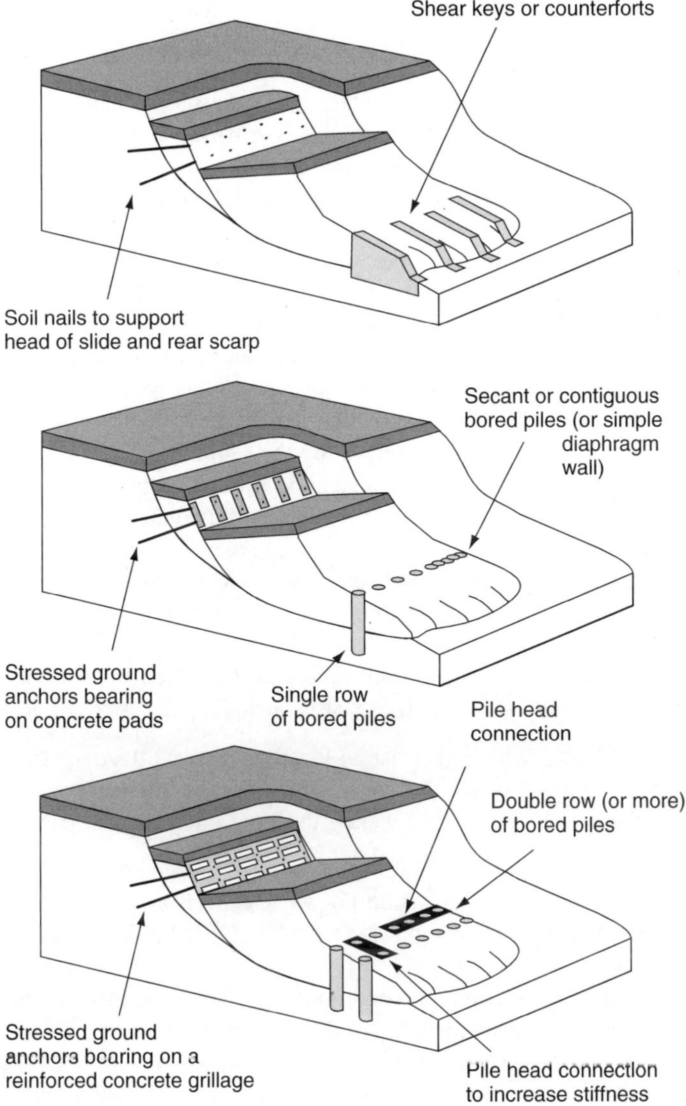


**Figure 19.21**  *Shear keys and piled systems are increasingly used as stabilization measures*

be immaterial, and the precise conditions which could cause failure to (re-)occur would be obvious. It is sadly the case that some engineers, as well as most clients, believe that remedial works for slope stabilization are good for all circumstances, and also for all time, without any maintenance.

This approach is best visualized in its entirety in the framework of a risk analysis (Fell and Hartford, 1997). If a remedial scheme does not offer a significant reduction in risk, then it is inappropriate.

## 19.9    Analysing Remedial Schemes

It is assumed that the reader is conversant with the basic principles of slope stability analysis as covered in innumerable elementary books on soil mechanics. In principle, the analyst considers the equilibrium of a mass of soil or rock which has a tendency to slide on a potential rupture surface. If the forces and moments which tend to cause movement (components of the weight of the sliding mass) exceed the resistance to sliding along the rupture surface, failure is to be expected, and the problem becomes one of consideration of the dynamics of movement, and where the mass comes to rest eventually. If, on the other hand, the resistance available is greater than that which is required to balance the destabilizing forces, then equilibrium is possible. The ratio of available forces (or moments) resisting and causing instability is expressed as a factor of safety.

The art of stability analysis nowadays lies in selecting the geological structure, appropriate parameters and soil conditions, and hydraulic regime for input into a 'black-box' commercial computer code, and the interpreting the results which come out of it. It was recognized long ago that the computational procedures of slope stability analysis are tedious and error prone, and thus the problem is an ideal application for the computer.

Conventional limit equilibrium stability analyses are at their best when used to study slides with curved shear surfaces, and water pressure distributions represented by a piezometric line within the soil or rock mass. They are used in two and three dimensions, and are applied to some slow flow problems. They throw up numerical instabilities in a number of well-understood cases, insight into which is given by Whitman and Bailey (1967) and Bromhead (1986, 1991). These cases include:

- piezometric conditions which give rise to locally negative effective stresses;
- shear strength properties which change rapidly along the slip surface;
- slip surface shapes which tend to separate the slide into different parts, with tension between them.

In addition, some methods of analysis do not work well, if at all, on some slip surface shapes: for example, Morgenstern and Price's method (1965) fails to converge for planar block slides.

The case of high local piezometric conditions might be experienced where there is a buried spring. A leaky site investigation borehole penetrating an artesian layer beneath a slide can have the same effect. So too can an underground drain, blocked at its exit. These cases are sketched in Figure 19.22. In some cases, the analysis can be made to work by simply taking wider slices in the area that is giving trouble, although this is not possible where the problem occurs at a location where the slip surface is significantly curved. If a wider slice is taken where the slip surface is planar, and the irregularities in ground profile are averaged out, piezometric conditions and variations of soil densities can work up to a point, although clearly when this procedure leads to a poor representation of actual soil conditions, errors will start to appear.

The problem of varying soil properties is one that was explored by Whitman and Bailey (1967). They found that a slip surface passing through a toe berm of granular material could give problems for some methods of analysis. The problem becomes particularly acute when the toe berm is of a material with a high angle of shearing resistance, especially relative to the rest of the slide, and the slip surface emerges upwards.

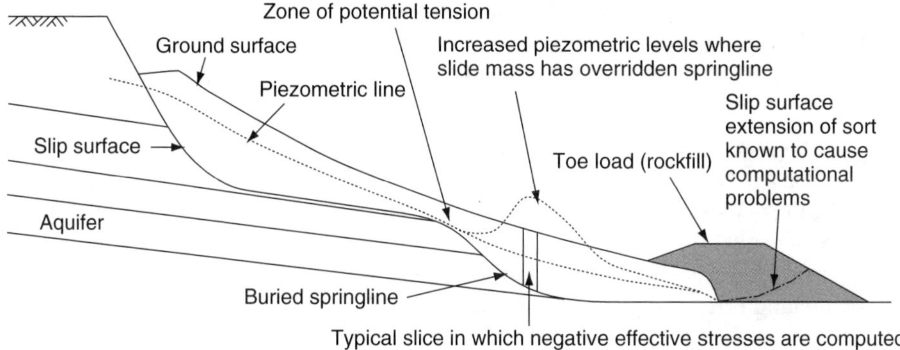


**Figure 19.22**   *Some of the situations that cause difficulties in limit equilibrium analyses*

Finally, a problem appears when exploring the stabilizing effect of very small changes in (say) porewater pressure in a large landslide. Then, the tolerance on convergence of the factor of safety may exceed the change in factor of safety calculated.

## 19.10   Simple Design Rules

It is when the decision is taken to at least consider the stabilization option that geotechnical investigations come to the fore. The primary tools of the geotechnical professional dealing with landslide hazard, properly investigated, are stability analysis and seepage (hydrological) modelling. Limit equilibrium stability analysis is best for computing the effects of toe loading and regrading, and of drainage. With appropriate modifications, limit equilibrium methods are adaptable to consider stressed ground anchors. It is least suited to assessing the effects of soil nails, dowels and other forms of passive structural intervention. Indeed, the notional factor of safety in these latter cases is not determined on the same basis as the former (mobilization of shear force on the rupture surface). Stability analyses work least well with rigid inclusions such as piles and shear keys spanning the shear surface. If standard analyses are done, inserting concrete properties in particular 'soil' zones, improbably large factors of safety are produced. Most software will do this without complaint. If the offending case is re-analysed with underriding or overriding slip surfaces, it is readily seen that this methodology is flawed. Piles and shear keys must be analysed using other methods (see below).

In my experience, rapid design assessments are best conducted from an initial evaluation of average levels of shear and effective normal stress in the landslide (Bromhead and Harris, 1986).

With the widespread use of computers, and in particular desktop computers, in geotechnical design offices, slope stability computations have become routine, and simple. There are, however, instances where the application of a few simple principles utilizing information routinely supplied by slope stability analysis programs (and equally as routinely ignored in the design office) can lead to rapid design decisions being taken. These can rule out much repetitive 'trial-and-error' work. These simple principles are described in detail in the Bromhead and Harris (1986) paper, with examples, and include:

Calculating stress levels in the landslide for correlation with laboratory test data (or indeed, for specification of test data at appropriate stress magnitudes)

(i)   Deciding what shear strength parameters give 'F = 1'
(ii)  Assessing whether drainage is a feasible measure
(iii) Location of cuts and fills on unstable slope, and approximate sizing of counterweight embankments
(iv)  Approximate calculation of anchor loads
(v)   Evaluating forces on walls in landslides.

These techniques are not intended to replace proper and systematic analyses, but to act in conjunction with them to speed up the design process.

Few procedures in geotechnical engineering, or for that matter other branches of civil engineering, are genuinely design methods. Engineering analysis is usually a matter of testing the feasibility or economy of a design concept. The evolution of a design is not therefore the direct result of the computational process, but is the result of a trial-and-error 'think of a concept and test it' procedure. Nowhere is this better shown than in slope stability analysis. By and large, slope stability analysis is undertaken with analytical computer software of a variety of levels of sophistication, but virtually all of this software postulates the existence of a 'design' or actual cross-section, for which it is desired to obtain some index of stability. This index might be the conventional factor of safety, or it might be the alternative critical seismic acceleration – see for instance, Sarma (1973). Generally, the index of stability is evaluated for a range of trial surfaces to find the most critical; or in cases where the position of an actual failure surface is known, a more restricted range of surfaces in the vicinity of the known failure surface is investigated. However, in no case does the method indicate what design slopes might be acceptable.

Even stability charts, which summarize the results of a number of slope stability analyses, are usually set up to yield a factor of safety: for example Bishop and Morgenstern's charts (1960). To evolve a design section requires several attempts until a satisfactory slope is obtained. The exception is in the case of the simple stability charts by Taylor (1937), from which a design slope can be obtained directly. However, any attempt to extend from a total stress analysis to one invoking effective stress behaviour (e.g. Spencer, 1967) by necessity adopts a recursive procedure since the behaviour of the seepage regime when the slope geometry is changed is not a linearly dependent function.

In the following, a number of simple but none the less useful techniques are presented, which speed up the process of finding the desired solution by eliminating many of the fruitless searching analyses. No claim is made that these eliminate the need for a final check analysis, nor that they are rigorous solutions (although some of them are). Their purpose is to assist in the evolution of the final design.

### 19.10.1   Calculating Stress Levels in the Landslide

The first step in many of the procedures outlined below is a stress analysis of the slide. In most methods of slope stability analysis, some stress analysis is done, but the details of this are usually lost on production of a 'factor of safety', although it is commonplace in non-circular slip analysis to be able to recover at least part of the stress analysis from the output. This takes the form of a table of 'interslice' forces, in many cases offered as a means by which the analyst may make a more or less subjective assessment of the 'reasonableness' of the solution in respect of two principal criteria:

- Is the location of the line of thrust within the section at all, or is it within the 'middle third' (both sorts of 'no-tension' criteria)?
- What is the mobilization of shear strength between the slices?

In the most straightforward procedure, the program will compute the shear and normal interslice force components explicitly (e.g. Morgenstern and Price's 1965 procedure), from which the equilibrium forces on the slip surface may be obtained. In some software these values form part of the printed output; in others, merely a summary, or, occasionally, nothing at all, is produced. The latter result is commonest with 'slip–circle' programs, where the output is extensive usually as a result of the analysis of a significant number of individual slip surfaces. Non-circular slip analysis programs usually have some output for the reasons discussed above. With commercial software, it is unlikely that the source code will be available for modification, and whatever output is obtained has to suffice.

If sufficient information is included in the output, then the levels of shear and normal stress acting on the slip surface can be derived from the summed force resultants and the length of slip surface affected. This simple expedient may not be a practical approach, however, and the following procedure can be used in its stead.

Three parameters define the stress conditions acting along the slip surface of a particular landslide. They are:

- The average shear stress ($\tau_{av}$);
- The average normal total stress ($\sigma_n$), or average normal effective stress ($\sigma'_n$);
- The average porewater pressure, or the ratio of average porewater pressure to average normal total stress ($u_{av}/\sigma_n$). The latter is termed the pore pressure ratio, $r_u$. This parameter may be used as a simple way of defining porewater pressure conditions in an analysis, or it may be one of the outputs.

To obtain the average shear stress acting along the slip surface, perform an analysis using a single cohesive shear strength parameter acting along the whole length of slip surface. Suppose this to be $c^*$. A corresponding factor of safety is obtained from the analysis, $F^*$. It will be found that the cohesion or shear strength(s) required to be available for a factor of safety of 1.0 is:

$$s = c^*/F^*,$$

and this is also the average shear stress ($\tau_{av}$) acting around the slip surface. Indeed, it can easily be shown to be so for a slip circle, although it is less clear in the case of a non-circular slip surface, and there may be a 2 to 3% difference between the 'average' shear stress computed by means of this rule, and that obtained by summing the shear forces and dividing by the appropriate length of slip surface. In some slip circle software, an overturning ($M_o$) or resisting ($M_r$) moment is output. The average shear stress can be obtained from this without an analysis with $c^*$, since:

$$M_r = FM_o = sLR,$$

in which $L$ is the length of slip surface, and $R$ is its radius.

However the average shear stress is obtained, it is thereafter used in the estimation of the average normal effective stress $\sigma'_n$ by means of an analysis involving an angle of shearing resistance $\phi'$ (with or without a corresponding $c'$) to produce a new factor of safety $F^+$, and then:

$$F^+ = (c' + \sigma'_n \tan \phi')/s$$

so that $\sigma'_n$ may be found from:

$$\sigma'_n = (sF^+ - c')/\tan \phi'$$

This gives the 'working' average stresses. One further parameter, the average total normal stress, can be obtained from the above equation for $\phi'$ by first re-analysing the section with zero porewater pressures. It will be found that a minimum of data manipulation is required, if within the data submitted to the program used it is possible to:

(a) set all $r_u$ values to zero; or
(b) take the unit weight of water to be zero.

The average porewater pressure ratio $r_u$ for the slip surface is the ratio of average porewater pressure to average normal stress (usually defined in terms of average vertical stress), and may be computed from the following:

$$r_u = (1 - (\sigma_n - \sigma'_n)/\sigma_n)$$

Sometimes, $r_u$ is available as an input parameter. Setting $r_u$ equal to zero removes porewater pressures; setting it equal to $\gamma_w/\gamma$ is equal to putting the piezometric line at ground level.

### 19.10.2   Finding Shear Strength Parameters for '$F = 1$' – the Essence of Back-Analysis

Shear strength parameters may be obtained in two ways: from the stress analysis outlined above, or by manipulations of the trial values of $c'$ and $\phi'$ and corresponding factor of safety. Note that there is an infinity of possible $c' - \phi'$ combinations to yield the same factor of safety, and it is commonplace (indeed correct in many cases) to assume $c' = 0$. In this situation, $\phi'$ for $F = 1$, often styled the *mobilized angle of shearing resistance*, $\phi'_m$, is simply

$$\phi'_m = \tan^{-1}(s/\sigma'_n);$$

it is also simple to show that, in an analysis using $\phi'$ alone to yield a non-unity factor of safety $F^+$,

$$\phi'_m = \tan^{-1}([\tan \phi']/F^+)$$

Were there a cohesion present in this analysis, one would also predict a mobilized cohesion $c'_m$ equal to $c'F$.

### 19.10.3  Answering the Question: Is Drainage Feasible?

The feasibility in principle of drainage as a remedial measure is indicated by the magnitude of $r_u$. Where $r_u$ is high, drainage can produce a large increase in stability, but where it is low, drainage is of proportionately less use. To estimate the maximum effect of drainage, that is when all porewater pressures have been removed, compute $F'$ from the original factor of safety $F_0$, where:

$$F' = \frac{s + \sigma_n \tan \phi'}{s + \sigma'_n \tan \phi'} F_0$$

or

$$F' = F_0/(1 - r_u)$$

In the case of a non-cohesive soil:

$$F' = F_0 \sigma_n / \sigma'_n$$

Obviously, if $F'$ is not as high as the desired factor of safety, $F_r$, drainage on its own is not feasible. Suppose that $F'$ exceeds the required factor of safety; what average drainage is required?

$$F_r = \frac{s + \sigma_{nr} \tan \phi'}{s + \sigma'_n \tan \phi'} F_0$$

from which we find the required average normal effective stress $\sigma'_n$ since the average drawdown of the piezometric line must be

$$(\sigma'_{nr} - \sigma'_n)/\gamma_w$$

Is this obtainable? More to the point, since a uniform drawdown is not usually achievable, with only small or perhaps even no drawdown at each end of the slip surface, so that the drawdown is distributed linearly or (as shown in Figure 19.23) parabolically, can twice or one-and-a-half times this be obtained as a maximum? (Note: the average of a parabolic distribution is 2/3 the maximum.)

### 19.10.4  Where to Place Counterweight Embankments, and How Big Should They Be?

The optimum location of a counterweight embankment is indicated by the 'neutral line' theory (Hutchinson, 1977, 1984b). The 'neutral line' is the trace, on a plan of a particular landslide, of a series of 'neutral points'. Each of these 'neutral points' is defined for a cross-section as the location on the ground surface (or on the slip surface perpendicularly beneath it) at which the imposition of a load has no influence on the factor of safety. Loads placed on each side of the neutral point will have a beneficial or detrimental

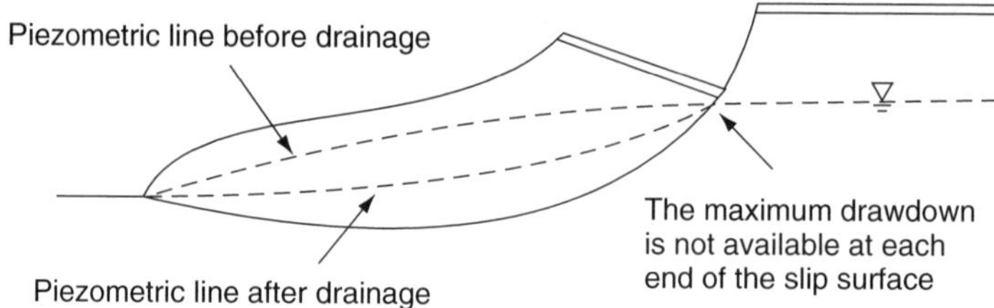



**Figure 19.23**  *Even in a simple case, the response to drainage across a slide is likely to be non-uniform, and the average effect is much less than the maximum effect*

effect on the factor of safety. Attention was drawn by Greco (1991) to the fact that the neutral line approach is only valid for slopes with pre-existing surfaces of rupture which continue to be followed: if the rupture surface migrates or is ill defined, the method may be misleading. However, the post-rupture strength of clays means that slides almost always follow the original rupture surface, and the method is of widespread applicability. Hutchinson (1983) describes the process of locating the rupture surface or slip surface of landslides in clays.

Counterweights also represent two extreme cases in terms of the generation and dissipation of porewater pressures under the applied loading. Hutchinson (1977) terms these respectively fully undrained and fully drained loading. An intermediate case, where either the amount of pore pressure generated is less than the amount of the total stress change (i.e. $\bar{B} < 1.0$), or where some consolidation has taken place from an initially higher generated porewater pressure, can be envisaged.

$\bar{B}$ relates to pore pressure generation by total stress change only. This theory relates to pore pressure change, by undrained loading, followed by drainage-related decrease in pore pressure. Hence the introduction of the new composite parameter $B^*$. Should the increase of porewater pressure in this intermediate case be represented by a parameter $B^*$ times the increase in vertical stress, the location of neutral points may be found from positions on the slip surface which satisfy the following equation:

$$\tan \alpha = (1 - B^* \sec^2 \alpha) \tan \phi' / F$$

in which $\alpha$ is the angle of inclination from the horizontal of the slip surface. $F$ is the target final factor of safety. This resolves a paradox in the original Hutchinson theory where the neutral points move differently if a load is applied in two parts or in one step.

There must be neutral points for drained loading ($B^* = 0$), undrained loading ($B^* = 1$) and intermediate points. The first step is to locate the neutral points. If more than one cross-section through the landslide has been investigated, we can draw the neutral line. This delineates feasible locations for the placement of fills, or for the excavation of cuts in the slope. Figure 19.24 shows this for a fairly simple slide. Figure 19.25 shows how a complex slide may have a number of neutral points.

If it is not possible to place fills where the neutral line delineated fill areas are shown, and at the same time it is not possible either to make cuts where the method indicates potential cut areas, then cut and fill solutions are not possible. However, when the neutral

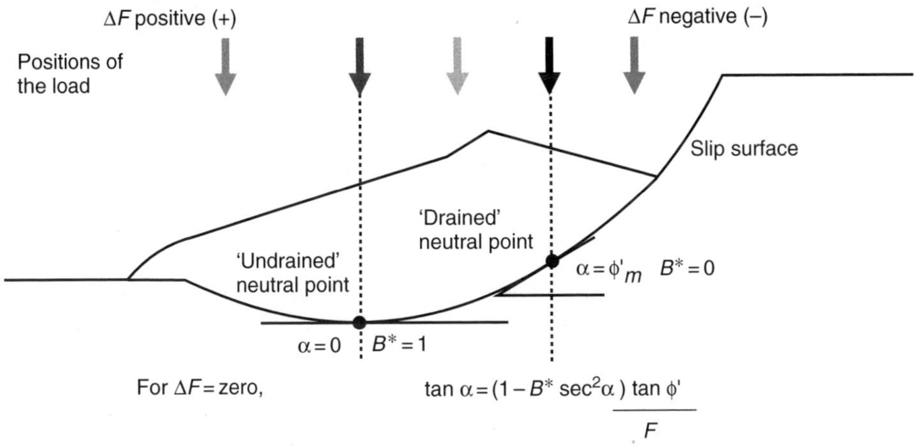



**Figure 19.24**  *Hutchinson's neutral line theory is illustrated with a simple rotational slide*

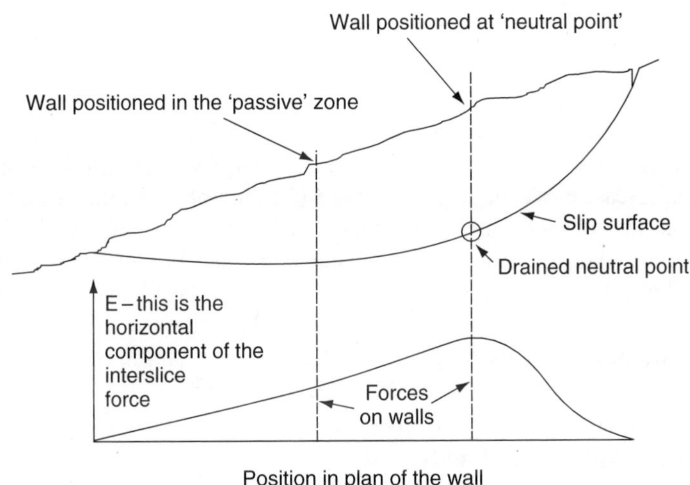



**Figure 19.25**  *Hutchinson's neutral line theory in practice may require the slide system to be subdivided into a variety of zones*

point theory indicates that cuts and/or fills might be permissible, there is the problem of deciding their approximate size. The following is a possible approach:

1. On the section, divide the slide into 'active' and 'passive' zones by a vertical through the appropriate neutral point. Measure areas in the active and passive zones ($A_a$ and $A_p$ respectively in Figure 19.26).
2. Express the ratio of these $R_{oa}$ as:

$$R_{oa} = A_p / A_a$$

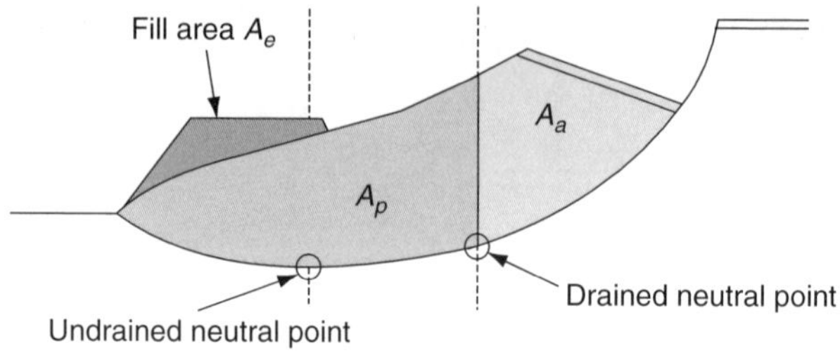


**Figure 19.26**    *Approximate sizing of counterweight embankments using area ratios can be improved considerably if calibrated by some actual analyses*

3. Following a stabilizing muckshift, the new area ratio $R_{1a}$ (defined in the same way as the original area ratio $R_{oa}$) should be increased in approximately the same proportion as the factors of safety, that is,

$$R_{1a}/R_{oa} = F_1/F_0 = (R_p + R_e)/R_p$$

Note: This is *very* approximate.

Naturally, in the commonest approach, which is to use toe loading without corresponding slide head cuts, the ratios can be evaluated in terms of the passive areas alone, since the active area is a common factor. Partial sums of slice weights can be substituted for cross-sectional areas, especially where these form part of the initial stability analysis output, without loss of 'accuracy', since this is only a rule of thumb. It should be noted that this does not take into account the greater unit efficiency of fill placed at the toe compared to fill placed nearer the neutral point, and so the likely sign of the error in using this approach can be inferred. Furthermore, the method is really only applicable where the earth-moving operation produces a small effect on the overall slope section, rather than inflicting gross changes on the slope geometry.

### 19.10.5    Forces on Walls in Landslides

A retaining wall in a landslide may have one of several roles:

- It retains fill, which in turn acts to resist movement: the wall has no structural loading specifically arising from its retaining action on the slide (Type 1).
- It is a continuous wall, positioned so as to resist the action of the slide. The loading on the wall arises from movement of the slide (Type 2).
- It is an intermittent structure, like a row of piers of a viaduct penetrating the slide mass, and acted on by forces as the slide moves past the structure. The structure may not prevent movement of the slide as a whole, but will impede its action locally (Type 3).

A Type 1 wall can be designed using ordinary earth pressure theories in most cases, or it may act as a wave protection structure if it retains the toe fill, stabilizing a coastal landslide.

Type 2 walls are more difficult. Essentially, they are passive structures, and can only be stressed up by some further movement of the slide. The key element in analysis and design is to estimate the forces on such a wall. As a first step, consider the forces between the slices at the intended wall position. Obviously these increase from zero at the toe of the slide to a maximum (which is found at the 'neutral point'), and then fall to zero at the head of the slide. If a wall at a certain position (Figure 19.27) is capable of resisting a force equal to the interslice force, then to all intents and purposes the soil below the wall could be removed, without in any way adversely affecting the stability of the given slide (but see below for other effects when the soil is removed!). A wall capable of resisting $X$ times the computed force would have a factor of safety of $X$ against its own failure. It does not raise the factor of safety of the slope, as conventionally calculated, but does increase its reserve of safety against being destabilized by some other means, for example an unloading at the toe, or a rise in the groundwater table. This approach leads to the design of walls much lower in the slide mass (where the forces are lower), and should be used with caution if the possibility of 'overriding' slip surfaces, as shown in Figure 19.17, is likely.

An alternative approach is to examine the 'shortfall' in resisting force, in the case of a slip circle by examining the relation:

$$\Delta F_{\text{required}} = \Delta M_r / M_o$$

and then taking $\Delta M_r$ as the moment to be resisted. Force equilibrium methods will obviously yield a shortfall in force from an equation analogous to this one. The provision

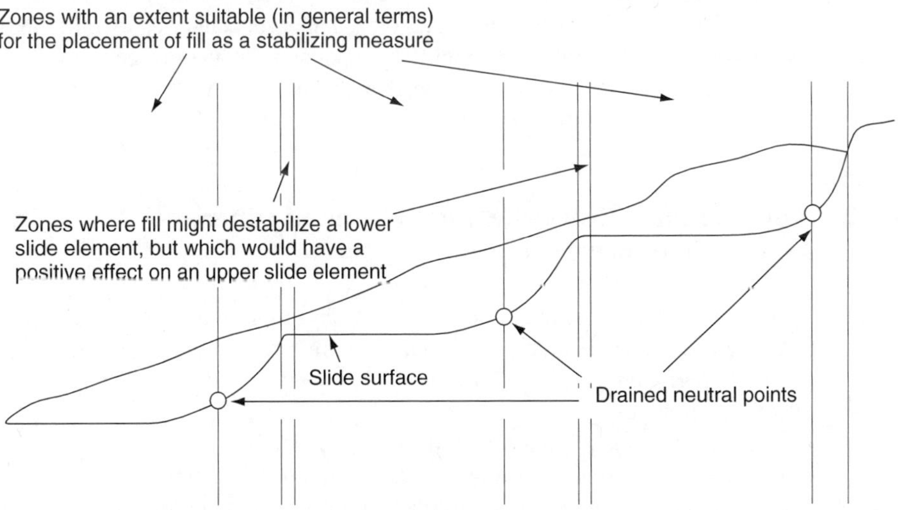


**Figure 19.27** *The 'interslice forces' represent how much force there is to resist if the slide mass downslope is removed. A wall inserted into a slide mass will increase the shear resistance overall. The improvement in factor of safety depends on the notional force the wall can carry*

of a retaining wall capable of supplying the shortfall increases the apparent factor of safety, and can be seen to improve the safety factor to any desired level. The optimum position for a wall designed on this basis is at, or slightly below, the neutral point.

Non-continuous (Type 3) structures pose yet another class of problem. It is possible to view them as rigid obstacles to the progress of the landslide, so that it slides or flows around them. The forces on the piers then amount to the maximum force that the slide can exert. Naturally, this cannot be more than the interslice force at the appropriate slide section. An alternative viewpoint is to see the piers as moving upslope through the slide, carving out a swathe of soil in a passive 'bow wave' as they do so. The forces on this passive bow wave can then be calculated from (passive) earth pressure theory, including the side friction on the passive wedges. Naturally, the sum total of these forces, averaged so as to apply to 1 metre width of the slide, cannot exceed the interslice force: it is perhaps easier to see why from this viewpoint – if they were to do so, the upper part of the slide mass would be carried uphill! An obvious impossibility, more so since the forces exerted on the slide by the piers are reactions to the loads that the slide, under the effects of gravity, applies to the piers.

## 19.11   Bioengineering

It was noted above that a good vegetation cover on a slope is most usually beneficial, since it moderates the infiltration characteristics, controls runoff erosion, and contributes through transpiration to reducing the soil moisture. There are innumerable cases of shallow slides being initiated after deforestation of slopes, and this supports that general conclusion. Conversely, tree root growth opens fractures in rock masses, and wind loading may complete the process of detachment of the root-opened rock and tree together into a rockslide or rockfall. On the whole, the weight of even a dense cover of trees is marginal or insignificant except for the shallowest of landslide types, and the same may be said of the overall effect of root reinforcement.

## 19.12   Assessing the Efficacy of Strategies, and of Different
##              Types of Remedial Measure

First of all, we must consider the overall efficacy of each of the four main strategies. We can start with the acceptance strategy. It is sometimes referred to as 'Do nothing'. This strategy appears superficially attractive at times of low landslide activity. Doing nothing represents least up-front cost.

In the UK, coast erosion is often a source of landslide activity. Coastal sites may have geological or other natural environment interests, which militate against any engineering intervention. Coastal processes including littoral drift, interruption of sediment budget and so on provide additional pressure against intervention. Relict landslide masses, sometimes extending for several kilometres along the coast, and up to 500 m from toe to crest, are found in several locations. Activity is low, with extremely slow to very slow movement (if any) recorded according to the Cruden and Varnes (1996) scheme (refer also to Varnes, 1958). Extremely rarely, episodes of moderate movement rates are experienced.

These landslides are subject to continual coastal erosion, sometimes causing recession of a coastal cliff, and where basal shear surfaces lie beneath sea level, foreshore lowering is an additional effect. These landslides are also subject to fluctuating groundwater levels as a result of cyclic weather variations (most notably seasonal variation).

### 19.12.1   Case 1: Folkestone Warren, Kent

A large coastal landslide system between the coastal towns and English Channel Ports of Folkestone and Dover is known as Folkestone Warren. The stratigraphy in this area is chalk overlying gault (an overconsolidated clay) overlying Folkestone Beds (a sand within the lower greensand). These beds dip gently coastwise to the east and slightly inland. Landsliding follows a bedding-controlled basal shear surface close to the base of the gault, and landslides occur from where the base of the gault descends to beach level up to where it disappears below the beach. Landsliding is then replaced as the primary mechanism by marine erosion at the base of the cliff and chalk falls and slides.

Coastal erosion downdrift of Copt Point (Figure 19.28) has formed a log-spiral bay, and it is believed (Hutchinson *et al.*, 1980) that construction and extension of the harbour in the nineteenth and early twentieth centuries has intercepted littoral drift and intensified erosion. This landslide system is of particular interest since it is crossed by the railway line from Folkestone to Dover, built in the 1840s. The railway was periodically affected by chalk falls from its rear scarp, mass movements of the whole complex (particularly in 1877, 1896 and 1915), and movements of the frontal margin of the slide complex, sometimes but not always affecting the railway. Following Hutchinson (1969), the massive movements of the whole complex are termed 'M'-type slides, and the rotational movements of parts of the coastal margin 'R'-type slides (Figure 19.29).

Landslide treatments in response to the elements of the problem are as follows. A seawall and groynes have been constructed to control erosion. These are regularly undermined and need repair. Rock armour is used now in preference to reinforced concrete seawall work. A toe-weighting embankment armoured with concrete was constructed in the 1950s. This has suffered distortion from landslide movements (and from impact by a shipwreck!). Hutchinson *et al.* (1980) consider that it improves the stability of the frontal margin of the Warren by 45% (maximum) and of the whole complex by 4–5% (maximum). A system of 22 drainage adits draws down the water table in the main slide complex. Water is discharged at top of seawall or top of toe-weighting level. Nothing has been done about chalk falls from the rear scarp, although a watch is kept.

The primary question is: 'Has the stabilization of the Warren been effective?', with supplementary questions: 'What elements of the scheme are best or worst?' and 'What else needs to be done?' With three M-type slides between 1846 and 1915, and none since, the remedial measures are apparently effective. However, movements of R-type slips are common, and these do affect parts of the track, and interfere with access roads to amenity sites on the Warren as well as access to the coast defence. Clearly, it is impossible to identify individual coastal erosion events responsible for instability. However, the extraordinary wet weather of 1915 is often cited as a trigger factor for the landslide events of that year, and so the weather record was examined.

The rainfall, shown for Cherry Gardens 1868–date, is extremely variable. Clearly, a study of the hydrology of the Warren needs to consider nett infiltration, not simply gross rainfall, but the two are related, and for simplicity, the gross rainfall is plotted on

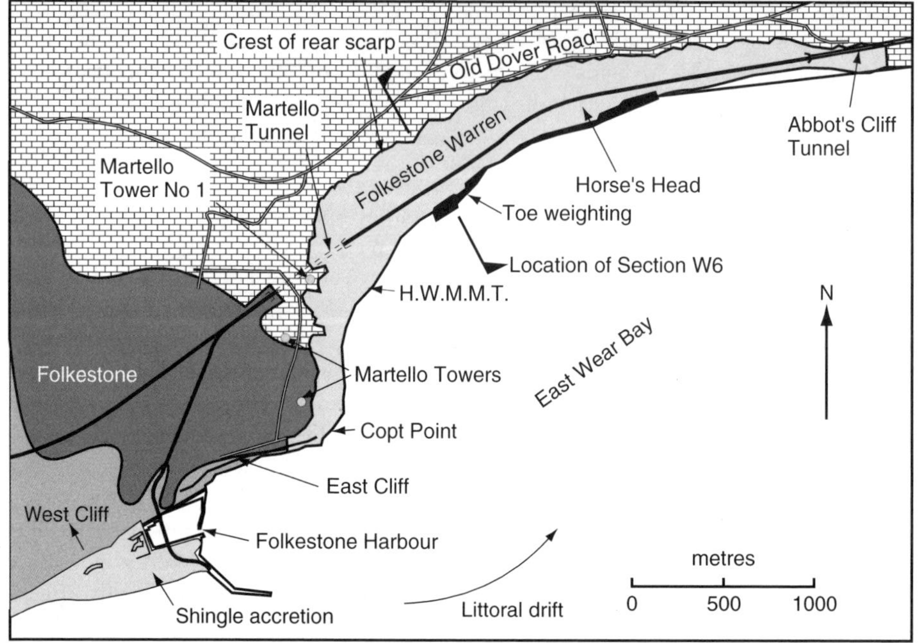



**Figure 19.28** *Map of the landslides at Folkestone Warren (see Hutchinson et al., 1980, reproduced by permission of The Geological Society)*

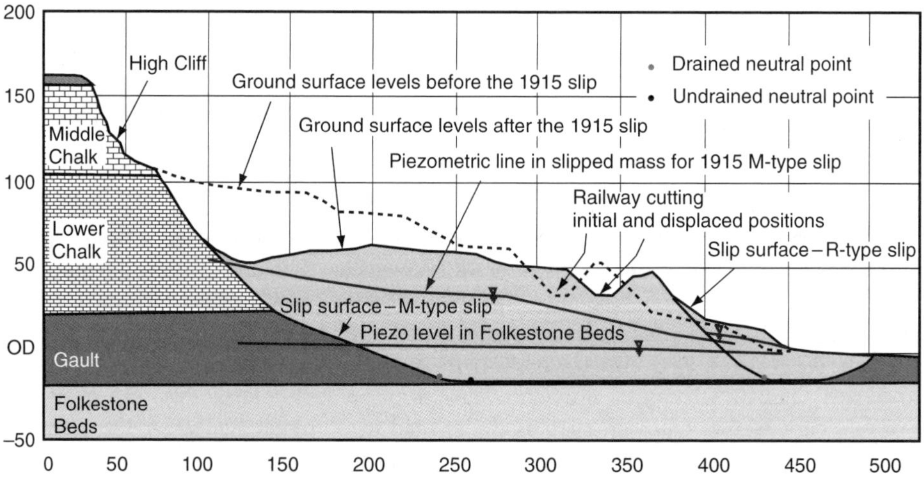



**Figure 19.29** *Cross-section W6 of the Folkestone Warren landslide (after Hutchinson et al., 1980, reproduced by permission of The Geological Society)*

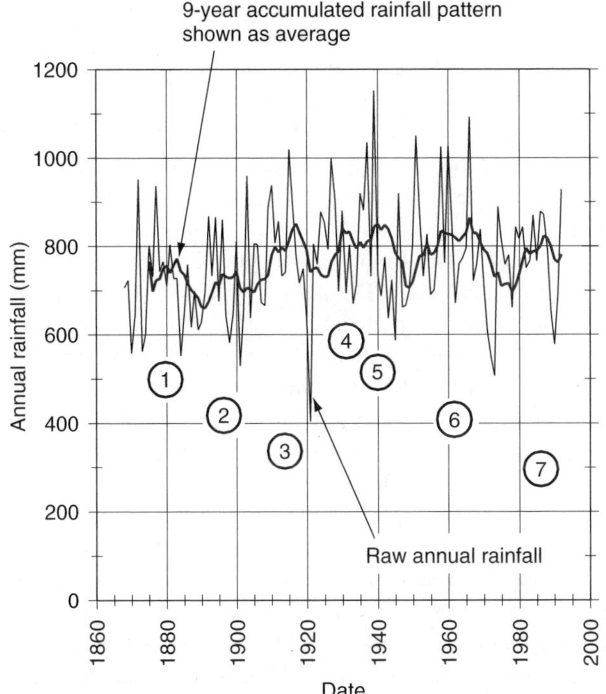



**Figure 19.30**  *Annual rainfall records for the Folkestone area, with the 9-year moving average superimposed. Landslide activity in the historic record is noted (nos 1 to 7)*

Figure 19.30. Folkestone Warren is too big to respond to just one day/week/month of rain – there simply isn't enough water to raise the groundwater table appreciably. This would need a run of wet years. It is, after all, the difference between mean rainfall and actual rainfall that raises the groundwater table. In order to gain some appreciation of the effects of runs of wet years, research has considered the optimum run length, but for the purposes of this assessment, 9-year cumulative rainfall has been considered (Ibsen, pers. comm.). This is plotted as the 9-year average on the graph to preserve the scale. A trend of increasing rainfall forms the background to the problem. Clearly, without coastal erosion, stability of this landslide system was going to reduce.

The 9-year cumulative rainfall shows clear peaks in 1877, 1896 and 1915 – years of M-type landslides. The 1915 peak contains exceptional rainfall in that year as well as a long-term running average more severe than the two preceding cases. It should not be surprising that there was a sequence of events from chalk falls to an M-type landslide in that year. It is noted that the slip in 1877 was an M-type landslide, although this was probably a chalk fall, based on the famous photo of Edward Watkins (Chairman of the railway company) on site during the remedial works. This could correspond to the peak rainfall of 1877. Associated ground movements might have followed in the succeeding years as the rainfall built up to an accumulated high.

The M-type landslide reactivation of 1896 occurred against a background of lower peak rainfall, and a lower accumulation. Cracking occurred in the Martello Tunnel on the western part of the Warren, so the tunnel portal was moved westwards and a cutting was opened out over the tunnel by the railway company. The rationale was to minimize the dig depth if a rescue was later needed, but to continue to provide the protection against chalk falls that the railway has when in tunnel. Spoil was placed seaward of this location.

In 1915, the rainfall had both a large long-term accumulation and a high peak. Both factors were greater than ever before in the record. It took some 8 years to reopen the line. Part of this time was spent constructing drainage adits. Adit construction had started earlier, but the 1915 slip overwhelmed the workings. Slips at the Warren were also noted in the records in 1937 and 1940. Both were R-type. The 1937 slip involved chalk spoil from the Martello cutting excavation. Detailed subsurface investigations commenced, and Terzaghi was consulted. The accumulated rainfall was on two occasions as severe as in 1915; the annual rainfall (especially in 1939) was worse. However, M-type movements did not occur. This is evidence that the initial adit construction was highly effective. Stabilization works were put into abeyance during the Second World War. Afterwards, a programme of works was undertaken. Toe weighting was constructed and the adit system made more comprehensive (Viner-Brady, 1955).

In the 1960s there was again high accumulation of rainfall, and high individual year peaks. Movements of the R-type occurred at Horsehead Point. Some small R-type landslide movements continued throughout the 1980s, when there was a run of years without a particularly dry year, so that a long-term accumulation of rain occurred without a high annual peak, and they were particularly severe in the winter of 2000–2001. Without remedial works we could easily have had 1915-scale movements in 1940, 1960/66, and movements similar to the 1877/96 events in 1980/85 at the very least. The conclusions are drawn as follows:

1. The adit construction after 1915 had a major impact on the stability of M-type landslides, but no effect on R-type landslides.
2. The toe weighting, with a small calculated improvement for M-type landslides and a big one for R-type slides, in actually ineffective for R-type slides.
3. The reason for this is probably that the adits leak water into R-type slides at a high elevation where the floor of the adit is cracked. Carrying drainage discharges across an active landslide system is poor practice, although unavoidable.
4. Seawall construction and rock armouring may 'hold the line' against erosion, but provides no positive improvement in stability.

### 19.12.2   Case 2: Small Chine, Isle of Wight

The second case is also a coastal landslide, and comes from my files, but it is on a much smaller scale than the Folkestone Warren landslide system. This is a small landslide on the southwest coast of the Isle of Wight (southern England), which has taken place in Wealden clays at a location known as Small Chine. The Wealden strata dip inland, and as marine erosion cut through a series of resistant sandstones, it exposed clay strata. Two clay strata are involved, which gives the landslide rupture surface a staircase form. Furthermore, the combination of marine erosion and landsliding was opening the pathway to the sea for the whole landslide system. This had the result of increasing the width of the landslide and the rapidity with which it sapped back inland towards the main coastal highway. The landslide is also bounded in part by the line of a fault.

The coastal highway under threat from the inland enlargement of the landslide system was built in the 1860s as a defensive work against a perceived invasion threat. In the 1930s, it came into the ownership of the local authority. It had already been breached and, to bring it up to even the standards of that time, some realignment of the route was undertaken, along with tarmacadam surfacing. Today, the road forms a tourist amenity. It is, however, under threat from cliff line retreat in a number of locations. Of these, the critical ones were associated with the section of road where it runs through a National Trust area near the crest of a high chalk cliff. A public inquiry was to be held into a proposed route change for the road. However, on the basis of retreat measurements, the road was liable to be breached at Small Chine before the public inquiry, or long before road realignment could take place.

Already, a gravel-filled cutoff trench drain was installed around the landslide, with a remote outfall. This had proved ineffective: no decrease in retreat rate was experienced. Trial pits showed why – there was no flow through the superficial soils towards the slide, which was driven by a combination of rainfall landing on its surface and marine erosion. A system of trench drains in the landslide itself was designed and installed in 1985. A 5-year target lifespan was required. It was the intention to provide sufficient increase in safety to cope with 5 years' retreat of the toe without reversion to instability. These new gravel-filled trench drains are shown in Figure 19.31.

In the event, the public inquiry was lost, and the works were abandoned. They did survive the extreme wet weather of 1988, itself the culmination of a run of wet years. However, this landslide system is small, and responds not to numerous years of accumulated rainfall, but to single-year (or shorter) periods of intense rainfall.

**Figure 19.31**   *Shallow drainage in the landslide at Small Chine, Isle of Wight. The view is to the northwest. Gravel filled drains are approximately 0.7 m wide*

The drainage system began to break up about 10 years after installation, and today is almost completely ineffective. Landslide movements undermined the rear scarp, and retreat of the scarp recommenced. Two consecutive retrogressive slides have occurred, the second taking the line of the ineffective cutoff drain as a preformed tension crack.

The efficacy of the works was thus as follows:

1. Cutoff drain. Completely useless, failed to even reduce recession rate of head of the slide.
2. Drains in the landslide. Worked as planned. Perhaps overdesigned, especially as coastline retreat was less than predicted; also survived particularly heavy rainfall years not anticipated in design. Thus lasted twice as long as desired.
3. When the toe erosion overcame the benefit from the drains, slide head retrogression restarted, and the cutoff drain served as a convenient tension crack. Soil nailing (not an option at the time of construction) would probably have proved more effective.

### 19.12.3   Case 3: Highlands Court, Southeast London

This landslide was triggered by excavations at the toe of a slope in London Clay for the mundane reason of increasing light to the lowest storey of an newly constructed apartment block. The slip undermined another apartment block upslope (Allison *et al.*, 1991).

Although the parties and their engineers cooperated in the early stages of investigation, litigation over the landslide put a stop to cooperation over the remedial scheme. The owners of the lower site had a piled scheme with a grid of bored piles installed in their site. This should have been sufficient to stabilize the whole slide. It did prevent movements towards the lower apartment block (Figure 19.32).

However, settlements at the head of the slide continued, further threatening the upper apartment block, and threatening to permit further retrogression upslope. The owners

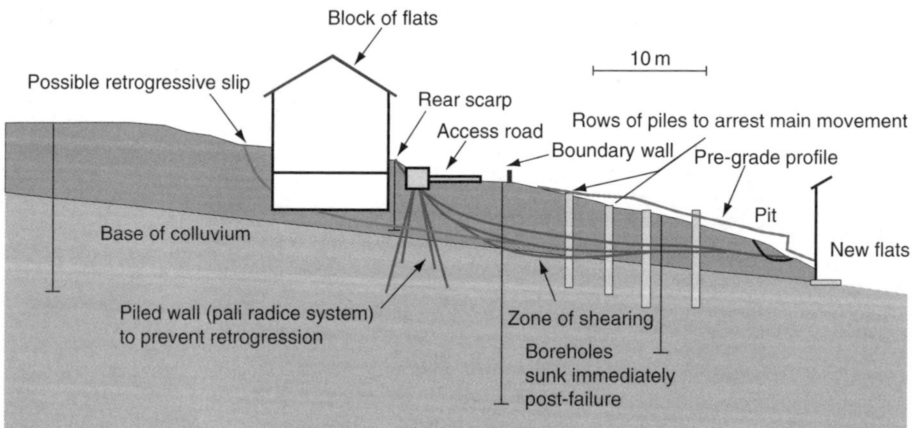


**Figure 19.32**  *Schematic cross-section of the Highlands Court landslide, S. London (see Allison et al., 1991, reproduced by permission of Thomas Telford Ltd)*

of the upper site needed to install their own combination of retaining structure and underpinning, based around a pali radice proprietary piling system. Eventually, in the absence of good vegetation cover on the lower slopes, the slide mass began to erode into a series of very small mudslides between the piles, and a system of 1 m high (approx.) retaining walls spanning between the piles was constructed. The differential movements in the landslide were a confirmation of the Picarelli *et al.* (1995) model for slow-moving landslides. The primary conclusion here was that even though calculations showed that the original piled scheme was satisfactory for the whole landslide, it did not meet the actual needs of at least half of the people whose properties were under threat.

## 19.13   Negative or Ineffective Treatment

Two small case records illustrate this. Engineer X was consulted as to the lifespan of a house at risk from coast erosion. He consulted old maps, and determined an average coastal cliff recession rate. On this basis, the house had a *c.* 200-year life expectancy. On the basis of his advice, the house was purchased. Within a few years, the cliff recession rate changed, and the house was damaged by extraordinarily rapid recession of the cliff. Engineer X is now being sued. The lesson is that the rate of natural processes can increase significantly, and, accordingly, assessments of risk based on past behaviour are necessarily imprecise.

Coastal stabilization at Charmouth involved stabilization of a coastal landslide by trench drains, the construction of a rock-armoured toe to prevent shoreline changes, and soil nailing to stabilize the rear scarp – the three elements installed in that sequence. During construction of the drainage system, formation of drain outfalls was prevented by high tides. Heavy rainfall filled the drain system, and led to failure involving retrogression of the rear scarp.

In this second case, the outfalls should have been constructed first. Alternatively, the soil nailing of the rear scarp might have prevented the retrogression of the slide into previously unfailed ground.

## 19.14   Conclusion

Slope stabilization by the construction of engineering works may be neither cost-effective nor strictly necessary. There are various of other methods of dealing with landslides that may be more appropriate in a given case. Where engineering works are undertaken, the engineer has an enormous range of options – a choice that may seem bewildering at first. The continual development of new materials and techniques adds to the list. However, in most practical situations the list quickly resolves itself into relatively few techniques. For example, a wholesale regrading of an unstable slope occupied by a historic monument which it is desired to preserve would not be an appropriate choice, and a scheme would be required which sympathetically treated the topography in the immediate vicinity of the monument.

Engineered remedial works should only be undertaken except in emergency following a full subsurface investigation which gives understanding of the internal structure and mechanics of the landslide in question.

The choice of treatment methodology for landslides can therefore be considered as a four-way choice, based on the component parts of the risk equation. Only one of these choices is stabilization. For non-stabilization options, the primary risk is that future landslide behaviour will differ markedly from past behaviour.

Small slopes can be stabilized to a factor of safety that entirely overcomes all negative effects, acting singly or in combination; large slopes cannot. We can never be entirely sure in a partial stabilization of a large slope that conditions in future (e.g. rainfall) will not exceed our expectations. There is also a converse problem that if conditions are better than expected – for example a period of drought – this may lead to otherwise necessary stabilization measures being deferred, or to the engineer being accused of overcaution.

Combinations of techniques may disguise the fact that one or more elements of the scheme is underperforming. This chapter also gives some guidance on approximate sizing of the remedial scheme based on simple parameters for the landslide. Most landslides which are amenable to stabilization by engineering works are rotational and translational slide types, especially those which are comparatively slow moving, or stationary and where reactivation is feared. Structural solutions usually rely on the landslide body retaining some coherence, and since they take time to construct and require safe access, these solutions are inappropriate to fast-moving landslides or flows. In these cases, catch or deflection structures may be used, or a different strategy altogether must be used to reduce landslide risk.

# References


Allison, J.A., Mawditt, J. and Williams, G.T., 1991, The use of bored piles and counterfort drains to stabilize a major landslip – a comparison of theoretical and field performance, in *Slope Stability Engineering – Applications and Developments* (London: Thomas Telford), 347–354.

Angelli, M.G., Gasparetto, P., Menotti, R.M., Pasuto, A. and Silvano, S., 1994, A system of monitoring and warning in a complex landslide in Northeastern Italy, *Landslide News*, No. 8. August, 12–14.

Ayres, D.J., 1985a, Stabilisation of slips in cohesive soil by grouting, *Technical Note 9, Failures in Earthworks* (London: Thomas Telford), 421–427.

Ayres, D.J., 1985b, Treatment of shallow underground fires, *Technical Note 23, Failures in Earthworks* (London: Thomas Telford), 470–472.

Baldwin, J.E., Donley, H.F. and Howard, T.R., 1987, On debris flow/avalanche mitigation and control, San Francisco Bay area, California, *Geological Society of America, Reviews in Engineering Geology*, **VII**, 223–236.

Barley, A.D., 1991, Slope stabilization by new ground anchorage systems in rocks and soils, in *Slope Stability Engineering – Applications and Developments* (London: Thomas Telford), 335–340.

Beles, A.A. and Stanculescu, I.I., 1958, Thermal treatment as a means of improving the stability of earth masses, *Géotechnique*, **8**, 158–165.

Bianco, B. and Bruce, D.A., 1991, Large landslide stabilization by deep drainage wells, in *Slope Stability Engineering – Applications and Developments* (London: Thomas Telford), 319–326.

Bishop, A.W., 1973, The stability of tips and spoil heaps, *Quarterly Journal of Engineering Geology*, **6**, 335–376.

Bishop, A.W. and Morgenstern, N.R., 1960, Stability coefficients for earth slopes, *Géotechnique*, **10**, 129–150.

Bromhead, E.N., 1984, An analytical solution to the problem of seepage into counterfort drains, *Canadian Geotech Journal*, **21**, 657–662.

Bromhead, E.N., 1986 and 1991, *Stability of Slopes* (Edinburgh: Blackie & Sons).

Bromhead, E.N. and Dixon, N., 1984, Pore water pressure observations in the coastal clay cliffs at the Isle of Sheppey, England, *Proceedings of the 4th International Symposium on Landslides*, Toronto, vol. 1, 385–390.

Bromhead, E.N. and Harris, A.J., 1986, Rapid design assessments from slope stability calculation results, in *Proceedings of the Conference on Computers in Geotechnical Engineering*, A.I.T. Bangkok (Amsterdam: Balkema).

Bromhead, E.N., Coppola, L. and Rendell, H.M., 1996, Stabilization of an urban refuse dump and its planned extension near Ancona, Marche, Italy, *Engineering Geology of Waste Disposal*, Geological Society Engineering Geology Special Publication No. 11, 87–92.

Byce, J., 1984, Rapid tunnelling solves Utah's landslide crisis, *Tunnels & Tunnelling*, September, 15–17.

Chandler, M.P. and Hutchinson, J.N., 1984, Assessment of relative slide hazard within a large, pre-existing coastal landslide at Ventnor, Isle of Wight, *Proceedings of the 4th International Symposium on Landslides*, Toronto, vol. 2, 517–522.

Cruden, D.M. and Varnes, D.J., 1996, Landslide types and processes, in *Landslides, Investigation & Mitigation*, Special Report 247, US Transportation Research Board.

Davis, G.M., Fort, D.S. and Tapply, R.J., 2002, Handling the data from a large landslide monitoring network at Lyme Regis, Dorset, UK, in R.G. McInnes and J. Jakeways (eds), *Instability, Planning and Management* (London: Thomas Telford), 471–478.

Department of the Environment (DoE), 1990, *Planning Policy Guidance: Development on Unstable Land, PPG14* (London: HMSO).

Dixon, N., Kavanagh, J. and Hill, R., 1997, Monitoring landslide activity and hazard by acoustic emission, *Journal of the Geological Society of China*, **39**(4), 301–327.

Ellis, I.W., 1985, The use of reticulated pali radice structures to solve slope stability problems, *Technical Note 11, Failures in Earthworks* (London: Thomas Telford), 432–435.

Fell, R. and Hartford, D., 1997, Landslide risk management, in D.M. Cruden and R. Fell (eds), *Landslide Risk Assessment* (Rotterdam: Balkema), 51–109.

Fookes, P.G. and Sweeney, M., 1976, Stabilisation and control of local rockfalls and degrading rock slopes, *Quarterly Journal of Engineering Geology*, **9**, 37–56.

Geomorphological Services Ltd (GSL), 1991, *Coastal landslip potential assessment: Isle of Wight Undercliff, Ventnor*. Final Report to the Department of the Environment.

Ginzburg, L.K., 1979, *Antislide retaining structures*, Moscow (in Russian).

Greco, V.R., 1991, Slope stabilization by means of cuts and fills, in *Slope Stability Engineering – Applications and Developments* (London: Thomas Telford), 421–426.

Hungr, O., Morgan, G.C., VanDine, D.F. and Lister, D.R., 1987, Debris flow defenses in British Columbia, *Geological Society of America, Reviews in Engineering Geology*, **VII**, 201–222.

Hutchinson, J.N., 1969, A reconsideration of the coastal landslides at Folkestone Warren, Kent, *Géotechnique*, **19**, 6–38.

Hutchinson, J.N., 1977, Assessment of the effectiveness of corrective measures in relation to geological conditions and types of slope movement, *Bulletin of the International Association Eng. Geol.*, **16**, 131–155.

Hutchinson, J.N., 1983, Methods of locating slip surfaces in landslides, *Bull. Assoc. Eng. Geol.*, **XX**(3), 235–252.

Hutchinson, J.N., 1984a, Landslides in Britain and their countermeasures, *J. Japan Landslide Soc.*, **21**, 1–25.

Hutchinson, J.N., 1984b, An influence line approach to the stabilization of slopes by cuts and fills, *Canadian Geotechnical Journal*, **21**, 363–370.

Hutchinson, J.N., 2001, Fourth Glossop Lecture. Reading the ground: morphology and geology in site appraisal, *Quarterly Journal of Engineering Geology and Hydrogeology*, **34**, 7–50.

Hutchinson, J.N., Bromhead, E.N. and Lupini, J.F., 1980, Additional observations on the Folkestone Warren landslides, *Quarterly Journal of Engineering Geology*, **13**, 1–31.

Hutchinson, J.N., Bromhead, E.N. and Chandler, M.P., 2002, Landslide movements affecting the lighthouse at St Catherine's Point, Isle of Wight, in R.G. McInnes and J. Jakeways (eds), *Instability, Planning and Management* (London: Thomas Telford), 291–298.

Jaeger, J.C., 1969, The stability of partly immersed fissured rock masses and the Vajont rock slide, *Civil Engineering and Public Works Review*, 1204–1207.

Kelly, J.M.H. and Martin, P.L., 1986, Construction works on or near landslides, *Proceedings of the Symposium of Landslides in South Wales Coalfield*, Polytechnic of Wales, 85–103.

Leadbeater, A.D., 1985, A57 Snake Pass, remedial work to slip near Alport Bridge, in *Failures in Earthworks* (London: Thomas Telford), 29–38.

McConnell, R.G. and Brock, R.W., 1904, *Report on the great landslide at Frank, Alberta*, Report for 1903, Department of the Interior, Ottawa, Canada, Part 8.

Morgenstern, N.R. and Price, V.E., 1965, The analysis of the stability of general slip surfaces, *Géotechnique*, **15**, 79–93.

Penman, A.D.M., 1985, The failure of Acu Dam. *Technical Note 6. Failures in Earthworks* (London: Thomas Telford), 414–416.

Picarelli, L., Russo, C. and Urciuoli, G., 1995, Modelling earthflow movement based on experiences, *Proceedings of the XI European Conference Soil Mechanics and Foundation Engineering*, **11**, 157–162.

Popescu, M.E., 1991, Landslide control by means of a row of piles, *Slope Stability Engineering – Applications and Developments* (London: Thomas Telford), 389–394.

Sarma, S.K., 1973, Stability analysis of embankments and slopes, *Géotechnique*, **23**, 423–433.

Siddle, H.J., 1984, South Wales spoil stabilisation tunnels in sandstone, *Tunnels & Tunnelling*, December 17–19.

Skempton, A.W., 1990, Historical development of British embankment dams to 1960, *Clay Barriers for Embankment Dams* (London: Thomas Telford), 15–52.

Skempton, A.W. and Coats, D.J., 1985, Carsington Dam failure, *Failures in Earthworks* (London: Thomas Telford), 203–220.

Skempton, A.W. and Hutchinson, J.N., 1969, Stability of natural slopes and embankment foundations, *Proceedings of the 7th International Conference on Soil Mechanics & Foundation Engineering, State of the art volume*, 291–340.

Spencer, E.E., 1967, A method for the analysis of the stability of embankments assuming parallel inter-slice forces, *Géotechnique*, **17**, 11–26.

Taylor, D.W., 1937, Stability of earth slopes, *Journal of the Boston Society of Civil Engineers*, **24**, 197–246.

Thomson, S. and Hayley, D.W., 1975, The Little Smoky landslide, *Canadian Geotechnical Journal*, **12**, 379–392.

Van Rose, S., 1983, *Earthquakes*. Institute of Geological Sciences, HMSO.

Varnes, D.J., 1958, Landslide types and processes, in E.B. Eckel (ed.), *Landslides in Engineering Practice*, US National Academy of Sciences, Highway Research Board, Special Report **29**, 20–47.

Varnes, D.J., and the Commission on Landslides and other Mass Movements, 1984, Landslide hazard zonation: a review of the principles and practice, *Natural Hazards* (Paris: UNESCO), **3**.

Vaughan, P.R., 1991, Mechanics of Landslides, *Slope Stability Engineering – Applications and Developments* (London: Thomas Telford), 1–10.

Vaughan, P.R., Dounias, G.T. and Potts, D.M., 1990, Advances in analytical techniques and the influence of core geometry on behaviour, *Clay Barriers for Embankment Dams* (London: Thomas Telford), 87–108.

Viner-Brady, N.E.V., 1955, Folkestone Warren landslips: remedial measures 1948–1950, *Proceedings of the Institute of Civil Engineers*, Railway paper No. 57, 429–441.

Watson, J.D., 1956, Earth movements affecting LTE railway in deep cutting east of Uxbridge, *Proceedings of the Institute of Civil Engineers*, **11**, 320–323.

Watson, P.D.J. and Bromhead, E.N., 2000, The influence of waste water disposal on slope stability, *Proceedings of the 8th International Symposium on Landslides*, Cardiff, June 2000 (London: Thomas Telford), 1557–1562.

Weeks, A.G., 1969, The stability of the Lower Greensand Escarpment in Kent, PhD Thesis, Surrey University.

Whitman, R.V. and Bailey, W.A., 1967, Use of computers for slope stability analysis, *Proceedings of the American Society of Civil Engineers, Journal of Soil Mechanics Division*, **93**, 475–498.

Wilson, R.L. and Smith, A.K.C., 1983, The construction of a trial embankment on the foreshore at Llandulas, *Shoreline Protection* (London: Thomas Telford), 223–233.

# PART 4

## 'END-TO-END SOLUTIONS' FOR LANDSLIDE RISK ASSESSMENT

# 20

# Towards the Development of a Landslide Risk Assessment for Rural Roads in Nepal

*David N. Petley, Gareth J. Hearn and Andrew Hart*

## 20.1 Introduction

Young, high fold mountain chains are probably the most dynamic geomorphological environments on earth. The rapid uplift that occurs in these environments is almost always associated with high rates of erosion. In many cases, landsliding is the critical mechanism by which this erosion is accomplished. Thus, in such environments landslides are an inevitable and necessary part of the natural landscape process system. However, these landslide cause significant problems in areas in which humans are located. Nepal, Bhutan and parts of India, which straddle the southern edge of the Himalayan massif (Figure 20.1), are countries with such problems. In the case of Nepal and India, and to a lesser extent Bhutan, these problems are being exacerbated by the relatively rapid processes of development that are occurring through, for example, the construction of rural access roads. Despite high levels of investment, the maintenance of these roads, and of other pieces of infrastructure, is proving a significant headache, primarily because of the occurrence of landslides.

As a result of this significant landslide problem, in recent years there have been increasing levels of interest in the development of landslide hazard and risk assessment techniques in the Himalayan region. Several approaches have been proposed, including the Landslide Hazard Evaluation Factor (LHEF) scheme of Anbalagan (1992a, 1992b) and Anbalagan and Singh (1996), and the mountain risk engineering approach of Deoja *et al.* (1991). These schemes have significant strengths and some success has been met

*Landslide Hazard and Risk*   Edited by Thomas Glade, Malcolm Anderson and Michael J. Crozier
© 2004 John Wiley & Sons, Ltd   ISBN: 0-471-48663-9

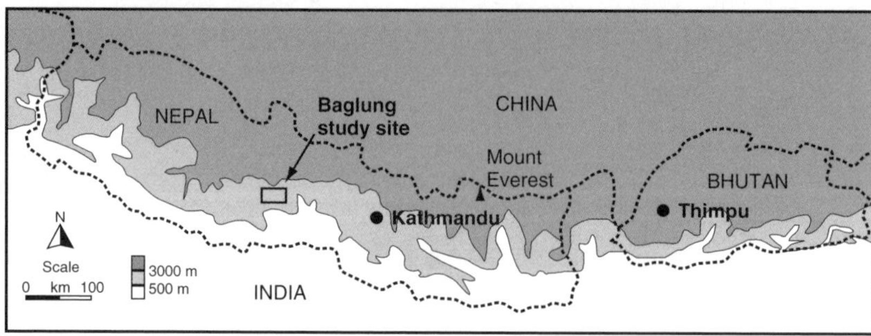


**Figure 20.1**  *Map illustrating the location and relief of Nepal and Bhutan, including the Baglung study site*

with their application within Nepal. However, in both cases the techniques are complex and detailed, requiring a high level of knowledge and skill for successful use. Whilst they are both undoubtedly useful for the detailed analysis of high-cost roads, their application in the construction of rural and agricultural roads has been limited. In addition, these techniques do not allow the analysis of true hazard (and thus true risk) as the calculation of landslide magnitude (i.e. volume and area affected and frequency of occurrence) is not established. Thus there is a need to develop a true landslide hazard and risk technique that can be applied in the development of low-cost roads in rural mountainous areas.

This chapter presents the initial results of the three-year Landslide Risk Assessment (LRA) in the Rural Access Sector project to develop such a simple scheme. The project has been based in Nepal and Bhutan and, although the results presented here concentrate primarily on the former, the scheme is designed to be equally applicable to the latter, and to similar areas. In this chapter we present a highly simplified scheme and demonstrate that in Nepal at least, a landslide hazard scheme that is based purely upon slope angle and geology can be surprisingly effective.

## 20.2  Landslide Hazards in Nepal

Nepal lies across the major boundary thrusts along which much of the deformation associated with the collision of India and Eurasia has been accommodated. These large, northward-dipping thrust plains have allowed the rapid and sustained uplift that has formed the young fold mountain chain of the Himalayas. The southern margins of both countries are situated on the Ganges Plain, the great northern plain of the Indian Subcontinent, and have elevations of just 100 m a.s.l. However, in Nepal the bulk of the terrain consists of steep, rugged mountains that of course include the highest elevations on earth. Indeed eight of the world's ten highest peaks, and nine of the fourteen peaks that exceed 8000 m, are located within Nepal. Despite this, the population density of Nepal is surprisingly high at 164 people km$^{-2}$ and growth rates are high at 2.37% in Nepal (Shrestha *et al.*, 1999). Unfortunately, there is a high level of poverty in Nepal, with 37.7% of the population lying below the $1 per day poverty line (UNDP, 2001).

Thus the combination of high population density, rapid population growth rates, low incomes and steep terrain with extreme relative relief means that both countries are

***Figure 20.2*** *Photograph showing an example of the impacts of a landslide on a low-cost road. The road at this site, located in the Baglung study area, is being damaged by a creeping landslide in the terraced colluvium that the alignment crosses. It is likely that reactivation of this slope has occurred as a result of loss of support to the slope caused by road construction*

vulnerable to the effects of landslides (Figure 20.2). Add to this the implications of a weak, young set of rocks; a monsoonal climate with long-duration, high-intensity rainfall; and the relatively high frequency of occurrence of seismic activity, and the significant problems associated with landslides in this region become clear. Indeed, on the southern flanks of the Himalayan Range, landslides represent the dominant hillslope process (Whitehouse, 1990). Within the Low and Middle Himalaya of Nepal landslides represent an important hazard, destroying roads and other infrastructure, farmland, and in some cases whole communities (Ives and Messerli, 1981; Schelling, 1988). This has been exacerbated in Nepal by the rapid development of infrastructure in the last 20 years, which has led to increased vulnerability to the effects of landslides. For example, according to government statistics, in Nepal the national road network increased from 3173 km in 1974 to 13 709 km in 1998. However, a significant proportion of these roads are damaged every year, and in some cases may be lost altogether, due to the impacts of landslides (Figure 20.3).

The increasing impact of landslides in Nepal was demonstrated by Hart *et al.* (2003), who compiled a database of landslide activity over a 30-year period. This clearly demonstrated an increasing level of loss associated with landslides in Nepal, with 185 recorded fatalities in 2001 and 342 recorded landslide fatalities in 2002. Our own data suggest that in 2003 the number of landslide fatalities in Nepal was 244, most occurring in a large

*Figure 20.3* Photograph from the Baglung study area showing substantial damage to a road caused by a translational landslide. The landslide moves c. 20 m in each monsoon season. Displaced sections of road can be seen downslope from the current alignment

rainstorm event on 30–31 July. The average number of fatalities in the period 1970–2000 was 65 per annum, but for the period 2000–2003 the average number has been 206 per annum.

## 20.3   Landslide Hazard and Risk Mapping in Nepal

As a result of the key problems associated with landslides in the Himalayas, some attempts have been made to develop landslide susceptibility and hazard schemes for Nepal. For example, landslide inventory mapping was conducted by Wagner (1983), by which the key factors (such as lithology, geological structure and slope angle) in increasing landslide susceptibility could be determined. A more advanced susceptibility technique was developed for the Midland zone of Nepal by Keinholz *et al.* (1983, 1984). Unfortunately, however, the authors admit that the final output is only applicable to a relatively small, albeit densely populated area, based upon the Kathmandu Valley.

A more comprehensive study was undertaken as part of a United Nations mountain hazard mapping study by Zimmerman *et al.* (1986) and White *et al.* (1987). Here a land systems approach based on detailed geomorphological mapping was used. In each land system, the likely range of hazards and their associated risks were highlighted. Although undoubtedly worthy, the aim of the study was not to determine the location of individual landslides and their associated risks. A range of other geomorphological mapping-based approaches have also been used, including studies by Laban (1979), Dhital *et al.* (1991), Dangol *et al.* (1993), Dixit (1994), and Thapa and Dhital (2000). In all cases the work essentially allowed the generation of landslide inventory maps rather than hazard maps.

A useful study was undertaken by Hearn (1987, 1997) in the Dharan–Dhankuta area, which had been affected by severe landslide problems triggered by monsoon rainfall. The distribution of landslides within a training area, determined through geomorphological mapping, was compared with rock type, slope angle, slope physiography, slope aspect and land use. Using a chi-squared analysis, relationships were tested, allowing the construction of a composite index of susceptibility that was then applied across the whole of the study area. Some considerable success was noted with this technique, although, interestingly, Hearn (1987) suggested that the technique yielded little that could not be achieved through conventional geomorphological mapping.

Empirical rating schemes to determine landslide susceptibility have been used with some success in Nepal, particularly along road alignments and for watershed management purposes. These have often been based upon the mountain risk engineering (MRE) scheme proposed by Deoja *et al.* (1991). In this scheme a wide range of landslide factors are analysed and given a rating, including material type, slope angle, soil thickness, land use, drainage, hydrogeology, geoindicators and geological structure. Whilst this is without doubt a successful and powerful technique, the amount of detailed measurement required precludes its widespread adoption in rural access schemes. A similar approach has been used by Anbalagan (1992a, 1992b) in proposing the Landslide Hazard Evaluation Factor scheme (LHEF). In both cases these schemes use summation of a score for a wide variety of factors. As such they are complex to use and require a high degree of specialist knowledge. It is also questionable whether summation is the most appropriate method of analysis.

Finally, there have been several attempts to use GIS to assess landslide susceptibility in Nepal using empirical rating techniques (e.g. Thapa and Dhital, 2000), multivariate statistics (Skirikar *et al.*, 1998; Dhakal *et al.*, 1999) or deterministic techniques (Joshi *et al.*, 2000). In all cases landslide magnitude and frequency have not been considered. In addition, the effectiveness of the approach has rarely been systematically appraised. Joshi *et al.* (2000) found, however, that their deterministic approach yielded generally good results, although problems were encountered in low-susceptibility zones.

Landslide risk mapping has rarely been undertaken in Nepal, although both JICA (1997) and Khanal (1999) did incorporate socio-economic factors into a GIS database. In addition, Dhital *et al.* (1991) attempted to estimate risk in terms of damage and maintenance cost estimates for road alignments.

Thus, although some considerable effort has been put into landslide susceptibility mapping in Nepal, this has rarely if ever been developed to produce true landslide hazard or risk maps. In addition, the techniques that have been developed have in general been complex and intricate. Although this might produce quite accurate outputs, the complexity of the schemes has in general precluded their adoption in infrastructure development projects, especially in low-cost agricultural roads.

## 20.4    A Simple Landslide Hazard and Risk Mapping Approach

During the LRA project, detailed research has been undertaken into the effectiveness of existing landslide susceptibility, hazard and risk schemes based upon a number of field sites located in a range of geological environments within both Nepal and Bhutan. Whilst the detailed results are beyond the scope of this chapter, we found that all of the schemes that we tested performed poorly in the field areas that we studied. To this end we have developed a number of alternative schemes tailored for different purposes. One of these schemes, described here, is a very simple formulation based purely upon the slope angle and geology. Although it has having lower accuracy than some other schemes that we are developing, this system has the advantage of being simple and quick to apply, even by practitioners with little training. The accuracy of the results is none the less quite good (see below). The technique has been developed to attempt to consider risk, albeit in a very simplistic way.

The simple approach reported here has been developed in conjunction with the ethos of low-cost road construction in Nepal in particular, which is heavily weighted towards community participation as a means of gaining ownership on a development scheme – a logical approach as most roads are funded through donors such as the Asian Development Bank, the World Bank, and nation-state donors such as the UK Department for International Development. The primary purpose of these roads is to allow poverty alleviation in rural areas through the provision of access to markets.

The low-cost, community basis of the construction of these roads precludes the application of highly technical landslide hazard and risk schemes. The financing of the roads does not provide for such analyses by dedicated experts, and in many cases no detailed analysis of landslide hazard is undertaken prior to corridor selection and determination of road alignment. In many ways this is important, as such a relatively simple approach to route selection renders the constructed road vulnerable to the effects of mass movements. Thus, in the development of a landslide hazard and risk scheme the aim has been to ensure that the outputs would be applicable at the local level by non-geotechnical specialists, who usually include the district road engineer. As such the scheme may not be ideal for high-cost road projects, and inevitably it is less precise than some other schemes. However, the results from our studies suggest that it provides a good level of information to the road planning and construction process.

The simple scheme reported here is based on a step-by-step analysis of landslide susceptibility, hazard and risk based on the best possible combinations of available data. The scheme is deliberately simple in its approach in order to allow its wider application. Whilst this may be seen as a disadvantage in terms of precision, our analyses have suggested that the loss of accuracy associated with the simple system for assessing susceptibility, for example, is acceptable. The scheme is based upon the use of three major steps: susceptibility mapping, hazard mapping and finally risk mapping (Figure 20.4). In susceptibility mapping, the location of areas liable to undergo landsliding are identified. In hazard mapping, the probability of failure is estimated using a number of techniques, and the runout distance, and hence the area likely to be affected, is determined. In risk mapping, the landslide hazard assessment is combined with an estimate of the vulnerability of the local infrastructure and population to produce an estimate of landslide risk.

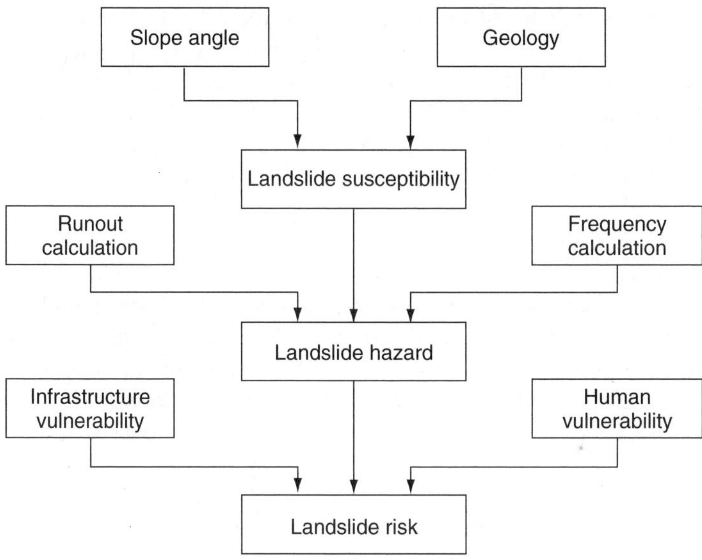


**Figure 20.4**   *Flow diagram illustrating the basic form hazard and risk analysis described here*

## 20.5   Landslide Susceptibility Assessment

The simple methodology described here has been developed using four field sites in differing geological and topographical regions of Nepal. To allow this development process to occur, a definitive landslide location map for each area was created through a combination of field mapping, aerial photograph analysis and satellite image interpretation using multispectral data obtained from the Landsat 7 ETM+, SPOT IV and IRS platforms. In one case, this was supplemented with 1 m resolution IKONOS panchromatic imagery. The outputs from these analyses are provided in Petley *et al.* (2002). This provided a benchmark against which the landslide hazard and risk assessment strategy could be developed. Detailed field mapping was then undertaken at 1:25 000 scale of geology (solid and superficial); land use; slope angle, slope aspect and relative relief, obtained using a DEM constructed from topographic data, infrastructure, and population patterns. A GIS was then used to determine the relative importance of each of these factors in estimating the susceptibility to failure based upon the definitive landslide map. Thus a simple factor-based system for assessing landslide susceptibility was developed.

The outcome of this analysis demonstrated that for the study areas the most important factors determining the susceptibility to landsliding were slope angle and geology, with all other factors proving to have low or insignificant control on stability. Thus a simple factor-based scheme was constructed using just these two factors. Although such a scheme may appear to be overly simplistic, it successfully predicts the occurrence of 70% of the slope failures in all study areas, performing better than many more complex schemes or indeed than existing techniques such as LHEF. An added advantage is the widespread availability of this information (in Nepal good-quality 1:25 000 scale topographic maps and 1:50 000 scale geology maps are currently being constructed with a nationwide coverage) and the simplicity of the application of the scheme. A factor-based scheme has the advantage of being simple to apply by non-specialists, quick to use, with a limited number of possible outputs, meaning that the results are easy to interpret. The disadvantages are the potential oversimplification of the results and the potential for ignoring key factors in particular cases. Of course, all schemes are only as good as the data upon which they are based. In areas of poor-quality baseline data this may be a key constraint.

The importance of geology and slope angle in the triggering of landslides has been noted previously. For example, the detailed analysis of Gerrard (1994) for a range of sites in both the eastern and the western Himalayas found that geology was the most important factor in determining landslide occurrence and suggested that 'sufficient relationships exist to place the most common rock types in order of susceptibility to landsliding' (p. 229). Interestingly, he noted that these relationships are consistent across the various tectonic domains through the Middle and High Himalayas. Similar observations have been made by for example Dangol (1994) and by Caine and Mool (1982). However, it must be pointed out that the detailed study of Brunsden *et al.* (1981) was much less conclusive, finding at best a weak correlation between rock type and landslide occurrence, except where only undercut slopes were considered.

Of course, it would be naïve to believe that a simple scheme based only on geology and slope angle will explain the occurrence of all landslides, a point noted by Gerrard (1994, p. 221): 'It would be dangerous to assume that landsliding in the Himalayas can be completely explained and predicted by the nature of the rock type.' None the less Gerrard (1994) also noted that 'with respect to specific rock types, the order from most susceptible to least susceptible to landsliding . . . is sufficiently consistent to be used as an initial assessment of landslide susceptibility'(p. 221).

In Tables 20.1 and 20.2 the form of this simple susceptibility mapping technique is given. The bin size for each class has been defined on the basis of a combination of expert judgement and trial and error. It must be stressed that this scheme is designed for application on low-cost roads in agricultural areas of less economically developed countries, and thus is probably not applicable in some other circumstances, especially where investment in the road is higher. Landslide susceptibility is determined by multiplying the two factors together. Whilst most previous susceptibility systems have worked on the basis of summation, multiplication is a more logical approach. For example, in an area in which slope angle is very low, landslide potential is also likely to be low regardless of geology or any other factor. A summative scheme may not be able to reflect this; a multiplicative scheme can so do. Scores are most logically defined in terms of terrain units (see Finlayson, 1984 for example), in which the landscape is divided into 'facets' of similar topography, geology, and so on. Boundaries between facets are defined by changes in materials, forms or processes. The terrain unit approach is extremely powerful for landslide hazard assessment as most slope failures respond to geomorphology, geology and human activity.

Once a score has been derived, it can be simply classified into areas of relative susceptibility; we suggest the scheme defined in Table 20.2, based upon the results from our project development areas. The uneven class sizes are needed due to the multiplicative nature of the scheme. With these classes, the probability of obtaining any

**Table 20.1**   *Ratings for the Himalayan landslide susceptibility assessment scheme*

| Lithology | Rating |
| --- | --- |
| Massive quartzite | 1 |
| Massive igneous rocks | 1 |
| Gneiss | 2 |
| Massive limestone | 2 |
| Quartzite and limestone with interbedded clay | 3 |
| Well-cemented sandstone with interbedded clay | 3 |
| Unweathered slate and phyllite | 4 |
| Unweathered schist | 4 |
| Poorly cemented sandstone with interbedded clay | 5 |
| Shale with interbedded clay | 6 |
| Highly weathered shale, phyllite, schist | 7 |

| Slope angle | Rating |
| --- | --- |
| 0–5° | 0 |
| 5–15° | 1 |
| 15–25° | 2 |
| 25–30° | 3 |
| 30–35° | 4 |
| 35–40° | 5 |
| 40–50° | 6 |
| >50° | 7 |

**Table 20.2**   *Multiplied scores and associated susceptibility levels for the Himalayan landslide susceptibility assessment scheme*

| Multiplied rating | Relative susceptibility |
| --- | --- |
| 0–3 | Very low |
| 4–7 | Low |
| 8–14 | Moderate |
| 15–24 | High |
| 25–49 | Very high |

given susceptibility score from a random selection of slope angle and geology ratings is approximately the same.

This simple scheme provides an estimate of relative susceptibility. In some cases this might be sufficient to allow the selection of a route corridor. For example, if a road is planned between two villages, the need may only be to select the safest possible alignment. On the other hand, if the exercise is undertaken as part of a feasibility appraisal, a determination of the actual levels of hazard may be necessary to determine whether a viable road can be constructed. Of course, if landslide risk levels are being evaluated, an assessment of hazard is required as an interim step.

## 20.6   Landslide Hazard Assessment

To determine landslide hazard, an assessment is needed of landslide magnitude (i.e. the surface area and depth of the landslide), its frequency/probability and likely runout distance. There is little doubt that this is the most difficult task in landslide hazard mapping in the Himalayas. For Nepal, we have undertaken a study of the residence times of landslide scars of different types in the landscape – that is, once formed, for how long a landslide feature will still be identifiable. This has been undertaken through the analysis of a series of epochs of aerial photographs. This analysis, in combination with our landslide distribution maps, has allowed an estimate to be made of the frequency of occurrence. Analysis of aerial photographs suggests that landslides remain resident in the landscape for a mean period of about 50 years. Of course some landslides are visible for much less time than this, especially where the failure is shallow and has occurred in agricultural terrain. In other cases, especially where the landslide is large and deep, the scar and debris may be visible for millennia. Thus the population of landslides seen in the landscape is representative of about 50 years of failures, providing an indication of frequency. We have used the average density across all of our field sites for each lithology/slope combination, and the 50-year residence time, to give an indication of the frequency of occurrence of failures.

A key aspect of the development of a landslide hazard assessment scheme is the consideration of the frequency of occurrence. To this end, we have also undertaken an analysis of the two environmental triggers that lead to instability in slopes in the Himalayas: seismic activity and rainfall. Taking seismicity first, the location of both countries on the Himalayan collision zone renders them both highly vulnerable to the effects of seismic activity – and there is little doubt that earthquakes are major triggers of landslides. Known seismic events since the eighteenth century are listed in Table 20.3; note that this list is not exhaustive, as many of the earlier events were probably not recorded. Bilham *et al.* (2001) analysed seismicity along the Himalayan chain and concluded that western Nepal appears to have built up an enormous amount of strain energy as a large earthquake has not occurred in this area for at least three centuries, and possibly as much as 700 years. Thus, based upon the 'seismic gap' theory, a major seismic event appears to be overdue in this area. In eastern Nepal and Bhutan the situation is less worrying, as the 1934 'Bihar–Nepal' earthquake is believed to have released much of the tectonic stress in this area. This earthquake is believed to have killed over 10 500 people, and to have triggered thousands of slope failures over an area of over 120 000 km$^2$ (Keefer, 1984). Thus a similar earthquake would not normally be expected in the foreseeable future, although Bilham *et al.* (2001) suggest that an event with magnitude greater than 7.0 could occur at any time.

Unfortunately, with the large seismic gaps that occur along the Himalayan chain the statistical analysis of seismic occurrence is unlikely to be meaningful (see Krinitzsky, 2002, for example). However, analyses for the Kathmandu Valley suggest a return period for a magnitude 8.0 earthquake of about 81 years. Such an event would lead to landsliding over an area of about 50 000 km$^2$ and would be devastating to any low-cost road as such structures cannot be designed economically to survive such an event. In the context of the hazard assessment for this project, which was designed for route selection for low-cost roads, the occurrence of such seismic events is effectively meaningless, therefore,

***Table 20.3***   *Significant earthquakes known to have affected Nepal and Bhutan (data from Bilham et al., 2001)*

| Date | Latitude (°E) | Longitude (°N) | Magnitude Ms |
|---|---|---|---|
| Jul. 15, 1720 | 29 | 77.5 | Unknown |
| Sept. 1, 1803 | 30 | 78.0 | 8.0 |
| Aug. 26, 1833 | 28 | 88.5 | 7.7 |
| Jan. 10, 1869 | 25 | 93.0 | 7.5 |
| Jun. 12, 1897 | 26 | 91.0 | 8.1 |
| Apr. 4, 1905 | 32.3 | 76.3 | 7.8 |
| Jul. 8, 1918 | 24.5 | 91.0 | 7.6 |
| Jul. 2, 1930 | 25.8 | 90.2 | 7.1 |
| Jan. 15, 1934 | 26.6 | 86.8 | 8.2 |
| Oct. 23, 1943 | 26.8 | 93.0 | 7.2 |
| Jul. 29, 1947 | 28.63 | 93.73 | 7.7 |
| Aug. 15, 1950 | 28.5 | 96.7 | 8.5 |
| Jan. 19, 1975 | 32.38 | 78.49 | 6.2 |
| Aug. 6, 1988 | 25.13 | 95.15 | 6.6 |
| Aug. 21, 1988 | 26.72 | 86.63 | 6.4 |
| Oct. 20, 1991 | 30.75 | 78.86 | 6.6 |
| May 22, 1997 | 23.08 | 80.06 | 6.0 |
| Mar. 29, 1999 | 30.41 | 79.42 | 6.8 |

as one could reasonably assume that any alignment would be destroyed. For this reason, therefore, our scheme has deliberately been designed to disregard the effects of seismicity, but the results should be considered in the context of significant levels of seismic hazard. Of course, further research is needed into this topic, and consideration of the likely location of seismically triggered landslides (close to the ridge-top rather than low in the slope for hydrologically triggered slides) has been built into the design guidelines that have also been derived from this project.

Hydrological triggers for landslides are supposedly easier to analyse. We have approached this issue through examination of the precipitation records for the study areas, which have been compared to a database of the dates of failures. However, we have been unable to come up with convincing correlations between the precipitation data and the known dates of landslides.

The second issue that must be addressed to compile a hazard map is that of area affected. In fact there are two interrelated issues here. First, we must consider the surface area that will become unstable in the initiation of a landslide. Second, we must consider the area that is likely to be engulfed by that landslide – that is, the runout area. Unfortunately neither is an easy problem with which to deal.

It is likely that once failure has been triggered, the landslide mass itself will include an area upslope of the area indicated as being unstable in the analysis, and also an area downslope and to each side as well. Thus, although an area upslope might have a lower susceptibility rating, a failure triggered from downslope might still cause movement at this point.

The following sections provide detail of a proposed method with which to deal with these problems to provide an estimate of the area to be affected by a landslide.

### Areas with a 'very high' susceptibility rating

Clearly, the whole of an area considered to be 'very high' susceptibility will also be 'very high' hazard. If upslope the susceptibility reduces, then the lowest part of this zone should also be considered to be 'very high' hazard. We propose the application of an upslope buffer of 50 m (horizontal distance) if the susceptibility is 'high', 10 m if the susceptibility is moderate or low. These buffers are based upon expert judgement.

### Areas with a 'high' susceptibility rating

Once again, the all areas considered to be 'high' susceptibility will also be 'high' hazard, unless they can be given a higher hazard class. If upslope the susceptibility reduces, then the lowest part of this zone should also be considered to be 'high' hazard. We propose the application of an upslope buffer of 10 m (horizontal distance), again based upon expert judgement.

### Areas with a 'moderate' susceptibility rating

All areas considered to be 'moderate' susceptibility will also be 'moderate' hazard, unless they can be given a higher hazard class. If upslope the susceptibility reduces, then the lowest part of this zone should also be considered to be 'moderate' hazard. We propose the application of an upslope buffer of 10 m (horizontal distance), again based upon expert judgement.

### Areas with a 'low' susceptibility rating

All remaining areas are given a 'low' or 'very low' hazard rating according to their susceptibility score.

## 20.7   Runout

Runout (the distance downslope that the landslide will affect) is deduced using the angle of internal friction ($\phi'$) of the materials involved (given in Table 20.4 and the angle ($\alpha$) of the slope below). In the case of 'high' and 'very high' susceptibility, the corresponding hazard class ('high' and 'very high' respectively) is expected to extend downslope for as far as $\alpha \geq \phi'$ plus an extra 20 m horizontal distance to allow the movement to stop. If, however, the downslope area has a higher hazard rating, then this area should not be reclassified. No runout calculation is necessary for the 'moderate' and 'low' susceptibility classes as it is assumed that failure will not occur on these slopes. Runout distance should be calculated for all areas downslope of 'high' and 'very high' susceptibility classes, but it is reasonable to assume that the lowest margin for which it needs to be determined is the main fluvial channel at the bottom of a major valley.

We are aware that the use of such a simple criterion for determining runout distance is overly simplistic, but it provides an approach that has some technical justification. Several studies have suggested a relationship between runout distance and the angle of internal friction of the landslide debris (Corominas, 1996; Kilburn and Sorenson, 1998; Griffiths *et al.*, 2002, for example).

**Table 20.4** *Angles of internal friction for runout calculation purposes*

| Material | $\phi'(°)$ |
|---|---|
| Basalt | 45 |
| Granite | 47 |
| Hard sandstone | 40 |
| Dolomite | 40 |
| Limestone | 37 |
| Quartzite | 35 |
| Gneiss | 35 |
| Poorly sorted colluvium | 34 |
| Shale | 32 |
| Slate | 32 |
| Soft sandstone | 30 |
| Clay with low organic content | 25 |
| Clay with very high organic content | 14 |

The simple approach adopted above could be supplemented with the use of runout curves. Care is needed here because, although considerable effort has been put into deriving such curves (see Corominas, 1996; Kilburn and Sorenson, 1998; Rickenmann, 1999; Griffiths *et al.*, 2002, for example) using parameters such as fall height, slope angle, volume and runout distance, the validity of many of these relationships has recently been questioned (Legros, 2002), with the only mathematically justifiable relationship between being that between volume and runout (see Hsu, 1975, for example). Legros (2002), using a large data set compiled by Hayashi and Self (1992), suggested that the following equation can be used to describe the relationship between landslide volume and runout length:

$$L_{max} = 15.6\,V^{0.39},\tag{1}$$

where $L_{max}$ is runout distance and $V$ is the volume of the landslide. This relationship provided an $R^2$ value of 0.91 for the extensive data set of Hayashi and Self (1992). Clearly this relationship is sufficiently strong to form the basis of a runout distance calculation, but it requires that a landslide volume be derived. This is problematic as it requires a calculation of both the surface area of the landslide and its depth, neither of which are easy. One solution to this problem is to estimate surface area using topographic controls, and to use an empirical relationship to determine depth, or to estimate depth and then use a similar relationship to determine surface area and hence volume. Neither approach is satisfactory. As a result, although attractive, the runout curve approach was deemed unsuitable for this study, and thus the angle of internal friction/slope angle approach have been used, with buffer zones to allow the movement to stop.

## 20.8  Risk Determination

Deriving a risk assessment from the hazard map is a relatively straightforward task assuming that the required data are available. As with all risk assessments, some thought is required as to the timescale over which the risk assessment is being undertaken.

We have used 20 years, which is the usual notional design life of a low-cost road in Nepal according to Department of Roads standards. Assuming that risk is taken to be hazard × vulnerability, the major requirement is an assessment of the vulnerability of the infrastructure and people. In our technique two risk maps are produced, one of which considers the vulnerability of the population, while the other considers the economic cost. We make no attempt to factor in aspects such as loss of trade as vulnerability estimates cannot be accurately determined for the area after the construction of the road. We also do not consider the effects of the relocation of the population once a road is constructed (many people move to live close to a newly built road). Finally, and perhaps most importantly, we have deliberately taken a very simplistic view of vulnerability. A truly rigorous assessment of vulnerability would consider the 'exposure' of the elements at risk to that particular hazard. Thus, for example, a slow-moving landslide might cause only minor damage to a structure, whereas a catastrophic slide might lead to complete loss. Thus vulnerability depends not just on the 'value' of the asset, but also on the nature of the hazard, and the ability of the infrastructure to withstand it. However, such assessments are complex and difficult to undertake, and are thus not suitable for a low-cost assessment. Thus we have assumed a 100% vulnerability – that is, that the event will lead to complete loss. Clearly this leads to an exaggerated assessment of risk, but one that is essentially a worst-case scenario. There are many disadvantages to this approach, but it does provide consistent and easily understood results.

To assess vulnerability, mapping of the human population and land use is needed. A substantial part of this can be undertaken using aerial photography, which allows evaluation of land use, building and road locations and so on. Mapping should include roads, tracks and paths; canals and other man-made hydrological systems; land use; houses and farm buildings; schools, health posts; administrative centres; and any industrial or commercial use of the land. Ground mapping should be used for truthing and verification purposes. Using these data, two vulnerability maps should be produced. One shows economic values created using local data for values of the land, roads, buildings and so on. The other shows vulnerability of the human population, based upon average numbers of occupants per house, or number of children in a school. Some double counting is inevitable (a child will be counted as both an occupant of a house and a school), resulting in an overestimate of risk. However, this seems to be the most reasonable way of determining a useful figure.

Having constructed these two vulnerability maps, the values can then be multiplied by those on the hazard map to produce two final risk maps – one showing risk to people over the 20-year period; the other showing economic risk over the same period. As noted above, because the vulnerability of any particular element to the hazard is considered to be 100% over the 20-year period, these risk maps are worst-case scenarios, consistently overestimating the true level of risk.


## 20.9   The Baglung Study Site

The effectiveness of this scheme can be demonstrated using a study area in Baglung District of central Nepal (Figure 20.1). Baglung District is located in the Middle Himalaya, a 10–35 km wide belt of schist, phyllite, gneiss, quartzite, granite and limestone belonging to the Lesser Himalayan Zone. The study area consists of a crescent-shaped section approximately 35 km in length and 5 km in width (Figure 20.5). The section follows the line of a planned low-cost road that runs from the town of Baglung in the east to a small settlement at Burtibang in the west. The eastern section follows the Kathe Khola river valley at an elevation of about 800 m, before the road climbs up over a saddle with a maximum elevation of 2100 m. Thereafter it descends into another valley with an elevation of about 1400 m. There are no existing roads within the study area, although there are footpaths as the area has extensive terraced farming. The range of the main ridges is about 2500 m.

The lower valley sides are over-steepened and in places undergoing active toe erosion, whereas the mid-valley sides are less steep and are characterized by thick soil deposits, particularly on the north side of the river. Subsidiary rocky ridges extend from upper to mid-valley sides and the soil is thinner at these locations. The upper valley sides close to the watershed ridge crests are steep, rocky and barren, with thin or absent soil deposits.

The western part of the site between Rageni and Kharbang lies within the watersheds of the Daram Khola and Gaudi Khola rivers. The valley floors in the upper part of the watershed are broad, with extensive alluvial flats, typically at an elevation of about 1200 m. The mid-valley sides are steeper than in the Kathe Khola catchment, and are characterized by thinner soils and many rocky, forested areas. The upper valley sides are very steep and high, up to about 3200 m elevation, and incorporate many very large rock cliffs and large dissected rocky bowl-shaped erosion source areas. Slope angles are typically in the range 15–55° across the study area in general (Figure 20.6).

Geologically, the area lies within the Lesser Himalayan Zone, a tectonic unit bounded by the Main Central Thrust (MCT) to the north and the Main Boundary Thrust (MBT) to the south. The strike of the principal thrust faults, fold axes and foliation is generally NW–SE, and this trend defines the orientation of the main valleys and watershed ridges within the area. Minor fold axes and faults have a NE–SW strike. Thrust and foliation dip is generally to the NE, and foliation dip is typically about 30–40°, though the dip angle and direction of foliation varies locally due to folding. The strata in the study area mainly comprise Late Pre-Cambrian metasediments of the Nawakot Group (Table 20.5, Figure 20.5).

Two rain gauges are located within this area, one on the eastern margin and one to the west. To the east the mean annual rainfall is 1939 mm, of which 1508 mm (77% of

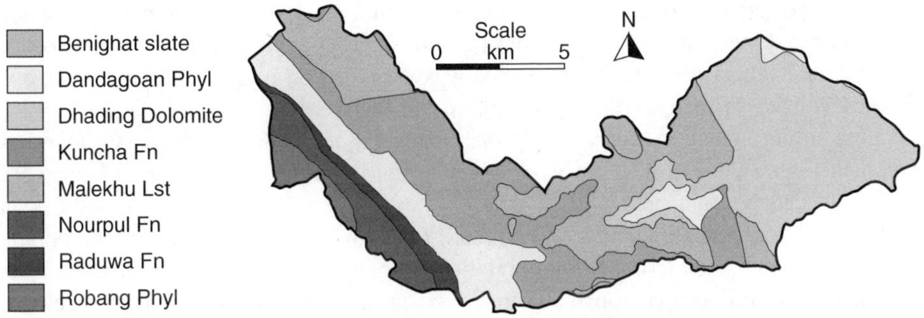


**Figure 20.5**  *Geological map of the Baglung study area*

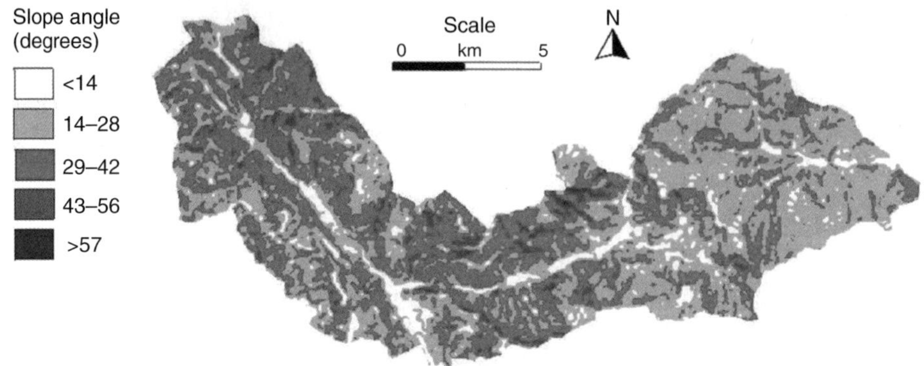



**Figure 20.6**   *Slope angle distribution map for the Baglung study area*

**Table 20.5**   *Principal geologic units in the Baglung study area*

| Unit | Lithology |
| --- | --- |
| Robang Phyllite | Phyllite with thin intercalations of quartzite |
| Malekhu Limestone | Dolomite and siliceous limestone and thin intercalations of slate |
| Benighat Slate | Weathered, calcareous laminated slates |
| Dhading Dolomite | Dolomites with some thin bands of slate |
| Nourpul Fm | Quartzite, phyllite and calcareous rocks |
| Dandagaon Phyllite | Phyllite with thin intercalations of quartzite |
| Raduwa Fm | Coarse crystalline garnet, biotite, muscovite schist and quartzite |
| Kuncha Fm | Phyllite with thin intercalation of quartzite |

the total) occurs during the monsoon, whilst to the west these values are 1921 mm and 1565 (81%) respectively. Caine and Mool (1982) proposed that a rainfall intensity of 100 mm in 24 hours typically triggers landslides within Nepal. In the east of the study area this amount is exceeded 0.51 times per year on average, whilst to the west this occurs 1.15 times per annum. However, our studies can find no correlation between landslide occurrence and the number of times that this threshold is exceeded in any year. So, for example, in Baglung in 1997 this threshold was exceeded on four occasions, but we can find no evidence of increased landslide activity in that year.

The resulting landslide susceptibility map using the methodology described above is shown in Figure 20.7. Susceptibility is given in terms of scale from very low to very high. This can be compared with a map of the distribution of mapped landslides in the Baglung field site (Figure 20.8), compiled from ground mapping, aerial photograph interpretation and satellite image interpretation. The comparison of the two maps shows that there is a good correlation between the occurrence of landslides and the areas of high and very high susceptibility. Statistically, this can be analysed using the chi-squared test (Table 20.6). The statistical data suggest that the technique has put 214 of the 231

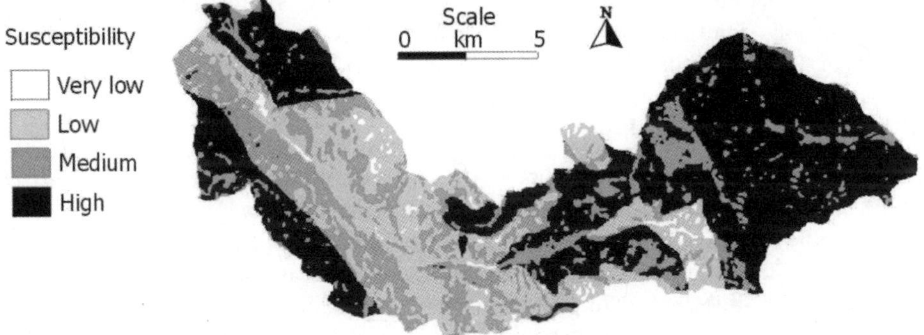


**Figure 20.7**   *Susceptibility rating for the Baglung study area*

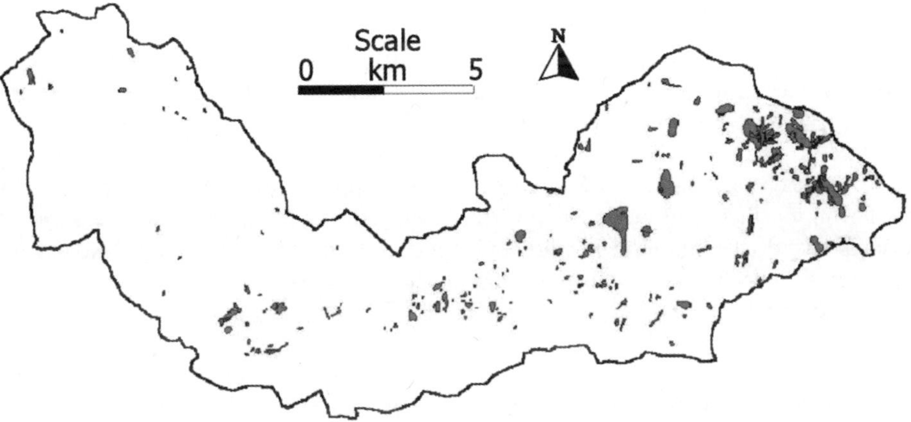


**Figure 20.8**   *Landslides distribution map for the Baglung study area*

**Table 20.6**   *Statistical analysis of the Baglung study area results*

| Susceptibility class | % area | Expected failures (for chi square) | Observed failures |
| --- | --- | --- | --- |
| Very low | 3.9 | 9 | 0 |
| Low | 29.8 | 69 | 17 |
| Moderate | 39.3 | 91 | 96 |
| High and v. high | 28.8 | 67 | 118 |
| Total | 100 | 1 | 231 |
| Chi-square value $[\Sigma((O-E)^2/E)]$ | 88.11 | Significance confidence level | 0.001 |

landslides in the moderate, high and very high categories, and that the results obtained are statistically significant, although of course the chi-squared test is not necessarily an indication that the technique is successful.

### 20.9.1    Landslide Hazard

For this study area we have determined landslide hazard purely in terms of the density of the landslides and the residence period of landslides in the landscape. We have not as yet been able to determine the runout distance for the areas of high and very high hazard. Here, we find that the density of landslides varies from zero in the very low susceptibility category to 0.78 slides/km$^2$ in the most susceptible hazard class. Our analysis of the return periods of landslides suggests that this population of landslides could be expected to occur in a 50-year period. We have therefore taken this to be the frequency of occurrence of these landslides, allowing a probability to be estimated. This allows the production of the hazard map in Figure 20.9.

### 20.9.2    Landslide Risk

To complete a true landslide risk map, a true landslide hazard map must be incorporated with a vulnerability map. Clearly, to date it has not been possible to generate a true hazard map for the Baglung area due to the problems associated with estimating runout distance. Thus it is not possible at present to generate a risk map. However, the social mapping has been conducted for the field area. In Figure 20.10, the land use for Baglung is shown, together with the location of the main settlements. In this case, land use has been simply divided into forest, cultivation, grazing land and areas for which no data are available. The location of the main settlements in the area is shown. To date no data have been available in the western part of the project area due to the occurrence of growing civil unrest during the 2001 field seasons.

A comparison of the landslide hazard map and the vulnerability map shows that a large number of communities are at risk in the eastern part of the study area, and also that this is an area of intense cultivation, meaning that there is a reasonable chance of economic losses from the destruction of farmland. This is an important consideration in

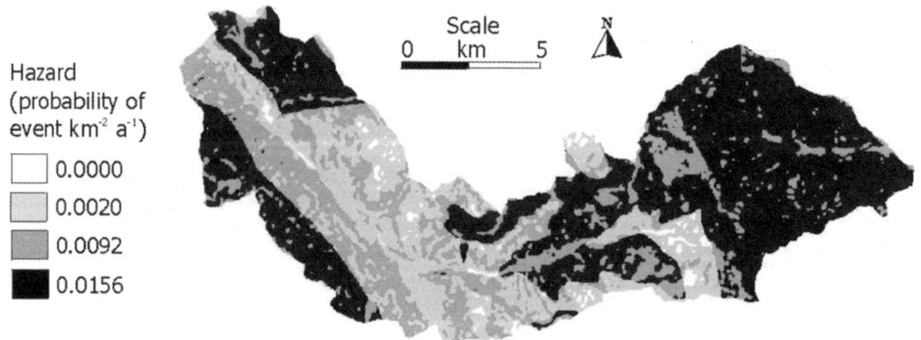


**Figure 20.9**  *Landslide hazard map for the Baglung study area*

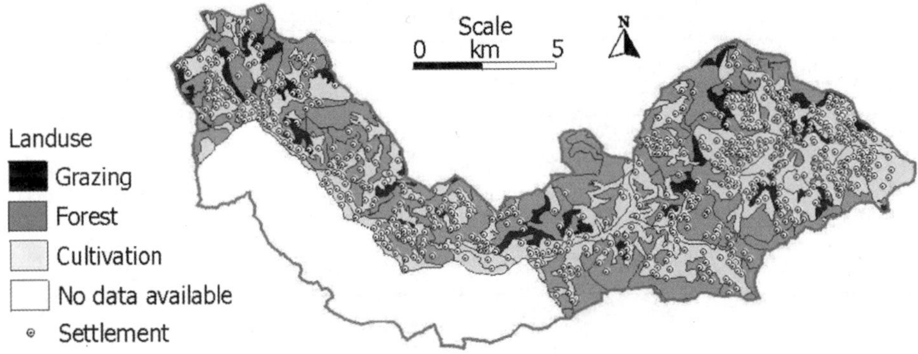


**Figure 20.10**   *Vulnerable elements in the Baglung study area*

this poor, subsistence-based rural region. As far as the road is concerned, the hazard map suggests that the selection of an optimum route in the east of the area is difficult due to the prevailing high hazard rating, but to the west the road should be built relatively low in the valley (but clearly above the level of the monsoon floods), where hazard is at its lowest. This is an interesting finding as there is considerable discussion in Nepal about the optimum routing in terms of valley wall or valley floor alignments.

## 20.10   Discussion

In this chapter a simple methodology has been proposed for determining the susceptibility of an area to landsliding based upon two very simple inputs. We acknowledge fully that this is not a definitive scheme and indeed that better, more accurate schemes can be devised. Indeed, our research is ongoing to try to develop an improved scheme for the Nepalese and Bhutanese Himalayas. However, the aim of the scheme described here was to allow rapid assessments of landslide susceptibility to be undertaken for rural access purposes. As mentioned in the introduction, these roads are built using low-cost principles without the use of highly qualified geoscientists. Landslide hazard assessment schemes are not used at present on such schemes because these assessment techniques are too complex to be used easily. The scheme presented here is deliberately simple to use to avoid these problems.

The indications from the study presented here, and from those in other areas of Nepal, are that the scheme is relatively successful in determining landslide susceptibility (Tables 20.6 and 20.7). Clearly further refinement is needed, but the technique appears to be promising. Of course all such schemes are dependent upon the quality of the information input. In much of the Himalayas the quality of both geological and slope information is quite low. However, compiling simple geology and slope maps remains a relatively simple task, and is one that may be necessary in determining a road alignment anyway.

In this chapter we have also described a technique by which this susceptibility map can be converted into a hazard map. In the case of the frequency of landsliding, we propose that the density of landslides in conjunction with the analysis of the residence

**Table 20.7**    *Density of landslides calculation*

| Susceptibility class | % area | Observed failures | Landslide density (slides/km$^2$) | Landslide hazard (events km$^{-2}$a$^{-1}$) |
|---|---|---|---|---|
| Very low | 3.9 | 0 | 0 | 0 |
| Low | 29.8 | 17 | 0.10 | 0.0020 |
| Moderate | 39.3 | 96 | 0.46 | 0.0092 |
| High and v. high | 28.8 | 118 | 0.78 | 0.0156 |
| Total | 100 | 231 | 0.44 | 0.0088 |

time of landslides within the landscape provides a mechanism by which this can be determined. Of course this technique is not fool-proof and indeed has some flaws, especially as any given period of 50 years is unlikely to be representative of the frequency of landslide-triggering events, especially if climate change is occurring. Given that there has not been a significant seismic event in western Nepal in the last 50 years, the data certainly do not represent seismic events. Finally, of course, the residence time of a landslide will vary according to land use, rock/soil type and precipitation frequency. As a result, this 50-year period may not be uniformly applicable.

However, we believe that this technique provides at least a way to get an indication of the frequency of events that is certainly no less reliable than other existing schemes, especially if used in conjunction with a more detailed analysis of triggering events, both seismic and precipitation.

In terms of the runout distance of a landslide, we advocate the use of a buffer approach together with an examination of the relationship between angle of internal friction and slope angle. The accuracy of such an approach is probably only moderate, and indeed some recent research has suggested that runout is related primarily to landslide volume rather than friction angle (Legros, 2002), although this is not yet widely accepted. However, determining the volume of a landslide is essentially impossible if a spatial approach to hazard is being used. Both the angle of internal friction and the slope angle are comparatively easily determined and it is intuitive that runout will in part be determined by their relationship.

In terms of risk mapping, this is completed by combining data on the vulnerability of the infrastructure with the hazard assessment. The actual production of the risk map is a relatively simple task, although the actual vulnerability mapping is laborious and slow.

## 20.11    Conclusion

The techniques that are described here appear to have been well received in Nepal and Bhutan, although further work is needed to refine them. We do not see the methodology presented here as the definitive technique for hazard and risk mapping in Nepal and Bhutan, but we do consider it to be an effective technique for at least gaining an impression of landslide susceptibility, hazard and risk in low-cost road planning. The use of geology and slope as the major inputs reflects previous research undertaken by, for example, Gerrard (1994), who demonstrated that these are the most important factors in determining landslide location. The Baglung example presented here illustrates the application of the scheme, although further work is needed in this area to allow true hazard and risk maps to be produced.


## Acknowledgements

This research was undertaken as part of the R7815 Landslide Risk Assessment in the Rural Access Sector project, funded by the UK Department for International Development, whose input is gratefully acknowledged. We would also like to thank the Department of Local Infrastructure Development and Agricultural Roads, Nepal and the Department of Roads, Bhutan. We would also like to acknowledge the assistance of Ivan Hodgson and Will Crick of Scott Wilson UK; Sushil Tiwari and Bhim Uppadaya of DoLIDAR Kathmandu; Prakash Lamichani, Durga Paudyal and Prakash Jha of Scott Wilson Kathmandu, and B.N. Upreti and Megh Raj Dhital of Tribhuvan University Kathmandu. All have made a very important contribution to the project.


## References

Anbalagan, R., 1992a, Terrain evaluation and landslide hazard zonation for environmental regeneration and landuse planning in mountainous terrain, in F. Bell (ed.), *Landslides* (Rotterdam: Balkema), 861–868.

Anbalagan, R., 1992b, Landslide hazard evaluation and zonation mapping in mountainous terrain, *Engineering Geology*, **32**, 269–277.

Anbalagan, R. and Singh, B., 1996, Landslide hazard and risk assessment mapping of mountainous terrains – a case study from Kumaun Himalaya, India, *Engineering Geology*, **43**, 237–246.

Bilham, R., Gaur, V.K. and Molnar, P., 2001, Himalayan Seismic Hazard, *Science*, **293**, 1442–1444.

Brunsden, D., Jones, D.K.C., Martin, R.P. and Doornkamp, J.C., 1981, The geomorphological character of part of the Low Himalaya of eastern Nepal, *Zeitschrift für Geomorphologie, Supplementband*, **37**, 25–72.

Caine, N. and Mool, P.K., 1982, Landslides in the Kolpu Khola drainage, Middle Mountains, Nepal, *Mountain Research and Development*, **2**, 157–173.

Corominas, J., 1996, The angle of reach as a mobility index for small and large landslides, *Canadian Geotechnical Journal*, **33**, 260–271.

Dangol, G., 1994, Geological database of the Jhiku Khola watershed and potential risk assessment study, in *Soil Fertility and Erosion Issues in the Middle Mountains of Nepal, Workshop. Proceedings of the Integrated Survey Division Kathmandu and Department of Soil Science*, University of British Columbia, Canada, 180–197.

Dangol, V., Shakya, U., Wagner, A. and Bhandari, A.N., 1993, A landslide inventory study (after the disaster of July 1993) along the Tribhuvan Highway, Central Nepal, *Bulletin of the Department of Geology*, Tribhuvan University, **3**, 59–69.

Deoja, B.B., Dhital, M., Thapa, B. and Wagner, A., 1991, *Mountain risk engineering handbook* (Kathmandu: International Centre for Integrated Mountain Development), 2 vol.

Dhakal, A.S., Amada, T. and Aniya, M., 1999, Landslide hazard mapping and the application of GIS in the Kulekhani watershed, Nepal, *Mountain Research and Development*, **19**, 3–16.

Dhital, M.R., Upreti, B.N., Dangol, V., Bhandari, A.N. and Bhattarai, A., 1991, Engineering geological methods applied in mountain road survey, *Journal of the Nepal Geological Society*, **7**, 49–67.

Dixit, A.M., 1994, Report of the landslide inventory survey in a part of Bajhang district, unpublished report, Department of Mines and Geology, Kathmandu.

Finlayson, A.A., 1984, Land surface evaluation for engineering practice: applications of the Australian PUCE system for terrain evaluation, *Quarterly Journal of Engineering Geology*, **17**, 147–158.

Gerrard, J., 1994, The landslide hazard in the Himalayas: geological control and human action, *Geomorphology*, **10**, 221–230.

Griffiths, J., Mather, A.E. and Hart, A.B., 2002, Landslide susceptibility in the Rio Aguas catchment, SE Spain, *Quarterly Journal of Engineering Geology and Hydrogeology*, **35**, 9–18.

Hart, A., Lamichhane, P., Jha, P. and Subba, M., 2003, The Incidence of Reported Landslide Activity in Nepal, *Proceedings of the PIARC Seminar, Kathmandu*, 232–240.

Hayashi, J.N. and Self, S., 1992, A comparison of pyroclastic flow and landslide mobility, *Journal of Geophysical Research*, **97**, 9063–9071.

Hearn, G.J., 1987, An evaluation of geomorphological contributions to maountain highway design with particular reference to the Lower Himalaya, unpublished Ph.D. thesis, University of London.

Hearn, G.J., 1997, Principles of low cost road engineering in mountainous regions, with special reference to the Nepal Himalaya, *Overseas Road Note*, **16**, Transport Research Laboratory, Crowthorne, UK.

Hsu, K.J., 1975, Catastrophic debris streams (Sturzstroms) generated by rockfalls, *Bulletin of the Geological Society of America*, **86**, 129–140.

Ives, S. and Messerli, B., 1981, Mountain hazards mapping in Nepal – introduction to an applied mountain research project, *Mountain Research and Development*, **1**, 223–230.

JICA, 1997, The development study in integrated watershed management in the western hills of Nepal, Japan International Cooperation Agency, unpublished report.

Joshi, J., Majtan, S., Morita, K. and Omura, H., 2000, Landslide hazard mapping in the Nallu Khola watershed, Central Nepal, *Journal of the Nepal Geological Society*, **21**, 21–28.

Keefer, D.K., 1984, Landslides caused by earthquakes, *Bulletin of the Geological Society of America*, **95**, 406–421.

Keinholz, H., Hafner, H., Schneider, G. and Tamrakar, R., 1983, Mountain hazards mapping in Nepal's Middle Mountains, *Mountain Research and Development*, **3**, 195–220.

Keinholz, H., Hafner, H. and Schneider, G., 1984, Stability, instability and conditional instability, *Mountain Research and Development*, **4**, 55–62.

Khanal, R.K., 1999, Study of landslide dynamics and management in Madi Watershed, the western hills and mountains of Nepal, unpublished report, International Centre for Integrated Mountain Development, Kathmandu.

Kilburn, C.R.J. and Sorensen, S.-A., 1998, Runout lengths of sturzstroms; the control of initial conditions and of fragment dynamics, *Journal of Geophysical Research, B, Solid Earth and Planets*, **103**, 17 877–17 844.

Krinitzsky, E.L., 2002, Epistematic and aleatory uncertainty: A new shtick for probabilistic seismic hazard analysis, *Engineering Geology*, **66**, 157–159.

Laban, P., 1979, *Landslide occurrence in Nepal*, HMG/FAO and UNDP, Department of Soil Conservation Report, Kathmandu.

Legros, F., 2002, The mobility of long-runout landslides, *Engineering Geology*, **63**, 301–331.

Petley, D.N., Crick, W.D.O. and Hart, A.B., 2002, The use of satellite imagery in landslide studies in high mountain areas, *Proceedings of the 23rd Asian Conference on Remote Sensing (ACRS 2002)*, Kathmandu, November 2002. Available online at: http://www.gisdevelopment.net/aars/acrs/2002/hdm/48.pdf.

Rickenmann, D., 1999, Empirical relationships for debris flows, *Natural Hazards*, **19**, 47–77.

Schelling, D., 1988, Flooding and road destruction in Eastern Nepal, *Mountain Research and Development*, **8**, 78–79.

Shrestha, N.R., Conway, D. and Bhattarai, K., 1999, Population pressure and land resources in Nepal: A revisit, twenty years later, *Journal of Developing Areas*, **33**, 245–268.

Skirikar, S.M., Rimal, L.N. and Jäger, S., 1998, Landslide hazard mapping of Phewa Lake catchment area, Pokhara, Central West Nepal, *Journal of the Nepal Geological Society*, **18**, 335–341.

Thapa, P.B. and Dhital, M.R., 2000, Landslide and debris flows of 19–21 July 1993 in the Agra Khola Watershed of Central Nepal, *Journal of the Nepal Geological Society*, **21**, 5–20.

UNDP, 2001, *Nepal human development report 2001*, UNDP, Kathmandu, Nepal.

Wagner, A., 1983, *The principal geological factors leading to landslides in the foothills of Nepal: a statistical study of 100 landslides*, HELVETAS Swiss Technical Cooperation report, Lausanne.

White, P.G., Fort, M. and Shrestha, B.L., 1987, Prototype 1:50 000 scale mountain hazard mapping in Nepal, *Journal of the Nepal Geological Society*, **4**(1–2), 43–53.

Whitehouse, I.E., 1990, Geomorphology of the Himalaya: a climatic–tectonic framework, *New Zealand Geographer*, **46**, 75–85.

Zimmerman, M., Bichsel, M. and Keinholz, H., 1986, Mountain hazards mapping in the Khumbu Himal, Nepal, *Mountain Research and Development*, **6**, 29–40.

# 21

# Quantitative Landslide Risk Assessment of Cairns, Australia

*Marion Michael-Leiba, Fred Baynes, Greg Scott and Ken Granger*

## 21.1  Introduction

This quantitative landslide risk assessment was part of a multi-hazard risk assessment of the Cairns area undertaken by the AGSO (now Geoscience Australia) *Cities Project*. All authors worked for AGSO at the time, either as employees or consultants. The material for this chapter was entirely sourced from Michael-Leiba *et al.*, 1999a, b, 2000, 2001; and Granger *et al.*, 1999. The *Cities Project* undertakes risk assessments aimed at reducing the risks posed by a range of geohazards to Australian urban communities. Cairns is the most northerly eastern Australian city, with a resident population of about 120 000. The positions of the localities mentioned in this chapter are shown in Figure 21.1.

The Cairns area has a wet tropical climate, with distinct wet and dry seasons. The heaviest rain occurs during the summer months. Rainfall is over 2000 mm per annum in the coastal corridor, influenced by the proximity of the 700 m high escarpment to the west; it increases to about 7000 mm on the summit of the range behind the escarpment.

Bedrock in the Cairns region consists of a more than 200 million-year-old sequence of folded and cleaved metamorphosed Silurian and Devonian sediments, and Permian and Permo-Triassic granite bodies. The prominent escarpments behind Cairns are believed to have been formed from a modified land surface more than 65 million years old, which was formerly a continental highland; the granite bodies probably formed the highest points on this land surface because of their resistance to erosion. Around 60 million years ago, the eastern part of the continental highland was rifted, leaving a steep eastern slope. This slope has been retreating since then to reach a position, about a million years ago, close to that of the present. Erosion has occurred most rapidly in the metamorphosed

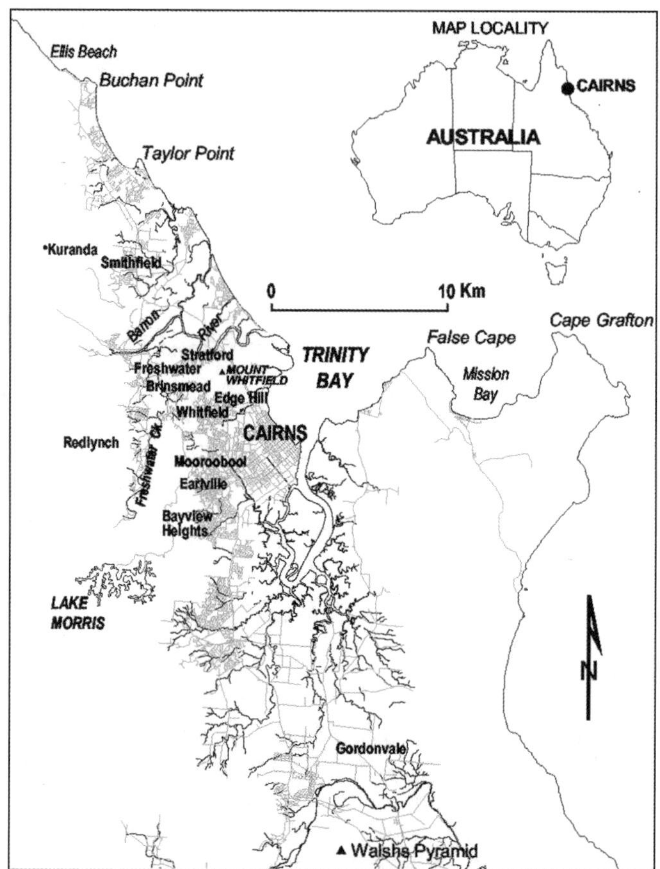


***Figure 21.1***   *Locality map*

sediments, leaving the granite as isolated hills and ranges. Whether or not there has been further uplift or faulting since the rifting around 60 million years ago is not known. There is no direct evidence that the Trinity Inlet–Mulgrave River corridor is a former rift valley, but this seems possible given the physiography and geological associations. The modern Cairns area consists of the plateau on which Kuranda is built, the rocks and landslide deposits of the escarpment, and the largely alluvial deposits of Trinity Inlet, the Barron River mouth, the Freshwater Valley and the coastal plain (Queensland Department of Mines, 1988).

The escarpment continues to be worn back by erosion and landslides. On the coastal plain, near the foot of the escarpment, a number of debris-flow fans and deposits are found. The geological history is summarized in Table 21.1.

The following information is from Granger *et al.*, 1999:

Recently, residential development in Cairns has spread to the northern beaches, the low hills and foot slopes of the Whitfield Range, and into the Mulgrave River valley towards Gordonvale. The population of Cairns has grown from 71 500 in 1983, to 128 000 at the 1996 census of whom almost 10 000 were overseas visitors.

***Table 21.1***   *Geological history*

| Age | |
| --- | --- |
| Holocene<br>  10 000 to present | Deposition of alluvium and marine sediments at or around<br>  current sea level |
| Pleistocene–<br>  Holocene | Deposition of alluvium, wind-blown deposits and colluvium on<br>  lower slopes and adjacent to streams |
| Pleistocene<br>  1.8 m y–10 000 yr<br>  before present | Deposition of alluvium, wind-blown deposits and colluvium as<br>  various terraces, fans, cones and aprons<br>Formation of basaltic lavas and scoria cones |
| Early Tertiary | Rifting of continental highland |
| Triassic–Permian | Granite intrusion |
| Devonian | Regional metamorphism, folding, cleavage |
| Silurian–Devonian | Deposition of muds and sands of Hodgkinson formation |

The city is the transport, logistic and administrative centre for an increasingly large hinterland which includes northern Queensland, Torres Strait, and significant mining operations in Papua New Guinea and the Indonesian Province of Irian Jaya. Cairns is now a major tourist destination, and the centre for a significant proportion of Australia's sugar industry and fishing operations in the northern Great Barrier Reef.

Extreme rainfall events are associated with tropical cyclones. Cairns comes under the influence of tropical cyclones on average at least once every two years, though 'direct hits' by severe tropical cyclones are not common.

Temperatures rarely exceed 35.0 °C or go below 10.0 °C for extended periods.

The natural vegetation of the area is a species-rich tropical rainforest. Extensive areas of this type still exist along the ranges and are now incorporated, under World Heritage listing, into the Wet Tropics Management Area. Rainforest grades to various forms of eucalypt-dominated forest or woodland and grassland in areas exposed to frequent burning, especially on the hill slopes. Most of the valley and coastal plain areas not occupied by urban development are under sugar cane cultivation or are covered by mangrove communities.

The objective of the landslide study was to provide information to the Cairns City Council on hazards, community vulnerability and risks for planning and emergency management purposes.


## 21.2   The Landslide Threat

In Cairns, rainfall intensities of sufficient magnitude to trigger slope failure have an average recurrence interval (ARI) of considerably less than one year, and landslides are not rare events.

The landscape around Cairns is dominated by a series of escarpments that are developing by scarp retreat by two main processes (Michael-Leiba *et al.*, 1999a):

- on steeper bedrock slopes, and bedrock slopes masked with a relatively thin mantle of broken rock and finer material, weathering and erosion leads to landslides (rock falls, rock slides, debris slides, and small debris flows confined to the slope). By this process rock and soil moves down slope under the influence of tropical rainstorms and gravity; and

- during the more extreme rainfall events, the combined effect of multiple landslides in the upper parts of gully catchments, and the remobilisation of accumulated debris in the major gully systems, periodically results in large debris flows. These can extend onto the depositional plains at the base of the bedrock slopes.

Debris flows are a type of landslide triggered by the action of torrential rain on loose material on a mountainside or escarpment. The boulders and finer material, mixed with water, flow down the slope as a torrent. The coarser material (the proximal part of the debris flow) is deposited near the base of the slope, while the finer material (the distal part of the debris flow) travels further as a flash flood across the floodplain. Debris flows can be highly destructive. One definite large debris-flow event described below, and two probable ones, have occurred in the Cairns region since European settlement of the area in 1876.

On 12 January 1951, a deluge of about 700 mm of rain in just under 5 hours triggered debris flows that affected 10 km of the Captain Cook Highway behind Ellis Beach. Huge quantities of debris were swept from the mountainside onto the road and over the precipice into the sea. Boulders up to 3 m long were hurled into the Pacific 'like marbles'. Large slabs of bitumen were tilted up from the road and landslide debris was piled up as high as 3 m. All culverts and inverts in this area were either damaged considerably or washed away entirely. The highway was not expected to carry normal traffic for at least two weeks (*Cairns Post*, 15 January 1951).

Landslides on hillslopes, most commonly batter failures, periodically block roads, particularly Lake Morris Road and Kuranda Range Road, and the Cairns–Kuranda railway has an even more spectacular history of dislocation by landslides. Instances of landsliding have been recorded in the established suburbs, either on cuts behind houses or road cuts or fills. Two houses have been destroyed and several building blocks written off as a result.

On 31 May 1900, the landslide with the fourth largest number of Australian landslide fatalities happened in Cairns. Five men were killed and one buried alive for 90 minutes when an 8 m deep tramway cutting they were constructing at Riverstone for the mill at Gordonvale caved in spontaneously. The location was in a river terrace in Gordonvale, 3 km WNW of Walsh's Pyramid.

## 21.3  Landslide Risk Assessment Methodology

The approach used in this study involved the following steps (Michael-Leiba *et al.*, 1999a):

- mapping, description and interpretation of the geology and geomorphology of the study area in as much local detail as possible, whilst ensuring uniform general coverage of the area. This required consideration of existing reports and publications, interpretation of aerial photographs, discussion with Cairns City Council, local consultants and the Cairns Historical Society, and, most importantly, ground inspection of areas, readily accessible from public land, where it was suspected that landslides of various forms may have occurred;
- definition of the slope processes, especially the landslide types, and their mode of occurrence;
- collection of information on process rates, from which landslide hazard may be assessed. This information essentially relates to how often landslides occur in the historical and geologic record, where they occur, and what appears to trigger landslide

events. Landslides were also logged following the passage of two tropical cyclones, Justin in 1997, and Rona in 1999;

- entry of the landslide location data and slope process interpretation into a GIS;
- use of shadow angles and a digital elevation model in the GIS to define areas which may be susceptible to debris flow runout;
- identification of GIS landslide hazard polygons;
- derivation of magnitude–recurrence relations for the landslide slope processes. Bureau of Meteorology rainfall intensity–frequency–duration curves were used to assess the mean recurrence intervals of rainfall events that triggered landslides. The total volumes of landslides triggered by each of these rainfall events were estimated;
- calculation of landslide hazard for the various polygons;
- assessment of the vulnerability of the elements at risk (people, buildings, and roads) to destruction by the various landslide hazards;
- development of GIS maps depicting specific risk of destruction for people, buildings, and roads, and of road blockage, assuming all landslide hazard polygons are developed (for subdivision, e.g. have a network of roads). These maps are useful as a planning tool; and
- development of GIS total risk of destruction maps for resident people, and buildings (houses and blocks of flats) for the developed parts of the landslide hazard polygons. Information on the elements at risk in each landslide hazard polygon was obtained by interrogating the GIS;
- assessment of the total risk for roads (running parallel to the escarpment) on hillslopes, in a 1 in 10 year rainfall event. This was tabulated as the length of road per 10 km covered by landslide debris, and the length of road destroyed by landslides per 10 km of road.

This approach is presented schematically in Figure 21.2. Peer review, reasonableness checks and consideration of the acceptability of risk estimates were also used as an essential part of the process, but for simplicity's sake were not included in Figure 21.2. In this report, the method used for calculating hazard from a magnitude recurrence relation graph is illustrated in Figure 21.3.

## 21.4  Slope Processes

Using geological and geomorphological observations and historical information, a regional map of landslide hazards in the Cairns area was produced. This map was entered into a GIS containing comprehensive information on buildings, roads and demography.

Two main landslide slope processes were identified. The first group are landslides which occur on slopes developed in weathered bedrock and thin colluvial cover. The second group, which had not previously been identified as a threat, are occasional large debris flows (containing boulders up to several metres in diameter) which extend from the major gully systems on to the plains and form fans at the base of the slopes.

These slope processes are illustrated in Figure 21.4. Definitions of geomorphic units (b1, b2, etc.) are presented in Table 21.2.

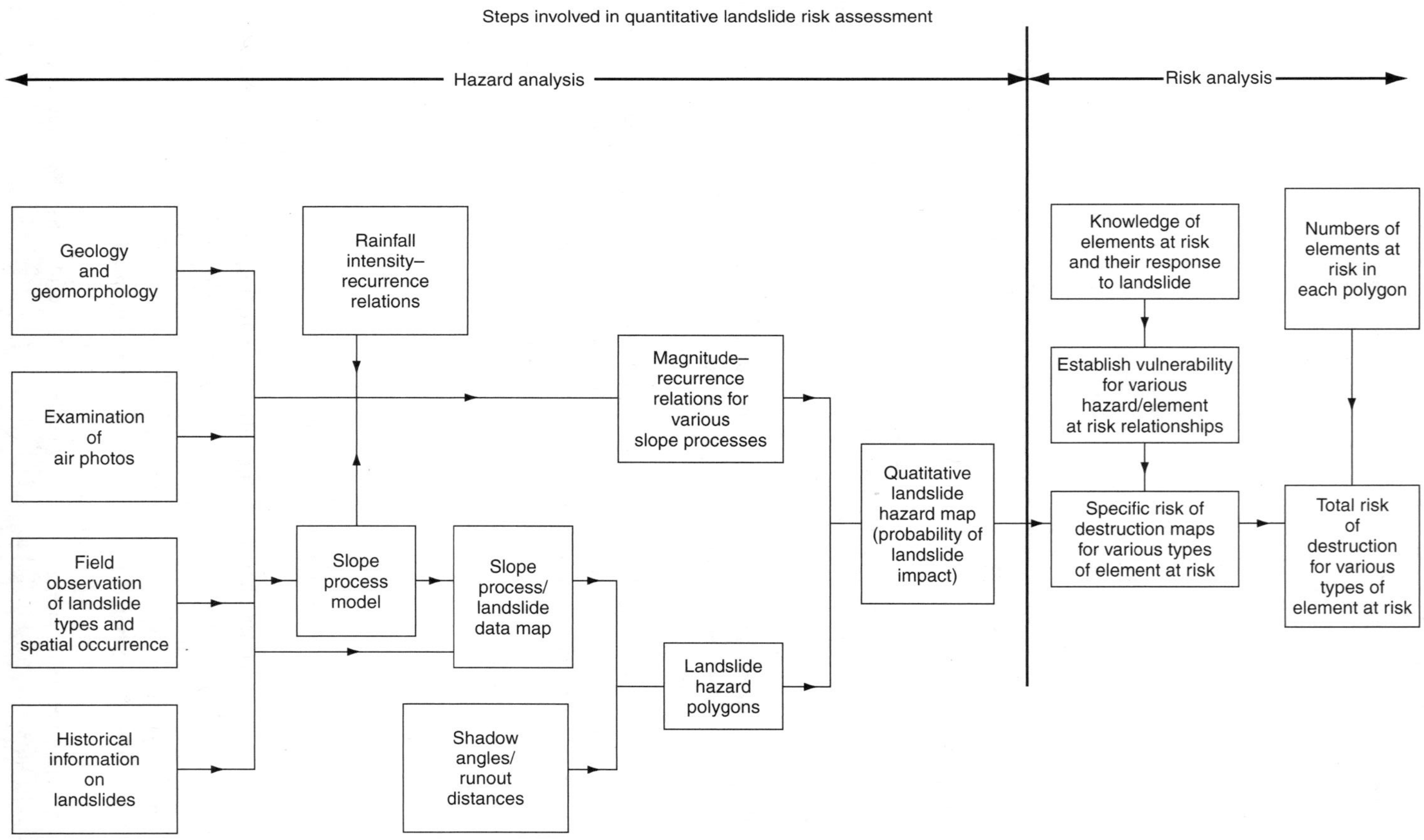

*Figure 21.2*  *Quantitative landslide risk assessment process*

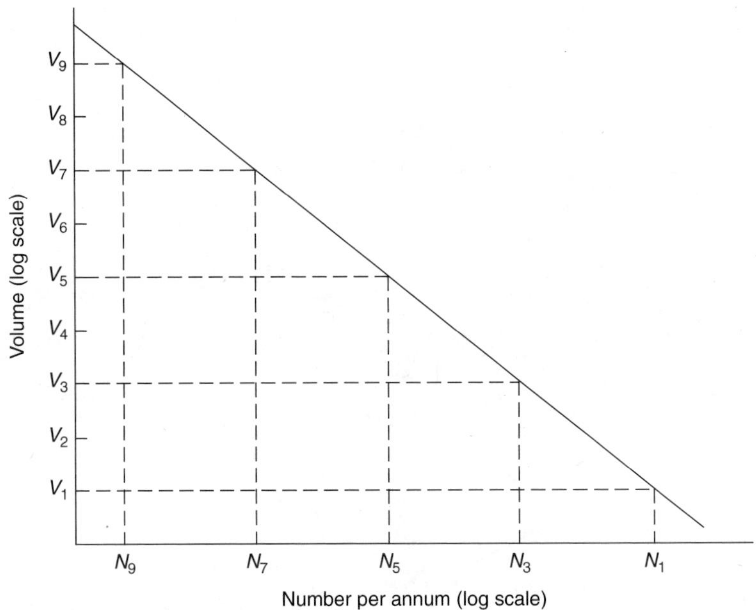


| Volume interval | Number of landslide events in volume interval | Mid-point on graph of volume interval | Area of landslide = vol./thickness= $v_2$/thickness, etc. | Probability of impact at a specified point in a polygon with area $A_{polygon}$, given that a landslide happens in the polygon | Hazard = annual probability of impact at a specified point |
|---|---|---|---|---|---|
| $V_1$ to $V_3$ | $N_1 - N_3$ | $V_2$ | $A_2$ | $P_2 = A_2 / A_{polygon}$ | $P_2 (N_1 - N_3)$ |
| $V_3$ to $V_5$ | $N_3 - N_5$ | $V_4$ | $A_4$ | $P_4 = A_4 / A_{polygon}$ | $P_4 (N_3 - N_5)$ |
| $V_5$ to $V_7$ | $N_5 - N_7$ | $V_6$ | $A_6$ | $P_6 = A_6 / A_{polygon}$ | $P_6 (N_5 - N_7)$ |
| $V_7$ to $V_9$ | $N_7 - N_9$ | $V_8$ | $A_8$ | $P_8 = A_8 / A_{polygon}$ | $P_8 (N_7 - N_9)$ |
| | | | | | Sum of this column = $H$ |

$H$ = hazard = probability per annum of impact of a landslide (volume in the range $V_1$ to $V_9$) at any point in the polygon

$$= P_2(N_1 - N_3) + P_4(N_3 - N_5) + P_6(N_5 - N_7) + P_8(N_7 - N_9)$$

$V$ = vulnerability = probability of an element at risk being destroyed by a landslide, if affected

$E$ = number of elements at risk in a polygon

Specific risk = $H*V$ = annual probability of a given element being destroyed by a landslide

Total risk = $H*V*E$ = number of elements per annum expected to be destroyed by a landslide

**Figure 21.3**   *Hazard assessment from magnitude–recurrence graph*

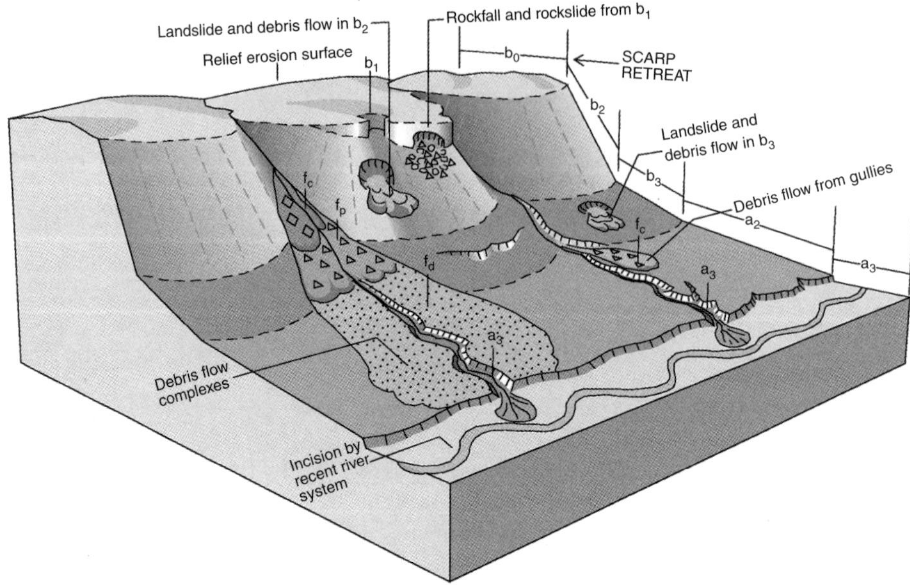


**Figure 21.4**   *Slope processes*

**Table 21.2**   *Geomorphological units and potential hazards*

| Unit | Landslide type |
| --- | --- |
| b0 – upper interfluves, creep slopes and convex creep slopes – remnants of Mesozoic peneplain | Landslide? |
| b1 – fall faces or steep slopes with cliff lines, developed in bedrock | Rockfall |
| b2 – transportational midslopes developed as ridges and gullies in bedrock | Rockfall, small landslides |
| b3 – bedrock footslopes, concave or planar deeply weathered bedrock sometimes locally covered with varying thickness of clayey colluvium from one to several metres thick | Rockfall, small landslides |
| fc – massive core debris-flow deposit with irregular lobate surface, numerous boulders greater than 1000 mm | Proximal debris flow |
| fp – proximal debris-flow fan/outwash with gentle undulating convex slope forms | Proximal debris flow |
| fd – distal outwash fan with uniform low-angle slopes | Distal debris flow |
| a2 – distal outwash fan with uniform low angle slopes and no obvious major debris-flow source (note that fd can grade into a2 or a2 alone can occur at the base of b3) | Distal debris flow |
| a3 – seasonal floodway incised into surface, possible transport corridor for debris flow | Distal debris flow grading into flood |

It is conceivable that locally some form of rotational or translational slides could happen in the accumulated fan debris where such deposits are themselves undercut by erosion. Small batter failures were observed in this material after Cyclones Justin and Rona. Such processes have not been considered further in this preliminary assessment.

Landslides in the walls of trenches and cuttings in the plains have also not been considered in this study, but it must be remembered that this sort of failure can, and does, cause fatalities.

No evidence of other forms of landslide was obtained during the study. In particular, there was no record of any deep-seated, slow-moving landslides having occurred. The landslides observed tend to be very shallow.


## 21.5   Landslide Hazard Assessment

### 21.5.1   Magnitude–Recurrence Relations

For the purpose of deriving recurrence intervals for landslide events of various minimum volumes, two processes were considered.

First were the landslides originating on the hillslopes and affecting the landslide hazard in the b1, b2 and b3 polygons. For the purposes of this regional pilot study, they were considered to have a uniform rate of occurrence in these polygons throughout the map area. The recurrence rates were derived from observations of landslides on roads and developed slopes and so are only applicable to hillslopes that are undergoing development. This was considered appropriate, as one of the uses of the maps is to be as a planning tool for the Council. Also, insufficient data are available for natural slopes to do quantitative work. The logarithm of the failure volume ($m^3$) per 10 km of the escarpment was used as a measure of landslide event magnitude. The length, 10 km, was chosen arbitrarily. A landslide event was taken to be the suite of landslides triggered by a single rainfall event. As a simplifying assumption, all parts of the b1, b2 and b3 polygons were assumed to be equally susceptible to slope failure and to be subject to the same recurrence relation.

The second process was the occurrence of debris flows large enough to run out on to the plains. Examples are those in the Ellis Beach area. Because of their runout, they may constitute a landslide hazard in flatter areas where landslides would not normally be considered a problem. For the 10 000 year rainfall event, these were assumed to occur in all gullies throughout the area. For smaller rainfall events, only one or two debris flows may occur in the map area. Their frequency was calculated for the whole of the map area, then reduced proportionately to consider only 10 km of escarpment. The logarithm of the failure volume ($m^3$) was taken as a measure of the magnitude of the debris-flow event (a suite of debris flows triggered by a single rainfall event).

The landslide magnitude was taken to be the logarithm of the landslide volume in $m^3$ per 10 km of escarpment. In summary, recurrence relations per 10 km of escarpment were established for:

- the total volume of a set of landslides triggered by a rainfall event, along roads and the railway up the escarpment (Figure 21.5a);

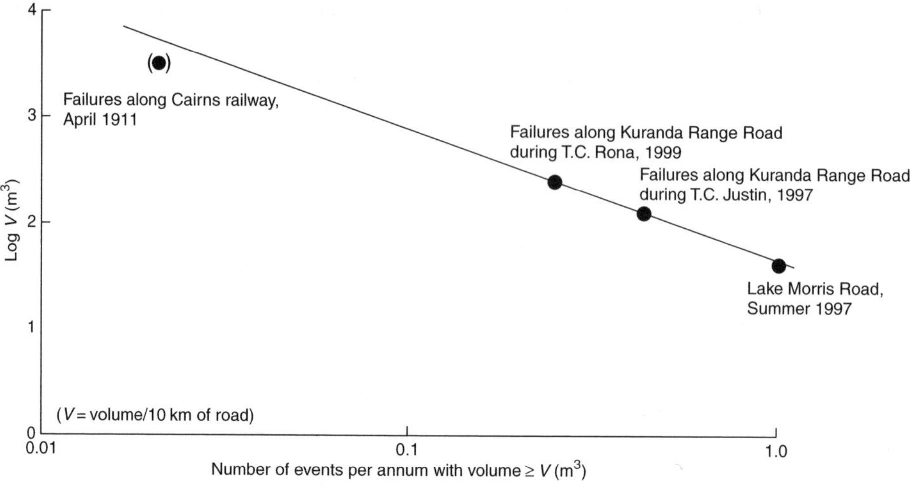



**Figure 21.5a**  *Recurrence of landslide along roads and railways along the escarpment*

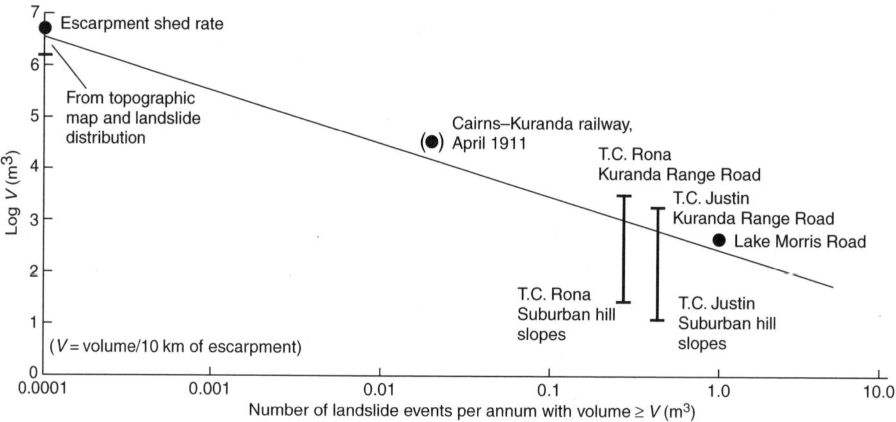



**Figure 21.5b**  *Recurrence of landslides on hillslopes*

- the total volume of a set of landslides triggered by a rainfall event on fully developed slopes (Figure 21.5b); and
- the total volume of a number of debris flows, triggered by a single rainfall episode, which extend on to the plain (Figure 21.6).

### 21.5.2  Shadow Angles

A landslide or debris flow originating in one geomorphic unit can extend and impact on the unit downslope. The extent of any impact may be conveniently defined by the

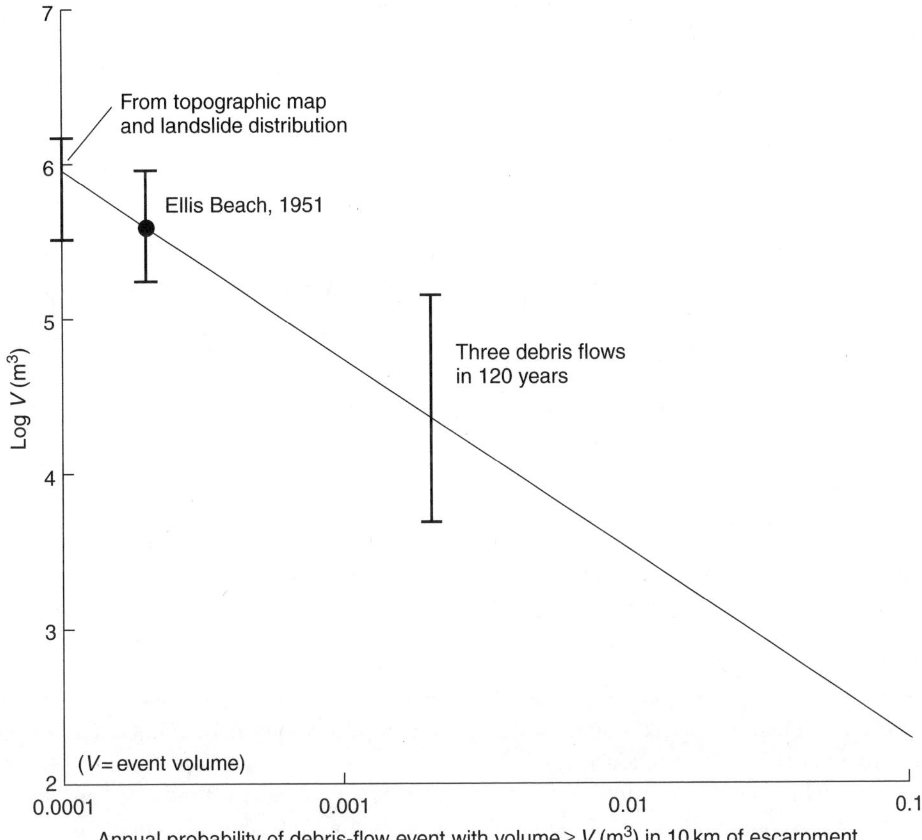


**Figure 21.6**   *Recurrence of debris flows*

shadow angle relevant to that process (Hungr, 1997; Wong *et al.*, 1997). In this study, the shadow angle is taken to be the angle between the horizontal and a line drawn from the limit of the proximal or distal part of the debris flow to the top of the escarpment or ridge crest (Figure 21.7).

The proximal portion is that part of the debris flow closest to the source of the landslide. It has a lumpy or convex surface and contains large boulders up to several metres in size. The distal portion of a debris flow is more gently sloping and contains the finer-grained sediments that are deposited further from the landslide source.

Based largely on field observations, shadow angles of 19° and 14° were chosen to represent the limits of the proximal and distal portions respectively of potential debris flows. Using a GIS, the extent of areas covered by these shadow angles was delineated, thus defining hazard zones on the gentle slopes below the gullies (Figure 21.8). These zones represent the limit to which a debris flow might conceivably extend were it to originate high in the catchment of a particular gully system. The polygons thus derived may be thought of as defining the extent of the debris-flow hazard.

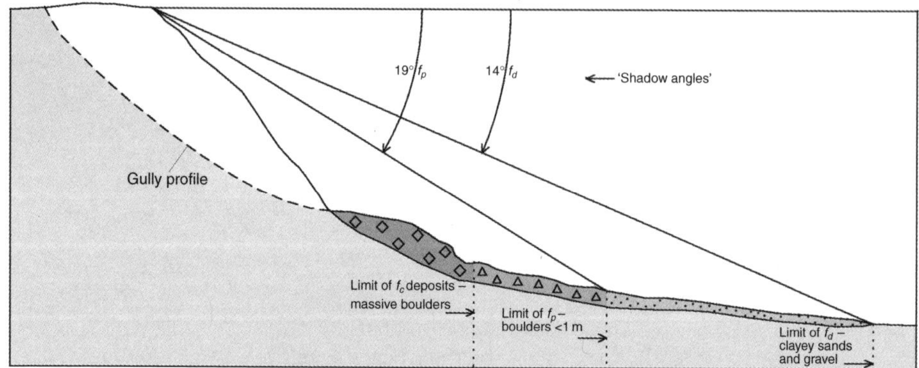


**Figure 21.7**   *The shadow angle concept*

### 21.5.3   Landslide Hazard Polygons

GIS polygons were used to delineate and characterize the areas that could be affected by landslides. Three main categories were chosen:

- hillslopes of bedrock, sometimes locally covered with colluvium (b1, b2 and b3);
- areas that could be affected by the proximal portions of debris flows (fc, fp, and parts of other geomorphic units of the gentle slopes, below b1/b2/b3, susceptible to proximal debris-flow runout); and,
- areas which could be affected by the distal portions of debris flows (fd, and parts of other geomorphic units of the gentle slopes, below b1/b2/b3, susceptible to distal debris-flow runout).

### 21.5.4   Landslide Hazard

The landslide hazard is the probability, $H$, per annum, of a point being affected by a landslide. For each polygon, the magnitude–recurrence relations were used to estimate $H$. The method is shown in Figure 21.3.

For points on the escarpment, the hazard occurrence probability is estimated to be 0.02% (an ARI of 6000 years), assuming that the slope is developed. Thus, for undeveloped parts of the escarpment, this figure predicts what the hazard would be *if* the slope were to be developed with a network of roads 100 m apart, parallel to the escarpment, without adequate mitigation measures being taken. The hazard would be expected to be considerably less on slopes developed with geotechnical consultation. It would also probably be less on undisturbed slopes. A simplifying assumption used was that the hazard is uniform at all points on a developed slope, whereas it could vary with distance from cuts or the edge of fills.

The hazard probability, rounded to one significant figure, in areas which may be affected by the proximal parts of debris flows is calculated to be 0.01% (an ARI of 8000 years), and for the distal parts of debris flows, 0.01% (an ARI of 9000 years).

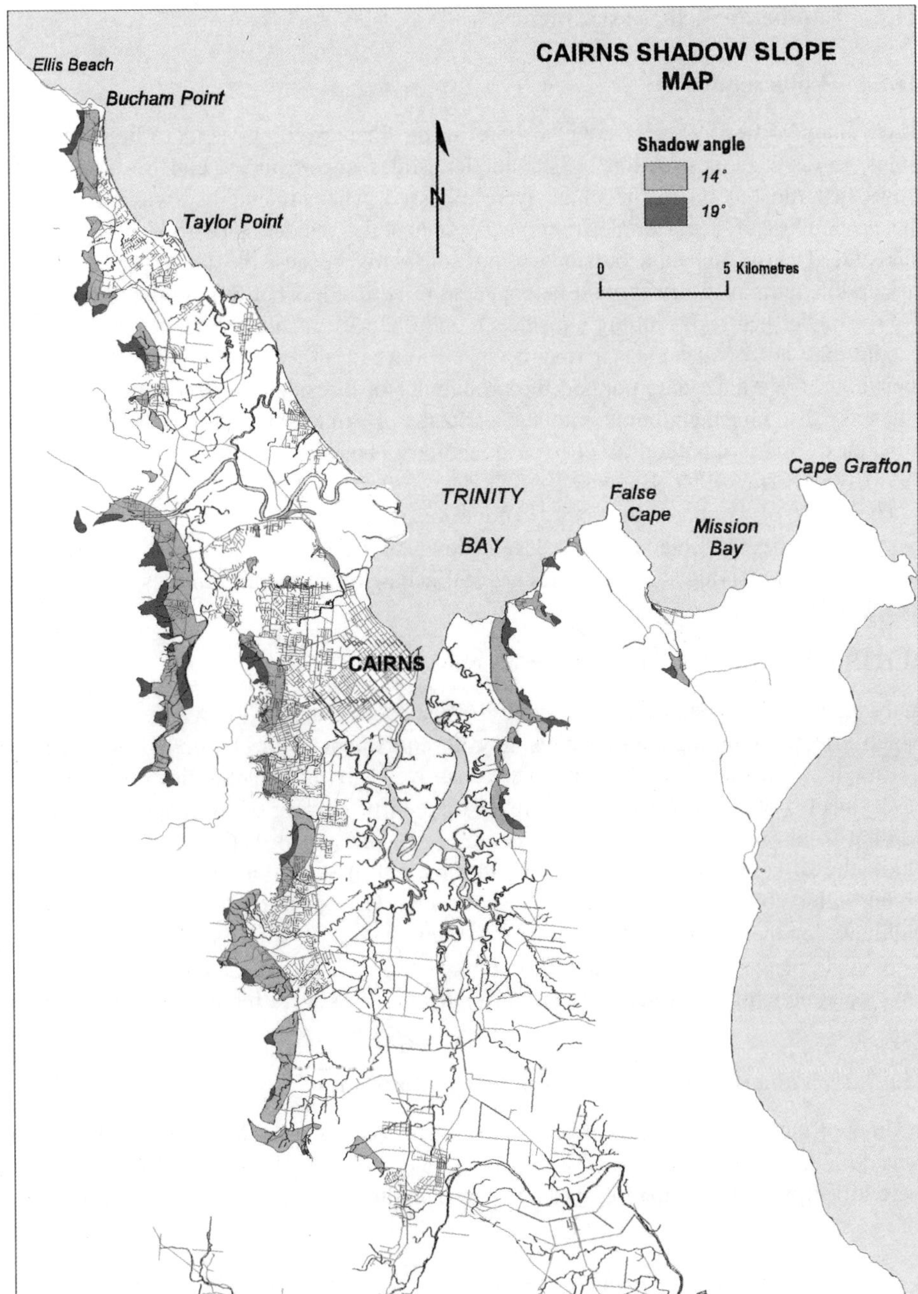



**Figure 21.8**    *Cairns shadow slopes map*

## 21.6   Landslide Risk Assessment

### 21.6.1   Vulnerability

The vulnerabilities ($V$) of resident people inside their dwellings, and to buildings and roads, to death or destruction by smaller landslides on hillslopes, and by large debris flows that run out on to the plain, were assessed. The vulnerability was taken to be the probability of death or destruction given that the residence or road was hit by a landslide. The risk to people outside was not considered because most people in landslide-susceptible parts of Cairns would be expected to be inside when it is raining heavily.

For smaller landslides hitting a residence on hillslopes, it should be noted that resident people may not be at home, or may be in a room not affected by a landslide that hits their house. We have incorporated the probability of a resident person being hit by the landslide that hits their home into the estimate of vulnerability for people living on hillslopes. This is common practice in quantitative landslide risk assessment (e.g. Fell, 1994; Wong *et al.*, 1997; Baynes, 1995).

The value of $V$ ranges between 0 (none destroyed) and 1 (all destroyed) with the type of landslide and element at risk, as shown in Table 21.3. The methods used and assumptions in arriving at these values are outlined below. The uncertainties in the figures in Table 21.3 are not known.

#### 21.6.1.1   Vulnerability of buildings on hillslopes

In the Australian landslide database and Cairns landslide database there were 24 landslides which affected buildings, and which may be equivalent to the b1/b2/b3 landslides in Cairns. In 11 of these, no buildings were destroyed, giving vulnerabilities of 0 in these cases. In 11 cases, all the buildings were destroyed (the number of buildings per landslide varying from one to three), so the vulnerability for these is 1.0. For the remaining two out of the 24 landslides, two out of 16 and three out of seven buildings were destroyed, giving vulnerabilities of 0.1 and 0.4 respectively. The weighted mean vulnerability of buildings for the 24 landslides is $V_s = 0.5$. This may be a conservative value, because there could be cases where buildings were impacted by landslides and not destroyed that have gone unreported. Assuming that only half the cases were reported in the database, $V_s = 0.25$.

#### 21.6.1.2   Vulnerability of resident people on hillslopes

In three of these 24 landslides, people were killed. At Walhalla in Victoria, both people were killed, giving a vulnerability of 1.0. At Coledale, NSW, two out of five people were killed, giving a vulnerability of 0.4. At Thredbo, NSW, 18 out of 19 people were

*Table 21.3*   *Vulnerability to destruction of people, buildings and roads*

| Unit | Resident people | Buildings | Roads |
|---|---|---|---|
| Hillslopes | 0.05 | 0.25 | 0.3 |
| Units susceptible to proximal debris flow | 0.9 | 1.0 | 1.0 |
| Units susceptible to distal debris flow | 0.05 | 0.1 | 0.3 |

killed, giving a vulnerability of 0.9. For the remaining 21 landslides, either no people were in the buildings, or else they were not killed, giving a vulnerability of 0. This gives a weighted mean vulnerability for the 24 landslides of $V_p = 0.1$. This value could be conservative because landslides which cause death are more likely to be reported than those which do not, and because fill failures were the cause of two of these landslides. Assuming that only half the cases of buildings being affected by landslides were reported in the database, then $V_p = 0.05$. This figure is in good agreement with the value suggested by Wong *et al.* (1997) for the vulnerability of a person in a building if debris strikes the building. As it is derived from figures relating to buildings affected by landslides in the database, it takes into account the fact that buildings may not be occupied all the time.

### 21.6.1.3    Vulnerability of roads on hillslopes

The data in the Australian landslide database are not detailed enough to calculate a vulnerability for roads. However, Cairns City Council informed us that Kuranda Range Road is totally blocked about once a year, but needs partly remaking no more than once in two years. This gives a vulnerability of no more than 0.5. Lake Morris Road gets totally blocked about three times a year, but needs partly remaking at most once every two years. This gives a vulnerability of at most 0.17. The mean of 0.5 and 0.17 is 0.3, so 0.3 was taken as the value of the vulnerability, $Vr$, of roads.

### 21.6.1.4    Vulnerability to large debris flows

There was insufficient information in the Australian Landslide Database to calculate vulnerabilities to large debris flows, so values were assumed from knowledge of the type of event and judgement (Michael-Leiba *et al.*, 1999a). As large debris flows are triggered by extreme rainfall events, resident people were assumed to be occupying their dwellings.

### 21.6.2    Specific Risk of Destruction

Specific annual risk of destruction is the probability per annum of a person, building or section of road at a given point in the Cairns area being destroyed by a landslide. The specific risks to individual people, buildings and roads in susceptible parts of Cairns, if the areas were to be developed, have been calculated from the equation

$$\text{specific risk} = H \times V$$

and mapped using the GIS. An example is shown in Figure 21.9.

The specific annual risk of destruction of individuals, buildings and roads, and of road blockage is shown in Table 21.4. The possible range in values, attributed to uncertainties in the recurrence relations, is given in brackets.

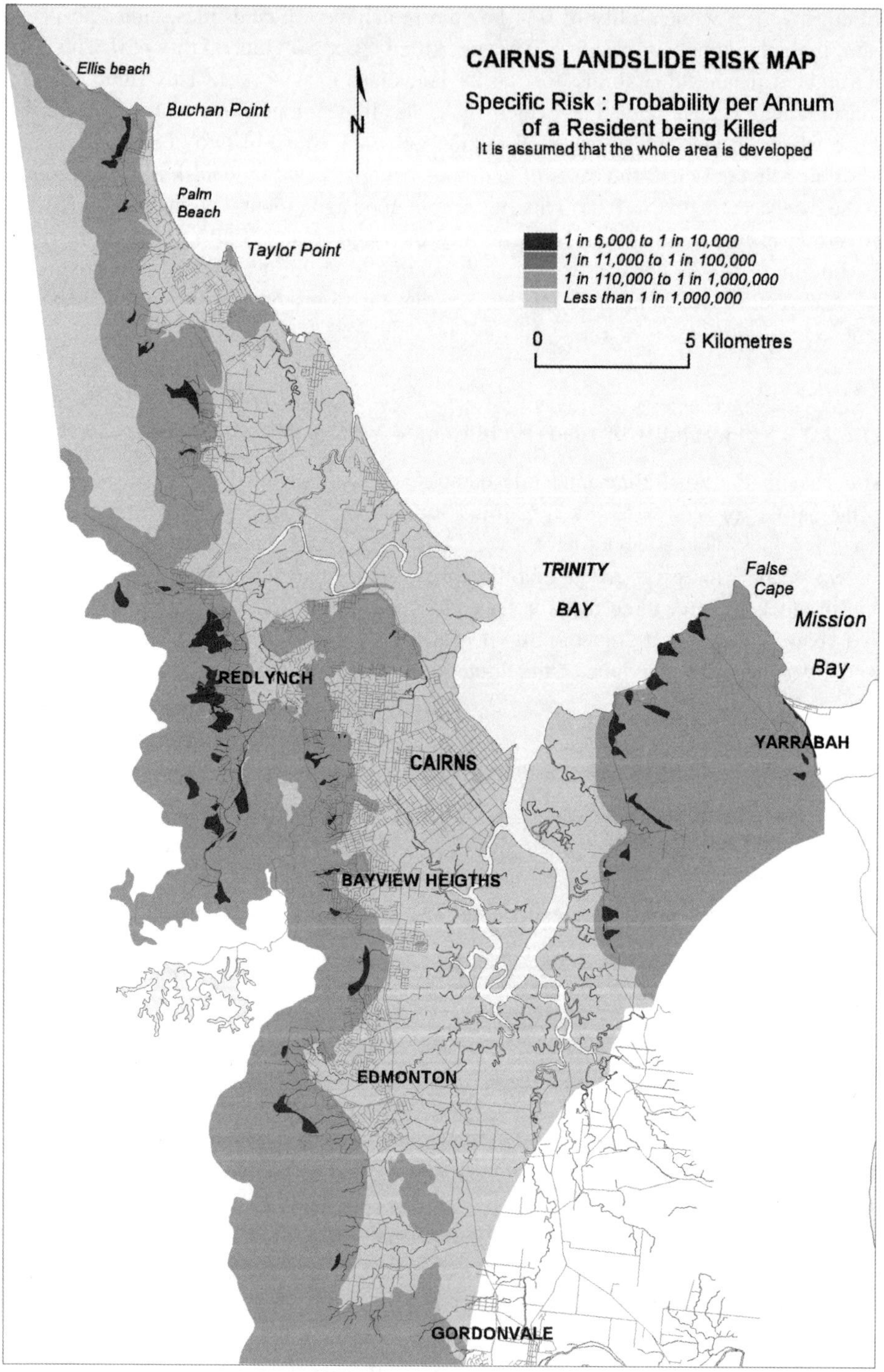

***Figure  21.9***   *Specific annual risk of fatality for resident people*

**Table 21.4**  *Specific annual risk*

| Unit | Death of resident people | Building destruction | Road destruction | Road blockage |
|---|---|---|---|---|
| Hillslopes | 0.0008%<br>1 in 100 000+<br>*(1 in 5 million<br>to 1 in 40 000)* | 0.004%<br>1 in 20 000<br>*(1 in 1 million<br>to 1 in 8000)* | 0.005%<br>1 in 20 000<br>*(1 in 1 million<br>to 1 in 8000)* | 0.02%<br>1 in 6000<br>*(1 in 300 000<br>to 1 in 2000)* |
| Units susceptible to proximal debris flow | 0.01%<br>1 in 9000<br>*(1 in 50 000<br>to 1 in 2000)* | 0.01%<br>1 in 8000<br>*(1 in 50 000 to<br>1 in 1000)* | 0.01%<br>1 in 8000<br>*(1 in 50 000<br>to 1 in 1000)* | 0.01%<br>1 in 10 000+<br>*(1 in 60 000<br>to 1 in 2000)* |
| Units susceptible to distal debris flow | 0.0005%<br>1 in 200 000<br>*(1 in 1 million<br>to 1 in 30 000)* | 0.001%<br>1 in 90 000<br>*(1 in 500 000<br>to 1 in 20 000)* | 0.003%<br>1 in 30 000<br>*(1 in 200 000<br>to 1 in 5000)* | 0.007%<br>1 in 10 000+<br>*(1 in 60 000 to<br>1 in 2000)* |

### 21.6.3  Total Risk of Destruction

Total risk is the number of elements at risk expected to be destroyed by a landslide in a given GIS polygon in a given period of time. The GIS polygons were interrogated to assess the nature and number of elements at risk ($E$). Maps which quantitatively depict the total risks per km$^2$ per 100 years for residential people and buildings in each GIS polygon in the currently developed parts of Cairns were constructed from the data for each polygon. These were based on the equation

$$\text{total risk} = H \times V \times E,$$

where $E$ is the number of houses and flats, or people living in houses and flats, in a polygon. The greatest total risk for buildings (houses and flats) is on the hillslopes, where it is estimated that a total of 13 buildings throughout the map area could be destroyed in 100 years, if no mitigation measures were taken. The highest total risk for people living in houses and flats is in the proximal parts of debris flows. It is estimated that a total of 16 people in the map area could die over 100 years in these areas.

### 21.6.4  Total Risk – Roads on Hillslopes

The hillslope failure rate along roads, in the first magnitude–recurrence relation, is the volume of landslide debris per 10 km of road. The recurrence relation was largely derived from failure volumes logged along roads up the escarpment after rainfall events.

The calculations of the total length of road covered by debris, and the length destroyed, in a 1 in 10-year ARI rainfall event in the Cairns local government area are shown in Table 21.5. The landslide volume per 10 km of road was read from the landslide recurrence graph. The landslide area was calculated assuming a mean depth of 1.5 m. The width was estimated by assuming that the landslide failure surface is square in plan. It was assumed that all the debris from the batter is dumped on to the road.

**Table 21.5**  *Total risk for roads on hillslopes for the 1 in 10-year rainfall scenario*

| Parameter | Value |
| --- | --- |
| Landslide volume ($m^3$/10 km of road) | 617 |
| Landslide area ($m^2$/10 km of road) | 411 |
| Maximum length of road covered by debris (m/10 km of road)[1] | 20 |
| Maximum length of road destroyed (m/10 km of road)[2] | 6 |
| Maximum total length (m) of road covered by landslide debris | 238 |
| Maximum total length (m) of road destroyed | 71 |

*Notes:*
[1] Length of road affected assumed equal to width of landslide.
[2] width $\times V$, where $V = 0.3$, vulnerability of roads.

In Cairns there are 119 km of roads on the hillslopes. It was estimated that a 14 m$^3$ landslide could block a 6 m wide road, and a 60 m$^3$ landslide could block a 10 m wide road, so that total road blockage could occur in the 1 in 10-year rainfall event. However, the estimates of road blockage and destruction are maximum values, because the batter failures could occur as a large number of tiny landslides, instead of a few larger ones, and not block the road completely.

## 21.7   Community Risk from Landslides

As part of risk evaluation in a regional risk assessment, one needs to consider the impact of a hazard not only on individual people and structures, but also on the resilience and viability of the community as a whole.

### 21.7.1   Building Destruction

The total risk of destruction by suburb (Figure 21.1) for all types of buildings is given in Table 21.6 for the ten suburbs with the greatest risk, in descending order of risk. These values do not compensate for the differing areas of the suburbs. Note that with good engineering practice, such as adequate drainage and retaining walls, commonly used in developing the hillslopes in Cairns, the actual number of buildings destroyed per 100 years would be expected to be considerably lower that that shown in the Table 21.6.

The parts of these suburbs that are at greatest risk of landslide are in the Freshwater Valley, the lower slopes of the coastal escarpment, or near the base of Mount Whitfield.

These are dormitory suburbs either on hillslopes, or near the base of the slopes in potential runout zones for large debris flows. The population of Cairns is expected to reach 187 000 by 2021 (DLGP, 2001). As the population grows, more development will take place in such areas, and the landslide risk will increase unless adequate mitigation measures are put in place at the time of development.

However, most of the critical facilities, such as hospitals and emergency services, essential to community recovery after a disaster are in the older, flatter parts of Cairns that are unlikely to be exposed to landslide.

**Table 21.6**  *Total risk of destruction by landslide of all types of buildings – estimated number destroyed per suburb per 100 years, if no mitigation measures taken*

| Suburb | Total risk |
| --- | --- |
| Redlynch | 6.0 |
| Mooroobool | 3.0 |
| Bayview Heights | 2.0 |
| Freshwater | 2.0 |
| Whitfield | 1.0 |
| Brinsmead | 1.0 |
| Smithfield | 1.0 |
| Stratford | 1.0 |
| Earlville | 0.9 |
| Edge Hill | 0.8 |

### 21.7.2   Isolation

Because the highways and railway that provide access to Cairns from the north and the Tableland pass through country with steep slopes, they may be blocked by landslides in the event of prolonged or intense precipitation. Outside the study area, the highways to the south may also be blocked by landslides. This makes the Cairns community particularly vulnerable to isolation by land.

The Cairns–Kuranda railway has a spectacular history of dislocation by landslides, the earliest of which was recorded during its construction in 1891. The most disruptive incident started on 15 December 1910, when a landslide at the Kuranda end of No. 10 tunnel partly closed the tunnel for more than two months. Several episodes of sliding occurred during this time (Broughton and Stephens, 1984). The line was cleared for goods traffic on 25 February 1911 and for all traffic on 6 March 1911 (*Cairns Post*, 27 February and 7 March 1911). Another disruptive episode started on 5 March 1954 when a large landslide in the Red Bluff area, with its head well above the railway, and toe well below it, blocked the railway until 22 April 1954 (A. Broughton, Cairns Historical Society, pers. commu., 1997).

### 21.7.3   Utilities

The Cairns water supply intake main crosses Freshwater Creek. Flash flooding in the creek, or debris flows, have the potential to disrupt the Cairns water supply by blocking the intake or destroying sections of the pipeline. There have been two instances in the twentieth century of the Cairns water supply intake main being broken by debris flows or flash floods.

In 1927 and again in 1984 or 1985, boulders smashed the water main at the No. 1 and No. 3 crossings respectively of Freshwater Creek. During the latter incident, the water supply pipeline slipped with a mudflow which took out the anchor blocks (Cairns City Council, 1927 and D. Gallop, Cairns City Council, pers. commu., 1997).

## 21.8   Discussion and Conclusions

### 21.8.1   Advantages of this Methodology

This approach to landslide risk assessment was adopted for the following reasons:

- It provides a rigorous, transparent, robust assessment methodology.
- It is based on an understanding of all the available observational data relating to geology and geomorphology, slope processes and the complex factors that control slope process rates.
- Because it is quantitative, it is more effectively communicated, and more effectively supports the development of management strategies to respond to, and to mitigate the landslide risks, in proportion to the absolute level of risk.
- Because it is quantitative, it allows comparison with other risks affecting the community. For example, the risks associated with, say, flood can be compared with those associated with landslide, and limited resources allocated in proportion to the level of each risk.

### 21.8.2   Limitations

There are some limitations that should be recognized but could not be dealt with in this reconnaissance-level study. The main ones are:

- the paucity of the data from which the landslide magnitude–recurrence relations were derived. As the error bars for the data points are, in some cases, more than two orders of magnitude, errors in the risk estimates may be large;
- the regional nature of this study. Mapping was at a reconnaissance level only;
- the assumption of a uniform process rate in time and across, and from top to bottom of, the entire escarpment profile, irrespective of local geomorphology, rock type, soil cover, position on the escarpment, or location of cuts and fills;
- the assumption that the shadow angles are uniform for all debris flows in the area;
- the assumption that vulnerability is independent of landslide magnitude;
- the assumption that debris-flow runout is not affected by the presence of large obstacles;
- the assumption that landslide intensity is uniform across a landslide;
- lack of discrimination between the effects of shorter-duration, higher-intensity rainfall events, of antecedent rainfall, and of longer-duration, lower-intensity rainfall events. The data tend to be skewed towards observations after tropical cyclones, which tend to be shorter-duration, higher-intensity rainfall events.

### 21.8.3   Acceptability of Specific Risk Estimates

From a consideration of the *de facto* record of acceptable and tolerable annual risk criteria deduced from questionnaires and land use planning documents for potential hazards such as dams, nuclear power stations and landslides, Fell and Hartford (1997) have suggested 1 in 1 million as a possible tolerable specific annual risk level for the average of persons at risk on both new and existing engineered slopes. They suggest 1 in 10 000 and 1 in 100 000 as the tolerable specific annual risk levels for the person most at risk on existing and new engineered slopes, respectively. For landslides on natural slopes the situation is less clear, but they think that the public may tolerate risks as high as 1 in 1000.

Table 21.4 gives estimates for the specific risk of fatality from landslides on hillslopes, and from the proximal and distal parts of debris flows. The specific risk for people living on the escarpment is acceptable, using Fell and Hartford's (1997) criteria, if the slopes are developed with appropriate landslide mitigation measures, as these may reduce the risk to around 1 in 5 million. If new developments took place on the escarpment without these mitigation measures, the risk could rise to about 1 in 40 000, which may not be considered tolerable under Fell and Hartford's (1997) criterion for newly developed slopes. It is possible that the specific annual risk of fatality assessed in this report for people living in areas susceptible to the proximal parts of large debris flows may be considered tolerable, provided that people are informed of the risk before they purchase property, and because the large debris flows in the Cairns area are a natural feature of the landscape. For the distal parts of debris flows, the risk is probably tolerable, considering the uncertainty in the estimates.

### 21.8.4  Acceptability of Total Risk Estimates

It is estimated that a total of 23 houses and/or blocks of flats could be destroyed, and 29 of their residents killed, by landslides in the Cairns area in a 100-year period, if no mitigation measures were taken. This assumes the present distribution of buildings and people. If the population continues to grow and spread into areas potentially exposed to landslide, these totals could be considerably higher.

Thirteen of these 23 buildings and eight of the estimated fatalities are on the hillslopes, and this toll could be reduced, and in many cases has been, possibly to zero, by appropriate mitigation measures with geotechnical consultation.

However, six of the 23 buildings and an estimated 16 of their residents could be destroyed over a 100-year period in areas susceptible to the proximal parts of large debris flows. A further four buildings out of the 23, and five of their residents could succumb to the distal parts of large debris flows. While risk could be mitigated by engineering works, such as levees, for the smaller of these debris flows, there is residual risk from rare, larger debris flows. This would be difficult, if not impossible, to mitigate cost-effectively.

# References

Baynes, F., 1995, *Landslide risk assessment, Flying Fish Cove, Christmas Island, Indian Ocean,* Golder Associates Report.

Broughton, A.D. and Stephens, S.E., 1984, A magnificent achievement. The building of the Cairns Range Railway, in A.D. Broughton and S.E. Stephens (eds), *Establishment of Trinity Bay* (Cairns: Historical society of Cairns, North Queensland), 24–36.

Cairns City Council, 1927, Minutes of meeting of Cairns City Council, February 1927 (unpublished).

DLGP, 2001, *Population trends and prospects for Queensland 2001 edition,* Planning Information and Forecasting Unit, Department of Local Government and Planning, Brisbane.

Fell, R., 1994, Landslide risk assessment and acceptable risk, *Canadian Geotechnical Journal,* **33,** 260–271.

Fell, R. and Hartford, D., 1997, Landslide risk management, in *Landslide Risk Assessment. Proceedings of the International Workshop on Landslide Risk Assessment*, Honolulu, Hawaii, USA, 19–21 February 1997 (Rotterdam: A.A. Balkema), 51–109.

Granger, K., Jones, T., Leiba, M. and Scott, G., 1999, *Community risk in Cairns. A multi-hazard risk assessment*, Cities Project, Australian Geological Survey Organisation.

Hungr, O., 1997, Some methods of landslide hazard intensity mapping, in *Landslide Risk Assessment. Proceedings of the International Workshop on Landslide Risk Assessment*, Honolulu, Hawaii, USA, 19–21 February 1997. (Rotterdam: A.A. Balkema), 215–226.

Michael-Leiba, M., Baynes, F. and Scott, G., 1999a, Quantitative landslide risk assessment of Cairns, *AGSO Record*, 1999/36.

Michael-Leiba, M., Scott, G. and Granger, K., 1999b, Community risk in Cairns. *Proceedings of Natural Disaster Conference*, Canberra, Australia.

Michael-Leiba, M., Baynes, F. and Scott, G., 2000, Quantitative landslide risk assessment of Cairns, Australia. *Proceedings of VIII International Symposium on Landslides*, Cardiff, Wales.

Michael-Leiba, M., Scott, G., Baynes, F. and Granger, K., 2001, Quantitative landslide risk assessment of Cairns, Australia, *Proceedings of 14th South East Asian Geotechnical Conference*, Hong Kong.

Queensland Department of Mines, 1988, *Cairns Region, Sheets 8064 & 8063 (part), 1:100 000 Geological map Commentary*.

Wong, H.N., Ho, K.K.S. and Chan, Y.C., 1997, Assessment of consequence of landslide, in *Landslide Risk Assessment. Proceedings of the International Workshop on Landslide Risk Assessment*, Honolulu, Hawaii, USA, 19–21 February 1997 (Rotterdam: A.A. Balkema), 114–119.

# 22

# The Story of Quantified Risk and its Place in Slope Safety Policy in Hong Kong

*Andrew W. Malone*

## 22.1  Introduction

Each summer rainstorms trigger many landslides in Hong Kong and slope safety is a matter of some public concern. The slope problems of this former British colony and, since 1997, Special Administrative Region of China are the product of adverse terrain, severe climatic conditions, restricted land supply and phenomenal population growth in the past 50 years. Pressure on land is acute, with 7 million people occupying an area of about $400\,km^2$, much of which is developed hillside. Rainfall in the urban areas ranges annually from about 1 m to 3.4 m and produces each year between about 80 and 800 reported 'landslides' (slides, flows, falls, retaining wall collapses, etc.). Nearly all are failures in *man-made* slopes – in cuttings and to a lesser extent in hillside embankments ('fill slopes') and retaining walls. In this chapter, the word 'slope' generally means 'man-made slope'. About 90% of the reported failures yield less than $50\,m^3$ of debris but many of these small collapses are still quite risky, being mobile and occurring next to roads and buildings. The degree of risk is indicated in the fatality figures. In the 15 years to 2002, 15 people were killed by landslides; but more than 450 people had been killed in the previous 35 years. This improvement, achieved despite population growth of about 1 million per decade, is the product of a control regime introduced progressively since the 1970s to satisfy society's rising demand for safety.

Hong Kong's population started to grow rapidly after the Second World War. Civil war in China led to mass immigration from the north, creating an acute housing shortage

*Landslide Hazard and Risk*   Edited by Thomas Glade, Malcolm Anderson and Michael J. Crozier
© 2004 John Wiley & Sons, Ltd   ISBN: 0-471-48663-9

and giving rise to squatter encampments throughout Hong Kong Island and the Kowloon Peninsula. The shantytowns were insanitary, insecure and prone to fires, flooding and landslides. Following a squatter fire in 1954 in which 54 000 people lost their homes, the government began to build low-cost multi-storey housing, mainly for the resettlement of squatters. In the following two decades the lower portions of the hillsides of northern Hong Kong Island and Kowloon became intensely urbanized, with massive earthwork terracing for buildings and roads. In that era controls were generally lacking in the construction industry and the quality of earthworks sometimes proved dangerously inadequate. A succession of multi-fatality landslides in the 1960s and 1970s, mainly mobile flows

***Figure 22.1***   *The landslide of 18 June 1972 at the Sau Mau Ping licensed area*

**Figure 22.2**   *The landslide of 18 June 1972 at Po Shan Road*

resulting from the collapse of man-made slopes, provoked public outcry. Two of the most serious occurred on the same day in 1972, killing 138 people: these landslides are illustrated in Figures 22.1 and 22.2. In response to each disaster the government introduced corrective measures corresponding to its perceived cause. Finally, in 1977, amidst public clamour for action, a reform-minded governor created a slope safety agency with broad regulatory and works functions.

Disaster and corrective response continued until a comprehensive slope safety regime eventually came into place. Under this regime, managed by the Geotechnical Engineering Office (GEO), earthworks are tightly controlled, slopes are regularly inspected and routinely maintained and defective slopes built before 1978 are being upgraded. The hillside shantytowns are being cleared and rainstorm emergency preparedness measures are in force. Education, information and community outreach programmes are all under way. The overall effectiveness of the slope safety effort is monitored through a landslide recording and investigation programme.

Such a slope safety regime is not cheap. Each year the government spends about US$88 million on routine slope maintenance, around US$115 million on upgrading its pre-1978 slopes and initiating statutory orders for the upgrading of privately owned slopes, and some US$25 million in recurrent costs to maintain the slope safety agency. US$228 million is a large annual outlay, amounting to about 0.73% of public expenditure, and to this must be added the non-government expenditure on upgrading and maintenance of privately owned slopes. This level of spending should be judged in its social context: Hong Kong has become a wealthy society (annual GDP per capita US$24 000) that has grown intolerant of landslide risk.

There are a number of other places in the world where landslides are becoming a significant public issue, with rising public expectation. Following public outcry after disaster, governments come under pressure to allocate resources to improve safety. Sooner or later they will face these two questions. Complete safety being unattainable, how safe is safe enough and what is an appropriate level of effort and expenditure on slope safety? How should the effectiveness of effort and expenditure on slope safety be measured?

## 22.2   Questions that Arise in the Public Policy Debate

### 22.2.1   What is an Appropriate Level of Effort and Expenditure on Slope Safety?

Public safety can be an emotive issue. All parties involved in public resource allocation (spending departments, the budget controller, the legislature, the public at risk, etc.)[1] stand to benefit from the application of logical processes to help decide appropriate levels of effort and expenditure by government on safety. What kind of rationales might apply? Starting at the most basic level, a government duty may exist under the law, for example the duty of landowner, employer or highway authority. But the legal minimum may not be enough to satisfy public expectation. In this situation the initial aim might be to devote the effort and expenditure needed to achieve a significant improvement in safety within an acceptable time. (Is there any reason to spend public resources on safety unless a significant improvement can be achieved in a timely manner? A government that merely wishes to reassure and be seen to be doing something so as to make people *feel* safer without making significant improvements may, or may not, depending on its longevity, come to regret having taken this approach.) There are also upper-bound rationales. Governments under pressure to spend may resort to citing capacity constraints to justify spending caps. More convincing justification for limited spending may be found through cost–benefit arguments (benefit disproportionate to cost) or relativity reasoning (safety levels lower than ours are accepted in similar societies elsewhere). In the following short account of Hong Kong's spending on slope upgrading we will see some of these rationales in play.

The programme to upgrade old man-made slopes has always taken the lion's share of slope safety funding in Hong Kong. In the Geotechnical Control Organisation's (GCO) early years spending on slope upgrading, including associated studies and landslide remedial works, was averaging some US\$7.5 million annually (0.25% of public expenditure). Yet each year only about 20–40 slopes, albeit large ones, were being upgraded out of an inventory of some 10 000 features, an unknown proportion of which were defective. Funding for slope upgrading had been deliberately limited in 1978 so as not to require the curtailment of public site formation work, which would have had an adverse effect on the housing programme and development in general in Hong Kong. The argument was that the resources of this section of the construction industry were fully stretched, as evidenced by the occurrence of delays and increasing prices.

---

[1] Public sector resource allocation in Hong Kong is controlled by the executive, essentially the civil service led by a chief executive (formerly governor) and his appointed cabinet, though public funding has to be passed by the Legislative Council, a representative unicameral chamber constituted partly by universal suffrage elections. The legislature's influence on the executive has changed markedly in the past 20 years, peaking in the mid-1990s.

With a very large financial commitment on slope works in prospect, it was not long before the question was being asked, 'How far should government go?' (Anon, 1982: 32). Concern was expressed that 'massive expenditure (on certain health and safety issues including slopes) should be at the mercy of public opinion, the sensational content of incidents, editorial opinion, departmental pressure and so on, with so little recourse to objective cost–benefit analysis and almost no overall control' (Anon, 1982: 33). In the years that followed, support waned within government for GCO's work. As early as 1984 it had to avert a demotion in the administration. During the 8 relatively dry years 1984 to 1991 its staffing level came under attack and spending on slope upgrading declined about 25% in real terms (reaching a low in 1991/92 of about 0.05% of public expenditure). It is perhaps relevant that only one fatal landslide had occurred during these years and this event, at a time of social protest in Beijing in 1989, received little publicity, increasingly critical of the government. Then in 1992 the climate of opinion dramatically reversed. The early 1990s brought a string of wetter than average years, with multiple-fatality landslides and accompanying publicity. These events created an irresistible demand for more resources to be put into slope safety. After 1994 GEO staffing was augmented significantly and annual spending on slope upgrading was increased about sevenfold in real terms during the next 7 years (reaching about 0.37% of public expenditure in 2002), the obstacle of GEO's internal capacity constraints having been surmounted by outsourcing work to consulting firms.

A cost–benefit argument was deployed as early as 1981 with the aim of justifying a limit on spending on one type of slope hazard-boulderfall. In deliberating on the potential hazard of boulder falls from the Conduit Cliffs, rocky bluffs which tower above high-rise buildings the Mid-levels area of Hong Kong Island (Sinclair, 1978), a government policy had been established in 1978 of giving low priority to preventive work to avoid harm from boulders falling from land owned by the government, except in cases of immediate and obvious danger. Among the various defences of this policy appears a claim that: 'the cost of such prevention tends to be out of all proportion to the death and injury it prevents' (Anon, 1982: 32). This is the first appearance of a cost–benefit argument in relation to slope safety policy in Hong Kong. The next appearance would be some 14 years later, as an integral part of the process known as 'risk management'.

### 22.2.2  How Should the Effectiveness of Effort and Expenditure on Slope Safety be Measured?

#### 22.2.2.1  When does the question arise?

In July 1994 a slope collapsed at Kwun Lung Lau in Hong Kong Island, killing five people and seriously injuring three others. There was a public outcry and the GEO was held to account. After such a disaster, when confidence in the safety agency is at its lowest ebb, the question arises: given this disaster, what, if anything, has been achieved by past effort and expenditure on safety? Nearly all the products of the slope safety effort are invisible, or at least go unrecognized, but at this crucial time the safety agency needs to be able to show tangible results to convince the sceptics that something of value has been achieved. No marks will be awarded for the mere consumption of resources. The agency that measures its output in terms of number of jobs completed or dollars spent may, when the crisis comes, find itself unable to prove its worth. So what can be done to prepare for the crisis? How can an improvement in safety be expressed as a number and measured?

Questions about the effectiveness of the GEO had been asked before 1994, in the wake of fatal landslides. In responding, attempts had been made to express the value derived from slope safety spending. By 1982 more than US$40 million had been spent on the upgrading of old slopes under the Landslip Preventive Measures (LPM) programme and a team of 30–40 professionals had been at work for some 4 years trying to ensure the safety of new site formations. These efforts were tested in 1982 by two severe rainstorms, when landslips killed 27 people, mostly in squatter areas. All the lives were lost in relatively small and isolated incidents. Unlike 1972 and 1976, in 1982 no major failure took many lives, and this was cited as evidence that slope safety efforts had been effective (Brand, 1983). By 1992 the slope safety office had been in operation for 15 years and nearly US$130 million had been spent on the LPM programme. Landslides during a rainstorm on 8 May had killed three people. In assessing the improvement in Hong Kong's landslide defences, it was argued that comparable rainstorms in 1982 and 1972 had caused many more deaths and created much greater disruption (Malone, 1992). These claims used the best indicators of the value of the effort and expenditure on slope safety that could be cited in 1982 and 1992, but, with hindsight, neither is seen to be fully persuasive.

Another occasion on which a query may be expected on the value of effort and expenditure on slope safety is when an application for public funds is being discussed with the budget controller or the legislature. Such a funding paper was being discussed by a Legislative Council sub-committee in 1993 when a member asked the deceptively simple question: 'How many lives will this one hundred million dollars save?' (note: HK$100 million). The GEO representative could say roughly how many old slopes could be upgraded for HK$100 million, or for how long this money would cover GEO's recurrent costs, but he could not begin to answer the member's question. Yet it is a fair question, for without an estimate of loss prevented (or similar measure), it is impossible to know whether spending on safety is doing any good. As we shall see below, it was to be another 4 years before a proper answer could be given to the legislator's question. By 1993, GEO had been bidding annually for funds for 16 years. It is curious that nobody had asked that question before.

### 22.2.2.2   Quantifying safety benefit

At the beginning of this section a number of rationales were suggested to help decide an appropriate level of effort and expenditure on slope safety (e.g. cost–benefit arguments). It is clear that some of these rationales cannot be employed unless we are able to define quantitatively and measure the state of safety. There are two ways to express the state of safety quantitatively. One uses only raw data (actual number of fatalities per year, cost of damage per storm, etc.). The other characterizes the state of safety by the theoretical quantity 'risk' (a mathematical concept whose definition, given later, is not to be found in the dictionary). 'Risk' is estimated through methodical calculations, called quantified risk analysis, yielding results such as 'expected number of fatalities per year'. We can quantitatively express the safety benefit of a programme in several different ways: number of lives or dollars saved, loss-rate reduction, risk reduction and so on. To deduce such safety benefits a comparison must be made of the situation with and without the safety regime. Various approaches can be envisaged. One way compares the detriment rate or risk level in one place with that in another, both places being closely similar except that one has a safety regime and the other does not. Another way compares the detriment rate or risk level in the same place at different times, before and after the implementation of the safety regime. The latter idea is illustrated in Figure 22.3, a sketch showing the notional effect of the various programmes of Hong Kong's slope safety system.

For such a 'with and without' exercise to be valid it will be necessary to show that 'like is being compared to like', that is, to isolate the effect of the safety regime from the effects all of the other variables that might influence safety. This will require an observational experiment, with careful preparation, data collection and interpretation. This kind of exercise has been attempted for the city of Angeles, USA. Analysis of landsliding during the 1969 rainstorms (Slosson, 1969) showed that sites developed prior to 1952 (pre-regulation) suffered a much higher failure rate than those developed after 1963 (regulation fully inforce).

The most satisfactory means of expressing the state of safety and the safety benefit of a programme is through the theoretical quantity 'risk'. But is it *feasible* to calculate landslide risk and use it for safety measurement? We will define 'risk' and consider the question of feasibility by examining the history of risk quantification in safety management in Hong Kong. A more general history of risk quantification is outside the scope of the chapter.

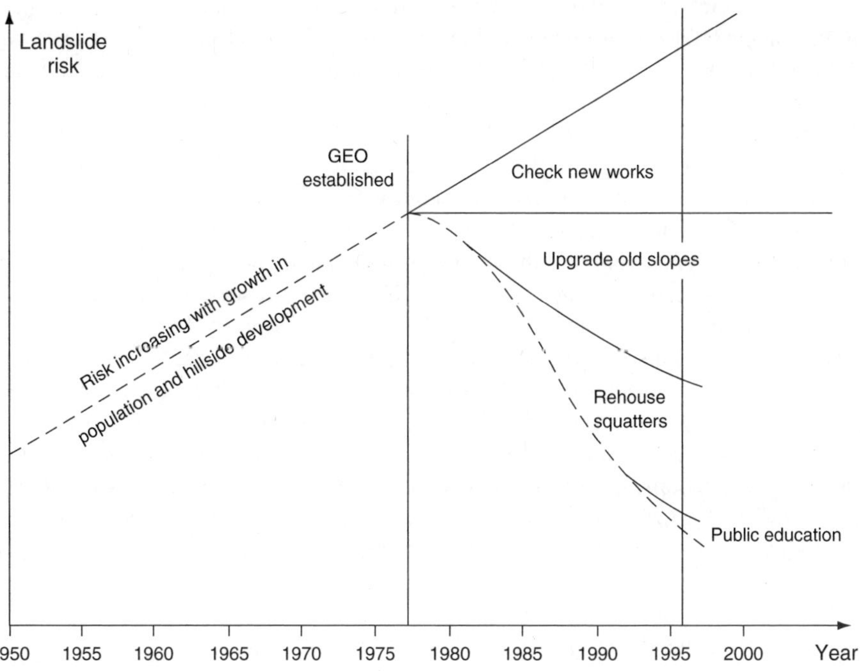


**Figure 22.3** *Postulated landslide risk trend in Hong Kong (Malone, 1998). Reproduced by permission of A.A. Balkema Publishers*

## 22.3   Peter Lumb's Contribution to Risk Quantification

The idea of applying risk thinking to slope safety emerges in the 1960s (Wu & Kroft, 1970) and was developed in the writing and practice of Peter Lumb. A pioneer of statistical analysis in geotechnical practice, Lumb taught at the University of Hong Kong between 1954 and 1986. Working as a geotechnical practitioner as well as a teacher, experience with practical soil engineering problems had led him to realize the importance of 'chaos, randomness and unforeseen eventualities' such that, for example, there would always be an unknown but finite probability that any steep slope, whether engineered or not, would collapse. Writing soon after the landslide disasters of 1972, he argues that in slope design, the acceptable failure probability should depend on the *consequences* of that failure (Lumb, 1975). The graver the consequences, the more certain we should be that failure will not occur, that is the lower should be the acceptable theoretical probability of failure. He was implicitly advocating the placing of a limit on 'risk' (meaning a quantity which combines the probability of occurrence of failure and the consequences of failure).

If risk could not be kept below an acceptable level, he argues, a slope works project should not go ahead: 'if the estimated . . . failure probability remains high then most attention should be given to ameliorating the consequences of failure . . . If, however, expected consequences are irreducibly grave then the project should be abandoned' (Lumb, 1975: 64). Probabilities and statistics are the tools to use when handling uncertainty and Lumb advocated the use of these methods in geotechnical design. The probability of slope failure would be calculated using applied mechanics (e.g. 'limit equilibrium analysis', by which the theoretical reserve of strength of a soil slope, its 'factor of safety', may be calculated) with probabilistic treatment (explained below). Consequence would be handled by means of utility theory, the economist's concept of true relative value of various choices.

In 'Statistical soil mechanics' (Lumb, 1983) the use is described of risk matrices (tables showing various hazards classified as to probability and consequence), methods are summarized for the quantification of probability and consequence of failure (separately and combined) and the concept is discussed of public tolerability in decision making about public safety, recognizing the notion of intolerable and negligible risk levels. We will see later that these ideas are at the heart of modern technological risk management.

Professor Lumb was an advocate of probabilistic thinking in design, though he was not to see a probabilistic approach to the design of slopes accepted into practice in Hong Kong. Foremost among the reservations expressed by Lumb's co-contributors to the *Geotechnical Manual for Slopes* (Geotechnical Control Office, 1979) was concern over the onerous task of measurement needed to compose the histograms of data and then their idealizations, called probability density functions. These are the curves, for example the bell-shaped curve, see Figure 22.4, which probability theory uses to characterize the variability of the theoretical slope stability parameters: soil strength properties, water pressures and geological–geometrical conditions. Such mathematical expression of variance is needed for probabilistic treatment.

Though the probabilistic approach was not generally adopted in local slope design practice, examples can be seen of the influence of risk thinking on technical policy. First, the *Geotechnical Manual for Slopes* (Table 5.2, 1st edn) stipulates a high factor of safety for slopes in high-consequence situations and permits a low factor of safety where the

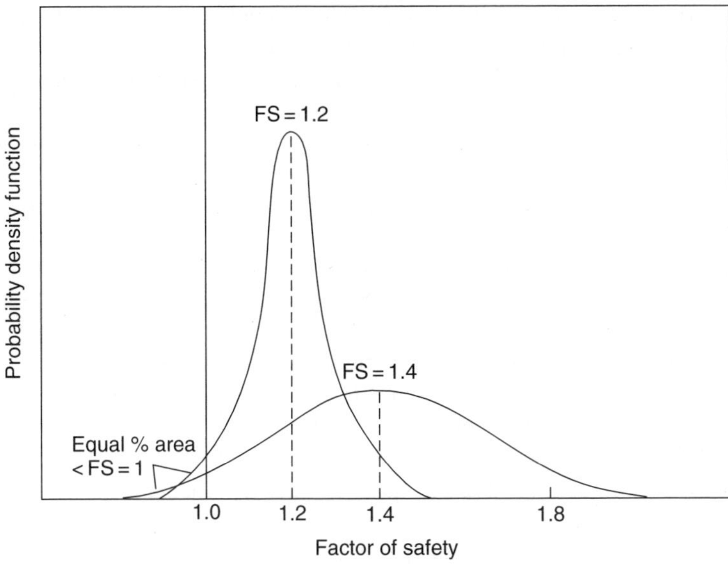


**Figure 22.4**  *The results of stability analyses of two cases with the same probability of failure*

consequences of failure are not so serious. Second, the rating system introduced in 1979 to form a waiting list of man-made slopes for stability study under the LPM programme, which built on an earlier rating system for masonry retaining walls (Binnie & Partners, 1978), relies on total scores which are the sum of a score for likelihood of failure and another for consequence of failure. Third, the revised *Geotechnical Manual for Slopes* (Geotechnical Control Office, 1984) permits a lower factor of safety for existing slopes than for new slopes, and this was justified using probabilistic thinking (Malone, 1985). It was argued that more reliable data are accessible for existing slopes than for slopes yet to be constructed, giving opportunity for reduced variance in the calculated factor of safety (a narrower and more peaked bell curve). If variance is reduced, it is possible to maintain the same theoretical probability of failure despite a lower (mean) factor of safety. This is illustrated in Figure 22.4, where two cases are shown, one having a high mean factor of safety and large variance and the other a lower mean factor of safety but reduced variance. Though their mean factors of safety differ, the cases have the same theoretical probability of failure, which is the area below the bell curve to the left of $FS = 1$ as a percentage of the total area below the curve.

As Lumb was writing 'Statistical soil mechanics', but in a quite different area of public safety, steps were being taken to apply formal risk management procedures in Hong Kong.


## 22.4  Managing the Safety of Major Industrial Hazards

In the 1980s a control regime was established in Hong Kong for managing safety in the vicinity of industrial sites containing substances with great destructive potential. Safety management procedures that had been devised in the nuclear power, oil and gas and chemical industries provided the framework of control.

By the early 1980s the coastal margin of the mountainous island of Tsing Yi had become home to oil terminals and other installations for the storage or manufacture of fuels and other chemicals. The $10\,km^2$ island also housed 50 000 people. Prompted by serious accidents elsewhere in the world involving volatile or toxic chemicals or flammable gases such as naphtha and liquefied petroleum gas (LPG), in 1982 the Hong Kong government commissioned a study on the threats to public safety arising from the storage of such substances on Tsing Yi. A quantified risk assessment procedure was employed to gauge the degree of threat at various locations on Tsing Yi and also to help the government decide whether these threats were excessive. The procedure involved the identifying of events that could give rise to hazardous incidents, such as leaks of inflammable gas from pipe valves, the examination of the potential consequences of such incidents and the estimation of their likely frequency of occurrence.

The theoretical quantity 'risk' would then be calculated using computer models developed by specialist firms of risk consultants. Mathematically, risk is the summation, for all conceivable hazardous incidents, of their combined probability and consequence. Two measures of risk were employed in the calculations: 'individual risk' and 'societal risk'. The former is expressed as the increase in the chance of death of a notional person at a given position. Individual risk (annual probability of death) may be shown in the form of a geographical distribution; see Figure 22.5, taken from a study described below (22.6.1). 'Societal risk' is the relation between the frequency of incidents and the resulting number of deaths (e.g. incident 1: LPG tank 1 explosion due to leak type A with 10 deaths, chance once in 10 000 years; incident 2: LPG tank 2 explosion due to leak type B with 1 death, chance once in 1000 years). Societal risk was represented as a frequency–number plot (*F–N* curve), a log-log plot of frequency, *F*, of *N* or more fatalities versus number of fatalities, *N*. An *F–N* curve is shown in Figure 22.6. Societal risk can also be expressed as an index called 'potential loss of life' (PLL), where PLL = the area under the *F–N* curve,

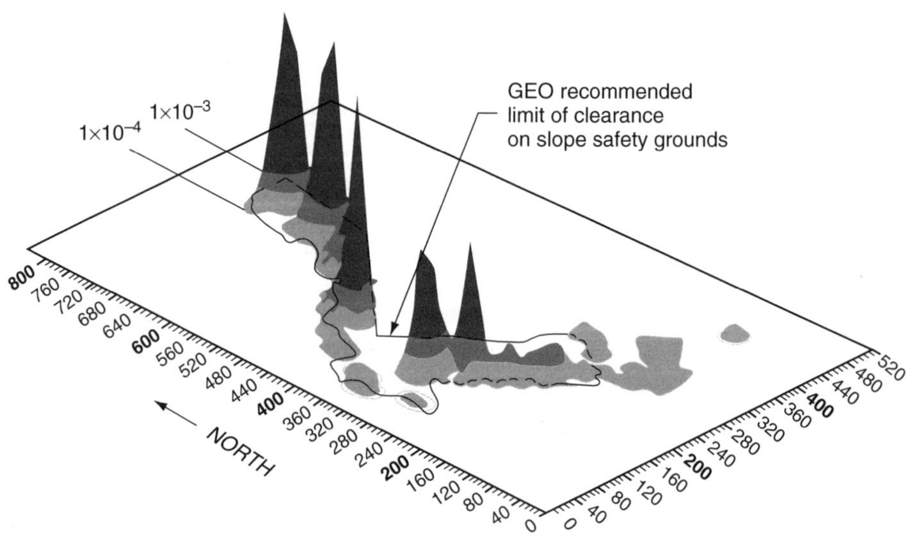


**Figure 22.5**  *'Individual risk' calculated by QRA for the Lei Yue Mun squatter area*

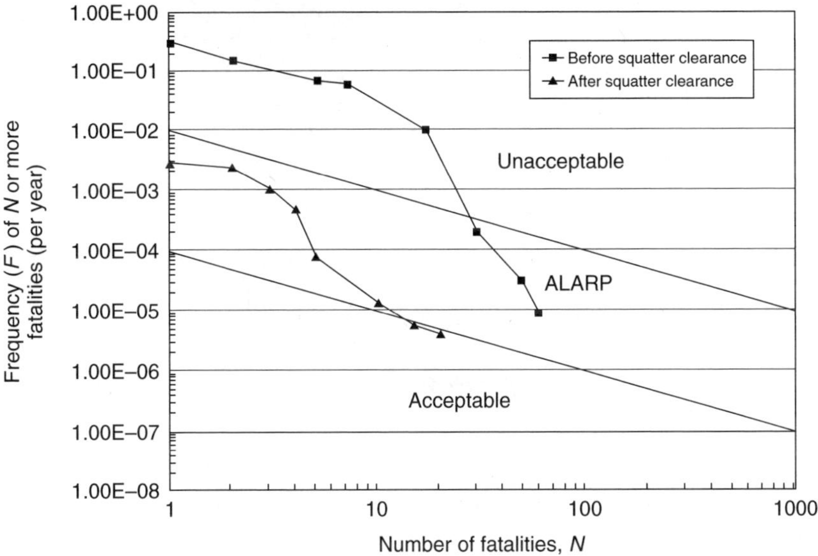


**Figure 22.6**  F–N *curves showing 'societal risk' calculated by QRA for the Lei Yue Mun squatter area (after Ho* et al., *2000). Reproduced by permission of CRC Press. © CRC Press*

$\Sigma F_1 N_1 + F_2 N_2 + F_n N_n \ldots$ The PLL index is the expected number of deaths per year caused by incidents at the installation. The estimation of risk using statistical or analytical modelling (risk analysis) and the evaluation of its significance became known as quantified risk assessment (QRA).

Having estimated the risks at several places on Tsing Yi, the numbers were compared with risk levels accepted elsewhere. In the light of these comparisons various town planning and land administration decisions were made by the government in 1984. The aim was to reduce risk to below a level considered unacceptable elsewhere. At one critical location in east Tsing Yi, where the Mobil oil terminal adjoined a high-rise housing estate, risk was found to be on the borderline of acceptability. Acting on the results of the risk assessment, planned housing expansion was abandoned and the safety of the oil terminal was improved. It was relocated some years later. The Mobil oil terminal case appears to be the first use of formal technological risk management procedures in government decision making in Hong Kong.

But Tsing Yi was not the only site of concern. In December 1986 an interdepartmental committee was set up in the government to coordinate action on the territory's 29 'potentially hazardous installations' (PHIs), installations storing defined quantities of certain hazardous chemicals, such as fuel gas, liquid oxygen and chlorine. The committee arranged for risk assessments to be carried out for each PHI. It also developed the 'risk criteria' to be adopted in Hong Kong, comprising an upper level indicating where risk becomes unacceptable and a lower level where it becomes broadly acceptable. The idea is illustrated in Figure 22.7. Where the risk level lay between the two limits, in the 'tolerable region', the risk criteria require that actions be put in hand to reduce risk to a level that is 'as low as reasonably practicable'. This is known as the ALARP requirement. Risk

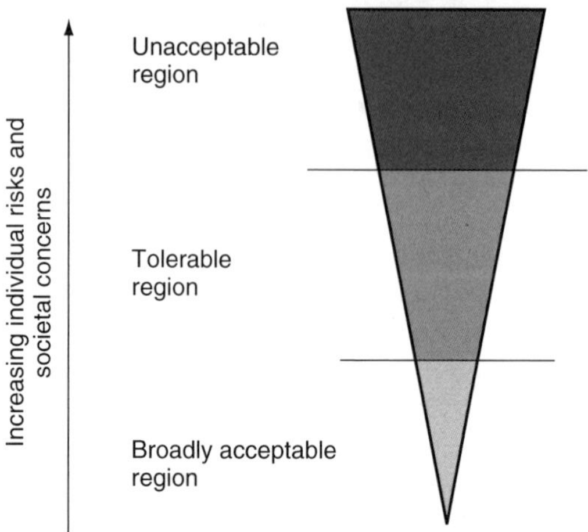



***Figure 22.7***   *HSE's tolerability of risk framework*

mitigation plans were implemented for each PHI and land use planning was thereafter managed to prevent increase in risk and if possible to reduce risk in the vicinity. The risk mitigation strategies have included PHI relocation, improvements to safety systems and constraints on residential development. Significant risk reduction has been achieved, but the risk management programme is an ongoing process because of population increase. A summary of the risk management process given in Figure 22.8.

The Hong Kong procedures generally follow the approach of the Health and Safety Executive (HSE), the UK occupational health and safety regulator, though Hong Kong lagged behind the UK in at least one respect – public consultation over PHI risks. By the late 1980s the HSE had embraced the notion of *public tolerability*: 'the opinion of the public should underlie the evaluation of risk' (HSE, 1988: 1), where ' "Tolerability" does not mean "acceptability". It refers to a willingness to live with a risk so as to secure certain benefits and in the confidence that it is being properly controlled'. The HSE published draft risk criteria in a consultative document in 1988 (ibid.) and following public consultation a further document was issued in 1992. The criteria include two risk limits: in suggesting the upper risk limit (the 'unacceptable' limit) for workers, the HSE chose a risk level comparable to those ordinarily accepted under modern conditions by workers in the UK (a risk of death of 1 in 1000 per annum). The upper risk limit for the public was set ten times lower. The lower limit (the 'broadly acceptable' limit) was selected to represent only a small addition to the ordinary risks of life (it was set at a risk of death of 1 in 1 million per annum). As safety regulator, the HSE aims to prohibit activities creating risk levels that are above the upper limit and requires risk levels between upper and lower limits to be 'reduced to ALARP'. But what does 'reduced to ALARP' actually mean?

In the United Kingdom the 'reasonable practicability' standard of care has been an injunction of safety law affecting employers since 1938, or earlier. The courts established that 'reasonably practicable' is a narrower term than 'physically possible' and allowed the employer, in his defence, to balance the quantum of risk against the time, trouble

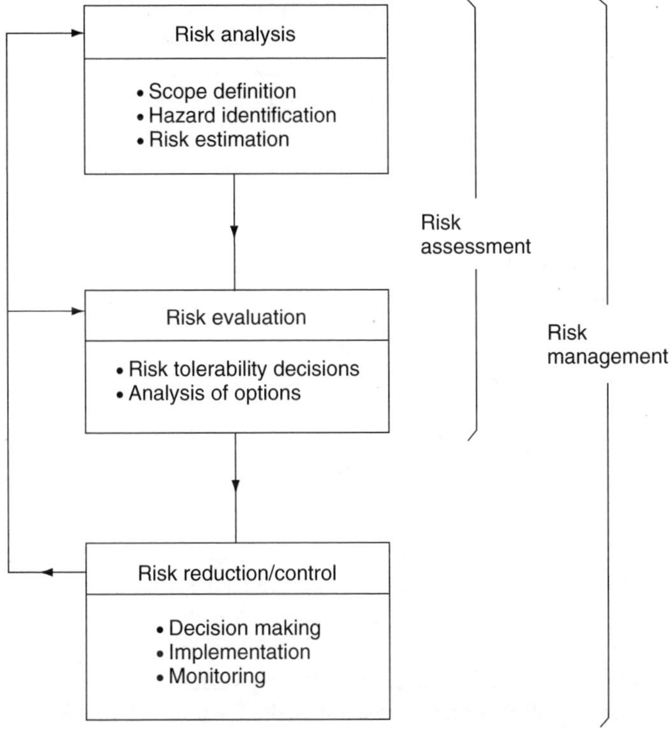



**Figure 22.8**  *The risk management process as applied in safety management*

and expense of the precautions needed to avert the risk: *Edwards vs the National Coal Board* [1949] 1 All ER 743. Accordingly, in regulating major industrial hazards the HSE expects duty holders to, among other things, reduce risk until a point is reached where the time, trouble or expense of a further increment of risk reduction becomes grossly disproportionate to its benefit. Once this point is reached, the regulator does not expect the employer to go any further. Sometimes cost–benefit analyses are needed to evaluate mitigation measures to reduce risk. When a risk reduction option produces the benefit of preventing fatalities, this requires putting a monetary value on achieving a reduction in the risk of death. The value of a statistical life used by HSE depends on the particular risk, higher values being regarded as appropriate for risks for which there is high public aversion, for example those that give rise to high levels of social concern. After 1987 it became practice for the Hong Kong regulator, in assessing PHI proposals, to require cost–benefit analysis of risk reduction proposals to help to show compliance with the ALARP duty. In these analyses the cost of risk reduction proposals is compared with the benefit of statistical lives saved, expressed as a reduction in the PLL index.

The HSE risk criteria were the subject of a public consultation exercise in UK in the late 1980s. Hong Kong's PHI risk criteria were not released to the public until the early 1990s (Planning Department, 1993; Wrigley and Tromp, 1994). Eventually it became normal practice to present the results of risk assessments to representative bodies, such as District Boards and the Legislative Council's Environmental Impact Assessment Panel.

In the 1990s these risk assessment methods came to be applied increasingly in other sectors in Hong Kong, for example transport of hazardous substances, railway safety and environmental hazards. The risk assessment process and criteria are now used under the Environmental Impact Assessment Ordinance for projects which manufacture, store, use or transport dangerous goods. A wide range of activities is covered; for example, even fireworks displays at the future Disneyland.

## 22.5    The Risk Management Tool in Slope Safety Policy in the 1990s

Slope safety management is another of the sectors in Hong Kong that took up risk assessment methods in the 1990s. As we shall see, risk management thinking was applied in reviewing slope safety policy in 1993 but it was not until 1995, as one of the consequences of a landslide disaster, that the risk management tool came to be recognized as an indispensable policy aid in slope safety. We will next look at how this came about.

At the behest of the PHI coordinating committee, risk specialists were employed to undertake risk assessments of Hong Kong's civilian explosives stores, which since 1991 had been under GEO management. In 1993 the Special Projects Division of GEO was tasked to study these risk assessments with a view to using a similar approach in slope safety. At around this time a number of staff members became aware of the work of Professor Robin Fell of the University of New South Wales on landslide risk assessment. The subject of risk quantification was established as a topic in the GEO Research and Development Programme in 1993 and the division began to utilize risk management thinking in its work.

Landslide statistics were used in 1993 in reviewing the government's slope safety efforts. Since 1984 the annual reports on rainfall and landslides had included data on the type of facility affected by landslides. Trends emerged from these data. The proportion of reported landslides affecting buildings was declining while that along roads was increasing. It was thought that these trends might be related to the policy of giving priority in the LPM programme to slopes affecting buildings rather than roadside slopes. This policy was reviewed. Statistical analyses of landslide and landslide casualty data for Hong Kong's roads for the period 1984–93 gave an annual fatality rate of $5 \times 10^{-5}$ (fatality per substandard road slope per year) (Wong and Premchitt, 1994). It was noted that this value is above the tentative acceptable risk levels for man-made slopes quoted by Professor Fell (Ho, 1993; Fell, 1994). The value also fell within the ALARP region for PHIs (Planning Department, 1993), requiring that all practical and cost-effective measures that can reduce risks should be sought. Estimates using failure rates for a November 1993 storm for one particular road, the South Lantau road, showed fatality rate to be within the ALARP region of the PHI societal risk criteria (Wong and Ho, 1994). These findings supported the case being made within GEO for a change in policy so as to include within the LPM programme the upgrading of defective slopes along any road with a high density of traffic.

The GEO was not alone in Hong Kong in its interest in applying risk methods in slope safety. In the early 1990s Dr Neil Kay at the civil engineering department of the University of Hong Kong was working to devise a cut slope design methodology using a risk approach. He too used historical failure data to determine frequency of failure (rather than attempting to derive probability of failure analytically, following Lumb) and in considering the need 'to determine the level of risk that is appropriate', reflects on the various approaches to deciding such risk criteria (Kay, 1994). Subsequently a comprehensive study of the GEO's landslide and rainfall records was carried out by Fell and Findlay of the University of New South Wales as consultants to GEO. Their study, undertaken between June 1994 and February 1995, provided a statistical analysis of landsliding and related losses, showed how such data might be used for the estimation of risk and summarized acceptable risk levels in various countries (Findlay and Fell, 1995).

### 22.5.1   Kwun Lung Lau Landslide of 23 July 1994

Risk quantification might just have remained a promising but little-known field of scientific enquiry in slope safety in Hong Kong had events not intervened. The power of risk management thinking as applied to slope safety policy was brought to the attention of a wide audience by Professor N.R. Morgenstern in his report to the Governor on the Kwun Lung Lau landslide (Morgenstern, 1994) and his subsequent evidence to the Kwun Lung Lau Select Committee of the Legislative Council (Anon, 1995). Slope safety had not gained such public prominence in Hong Kong since the 1970s, but grave concern was understandable in this situation. Not only had five people been killed in the landslide (Figure 22.9) and three other people seriously injured, but also some 3900 people had been evacuated from their homes during the night for fear of collapse of the high-rise buildings they occupied. Rescue efforts in the landslide debris continued for days, constantly televised. The GEO found itself overwhelmed by a sudden massive demand for information from the world news media and the local press and politicians. In responding to enquiries it was unable to convince the sceptics that its efforts generally in Hong Kong since 1977 had improved slope safety. Public censure followed when it became known that GEO had earlier investigated the stability of the collapsed slope and had concluded that its factor of safety was adequate. The political parties, then campaigning for District Board elections in September of that year, took up the cudgels on behalf of the victims and their candidates in hilly districts put slope safety at the top of their electoral agenda. Political party views expressed on slope safety were diverse, revealing both insight and serious misunderstandings. Views ranged from mild comment to extreme opinion, from a request for assurances that the government was 'doing its best' to improve slope safety to a demand that landslides be eradicated. But all parties were united in calling for a substantial injection of resources into slope safety. In October the Legislative Council voted to create a Select Committee (modelled on the US Congressional Committees) to enquire into the circumstances of the landslide and related issues: only the second Select Committee in Hong Kong's 150-year colonial history.

Drawing on risk management thinking, Morgenstern advanced ideas into the political arena that had not been heard locally before, at least not in relation to slope safety: 'It is important for any community to have some sense of its goals with regard to risk management. One cannot simply say that the goal is to reduce risk to zero. This is both unrealistic and undesirable' (Morgenstern, 1994: 18). Speaking to the Select Committee, he said 'you are within a range in which you can reduce risk further by further allocation of resources. You are not in a range yet in which your allocation of resources becomes

***Figure 22.9***   *The landslide of 23 July 1994 at Kwun Lung Lau*

unreasonable, that is to say close to zero risk in which you consume infinite resources . . . So you are in a range where it really is a matter of public policy for you to decide risk-taking versus resource allocation' (Anon, 1995:33, lines 3–6). It is clear that members of the Select Committee understood the main points being made.

The Kwun Lung Lau disaster brought to light weaknesses within the slope safety system and prompted deep introspection on the part of the senior management of GEO. Encouraged by Morgenstern's use of risk management thinking, the search for useful concepts and tools led us back into dialogue with the chemical engineering risk specialists and thence to the process safety and loss prevention literature. In due course the ideas that took form during this period were to help bring about a profound change in thinking on slope safety.

Chief amongst these was the idea of quantified risk trend as an index of safety. Risk expressed as a number would provide the much-needed measure of safety, with which we might discern trends, gauge the effect of our work, defend our achievements and set goals. We expected safety to have deteriorated post-war with population growth and then, from 1977, to have progressively improved as the slope safety system took effect. Expressed graphically, the safety record should appear as in Figure 22.3. This sketch shows the notional improvement since 1977 attributable to elements of the slope safety system. If we could quantify risk, the risk management process would become available as a policy aid, just as it had before when safety on Tsing Yi became an issue in 1982. By showing that risk remained above an acceptable level, and that spending on risk reduction remained cost-effective, we would be able to defend continued outlay on slope safety when support waned, as it had in the 1980s. By estimating the risk reduction likely to be brought about by further increments of slope upgrading work we might be able to provide some justification for our funding applications in terms of potential benefits. These aspirations were discussed within the GEO and then documented (Malone, 1996).

The second useful concept to come out of this period of thinking was the idea of the slope safety regime as *a system*. This leads to the perception of landslides as defects in the system, every one worthy of investigation and capable of revealing areas for improvement of the system.

Their interest having been aroused, the risk specialists were keen to try formal quantified risk assessment (QRA) for landslide risk, believing that they had the tools and that sufficient accuracy was achievable. They reminded us that landslide probability might be estimated not only from statistical treatments relying on slope stability theory, following Professor Lumb, but also directly from failure data or by judgement (belief). For this purpose use of archived failure data was the norm in QRAs for major industrial hazards and was promising, they felt, for slope safety applications, landslides being much more frequent in Hong Kong than were the hazardous events in the industries with which they were familiar (A.B. Reeves, pers. comm., 1995). The risk experts cast doubt over treating fatality statistics as a measure of risk, pointing out that historical casualty data failed to reflect the 'near misses' which we knew were common and felt instinctively must not be left out of the account (A.B. Reeves, pers. comm., 1995). The way forward was clear. To get the benefits of the risk approach, GEO would need to equip itself to undertake formal landslide QRA modelled on the approach used for major industrial hazards.


## 22.6   QRA Research Programme

Following dialogue with the local risk consultants and after discussion with members of the Slope Safety Technical Review Board (Professor Robin Fell, Sir John Knill and Professor Morgenstern) at its first meeting in July 1995, the Special Projects Division of GEO presented a 2-year R&D programme on quantitative risk assessment (QRA). Its objectives included: to determine individual and societal risks, to establish risk tolerability criteria, and to examine risk mitigation options. After preliminary research to compile a database of failures, the plan was to carry out risk assessments for the territory as a whole ('global' QRAs). Risk from man-made slopes, natural terrain landslides and boulder fall would be studied. Risk consulting firms would undertake the QRAs working closely with the GEO research managers and landslide specialists. Social scientists at local universities would input to projects on public perception and tolerability of risk.

The QRA research programme quickly produced results. In due course the 2-year programme was extended and by the end of the decade much of significance had been achieved, as we shall see in the following account, which concentrates on results rather than on methods. Ho *et al.* (2000) and the literature cited below give a more detailed account of the GEO's QRA work with some limited description of methodology. Bibliographic information is available at http://hkss.cedd.gov.hk/hkss/eng/studies/qra, a web page linked to the Hong Kong Slope Safety website. For details of the QRAs carried out, reference must be made to the unpublished reports cited in these bibliographies.

### 22.6.1   The Lei Yue Mun Risk Assessment Exercise

Because our first interest was in global risk quantification for policy purposes, site-specific QRA had been given a low priority in the research programme. But events were to dictate otherwise. In August 1995 debris slides and rockfalls from abandoned quarry faces at Lei Yue Mun, East Kowloon, deposited about $800\,m^3$ of debris on to squatter dwellings beneath. Luckily there were no casualties, but there was severe property damage. On inspection, hundreds of dwellings were judged to be in danger from further landslides. The customary criteria for safety clearance of squatter huts (GEO's squatter clearance 'Method Statement') were fully satisfied for more than half of the 25 ha site, but the residents refused to move out. The situation had reached stalemate between the government, which was contemplating eviction by force, and the residents. At this point a government secretariat policy officer, an engineer with experience of safety management in the gas industry, suggested that a QRA might help in decision making. The firm Atkins Haswell was commissioned and the job was done within a few weeks. The study, outlined by Hardingham *et al.* (1998), provided a geographical distribution of individual risk (Figure 22.5) and estimates of societal risk (Figure 22.6). Evaluation against the PHI risk criteria (Planning Department, 1993), referred to in 22.4, showed that individual and societal risk levels in the GEO clearance zone were unacceptable. As the squatters were occupying their dwellings voluntarily, the risk consultants felt that individual risk of $1 \times 10^{-4}$ (annual probability of death) would be a more appropriate unacceptable criterion than the $1 \times 10^{-5}$ applicable to PHIs, but even this level is exceeded generally within the GEO clearance zone (Figure 22.5).

The results provided more than just a benchmarking test for the safety clearance criteria of the squatter Method Statement. Calculations also indicated that when the residents had left the GEO clearance zone, societal risk would fall to levels within the ALARP region and the broadly acceptable risk region (Figure 22.6). Cost–benefit calculations showed that the costs of any additional rehousing of squatters from elsewhere in the village would be grossly disproportionate to the associated benefits. This finding would have provided a supporting argument for a decision to take no further action.

The Lei Yue Mun risk assessment included the first formal landslide QRA to be carried out for a Hong Kong site. (The first landslide QRA to be published appears to be that reported by Morgan *et al.*, 1992.) The Lei Yue Mun QRA uses site-specific historical failure data to estimate landslide frequency (115 previous failures identified from aerial photographic interpretation, GEO records and walkover survey) and employs a simple consequence model. The consequence model recognizes three hazard groups defined in terms of type of failure, debris travel distance and proximity of dwellings. Judgements were needed from landslide and risk specialists, for example about vulnerability (probability that death results from impact of specified debris), the temporal presence (day/night) of people in dwellings, chance of fire and so on. Decision making generally followed the HSE's tolerability of risk rationale, broadly as it had been applied to PHIs in Hong Kong for 10 years. The study was carried at a cost of about US$25 000 and was completed in November 1995.

Despite the evident uncertainties surrounding the models and parameter values assumed, the Lei Yue Mun exercise convinced the writer that risk management with QRA using the HSE tolerability of risk rationale could be a powerful aid to decision making in slope safety at the site level.

## 22.6.2   Preliminary Studies – Feasibility Study for QRA of Boulder Fall Hazard, Consequence Classification System for Roads, Database of Landslides and Casualties at man-made Features

While the Lei Yue Mun study was under way, other interesting results were coming out of the QRA research programme. By December 1995 the firm ERM-Hong Kong Ltd had completed the first phase of the Feasibility Study for QRA of Boulder Fall Hazard in Hong Kong and concluded that a global (territory-wide) QRA of boulder falls was feasible (Reeves *et al.*, 1998). In 1996 a consequence classification system for roads based on quantified risk considerations was devised (Wong *et al.*, 1997) and a database of landslides and landslide casualties at man-made features was completed, using information from the thousands of documented cases that had accumulated in the GEO files since systematic recording of landslides began in 1978 (Chan *et al.*, 1998).

In the hope of revealing trends, risk consultants DNV drew on this database to produced graphs of 'rolling average' fatality rate for Hong Kong as a whole for the period 1948–95. The results, extended to 1996, are shown in Figure 22.10: each point is the average of annual fatalities over a 15-year period (i.e. the given year and the previous 14 years). Clear trends emerged: a rise in fatality rate until the 1970s, when the 15-year average annual fatality rate peaked at about 20 per year, followed by a sustained decline to around three per year in the 1990s. The 10-year and 20-year averaging plots show a similar trend. Does this provide proof of improved safety? Is like being compared to like (22.2.2.2)? This pattern corresponds well with the postulated risk trend (Figure 22.3), but the decline might alternatively be caused by factors other than the safety regime. But between 1977 and 1995 no matching abatement in rainfall had occurred and Hong Kong's

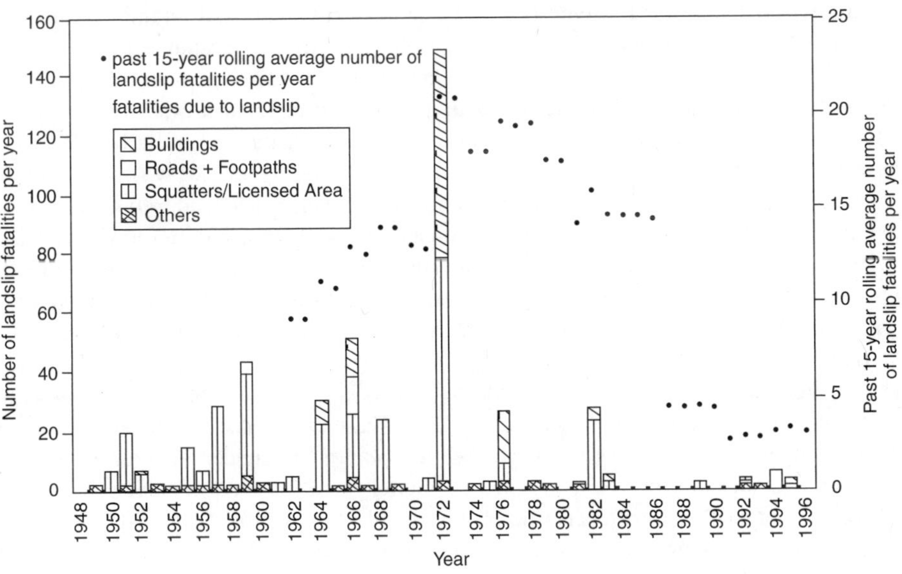


***Figure 22.10***  *Landslide fatalities in Hong Kong – annual and past 15-year rolling average (Malone, 1998). Reproduced by permission of Swets and Zeitlinger Publishers*

population had increased by 38%: these factors could not explain the trend. It may be argued that squatter clearance in this period, which was accompanied by a steep fall in squatter deaths due to landslides, would have happened anyway with rising prosperity, immigration control and sustained production of public housing. This is likely, but the GEO's squatter programme advanced by many years the selective re-housing of those squatters living on dangerous hillsides. For want of alternative plausible explanation, the decline in landside fatality rate was attributed to the introduction of the slope safety system (Malone, 1998). Since these results were published, the 15-year average annual fatality rate has declined to about one, despite a number of very wet years and continued growth in population.

Thus by 1996 we were in a position to express the state of safety as a number, using fatality data. The trends that emerged from the statistical smoothing would allow us to measure safety improvements and gauge the value of spending on safety. However, we remained mindful of the weakness of raw casualty statistics as a safety index, as compared to the quantity 'risk' (22.5.1).

### 22.6.3   Preliminary global risk assessment for the pre-1978 man-made slopes

Recognizing the need to express the state of safety in terms of risk, work began on a global risk assessment once the failure database had been compiled and a consequence classification had been devised. The class of slopes chosen for the first risk assessment was those built before controls became (relatively) effective. The cutoff date of 1978 was chosen for the purpose of distinguishing between 'old' and 'new' slopes. Wong and Ho (1998) outline the methodology they used for the global risk assessment, which was carried out with the help of the firm DNV. Failure frequencies for each of the specific hazards (failures in given types of feature – fills, cuts and retaining walls – of given mechanism, of given volume range) were assigned by reference to past failure rates, as evidenced in the failure database, the numbers of slopes in each category being obtained from the slope catalogue. An example of a specific hazard is, for example, a fill-slope liquefaction failure, debris volume $200$–$1000\,\mathrm{m}^3$. Each specific hazard has its corresponding frequency of occurrence.

The approach to consequence assessment was a 'generalized consequence model' (Wong *et al.*, 1997) which built on the consequence classification system for roads devised in 1996. The key factors taken into account include the area affected by the debris, the types of facilities affected and the spatial and temporal distribution of the population at risk. Consequence is expressed in terms of PLL (i.e. the expected number of deaths per year). The basis of the approach is to apply site-specific factors to standard PLLs. These were estimated for landslides $10\,\mathrm{m}$ wide and $50\,\mathrm{m}^3$ in volume affecting various types of facilities ('facility groups'), see Table 22.1. These 'reference PLLs' are calculated assuming the facility is at the worst possible location but under average occupation conditions. In the generalized consequence model the PLL for the standard landslide for the type of facility threatened (the 'reference PLL') is factored for scale, depending on the width and volume of the landslide considered, and for vulnerability, depending on the estimated likelihood of death due to the standard landslide affecting the *particular* facility. The calculation procedure for a single slope is given in Table 22.2. In the global QRA, societal risk is the sum of all of the site-specific risk.

**Table 22.1**   *Grouping of facilities for the global QRA study*

| Facility group no. | Facilities affected | Expected number of fatalities given a standard landslide |
|---|---|---|
| 1(a) | Buildings with a high density of occupation or heavily used (residential buildings, hotels, schools, markets, etc.) | 3 |
| 1(b) | Very heavily trafficked roads & footpaths, bus shelters, railway platforms, etc. | 3 |
| 2(a) | Buildings with a low density of occupation or lightly used (sports hall, churches, etc.) | 2 |
| 2(b) | Heavily trafficked roads & footpaths, railways, construction sites, etc. | 1 |
| 3 | Moderately trafficked roads & footpaths, densely used open spaces (playgrounds, car parks, etc.) | 0.25 |
| 4 | Roads & footpaths with low traffic density, lightly used open air recreation areas, etc. | 0.03 |
| 5 | Roads with very low traffic density, country parks, etc. | 0.001 |

**Table 22.2**   *Results of consequence assessment for the Fei Tsui Road landslide*

| Facility affected | Facility group no. (Reference PLL) | Vulnerability to death if debris impacts | Scaling factor for actual size of landslide | PLL |
|---|---|---|---|---|
| Open space | Group no. 5 (0.001) | 0.95 | 90 m/10 m = 9 | 0.01 |
| Fei Tsui Road | Group no. 3 (0.25) | 0.85 | 90 m/10 m = 9 | 1.91 |
| Baptist Church & kindergarten | Group no. 1 (3 × 2) | 0.17 | 20 m/10 m = 2 | 2.04 |
| Playground | Group no. 4 (0.03) | 0.15 | 50 m/10 m = 5 | 0.02 |
| | | | Total PLL = | 3.98 |

The preliminary global QRA gave a PLL of about 11 deaths per year. The number represents the risk level for the years 1984–95 overall, this being the window for collection of failure data. The PLL reflects the rainfall and population in these years and the state of effectiveness of the slope safety regime. The value of 11 deaths compares with about 3 deaths from the fatality data (Figure 22.10).

The preliminary global risk assessment gave risk per feature in the ratio 3:1:1 (cuts, fills, retaining walls) and a risk apportionment with cuts creating about 75% of the total risk. As indicated in Figure 22.11, around 55% of the total risk is derived from the 10% of the population of slopes that affects facilities in the highest consequence group: this group includes very busy roads and buildings with a high density of occupation or heavily used (see Table 22.1). The lowest consequence group includes country parks and roads with very low traffic density. The average risk levels of the highest and lowest consequence groups differ by three orders of magnitude. These results gave useful pointers to the best use of scarce resources for risk mitigation.

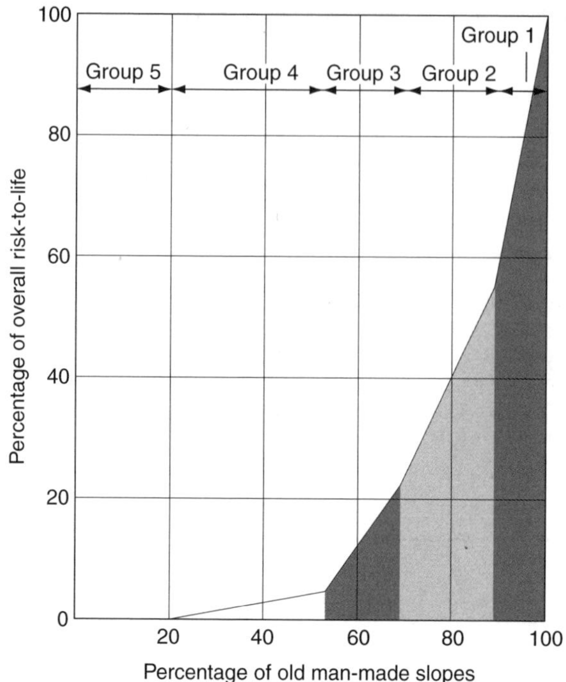



**Figure 22.11**   *Percentage of risk by facility group (Wong and Ho, 1998). Reproduced by permission of A.A. Balkema Publishers*

### 22.6.4   Site-specific QRA for the Fei Tsui Road Landslide

A large cutting at Fei Tsui Road, Chai Wan, had suffered a sudden massive collapse during a rainstorm on 13 August 1995. The debris, about 14 000 m$^3$ in volume, moved rapidly across open ground and the road, killing one person and injuring another. The collapse had occurred in a densely populated district of Hong Kong in the early hours of the morning and one would judge that the consequences would probably have been worse had the collapse occurred during the daytime. Calculations using the generalized consequence model reinforced this judgement, showing a PLL of about four fatalities (Wong *et al.*, 1997). The calculations are summarized in Table 22.2.

### 22.6.5   Preliminary Global QRAs of Boulder Fall Hazard and Earthquake-induced Landslides

At an early stage in the research programme preliminary global QRAs were carried out for boulder fall hazard and earthquake-induced landslides in Hong Kong. Completed in 1996, the results of the boulder study (Reeves *et al.*, 1998) showed that the 'historical risk level' from boulders (i.e. raw statistical data, not risk from QRA) between 1926 and 1995 was at least one order of magnitude lower than that from the pre-1978 man-made slopes. In the same study a preliminary site-specific QRA was carried out for four areas. Boulder fall risk appears to be at least two orders of magnitude below risk from man-made slopes:

pedestrians and squatters are the two most vulnerable groups.The earthquake-induced landslide QRA (Wong and Ho, 1999) compared the risk from earthquake-induced landslides at old and new slopes. The analysis showed that the risk from earthquake-induced landslides at man-made slopes designed for or upgraded to current geotechnical standards is one to three orders of magnitude lower than the risk posed by rain-induced landslides at old slopes.

These early QRAs gave us confidence that the approach was feasible and the results useful, and with the encouragement of the Slope Safety Technical Review Board, the two-year R&D programme was extended. Between 1996 and the end of 2001, work was carried out on risk tolerability criteria and two more global QRAs were completed: an assessment of global landslide risk reduction achieved by the LPM programme from 1977 to 2000, and a study of global risk from natural terrain landslides.

### 22.6.6 Studies on Risk Tolerability Criteria

As part of the QRA research plan the firm ERM was commissioned in 1997 to develop risk tolerability criteria for landslides and boulder falls from natural terrain in Hong Kong. After reviewing criteria used for major industrial hazards, dams and railways in several European countries, Australia and Canada, along with the local criteria for PHIs, dangerous goods transportation and railways, the study recommended adoption on a trial basis of Hong Kong's existing PHI criteria (22.4), with modifications. The exceptions were, with respect to societal risk, that no requirement was seen for a 'broadly acceptable' region on the $F$–$N$ curve (the 'acceptable' region of Figure 22.6) and the limit of tolerability for the number of fatalities from a single event was extended from 1000 to 5000. These recommendations became interim GEO guidelines for use in evaluating the results of natural terrain QRAs, published as GEO Report No. 75 (ERM, 1998).

### 22.6.7 QRAs of Global Risk Reduction Due to the Slope Safety System

The third objective of the QRA research programme was to examine risk mitigation options, such as those shown in Figure 22.3. Calculations had been made to estimate the costs and benefits of each of the elements shown on the figure using graphs of 'rolling average' fatality rate (Malone, 1998). The benefits were expressed as lives saved. The procedure involved extrapolating the pre-1978 fatality rate trend shown in Figure 22.10 (the 15-year rolling averages) and attributing to the slope safety system the difference between the extrapolated fatality rates and the actual fatality rates. Reasonable assumptions were made about the direct cost of the slope safety system, which was charged to life saving only (economic benefits were ignored). The crude estimates suggest that some elements of the safety system had been rather more cost-effective than others and that to the end of 1996 each life saved had cost about US$2.5 million to save.

How might such a number be judged? Two frames of reference exist. First, 'cost per life saved' figures may be viewed in the context of HSE's tolerability framework and ALARP. The figure of US$2.5 million is lower than the values of statistical life normally assumed in cost–benefit analyses conducted as part of risk assessments for technological hazards in Hong Kong practice. Society is indirectly paying much more than US$2.5 million for a statistical life saved in the regulated risk sectors e.g. PHIs, railways, because the regulators require that risk is reduced until the incremental cost of further risk reduction becomes not just equal to, *but grossly disproportionately more than, its incremental benefit*. Second, the figure of US$2.5 million is also well below the range of 'values of preventing a statistical fatality' from 'willingness to pay' estimates elicited from people attending focus groups (Social Sciences Research Centre, 1998). Therefore, on this framework of justification, there was a case in 1998 for continuing investment of effort and expenditure in landslide risk reduction in Hong Kong.

Costing around US$115 million annually at the present time, the Landslip Preventive Measures (LMP) programme for the upgrading of defective pre-1978 ('old') man-made slopes is by far the most expensive element of the safety regime. A QRA was conducted in 2000 to examine its effectiveness. The methodology followed the approach used for the preliminary global risk assessment for the pre-1978 man-made slopes described above (22.6.3). The study concludes that, up to the end of September 2000, the LPM programme had reduced landslide risk due to the pre-1978 slopes to about 50% of its level in 1977 (Cheung and Shiu, 2000). This was a noteworthy finding, giving for the first-time quantitative evidence that the US$650 million LPM programme 1977–2000 was actually having a significant impact on public safety.

The study also showed that a small but not insignificant proportion of the risk from the batch of slopes remains *after* treatment of the batch under the LPM programme because of the apparent inability of treatment under the programme to reduce the rate of failure of slopes greater than 20 m high. The main reason for the collapses of engineered slopes appears to be the failure to detect past slope instability and geological/hydrogeological conditions and past slope instability having a seriously adverse effect on stability. This indicates the need for improvements in site investigation to enhance the reliability of the programme.

Along with the assessment of the effectiveness of the past LPM programme, an attempt was made using QRA to predict its future impact. Calculations based on the preliminary global risk assessment for the pre-1978 man-made slopes described above (22.6.3) showed that with the slope safety system now in place, the risk from old man-made slopes will be reduced by the year 2010 to less that 25% of that in 1977 (Ho *et al.*, 2000). Such estimates of risk reduction were part of the case supporting a funding application made in 1998 for a 10-year extended LPM programme (2000–2010). The application was successful. The government has now adopted the goal 'to further reduce the landslide risk from old man-made slopes to below 25% of the 1977 level by year 2010' (Anon, 2000). To the best of the writer's knowledge this is the first time that quantified landslide risk has been cited in a governmental policy goal.

### 22.6.8    Studies of Risk from Landslides on Natural Terrain

It is estimated that, on average, just over 300 landslides occur on undeveloped terrain annually in Hong Kong, of which well below 1% reach the developed area. Some of these landslides are rockfalls and boulder detachments on natural terrain. Measures to counter such failures had always been a very minor part of the slope safety system, in keeping with the Conduit Cliffs 'boulder policy' decision referred to earlier (Sinclair, 1978). A few of the 300 landsides give rise to channelized debris flows. The Tsing Shan debris flow of 1990 (Figure 22.12) was the largest such event in recent decades. It deposited a total debris volume of about 26 000 m$^3$ and its debris trail extended for 1035 m, just reaching the urban margin. The Tsing Shan event delivered a major jolt,

**Figure 22.12**   *The debris flow of 11 September 1990 on Tsing Shan*

prompting a review of natural terrain safety policy. It was to be 12 years before a new strategy for natural terrain risk would be promulgated. However, the policy review when its time came was able to take advantage of risk management thinking. QRA studies, both global and site-specific, were carried out in the course of its development.

### 22.6.8.1   The global risk from landslides on natural terrain

Relying on high-level aerial photographs taken between 1945 and 1994, an inventory of landslides on natural terrain (the NTLI) was compiled as the database for subsequent studies into the distribution of these events. Staff of the New South Wales Department of Land and Water Conservation carried out the air photo interpretation in 1995 and 1996 under a consultancy agreement with GEO. With these data available, a preliminary study could proceed into the global risk from landslides on natural terrain (Sun and Evans, 1999). Average annual frequencies were estimated from 30 years' worth of data comprising the recent landslides in the inventory (recent at the time of photography). Consequence modelling followed the approach used for the preliminary global risk assessment for the pre-1978 man-made slopes described above and assuming the 1994 development situation. The calculated PLL best estimates range from 0.07 to 0.18, which may be compared with a PLL of about 11 for the pre-1978 man-made slopes (22.6.3).

Judged simply on the relative numbers of expected deaths, the results support the priority given since 1978 to risk from man-made slopes as compared to risk from natural hillsides. However, the natural terrain problem was shown to require some attention, as the $F\text{--}N$ curves for global societal risk are higher than many published acceptable risk criteria. Also there remains the unquantified risk from very rare natural landslides not represented in the NTLI but evident in the geological record.

### 22.6.8.2   QRA methods for site-specific assessment of landslide risk in natural terrain

A study was undertaken in 1998 and 1999 intended to develop QRA methods for site-specific assessment of landslide risk in natural terrain. Halcrow Asia Partnership carried out the work and part of it is summarized by Moore *et al.* (2001). Two undeveloped hillsides were studied, one 75 ha and the other 18 ha in area, on which open hillside landslides (i.e. not channelized), coastal landslides and potential rock- and boulder falls were recognized. Evidence was also seen of channelized debris flows and possible large-scale landslides. Aerial photographic interpretation, documentary data collection and field observations allowed 298 and 62 individual landslide features respectively to be recorded. Hazard models were developed for each of the landslide types, comprising data for frequency of occurrence, volume and debris runout probability. All numerical estimates were assigned an upper- and lower-bound value as a measure of uncertainty. Two kinds of consequence model were devised: the catchment scale model and the site-specific model. (In the terminology used here the former is still a 'site-specific' QRA.) The former is characterized by population densities according to the Outline Zoning Plan plus clusters representing hypothetical buildings and a major road. The site-specific models included a hypothetical three-storey house, multi-block mass housing and a major road. Vulnerability and temporal presence factors were applied for indoor and outdoor populations.

The differences between risk numbers calculated using upper- and lower-bound values revealed a significant degree of uncertainty. It amounted to one or two orders of magnitude for open-hillside landslides and about three orders of magnitude for channelized debris flows and large-scale landslides. The study proceeded to develop a QRA approach for development at the foot of open hillsides but not for channelized debris flows and large-scale landslides because of excessive uncertainty.

Natural terrain QRA is proving useful in site-specific decision making in other applications, for example in helping to judge whether transportation facilities need to be protected from landslides or if squatter settlements threatened by landslides on natural hillsides ought to be cleared. The latter problem is exemplified at Pa Mei Shan Ha squatter village in Tung Chung, Lantau, which is threatened by potential shallow debris flows and slides in the hillside overlooking the settlement. A rudimentary QRA was carried out, as outlined by Ho and Wong (2001). Their hazard model took a $100\,\text{m}^3$ standard landslide crossing a 10 m wide section of village boundary as the 1 in 100 years' probability event. The consequence model assumed debris travel to 10 m beyond the boundary and incorporated vulnerability and temporal presence factors. Simple calculations sufficed. Individual risk estimated for the various houses ranged from about $5 \times 10^{-5}$ to around $6 \times 10^{-4}$, which may be compared with the unacceptable limit of $1 \times 10^{-4}$ taken for Lei Yue Mun squatter village (22.6.1). It was decided to recommend voluntary safety evacuation. In another application, the use of QRA and cost–benefit analysis to evaluate the need for mitigation measures against natural terrain landslide risk for a water supply system (reservoirs, pipelines and an access road) serving the Hong Kong Disneyland development is described by Pinches *et al.* (2002). The risk was found to be too low to merit carrying out any risk mitigation measures on purely cost–benefit considerations.

### 22.6.8.3  Strategy for management of natural terrain landslide risk

In 2002 the Hong Kong government published its new strategy for management of natural terrain landslide risk (GEO, 2002: 2). The strategy 'aims at keeping the natural terrain landslide risk to a level that is as low as practically achievable'. In dealing with natural terrain risk affecting existing developments, a 'react to known hazards' principle has been adopted, modelled on the 'boulder policy' (22.2.1). Decision making about the need for risk mitigation is sometimes assisted by site-specific QRA and cost–benefit analysis, some details of which are given in GEO Report No. 75 (ERM, 1998).

### 22.6.9  Other QRA work up to the end of 2001

At least five more site-specific QRAs have been carried out in Hong Kong, to the end of 2001, and the tool has been used for risk-based priority ranking for old fill slopes and busy roads with a history of landslides.

## 22.7  Summary of Applications of QRA in Slope Safety Policy in Hong Kong

**Technical Policy**

1. The Lei Yue Mun risk assessment exercise provided a benchmarking test for the conventional squatter safety clearance criteria. The two methods gave similar results.
2. The findings of the QRA of earthquake-induced landslide risk lend support to the regulatory policy prevailing since 1978 of requiring no general allowance for earthquake loads in slope stability design.
3. The QRA to examine the effectiveness of the LPM programme showed that a not insignificant proportion of the risk from the batch of slopes treated remains after treatment. This points to the need for improvements in site investigation to improve the reliability of the programme.

**Resource Allocation**

4. Calculations of landslide fatality rate supported the case made in 1994 for a change in policy so as to include within the LPM programme slopes along all roads with high traffic density.
5. The preliminary global risk assessment for the pre-1978 man-made slopes showed risk per feature to be in the ratio 3:1:1 (cuts, fills, retaining walls) and that cuts create 75% of the total risk. About 55% of the total risk is derived from the 10% of the population of slopes affecting facilities in the highest consequence group. The results gave valuable pointers to the best use of scarce resources in risk mitigation.

6. The results of a study of global risk from natural terrain landslides support the policy of having given a higher priority to the upgrading of man-made slopes than the mitigation of risk from natural hillsides. However, the natural terrain problem was judged to require some attention, as the $F$–$N$ curves for global societal risk are higher than many published allowable risk criteria.
7. Under the new strategy for management of natural terrain landslide risk (GEO, 2002), decision making about the need for mitigation of risk affecting existing developments is sometimes assisted by QRA and cost–benefit analysis.
8. Calculations using the preliminary global risk assessment for the pre-1978 man-made slopes indicate that with the slope safety system now in place, the risk from old man-made slopes will be reduced by the year 2010 to less that 25% of that in 1977. These estimates of risk reduction were part of the case supporting the funding bid in 1998 for a 10-year extended LPM programme (2000–2010).

**Effectiveness of Spending**

9. The QRA to examine the effectiveness of the LPM programme showed that, up to the end of September 2000, the programme had reduced landslide risk due to the pre-1978 slopes to about 50% of its level in 1977.

**Risk-based Safety goal**

10. The government has adopted the objective 'to further reduce the landslide risk from old man-made slopes to below 25% of the 1977 level by year 2010'. To the best of the writer's knowledge this is the first time that quantified landslide risk has been cited in a governmental policy goal.

This summary of results completes the brief history of risk thinking in safety management in Hong Kong. We are now ready to return to the questions that become important in the resource allocation debate wherever landslides are a public issue.

## 22.8   Discussion on Risk Quantification in Slope Safety Policy

Following public outcry after disaster governments come under pressure to improve safety. It was suggested earlier that, sooner or later, they will face these questions: Complete safety being unattainable, how safe is safe enough, and what is an appropriate level of effort and expenditure on safety? How should the effectiveness of effort and expenditure on safety be measured?

To summarize the position reached earlier (22.2.2.2), some of the suggested rationales to help decide an appropriate level of effort and expenditure on slope safety (e.g. cost–benefit arguments) cannot be employed unless we are able to define quantitatively and measure the state of safety and safety benefit. Of the two ways of expressing state of safety and safety benefit as a number, detriment rate (e.g. Figure 22.10) and the theoretical quantity 'risk' (e.g. Figures 22.5 and 22.6 and Table 22.2), the latter is technically preferred. But is its use feasible?

As we have seen in the foregoing account of the GEO QRA research programme, it was found to be feasible in Hong Kong, with its 20 year landslide database, to express slope safety as a number, in terms of risk, and to use this quantity to measure safety benefit. Thus it was shown that, up to the end of September 2000, the programme had reduced landslide risk due to the pre-1978 slopes to about 50% of its level in 1977 (Cheung and Shiu, 2000). The risk tool was also judged to be sufficiently trustworthy to be used by the government to define the safety goal relating to old man-made slopes: 'to further reduce the landslide risk from old man-made slopes to below 25% of the 1977 level by year 2010' (Anon, 2000). The adoption of such a goal decides in effect an appropriate level of effort and expenditure on slope safety, and the two matters (goal and resources) would be decided together in the resource allocation process.

Once a safety goal is adopted, benefits will flow to all parties involved in the public resource allocation process (safety agency, budget controller, legislators, the public at risk, etc.). In principle these benefits include the following:

(i) Reduced scope for misunderstanding. One understanding of the word 'safety' is 'freedom from danger or risks' (*Concise Oxford English Dictionary*). If no goal is stated, it may be thought that the achievement and preservation of a state of perfect safety (zero risk) is the implicit goal of a safety regime. This was the understanding voiced by some members of the Legislative Council in 1994. By adopting an explicit safety goal that is plainly not 'freedom from danger or risks', for example 'to reduce risk to below 25% of the 1977 level by year 2010', the scope is reduced for misunderstanding over the aim of spending on safety.

(ii) Holding the safety agency to account. When a safety goal has been adopted, progress towards the goal will be monitored in terms of safety benefit. The safety agency is thereby more readily accountable. When called for after a disaster, tangible results will be available to show what of value has been achieved by past effort and expenditure.

(iii) Showing that continued effort is needed. When support wanes, the demonstration that significant risk remains will confirm the need for continued effort and expenditure.

(iv) Best use of resources. When a safety goal has been adopted, it may be expected that applications for public funds for safety work will be required to state the contribution to be made by the work included in the application to the achievement of the overall goal. Thus it is likely that, all else being equal, the more cost-effective items will be selected in the resource allocation process in preference to the less cost-effective items.

(v) Justifying public expenditure. In defending spending decisions the resource allocator, whether the executive or the legislature, can cite the associated expected safety improvement.

Some aspects of the feasibility of the use of 'risk' in slope safety policy have yet to be tested in Hong Kong. The acid test of the use of 'risk' to express safety benefit and define goals will come in the wake of disaster. Risk numbers will then be employed to make safety improvement demonstrable and clearly intelligible so as to convince the sceptic that something of value has been achieved. The safety agency will be called upon to explain to the lay person how the risk numbers are derived (QRA). This will be a challenge: 'risk' is a mathematical concept beyond the reach of many people. Raw statistics have the advantage over 'risk' as an expression of safety that their derivation is rather more easily understood.


## 22.9   Conclusions

Professor Peter Lumb introduced aspects of risk thinking to the local geotechnical profession in the 1970s and use was made of these concepts in early geotechnical policy making. By the late 1980s the formal risk management process had been fully developed and employed in Hong Kong for the control of land use and population in the vicinity of potentially hazardous installations. GEO began to use the risk management process in slope safety policy in 1993. But it was only after the Kwun Lung Lau disaster in 1994 that the full potential of the risk management tool came to be appreciated. If risk could be expressed as a number, it would provide the much-needed index of safety through which we might discern trends, see the effect of our interventions, defend our achievements and set goals. If we could quantify risk, the risk management process would become available as policy aid, as it had before when the Hong Kong government was faced with a difficult land use problem involving a serious safety threat from highly explosive chemicals.

Having decided to develop capability in landslide QRA, rapid advance was possible by bringing in risk experts from the hazardous industries to work with the GEO's research managers and landslide experts and by following the risk management procedures for potentially hazardous installations that had already been fully agreed by the local regulator. Useful progress has since been made in policy application of QRA and the risk management process in slope safety. The results of global risk assessments have lent support to aspects of technical policy concerning slope design and squatter safety criteria, helped in decision making about mitigation of natural terrain risks and confirmed the effectiveness of LPM spending.

But perhaps most significantly, risk methods have been used to define the safety goal relating to old man-made slopes. Looking back, it seems surprising, given the tight rein on public spending in Hong Kong, that for its first 20 years GEO had no safety goal, that is, one expressing safety improvement in numbers. But then it did not feel the need for a safety goal for its own purposes until the landslide disaster of 1994, when the dangers of operating an expensive safety system without safety goals became apparent.

Along with its chronology of landslide risk quantification in Hong Kong, this chapter has shown how questions of importance in the resource allocation debate may be addressed using risk concepts. Through risk we are able to express the state of safety as a number and measure safety improvements in terms of risk reduction. Once we can quantify safety benefit we are able to gauge the effectiveness of investment in safety. There are good reasons for hazard-prone communities to adopt risk-based safety goals, or at least goals based on loss statistics (e.g. casualties), when they begin to spend resources to counter the risk. How else is success to be defined in the intangible world of safety? Once a safety goal had been adopted, an appropriate level of public effort and expenditure on slope safety had been decided for the time being. Adoption of such a safety goal will in principle bring benefits to all of the parties involved in safety.

## Acknowledgements

I wish to thank Leung Ka Fai for preparing the line drawings and the GEO for providing Figures 22.9 and 22.12.

## References

Anon., 1982, Is Hong Kong spending wisely on safety? Report of HKIE Professional Practice Board Forum December 9 1981, *Hong Kong Engineer*, **10**(4), 31–33.

Anon., 1995, Verbatim transcript of Legislative Council of Hong Kong Select Committee on Kwun Lung Lau Landslip and Related Issues. Public Hearing on 23 January 1995 p. 33, lines 3–6.

Anon., 2000, Slope Safety in Hong Kong, *Hong Kong Information Note*, Information Services Department, HKSAR Government, June, p. 6.

Binnie & Partners, 1978, Landslide Study Phase II C, Report on the Caine Road Area, unpublished.

Brand, E.W., 1983, How safe are Hong Kong's slopes? Press Release, 9 May. Text of a Speech to the Rotary Club of East Tsimshatsui by the head of the Geotechnical Control Office.

Chan, W.L., Ho, K.K.S. and Sun, H.W., 1998, Computerized databases of landslides in Hong Kong, in K.S. Li, J.N. Kay and K.K.S. Ho (eds), *Slope Engineering in Hong Kong* (Rotterdam: Balkema), 213–219.

Cheung, W.M. and Shiu, Y.K., 2000, Assessment of global landslide risk posed by pre-1978 man-made slope features: risk reduction from 1977 to 2000 achieved by the LPM Programme. Special Project Report SPR 6/2000, Geotechnical Engineering Office, Summary in *GEO News* Issue No. 228 march 2001, 8, 9.

ERM-Hong Kong Ltd, 1998, *Landslides and Boulder Falls from Natural Terrain: Interim Risk Guidelines*, GEO Report No. 75.

Fell, R., 1994, Landslide risk assessment and acceptable risk, *Canadian Geotechnical Journal*, **31**, 261–272.

Findlay, P.J. and Fell, R., 1995, A Study of Landslide Risk Assessment for Hong Kong, Report prepared for the Geotechnical Engineering Office, Hong Kong, unpublished.

Geotechnical Control Office, 1979, *Geotechnical Manual for Slopes*, Geotechnical Control Office.

Geotechnical Control Office, 1984, *Geotechnical Manual for Slopes*, 2nd edn, Geotechnical Control Office.

Geotechnical Engineering Office, 2002, *Management of natural terrain landslide risk*, Information Note 4/2002.

Hardingham, A.D., Ho, K.K.S., Smallwood, A.R.H. and Ditchfield, C.S., 1998, Quantitative risk assessment of landslides – a case history from Hong Kong, in K.S. Li, J.N. Kay and K.K.S. Ho (eds), *Slope Engineering in Hong Kong* (Rotterdam: Balkema), 145–151.

HSE (Health & Safety Executive), 1988, revised 1992, *The tolerability of risk from nuclear power stations* (London: HMSO).

Ho, K.K.S., 1993, Landslide risk assessment and acceptable risk level. Report of a lecture by Professor Robin Fell to HKIE Geotechnical Division, 22 July 1993, *Hong Kong Engineer*, **21**(12), 11–14.

Ho, K., Leroi, E. and Roberds, B., 2000, *Quantitative Risk Assessment: Application, Myths and Future Direction. GeoEng2000*, Volume 1: Invited Papers (Technomic Publishing Company), 269–312.

Ho, K.K.S. and Wong, H.N., 2001, Application of quantitative risk assessment in landslide risk management in Hong Kong, in K.K.S. Ho and K.S. Li (eds), *Proceedings of the 14th South East Asian Geotechnical Conference, Hong Kong* (Rotterdam: Balkema), vol. 1, 123–128.

Kay, J.N., 1994, An explicit risk based approach to slope stability, *Proceedings of the International Conference on Safety, Economy and Reliability in Marine Engineering and Construction*, Hong Kong.

Lumb, P., 1975, Slope failures in Hong Kong, *Quarterly Journal of Engineering Geology*, **8**(1), 31–65.

Lumb, P., 1983, Statistical soil mechanics, in *Proceedings of the 7th Asian Regional Conference on Soil Mechanics & Foundation Engineering*, Haifa, vol. 2, 67–81.

Malone, A.W., 1985, Factor of safety and reliability of design of cuttings in Hong Kong, *Proceedings of the 11th International Conference on Soil Mechanics & Foundation Engineering*, San Francisco, vol. 5, 2647.

Malone, A.W., 1992, Text of a Speech to the Y's Men's Club of Hong Kong, 1 October.

Malone, A.W., 1996, Thoughts on a new landslide risk management strategy, Discussion Note No. 1/96, Geotechnical Engineering Office, unpublished.

Malone, A.W., 1998, Risk management and slope safety in Hong Kong, in K.S. Li, J.N. Kay and K.K.S. Ho (eds), *Slope Engineering in Hong Kong* (Rotterdam: Balkema), 3–17.

Moore, R., Hencher, S.R. and Evans, N.C., 2001, An approach for area and site-specific natural terrain hazard and risk assessment, in K.K.S. Ho and K.S. Li (eds), *Proceedings of the 14th South East Asian Geotechnical Conference, Hong Kong* (Rotterdam: Balkema), vol. 1, 155–160.

Morgan, G.C., Rawlings, G.E. and Sobkowicz, J.C., 1992, Evaluating total risk to communities from large debris flows, *Geotechnique and natural hazards. GeoHazards '92* (BiTech Publishers Ltd), 225–236.

Morgenstern, N.R., 1994, *Report on the Kwun Lung Lau Landslide of 23 July 1994*, vol.1, Causes of the landslide and adequacy of geotechnical practice in Hong Kong.

Pinches, G.M., Smallwood, A.R.H. and Hardingham, A.D., 2002, The study of natural terrain hazard of Yam O, Lantau, in *Natural terrain – A constraint to development?* Institution of Mining & Metallurgy, Hong Kong Branch, 207–221.

Planning Department, Hong Kong Government, 1993, *Hong Kong Planning Standards and Guidelines*, Ch. 11.

Reeves, A.B., Chan, H.C. and Lam, K.C., 1998, Preliminary quantitative risk assessment of boulder falls in Hong Kong, in K.S. Li, J.N. Kay and K.K.S. Ho (eds), *Slope Engineering in Hong Kong* (Rotterdam: Balkema), 185–191.

Sinclair, K., 1978, Danger cliff plan shelved, *South China Morning Post*, Sunday 4 June.

Slosson, J.E., 1969, The role of engineering geology in urban planning, in *The Governor's Conference on Environmental Geology*, Colorado Geological Survey Special Publication 1, 815.

Social Sciences Research Centre, 1998, *Public Perception and the Tolerability of Landslide Risk*, Social Sciences Research Centre, The University of Hong Kong.

Sun, H.W. and Evans, N.C., 1999, The average annual global risk from natural terrain landslides in Hong Kong in 1994, Technical Note TN 5/99, Geotechnical Engineering Office, unpublished.

Wong, H.N. and Ho, K.K.S., 1994, The 5 November 1993 Lantau landslip study – general report on landslips at man-made features. Special Projects Report SPR 7/94. Appendix A. Geotechnical Engineering Office, unpublished.

Wong, H.N. and Ho, K.K.S., 1998, Overview of risk of old man-made slopes and retaining walls in Hong Kong, in K.S. Li, J.N. Kay and K.K.S. Ho (eds), *Slope Engineering in Hong Kong* (Rotterdam: Balkema), 193–200.

Wong, H.N. and Ho, K.K.S., 1999, Preliminary quantification of risk of earthquake-induced failure of man-made slopes in Hong Kong, *Geotechnical Risk Management, Proceedings of the HKIE Geotechnical Division Annual Seminar*, 67–76.

Wong, H.N., Ho, K.K.S. and Chan, Y.C., 1997, Assessment of consequence of landslides, in D.M. Cruden and R. Fell (eds), *Landslide Risk Assessment, Proceedings of the International Workshop on Landslide Risk Assessment, Honolulu, Hawaii* (Rotterdam: Balkema), 111–149.

Wong, H.N. and Premchitt, J., 1994, Review of aspects of the Landslip Preventive Measures Programme, Special Projects Report SPR 5/94, Geotechnical Engineering Office, unpublished.

Wrigley, J. and Tromp, F., 1994, Risk management of major hazards in Hong Kong, in *Proceedings of the Conference on Integrated Risk Management*, University of New South Wales.

Wu, T.H. and Kraft, L.M., 1970. Safety analysis of slopes, *J. Soil Mech. Found. Div. ASCE*, vol. 96 SM2, 609–630.

# 23

# Rockfall Risk Management in High-density Urban Areas. The Andorran Experience

*Ramon Copons, Joan Manuel Vilaplana, Jordi Corominas, Joan Altimir and Jordi Amigó*

## 23.1 Introduction

The Principality of Andorra is a mountainous country located in the central Pyrenees between France and Spain. The capital is Andorra la Vella, which is situated at the bottom of a valley of glacial origin. It is surrounded by mountains that exceed 2000 a.s.l. One of these mountains is the Enclar massif, the southeastern face of which is known as Solà d'Andorra, a steep rock slope with considerable rockfall activity (Figures 23.1 and 23.2). This slope is made up of granodiorite and hornfels outcrops traversed by joint sets that are mainly responsible for the instability of the slope. The joint spacing is closely related to the lithology. According to our observations, the granodiorites are less densely fractured than the hornfels. In the scree deposits accumulated at the foot of the slope, the granodiorite rock blocks display a volume (0.5 to 15 m$^3$) greater than that of the hornfels (0.02 to 1 m$^3$).

Given the lack of available land for building, Solà d'Andorra is subjected to intense urban pressure. Rapid development in the 1970s led to the construction of buildings in areas exposed to rockfall, that is, the districts of La Margineda, Santa Coloma and downtown Andorra la Vella (see Figure 23.1). In recent years, rockfalls have damaged a number of buildings (Figure 23.3). The most significant event occurred on 21 January 1997 when a block 25 m$^3$ (Figure 23.4) in volume impacted against the Bon Repòs building in the district of Santa Coloma, causing considerable damage and injury to one person.

*Landslide Hazard and Risk*   Edited by Thomas Glade, Malcolm Anderson and Michael J. Crozier
© 2004 John Wiley & Sons, Ltd   ISBN: 0-471-48663-9

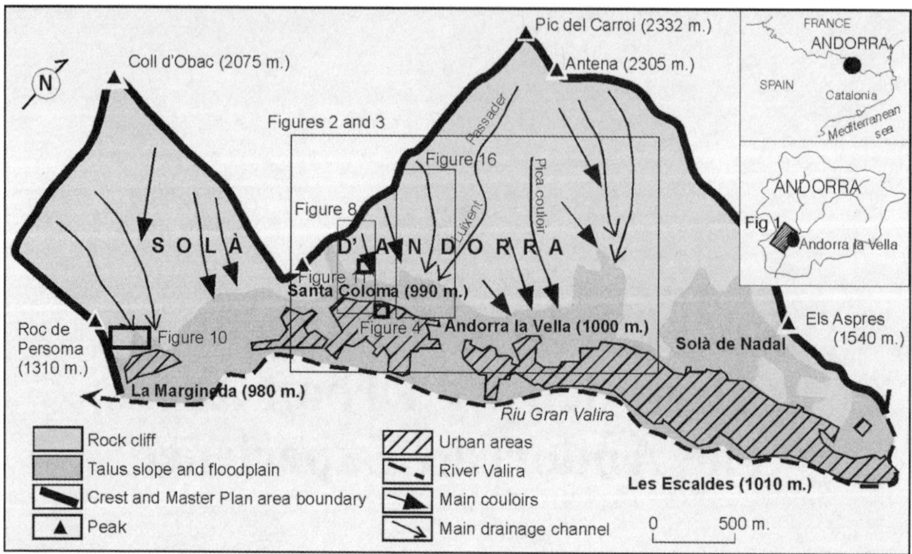



**Figure 23.1**  *Location and detail of the study area, the town of Andorra la Vella and Solà d'Andorra, in the Principality of Andorra*

**Figure 23.2**  *General view of Solà d'Andorra la Vella*

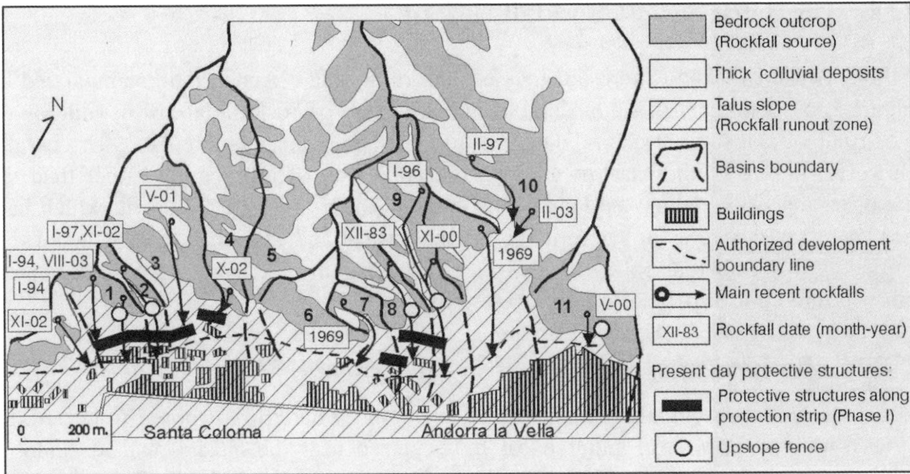



**Figure 23.3** *Detail of the most urbanized sector, including a record of the most recent rockfall events. Numbers show the main rockfall basins (1: Forat Negre, 2: Alzina couloir, 3: Boneta couloir, 4: Cirera couloir, 5: Luixent Passader channel, 6: Basera Mateu wall, 7: Ramenada couloir, 8: Coll d'Eres couloir, 9: Pica couloir, 10: Pedra Plana couloir, 11: Corbs wall)*

**Figure 23.4** *Boulder of 25 m$^3$ in the garage of the Bon Repòs building. This block crashed into the building causing one casualty in January 1997*

## 23.2   The Battle against Rockfall Hazard

All the measures taken to forestall rockfall hazard at Solà d'Andorra are summarized in Figure 23.5. A number of studies and technical reports on rockfall hazard in addition to the installation of some structural measures had been undertaken before the 1997 rockfall. However, hazard studies had focused on the collection of information rather than on the provision of guidelines on land use at Solà d'Andorra. The 1997 event, which had become a major cause for concern for the authorities and the inhabitants, marked the beginning of a coherent policy of rockfall risk management. This event also prompted other hazard control policies in the whole principality.

### 23.2.1   Measures Taken before the Rockfall Event of 1997

Before 1997, diverse rockfalls had focused attention on buildings exposed to hazard. In 1969 a rockfall of several hundreds of $m^3$ occurred near the old hospital at Andorra la Vella, but did not cause any human loss or material damage. In 1983 several rock blocks exceeding $1\,m^3$ damaged two new buildings and one under construction. Despite extensive material damage, there were no victims. In 1994 another rockfall impacted against some warehouses and buildings under construction. In 1996 some rock blocks reached farmlands and a promenade used for recreation.

Earlier attempts to manage the hazard suffered from a lack of studies on spatial distribution. In June 1980 an ordinance prohibited construction in areas exposed to snow avalanches, rockfall and torrential activity.

In 1989 a landslide-flooding hazard map at 1:25 000 scale, which covered the whole country, was completed. A geomorphological approach was used to prepare the map and four hazard categories were considered (Corominas *et al.*, 1990). Despite its relatively small scale, the authorities used the map for issuing building permits.

In 1994 more detailed hazard zoning was carried out (1:10 000 scale) in Andorra la Vella. However, this report was merely informative given that neither protective measures nor specific restrictions for land development were proposed.

### 23.2.2   Policy of Rockfall Risk Management at Solà d'Andorra

The impact of the 1997 rockfall event prompted the local authorities (Municipality of Andorra la Vella) and the government (Andorran Ministry for Land Use and Public Works) to adopt a specific policy on rockfall risk management at Solà d'Andorra. The first task of this policy was the Rockfall Risk Management Master Plan, which was completed in May 1998 (Figure 23.5) (Corominas *et al.*, 2003b). The Master Plan included detailed studies on rockfall hazard assessment (the methodology of which is discussed below), simple guidelines for issuing building permits and for protecting buildings exposed to hazard.

After the implementation of the Master Plan it was necessary to adopt structural measures to mitigate the risk. These measures were implemented by the Rockfall Risk Mitigation Plan. This Plan made provision for protective design studies and for the installation of structural measures such as dykes and fences to protect exposed buildings. A methodology known as 'Eurobloc' (Copons *et al.*, 2001a, b) was devised to afford protection in terms of typology, maximum absorbable energy and height.

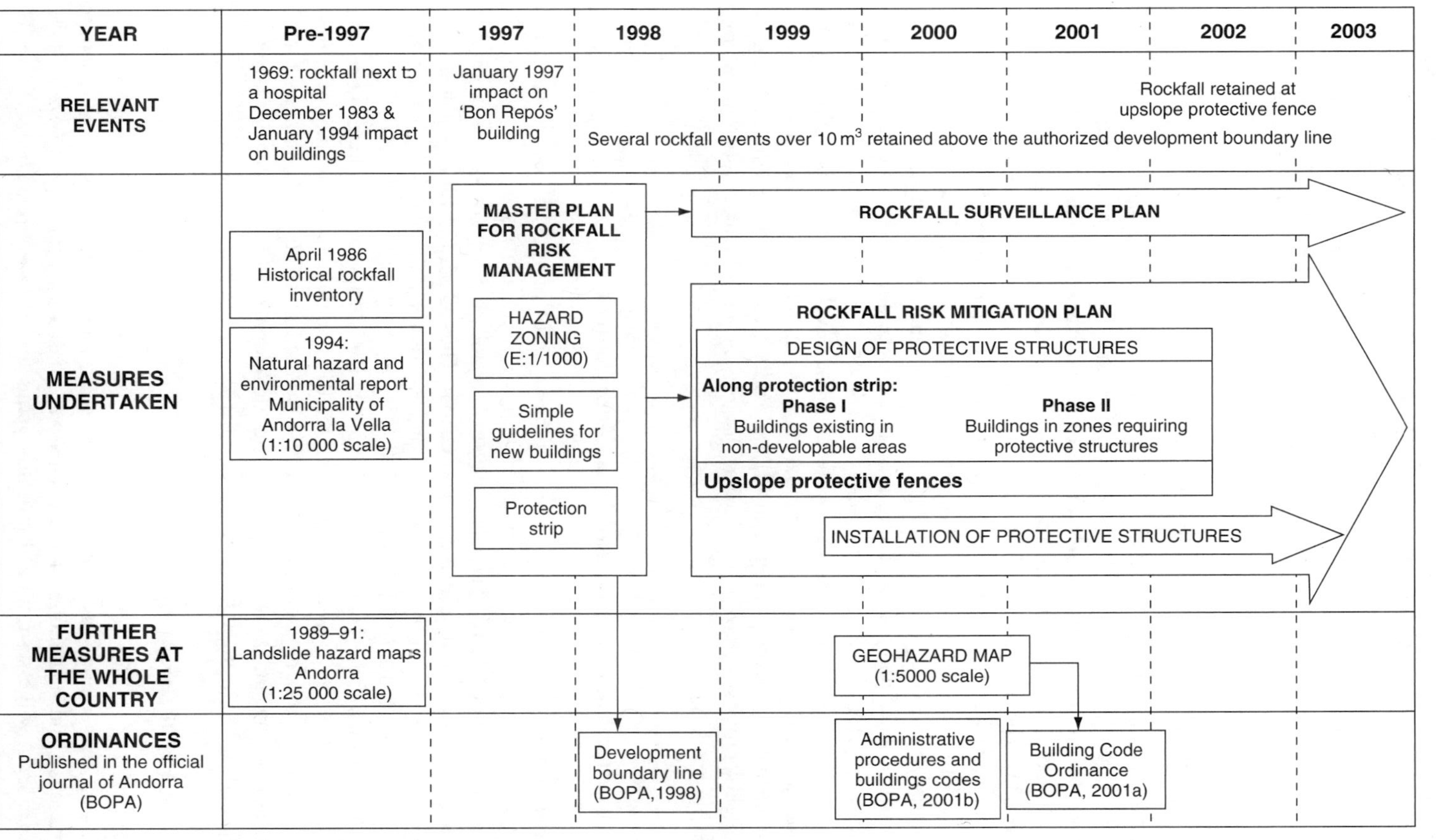


***Figure 23.5*** *Main rockfall events, measures undertaken and ordinances established at Solà d'Andorra*

In parallel to the Mitigation Plan, a Rockfall Surveillance Plan of the rock slope was implemented at Solà d'Andorra. The reason for this was twofold: (i) since the protective works were conceived for relatively small rockfall events, the Surveillance Plan seeks to forestall potentially large rockfall events (those exceeding thousands of $m^3$); (ii) new rockfall events are identified and described, enabling us to validate and update the rockfall trajectography models used to design the protective structures (Amigó *et al.*, 2001). These new events are stored in a rockfall database.

### 23.2.3   Other Measures Taken in the Principality of Andorra

A number of studies were carried out (Escalé, 2001): the snow avalanche inventory map (1988–99) (see Ferrer *et al.*, 2000), the study of the flood hazard for the whole river network of Andorra (1999–2002), and the Geotechnical and Landslide Hazard Zoning Map of Andorra (1999–2001) (henceforth Geohazard Map) at 1:5000 scale (Corominas *et al.*, 2003a). The most striking feature of the latter study was that, besides the landslide hazard map, a specific administrative procedure was established for issuing building permits (Corominas *et al.*, 2002). This procedure, which also applied to Solà d'Andorra, was published in the Official Journal of Andorra in June 2001 (BOPA, 2001a).

The Urban and Land Use Planning Law of 29 December 2000 (BOPA, 2001b) was the decisive factor in regulating buildings at risk. In accordance with this law (Escalé, 2001): (a) zones exposed to natural hazard cannot be developed; (b) urban planning must take into account zones exposed to natural hazards; (c) the government of Andorra will be responsible for hazard zoning; and (d) the Andorran government will determine the type of protective structure at sites where hazard can be mitigated and reduced to an acceptable level.

## 23.3   Master Plan for Rockfall Risk Management

The Master Plan was conceived as a preliminary tool of rockfall risk management, albeit not the definitive one, to be implemented within a relatively short period in order to forestall rockfall hazard. The only way to forestall this hazard was to prohibit the construction of new buildings in areas exposed to it. Thus all permits for new constructions were suspended in the period between the Bon Repòs rockfall of January 1997 and the completion of the Master Plan in May 1998.

The Master Plan had a dual role: (i) to demarcate the areas where development could be authorized; and (ii) to provide guidelines on the development and protection of exposed buildings. The methodology of the Master Plan was as follows: (i) to undertake a rockfall hazard assessment and zoning; (ii) to restrict development in accordance with hazard zoning; and (iii) to propose protective measures.

### 23.3.1   Rockfall Hazard Assessment and Zoning

The first step of the Master Plan was to carry out the rockfall hazard zoning at Solà d'Andorra. The zoning was made at 1:1000 scale and the criteria used to determine the frequency and magnitude of the events were based on geomorphic (size of fallen boulders, silent witnesses etc.) and trajectographic analyses (Figure 23.6).

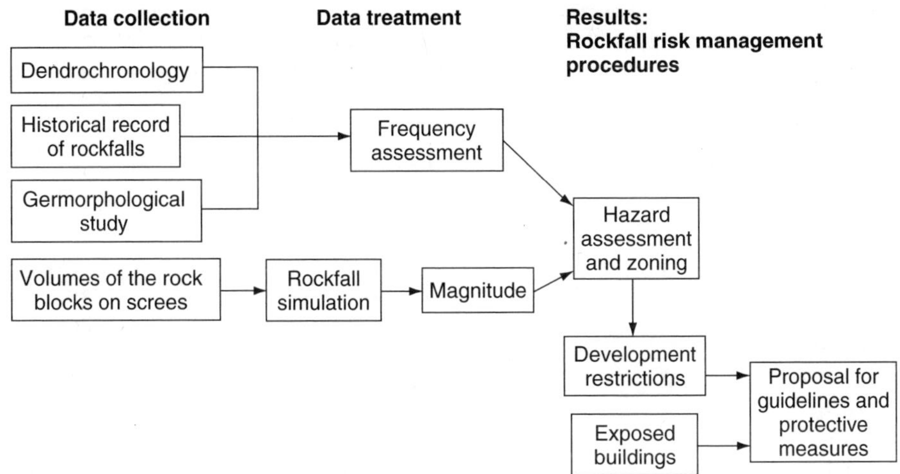


**Figure 23.6**  *Methodology followed by the Master Plan*

### 23.3.1.1  Frequency assessment of rockfalls

Rockfall frequency was evaluated using an inventory of historical rockfalls and eye-witness accounts, a dendrochronogical study of damaged trees, and a geomorphological study of rockfalls.

The historical record was an inventory of recent rockfalls obtained from questionnaires and consultations of records. The questionnaires, which were filled in by eyewitnesses, together with the information obtained from historical documents, reports and newspapers, enabled us to make an inventory of a large number of recent rockfalls. These rockfalls and the high population density yielded a great deal of evidence. A dendrochronological study (Moya, 2002) of trees damaged by the impact of rock blocks was also carried out. The localities selected for dendrochronological analysis were situated at the couloir outlets and at the foot of the rock walls. This study was carried out in the urban areas given the complexity of the dendrochronological techniques. The geomorphological study of rockfalls consisted in identifying and mapping the scars, the rockfall trajectories and the fallen boulders in the field. This enabled us to make an inventory of the rockfalls without specifying the date. Rockfall events detected by geomorphological techniques were occasionally related to those identified by historical and dendrochronogical methods. Moreover, the geomorphological study allowed us to identify older rockfalls without any historical reference. These older rockfalls had been colonized by vegetation and their mapping was often difficult. Thus the geomorphological study enabled us to obtain a record of rockfalls that occurred at greater intervals.

The historical record allowed us to determine the frequency of rockfalls that could affect urban areas. Moreover, the geomorphological record helped us to make a rough estimate of the frequency in non-urban areas. The dendrochronological study yielded a higher frequency of rockfall events given that it also included the rockfalls retained before reaching urban areas (Table 23.1). These studies showed that the rockfall frequency was much higher at the foot of the rock slope at Solà d'Andorra (obtained

**Table 23.1**   *Estimated frequencies obtained from both historical and dendrochronological records at different locations at Solà d'Andorra (see Figure 23.1)*

| Site | Estimated frequencies (in years) from: | |
| --- | --- | --- |
| | Dendrochronological records | Historical records |
| Forat Negre | 2 to 5 | 7 to 9 |
| Alzina couloir | 2 | 2 |
| Boneta and Cirera couloirs | 3 | 10 |
| Luixent Passader channel | 5 to 15 | Tens of years |
| Ramenada couloir | 1 to 4 | 30 |
| Coll d'Eres couloir | – | 20 |
| Pica couloir | – | 20 |

from dendrochronogical records) than in the recently urbanized areas (based on historical records).

### 23.3.1.2   Magnitude and runout assessment of rockfalls

The magnitude, which is regarded as the capacity to cause damage, was expressed as energy (kilojoules). This magnitude was determined in two steps. First, a statistical analysis of the volumes of the rock blocks forming the scree deposits was carried out, yielding a distribution of volume frequency (Copons *et al.*, 2000b). Second, a calculation of the energies of rock blocks was made. This step was carried out by means of the 'Eurobloc' rockfall simulation code (López *et al.*, 1997; Copons *et al.*, 2000a). The simulation process considered different rockfall scenarios, each with a different volume. The volumes to be simulated were selected on the basis of the statistical analysis of the volumes of accumulated blocks. The volumes measured were grouped into three volumetric classes, each with its own volumetric distribution as shown in the discriminant analysis in Figure 23.7. The volumes represented by the simulation were obtained from the highest percentiles (90 and 99%) of each volumetric class so that the volume chosen represented most of the blocks in the terrain. Therefore the volumes of 0.5, 1, 2.5, 5 and $10\,m^3$ were considered the most suitable for the different scenarios to be simulated (Copons *et al.*, 2000a).

The 'Eurobloc' simulation code is an analytical and numerical rockfall simulation model based on dynamics and kinematics. The code provides data on the trajectories, rebound heights and the energies of the blocks along the slope. The inputs are the topography (TDM), the starting points, the block volumes (in our case 0.5, 1, 1.5, 5 and $10\,m^3$) and the slope characteristics (terrain typology, rugosity and vegetation cover) (Krummenacher and Keusen, 1997). The starting points correspond to the scars of former rockfalls, interpreted in most cases from a geomorphological study. Terrain typology, rugosity and vegetation cover are obtained directly from terrain mapping. The results of the model are a three-dimensional representation of the block trajectories, energies and rebound heights on a map at the selected scale, which in the case of the Master Plan was 1:1000 (Figure 23.8). The simulated rockfalls are usually calibrated with real data from recent rockfalls (size of the detached blocks, trajectories, maximum reach, etc.) (Copons *et al.*, 2000a).

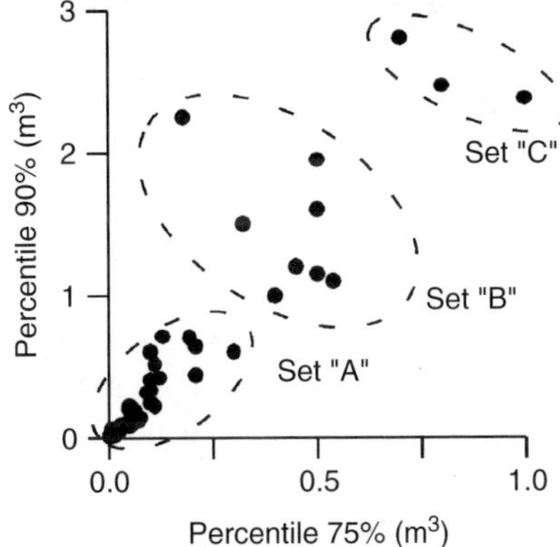



**Figure 23.7** *Ratio between 75 and 90 percentiles of volumes of fallen blocks measured in different sampling plots. The analysis shows the presence of three volume sets. Each dot corresponds to a sampling plot*

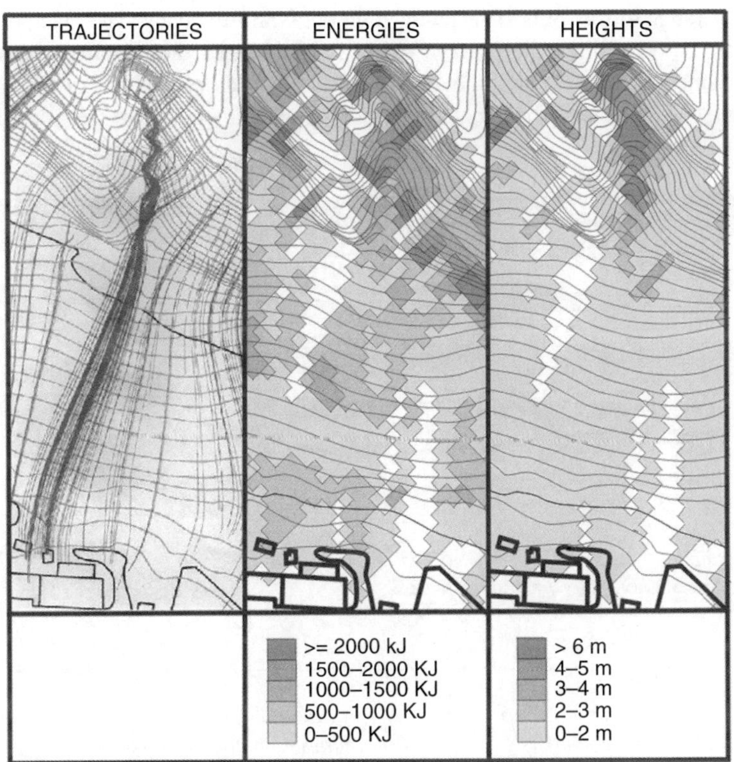



**Figure 23.8** *Numerical simulation of the rockfall process. The example was performed with blocks of 1 m³*

### 23.3.2   Restrictions on Development at Solà d'Andorra

The lack of developable land, its high cost, and the rapid urban growth of the town have led to the construction of buildings exposed to rockfall hazard. Thus it goes without saying that the restrictions on development have aroused strong feelings among the landowners and the authorities. Moreover, there was a pressing need for mitigating rockfall risk. Restrictions on building were the first measures to be adopted.

Trajectography and frequency studies were used to define an upper boundary known as the development boundary (Figure 23.9), above which the energy of the fallen blocks was too high to be retained by protective structures. A lower boundary, known as the development boundary without restrictions, was also defined to indicate the area where fallen rocks could be retained by protective structures (Figure 23.9). Building without the need for protective structures was permitted in areas located below this lower boundary not reached by rockfalls. The demarcation of the upper boundary on the topographic map was carried out taking into account the frequency and magnitude of the most usual rockfalls. The upper and lower boundaries coincided with the lines between the zones of high and medium hazard, and medium and low hazard, respectively (Lateltin, 1997; Raetzo *et al.*, 2002).

Given the high frequency of rockfalls in almost all the sectors of the study area, zones suitable for building had to meet two conditions. The first condition was that the area had to be out of range of rockfalls. This reduced the rockfall frequency. The reach probability, which was expressed as a percentage of the fallen blocks that reached a hypothetical horizontal line on the slope, was obtained from the trajectory analysis of the simulation. In our case, the protected urban zones were only reached by less than 1% of the simulated

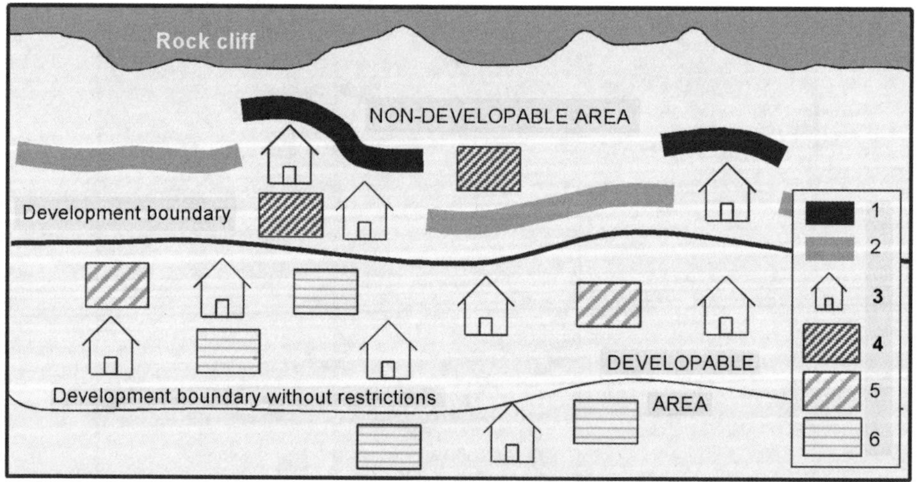


*Figure 23.9*  *Development map applied by the Rockfall Master Plan. (1 and 2: Protection strip. Installation of protective structures for buildings in the non-developable area and in the developable area, respectively. 3: Buildings. 4: Plots of land where building is prohibited (case one). 5: Developable plots of land where protective structures are necessary (case two). 6: Developable plots of land exempted from protective structures (case three))*

blocks. The second condition was that the magnitudes or energies of the rock blocks had to be low enough to be intercepted by the usual protective structures. A large protective structure such as an earth barrier is effective in intercepting rockfalls of up to 10000 kJ since higher energies are regarded as non-manageable (Corominas *et al.*, 2003a).

It should be pointed out that large rockfalls corresponding to rock slabs (thousands of m$^3$ in volume) cannot be retained by structural defences. However, it should be noted that such rockfall volumes are infrequent at Solà d'Andorra. Furthermore, only one rock avalanche (millions of m$^3$ in volume) has been identified in the River Valira Valley since the Pyrenean deglaciation 30 000 years ago (Bordonau *et al.*, 1993). Consequently, neither rock avalanches nor rock slabs were considered when demarcating the upper development boundary.

The lower boundary had to be located out of range of rockfalls, that is, at a distance of more than 100 m from the upper boundary line. The results of the simulation showed that all the simulated blocks came to rest above the lower boundary line. On the other hand, the maximum reach for large rockfalls (thousands of m$^3$) was derived from the empirical tables of the reach angle (Corominas, 1996). The results showed that the majority of large rockfalls did not cross the lower boundary line.

A number of guidelines were issued for new buildings. These prohibit building above the upper boundary line (case one in Figure 23.9). On the other hand, protective structures must be erected for new buildings located between the upper and lower boundary lines (case 2 in Figure 23.9).

### 23.3.3   Proposal for Protective Measures

The proposed zoning for development also established a strip of land containing the protective structures, known as the *protection strip*, located immediately above the upper boundary line. This protection strip was designed to forestall the situation where building proprietors could install their own protective structures in the developable zone.

All the protective structures had to be located along the protection strip regardless of the distance from the new buildings to be protected (case 2 in Figure 23.9). However, some new buildings are exempted from the obligation to install protective structures along the protection strip. This occurs when these buildings are located immediately below older ones or when they are built below the protective structures already in place (case 3 in Figure 23.9). Furthermore, all new buildings located below the lower boundary line are exempted from protective structures (case 3 in Figure 23.9).

The guidelines on development were published without delay in the official journal of Andorra in 1998 (BOPA, 1998). Thus the Master Plan passed from being an informative study to being a useful tool of land use planning. Subsequently, in 2001, the guidelines drawn up by the Master Plan were replaced by the Building Code Ordinance (BOPA, 2001b), using the same boundary lines (Corominas *et al.*, 2003a).

## 23.4   Rockfall Risk Mitigation Plan

The Master Plan revealed the existence of some buildings that should not have been authorized given their location above the development boundary line. These buildings had to be afforded immediate protection. Moreover, suitable protective structures had to be designed along the protection strip. The design of these structures and their installation were included in the Rockfall Risk Mitigation Plan (henceforth Mitigation Plan).

Protective structures were proposed at two sites: along the protection strip defined in the Master Plan (see Figure 23.3 and 23.10), and further upslope close to the couloir outlets (Figures 23.3 and 23.4). In the latter site, the protective structures were located in sectors with the highest number of rockfall paths. Several selected locations coincided with the couloir outlets above the talus deposit apex.

The purpose of the protective structures along the protection strip is to prevent the rockfall from reaching the protected buildings. Moreover, the protective structures at the couloir outlets are used to retain smaller rockfalls and to reduce the energy of the larger rockfalls before they reach the protection strip. Thus the protective structures at the couloir outlets are complementary to those located along the protection strip. These structures significantly minimize the hazard level and, hence, the risk in populated areas.

Given the wide extension of the exposed area, the protective structures were designed in two phases taking into account the number of buildings exposed to hazard (Figure 23.5). In the first phase, between June 1998 and April 1999, they were designed for buildings located above the development boundary line (Figure 23.9). In the subsequent phase, between February 2000 and June 2001, the protective structures were designed for buildings located below this line (Figure 23.9). In the first phase, 1365 m of the 7.5 km length of the slope were considered, whereas 3100 m were contemplated in the subsequent phase. The rest of the talus slope, 3000 m in length, corresponded to non-urban areas where development was not expected in the short term. Since April 1999 the authorities

**Figure 23.10**  *Protective fences in the protection strip defined in the Master Plan*

**Figure 23.11** *Protective fence located upslope at a couloir outlet. This fence was damaged by a rockfall event in November 2002*

have been installing protective structures for buildings located above the development boundary line.

The design of the protective structures raised two important questions: (i) how can rockfall risk be quantified? and (ii) how much risk do the protective structures reduce? The methodology developed at Solà d'Andorra to answer these questions is presented below.

Given the wide extension of the rock cliff, its inaccessibility and the presence of a large number of potential rockfall sources, removal of the unstable blocks or anchoring rock masses at the cliff face were not considered feasible.

### 23.4.1 Design of Protective Structures along the Protection Strip: The 'Eurobloc' Methodology

The lack of standardized design criteria for protective structures and the need for these studies at Solà d'Andorra gave rise to an original methodology known as 'Eurobloc' (Copons *et al.*, 2001a, 2001b). 'Eurobloc' designs protective structures taking into account the residual risk of the protected building. The residual risk is the risk that remains in the exposed building after the installation of protective structures. Given that the buildings are 'elements at risk', the ideal protection is when buildings have a residual risk equivalent to 'tolerable risk' (Fell and Hartford, 1997) or 'acceptable risk' (Cruden and Fell, 1997; IUGS Working Group, 1997). The quantitative assessment of residual risk allowed us to select the most suitable protection in terms of maximum absorbable energy and height.

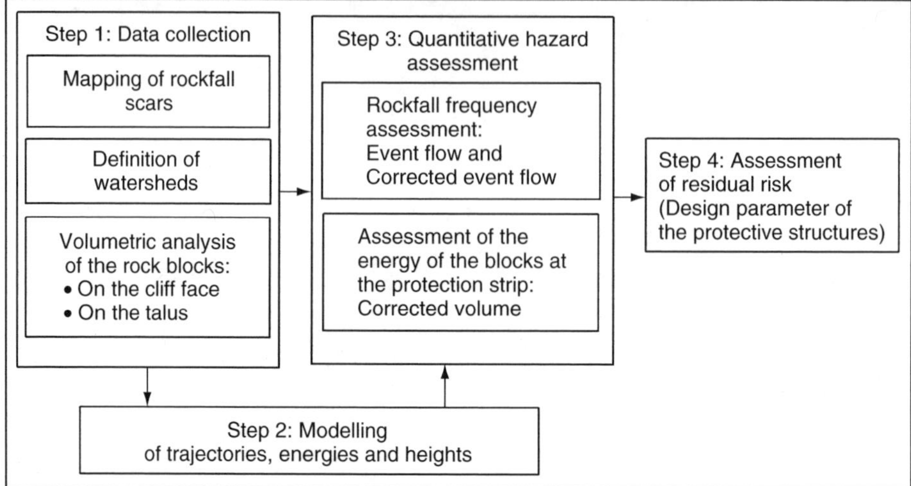

*Figure 23.12* *The methodology used for the design of protective structures in the protection strip*

The 'Eurobloc' methodology consists of four well-defined steps (Figure 23.12). Step one involves the collection of data in the field. Step two is the rockfall simulation process. Step three consists in determining the frequency of rockfall and the energy of rock blocks along the protection strip. Step four quantifies the residual risk of the protected buildings as a function of the energy of the rock blocks, the rockfall frequency and the protection characteristics.

### 23.4.1.1 Estimate of rockfall frequency

A number of basins, similar to hydrographic ones, were considered on the rock slope under study. The energy and frequency of the rockfalls were studied in each of these basins. The rockfall frequency along the protection line not only depends on the characteristics of the source area (distribution of joint sets, lithologies, etc.), but also on the rockfall trajectories (bedrock morphology, presence of obstacles, etc.). Depending on the characteristics of the trajectory, the rockfall can acquire different energies and reaches. Thus each basin displays its own rockfall frequency along the protection strip.

In the absence of detailed rockfall records, which can help us to determine the quantitative frequency, other criteria were used. These were the number of rockfall scars and the potentially movable rock blocks, known as the *rockfall indicators in starting zones* (henceforth starting indicators).

In each basin, the relationship between the number of starting indicators and the length of the protection strip establishes a parameter known as the *event flow*. This parameter is defined as the number of starting indicators for each 100 m of the protection line located below.

The event flow is a rough estimate of the number of rockfalls along the protection strip during the last few thousand years. The indicators provide a maximum frequency given that not all the rock blocks reach the bottom of the valley. It is possible to determine a more accurate frequency with the aid of rockfall trajectographic models. The iterative calculation of the trajectories allowed us to determine the percentage of rock blocks that reach the protection strip from different starting points. The result of multiplying the event flow by this percentage yields the *corrected event flow*.

Although the inventory of rockfall scars presents a rock slope that exerts considerable influence on the value of the event flow, it only represents a given period of time in the geomorphological history of the slope. It should be borne in mind that more recent rockfalls could erase the morphological traces of older rockfalls at similar starting points. Thus the inventory of rockfall scars corresponds to the last few thousand years. Traces of rockfalls before 10 000 BP would probably have been obliterated by subsequent rockfalls. Furthermore, these older rockfalls could have occurred under climatic conditions different from the present ones.

### 23.4.1.2   Energy of rock blocks along the protection strip

The calculation of the energy of rock blocks was carried out in two steps: (i) the determination of the rock block volumes that could reach the protection strip, and (ii) the determination of the block energy by simulation.

The volumes of the detached rock blocks were obtained in two ways: by direct measurement of rock blocks accumulated in the scree deposits, and by estimating the size of rock blocks that could fall on the rock slope (Copons *et al.*, 2000b). The latter volumes were determined by a statistical study of the joint spacing in the different joint sets (Copons *et al.*, 2000b) given that these joint sets are responsible for decomposing the rock mass into polyhedral elements (Jaboyedoff *et al.*, 1996; Rouiller and Marrow, 1997; Rouiller *et al.*, 1998). The joint planes of the rock usually correspond to the faces of the polyhedrons of the rock blocks and the spacing of the joint planes is proportional to the sides of the rock bodies that become detached.

The rock blocks that reach the protection strip have volumes similar to those of the blocks located below the strip, that is, in the plots to be protected. Thus, in order to measure the rock blocks in the runout zone, it is necessary to focus our attention on the scree deposits, especially those below the protection strip. However, the high degree of urbanization and the considerable forestation at the screes below the protection strip prevented us from pinpointing with accuracy the rock block volumes. Moreover, most of the largest blocks had been removed, leaving only remnants in parks or in plots reserved for buildings. Thus most of the rock block volumes were measured above the protection strip but in few cases some blocks were measured below this strip.

The impossibility of detecting a sufficient number of rock blocks below the protection strip plus the presence of some large rock blocks midway up the talus slope prompted us to study the variations of the rock block volumes on the slope (Copons *et al.*, 2001a). These variations in volume showed a tendency to increase downhill as on the talus slopes (Evans and Hungr, 1993), although the volumetric distribution of some rock blocks measured below the protection strip resembled those measured immediately above. The statistical results of the block volumes measured above were corrected given the tendency of the volume to increase downhill. The resulting volumetric distribution had to fulfil two conditions: (i) be slightly higher than the volumetric distribution calculated at the screes above the protection strip; and (ii) be considerably lower than that calculated at the rock slope. The resulting volumetric distribution is known as the *corrected volume*.

The corrected volume (*Vc*) would be the result of weighting the volumetric distribution calculated at the rock slope (*Vr*) and that measured at the scree deposits (*Vs*). In this case and in accordance with the heuristic criteria, Copons *et al.* (2001a) proposed the following equation:

$$Vc = \frac{4Vs + Vr}{5}$$

The constants of this equation were estimated by weighting the largest volumes at the scree deposits and at the rock slope to obtain the volumes located below the protection strip. Subsequently, the constants were verified bearing in mind the few fallen blocks that had recently reached the protection strip. At present, the volumes of these fallen blocks resemble those of the corrected volume, which corroborates the reliability of the estimation of constants.

On the other hand, simulation can determine the distribution of energies and heights of rock blocks as a function of the volumes (in this case 0.5, 1, 2.5, 5 and $10\,\mathrm{m}^3$) along the protection strip. This calculates the percentage of simulated blocks retained by protective structures in accordance with their maximum absorbable energy and height. Figure 23.13 shows a graph of the energy distribution of different volumes of blocks along the protection strip. A vertical straight line on the graph represents the maximum absorbable energy of protection, which is 2000 kJ in Figure 23.13. By varying the maximum absorbable energy of the protection, and therefore the position of the straight line of the graph, it is possible to vary the percentages of the volumes retained. In the above example, the protection of 2000 kJ retains all the blocks of 0.5 and $1\,\mathrm{m}^3$ volumes since they all yield a lower energy. Moreover, volumes of 2.5, 5 and $10\,\mathrm{m}^3$ are retained with a percentage of 95.7%, 88.4% and 60.6%, respectively.

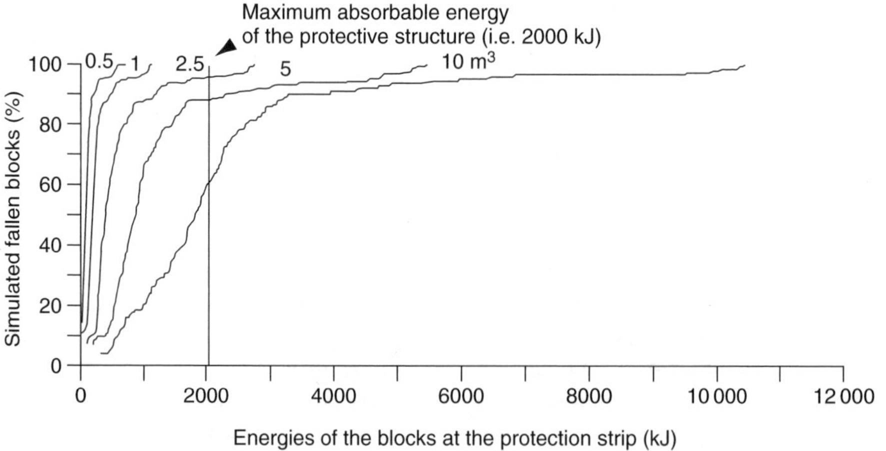


**Figure 23.13** *Range of energies of blocks (0.5, 1, 2.5, 5 and 10 m³) in the protection strip, obtained with the numerical model. Plot of the maximum absorbable energy that allows us to determine the percentage of simulated blocks retained*

### 23.4.1.3  Estimate of residual risk

The determination of the residual hazard was performed in two steps. The first step consisted in the determination of the percentage of rock blocks retained before reaching the protected areas. This percentage is a parameter known as the *degree of protection*. The second step consisted in the calculation of the *residual risk* remaining in the protected buildings, using the degree of protection and event flow parameters.

The degree of protection is the result obtained by multiplying the corrected volume and the percentage of simulated rockfall retained before reaching the protected areas. The resulting percentage consists of the corrected volumes retained by the protective structures plus those of the rock blocks that come to rest naturally. In our case, the ideal value of the degree of protection always exceeded 99%, although other authors regard 99.9% as an acceptable value (Rouiller *et al.*, 1998). In the example of Figure 23.13, simulated volumes of 0.5, 1, 2.5, 5 and 10 m$^3$ represent 49.6%, 33.9%, 14.5%, 1.47 % and 0.51% of the corrected volumes, respectively. The result obtained by multiplying these percentages and those of the simulated rock blocks retained before reaching the protected areas offers a degree of protection that is slightly higher than 99% (see Table 23.2).

The residual risk is the result obtained by multiplying the percentage of the corrected volume blocks that reach the protected areas, and the event flow parameter. The event flow parameter is the maximum estimate of the frequency along 100 m of the protection strip. The residual risk is an estimate of rockfall frequency below the protection strip after the installation of protective structures. All rock blocks breaking through the protective structures and impacting against buildings are regarded as negative, regardless of the resulting energy. As in the event flow, the residual risk is expressed by events that cross the protection strip every 100 m.

The determination of the residual risk in different areas showed that an acceptable risk in areas destined for residential buildings is one event per 100 m. It was not possible to attain this value of residual risk by means of conventional protection in non-developable areas. Bearing in mind that the event flow parameter represents the last thousand years, a residual risk of one event per 100 m in the protected zone would be equivalent to $10^{-5}$

*Table 23.2*  *Obtain the degree of protection by adding the retained corrected volume percentages*

| Simulated volumes | Corrected volumes (%) | Retained simulated blocks (%) | Retained corrected volumes (%) |
|---|---|---|---|
| 0.5 | 49.6 | 100 | 49.6 |
| 1 | 33.8 | 100 | 33.8 |
| 2.5 | 14.7 | 95.7 | 14.1 |
| 5 | 1.4 | 88.4 | 1.2 |
| 10 | 0.5 | 60.6 | 0.3 |
| | | | **99%** |

Note:
Percentages in column four are the results obtained by multiplying the percentages of the corrected volumes (second column) and those of the retained simulated blocks (third column) for every simulated volume (from 0.5 to 10 m$^3$).

and $10^{-6}$ events per metre and year. These values are regarded as an acceptable risk for buildings in the literature (Fell and Hartford, 1997).

### 23.4.2   Location and Design of Upslope Protection

Given that the upslope protection is complementary to the structures along the protection strip, more attention was focused on location than on design. The most suitable location had to fulfil two conditions. The first was that the location should intercept as many rockfalls as possible. The second was that the protection had to be situated in places where the rock blocks presented less energy and rebound height. The method of selecting the location had to take into account the results of the simulation. The selection was carried out by identifying the sectors where the trajectographic map presented a high density of rockfall trajectories. Then, definitive locations were selected using the profiles of the slope, at 1:1000 scale, representing the maximum energies and rebound heights of the simulated blocks (Figure 23.14).

The most suitable protective structures were 'woven wire-rope nets' (Wyllie and Norrish, 1996) or 'dynamic fences'. Given that the selected locations were usually inaccessible and the space available for the installation was limited, dynamic fences of 500, 750, 1000, 1500 and 2000 kJ of maximum absorbable energy with heights of 4 and 6 m were selected for upslope protection. Fences of greater energies and heights were more difficult to install and required more space. The method of selecting the energy and

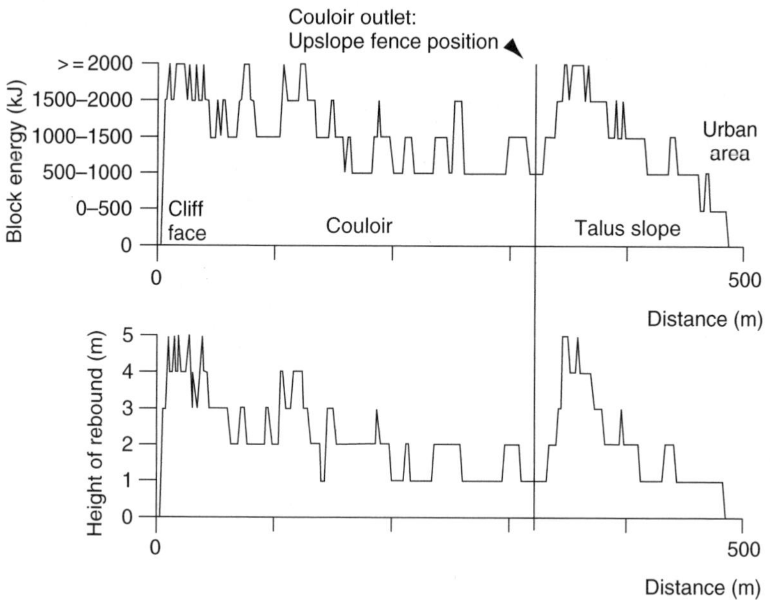


**Figure 23.14** *Example of energies and rebound heights of block profiles along a couloir, from the cliff face to the talus slope. Such profiles allow the identification of the most suitable location of the upslope protective fence (taking into account the lowest energies and rebound heights)*

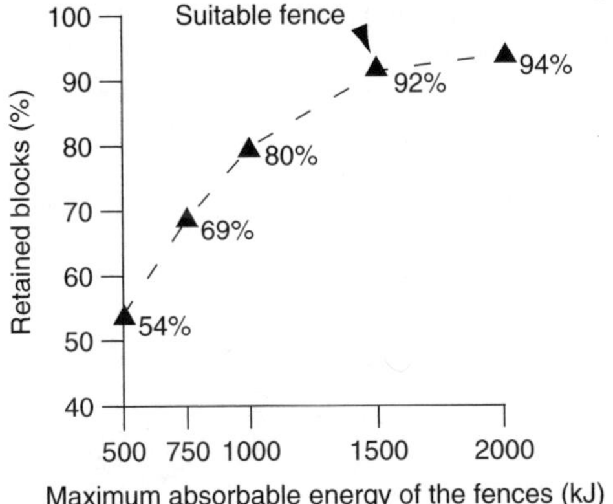



**Figure 23.15** *Plot showing the percentage of simulated blocks retained by protective fences placed at the couloir outlets, with different absorbable energies. The arrow indicates the most suitable design energy of the upslope protective fence (1500 kJ in the example)*

height consisted in using the simulated energies and rebound heights in the location of the protection. The optimum energy of suitable protection is that for which all the fences of maximum absorbable energy do not show a significant improvement in the percentage of retained blocks (Figure 23.15).

## 23.5 Rockfall Surveillance Plan

The rockfall risk management policy at Solà d'Andorra was completed with the Rockfall Surveillance Plan, which was implemented immediately after the Master Plan in 1998. The Surveillance Plan had two main objectives: (i) to identify and document the rockfalls occurring over the years; and (ii) to predict the occurrence of large rockfalls (thousands of m³) for which structural protection is not possible. Identification was carried out by surveying the foot of the slope. Nevertheless, the most useful information is that supplied to the emergency services (fire brigade, police) by eyewitnesses. Thus 16 rockfall events were documented from 1999 to 2003. Prediction was made with additional reconnaissance flights by helicopter. Given that small rockfalls could anticipate the occurrence of larger ones as in the case of Randa in Switzerland (Schindler *et al.*, 1993), subsequent events produced during a relatively short interval in the same sector could be a sign of larger rockfalls.

The Surveillance Plan covers 3 km² of the 5.2 km² of the surface of the rock slope at Solà d'Andorra, where rockfalls could reach urban areas.

### 23.5.1 Documentation of Rockfalls

Each identified rockfall is studied in detail and documented immediately. The documentation considers (a) a detailed study of the rockfall scar; (b) the relationship between the data of the rockfall and the possible triggering factors; and (c) mapping of rockfall trajectories and detached blocks.

The rockfall mapping is compared with the results of the simulation (Figure 23.16) from earlier hazard zoning and protection studies (Amigó *et al.*, 2001). Volumes of detached blocks are used to verify the aforementioned constants in the equation to obtain the corrected volume.

**Figure 23.16**  *Validation of the simulated trajectories (thin lines) compared with trajectories followed by rockfall events (in this example, a rockfall of May 2001)*

### 23.5.1.1   Rockfall scars

Aspects related to the mobilization of the rock mass are observed on the rockfall scar. These features include initiation mechanisms (slide, topple, etc.), the geometry of the mobilized block, possible presence of water, weathered soils in the joints, tree roots and other unstable rocks.

### 23.5.1.2   Triggering factors

In an attempt to identify the factors triggering the rockfall, meteorological conditions as well as possible earthquakes before the rockfall event are taken into consideration. The most common factors triggering rockfalls are rain and temperature, and the effect of vegetation. Many rockfalls occur between the months of November and February when the rains are accompanied by relatively low temperatures though not necessarily below 0°C (e.g. Servei Geologic de Catalunya, 1982). However, a distinct cause–effect relationship between rains and rockfalls cannot be established. On the other hand, between the months of April and May, rockfalls of medium to small size (less than $10\,\text{m}^3$) take place without prior rain or associated earthquakes. Given the presence of tree roots (mainly evergreen oaks and pines) on many scars of small rockfalls, it can be assumed that these contribute to the instability of rock blocks, especially during the months of maximum growth (April and May). Thus the wedge effect of the tree roots could also be regarded as a triggering mechanism of small rockfalls.

### 23.5.1.3   Rockfall mapping

Detailed mapping of rockfalls includes the location of the source, the trajectories and the position of fallen blocks. These data are essential for the verification of the results of the simulation model. Trajectories of the rockfalls are reconstructed from the damaged vegetation and from the impacts on the bedrock and on blocks of the talus slope. When the blocks reach a forested area, the rebound heights are also estimated by measuring the impact heights on the tree trunks. Finally, the fallen blocks can be identified without difficulty because of their broken surfaces. These blocks are mapped in detail and their volumes and morphologies are measured.

The trajectories and the maximum reach of the detached rock blocks are compared with the results of the simulation (Figure 23.16). Given that the maximum reach of the rock blocks depends on their energy, the comparison of the real and simulated maxima enables us to verify the results of the energy calculations of the simulation and the validity of the protection designs (Amigó *et al.*, 2001).

### 23.5.2   Inspection of Rock Slope by Helicopter: Possible Prediction Strategy of Large Rockfalls

Reconnaissance flights by helicopter over the rock slope are carried out at Solà d'Andorra annually between December and February. The aim of these flights is to detect small morphological changes in the rock slope (unusual concentration of small rockfalls, growth of cracks, etc.) which could be interpreted as precursory features of larger rockfalls.

Although the reconnaissance covers the whole outcrop, the large extent of the area does not allow us to study it with the same degree of detail. Therefore, some points or areas of the rock slope with a greater propensity to trigger large rockfalls were selected.

The criteria for this selection include the presence of rock blocks separated by joints or open cracks in the rock massif, recent scars, zones with a high density of small scars and so on. An inventory of these points and areas with the highest probability of triggering larger rockfalls was prepared after the first reconnaissance flight, and is used as a reference for subsequent flights. The photographs taken during different flights are used to identify new rockfall scars, to monitor old scars and to detect morphological changes in the cliff face. Evidence of potentially large rockfall events has not been observed to date.

## 23.6   Final Remarks

The significance of rockfall activity, the high number of buildings at risk and the considerable investment in risk management combine to make Solà d'Andorra a unique example in the world. In this regard, a large number of technical reports have enabled us to make headway in rockfall hazard zoning and design of protective structures.

The Rockfall Risk Management implemented at Andorra la Vella has succeeded in curbing urban growth in a zone highly exposed to rockfall hazard, and in minimizing the risk to the buildings exposed to rockfall.

This policy implemented the following key measures:

- The Rockfall Risk Management Master Plan. This was designed to identify the urban areas exposed to hazard, to establish a boundary line to prohibit development and to provide guidelines for the protection of new buildings exposed to hazard.
- The Rockfall Risk Mitigation Plan. This sought to devise the most suitable protective structures considering the residual risk to the protected buildings, and to provide protective structures for the most exposed buildings.
- The Rockfall Surveillance Plan. This entailed (i) documentation of new rockfall events to verify the results of the rockfall numerical models, and (ii) a strategy to forecast large rockfalls (thousands of m$^3$) for which structural protection is not available.

This policy was complemented by the Building Code Ordinance (BOPA, 2001a), which includes an administrative procedure for issuing building permits in areas exposed to hazard.

Spatial restriction and building codes implemented by the Master Plan are currently in force for new developments. A large part of the Mitigation Plan has been completed. Approximately half of the protective structures, which protect buildings in non-developable areas, are already in place, thus improving the safety of the local inhabitants. The Surveillance Plan is in operation and it is hoped that it will continue to be operational in the future. Awareness of rockfall hazard has been raised by keeping the public informed of the works undertaken.

Rockfall hazard zoning at a detailed scale (1:1000 scale), implemented by the Master Plan, is indispensable for issuing building permits. Our experience shows that a scale less detailed than 1:1000 or 1:2000 is unsuitable for this purpose.

The Surveillance Plan has allowed us to better understand rockfall dynamics in the study area and to upgrade procedures for a quantitative assessment of the residual risk. At present, the data obtained from the Surveillance Plan have not been used to update the results of hazard zoning and protective structure design at Solà d'Andorra. However, we have a database to implement our work at other sites.

The risk management policy implemented at Solà d'Andorra has been satisfactory to date. The success of this management offers considerable hope for its application in other similar scenarios.

## Acknowledgements

The Andorran Ministry for Land Use and Public Works (Andorran government) and the local administration of Andorra la Vella supported the works undertaken at Solà d'Andorra. George von Knorring reviewed the English style of the text.

## References

Amigó, J., Altimir, J. and Copons, R., 2001, Verificación de los resultados del modelo de simulación Eurobloc a partir de casos reales de caídas de bloques rocosos, *V Simposio Nacional sobre Taludes y Laderas Inestables*, Madrid, vol. **2** (in Spanish), 653–663.

Bordonau, J., Vilaplana, J.M. and Fontugne, M., 1993, The glaciolacustrine complex of Llestui (Central Southern Pyrenees): a key-locality for the chronology of the last glacial cycle in the Pyrenees. *C.R. Acad. Sci. Paris*, vol. 316, Série II, 807–813.

BOPA, 1998, Decret relatiu a la caiguda de blocs rocosos a la Solana d'Andorra la Vella i Santa Coloma, *Butlletí Oficial del Principat d'Andorra*, **10**(24) (in Catalan), 750–758. http://www.bopa.ad.

BOPA, 2001a, Decret d'aprovació del Reglament per a la realització de treballs o activitats que modifiquin l'estat actual del terreny, *Butlletí Oficial del Principat d'Andorra*, **13**(71) (in Catalan), 1950–1960, http://www.bopa.ad.

BOPA, 2001b, Llei General d'Ordenació del Territori i Urbanisme (LGOTU), *Butlletí Oficial del Principat d'Andorra*, **13**(10) (in Catalan), 123–175, http://www.bopa.ad.

Copons, R., Altimir, J., Amigó, J., Díaz, A. and Vilaplana, J.M., 2000a, Eurobloc: Un modelo de simulación de caída de bloques y su máxima adaptación a la realidad, *Geotemas*, **1** (in Spanish), 219–222.

Copons, R., Vilaplana, J.M., Altimir, J. and Amigó, J., 2000b, Estimación de la eficacia de las protecciones contra la caída de bloques. *Revista de Obras Públicas*, **3394** (in Spanish), 37–48.

Copons, R., Altimir, J., Amigó, J. and Vilaplana, J.M., 2001a, Estudi i protecció enfront les caigudes de blocs rocosos a Andorra la Vella: Metodologia Eurobloc, *Primeres Jornades del CRECIT (Institut d'Estudis Andorrans), La gestió dels riscos naturals, Andorra la Vella* (in Catalan), 134–149, http://www.iea.ad/crecit/primeresjornades.html.

Copons, R., Altimir, J., Amigó, J. and Vilaplana, J.M., 2001b, Metodologia Eurobloc para el estudio y protección de caídas de bloques rocosos. Principado de Andorra, *V Simposio Nacional sobre Taludes y Laderas Inestables, Madrid*, vol. **2** (in Spanish), 665–676.

Corominas, J., 1996, The angle of reach as a mobility index for small and large landslides, *Canadian Geotechnical Journal*, **33**, 260–271.

Corominas, J., Esgleas, J. and Baeza, C., 1990, Risk mapping in the Pyrenees area: a case study, *Hydrology in Mountainous Regions II*. IAHS Publication 194, 425–428.

Corominas, J., Copons, R., Vilaplana, J.M., Altimir, J. and Amigó, J., 2002, Hazard assessment and management experience in the Principality of Andorra, International Conference on Instability Planning and Management, Ventnor, Isle of Wight, UK. R.G. McInnes and J. Jakeways (eds), *Instability Planning and Management* (London: Thomas Telford), 687–694.

Corominas, J., Copons, R., Vilaplana, J.M., Altimir, J. and Amigó, J., 2003a, Integrated landslide susceptibility analysis and hazard assessment in the Principality of Andorra, *Natural Hazards*, **30**(3), 421–435.

Corominas, J., Copons, R., Vilaplana, J.M., Altimir, J. and Amigó, J., 2003b, in press, From landslide hazard assessment to management, The Andorran experience, *International Conference on Fast-Slope Movement, Prediction and Prevention for Risk Mitigation, IC-FSM2003*, Naples.

Cruden, D. and Fell, R. (eds) 1997, *Landslide risk assessment, Proceedings of the Workshop on Landslide Risk Assessment*, Honolulu, Hawaii, USA, 19–21 February 1997 (Rotterdam and Brookfield, USA: A.A. Balkema).

Escalé, J., 2001, La nova llei d'ordenament territorial a Andorra, *Primeres Jornades del CRECIT (Institut d'Estudis Andorrans), La gestió dels riscos naturals, Andorra la Vella* (in Catalan), http://www.iea.ad/crecit/primeresjornades.html.

Evans, S.G. and Hungr, O., 1993, The assessment of rockfall hazard at the base of talus slopes, *Canadian Geotechnical Journal*, **30**, 620–636.

Fell, R. and Hartford, D., 1997, Landslide risk management, in D. Cruden and R. Fell (eds), *Landslide Risk Assessment* (Rotterdam: A.A. Balkema), 51–109.

Ferrer, P., Furdada, G. and Vilaplana, J.M., 2000, Organización e Informatización del Catastro de aludes de Andorra, *Geotemas*, **1**, 173–176.

IUGS Working Group on Landslides, 1997, Quantitative risk assessment for slopes and landslides, in D. Cruden and R. Fell (eds), *Landslide Risk Assessment* (Rotterdam: A.A. Balkema), 3–12.

Jaboyedoff, M., Philippossian, F., Mamin, M., Marro, Ch. and Rouiller, J.D., 1996, *Distribution spatiale des discontinuités dans une falaise, Approche statistique et probabilistique*, PNR 31, Hochschulverlag AG an der ETH Zürich (in French).

Krummenacher, B. and Keusen, H.R., 1997, Steinschlag-Sturzbahnen – Modell und Realität, *Instabilités de falaises rocheuses, chutes de blocs at ouvrages de protection, Steinschlag*, Publications de la Société Suisse de Mécanique des Sols et des Roches (in German), 17–23.

Lateltin, O., 1997, *Recommandations: prise en compte des dangers dus aux mouvements de terrain dans le cadres des activités de l'aménagement du territoire*, OFAT, OFEE and OFEFP, Switzerland (in French).

López, C., Ruíz, J., Amigó, J. and Altimir, J., 1997, Aspectos metodológicos del diseño de sistemas de protección frente a las caídas de bloques mediante modelos de simulación cinemáticos, *IV Simposio nacional sobre taludes y laderas inestables*, Granada, vol. **2** (in Spanish), 811–823.

Moya, J., 2002, *Determinación de la edad y de la periodicidad de los deslizamientos en el Prepirineo Oriental*, PhD. thesis, UPC, Barcelona (in Spanish).

Raetzo, H., Lateltin, O., Bollinger, D. and Tripet, J.P., 2002, Hazard assessment in Switzerland – codes of practice for mass movements, *Bulletin of Engineering Geology and Environment* **61**, 263–268.

Rouiller, J.D. and Marro, Ch., 1997, Application de la méthodologie 'MATTEROCK' à l'évaluation du danger lié aux falaises, *Eclogae Geol. Helvetica*, **90** (in French), 393–399.

Rouiller, J.D., Jaboyedoff, M., Marro, Ch., Philippossian, F. and Mamin, M., 1998, *Pentes instables dans le Pennique valaisan, Rapport final du Programme National de Recherche PNR 31/CREALP 98*, vdf Hochschulverlag AG, Zürich (in French).

Schindler, C., Cuénod, Y., Eisenlohr, T. and Joris C.L., 1993, Die Ereignisse vom 18. April und 9. Mai 1991 bei Randa (VS) – ein atypischer Bergsturz in Raten, *Eclogae Geol. Helvetica*, **86**(3) (in German), 643–665.

Servei Geològic de Catalunya, 1984, *Efectes geomorfològics dels aiguats del novembre de 1982*. Informes no. 1. Departament de Política Territorial i Obres Públiques, Barcelona (in Catalan).

Wyllie, D.C. and Norrish N.I., 1996, Stabilisation of rock slopes, in A.K. Turner and R.L. Schuster (eds), *Landslides: investigation and mitigation*, Special Report 247, Transportation Research Board, National Research Council, Washington, DC, 474–504.

# 24

# Landslide Risk Assessment in Italy

*Marino Sorriso-Valvo*

## 24.1 Introduction

Landslides have always been a major problem for land management in Italy. Much attention has been devoted to this problem and to geomorphological, hydraulically triggered disasters in general, especially since the beginning of the twentieth century (Almagià, 1910; Catenacci, 1992; Guzzetti, 2000). In the period after 1860, some order was brought into the study of the problem with the establishment of the Civil Engineer Corps (Genio Civile). From historical records, it appears that in the twentieth century extreme rainfall events hit Italy at 5- to 10-year intervals, triggering numerous landslides all over the country. As regards landslides triggered by natural events, the most catastrophic year was probably 1951, when violent storms caused floods and landslides resulting in hundreds of casualties in northern and southern Italy.

Landslide incidence is very high. The exact number of cases is still unknown because certain parts of Italy have not yet been studied. However, the number of the various types of mass movement of some relevance amounts to tens of thousands. The percentage of total area affected is given as slightly more than 10%, but this is possibly an underestimation. In some extended areas where particularly weak rocks crop out, the percentage has been estimated to reach 95% (Sorriso-Valvo, 1993; Conaker and Sala, 1998). Figures 24.3 to 24.9 illustrate some examples of mass movement in Italy.

Disasters remain in the memory and in the tradition of the people. The persistence of this memory maintains awareness of the threat posed (Sorriso-Valvo, 1988); there are also historical records, where large-scale landslides are frequently recorded, though the smaller ones tend to be omitted, even if they caused fatalities (Guzzetti, 2000).

*Landslide Hazard and Risk*   Edited by Thomas Glade, Malcolm Anderson and Michael J. Crozier
© 2004 John Wiley & Sons, Ltd   ISBN: 0-471-48663-9

In Italy disastrous landslides are common: in the nineteenth and twentieth centuries, different types of large-scale landslides occurred involving large-scale loss of life, such as those at Mount Antelao (Veneto, 1814, 1925, Figure 24.8), Ancona (Marche, 1856, 1921, 1982); Costiera Amalfitana (Campania, 1910, 1924, 1953); South Calabria (1951, 1953); Vajont (Veneto, 1963); Pisticci (Lucania, 1968); Stava (Veneto, 1985); Valtellina (Lombardia, 1987); Sarno (Campania, 1998). The death toll in the twentieth century alone is estimated to be more than 5600; from 1410 to 1999, 996 recorded landslides occurred causing 12 412 victims (Guzzetti, 2000).

Many landslides, some of which are not included in the Guzzetti (2000) records because of difficulties in separating the victims from landslides from those caused by earthquakes, occurred during or soon after large earthquakes (Casamicciola, 1883; Calabria, 1905; Messina, 1908; Friuli, 1976; Irpinia, 1930, 1980); the most disastrous combined effects of earthquakes and consequent landsliding occurred in 1783 in southern Calabria, where more than 300 lakes were formed by hundreds of huge landslides that dammed the streams. Entire villages disappeared under gigantic landslides; there were more than 30 000 fatalities, but it is impossible to know how many of these were due to landslides only. After the earthquakes, there was an outbreak of malaria, and the fatalities were comparable in number to those killed by earthquakes and landslides directly (De Dolomieu, 1785).

The greater part of the economic cost of landslides in Italy, however, is due essentially to the 'normal-scale' phenomena that affect the country every year. Delmonaco *et al.* (2002) assessed landslide susceptibility at the national scale, and found that most of Italy (roughly 60%) has an intermediate to very high degree of susceptibility to landslides; exceptions are most of the lower Po Valley, Apulia, and parts of Sardinia, Lazio, Sicily and Friuli.

After World War II, the attention paid in Italy to landslides and floods (split data are not available) can be measured by the amount and cost of legislation between 1945 and 1990. These data are shown in Table 24.1.

It was only after the flood in Florence in 1967 that Italian governments made a serious legislative and financial effort (Table 24.1) and decided to set up a committee to analyse the problem of geomorphological hazards in depth. It produced a report that was the starting point for a new approach to geomorphological disasters in Italy – the De Marchi Report.

Though a valuable document, this report was responsible for a semantic mistake, using the adjective *hydrogeologic* instead of *hydro-geologic* or *hydro-geomorphologic*, as a comprehensive term to describe the disasters caused by water on the surface of the earth. Indeed, there is no one generally adopted term to indicate the complex of phenomena of mass movement, erosion and flooding. All of these, in fact, represent different aspects of the same problem – soil conservation and land management. The use of the specialized

**Table 24.1**  *Legislation and cost of interventions to reduce landslides and flooding in Italy (Catenacci, 1992, reproduced by permission of Agenzia per la Protezione dell'Ambiente e per i Servizi Tecnici)*

|  | 1945–50 | 1951–60 | 1961–70 | 1971–80 | 1981–90 |
|---|---|---|---|---|---|
| Number of legislative acts | 4 | 11 | 10 | 10 | 56 |
| Funds (million euro) | 50 | 1600 | 8500 | 2500 | 4100 |

term instead of a general one is still widespread in the Italian lexicon, even in scientific texts (Sorriso-Valvo, 2002). This problem is mentioned here in order to make non-Italian readers aware of what is really meant by '*rischio idrogeologico*' (hydrogeological risk).

Crescenti (1998) observed that, especially in technical reports and in some documents of a legal character, another Italian term is used in a way that may convey a different idea with respect to what is really intended: *pericolosità*, which is translated as hazard, but whose real meaning may be different. Indeed, in most of these works, the evaluation of *pericolosità* is performed by combining the frequency of recurrence with the magnitude of phenomena, obtaining highest values for most frequent and most intense ones. The difference is quite significant, since hazard, as obtained from probabilistic models, is always shown to be in inverse proportion to the magnitude of the phenomenon.

In this section, the terminology adopted in this chapter is briefly illustrated. Except for 'danger', all the other terms are used with the meaning currently used in the literature (e.g. Canuti and Casagli, 1994). In order to avoid confusion when referring to Italian works, the correct terms will be used in this text, and any differences with Italian terms will be described.

**Susceptibility**: proneness to a given phenomenon. For landslides, it includes the territorial or site characteristics (rock type and structure, geological setting, slope angle and form, groundwater conditions, vegetation cover) but does not include the external causative factors and the time dimension.

**Magnitude**: parametric expression of the intensity of a given event.

**Hazard**: the probability that a given event of given magnitude will occur within a fixed time interval.

**Danger**: measurement based on a rank scale of the probability of occurrence of an event, combined with its magnitude and damage potential. In this chapter, it may correspond to Italian *pericolosità*.

**Danger or hazard indicators**: elements used to define the degree of danger or hazard.

**Vulnerability**: percentage of loss of value of a given good, or of number of lives, when affected by a given event of given magnitude.

**Risk**: hazard times vulnerability times value of good at risk. It is expressed in terms of money or loss of lives. Normally the fixed time interval for hazard is one year.

**Specific risk**: risk relative to a given type of phenomenon.

**Total risk**: risk due to a given causative event (an earthquake, a storm).

## 24.2   Evaluation of Landslide Hazard – Some Shortcomings

One crucial point when dealing with the landslide hazard of natural slopes is that its evaluation is very complex and difficult, if not impossible, because probabilistic models of recurrence–magnitude relationships cannot be established due to the lack of historical records and difficulties in defining the magnitude of phenomena. Sorriso-Valvo (2002), Guzzetti *et al.* (1999) and other authors all agree on the difficulty of this task.

The interested reader is referred to the literature on this topic, and, of course, to the relevant chapters in this book.

For the single event, geotechnical slope models provide a much better definition of the hazard, especially for artificial slopes, whose internal structure is much better known than that of natural slopes (Aleotti and Chowdhury, 1999). Hazard assessment for a single event, however, is very expensive, and an extended study based on complete slope modelling is not economically viable for extended areas. Notwithstanding this, the path to a sounder definition of the landslide hazard probably lies in an approach that combines some simplified single-event procedures with statistics-based spatial analysis adopted for diffuse events.

Here I recall a few points with regard to diffuse events, as this is the most critical area due to the need to enforce recent legislation. The following points help to explain why the assessment of landslide hazard, except for the case of artificial slopes, lags behind the evaluation of other hazards, for example floods or earthquakes.

1. Rainfall is by far the major cause of landslides, both as a predisposing and a trigger factor. However, the effects of rainfall are influenced or controlled by a number of other factors (e.g. tectonic activity, geological structure, local climate, other physical morphogenetic processes, biological activity, non-biological human activity, and so forth) so that understanding the relationships between all these is extremely complex, and one landslide is different from another.

   The above holds true for all inhabited or potentially inhabited areas of the earth, but landslides can also occur in deserts and arid zones, and rainfall may be a factor, though not necessarily the only or even the most important one.

2. Seismic shaking is the second most important cause of landslides. Several studies have found a number of common characteristics of earthquake-triggered landslides worldwide (among others Agnesi *et al.*, 1983; Keefer, 1984; Gringeri Pantano *et al.*, 2002; Sorriso-Valvo, 1986; Updike *et al.*, 1988). Recent studies present methods for the evaluation of hazard from seismically triggered landslides (e.g. Jibson *et al.*, 2000).

3. Volcanic activity is another trigger factor for landslides (e.g. Schuster, 1984; Kadomura *et al.*, 1983). Though active volcanoes are not usually found in inhabited areas, those that do pose a very serious threat to life and property.

4. Although evidence for the contemporaneous and complex interplay of different factors in causing landslides has long been recognized (e.g. De Dolomieu, 1785), the multiple sources of landslide hazard are generally underestimated.

5. As mentioned before, time series of trigger events related to landslides of known magnitude are scarce, even though detailed rainfall time series are available. The problem is that distribution is very different, even for the same rainfall, from one site to another, and thus the rainfall data of the actual landslide site should be used, but, in some papers, correlations are made between landslide movements and (non-specific) rainfall series (Sorriso-Valvo *et al.*, 1994).

6. Another problem is the definition of the magnitude of a landslide. For landslides there are various proposals (Aleotti and Chowdhury, 1999; Sorriso-Valvo, 2002) but general agreement is lacking. Canuti and Casagli (1994) proposed that the magnitude of landslides be measured on the basis of their velocity. Guzzetti *et al.* (1999) suggest the adoption of different criteria for the evaluation of magnitude for the different types of landslides, for example volume, velocity, momentum or kinetic energy. This method is

perhaps more suitable for a description of the process, but if the concept of magnitude has to convey the idea of danger, velocity should be included in the criteria to express the event accurately.

For diffuse landslides, the concept of magnitude should include the spatial dimension. In this case magnitude should refer to all the triggered events, for example the whole volume or the whole loss of energy.

7. Every type of mass movement responds differently to the different predisposing and trigger factors. Thus, for each type of mass movement a different type of model for evaluating the hazard must be adopted. As several types of mass movement may coexist over a given territory, a hazard evaluation for any given type must be performed, making the whole task complex, time-consuming and difficult to handle without the help of a GIS.

8. Only a few types of mass movement can be considered as stationary processes, for example translational slide as long as the displacement of the mass does not imply relevant changes in the conditions along the shearing surface; or a recurrent debris flow as long as the overall morphology of the involved slope or area does not change significantly.

For diffuse landslides, one could consider a territory where geological, geomorphological and climatic conditions do not change within the considered time span as an invariant system upon which a number of processes occur that on the whole can be regarded as a single stationary process.

9. With diffuse landslides there are problems with mapping. Guzzetti *et al.* (2000) showed that landslides can be mapped cost-effectively by means of aerial photo interpretation and field surveys, but the degree of uncertainty still remains high, especially if the field survey is not carried out meticulously.

Based on my experience, however, it is worth reconsidering the practicality of landslide hazard evaluations. Indeed, in my experience, no reliable hazard assessment has ever been achieved. At the present time, given the general lack of confidence in hazard evaluation, it is wiser either to abandon or leave unused landslide-affected territory, or to adopt measures to eliminate or at least reduce hazard.

On the research front, instead, one can be optimistic, as most of the above-mentioned shortcomings are close to being overcome by the use of new technology, such as artificial intelligence methods (Fernández-Steeger *et al.*, 2002; Gomez and Kavzoglu, 2003).


## 24.3   The Evaluation of Diffuse Landslide Hazard in Italy – a Historical Overview

The following few examples are briefly presented in order to give an idea of the present state of the art for diffuse landslide hazard (danger) evaluation in Italy.

Pioneer works in Italy date back to the 1970s (Carrara, 1983; Carrara *et al.*, 1978, 1982a, 1982b, 1992). The first known risk evaluation is in Carrara *et al.* (1982a), where the interference of mass movements with land use is depicted by means of a simple land classification procedure. This study contains one of the earliest examples of the use of computer-assisted techniques for land evaluation.

These studies were carried out within the framework of the Progetto Finalizzato 'Conservazione del Suolo' ('Soil Conservation Project'), a scientific initiative under the auspices of CNR (National Research Council) that financed tens of teams from universities, research agencies, public and private bodies over a period of 5 years, with aim of bringing up to date scientific knowledge of the country's hydrogeomorphological hazards. The results were good, albeit incomplete. At the beginning of 1990s, the Dipartimento per la Protezione Civile (Civil Protection Department, a government agency) reached an agreement with CNR, on the basis of which three national research groups were set up for the study of volcanic, seismic and hydrogeomorphological hazard. The last of these is still active, and includes more than 70 teams working on floods, landslides and erosion. This experience of coordinated nationwide research into the problem of landslides involving virtually all universities and research agencies is unique on the international scene.

In these pioneer studies, evaluation procedures were based on statistical analysis. The best results were obtained by means of multivariate statistical analysis (Carrara *et al.*, 1982a, 1992; Carrara, 1983), even though, in spite of the high significance level of statistical procedures (Carrara *et al.*, 1982b), in many cases the scores for correct classifications were only slightly higher than what would have been expected by chance. For example, Carrara *et al.* (1982b), by means of discriminant analysis, classified 73% of the land unit parcels of the River Ferro basin correctly as stable/unstable, while the expected percentage of cases correctly classified by chance is 50%. The best result was obtained by Carrara *et al.* (1982a), who, by means of the same procedure, assigned the various landslides to the four appropriate zones. Cases correctly assigned were from 49.7% to 87.9% (vs 25%). These scores, however, remain lower than 95%, that is, statistical certainty. In other words, these procedures are suitable for assessing the influence of some territorial characteristics, on the characteristics and territorial distribution of landslides, but cannot be used, so far, for sound land classification in terms of landslide hazard.

Statistical results of the relationships between landslide occurrence and land features can also be used for land classification in a much easier way than complex multivariate statistics. For example, in the $750\,\mathrm{km}^2$ extended area of the National Park of Aspromonte, sites with differing degrees of danger due to debris flows (of various sizes) and soil slips were surveyed by means of simple statistical analysis (Sorriso-Valvo, 2001). The probability is obtained from the observed relative frequency of occurrence of any given type of landslide over any given type of rock, according to the slope gradient. Class limits of slope gradient are different for different types of rock and the maximum frequency of landslides occurs on intermediate slope gradients. The slope gradient was derived by a 40 m square-grid DEM. The ranking of danger ($D$) was divided into five classes between 0 and 4 in order to match the ranking of hazard stipulated in the recent legislation (see section 24.1). As an example, the classification of landslide danger for a small part of the Aspromonte National Park is shown in Figure 24.1 (see also Colour Plate section, Plate 11).

Recently, for land classification based on the occurrence of one modality of a dichotomous variable (i.e. landslide/non-landslide), logistic regression has been adopted in order to avoid unrealistically negative values for probability of occurence (Bernknopf *et al.*, 1988). This procedure was used in a few case studies in Italy (Carrara *et al.*, 1992) and is being tested in current case studies in southern Italy.

When dealing with the assessment of characteristics of a system of elements that are not measurable on parametric scales (e.g. rock type, local geomorphological setting, aspect, vegetation cover, land use and so forth), the human brain is much more effective than a computer. This can be simulated by means of artificial intelligence methods. One of

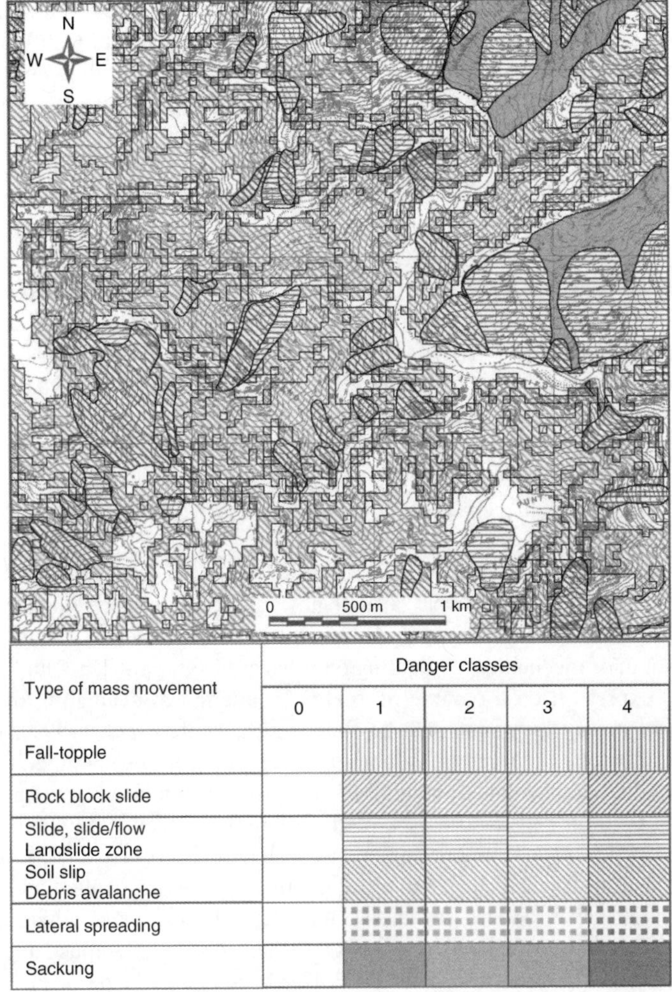



| Type of mass movement | Danger classes | | | | |
|---|---|---|---|---|---|
| | 0 | 1 | 2 | 3 | 4 |
| Fall-topple | | | | | |
| Rock block slide | | | | | |
| Slide, slide/flow Landslide zone | | | | | |
| Soil slip Debris avalanche | | | | | |
| Lateral spreading | | | | | |
| Sackung | | | | | |

**Figure 24.1**  *Part of land classification of* Parco Nazionale dell'Aspromonte *(Aspromonte National Park, South Calabria) as regards landslide danger, for all the different types of mass movement. Cell size of square-grid DEM is 40 m. In case of superposition of different danger classes, the highest one is shown in the display. The same map can be displayed for each layer (type of mass movement). From Sorriso-Valvo (2001), reproduced with permission of* Ente Parco Nazionale dell'Aspromonte. *(See also Plate 11)*

these is the neural networks (NN) method (Aleotti and Chowdhury, 1999), recently used for forecasting the evolution of lakes created by landslides (Casagli and Ermini, 2001).

As regards rainfall as a trigger of landslides, the Italian literature on the subject is too extensive to give an exhaustive picture here. Nevertheless, it is worth referring to the FLaIR model (Sirangelo and Versace, 1992; Sirangelo *et al.*, 1996; Versace *et al.*, 2000), a sophisticated model for the Forecasting of Landslides Induced by Rainfall. Forecasting can be considered as a short-term, conditioned hazard evaluation made on the basis of probability of occurrence. The conditioning event is the trigger event – rainfall. FlaIR is most suitable for single case studies; it represents a generalization of almost all the known empirical forecasting models. The advantage of FLaIR is essentially in its standardization of procedures, especially important for warning systems. The model has sufficient flexibility and reliability to deal with a large number of types of landslides. There is still the limitation that, especially when one single mobilization event is known, the landslide should be a stationary phenomenon, which is not generally the case. This model, however, can represent the link in the chain that connects the procedures for the assessment of hazard for single cases with those for diffuse landsliding, as the information required for each case needs to be precise but is limited to a few parameters only.

In Italy, special attention has been devoted in the last two decades to large-scale landslides and deep-seated gravitational slope deformations. Two types of dangerous phenomena are included in this category (Sorriso-Valvo, 1984a, b; Oyagi *et al.*, 1994): large-scale, often extremely rapid, complex landslides of the rock slide–debris avalanche (or rock avalanche) type, and the deep-seated rock creeping (Sackung and lateral spreading). A Sackung may be the long-lasting creep deformation that represents the preparatory stage of a catastrophic collapse (Guerricchio and Melidoro, 1973; Emery, 1978; Oyagi *et al.*, 1994).

A national study group has been active for more than 10 years (Sorriso-Valvo, 1984b; Crescenti and Sorriso-Valvo, 1995), so that now most researchers and technicians are aware of the presence and the peculiarities of these phenomena. As little is known, however, about their mechanics, evaluating their hazard is very difficult, and it is virtually impossible to express it on a parametric scale. An attempt at evaluating the hazard for this category of mass movements can be found in Antronico *et al.* (1999), who studied the various phenomena observed in Calabria (southern Italy). They adopted three specific risk classes on the basis of the summation of specific and standardized danger indicators.

In the field of remote sensing, the new methodologies and technology being developing in Italy have opened up promising perspectives. So far, however, this promise has not been translated into practical use. Of some interest for hazard evaluation and warning is the differential interferometry (DIFSAR) technique, by means of which variations in elevation of one given point can be measured. Such differences must not exceed one wavelength (from a few mm to a few cm). This technique can be used for surveying and monitoring small precursory ground and/or building movements, over vast areas with low gradient and suitable aspect, which may be associated with imminent landslide movements. Problems arise from the fact that not all sites may be suitable, and movements may result from other causes, such as variations in groundwater levels and tectonics (Spacebel *et al.*, 2001).


## 24.4    Italian Laws for Downgrading Landslide Hazard

Due to the high incidence of landslides, and of hydrogeomorphological events in general in Italy, various attempts have been made to reduce the risk to people and property, often as an emotive response to major disasters. A more rational legislative approach was adopted, however, in the second half of the 1990s, when several decrees were passed and eventually a comprehensive law was enacted.

The following section gives an outline of the more relevant and recent legislation.

### 24.4.1   Law No. 183, 1989

This is the first comprehensive law issued in Italy on land management and soil protection. It enforces rules for the reorganization of soil defence (within the widest sense of the term). It is aimed at ensuring the protection and reclamation of land and water, the use and management of water supplies for social and economic development, and the protection of the environment. The river basin is the reference territorial entity.

The main tool for the fulfilment of these requests is the 'Basin Plan', a comprehensive planning act. A Basin Authority (to be set up) is charged with implementing such a plan.

Only a small part of the activity prescribed by the law had been accomplished in May 1998, when a number of killer mudflows occurred in the Sarno area, near Naples (see section 24.4.3).

As a result, the government decided to pass a much more stringent measure, decree no. 180.

### 24.4.2   Law No. 493, 1993

This law sets up the Basin Authorities throughout the country, defining their role and obligations. The Basin Authority (BA) oversees other agencies, including local government, as regards land planning and management. There are three types of Basin Authority, depending on their territorial setting:

 (i) national BA: these are the most important ones, as they concern large river basins and extend over different regions;
 (ii) interregional BA: they concern one or more small river basins and extend over two regions or more;
(iii) regional BA: they concern several small river basins, as in the case of Calabria and Sicily, and are normally confined to one region.

The work of Basin Authorities is carried out through the 'Basin Plan', a comprehensive programme of land management projects.

### 24.4.3   Decree No. 180 of 11 June 1998, 'Sarno', and Law No. 267 of 3 August 1998

In May 1998 a violent storm in Campania triggered tens of mudflows, some of which were of considerable size. Fatalities amounted to more than 160. Flooding also occurred but its effects were limited.

Under pressure from public opinion, a government decree was quickly enforced to deal with the problem of hydrogeomorphologic disasters. Moreover, stringent deadlines were laid down before which all the Basin Authorities of Italy had to:

1. complete the reconnaissance of geological–hydraulic phenomena threatening property and lives in Italy;
2. evaluate the hazard levels in correspondence to the property at risk;
3. assess the risk;
4. enforce urgent rules for land use and management, including urban plans.

Taken as a whole, these acts constitute the 'Piano di Assetto Idrogeologico', that is, the Hydrogeomorphological Setting Plan (HSP) which will be illustrated in section 24.4.4.

In other words, this decree set a definitive deadline for the practical application of the 1989 law (no. 183) that had not been previously applied. This legislation was confirmed by another law (no. 267) on 3 August 1998, which contained technical directions. Local authorities were allowed funding for interventions only if they were in accordance with the schedule laid down by the law.

The aim of such legislative acts was important and laudable, except for the fact that the deadline posed was so stringent that reliability was sacrificed. In fact, all the procedures had to be completed in less than two years, even in areas where there was limited knowledge on geomorphological phenomena.

There was heated debate at conferences on this decree. Most of the well-informed researchers cast doubt on the outcome; however, others maintained that it was important to rise to the challenge, and the adventure went on.

Unfortunately another disaster occurred in September 2000 in Calabria, where after a violent summer storm, several small landslides along the Ionian coast of southern Calabria caused widespread damage and, more seriously, a flood killed 19 people on a camp site on the bankfull flood bed of the Soverato torrent.

Politicians reacted again by enforcing even more stringent terms for the accomplishment of the work required by the 1998 legislation, creating more and more problems for those regions where the work was already behind schedule. Note that the regions involved are those that have the most need for funding for structural and non-structural interventions.

The work was demanding, but by 31 December 2001, a number of Basin Authorities had completed the requirements laid down in the decree (no. 180), and most of the remaining ones finalized their task in the following months.

### 24.4.4   The Hydrogeomorphological Setting Plan (HSP)

The HSP consists in land classification on the basis of risk from hydrogeomorphological hazards. It is the most important part of the whole procedure set up by the legislation on this issue. It is the preliminary, basic part of the Basin Plan. The agency in charge of implementation is the Basin Authority (BA); however, local authorities (from the municipal to regional level) must comply with it or submit their technical observations and proposals for modifications that must be approved by BA. After this procedure, the HSP is put into action and all agencies must conform to it. However, as knowledge progresses and situations change, the HSP is subject to continuous upgrading and updating.

### 24.4.4.1   Aims of the Hydrogeomorphologic Setting Plan

The aims of the HSP are to start a rational planning process and to establish an appropriate and effective soil defence policy at the basin/regional level.

Due to delays across the country in the application of laws to reduce hydrogeomorphological hazard, there is an urgent need for prompt action; thus the first stage of application of HSP entails:

- identifying priority sites/areas to be restored to sustainable land use;
- evaluating the hazard, identifying and evaluating the elements at risk, and risk assessment. Risk is ranked in four grades (R1–R4) that can be summarized as follows:
  - R1 – Low risk – possibility of limited damage to persons and property;
  - R2 – Intermediate risk – possibility of minor damage to buildings, communication lines and lifelines, people affected but do not lose their capacity to react;

- R3 – High risk – serious threat to public safety; damage to constructions, communication lines and lifelines that put them out of use; interruption of economic and industrial activity; heavy damage to natural resources;
- R4 – Very high risk – possibility of loss of human life; irreparable damage to buildings, communication and lifelines; permanent stop of productive activity.

- implementing interventions and monitoring their effects;
- updating the programmes on the basis of the observed effects and the new demands.

In the subsequent phase, risk evaluation is extended to all urban and industrial areas. This is to be accomplished by means of:

- GIS and database implementation;
- surveys and mapping of mass movement, including avalanches, at a scale not less than 1:25 000;
- evaluation of the possible extent of damage of different types of phenomena;
- evaluation of the hazard and vulnerability of elements at risk, depending on the different types of mass movement;
- assessment of risk from landslides to elements at risk, depending on the type of landslide.

No precise methods are indicated for the assessment of hazard, vulnerability and risk. It will be shown in the following section that this has resulted in a rather uneven classification at the national scale.


## 24.5   Enactment of Law 267/1998 – Some Case Studies

The application of law 267 of 1998 required a great effort. However, almost all the Italian Basin Authorities have completed the implementation of their HSP. Few of them, however, have made the relative documentation easily accessible; in addition, not all the information is available, as most of Basin Authorities do not illustrate the methods applied, but just the results. Therefore, only a few cases selected from those for which information is readily available will be illustrated and discussed in this chapter. It is worth stressing here that, as regards the aims of this chapter, the focus is on the methodology rather than the results. Furthermore, due to the differences in territorial competence of the different types of basin authorities (see section 24.4.2), the various studied territories are differently defined (one or more river basins, a single region, or a part of it).

The contents of the various HSPs, in general, are similar, as they are based on the rules stipulated in the legislation. There are, however, some differences, which can sometimes alter the final picture significantly, due to the differences in environmental conditions, the methodology adopted, and initial background knowledge of the area in question. As a consequence, the conclusions of each HSP tend to vary considerably.


### 24.5.1   The Calabria Region

Calabria is the southernmost stretch of the Italian peninsula. It extends over an area of *c.* 15 000 km². Its southern and central parts take the form of an arc-shaped, narrow land with high relief and rugged morphology, with a geology characterized by an allochtonous crystalline core bearing, at sites, sedimentary covers ranging in age from Mesozoic to Palaeogene.

In the northern part, there are outcrops of extensive carbonate sediments bearing thick layers of low-grade metamorphic rocks, volcanites and clay-rich sediments. These terranes are tectonically covered by flysch and tectonic melange formations Mesozoic–Tertiary in age. These are also to be found beneath the crystalline tectonic units of central Calabria, where they crop out in scattered tectonic windows.

Upon the crystalline and carbonate units, Neogene sediments were deposited in several basins.

Quaternary block faulting gave the territory its present aspect, with the morphology strongly controlled by tectonics.

The typical high and seasonal nature of the rainfall, with a wet period in winter, together with the intense seismicity, are the two major triggers of landslides. For more details refer to Sorriso-Valvo (1984a, 1993) and Sorriso-Valvo and Tansi (1996).

Due to its morphology, there are no large river basins in Calabria. The largest is that of the river Crati, which extends over an area of only $c$. $600\,km^2$. For this reason there are no single-river BAs, but one single regional BA that includes almost all the surface of Calabria, except for a few hundred $km^2$ that belong to three river basins that extend over a part of another region. These latter basins, therefore, are under the control of other BAs. Calabria was one of the first regions to fulfil the requirements of the legislation regarding BAs. All the following information regarding Calabria's HSP comes from Autorità di Bacino della Calabria (2001).

The procedures for implementing the HSP in Calabria for landslides were those indicated in the legislation. The technical procedures were elaborated in cooperation with the Institute of Research for Hydrogeological Protection (IRPI) of the National Research Council (CNR) and other consultants. The work was performed by a team appointed by the Basin Authority.

At the beginning of the process of implementation of the HSP, the GIS and relational database contained:

- literature and archive-derived data:

    - on mass movement
    - on damage
    - on countermeasures undertaken;

- geological, geomorphological, land use, urban e-maps;
- urban plans at different levels;
- local regulations.

During the work, the following data sets were stored and filed:

- Mass movement mapping at 1:10 000 scale.
- Implementing the landslide maps in the GIS.
- Compilation of a field form for each landslide. The form includes more than 100 variables concerning landslide and slope morphometry, slope geology, landslide characters, land use, damage, countermeasures, historical information, and so forth. The data are filed in a relational database linked to a the regional GIS.

A great deal of data was available from studies made over the last 30 years by research institutions, such as the IRPI of the CNR, the University of Calabria, and others. As these studies had been made for different purposes using different scale maps, it was necessary to standardize the information by means of aerial photo interpretation and field work.

The very stringent legal guidelines, however, made it necessary to consider even those areas where there is only slight evidence of the presence of mass movement. This choice was reinforced by the fact that the law made it very clear that doubtful cases should be included in the interests of safety. The result is the inclusion of a number of landslides which are redundant with respect to the real situation on the ground. However, as the HSP is to be continuously updated, this problem can be addressed at a later stage.

Landslide hazard was evaluated by means of a rather simple procedure, consisting of:

- Ranking danger by activity (five classes);
- Ranking magnitude by type of mass movement and velocity (four classes);
- Combining activity and magnitude in five hazard levels (two levels for Sackung and Jahn's Lateral Spread).

Table 24.2 shows the ranking of the *pericolo* ('danger') whose grade is based here on recurrence only. Thus the term *pericolo* does not correspond to 'danger' as used in this paper (see section 24.1), as it does not include magnitude. Recurrence was ranked on the basis of historical records (if available), multitemporal aerial photo interpretations and field evidence.

For the evaluation of danger (called *Indice di Pericolosità* = Hazard Index), the magnitude of events is also introduced in a two-entry matrix. To this end, a simple ranking of magnitude, expressed by the velocity of the process, was adopted, as shown in Table 24.3.

**Table 24.2**  *Ranking of* pericolo *levels*

| | Degree of activity | | | | |
|---|---|---|---|---|---|
| | Active, suspended | | dormant | Inactive | |
| Recurrence interval (years) | 1–10 | 10–100 | >100 | Undefined | Undefined |
| *Pericolo* rank | Very high | High | Intermediate | Low | Null |
| *Pericolo* class | D4 | D3 | D2 | D1 | D0 |

**Table 24.3**  *Ranking of the magnitude of landslide types*

| Type of mass movement | Magnitude rank (range) |
|---|---|
| Fall | V4 |
| Rapid flow | V4 |
| Pure sliding | V2 |
| Flow (earth, mud) | V2/V3 |
| Slide-flow | V2/V3 |
| Slow deformation | V1 |

Next, the two-entry matrix was created, assigning to the different combinations of activity-based danger and magnitude a 'Hazard Index' (IP) ranking. Hazard Index, in this case, corresponds to danger. The danger assessment so obtained includes, very roughly, the temporal dimension (recurrence, as shown in Table 24.2), but this cannot be considered sufficient for a true hazard evaluation in terms of calculated probability. However, with the data sets and working time available for the fulfilment of the requirements of the legislation, it was not possible to use a sounder procedure.

The Hazard Index (IP) matrix for landslide phenomena is shown in Table 24.4. The same for deep-seated gravitational slope deformations is shown in Table 24.5.

Using danger instead of hazard does not fulfil the legislation that requires the evaluation of hazard; however, danger conveys better the idea of expected effects.

Vulnerability was not assessed. Only two classes of property at risk were defined: urban areas and main lifelines. Then, risk was ranked by combining the four classes of IP with the two classes of property at risk.

The results of the analysis are summarized in Table 24.6, where the number of observed landslides, and the total surface of phenomena, are shown for each province in Calabria (a) and for each landslide type (b). Landslide incidence is not given because only part (roughly 40%) of the whole territory has been analysed.

### 24.5.2    The Lazio Region

Assessment using the methodology indicated by the 1998 decree (no. 180) was first applied in the Lazio Region (Prestininzi, 2000).

This region, whose capital is Rome, is characterized by the carbonatic sediments of the Apennine mountain range in the southern and eastern parts; by Tertiary and

**Table 24.4**  *Two-entry landslide danger matrix (IP = hazard index in Autorità di Bacino della Calabria, 2001)*

| Magnitude | Danger | | | | |
|---|---|---|---|---|---|
| | D0 | D1 | D2 | D3 | D4 |
| V1 | IP0 | IP0 | IP1 | IP2 | IP2 |
| V2 | IP0 | IP1 | IP2 | IP3 | IP4 |
| V3 | IP0 | IP1 | IP2 | IP4 | IP4 |
| V4 | IP0 | IP2 | IP3 | IP4 | IP4 |

**Table 24.5**  *Two-entry matrix of danger for Sackung and Jahn's lateral spread*

| Type | Activity | |
|---|---|---|
| | Active, suspended | Dormant or quiescent |
| Sackung | IP3 | IP2 |
| Jahn's lateral spread | IP2 | IP1 |

**Table 24.6**   *(a) Regional frequency and surface extent of landsliding in the provinces of Calabria (b) according to the different landslide types and undifferentiated grouping (landslide zone)*

(a)

| Province | Catanzaro | Crotone | Vibo Valentia | Reggio Calabria | Cosenza | Totals |
|---|---|---|---|---|---|---|
| No. | 1711 | 322 | 743 | 1914 | 3238 | 7928 |
| Area (km$^2$) | 128.640 | 25.259 | 53.717 | 115.947 | 312.552 | 636.115 |

(b)

| Type | Flow | Rapid flow | Complex | Fall | Dgsd | Slide | Landslide zone |
|---|---|---|---|---|---|---|---|
| No. | 123 | 79 | 831 | 111 | 36 | 4229 | 2575 |
| Area (km$^2$) | 6.297 | 1.682 | 61.595 | 2.910 | 59.173 | 154.466 | 354.123 |

*Note:*
Dgsd stands for deep-seated gravitational slope deformations (Sackung and Jahn's lateral spread).

Quaternary sedimentary rocks, mainly represented by sandy and clayey deposits, and by allochtonous clayey melange; and by Tertiary and Quaternary inactive volcanoes and related volcaniclastic deposits, with typical crater lakes (Ogniben *et al.*, 1975). The morphological relief is in general low to intermediate, as well as the slope gradient, except the more rugged rocky areas of the Apennine chain. Most of the territory falls within the basin of the River Tevere.

This early study was not performed or commissioned by the BA, but by the regional government. However, it served as a methodological reference for the subsequent work done by BAs all over the country, even though each has adapted the methodology to local characteristics and, in some cases, introduced additional procedures (see section 24.5.5).

The work is essentially based on landslide maps. For this reason, 'first motion' landslides were not taken into consideration, as additional information was required (section 24.2).

Present phenomena are classified according to the type of mass movement (including deep-seated gravitational slope deformations), and their degree of activity (active, dormant, non-active). Landslides are classified according to the classification of Cruden and Varnes (1996).

Data on mass movement were gathered from a great number of documents, including urban plans and archive material produced by research or local government agencies.

The data set was filed in a GIS for a territorial analysis. For each event mapped, a form was compiled that included a large number of variables, all stored in a database.

The landslide map was made at a scale of 1:25 000 by means of aerial photo interpretation and field work.

Only zones R3 and R4 have been mapped. Hazard has been assumed to be 1 (existent landslide) or 0 (non-existent landslide), thus taking the form of a dichotomous variable. No real evaluation of hazard has been performed, essentially because the temporal dimension is lacking. As regards zones R4 and R3, the risk map is obtained by combining the dichotomous hazard map with that of the elements at risk. Of course, they appear only in the areas where landslides interfere with elements at risk.

This work uncovered 17 327 phenomena (including mapped and non-mapped landslides), 2507 of which were previously known. This sevenfold ratio is a clear indication of the need for such mapping to improve our territorial knowledge. Most of the mapped landslides (*c.* 8600 over *c.* 10 400) were in an active state.

### 24.5.3    The Marecchia and Conca Rivers

The Marecchia and Conca Rivers flow on the western side of the Apennine chain, across the Emilia-Romagna, Marche and Toscana regions. The area actually includes seven river basins, covering an area of 1333.8 km$^2$. The largest river basin (Marecchia) covers 610 km$^2$.

All the following information is taken from the report of the HSP of the Interregional Basin Authority of the Marecchia and Conca Rivers (Bruschini *et al.*, 2001).

The geology of the whole area is highly complex due to the tectonic crustal deformation, sintectonic and late tectonic sedimentation, which all occurred during the Alpine Orogeny and are still ongoing. At present, the most ancient flysch and tectonic melange-bearing units are overthrust upon the more recent autochtonous sedimentary units; this overthrust occurred over an area of more than 200 km from SW to NE; during this orogenic transport, sediments were deposited over the allochtonous nappes.

The prevalence of pelitic components in the different units, and their structural complexity, make the slopes intrinsically susceptible to landslides and erosion.

The landscape of the area is characterized by a wide variety of forms, from gently rolling hills to abrupt limestone and sandstone cliffs. Several villages lie on top of rocky layers capping softer sediments, that is, one of the classic structural conditions for landslides. Consequently a large amount of historical data on landslide mobility has been collected over the last three or two centuries. Besides rainfall, earthquakes are one of the most frequent causes of landslide reactivation.

The HSP report on the Marecchia and Conca basins contains a large section devoted to the description of the environmental conditions of the studied area. One section is devoted to the geomorphological and hydraulic problems found at several sites. The general plan for the downgrading of risk is outlined; then the financial requirements, the monitoring and checking of the work and the procedures for the management of the HSP are illustrated in detail.

With respect to mass movements, information was taken from already existing documents produced by regional agencies. These documents (landslide inventory maps) needed only a little updating and, of course, some standardization because of differences in landslide classifications. The landslide inventory covered the whole surface of the Marecchia and Conca area; thus it was possible to measure landslide incidence. Values for the different basins are shown in Table 24.7.

More detailed study has been devoted to landslides that have or are likely to cause damage. Inventory mapping was carried out (1:25 000 scale) and a database implemented using the procedures of the national landslide inventory set up by the National Geological Survey (IFFI Program). The landslide classification adopted was that of Varnes (1978), modified along the lines of WP/WLI (1990, 1991, 1993) and Cruden and Varnes (1996). It is noteworthy that in this area the rapid flows were considered a separate category of phenomenon, the same as in Calabria. There are only 303 of these damaging or potentially damaging phenomena; 62% of them are of slide type, 32% are flows, 6% are falls; 78%

**Table 24.7**   *Landslide incidence in the drainage basins of the Marecchia and Conca river area*

| Basin | Sub-basin | Area (km$^2$) | Landslide incidence (%) |
|---|---|---|---|
| River Marecchia, mountain section | | 446.4 | 44.5 |
| R. Marecchia, low hill section | | 154.4 | 2.9 |
| Torrent Conca | | 175.2 | 34.0 |
| T. Tavollo | | 28.1 | 5.3* |
| T. Ventena | | 45.5 | 16.0 |
| T. Melo | | 51.7 | 15.0 |
| T. Uso | | 140.6 | 14.0 |
| T. Foglia | | 65.6 | 31.1 |
| R. Savio | T. Fanante | 64.5 | 41.0 |
| T. Marano | | 83.2 | 8.0 |
| All | | 1395.8 | 27.2 |

*Note:*
* = for the part in the Emilia-Romagna region only. The total surface is not the same as that of the Basin Authority area because the administrative boundaries of the Basin Authorities do not correspond with the watersheds in places.

are active (or recently active) and 22% are dormant. Landslide surface, area (volume for falls), movement type, velocity, and state and distribution of activity (WP/WLI, 1993) are the relevant elements for the definition of the hazard of the landslides. Two-entry matrices adopted for hazard ($H$) evaluation are shown in Tables 24.8 to 24.10.

To assess the expected damage, the two entries are the values of elements at risk and vulnerability (V0 to V4, from potential to not recoverable loss, respectively). Elements at

**Table 24.8**   *Hazard index P*

| Activity | Movement type | | |
|---|---|---|---|
| | T1 | T2 | T3 |
| Stabilized | P0 | P0 | P0 |
| Dormant | P1 | P1 | P2 |
| Active | P1 | P2 | P3 |

*Note:*
T1 = flow and lateral spread: T2 = slide; T3 = topple, fall and fast landslides. T1 becomes T3 if the velocity is high.

**Table 24.9**   *Hazard index M, obtained combining the hazard index P with the activity distribution*

| Activity distribution | Index P | | | |
|---|---|---|---|---|
| | P0 | P1 | P2 | P3 |
| Expanding | M0 | M0 | M1 | M2 |
| Constant | M0 | M1 | M2 | M3 |
| Diminishing | M0 | M2 | M3 | M4 |

**Table 24.10**   *Hazard*

| Area (ha) (Volume (m³) for T3 types) | Index M | | | | |
|---|---|---|---|---|---|
| | M0 | M1 | M2 | M3 | M4 |
| $A < 0.5; V < 1$ | H0 | H1 | H1 | H2 | H2 |
| $0.5 > A < 3$ | H1 | H2 | H2 | H3 | H3 |
| $A > 3; V > 1$ | H2 | H3 | H3 | H4 | H4 |

*Note:*
For T3 landslides (falls, topples and fast events) the volume only is considered; thus values of H must be considered in the first and third rows only.

risk are ranked in four groups (from E1 to E4) with increasing value. Ranking is relative, neither real nor percentage values are given, and human lives are not explicit.

The expected damage matrix is shown in Table 24.11. Values range upwards from D0 to D4.

Finally, the relative risk (RR) is determined by crossing $H$ with $D$, as shown in Table 24.12. RR ranges from RR1 (low risk), to intermediate (RR2), high (RR3), very high (RR4). Notice how zero risk is not considered, as this work was done for well-known landslides only, thus it covers only a small part of the Basin Authority area.

Considering the 303 events for which the risk assessment has been performed, the percentages in terms of surface for the four relative risk ranks were the following:

$$RR1 = 3\%; RR2 = 43\%; RR3 = 36\%; RR4 = 18\%.$$

**Table 24.11**   *Expected damage (D) of elements at risk*

| Relative value of elements at risk | Vulnerability | | | | |
|---|---|---|---|---|---|
| | V0 | V1 | V2 | V3 | V4 |
| E1 | D0 | D1 | D1 | D2 | D2 |
| E2 | D1 | D1 | D2 | D3 | D3 |
| E3 | D1 | D2 | D3 | D3 | D4 |
| E4 | D2 | D2 | D3 | D4 | D4 |

**Table 24.12**   *Relative risk according to the hazard and expected damage ranking*

| Hazard | Expected damage | | | | |
|---|---|---|---|---|---|
| | D0 | D1 | D2 | D3 | D4 |
| H0 | RR1 | RR1 | RR1 | RR2 | RR2 |
| H1 | RR1 | RR1 | RR2 | RR2 | RR2 |
| H2 | RR1 | RR2 | RR2 | RR3 | RR3 |
| H3 | RR2 | RR2 | RR2 | RR3 | RR4 |
| H4 | RR2 | RR2 | RR3 | RR4 | RR4 |

### 24.5.4 The River Adige

The River Adige basin (Autorità di Bacino Nazionale del Fiume Adige, 2001) drains from the Dolomites (Central Alps) to the Adriatic Sea, an area of $c$. 12 200 km$^2$. Its headwaters are in the Alto Adige (South Tyrol) sub-region; the watershed lies at elevations ranging from 2500 to 3500 m a.s.l. It enters the Trentino sub-region and then, downstream, the Veneto Region. It is a mountain river basin for 3/4 of its length and for 4/5 of its area.

In the high mountain reaches, the valley bottoms lie at elevations between 1300 and 1500 m a.s.l.; these figures descend to 240 m at Bolzano and 190 m at Trento, $c$. 240 km from the mouth of the Adige River to the Adriatic Sea.

The geology of the Adige basin has been divided into three zones according to the tectonostratigraphic setting, and precisely:

1. Pennidica zone
2. Austroalpino zone
3. Southern Alps zone.

The Southern Alps zone is separated from the Austroalpino zone to the north by the Linea Insubrica, a well-developed strike-slip fault. Many more tectonic lines cross the area whose geological units have been affected by the complex and pervasive deformations of two orogenic periods (Ercinian in the Palaeozoic era and Alpine from the Cretaceous to the present).

The Pennidica zone consists of three tectonic complexes of high-grade metamorphic rocks (gneiss, garnet gneiss, mica-schist, quartzite, marble) and ophiolite-bearing schist. Where metamorphic units contain low-grade or degraded levels, deep-seated gravitational slope deformations develop.

In the Austroalpino zone, there is a metamorphic basement made of units with different metamorphic grades. This Palaeozoic basement is overlain by a Mesozoic cover of different units that includes low-grade metamorphic rocks, volcanites, magmatites and sedimentary rocks.

The Southern Alps are characterized by the Mesozoic–Eocene dolomite and limestone of the Dolomite Mountains. They include the volcanic units, the plutonic rocks and the effusive Permian porphyry banks. All these are hard rocks that make a typical contrast with the much softer units of volcanic and sedimentary origin.

Not surprisingly, the hardness of the rocks determines the typology of mass movement that develops on the steep slopes of the mountains.

In the deglaciated valleys, the removal of the confining pressure at the side of the deep valleys has resulted in the development of large-scale landslides and deep-seated gravitational slope deformations. The bottom of these valleys is covered with thick moraines that constitute an enormous bank of material available for torrential transport and the development of debris flows.

The drainage network is well developed, adapted to structures in the mountain ranges, even though the high and intermediate reaches of the valleys have features of former glacial valleys.

In the highest reaches, glaciers cover an area of $c$. 200 km$^2$. Seasonal snow cover extends over a much larger area ($c$. 3000 km$^2$), with the hazard of avalanches.

The implementation of the HSP in this area encountered a number of problems due to the overlapping of several local authorities, even though the Basin Authority should be in overall control. As a result of these constraints, only partial documentation has been produced. It will be shown in the following section that, currently, each document has been made using a slightly different procedure.

In compliance with the legislation, in the part of the River Adige basin extending over the provinces of Trento and Bolzano, the areas with very high, high, intermediate, and low landslide risk were studied and mapped; data on each phenomenon were stored in a database. Individuation and data derive from land survey archives available for the entire Adige basin. This condition, which is common for the other basins in northern Italy, rendered possible the full accomplishment of the requirements of the 1998 law (no. 267). In this territory, 415 major landslides have been found. Forty-three per cent are falls, 15% slumps, 12% are complex phenomena, and 8% are flow (mainly debris flow) phenomena (Autorità di Bacino Nazionale del Fiume Adige, 2001).

This procedure for hazard evaluation was carried out only for phenomena threatening public and private property, and valuable environmental sites.

After the reconnaissance phase, aerial photo interpretation and field work was undertaken to standardize all the information gathered.

The report by the Basin Authority considered the following elements: the threatened zone for each phenomenon; the mechanical elements of the landslide, such as volume and velocity; the reconnaissance of the elements at risk; and assessment of the expected damage for each type of element at risk depending on the intensity of the event.

The mapping of landslides for the R4 and R3 zones was performed on maps at a scale of 1:10 000. From the Basin Authority report, it is clear that an attempt was made to avoid over-identification, that is, the mapping of non-existent landslides. This was done in order to avoid creating too many restrictions in an area where the land available for urban development is rather scarce.

Maps so obtained were compared with existing maps of the involved provinces, except for Bolzano, that is, the inner high mountain zone. In the case of discrepancies the elaboration was checked again. If any discrepancy still persisted, the results of the analysis conducted by the Basin Autority were eventually applied.

No details are given in the Report about the procedure adopted for each different step. However, one can imagine that the level of data elaboration is rather basic, given that the general level of the data collected was unavoidably not very detailed.

In the province of Bolzano 39 areas were studied, 23 in the province of Trento, and none in the Veneto region. These are well described in the archives.

The high-risk areas were also mapped through aerial photo interpretation in addition to the information available in the public archives.

Areas of intermediate risk were identified on the basis of geomorphological analysis and historical records. This group includes mostly old landslides on which corrective measures had been previously carried out to render them safe.

The most frequent cases are low-risk areas, as shown in Table 24.13. They have not been mapped, however, because information on these phenomena is very uneven and generally limited to indications of their location, except for the cases in which intervention had been carried out in the past.

In the Basin Authority Report, there is, to date (7 December 2002), a summary of the events regarding the Veneto region only, that is, the lower portion of the basin, where no very high-risk landslides have been found.

**Table 24.13**   *Number of cases of each risk level*

| Risk level | Reckoned cases |
| --- | --- |
| R4 | 62 |
| R3 | 68 |
| R2 | 110 |
| R1 | 647 |
| All levels | 887 |

Besides classification according to landslide risk, a general plan regarding counter-measures for this area was produced and adopted.

Virtually all kinds of corrective measures, including those aimed at reducing erosion, were taken into consideration, depending on the different situations.

Natural engineering techniques were used wherever feasible, in order to preserve the high naturalistic and landscape value of the whole area of the river Adige basin.

The total estimated cost for the R3 and R2 zones (178 in total, see Table 24.12) was over 2.9 million euro. No estimate for the cost of intervention is provided for R4 zones, but it would be higher, even though the cases are much less numerous (62, see Table 24.12).

The cases above do not include debris flows. These actually represent the largest number of mass movement phenomena, and they generally fall into the category of mass transport events. According to the report by Autorità di Bacino Nazionale del Fiume Adige (2001), they usually develop where large amounts of debris are available, where slope gradients are high, when rainfall and/or thawing snow occur with great intensity. Such conditions are widespread mostly in the high mountain reaches. Most of the events occur in summertime, when rainfall is very intense (between 30 and 80 mm/day).

These phenomena are very dangerous and have been studied by a number of researchers, from different points of view: the mechanics of the processes, hazard evaluation, and the possible techniques for effective intervention.

Particular attention was paid to these events. Preliminary reconnaissance work, based on the available archive data for the last 300 years, was conducted and the events were mapped on a 1:300 000 scale. For the Bolzano province, an archive is available from which relevant information was extracted.

A study of countermeasures undertaken in the last 100 years was conducted and these interventions were mapped.

Finally, all the torrents that were identified as potentially or actually affected by debris flows, on the basis of historical analysis, aerial photo interpretation and expert reports, have been mapped. All mapped information was georeferenced in a GIS.

The comparison of zones classified as affected by debris flows, with the territorial and temporal distribution of corrective measures, has given us a picture of how the downgrading of the risk of these phenomena has been effective in the study area. It appears that the most effective interventions were undertaken in the second half of the twentieth century.

In conclusion, the HSP of the River Adige, for landslide hazards, includes:

(a) mapping and land use restrictions for R2 and R3 risk levels for well-known landslides and debris flows;
(b) the rules for use in R2 areas, based essentially on current urban plans;
(c) the mode of local government participation in the checking of, and modifications to, the plan itself.

### 24.5.5    Northwest Campania

The Northwest Campania Basin Authority covers an area of *c.* 1500 km$^2$ that has peculiar geological and morphological characteristics, including volcanic, alluvial and limestone areas. Volcanic areas are on the northern side of Mount Vesuvius, the Naples area, the Campi Flegrei zone, and the islands of Ischia and Procida; the alluvial plain surrounds Mount Vesuvius and extends eastwards towards the western slopes of the Apennine mountain range; the limestone area is found on the western parts of the Apennine range.

Vesuvius and Campi Flegrei are different types of volcanic structures: Vesuvius is a strato-volcano well known for its extremely destructive eruptions in ancient times; the Campi Flegrei zone is less well known, but it is an area scattered with dozens of explosive volcanic structures; Ischia is a volcanic island with fumaroles. The result of the explosive activity of these volcanoes is the mantling of almost all the Northwest Campania Basin Authority territory with volcaniclastic deposits (pumice, ash, ignimbrite, and so on) from different periods of Late Quaternary. This cover lies upon every kind of relief and every type of rock. Its emplacement came from fallout after Plinian explosions, or *nuée ardent*. These deposits may reach a thickness of hundreds of metres, are rather unstable and have caused several mud and debris flows, often with a large number of casualties. The most recent event occurred in May 1998, when more than 270 people died in the Sarno area when huge mudflows were triggered by extremely heavy rainfall. This was the event that pushed the Italian government to pass the 1998 decree (no. 180). Probably because of this dramatic situation, the study on landslide risk conducted in this area (Autorità di Bacino Nordoccidentale della Campania, 2002) investigates debris flows thoroughly, but treats rockfalls and topples only briefly, and completely ignores other types of landslides, even though they are present in the area (Vallario, 2001).

However, mudflows are treated with reference to the possibility of onset (susceptibility) in the source areas, and the possible spread of the landslide tongue.

The first item is addressed by adopting an empirical formula whose general expression is:

$$S = A^*B^*G^{(1+K+\cdots+N)}, \tag{1}$$

where $S$ is the susceptibility to mudflows, $A$, $B$, $K$, ... $N$, are local geomorphic or land-related parameters, and $G$ is the slope gradient at the source area of the mudflow (normally these phenomena initiate as a mud- or earthslide).

For the whole study area, the following formula was adopted:

$$S = L^*Dc^*G^{(1+T+Dr)}, \tag{2}$$

where $L$ is land use, $Dc$ is the distance to the cliff, $T$ is the thickness of the pyroclastic cover, and $Dr$ is the distance to the mountain road. The report does not explain the criteria for the selection of the parameters in formula 2. However, all these formulae, being in the form of regression first-order polynomials, are probably the result of statistical analyses. By means of statistical analysis, in fact, relevant parameters are considered or discarded for different sub-zones. The criteria for determining the relevant parameters, however, are not illustrated. For example, for one part of the territory, the parameters $Dc$ and $Dr$ do not influence the territorial distribution of $S$, thus formula (2) takes the form:

$$S = L^* G^{(1+T)}, \tag{3}$$

while in another sub-zone only $Dc$ is not relevant; thus formula (2) takes the form:

$$S = L^* G^{(1+T+Dr)}. \tag{4}$$

In another sub-zone (Campi Flegrei), formula (2) becomes:

$$S = G^{(1+T)}. \tag{5}$$

The variables are measured on a conventional parametric scale. It can be seen that the principal controlling variable is the slope gradient, and that other parameters are increasing or decreasing factors of the base or of the exponent of the power. In general $L$ is assigned values that range from 0.0001 for stable meadows, to 1.5 for deciduous forest.

It seems clear that there is likely to be a correlation between $G$ and $L$ (higher values are assigned to types of land use typical of steep slopes, where other types of cultivation are not feasible); thus a similar result would be attained if the variable $G$ alone was adopted. A more useful approach seems to be, instead, the introduction of $T$ and, probably, $Dc$ and $Dr$. From regression analysis, it seems that correlation is inverse in the measured range 1 to 49 m and follows a negative-log law:

$$F = -1.802(\ln D) + 7.3515 \qquad (r^2 = 0.95), \tag{6}$$

where $F$ is the frequency (%) of landslides for each $D$ class, $D$ can be either $Dc$ or $Dr$.

Formula (6) confirms the field observations made by me in the Sarno area after the 1998 disaster, and confirmed by other researchers, that several (almost all in the observed area) landslides initiated just upstream of natural cliffs or road cuts, indicating that the initial motion involved only a few m$^3$ of material, but the phenomena evolved into huge mudflows because of undrained overload of the soaked pyroclastic cover downstream, gathering more and more material lower down the slope.

The extent of the potential phenomena was determined by adopting values for the 'reach angle' ($= \tan^{-1} Dh/L$, where $Dh$ is the height and $L$ is the horizontal length of the mudflow) of 28° or 21° (according to the different sub-zones) for landslides on open slopes (i.e. non-conservative energy conditions in the sense of Nicoletti and Sorriso-Valvo, 1991), and 18° for channelled landslides (conservative conditions).

The landslide risk is evaluated by means of a two-entry matrix that considers in a rank scale both hazard (P3 > P2 > P1) and potential damage (D1 < D2 < D3 < D4),

**Table 24.14**  *Landslide danger (risk in the original text) matrix*

|      | P3   | P2   | P1   |
|------|------|------|------|
| D4   | R4   | R4   | R3   |
| D3   | R4   | R3   | R2   |
| D2   | R3   | R2   | R1   |
| D1   | R2   | R1   | R1   |

the latter being the product of the vulnerability (in %) and the value (in rank scale) of the elements at risk. On the basis of the procedure described above, what is here called 'hazard' is actually 'susceptibility'; thus the risk is actually an expression of danger. The key for classification is given in Table 24.14.

By means of the above methodology, which uses separate regression formulae for the different environments, the whole territory has been classified in terms of landslide susceptibility. As an example, a small part of it is shown in Figure 24.2 (see also Colour Plate section). Note that the zones of possible invasion by debris flows appear little extended, despite the fact that all the casualties and the largest part of the damage caused by the1998 Sarno disaster occurred in these zones (Brancaccio *et al.*, 2000).

**Figure 24.2**  *Example of land classification from map of the relative hazard (susceptibility) for trigger of and invasion by landslide in the HSP of* Autorità di Bacino Nordoccidentale della Campania *(2002), South Italy. Reproduced with permission of* Autorità di Bacino Nordoccidentale della Campania. *(See also Plate 12)*

**Figure 24.3** *Rockfall near Corleone, Sicily. The cliff is c. 30 m high. The rock is jointed and faulted sandstone*

**Figure 24.4** *The large-scale Vallone Colella landslide complex, Aspromonte, Calabria. The slope is made of gneiss. The mass movement is of the rockslide–debris-flow type. The landslide on the left is more than 1.5 km wide and more than 1 km high. In the foreground (light grey tones) several smaller landslides make the slope similar to badlands. This huge landslide complex is causing problems downstream because of excessive aggradation of the riverbeds. For scale, the trees on the skyline and on the right of the photo are 20 to 30 m high pines*

**Figure 24.5** *Upper part of the Monte Antelao (Alps, Veneto region). Pinnacles are limestone blocks spread by a long-lasting rock block slide. Debris flows originated from this area that caused hundreds of fatalities in the past*

## 24.6  Discussion

Before 1998, little had been done in Italy to fulfil the legislation regarding landslide risk. Thereafter a series of very stringent laws on land classification were implemented, even though they did not cover the whole territory and lacked procedural uniformity.

These gaps are striking if we consider the real needs of a country where natural hazards such as landslides are both frequent and widespread. This is despite the fact that, as regards other indicators of development, Italy has one of the highest social and economic standards in the world.

The Basin Authorities have been charged with the implementation of the classification of landslide, as well as flood risk. As local authorities, however, are given the right to legally oppose these classifications, the situation is under debate in some parts of the country.

The deadlines for fulfilment of the legal requirements were very stringent. In other words, Basin Authorities had to carry out, in less than two years, work that had remained neglected for more than a decade with respect to the 1989 law (no. 183) that, in its turn, had only been enacted after more than a century from the foundation of the Italian state (1860; unification and Rome declared capital in 1870).

The need for land classification on the basis of hydrogeomorphological risk became even more urgent after the September 2000 disaster in Calabria, where 19 people died when a camp site, located on the bankfull flood bed of the Soverato Torrent, was washed away in a flood.

Deadlines have to be met because, without HSP approval, there is no access to financial resources for countermeasures.

**Figure 24.6**  *Rockslide–debris flow near Piminoro, Aspromonte (Calabria). The landslide body is* c. *250 m long and 120 m wide (horizontal distances); the thickness is* c. *35 m. The artificial tunnel is working well in sheltering the important highway from the debris. Slope carved in gneiss*

In some of the areas (e.g. almost all the north of Italy), where the new studies could benefit from previous ones, this task could be achieved with relative ease, within the time allowed. On the other hand, in the areas where there was little and/or very uneven knowledge, there was a higher probability that the work would be flawed due to haste and the lack of previous knowledge.

In an area like Calabria, for example, doubtful cases have been, for the moment, classified as landslide-affected, in the interests of caution. This means that a larger portion of the territory than is actually necessary has been 'frozen' until more detailed studies clarify the situation.

Part of the scientific community was well aware of this fact, but the government went on its way, paying little attention to the arguments of most of the experts.

Another shortcoming that can be seen clearly in the few examples described above is the variety of methods used to classify the territory on the basis of landslide hazard, vulnerability and risk. This has led to the problem of disparities from one Basin Authority

***Figure 24.7*** *Rockslide slide–earth flow in the Licetto river basin (Calabria). The landslide occurred in 1978, involving Miocene sandstones lying on Paleozoic phyllites. It destroyed a road in the scar area and now threatens the buildings on the apex of the fan at the mouth of the canyon. The dangerous practice of building on fans is common all over Italy*

to another, in spite of the apparently similar classification criteria of risk in five R levels, including zero risk.

The practical impossibility of defining a real hazard for landslides (because probability distribution models are not available for most landslide types) has induced almost all operators to label 'hazard' what actually is 'danger'. Indeed, in these studies, hazard always incorporates the magnitude of the events, while hazard should be evaluated for any given level of magnitude. Magnitude should also be included in the reference criteria for assessing vulnerability, yet in the procedures described in this chapter vulnerability refers to the type of landslide only.

In the case of Northwest Campania, danger is called 'relative hazard'. In this case, too, the use of the term 'hazard' is incorrect, as it should include the temporal dimension. In addition, susceptibility is directly correlated with the danger level, not with the probability of occurrence for phenomena of a given magnitude ($S$ increases with so-called hazard, instead of decreasing with higher magnitudes).

In the case of the Lazio Region, the approach is based only on the spatial dimension. Indeed, hazard is considered as a dichotomous variable that can assume the value of 1 (existence), or 0 (non-existence) of a landslide, but for hazard in the true sense of

**Figure 24.8**  *Tongue of the Maratea landslide (Lucania, South Italy). The landslide is a large-scale, pre-Holocene earthflow involving phyllites and schists. The tongue extends below the sea, making the irregularly convex coastline. This landslide moves of some cm per year and it is causing problems for the city of Maratea (part of which is visible in the photo), to a highway, to an important railroad, to a tourist port, and to other lifelines*

**Figure 24.9**  *Tessina landslide. A large earthflow threatens the small village of Tessina (Alps), defended with the high wall of concrete blocks visible in the photo. The village has been evacuated for some months. The landslide tongue is in the foreground, moving to the right. This landslide is becoming quiet but reactivation is possible, and consequently a sophisticated system for monitoring the landslide movements and avoiding excessive accumulation of debris in the proximity of the village was designed and implemented by experts from CNR–IRPI (Padova Section)*

the term (that includes time and magnitude variables), dichotomous distribution is not sufficient, unless its probability is included.

The case of the River Adige Basin Authority offers a good example of difficulties posed by the complexity of the laws on risk assessment. Besides arguments over territorial competence between regional authorities and the Basin Authority, it is evident the technicians of the Basin Authority decided to avoid the risk of overextending the restricted areas by not assigning risk level R1 to the uncertain zones and deciding not to map them at all. This procedure may appear to be an unacceptable short cut, but it is actually a valid way of solving the problem of uncertain zones, as it is quite impossible to assess landslide hazard in these zones given the data and time constraints; on the contrary, where all the elements are clear the assessment of hazard is easier.

The state of the art of research on the assessment of landslide hazard in Italy is at a comparable level to that of other countries (Aleotti and Chowdhury, 1999; Sorriso-Valvo, 2002), and is, on average, slightly ahead with respect to its practical application. There is widespread, albeit uneven, use of innovative methods for evaluating the different natural components of hazard. Difficulties remain, however, especially for diffuse phenomena.

The HSPs discussed here do not consider the combination of causative factors in mass movement at all. For instance, it is an important omission that the effects of earthquakes, and their combination with rainfall, are not considered in a country with such high seismicity as Italy. In some cases (e.g. the Calabria Region), earthquakes are mentioned as trigger factors, but no analysis is performed using causative and multi-causative models.

The application of the HSP was, from the scientific point of view, a missed opportunity. It usually happens that science makes great progress during difficult periods (such as wars), because more knowledge means the possibility of overcoming difficulties. In this case, the very stringent deadlines laid down by the legislation did not give the scientific community the chance to really address the needs of the country. The timing was too much in contrast with what researchers were used to, so no great progress was made in the methodology for assessing landslide hazard. Rather, there was a generalized low-level application of very basic procedures, except in very few cases (such as the approach by the Northwest Campania Basin Authority).

Probably this is not true for all the zones studied, and in some of the as yet unpublished work some very innovative and successful methods may well have been used that make the precise evaluation of landslide hazard possible. If this is the case, it will be welcome.

To sum up, there is general uniformity in the mapping of landslide hazard in all HSP work; this is less so in risk maps (due to procedural differences). However, this uniformity is not consistent, because a relevant part of it is only on the graphics side. This can be easily inferred by comparing the different procedures adopted to evaluate the landslide hazard in the few Hazard Setting Plans briefly illustrated in section 24.5. Such differences are indeed found from one HSP to another. In addition, as many other HSPs have been performed (more than 30 overall), such differences are expected to become even greater than shown in this chapter. A special task force at the Ministry of the Environment is trying to overcome this crucial shortcoming.

Part of the fault is in the lack of precise directions in the legislation. However, researchers are also partly to blame because they were unable to convince the Basin Authorities of the need for a sounder, more scientific study, which would have put the country in a better position to cope with critical developments in the future. Unfortunately, the fear of missing out on the funding was stronger than good reasoning, and those researchers that had foreseen the low level of the final result were proved right.


## 24.7   Perspectives

Landslide hazard evaluation in Italy is a many-faceted activity resulting from a complex history whose principal mistake was to charge research agencies with performing the work that surveyors should have done. This occurred because surveying departments were too small or non-existent. At present, some regions do have their well-staffed surveying departments, but most of them do not.

The work of the HSP has been done by research agencies, by region or Basin Authority surveys with or without the assistance of research agencies, by private companies. Unfortunately, it was not possible to collect any of the HSP data done by private companies.

Due to this situation, the overall picture at present varies from zone to zone, and perspectives for future development and homogenization are rather vague:

1. Surveying agencies should now have the task of comparing their work in order improve procedures for evaluating landslide hazard and risk. This is not an easy goal to achieve, but it is vital, otherwise the ultimate aims of the legislation to evaluate landslide risk on a national scale will remain unmet.
2. Research agencies should abandon the role of substitute for surveying departments, and concentrate their efforts on finding sounder, more reliable and, possibly, cheaper procedures for the evaluation of landslide hazard, for both diffuse and single cases. Causative factors should be incorporated in hazard and, possibly, predictive models. At least rainfall, ice and snowmelt, earthquakes and human activity should be incorporated in such models.

    Research agencies should also transfer their new know-how to surveying departments, and work with them to monitor progress.
3. Public financing bodies (including national and local governments) should provide enough financial support to scientific agencies, which are still considerably underfunded, even after the application of recent legislation. To do this, they should extend for as long as necessary (at least a decade) the deadlines for results, as it is widely recognized that results in the field of mass movements can only be considered reliable after a long-term study. In the case of the 1998 decree no. 180 and subsequent legislation, exactly the opposite has happened. Last, but not least, public financing bodies should refrain from promoting applied research only, despite the fact that their political masters continuously declare that basic (or knowledge-driven) research is essential for the development of science.

Laws provide funds for further developments of HSP and complete the Basin Plans (section 24.4.1). Notwithstanding the above-mentioned shortcomings, the current laws have had the great value in posing the problem of hydrogeomorphological hazard and the need for efficient and sustained assessment, for the first time in actions, rather than words, and this new political direction must be encouraged to move towards a more rational approach, which is less dependent on crisis intervention.

In conclusion, under pressure from new legislation, the various HSPs have been produced as documents that can, and in most cases will have to be, be updated and upgraded. This gives the scientific community and land managers a further opportunity to upgrade the state of the art and the efficiency of landslide hazard assessment in Italy.

# References


Agnesi, V., Carrara, A., Macaluso, T., Monteleone, S., Pipitone, G., Sorriso-Valvo, M., 1983, Elementi tipologici e morfologici dei fenomeni di instabilità dei versanti indotti dal sisma del 1980 (alta valle del fiume Sele), *Geologia Applicata e Idrogeoogia*, **18**(1), 309–341 (in Italian).

Aleotti, P. and Chowdhury, R., 1999, Landslide hazard assessment: summary review and new perspectives, *Bulletin Eng. Geol. Env.*, **58**(1), 21–44.

Almagià, R., 1910, Studi geografici sulle frane in Italia, *Memorie Società Geografica Italiana*, vol. 14, Roma (in Italian).

Antronico, L., Gullà, G., Sorriso-Valvo, M. and Tansi, C., 1999, Grandi frane e deformazioni gravitative profonde di versante: un possibile approccio per la prevenzione ed alcuni approfondimenti di studio mirati alla previsione, *Atti del Convegno dei Lincei*, **154**, 119–126 (in Italian).

Autorità di Bacino della Calabria, 2001, *Piano-stralcio per l'Assetto Idrogeologico – Relazione*. Available on www.autoritadibacinocalabria.it (in Italian).

Autorità di Bacino Nazionale del Fiume Adige, 2001, *Piano Stralcio per la Tutela dal Rischio Idrogeologico – Individuazione e perimetrazione delle aree a rischio idraulico, da frana e da colata detritica – Relazione Illustrativa di Sintesi*. Available on http://www.adbpo.it (in Italian).

Autorità di Bacino Nordoccidentale della Campania, 2002, *Piano Stralcio per l'Assetto Idrogeologico, Relazione Generale*, vol. 2, S.El.C., Firenze (in Italian).

Bernknopf, R., Campbell, R.H., Brookshire, D.S. and Shapiro, C.D., 1988, A probabilistic approach to landslide hazard mapping in Cincinnati, Ohio, with applications for economic evaluation, *Bulletin Eng. Geol.*, **25**(1), 39–56.

Brancaccio, L., Cinque, A., Russo, F. and Sgambati, D., 2000, Le frane del 5–6 maggio 1998 sul gruppo montuoso Pizzo d'Alvano (Campania): osservazioni geomorfologiche sulla loro distribuzione e sulla dinamica delle connesse colate, *Quaderni di Geologia Applicata*, 7–1, 5.36 (in Italian).

Bruschini, M., Giovagnoli G. and Minarelli E., 2001, *Progetto di Piano Stralcio per l'assetto idrogeologico – Relazione*. Available on www.bacino-adige.it (in Italian).

Canuti, P. and Casagli, N., 1994, Considerazioni sulla valutazione del rischio da frana, in CNR-GNDCI – Regione Emilia- Romagna, *Fenomeni franosi e centri abitati; Atti del convegno, Bologna, 27 Maggio 1994*. CNR-GNDCI (846) (in Italian).

Carrara, A., 1983, A multivariate model for landslide hazard evaluation, *Mathematical Geology*, **15**, 403–426.

Carrara, A., Cardinali, M. and Guzzetti, F., 1992, Uncertainty in assessing landslide hazard and risk, *ITC Journal*, 1992–2, 172–183.

Carrara, A., Sorriso-Valvo, M. and Reali, C., 1982a, Analysis of landslide form and incidence by statistical techniques, Southern Italy. *Catena*, **9**(1/2), 35–62.

Carrara, A., Catalano, E., Reali, C. and Sorriso-Valvo, M., 1982b, Computer-assisted techniques for regional landslide evaluation, *Studia Geomorphologica Carpatho-Balcanica*, **15**, 100–113.

Carrara, A., Catalano, E., Sorriso-Valvo, M., Reali, C. and Osso, I., 1978, Digital terrain analysis for land evaluation, *Geol. Appl. Idrogeol.*, **13**, 69–127.

Casagli, N. and Ermini, L., 2001, Modellizzazione della genesi e del collasso di sbarramenti fluviali da frana tramite l'utilizzo di una rete neurale, *Mem. Soc. Geol. It.*, **56**, 31–40 (in Italian).

Catenacci, V., 1992, Il dissesto geologico e geombientale in Italia dal dopoguerra al 1990. *Memorie Descrittive della Carta Geologica d'Italia, Servizio Geologico Nazionale*, **47** (in Italian).

Conaker, A.J. and Sala, M. (eds), 1998, *Land Degradation in Mediterranean Environments of the World*. (Chichester: John Wiley & Sons).

Crescenti, U., 1998, Il rischio da frana: appunti per la valutazione, *Quaderni di tecniche di Protezione Ambientale*, **5**(2), 87–100 (in Italian).

Crescenti, U. and Sorriso-Valvo, M., 1995, IV Seminario del gruppo DGPV – Presentazione, *Memorie Società Geologica Italiana*, **50**, 5–7 (in Italian).

Cruden, D.M., and Varnes, D.J., 1996, Landslide types and processes, in A.K. Turner and R.L. Schuster (eds), *Landslides: Investigation and Mitigation*. Transportation Research Board, Washington, DC.

De Dolomieu, D., 1785, *Memoria sopra i tremuoti della Calabria*, Librai Francesi rimpetta S. Angelo a Nido, Napoli (in Italian).

Delmoneco, G., Leoni, G., Margottini, C., Puglisi, C., 2002, La Propensione del territorio italiano ai fenomeni franosi, *Italian Journal of Engineering Geology and Environment*, **1**(1).

Emery, J.J., 1978, Simulation of slope creep, in B. Voight (ed.), *Rockslide and Avalanche, 1, Natural Phenomena*, Elsevier, 669–693.

Fernández-Steeger, T.M., Rohn, J. and Czurda, K., 2002, Identification of landslide areas with neural nets for hazard analysis, in J. Rybář, J. Stemberk and P. Wagner (eds), *Landslides* (Lisse: Swets & Zeitlinger), 163–168.

Gomez, H. and Kavzoglu, T., 2004, Modelling landslide hazard using artificial neural networks in Jabonosa River Basin, Venezuela, *Engineering Geology*, in press.

Gringeri Pantano, F., Nicoletti, P.G. and Parise, M., 2002, Historical and geological evidence for seismic origin of newly recognized landslides in Southern Sicily, and its significance in terms of hazard, *Environmental Management*, **29**(1), 116–131.

Guerricchio, A. and Melidoro, G., 1973, Segni premonitori e collassi delle grandi frane nelle metamorfiti della valle della Fiumara Buonamico (Aspromonte, Calabria), *Geol. Appl. Idrogeol.*, 8–II, 315–346 (in Italian).

Guzzetti, F., 2000, Landslide fatalities and the evaluation of landslide risk in Italy, *Engineering Geology*, **58**, 89–107.

Guzzetti, F., Carrara, A., Cardinali, M. and Reichenbach, P., 1999, Landslide hazard evaluation: a review of current techniques and their application in a multi-scale study, Central Italy, *Geomorphology*, **31**, 181–216.

Guzzetti, F., Cardinali, M., Reichenbach, P. and Carrara, A., 2000, Comparing landslide maps: a case study in the upper Tiber river basin, Central Italy, *Environmental Management*, **25**(3), 247–263.

Jibson, R.W., Edwin, L.H., Michael, J.A., 2000, A method for producing digital probabilistic seismic landslide hazard maps, *Engineering Geology*, **58**, 271–289.

Kadomura, H., Okada, H., Imagawa, T., Moriya, I. and Yamamoto, H., 1983, Erosion and mass movements on Mt. Usu accelerated by crustal deformation that accompanied its 1977–1982 volcanism, *Journal of Natural Disaster Science*, **5**(2), 33–62.

Keefer, D.K., 1984, Statistical analysis of an earthquake-induced landslide distribution – the 1989 Loma Prieta, California event, *Engineering Geology*, **58**(3–4), 231–249.

Nicoletti, P.G. and Sorriso-Valvo, M., 1991, Geomorphic controls of the shape and mobility of rock avalanches, *Geological Society of America Bulletin*, **103**, 1365–1373.

Ogniben, L., Parotto, M. and Praturlon, A., 1975, Structural model of Italy, *Quaderni de 'La Ricerca Scientifica'*, **90**, CNR.

Oyagi, N., Sorriso-Valvo, M. and Voight, B., 1994, Introduction to the special issue on deep-seated landslides and large-scale rock avalanches, *Engineering Geology*, **38**, 187–188.

Prestininzi, A., 2000. *La valutazione del rischio da frana: metodologie e epplicazioni al territorio della Regione Lazio.*, Dipartim. Scienze della Terra, Univ. di Roma 1, p. 54, Duògrafi, Roma (in Italian).

Schuster, R.L., 1984, Effects of landslides and mudflows associated with the May 1980 eruption of Mount St. Helens, Nerthwestern U.S.A., *Journal of Japan Landslide Society*, **21**(3), 1–10.

Sirangelo, B. and Versace, P., 1992, Modelli stocastici di precipitazione e soglie pluviometriche di innesco dei movimenti franosi, *Atti del XXIII Convegno Nazionale di Idraulica e Costruzioni Idrauliche*, **4**, 361–373 (in Italian).

Sirangelo, B., Iiritano, G. and Vesace, P., 1996, Il preannuncio dei movimenti franosi innescati dalle piogge. Valutazione della probabilità di mobilizzazione in presenza di indeterminatezza nell'identificazione del modello FLaIR, *Atti del XXV Convegno Nazionale di Idraulica e Costruzioni Idrauliche*, **3**, 378–391 (in Italian).

Sorriso-Valvo, M., 1984a, Deep-seated gravitational slope deformations in Calabria (Italy), *Communications du Colloque 'Mouvements de Terrains'*, Série Documents du B.R.G.M., **83**, 81–90.

Sorriso-Valvo, M. (ed.), 1984b, Atti del I Seminario del gruppo informale del CNR 'Deformazioni Gravitative Profonde di Versante', *Bollettino Società Geologica. Italiana*, **193**, 667–729 (in Italian).

Sorriso-Valvo, M., 1986, Landslide activity in the area of the El-Asnam 1980 earthquake (Algeria), *Geologia Applicata e Idrogeologia*, **21**(2), 291–304.

Sorriso-Valvo, M., 1988, Le catastrofi naturali e l'Uomo, in A. Celant and P.R. Federici (eds), *Nuova città, nuova campagna, spazio fisico e territorio, Atti 24$^{mo}$ Congresso Geografico Italiano*, Patron, Bologna (in Italian), 569–575.

Sorriso-Valvo, M., 1993, The Geomorphology of Calabria – a sketch, *Geografia Fisica e Dinamica Quaternaria*, **16**, 75–80.

Sorriso-Valvo, M., 2001, Pericolo di Frana – Piano-stralcio Assetto Idrogeologico del Parco Nazionale dell'Aspromonte, Parco Nazionale dell'Aspromonte, Gambarie (unpublished report, in Italian).

Sorriso-Valvo, M., 2002, Landslides: from inventory to risk, in J. Rybář, J. Stemberk and P. Wagner (eds), *Proceedings of the First European Conference on Landslides*, Prague, Czech Republic, 24–26 June 2002 (Lisse: Balkema), 79– 103.

Sorriso-Valvo, M., Agnesi, V., Gullà, G., Merenda, L., Antronico, L., Di Maggio, C., Filice, E., Petrucci , O. and Tansi, C., 1994, Temporal and spatial occurrence of landsliding and correlation with precipitation time series in Montalto Uffugo (Calabria) and Imera (Sicilia) areas, in R. Casale, R.F. and Flageollet, J.C. (eds), *Temporal Occurrence and Forecasting of Landslides in the European Community*, Final Report, vol. II, European Commisison, Science Research Development, Programme EPOCH, Brussels, 823–869.

Sorriso-Valvo, M. and Tansi, C., 1996, Grandi frane e deformazioni gravitative profonde di versante della Calabria. Note illustrative della carta al 250.000, *Geografia Fisica e Dinamica Quaternaria*, **19**, 395–408 (in Italian).

Spacebel, Planetek, T.R.E., CNR-IRPI CS, University of Calabria, 2001, *MASMOV. Final Report*. ESA.

Updike, R.G., Egan, J.A., Moriwaki, Y., Idriss, I.M. and Moses, T.L., 1988, A model of earthquake-induced translatory landslides in Quaternary sediments, *Geological Society of America Bulletin*, **100**, 783–792.

Vallario, A., 2001, *Il dissesto idrogeologico in Campania*, CUEN, Napoli (in Italian).

Varnes, D.J., 1978, Landslide types and processes, in R. Schuster and D. Krizek (eds), *Landslides, Analysis and Control*. Transportation Research Board, Special report 176, Washington, DC, 11–33.

Versace, P., Sirangelo, B. and Iiritano, G., 2000, Soglie pluviometriche di innesco dei fenomeni franosi, *L'Acqua*, 2000 (3), 113–130 (in Italian).

WP/WLI, 1990, A suggested method for reporting a landslide, *Bull. Int. Ass. Eng. Geol.*, **41**, 5–12.

WP/WLI, 1991, A suggested method of a landslide summary, *Bull. Int. Ass. Eng. Geol.*, **43**, 101–110.

WP/WLI, 1993, A suggested method for describing the activity of a landslide, *Bull. Int. Ass. Eng. Geol.*, **47**, 53–57.

# 25

# An Initial Approach to Identifying Slope Stability Controls in Southern Java and to Providing Community-based Landslide Warning Information

*D. Karnawati, I. Ibriam, M.G. Anderson, E.A. Holcombe, G.T. Mummery,
J.-P. Renaud and Y. Wang*


## 25.1 Landslide Problems in Indonesia and the Socio-economic Cost

Two key parallel processes are taking place in terms of landslide risk and hazard assessment. The first is the increasing development of a very restricted number of numerical codes that allow a holistic assessment to be made of the processes involved in slope instability (Anderson *et al.*, 1997, 1998). The second is the pressing need to communicate such findings in a manner that is capable of being understood (and acted upon) by those most at risk. Nowhere is this a more pressing issue than in those parts of the world where such 'at risk' persons are among the poorest and where lives are regularly lost on a scale that most countries would find unacceptable. Currently, the use of some form of prioritization system which ranks slopes according to landslide risk is a commonly used empirical approach. The Hong Kong Geotechnical Control Office established the New Priority Classification Scheme (NPCS), which scores slope attributes in terms of instability risk and consequence of failure (Geotechnical Engineering Office, 1998). Such an essentially observational approach allows a rapid assessment of risk


*Landslide Hazard and Risk*   Edited by Thomas Glade, Malcolm Anderson and Michael J. Crozier
© 2004 John Wiley & Sons, Ltd   ISBN: 0-471-48663-9

to be made that can then be easily communicated to relevant agencies and the public. However, Dai *et al.* (2002) note that new methodologies are required to develop a better understanding of landslide hazard and to make rational decisions regarding landslide risk. The goal of course is one of rapid assessment, effective management and clear communication.

To advance towards this goal, this chapter seeks to review the opportunities that are now available for using a physically based model of slope stability to generate appropriate public awareness documentation. Attempting to establish 'end-to-end' connectivity, starting with field evidence of major landslide problems, and connecting with parameterization of geotechnical, hydrological and vegetation components, selection of a modelling platform, development of appropriate scenario simulations and the presentation of results in a public awareness format, is an important framework that needs a coherent and concentrated research effort. This chapter is an initial attempt at such connectivity. Such a development is seen to be at the heart of both assessing and communicating landslide hazard. In time, it is to be expected that such comprehensive connectivity will yield major developments in both model platforms and public awareness formulations.

Rain-induced landslides represent a major hazard in Indonesia. Between November 2000 and February 2003, 291 people lost their lives, 135 people were reported missing, thousands of people lost their homes and 657 ha of land were destroyed (Table 25.1, Figure 25.1). The total financial loss was estimated to be about US\$1 million. To avoid such substantial losses, landslide occurrence should be prevented or minimized by developing an appropriate warning system. Having regard to the socio-economic conditions of the landslide vulnerable areas in Indonesia, a low-cost, simple community-based slope management and warning system is required.

The high socio-economic costs to which we have referred highlight the importance of a new approach for developing a more refined warning system and slope protection management in rural areas, where the community does not have easy access to education and social welfare. The first part of this chapter briefly assesses Indonesian landslide conditions. These conditions are critical to basic considerations necessary to develop a socially appropriate warning and slope protection system. A numerical hydrology–slope stability model, parameterized by the slope conditions, is reviewed, and then developed and applied to type slope conditions. From these numerical experiments, thresholds of rainfall and porewater conditions are identified which are formulated and presented in a sufficiently simplified manner, making them appropriate for local community needs.

## 25.2   Slope and Landslide Conditions in Southern Java

Many slopes in south Java (Figure 25.1) that are vulnerable to landslides are formed by andesitic breccia and covered by sandy or silty clay with slope inclinations typically being steeper than 20° (Saroso, 1992; Karnawati, 1996, 1997, 2000). In 2000, 113 landslides occurred in the andesitic breccia slopes of Menoreh Mountain. Most of these failures occurred on slopes of 30° or more during, or immediately following, heavy rainfall events, certain of which exceeded 600 mm in five days in November 2000. Furthermore, by the end of 2002, 50 other landslides on the slopes of Sumbing, Merbabu and Merapi volcanoes were recorded. Many of them also occurred on 30° slopes (or steeper), formed by andesitic breccia and overlain by sandy clay. It was obvious that all the landslide

**Table 25.1** *Landslide socio-economic cost 2001–2003 in Indonesia*

| No. | Landslide location | Date | Socio-economic cost | |
|---|---|---|---|---|
| | | | Deaths | Other |
| 1. | Kotagede, Yogyakarta | January 2001 | – | 10 houses damaged |
| 2. | Sangihe Island, Province of North Sulawesi | January 2001 | 32 people dead 21 missing | |
| 3. | Lebak Regency, Banten Province | 8 February 2001 | 94 people dead | 1421 people and 389 houses threatened<br><br>61 houses damaged |
| 4. | Nias Island | 30 July 2001 | 50 people dead 114 people missing | 325 houses damaged |
| 7. | Penusupan Village, Sruweng District, Kebumen Regency | 4 October 2001 | 9 people dead | |
| 8. | Mangunweni Village, Ayah District, Kebumen Regency | 23 October 2001 | – | 35 houses burried |
| 9. | Ngelo and Purwodadi villages, Kajoran District, Magelang Regency | 19 November 2001 | – | 5 houses and 1 mosque damaged, tens of houses in Nglegok village downslope areas threatened |
| 10. | Kedungrong, Purwoharjo Village, Samigaluh District, Kulon Progo Regency, Yogyakarta | 20 November 2001 | 7 people dead | |
| 11. | Klepu, Banjararum Village. Kalibawang District, Kulon Progo District | 20 November 2001 | 2 people dead | |

**Table 25.1**  (Continued)

| No. | Landslide location | Date | Socio-economic cost | |
|---|---|---|---|---|
| | | | Deaths | Other |
| 12. | Ciliwung Valley, Jakarta | January 2002 | 4 people dead | 3 houses destroyed |
| 13. | Pacet Regency, Mojokerto Regency East Java | 11 December 2002 | 26 people dead<br>14 people missing | One tourism site (hot spring) destroyed and should be closed |
| 14. | Kalimatan mining sites | January 2003 | 5 people dead | |
| 15. | Bandung, West Java (mining site) | January 2003 | 4 people dead | |
| 16. | Kedungora District, Garut Regency, West Java | February 2003 | 21 people dead | |
| 17. | Blambangan Village, Banjarnegara Regency, Central Java | February 2003 | – | 16 houses destroyed |
| 18. | Kokap District, Kulon Progo Regency, Yogyakarta | February 2003 | – | 4 houses buried |
| 19. | National Park — Bahorok, North Sumatra | 2 November 2003 | 151 people dead<br>100 people missing | One tourist village buried |
| 20 | Plipir Village, Purworejo Regency, Central Java. | February 2004 | 7 people dead | 3 houses destroyed |
| 21 | Gowa Regency, South Sulawesi | 27 March 2004 | 32 people dead | 3 km road, 12 houses & 430 ha of land buried US$2.21 million |
| **TOTAL** | | | **444 people dead**<br>**246 people missing** | |

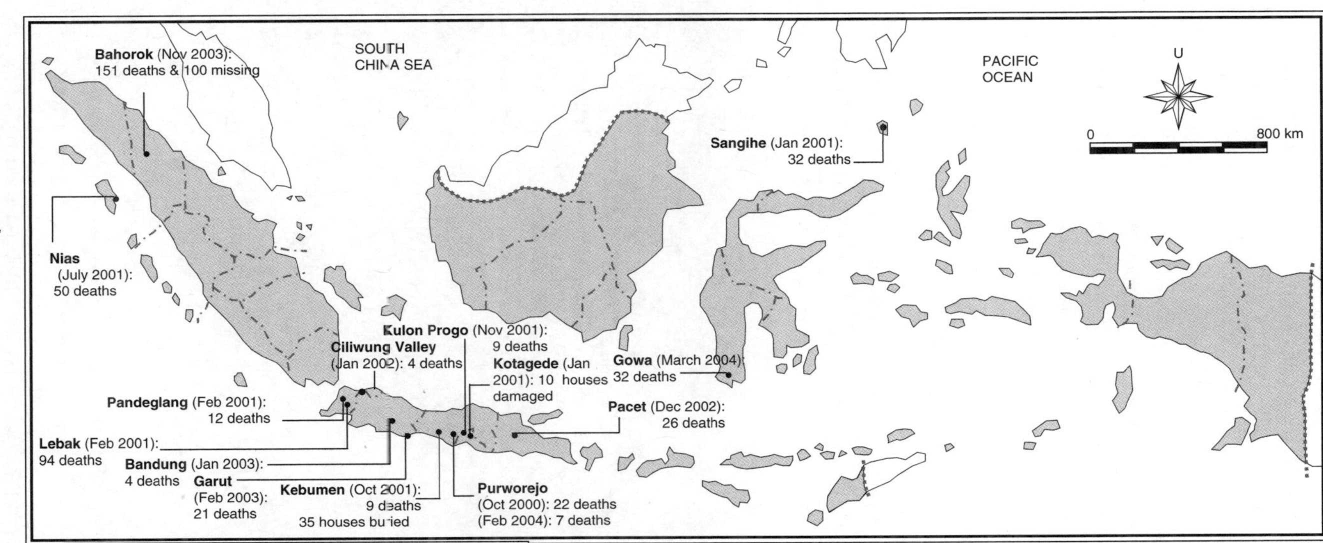


*Figure 25.1*    Main landslide occurrences in Indonesia in the period 2000–2004

events in those volcanic areas were during or immediately after rainfall events, where the average monthly rainfall approximates 250 mm, with the maximum daily rainfall exceeding 100 mm.

An important factor in terms of social awareness needs is not just the infiltration trigger, but the resulting speed of slip mass movement; in many cases this was as much as 25 m/minute, causing considerable loss of life (Table 25.1). Figure 25.2, for example, shows a debris slide on a 45° slope of sandstone covered by sandy clay in Penusupan Village, Regency of Kebumen in the south of central Java (4 October 2001) in which nine people were buried.

Slow mass movement, or creep (rates <1 m per year of lateral movement), occurs in several places in southern Java (such as in Semampir and Seling in Kebumen Regency, and on the provincial highways in Grobogan and Ngawi Regencies). Creep occurs on gentle slopes of around 15°. These gentle slopes typically comprise silty or clayey soil, underlain by low-permeability clay (such as montmorillonite) with paddy fields being the most common land use. On such slopes, whilst there is generally no loss of life, many houses are destroyed, because creep processes commonly occur over relatively large areas (in excess of 1 ha).

It was also evident that deforestation has an important role in increasing the number of rain-induced landslides. In December 2002 a debris flow occurred in Pacet (on the

**Figure 25.2**  *Debris slide on 45° slope of sandstone covered by sandy clay in Penusupan Village, Regency of Kebumen in the south of central Java (4 October 2001; 9 people buried)*

northwest slopes of Welirang Volcano, southeastern Java). Before this debris flow, there had been several days of rain, with intensities exceeding 50 mm/day. Deforestation in this area had reached 50%, equating to 195.8 ha of pine forest. Twenty-six people were buried instantly by the debris flow and 14 others were reported missing. In February 2003 another debris flow occurred on the slope of Mandalawangi Volcano at Garut Regency, in the south of west Java during several days of heavy rain (when rainfall intensities exceeded 100 mm/day). Twenty-two other people lost their lives and, again, extensive deforestation had taken place on the upper slopes of this volcano.

It is evident that vegetation cover on such slopes has an important role in controlling rain-induced landslides, and it is this proposition that will be further investigated through simulation modelling conducted in this chapter.

Field evidence of this form was used to parameterize three 'typical' slope types for southern Java (Figure 25.1). Three typical slopes of 15°, 30° and 45° were selected for numerical simulations (Table 25.2). These slopes were parameterized by appropriate geotechnical, hydrological and vegetation characteristics typical of andesitic breccia overlain by sandy clay (Table 25.3). This configuration is one of the most common rock and soil types on slopes in southern Java which are prone to landslides.

Five vegetation types were selected for the simulation experiments. Pine is one of the most common vegetation covers for the steeper slopes in Java, and consequently was selected as the vegetation cover for the 45° slope category (Figure 25.3a). Elsewhere, on 30° slopes, bamboo plantations are common. As well as bamboo, currently there is a emphasis on cultivating vertiver grass on such slopes for both protection and commercial reasons. Thus both bamboo and vertiver vegetation covers were applied to the 30° slope simulation. In the context of the shallower 15° slopes, paddy fields and cassava (Figure 25.3b) were selected as appropriate vegetation covers.

## 25.3 Current Inadequate Mapping of Landslide Risk and the Need for Improved Public Awareness

Surface and subsurface site investigations have been carried out to assist in the prevention and minimization of landslide consequences. Engineering geological mapping and landslide susceptibility mapping have been undertaken and maps prepared at scales ranging from 1:5 000 000 to 1:100 000. Other methods of mapping, such as conventional field mapping, statistical analyses, the use of remote sensing imagery and geographical information systems have also been implemented.

It could reasonably be expected that the results of these investigations and mapping techniques might provide the basis for an adequate landslide early warning system. However, such mapping methods do not seem an effective method of improving public awareness of landslide risk; as a consequence fatalities continue to occur as a direct result of landslide occurrence. Poor community access to, and understanding of, the mapped information (or investigation results) is seemingly always the problem. Most of the mapped results are too technical for the relevant social environments, and thus prove difficult for communities to understand.

A further problem is the accuracy of the mapped information. Most landslide sites are individually less than 1 ha in spatial extent, but their occurrence is widespread. Each landslide site will rarely be detected at a mapped scale of 1:100 000. However,

**Table 25.2** *Slope and landslide conditions analysed*

| No. | Field conditions | | | Simulated conditions | | |
| --- | --- | --- | --- | --- | --- | --- |
| | Slope conditions | | Type of movement | Impact | Slope inclination | Stratigraphic model | Engineering properties |
| | Slope inclination | Stratigraphic model | | | vegetation cover | | |
| 1 | Gentle slope ($\leq 20°$) See Figure 25.1 | Thick ($>2$ m) sandy clay, sandy silt, silty clay or clay of residual soils (commonly montmorillonite type) sometimes underlined by montmorillonite claystone or impermeable layer dipping towards slope inclination | **Slow movement** (creep) $<1$ m per year. Normally Initiated by slump at upper slope locations | • Structural damage; no loss of life <br> • Occurs over large area (larger than 1 ha), indicated by cracks in soil and structures, ground subsidence, and/or inclined trees/piles | 15° <br> • No vegetation cover <br> • Paddy <br> • Cassava | (a) Sandy clay as slope cover material <br><br> (b) Andesitic breccia impermeable underlying layer | $C' = 5\,kPa; \varphi' = 29°;$ $Ks = 5.49 \times 10^{-6}\,m/s;$ $e = 1.8; \theta s = 0.456;$ $\gamma s = 17.97\,kN/m^3;$ $\gamma d = 15.8\,kN/m^3$ <br><br> $C' = 13\,kPa; \varphi' = 33°;$ $Ks = 3.5 \times 10^{-7}\,m/s;$ $e = 0.57; \theta s = 0.38;$ $\gamma s = 18.97\,kN/m^3;$ $\gamma d = 18\,kN/m^3$ |
| 2 | Steep slope ($>20-30°$) See Figure 25.2 | • Thick ($>3$ m) loose sandy clay, sandy silt, silty clay or clay of residual soils underlined by denser soil or rock mass | **Rapid movement** ($>25$ m/minute) with deep slide or slump. This may develop as debris or earthflow | • Frequently causes loss of life and/or damaged structures <br> • Occurs in localized area ($\leq 1$ ha), but moving material may flow downslope <br> • Sometimes the moving material impounds river flow, causing flooding or downslope debris flow | 30° <br> • No vegetation cover <br> • Bamboo <br> • Vertiver | (a) Sandy clay as slope cover material <br><br> (b) Andesitic breccia impermeable underlying layer | $C' = 5\,kPa; \varphi' = 29°;$ $Ks = 5.49 \times 10^{-6}\,m/s;$ $e = 1.8; \theta s = 0.456;$ $\gamma s = 17.97\,kN/m^3;$ $\gamma d = 15.8\,kN/m^3$ <br><br> $C' = 13\,kPa; \varphi' = 33°;$ $Ks = 3.5 \times 10^{-7}\,m/s;$ $e = 0.57; \theta s = 0.38;$ $\gamma s = 18.97\,kN/m^3;$ $\gamma d = 18\,kN/m^3$ |

| 3 | Very steep slope ($>45°$) See Figure 25.3 | • Thick ($>3$ m) and loose sandy clay, sandy silt, silty clay or clay of residual soils underlined by denser soil or rock mass | **Rapid movement** (being more than 25 m/minute) with shallower slide or slump. This may also develop as debris or earthflow | • Frequently causes loss of life and/or damage to structures<br>• Relatively localized area of movement ($\leq 1$ ha), but moving material may flow downslope<br>• The moving material can block river flow and cause floods or downslope debris flow | 45° <br>• No vegetation cover<br>• pine | (a) Sandy clay as slope cover material | $C' = 26\,\text{kPa}$; $\varphi' = 29°$; $Ks = 5.49 \times 10^{-6}\,\text{m/s}$; $e = 1.8$; $\theta s = 0.456$; $\gamma s = 17.97\,\text{kN/m}^3$; $\gamma d = 15.8\,\text{kN/m}^3$ |
| | | | | | | (b) Andesitic breccia impermeable underlying layer | $C' = 26\,\text{kPa}$; $\varphi' = 33°$; $Ks = 3.5 \times 10^{-7}\,\text{m/s}$; $e = 0.57$; $\theta s = 0.38$; $\gamma s = 18.97\,\text{kN/m}^3$; $\gamma d = 18\,\text{kN/m}^3$ |

*Notes:*
$C'$ = effective cohesion; $\varphi'$ = internal friction angle; $Ks$ = saturated hydraulic conductivity; $e$ = void ratio; $\theta s$ = saturated volumetric moisture content being equal to effective porosity; $\gamma s$ = saturated unit weight; $\gamma d$ = dry unit weight.

**Table 25.3**   *Vegetation parameters for the simulated slopes*

| Parameter | Pine | Vetiver | Bamboo | Cassava | Paddy (dry)* |
|---|---|---|---|---|---|
| Root tensile strength kNm$^{-2}$ | $\tau_r = 18$ | $\tau_r = 12$ | $\tau_r = 10$ | $\tau_r = 10$ | $\tau_r = 8$ |
| Vegetation cover % | $v_c = 36.1$ | $v_c = 45$ | $v_c = 30$ | $v_c = 45$ | $v_c = 55$ |
| Leaf area index m$^2$m$^{-2}$ | $lai = 13.1$ | $lai = 1.5$ | $lai = 0.95$ | $lai = 5$ | $lai = 0.95$ |
| Aerodynamic resistance sm$^{-1}$ | $r_a = 7.5$ | $r_a = 40$ | $r_a = 7.5$ | $r_a = 40$ | $r_a = 40$ |
| Canopy resistance sm$^{-1}$ | $r_c = 70$ | $r_c = 70$ | $r_c = 120$ | $r_c = 80$ | $r_c = 50$ |
| Maximum transpiration ms$^{-1}$ | $T = 3.10^{-7}$ | $T = 4.5.10^{-7}$ | $T = 5.50 \times 10^{-7}$ | $T = 3.10^{-7}$ | $T = 5.1 \times 10^{-7}$ |
| Root depth/lateral extent m | $R_d = 4,$ $R_l = 1$ | $R_d = 3,$ $R_l = 0.5$ | $R_d = 0.5,$ $R_l = 1$ | $R_d = 0.40,$ $R_l = 0.25$ | $R_d = 0.3,$ $R_l = 0.2$ |
| Vegetation surcharge kNm$^{-2}$ | $S_w = 3$ | $S_w = 0.25$ | $S_w = 2$ | $S_w = 0.5$ | $S_w = 0.2$ |
| Root area ratio % | $RAR = 30$ | $RAR = 10$ | $RAR = 25$ | $RAR = 15$ | $RAR = 5$ |

*Paddy field (wet) simulated as water pond or water head of 20 cm on slope surface

inside a given 1 km$^2$ area (which for the scale reasons just outlined is likely to be considered as a very low-risk landslide zone), there might be several highly vulnerable sites, each of which covers an area of 50 m × 50 m or less. The communities living in those sites will thus consider they live in a safe zone, as defined by the currently available mapped information. Clearly, an inappropriate map scale for landslide risk may result in communities receiving misleading information. Furthermore, this is not a static mapping problem. Dynamic changes to risk occur since the vulnerability of specific zones may change from year to year in response to changes in land use, for example urbanization and deforestation.

Slope monitoring systems have been installed at several sites where there are important structures, such as highways and dams. These systems include inclinometers, strain-gauge, tilt-meters and extensometers. For the communities in landslide-vulnerable areas such equipment is expensive and complicated not only in terms of installation, but also in terms of operation and maintenance. Moreover, in order to provide an effective and accurate landslide warning system, continual analyses of the monitoring results are required. All of these factors combine to render such instrumentation methods unsuitable for the relevant local communities.

Learning from those experiences, it is evident that a different approach to developing an *appropriate* landslide warning system is required. Such a system should be as simple and community-friendly as possible so that it enables the community to provide, operate and maintain the system by themselves. A further important point in the context of Indonesian landslide occurrence is that it is much more preferable to establish a methodology which addresses issues of slope maintenance (such as appropriate land use) since post-failure

**Figure 25.3a** *Pine commonly cultivated on steep slope ($\geq 30°$) such as on this slope of sandy clay underlain by andesite in Pacet area, Mojokerto Regency, in the south of east Java (photo taken on 13 December 2002, 2 days after debris flow occurred at downslope where 26 people were buried). A slump occurred on this slope where there was no pine cover, indicated by the appearance of a curve-crack (with horse-shoe shape) in the foreground of pines*

rehabilitation is expensive, politically sensitive and totally disruptive in a social context.

This chapter seeks to address issues relating not only to triggering rainfall events but also the slope protection measures which could be developed in a bioengineering context. As we have noted, it is apparent that most of the areas vulnerable to instability are located on sloping agricultural land and for which improved knowledge of the relevant bioengineering controls is clearly of significance.

To overcome the difficulty of disseminating information relating to landslide risk in poor communities it is possible to attempt to increase public awareness at an appropriate level using carefully designed leaflets or posters. The currently available leaflets (Figure 25.4a) show typical slope types that are prone to landslides, as well as potential indicators of incipient instability. Additionally, such documentation shows simple precautions that the individual landowner or home owner can take to minimize landslide risk. From trialling such an approach in Java it was apparent that this dissemination method is effective, especially for the more vulnerable landslide areas in Java. This approach is seen as important in such developing areas – Holcombe *et al.* (2003), for example,

**Figure 25.3b**   *Cassava, particularly common in the south of central Java, planted on slope with other vegetation types (site: Magelang Regency, October 2002)*

have designed an analogous document for the Caribbean (Figure 25.4b) where similar conditions exist to those in Java, albeit at a smaller scale.

However valuable such an approach has been shown to be, more detailed information on the likely triggering rainfall characteristics which could be used to alert the community to the onset of instability is the next step in refining the public awareness methodology. Additionally, information concerning appropriate bioengineering species for slope stability 'enhancement' could be very beneficial in such environments. The modelling enhancement we propose in this chapter seeks to augment the generalized public awareness campaign with the critical information on triggering rainfall and species recommendations for given slope types.

## 25.4   Using a Combined Hydrology and Stability Model (CHASM) to Estimate Instability Triggering Conditions

A suitable platform for assessing critical triggering rainfall and the role of different vegetation covers on slope stability is the Combined Hydrology and Stability Model (CHASM). This model has been fully outlined elsewhere (Anderson, 1990; Anderson and Lloyd, 1991) and is available in both Windows (www.chasm.info) and UNIX form. The model structure is described briefly below and is illustrated in Figure 25.5; the basic equation set is given in Appendix 1.

**Figure 25.4a** *Community publicity material from Indonesia to both sensitize and inform local communities of the risks associated with landslides and the simple precautions that can be taken to reduce landslide risk*

CHASM is a physically based two-dimensional combined soil hydrology–slope stability model that allows simulation of changes in porewater pressures in response to individual rainfall events, and considers their role in maintaining slope stability (Anderson *et al.*, 1988). The procedure adopted to model the hydrological system is a forward explicit finite difference scheme. The hydrology model simulates flow over and through the discretized slope by moving water between adjacent cells. Rainfall is allowed to infiltrate the top cells governed by the infiltration capacity. Vertical flow through each column within the unsaturated zone is calculated using the Richards equation. This is solved in explicit form with the unsaturated conductivity defined by the Millington–Quirk (1959) procedure. Flow between columns is modelled in one dimension using the Darcy equation for saturated flow.

The techniques for stability assessment used in the model described here are Bishop's simplified circular method (Bishop, 1955) and Janbu's non-circular method (Janbu, 1954). These methods are used to determine the effective stress along the failure surface, the downslope shear force and the ratio between these two. The latter, termed the factor of safety (FS), provides a measure of the relative stability of the slope. On each hour of the simulation, soil-water pressures are calculated for each cell in the profile using prevailing moisture contents and the suction–moisture curve appropriate for that cell. The porewater pressures are then incorporated directly into the Mohr–Coulomb equation for soil shear strength. By calculating both the effective shear stresses along the failure surface and the soil shear strength, it is possible to determine the FS. For each hour of the simulation,

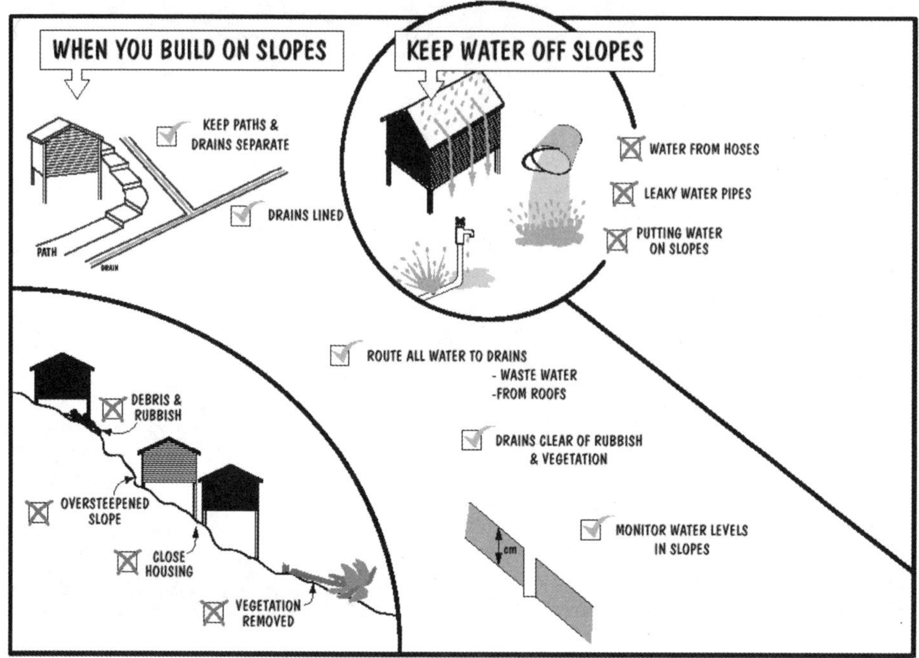



**Figure 25.4b**  *Public awareness material developed for the St Lucian Government in the context of providing advice on five simple steps to improve slope hydrology and reduce landslide risk (Holcombe et al., 2003)*

the minimum FS is obtained for the slope, with any temporal variations arising from dynamic hydrological responses (Wilkinson *et al.*, 2002).

In hydrological terms, vegetation affects slope hydrology and slope stability through three mechanisms. The first of these is interception. The modelling procedure here depends on the type of vegetation. Dense stands of tall grass can be flattened by intense rainfall to form a semi-permeable barrier. This is demonstrated by results from experimental plots in Hong Kong where between half and three-quarters of all rain formed runoff without the infiltration capacity of the soil having been reached (Lamb and Premchitt, 1990). This process is incorporated into CHASM by simply reducing the hourly rainfall intensity applied to the surface of the slope, according to the depth of grass. For interception by trees, the structure of the canopy is described by the free throughfall coefficient, the stemflow-partitioning coefficient, the canopy storage capacity and the trunk storage capacity (Rutter *et al.*, 1971; Valente *et al.*, 1997). The model estimates throughfall, stemflow and interception loss from input rainfall and meteorological data (Figure 25.5b).

The second hydrological mechanism relates to evapotranspiration and root water uptake. In order to calculate this, the Penman–Monteith equation for evapotranspiration is adopted (Monteith, 1973). This is coupled with the spatial distribution of the roots and the soil moisture content to determine the rate of water uptake from each portion of the soil (Feddes *et al.*, 1978; Wilkinson *et al.*, 1998). The final hydrological effect is concerned

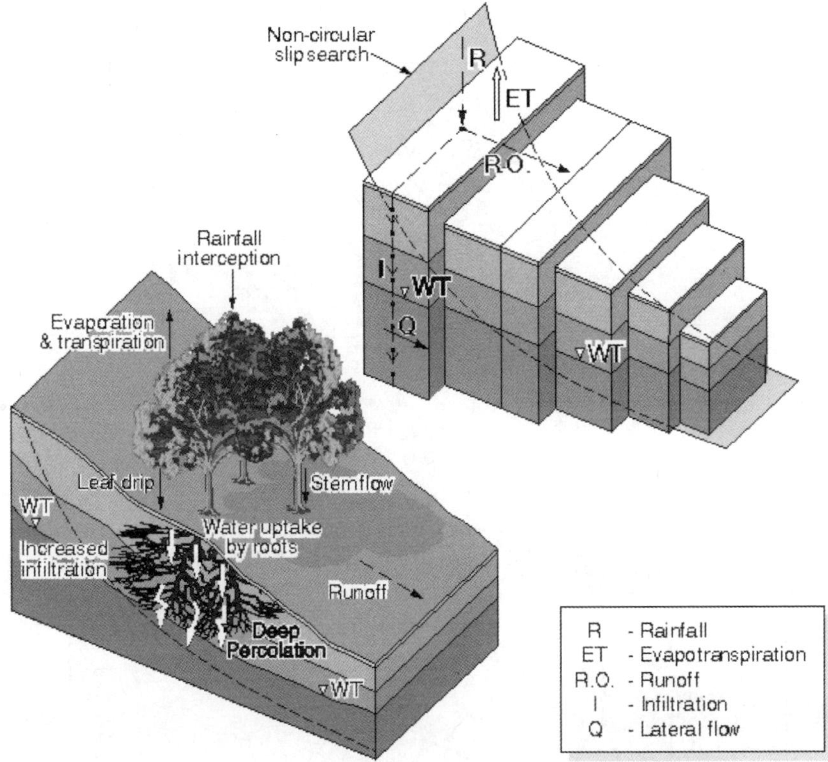



**Figure 25.5a**   *Schematic outline of CHASM model framework*

with the increase in soil permeability as a result of the root network. The magnitude of this effect is determined by an empirical equation relating the root area ratio to the saturated hydraulic conductivity (Collison, 1993).

The mechanical effects of vegetation include both root reinforcement and vegetation surcharge. Root reinforcement is represented by an increase in the effective cohesion in equation (7), Appendix 1. This is determined according the root tensile strength and the root area ratio of the root network (Wu *et al.*, 1979). The added surcharge is calculated according to the unit weight of the vegetation and is distributed through the soil profile according to the vertical distance from the ground surface.

## 25.5   Evidence of CHASM Model Performance and Capability in Assessing the Net Effect of Bioengineering Intervention

CHASM has undergone 'validation' in Malaysia, Hong Kong and New Zealand. Porewater pressure data collected from the Kuala Lumpur–Karak Highway in Malaysia provided an appropriate means of evaluating the two-dimensional finite-difference soil hydrology

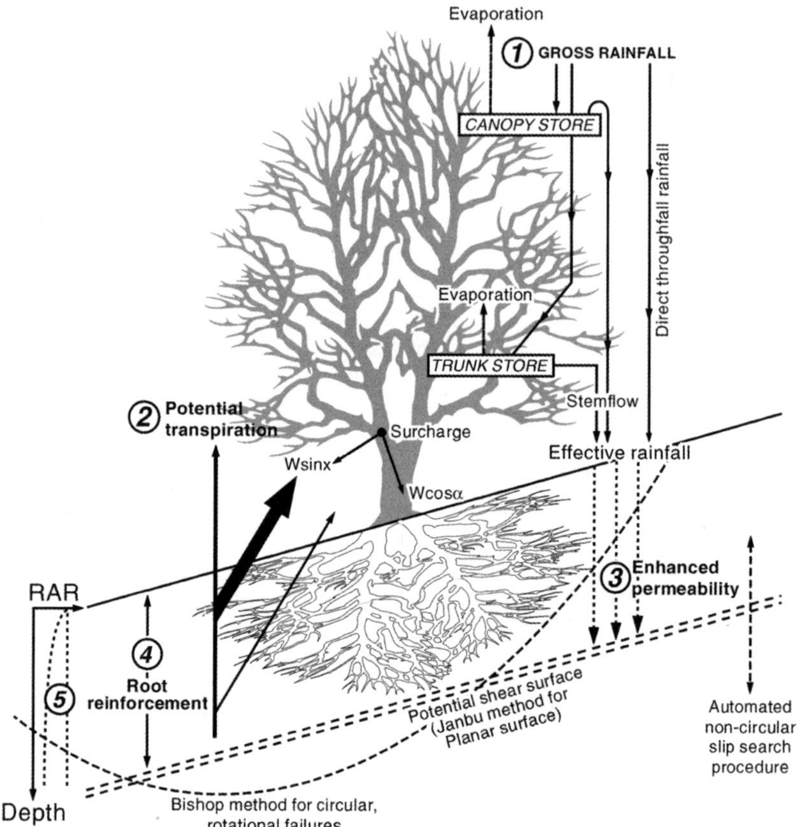



***Figure 25.5b*** *Schematic representation of the integrated vegetation–slope model (1 Rutter model: Rutter* et al., *1971; 2 Penman–Monteith: Monteith, 1973; 3 Permeability model: Collison, 1993; 4 Root reinforcement: Wu* et al., *1979; 5 Water uptake model: Feddes* et al., *1978)*

model (Anderson *et al.*, 1988; Anderson and Kemp, 1991). The model has also been used in a comparison of model results with a large sample of known stable and failed slopes in Hong Kong. In this validation exercise, 77% of failed slopes and 68% of stable slopes were correctly predicted by the model (Anderson, 1990). Finally, the vegetation model has been applied and validated in the Hawke's Bay region of the North Island of New Zealand. The sedimentological, historical and contemporary records here allowed a multi-level approach to model validation, which proved to be successful. On the basis of the validation exercises described above, it is feasible to produce charts for the rapid assessment of stability conditions. Anderson and Lloyd (1991) successfully used this technique to produce design charts for investigation into the stability and design of cut slopes in the tropics. In this example, each vegetation model component may be represented in terms of its likely parameter range and, when combined with the other system components, be used to produce a matrix of model simulations.

The design chart approach enables:

1. Determination of the net effect of vegetation on slope stability under a range of vegetation–slope conditions, thus providing a rapid bioengineering analysis tool.
2. Preliminary investigation into vegetation change scenarios. The charts may assist modelling growth and decay cycles associated with timber harvesting.
3. Direction of the types of investigation undertaken for a particular site. This may include more rigorous numerical approaches and simulations or different strategies for instrumentation and remediation.

Given the clear trend in the design chart, ranging from negative in the top left-hand corner to positive in the bottom right-hand corner, it is possible to delineate zones of vegetation effects into very strong, significant and negligible effects (Figure 25.6). CHASM modelling shows, for the example presented, that clear differentiation is possible between those slope topographic, geotechnical and hydrological characteristics that result in bioengineering having a net *negative* effect and those resulting in a net *positive* effect. In particular, there are especially critical elements of the result domain that illustrate the sensitivity that root area ratio and soil hydraulic conductivity play in establishing threshold bioengineering net benefits in relation to slope stabilization.

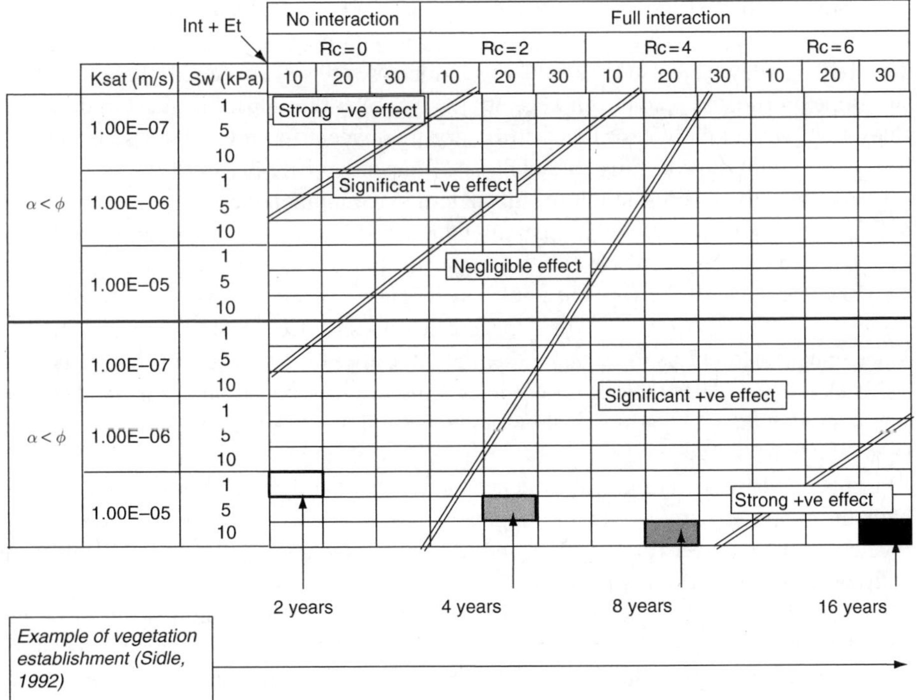


**Figure 25.6** *Delineating vegetation–slope effects for rapid assessment (α = slope angle, ϕ = angle of internal friction, Ksat = saturated hydraulic conductivity and Sw = vegetation surcharge, Rc = added root cohesion, kPa and Int + Et = % rainfall lost from interception and evapotranspiration)*

Against this background it is felt that CHASM can be used in numerical experiments to determine critical instability triggering conditions for type slopes in Indonesia and as a means of estimating appropriate bioengineering practices appropriate to improve stability. Critically, such information needs then to be formulated in a manner that is readily communicated to the poor communities at greatest risk. This is the basis and focus of the analysis that now follows.

## 25.6   Numerical Simulations

As discussed above, the numerical analysis relates to the stability study of three slope profiles with specified geotechnical characteristics, hydrological initial conditions and different types of vegetation cover, in response to four typical rainstorms. The analyses undertaken and presented below relate to assessments of each slope's factor of safety under conditions defined for 'constant vegetation and soil parameterization'. The geometrical profiles of the slopes used were: 15°, 30° and 45° slope angles at heights of 30 m, 40 m and 50 m respectively.

From a geotechnical point of view, the three slopes were considered as comprising two soil types, a sandy clay overlying a less permeable soil (see Table 25.2). Each simulation was conducted for 300 hours, starting with 120 hours of rain of selected intensity, followed by a dry period of 180 hours. Four rain scenarios with constant intensities of 160, 100, 50 and 20 mm/day were used. Several vegetation cases were considered; these were first bare slope profile and second full vegetation cover over the slope. For the 15° slope, the paddy field was initially considered to be dry (equivalent to uniform grass cover), and then as wet (with the soil fully saturated, and a detention capacity of 20 cm).

The numerical scenarios and the results in terms of minimum factor of safety are shown in Table 25.4. For the 15° slope, the simulation results show that instability problems occur only in the vegetated case with wet paddy field, if the initial groundwater level is higher than 50% of the top soil thickness (Figure 25.7 and Table 25.4). If the slope is non-vegetated, or vegetated with a species that doesn't need complete irrigation, such slopes remain stable. It is to be noted that the intensity of the rainstorm is less important in this case – the water table level dominates the stability conditions. This is of course consistent with the fact that for shallow slopes the soil water potential dominates the total potential conditions (since the elevation potential is relatively less significant).

For the 30° case the numerical evidence shows that the dynamic hydrology is critical (both the storm dynamics and the vegetation as it affects the porewater pressures). Vegetation, both vetiver (Figures 25.8 and 25.9 and Table 25.4) and bamboo is sufficient to provide increased stability in this case.

For the 45° slope, the vegetation becomes the critical factor in stability terms, beside the water level (Figure 25.10 and Table 25.4). The simulations demonstrate that the slope could be unstable, even in the case of very low rainfall intensity, but that stability could be restored by the occurrence of pine trees on the slope (Figure 25.9). In the case of this steeper slope, geometry and geotechnical properties become a dominant factor – thus the increase in cohesion afforded by vegetation is significant.

From the simulations, we see the progressive shift in process controls on slope stability from water table and soil water potential domination (15° slopes), through dynamic

*Table* **25.4** *Influence of vegetation cover on the factor of safety for 15°, 30° and 45° slopes in response to a 5 day rainstorm event (300 hrs simulation time)*

| Slope angle | | FOS for slope profile, rainstorms and vegetation type | | | | | | | | | | | |
| --- | --- | --- | --- | --- | --- | --- | --- | --- | --- | --- | --- | --- | --- |
| | | 15° | | | | 30° | | | | 45° | | | |
| Rainfall (mm/day) | | 160 | 100 | 50 | 20 | 160 | 100 | 50 | 20 | 160 | 100 | 50 | 20 |
| Vegetation type | | | | | | | | | | | | | |
| Non-vegetated | | 1.166 | 1.235 | 1.334 | 1.423 | 0.916 | 0.971 | 1.017 | 1.046 | 0.861 | 0.912 | 0.954 | 0.981 |
| Paddy field | dry | 1.456 | 1.452 | 1.450 | 1.454 | – | – | – | – | – | – | – | – |
| | wet | 0.977 | 0.977 | 0.977 | 0.977 | – | – | – | – | – | – | – | – |
| Cassava | | 1.456 | 1.457 | 1.459 | 1.461 | – | – | – | – | – | – | – | – |
| Vetiver | | – | – | – | – | 1.064 | 1.061 | 1.063 | 1.067 | – | – | – | – |
| Bamboo | | – | – | – | – | 1.057 | 1.051 | 1.056 | 1.063 | – | – | – | – |
| Pine | | – | – | – | – | – | – | – | – | 1.029 | 1.031 | 1.030 | 1.017 |

*Notes:*
for 15° slope, initial water table set at 75% of soil depth;
for 30° slope, initial water table set at 25% of soil depth;
for 45° slope, initial water table set at 10% of soil depth.

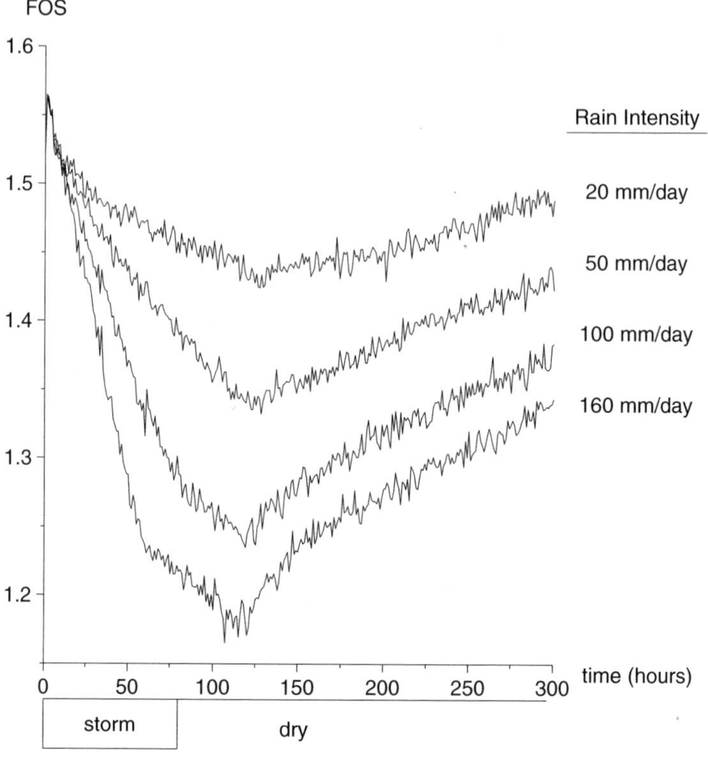


**Figure 25.7**   *Factor of safety for 15° non-vegetated slope in response to different rainstorm intensities*

hydrological controls (storm type and pore pressure changes due to vegetation – 30° slopes) to geotechnical and slope geometry dominance (45° slopes).

## 25.7   A Case Study Examining Process Failure Mechanisms – Menoreh Mountains, Southern Central Java

The simulation results for the 30° slope closely correspond to several landslide events occurring in the Menoreh Mountains in southern central Java. The largest landslide was in Kemanukan Village, where 22 people died; it occurred on 9 November 2000 during the third day of heavy rainfall, when the daily rain intensity exceeded 210 mm.

The failed slope comprised andesitic breccia covered by sandy clay (Figure 25.11) the geotechnical properties of which are illustrated in Table 25.5. The field–slope stratigraphy and corresponding engineering properties relate very closely to the slope configuration used in the simulated analysis above for the 30° slope.

Figure 25.12 shows the precipitation at the landslide area. Rainfall accumulation for the 10-day period 2–11 November amounted to 621 mm and thereby corresponds closely to the simulated rainstorm used above, with an average rain intensity of 100 mm/day. Engineering geological investigation confirmed that the Kemanukan landside was induced

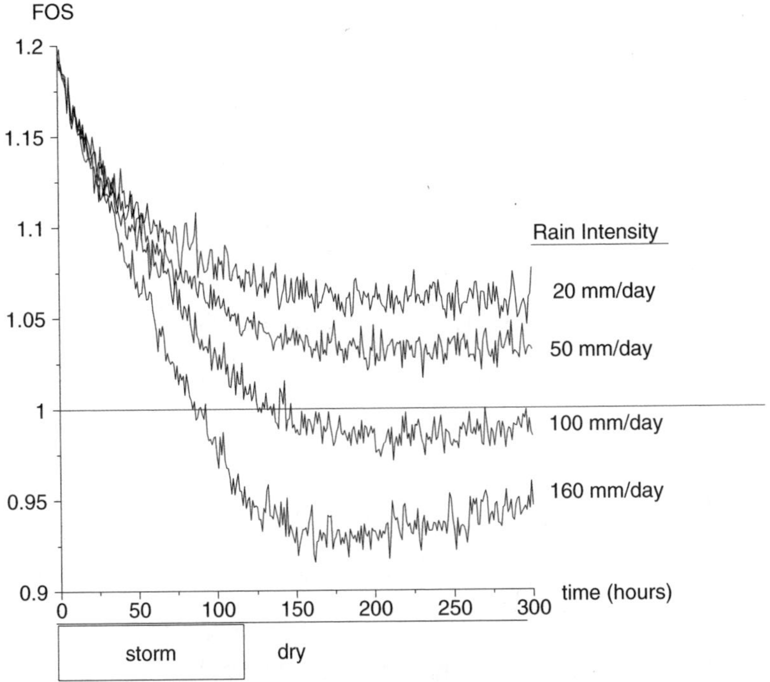



**Figure 25.8**   *Factor of safety for 30° non-vegetated slope in response to different rainstorm intensities*

by infiltrating rainwater in the sandy clay that cannot penetrate into the more impermeable andesitic breccia layer. The flow of mud from the failed slope indicated that the water content in the sandy clay was sufficiently high to create high water pressures in the slope. This apparent mechanism of slope failure conforms to the simulation results for the 30° slope category above (Figures 25.8 and 25.9). This slope profile shows that at the time of failure there was a significant groundwater table rise (being some 6 m above the initial level), as a response to infiltration on the non-vegetated slope. The moisture profile for the non-vegetated slope was initially in the range of 0.28 to 0.35, but increased sharply to the range 0.39 to 0.42 by the end of rainstorm (at hour 120).

However, we should note that the cohesion of the Kemanukan slope is somewhat higher than the simulated slope conditions used in the analysis above. Applying this higher field-cohesion value (12.74 kPa), the simulated slope remained stable, with a safety factor of 1.05 despite the 5-day, 100 mm/day rainstorm. In fact, the actual slope in Kemanukan failed under similar rainstorm characteristics. The slightly higher position of the initial water table in the Kemanukan slope than that in the simulated slope may be responsible for this difference. The initial water table level of the simulated slope was 25% of the total thickness of permeable sandy clay, being at a level about 8 m above the surface of impermeable layer (the total thickness of sandy clay was some 30 m).

Another simulation with the same slope angle and stratigraphy as those of Kemanukan slope was then implemented to check this proposition. With the initial groundwater table of 33% of the thickness of sandy clay layer (10 m above the impermeable layer of

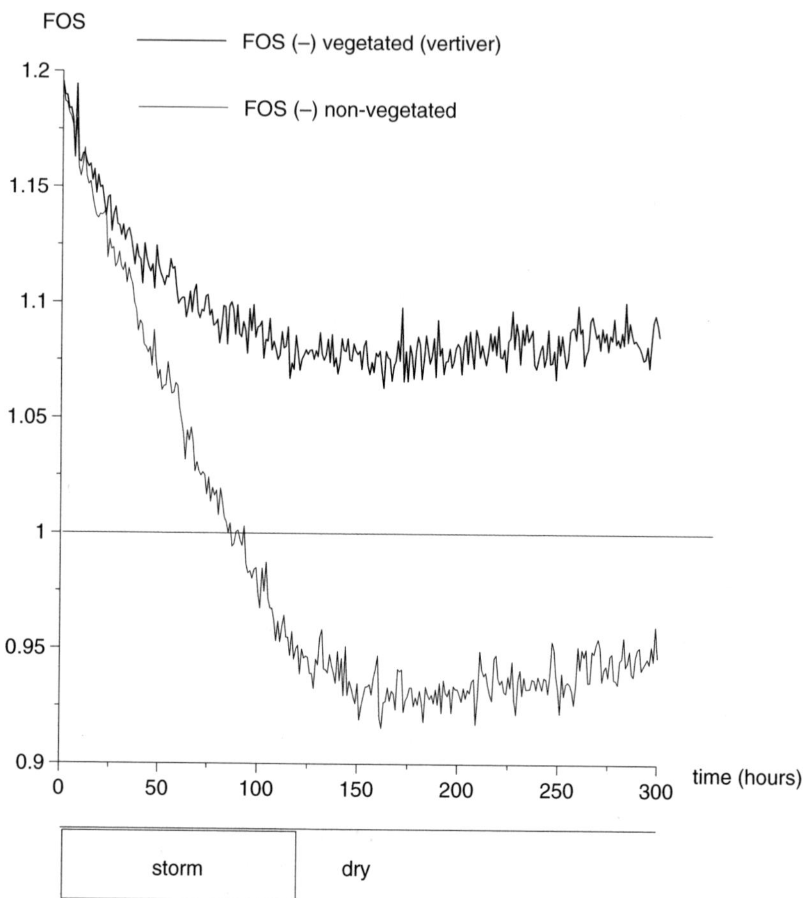


**Figure 25.9** *Influence of vegetation (vetiver) cover on factor of safety for 30° slope in response to 5 days of 160 mm/day rain intensity*

andesitic breccia), the simulation gave a safety factor of 1.04 before the rainstorm was applied. Thus the slope had been in critical condition before the rainstorm. This factor of safety then declined to $F = 0.99$ after 5 days of rain (intensity of 100 mm/day). This last simulation seems to be more realistic for the Kemanukan slope.

From the simulations we can establish that it was very likely that the initial field water table had risen to some 33% of the thickness of the sandy clay layer in the middle of wet season, just before the period of the rainstorm. The wet season in that year actually had started in September. The antecedent rainfall (accumulated rains occurring from September to early November) thus played a significant role in establishing the critical water table in the slope, just before the rainstorm started.

Vegetation cover was also shown by this Kemanukan slope simulation to play an important role in improving and maintaining slope stability. The simulation showed that the safety factor increased from 0.99 to 1.07 under the protection of vegetation cover (despite the 5-day, 100 mm/day rainstorm). Vegetation cover can provide additional

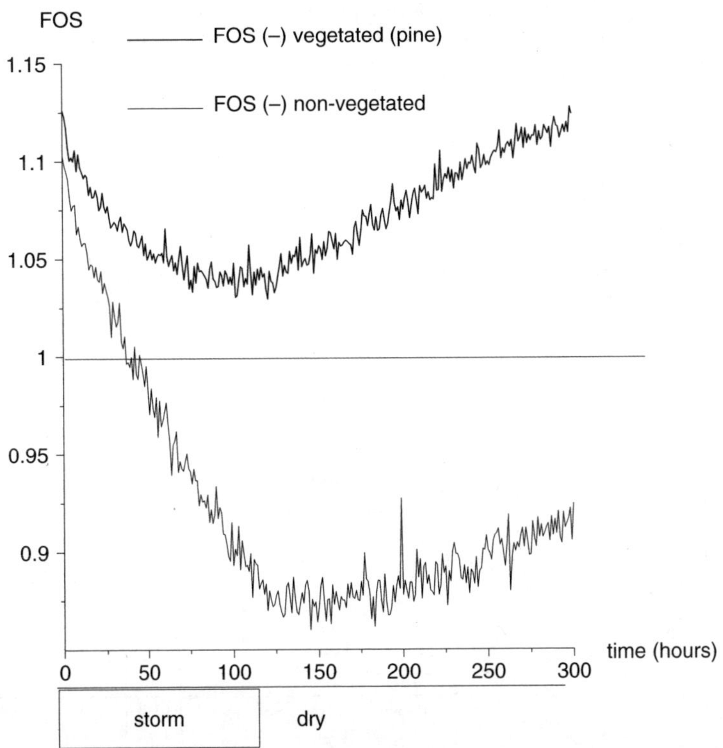


**Figure 25.10** *Influence of vegetation (pine) cover on factor of safety for 45° slope in response to 5 days of 160 mm/day rain intensity*

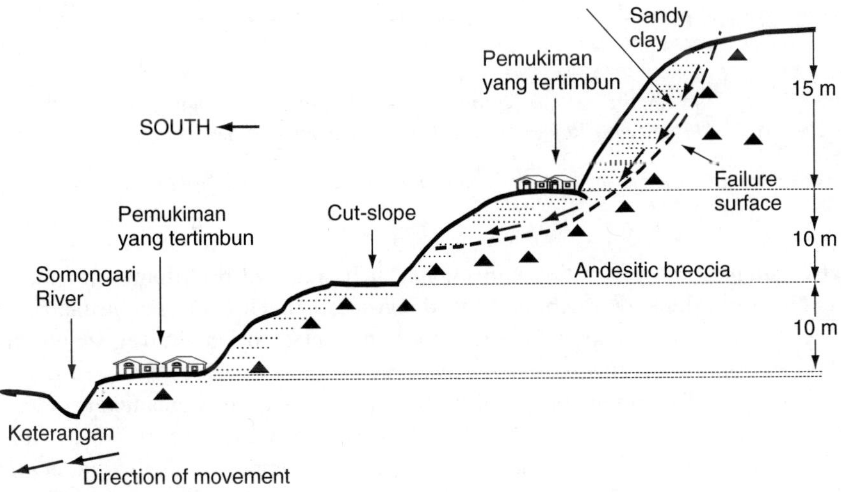


**Figure 25.11** *Schematic diagram of landslide on 30° slope of andesitic breccia covered by sandy clay at Kemanukan Village, where 22 people died on 9 November 2000*

**Table 25.5**   *Engineering properties of 30° Kemanukan slope*

| Properties | Matrix of andesitic breccia | Sandy clay |
| --- | --- | --- |
| $\rho s(KN/m^3)$ | 17.00 | 16.11 |
| $\rho d(KN/m^3)$ | 12.90 | 11.05 |
| Gs | 2.58 | 2.58 |
| Void ratio | 1.01 | 1.03 |
| K (m/s) | $3.42$ *to* $5.2.10^{-8}$ | $1.57^{-7}$ *to* $6.31\ 10^{-5}$ |
| C′ (kPa) | 35.00 | 12.74 |
| $\varphi'(°)$ | 45.00 | 26.84 |

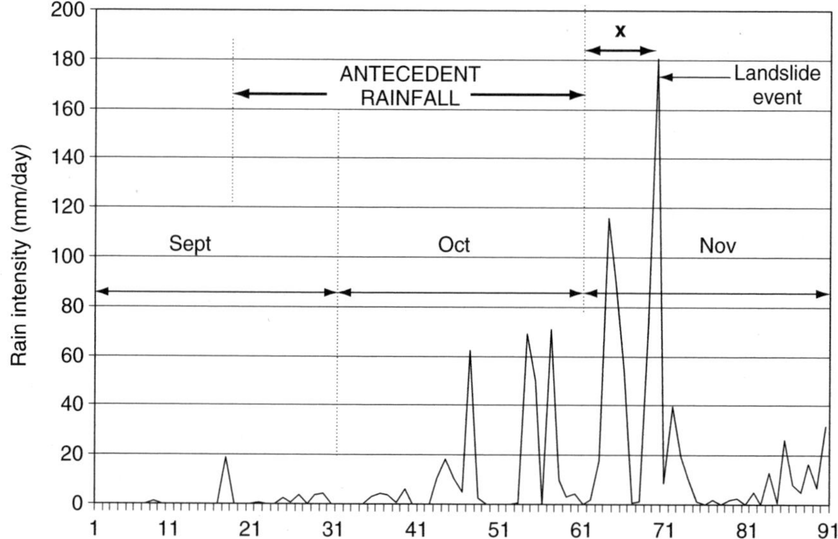


**Figure 25.12**   *Rain record data in Sermo, nearby Kemanukan Village (Directorate of Irrigation, Dept of Settlement and Regional Infrastructures, Yogyakarta Regency, Indonesia)*
Notes:
*x= one period of triggering rainstrom (2–11 Nov. 2000). Antecedent rainfall is about 10.8 mm/day lasting for 45 days.*

effective strength; however, this protection will be reduced by trimming of the slope, especially at the slope toe, such as for road construction. This was a contributory factor causing this slide on a slope which had been 'protected' by mixed vegetation, mainly bamboo, coconut trees, grasses and cassavas.

In conclusion, the simulations indicate that the landslide in Kemanukan is likely to have been significantly controlled by the slope geometry (steepness) and the relevant geotechnical properties. The dominant control, however, appears to be the initial groundwater table position and the subsequent rise in response to the rainstorm of 100 mm/day lasting for 5 days. Critically, the existence of vegetation cover can also improve the slope stability.

The critical groundwater table level on such slopes appears to approximate one-third of the thickness of the permeable soil or about one-third of slope height. If the water table in the slope has reached this 'critical level', the slope will fail without rainfall, unless, that is, there is vegetation cover providing additional 'effective' strength. However, when this critical groundwater level has not yet been achieved, the critical rainstorm suggested by the simulations is about 70 mm/day, as illustrated in Figure 25.12. The analysis of this slide is thus entirely consistent with the results shown in Table 25.4 for the 30° slope category in terms of process controls on stability and the identification of potential critical thresholds.

## 25.8   Refining the Public Awareness Documentation

We can structure the scenario modelling results to develop a set of structured 'decision' diagrams to present the dominant controls on stability for the three slope angles investigated. Figures 25.13–25.15 show the logic structure developed. The mapped thresholds are structured in terms of initial groundwater level and then discrimination is afforded using vegetation type. The resultant decision diagram for the three slope groups thereby categorizes the stability of the slopes in a manner that can, we believe, be sensibly converted to documentation that residents can both understand and act upon when appropriate. For this region of southern Java Figures 25.13–25.15 provide clear guidance and refinement of the pre-existing publicly available information which gives general guidance only (Figure 25.4a).

Figure 25.16 is an initial attempt at refining the documentation so that the results for the 45° slope category can be presented in a graphical form capable of being interpreted by residents in Java. Having made the connectivity from general landslide problem areas (Table 25.1), through parameterization (Table 25.3), model platform (Figure 25.5 and Appendix 1), scenario modelling (Table 25.4) resultant decision structures (Figures 25.13–25.15), a sample public awareness brochure has been configured

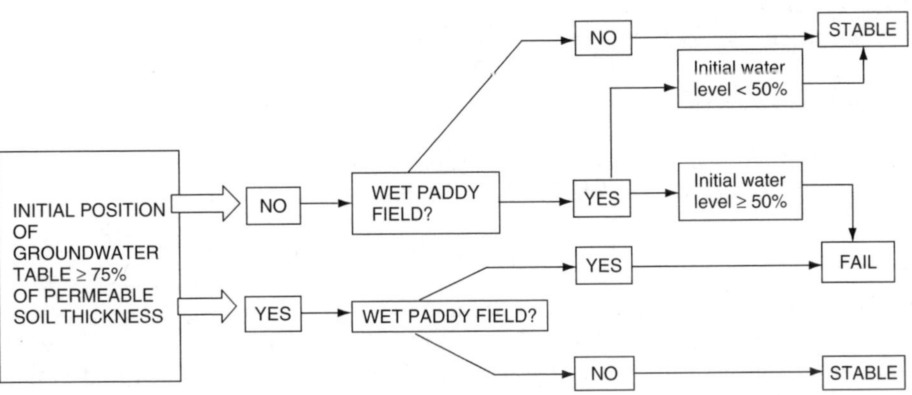


**Figure 25.13**   *Analytical assessment of the influence of initial position of groundwater table and vegetation cover on slope stability for 15° slope*

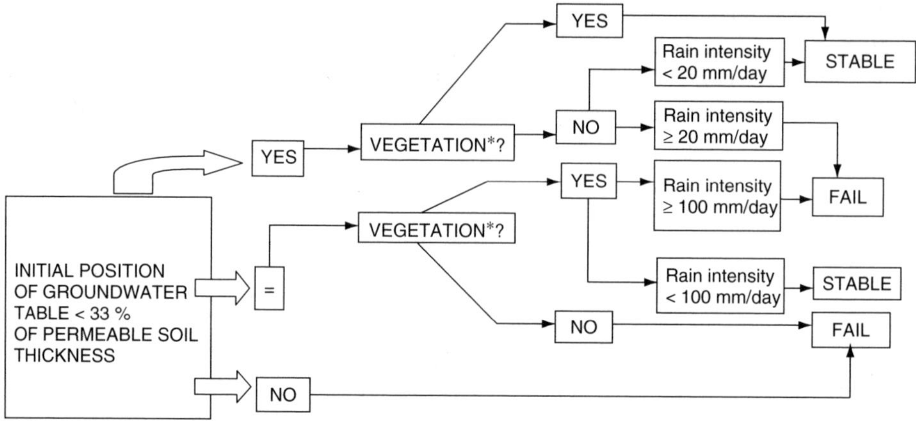


**Figure 25.14**  *Analytical assessment of the influence of initial position of groundwater table, vegetation cover and rain on slope stability for 30° slope*

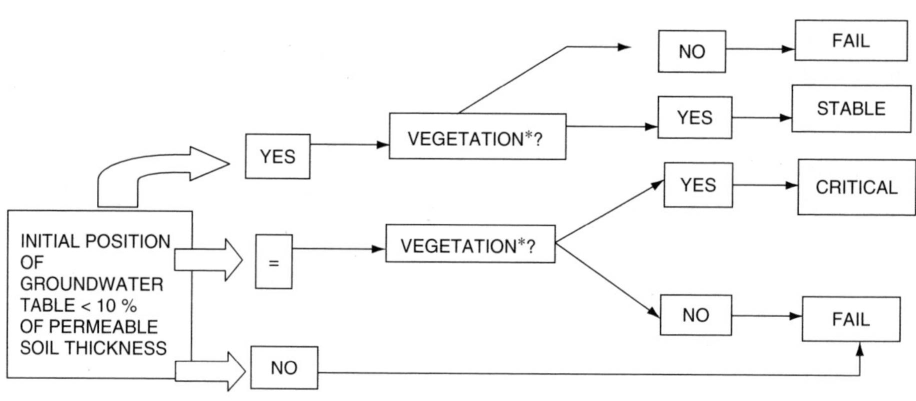


**Figure 25.15**  *Analytical assessment of the influence of initial position of groundwater table, vegetation cover and rain on slope stability for 45° slope*

for one of the slope groups by way of illustration (Figure 25.16). This general approach has considerable scope for refining the level of public awareness in those areas of the world at greatest risk from landslide hazard. Integration of this type will lead to more refined physically based schemes providing a numerical platform that is analytically robust and which, allowing for uncertainty, is none the less capable of providing specific 'best-estimate' decision structures upon which residents can act.

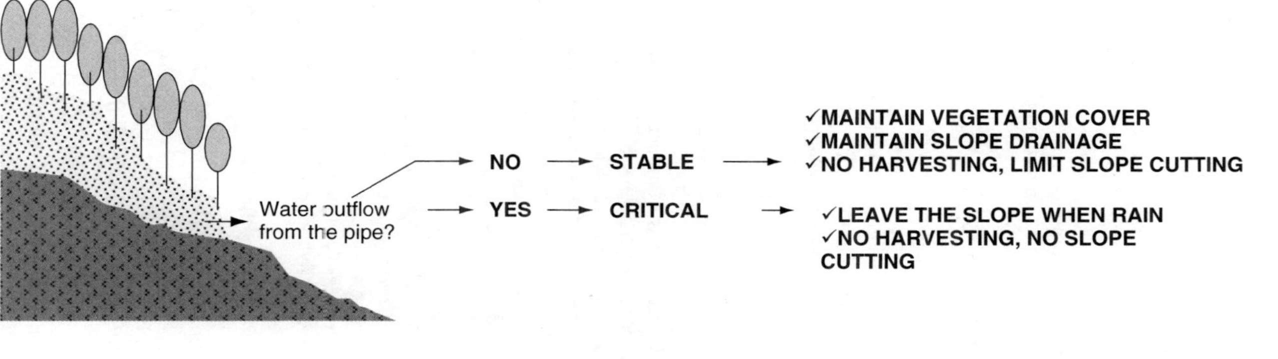
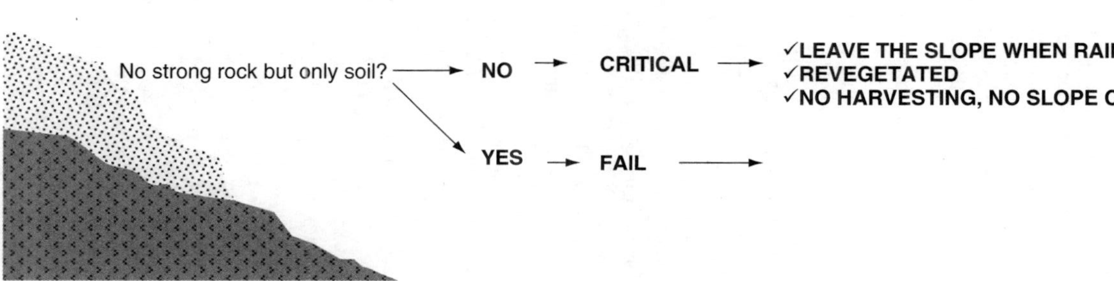


**Figure 25.16** *An initial attempt at structuring public awareness documentation which utilizes the CHASM modelling results for the 45° slope*

# References


Anderson, M.G., 1990, *A Feasibility Study in Mathematical Modelling of Slope Hydrology and Stability*, Geotechnical Control Office Civil Engineering Services Department, Hong Kong, Report CE 23/90.

Anderson, M.G. and Kemp, M.J., 1991, Towards an improved specification of slope hydrology in the analysis of slope instability problems in the tropics, *Progress in Physical Geography*, **15**(1), 29–52.

Anderson, M.G. and Lloyd, D.M., 1991, Using a combined slope hydrology–stability model to develop cut slope design charts, *Proceedings of the Institution of Civil Engineers*, **91**(2), 705–718.

Anderson, M.G., Kemp, M.J. and Lloyd, D.M., 1997, *Hydrological Design Manual for Slope Stability in the Tropics*, Transport Research Laboratory.

Anderson, M.G., Kemp, M.J. and Lloyd, D.M., 1988, Applications of soil water finite difference models to slope stability problems, *Proceedings of the 5th International Symposium on Landslides*, vol. 1, Lausanne, 525–530.

Bishop, A.W., 1955, The use of the slip circle in the stability analysis of slopes, *Géotechnique*, **5**(1), 7–77.

Collison, A.J.C., 1993, Assessing the influence of vegetation on slope stability in the tropics, Ph.D. Thesis, University of Bristol.

Coulomb, C.A., 1776, Essai sur une application des règles des maximis à quelques problèmes de statique relatifs à l'architecture, *Memorandum Academie Royale, Pres. Div. Sav*, **7**, 343–382.

Dai, F.C., Lee, C.F. and Ngai, Y.Y., 2002, Landslide risk assessment and management: an overview, *Engineering Geology*, **64**, 65–87.

Feddes, R.A., Kowalik, P.J. and Zaradny, H., 1978, Simulation of field water use and crop yield. Simulation Monographs. Pudoc. Wageningen.

Geotechnical Control Office, 1980, CHASE Cutslopes in Hong Kong: Assessment of stability by empiricism. Public Works Department, Hong Kong.

Geotechnical Engineering Office, 1998, Landslides and boulder falls from natural terrain: interim risk guidelines. GEO Report 75, Geotechnical Engineering Office, Hong Kong.

Holcombe, E.A., Ellis, D., Tooby, J. and Anderson, M.G., 2003, Public awareness documentation to improve slope stability conditions, Poverty Reduction Fund, St Lucia, unpublished MoSSaiC Management report 3.

Janbu, N., 1954, Application of composite slip surface for stability analysis, *Proceedings of the European Conference on the Stability of Earth Slopes*, **3**, 43–49.

Karnawati, D., 1996, Mechanism of rain-induced landslide in allophonic and halloysitic soils in Java, Ph.D Thesis, Dept. of Earth Sciences, Leeds University, UK.

Karnawati, D., 1997, Natural slope failure on weathered Andesitic Breccia in Samigaluh Area, Indonesia, *The 4th International Conference on Case Histories in Geotechnical Engineering*, Missouri, USA.

Karnawati, D., 2000, Assessment on mechanism of rain-induced landslide by slope hydrodynamic simulation, *GeoEng2000, An International Conference on Geotechnical and Geological Engineering*, Canberra, Australia.

Lamb, T.S.K. and Premchitt, J., 1990, Rainfall-runoff on slopes 1984–1988. Civil Engineering Services Report Hong Kong, Special Project Report I 1/90.

Millington, R.J. and Quirk, J.P., 1959, Permeability of porous media, *Nature*, **183**, 387–388.

Monteith, J.L., 1973, *Principles of Environmental Physics* (London: Edward Arnold).

Nash, D., 1987, A comparative review of limit equilibrium methods of stability analysis, in M.G. Anderson and K.S. Richards (eds), *Slope Stability* (Chichester: John Wiley & Sons Ltd), 11–73.

Richards, L.A., 1931, Capillary conduction of liquids in porous mediums, *Physics*, **1**, 318–333.

Rutter, A.J., Kershaw, K.A. and Robins, P.C., 1971, A predictive model of rainfall interception in forests. I. Derivation of the model from observation in a plantation of corsican pine, *Agricultural Meteorology*, **9**, 367–384.

Saroso, B.S., 1992, Ancaman gerakan tanah pada jaringan jalan di Jawa Barat. Paper presented on the Seminar Aplikasi Sistim Informasi Geografi untuk Jaringan Jalan di Indonesia (unpublished).

Sidle, R.C., 1992, A theoretical model of effects of timber harvesting on slope stability, *Water Resources Research*, **28**, 1879–1910.

Valente, F., David, J.S. and Gash, J.H.C., 1997, Modelling interception loss for two sparse eucalypt and pine forests in central Portugal using reformulated Rutter and Gash analytical models, *Journal of Hydrology*, **190**, 141–162.

Wilkinson, P.L., Brooks, S.M. and Anderson, M.G., 1998, Investigating the effect of moisture extraction by vegetation upon slope stability: further developments of a Combined Hydrology and Stability Model (CHASM), *British Hydrological Society International Symposium on Hydrology in a Changing Environment. Theme 4: Hydrology of Environmental Hazards*, 165–178.

Wilkinson, P.L., Anderson, M.G., Lloyd, D.M. and Renaud, J.-P., 2002, Landslide hazard and bioengineering: towards providing improved decision support through integrated model development, *Environmental Modelling and Software*, **17**, 333–344.

Wu, T.H., McKinnell III, W.P. and Swanston, D.N., 1979, Strength of tree roots and landslides on Prince of Wales Island, Alaska, *Canadian Geotechnical Journal*, **16**(1), 19–33.

Wu, W. and Sidle, R.C., 1995, A distributed slope stability model for steep forested basins, *Water Resources Research*, **31**, 2097–2110.

# Appendix 1: Principal Equation Set for CHASM

## 1. Richards Equation (1931)

$$\frac{\partial \theta}{\partial t} = -\frac{\partial}{\partial z}\left(D\frac{\partial \theta}{\partial z}\right) - \frac{\partial K}{dz}$$

$\theta$ = volumetric moisture content ($\mathrm{m^3\,m^{-3}}$)
$t$ = time (s)
$z$ = vertical depth (m)
$D$ = hydraulic diffusivity ($\mathrm{m^2\,s^{-1}}$)

## 2. Millington–Quirk Equation (Millington and Quirk, 1959)

$$K_i = \frac{K_s(\theta_i/\theta_s)^p \sum\limits_{j=i}^{m}\left((2j+1-2i)\psi_j^{-2}\right)}{\sum\limits_{j=1}^{m}\left((2j-1)\psi_j^{-2}\right)}$$

$p$ = pore interaction term
$K_i$ = unsaturated conductivity ($\mathrm{ms^{-1}}$)
$K_s$ = saturated conductivity ($\mathrm{ms^{-1}}$)
$\theta_i$ = unsaturated moisture content ($\mathrm{m^3\,m^{-3}}$)
$\theta_s$ = saturated moisture content ($\mathrm{m^3\,m^{-3}}$)
$\psi_j$ = suction value at moisture content $\theta_i$(m)
$m$ number of equal increments of $\theta$ from $\theta = 0$ to $\theta = \theta_s$
$j, i$ summation indices

## 3. Mohr–Coulomb equation (Coulomb, 1776)

$$s = c' + (\sigma - u)\tan\phi'$$

$s$ = soil shear strength $(kNm^{-2})$
$c'$ = effective soil cohesion $(kNm^{-2})$
$\phi'$ = effective angle of internal friction (degrees)
$\sigma$ = total normal stress $(kNm^{-2})$
$u$ = porewater pressure $(kNm^{-2})$

## 4. Bishop and Janbu Stability Equations (Bishop, 1955)

Bishop:

$$FOS = \frac{\sum_{i=0}^{n}(c'l + (P - ul)\tan\phi')}{\sum_{i=0}^{n} W\tan\alpha}$$

where

$$P = \left[W - \frac{1}{FS_0}(c'l\sin\alpha - ul\tan\phi'\sin\alpha)\right]\big/ m_\alpha$$

and

$$m_\alpha = \cos\alpha\left(1 + \tan\alpha\frac{\tan\phi'}{FS_0}\right)$$

$n$ = number of slices
$FS$ = factor of safety
$c'$ = effective soil cohesion $(kNm^{-2})$
$l$ = slice length (m)
$\alpha$ = slice angle (degrees)
$u$ = porewater pressure $(kNm^{-2})$
$\phi'$ = effective angle of internal friction (degrees)
$W$ = weight of the soil $(kNm^{-2})$
Janbu:

$$FOS = f_0\frac{\sum_{i=0}^{n}(c'l + (P - ul)\tan\phi')\sec\alpha}{\sum_{i=0}^{n} W\tan\alpha}$$

where $f_0$ is a correction factor to account for interslice shear forces.

## 5. Interception Equation (Rutter *et al.*, 1971; Valente *et al.*, 1997)

$$(1 - p - p_t)\int Rdt = \int Ddt + \int Edt + \Delta C$$

$$p_t\int Rdt = S_f + \int E_t dt + \Delta C_t$$

$S$ = canopy storage capacity (m)
$S_t$ = trunk storage capacity (m)

$p$  = free throughfall coefficient
$p_t$  = stemflow partitioning coefficient
$R$  = gross rainfall intensity $(\text{ms}^{-1})$
$D$  = drainage rate from the canopy $(\text{ms}^{-1})$
$E$  = evaporation rate of water intercepted by the canopy $(\text{ms}^{-1})$
$\Delta C$ = change in canopy storage (m)
$S_f$  = stemflow (m)
$E_t$  = evaporation rate of the water intercepted by the trunks $(\text{ms}^{-1})$
$\Delta C_t$ = change in the trunk storage (m)

## 6. Penman–Monteith Equation (Monteith, 1973)

$$E_p = \frac{\Delta R_n + \rho c_p VPD/r_a}{\lambda[\Delta + \gamma(1 + r_c/r_a)]}$$

$E_p$  = potential evapotranspiration rate $(\text{ms}^{-1})$
$r_a, r_c$ = aerodynamic and canopy resistance, respectively $(\text{sm}^{-1})$
$\Delta$  = slope of the saturation vapour pressure–temperature curve $(\text{kgm}^{-3}\text{K}^{-1})$
$VPD$ = vapour pressure deficit $(\text{kgm}^{-1}\text{s}^{-2})$
$c_p$  = specific heat of air $(\text{Jkg}^{-1}\text{K}^{-1})$
$R_n$  = net radiation $(\text{Wm}^{-2})$.

## 7. Root Reinforcement Equation (Wu *et al.*, 1979; Wu and Sidle, 1995)

$$\Delta c' = c'_R = t_R(\cos\theta \tan\phi + \sin\theta)$$

$c'$ = effective cohesion $(\text{kNm}^{-2})$
$c'_R$ = effective cohesion attributed to the root network $(\text{kNm}^{-2})$
$\theta$ = angle of shear rotation (degrees)
$\phi$ = angle of internal friction (degrees)
$t_R$ = average tensile strength of the roots per unit area of soil $(\text{kNm}^{-2})$

*Source*: Wilkinson *et al.* (2002).

# PART 5
## SYNOPSIS

# 26

# Landslide Hazard and Risk – Concluding Comment and Perspectives

*Thomas Glade and Michael J. Crozier*

## 26.1  Introduction

There is no doubt that the prerequisite for establishing a rational programme for managing risk is an objective assessment of risk and its careful evaluation in terms of community perceptions and aspirations. The research challenges required to achieve these objectives are manifest. First, there is still a poorly developed ability to comprehensively estimate landslide hazard. It is useful but not sufficient to represent the threat by identifying areas that show a range of susceptibility to landslide occurrence. Research needs to employ a range of modern earth science techniques in order to reveal the type of landslide that is likely to occur and the frequency of its occurrence. Furthermore, the expected landslide impact characteristics and intensity need to be clearly stated. How fast, how deep, how disrupted and how far will the material move?

There is a demand for risk assessments over a broad range of scales. The scale of investigation warranted is determined by a number of factors. One of these is the value and vulnerability of what is at risk. For example, a threatened school or hospital would demand a much more detailed assessment than a nature reserve or a parking lot. What is more, if there is a specific locational advantage for a particular site (in terms of economic or other values) compared to other sites, this may be sufficient to initiate detailed site investigation. However, the cost of investigation or the anticipated risk may outweigh the original site advantage and lead to the investigation of alternative sites.

*Landslide Hazard and Risk*   Edited by Thomas Glade, Malcolm Anderson and Michael J. Crozier
© 2004 John Wiley & Sons, Ltd   ISBN: 0-471-48663-9

If an area under investigation shows an unacceptable benefit/risk assessment and there are no obvious alternatives, then potential risk treatment measures need to be addressed and assessed from a feasibility and cost–benefit perspective. In some instances, there may be no 'site elasticity'. In other words, an established element under threat may gain its sole value by association with the site. In this case, the cost of remedial measures must be weighed against the uniqueness of the site values. For example, if it is unacceptable to shift the 'Leaning Tower of Pisa', the cost of risk reduction measures becomes less important in the ultimate decision. In reality, many elements at risk are essentially locationally fixed and have very little site elasticity. Main road and rail links in mountainous country are examples of where there are few options to relocate and engineering solutions rather than avoidance measures are required.

Broader-scale investigations are generally driven by longer-term planning horizons. With regional assessments, for example, the immediacy of the threat is less of a concern than the characterization and anticipation of future problems. Planning authorities want to know where problems might arise within their jurisdiction and what they can do to prevent them. These demands may be satisfied by the characterization of the region in terms of relative susceptibility to landslides. But ultimately, few rational planning decisions can be made without a clear statement of the hazard and risk. The challenge for scientists is to provide land managers with a statement not only of the frequency of landsliding expected at a site but also of the nature or the failure and its intensity.

Whereas certain economic units (for example, buildings and lifelines) can be readily identified as elements at risk, their vulnerability in the face of different types of landslide is currently poorly understood. A standardized and acceptable categorization of vulnerabilities with respect to landslide activity is a task yet to be achieved. An understanding of vulnerability requires a concerted effort of post-event record keeping using standardized protocols of damage assessment. There are, however, many other values (e.g. environmental quality) and activities in society that are also threatened or affected by landslides but which are much more difficult to represent in economic terms or are not readily assessed in terms of vulnerabilities. This aspect of risk estimation remains a challenge to economists, accountants, valuers, sociologists and psychologists. Despite their elusive quality, the representation of these elements at risk and their vulnerabilities are critical in characterizing loss and risk.

The derivation of a carefully crafted statement of risk is only the starting point for management. Essentially, it is an expert's representation of the problem. Communities or individuals are then faced with judging this statement before any decision to respond is considered.

The response to a calculated level of risk represents a value judgement. It is a balance of the risk and benefits of being exposed to this particular risk. Furthermore, risk reduction needs to be judged in terms of the costs involved compared with the benefits accrued. All these aspects are culturally, ethnically and value based. No one solution will 'fit all'. There is a timely warning here for the imported experts who deliver judgements for a foreign community. Decisions on risk acceptance or the acceptability of risk reduction measures must remain with the affected communities.

## 26.2   Hazard Assessment

An estimation of hazard is the fundamental starting point for any risk analysis. Landslide management decisions are strongly dependent on information on frequency, impact characteristics and magnitude of failures. Currently, the assessment of the probability of occurrence of a damaging event is being achieved by careful scientific appraisals of:

- monitored and modelled kinematics of active landslides;
- recorded past occurrences;
- spatial attributes of the terrain, including factors known to promote instability; identified by, for example, field surveys, remote sensing or GIS;
- the use of new dating techniques and the establishment of protocols for recording in a systematic fashion past landslide behaviour.
- computer simulation models of landslide occurrence under triggering conditions of rainfall and earthquakes (e.g. some using physical process-based estimations of unsaturated and saturated flow condition within the soil);
- runout models.

Depending on investigation scales, this information can be portrayed at specified levels of accuracy. For example, to determine frequency and magnitude of single landslides, detailed site investigations are crucial, but are commonly costly and time-intensive. On the other hand, spatial landslide hazard analysis, using different methods (e.g. qualitative or quantitative/mathematical approaches) depends very much on the availability of existing spatial information (e.g. geology, soil, vegetation, topographic information). In particular, spatial landslide inventories contain data indispensable for identifying critical stability factors or for verifying results of prior modelling.

## 26.3   Elements at Risk

Numerous approaches are available to assess different elements at risks. If the objective is to represent economic or human elements, the options are to use either existing databases, or, in their absence, to carry out specific surveys. In many jurisdictions most of the economic and human information is already available from government or private (e.g. insurance companies') databases. In the absence or lack of availability of specific data on the value of the elements at risk, reasonable assumptions can be made. However, it is a complex task, requiring considerable expertise, to assign average monetary damage values to classes of elements at risk (e.g. industrial district, residential area, agricultural regions, lifelines such as roads, railways, water supply, sewage, etc.). In addition, economic concerns are only one aspect of the problem; social and psychological consequences of landslide occurrences should also be taken into account when assessing elements at risk. Thus elements at risk can be determined by:

- establishing post event damage recording protocols;
- detailed field surveys including extensive questionnaires to assess object-based information; and
- deriving regional information from other sources such as official statistics.

The choice of assessment techniques is strongly dependent on the study's framework; however, it might also be influenced by administrative and legislative constraints.

Future efforts for establishing what is at risk need to be directed to the chains of consequences instigated by landslide occurrence. These chains of events may be extremely complex and attenuated. One salutary example illustrates the ramifications of government policy direction on land management. In the 1980s, the New Zealand government decided to address its negative balance of payments situation by attempting to increase its sheep and beef exports. Incentives were provided to farmers to clear scrub and forest from the steep land in order to provide more pasture for stock grazing. This process destabilized the slopes and landslides soon followed, destroying farm infrastructure and releasing sediment to the drainage system. Landslide erosion resulted in both enhanced runoff and downstream channel aggradation. The net effect was loss of primary productivity on the steep slopes, burial of high productive soils on the lower slopes, destruction of farm infrastructure and enhanced flooding downstream. Clearly this example shows that elements of risk from landsliding can be recognized well beyond the immediate on-site damage. Risk management procedures need to recognize all the possible consequential physical impacts.

Elements that are immediately and directly affected by landslide activity are easily recognized. However, there is yet a greater challenge to recognize and account for the less easily identifiable consequences of landslide impact, such as community viability, cultural values, and physical and mental well-being.

## 26.4   Vulnerabilities

The determination of vulnerability in landslide risk analysis focuses in particular on two aspects of vulnerability: economic loss and vulnerability with respect to life 'restrictions' (e.g. inquiry, death). When calculating risk, vulnerability needs to be registered against some aspect of the hazard (e.g. landslide volume) and qualified with respect to the element at risk (e.g. type of building). In terms of earthquake hazard, for example, this aspect of risk assessment is well established. In this instance, records are kept for the purpose of establishing the damage ratio (vulnerabilities) of different types of structures in particular isoseismal (earthquake shaking) bands. For example, the records might establish that within a zone where shaking intensities are registered as MM VII, brick houses may suffer destruction equivalent to 30% of their value, that is, a vulnerability of 0.3. Although in this book we have provided some examples of vulnerability with respect to landslide activity and elements at risk, they are largely educated guesses. There needs to be a much more concerted effort of recording and research to establish reliable vulnerability factors for combinations of elements at risk and types of landslides. There are even greater challenges to characterize vulnerabilities for the less tangible elements at risk. These less easily identifiable elements include the political, social, psychological and community vulnerabilities, to name only a few (see Winchester, 1992 or Blaikie *et al.*, 1994 for more details).

## 26.5   Conclusions and Perspectives

As population grows, there is inevitable pressure on the land resource. Initial settlement generally takes the best and safest places and subsequent settlement the more marginal sites. At the same time, increased urbanization, with its inevitable reliance of lifelines for survival, introduces a new set of risks. Inevitability, more of the population becomes at risk. As marginal sites are occupied there is a need to both inform those affected of the risk and to devise means for reducing the level of risk. It is increasingly important that the land occupier and the consent authorities are aware of the implications of increasing exposure to hazard.

Within this book the contributors have highlighted certain aspects of the landslide risk system that are still poorly understood or that require a better base of information before reliable risk assessments can be made. This does not detract from the importance of conventional field investigation, laboratory testing and stability analysis. Landslides will always need to be mapped, their subsurface and subsurface structure will need to be examined in the field, and their behaviour will need to be modelled. Some of those areas where future effort could usefully be directed to enhance our understanding of hazard include:

- The prediction of geometry and behavioural characteristic of first-time landslides.
- The further refinement of models capable of linking climate, slope hydrology and stability.
- Determination of sensitivity of different landslide types to changing boundary conditions such as climate and hydrology.
- Forecasting landslide movement: initial occurrence, reactivation, including the identification of precursors of failure.
- Determining the factors that control the mode of movement of existing landslides, especially the transition from creeping to catastrophic rates.
- The refinement of the relationship between geomorphic and geotechnical site models.
- Further refinement of landslide-triggering models based on rainfall and seismic shaking by taking into account preconditions such as availability of material, antecedent hydrology and exhaustion of available sites through previous landslide activity (event resistance).
- Application of developments in dating techniques including luminescence and cosmogenic techniques to refine frequency of movement.
- Ground truthing of airborne remote sensing techniques for stability assessment.

It is also evident from the contributions to this book that carefully kept records of past failures are indispensable to hazard analysis, particularly at the regional scale. Such inventories can be used as input data for the direct calculation of landslide susceptibility. In addition, if there is temporal and magnitude information available in the inventory, the probability of landslide occurrence of a given magnitude, in a specific time period and within a predefined location can be estimated. Thus the landslide hazard can be estimated. Another application of landslide inventories is their use for verification and validation of calculated susceptibility or hazard. If inventories need to be used for both analysis and validation of results, the data sets can be split in two groups, one for model development and one for validation (e.g. Chung and Fabbri, 1999). Major issues for comprehensive spatial analysis include:

- The implementation and maintenance of spatial data sets, including the use of information supplied by other disciplines (e.g. historical geography, archaeology, etc.).
- Whenever possible, the differentiation of these inventories by landslide type and magnitude.
- The increased usage of remote sensing techniques (both airborne and satellite imagery) for landslide detection and observation.

- Further improvements in spatial modelling techniques, including physically based modelling and numerical modelling approaches.
- The development of methodological concepts and definition of accuracy limits for landslide hazard and risk analysis for different scales, ranging from <1:10 000 to >1:750 000.
- The comparison of the current results from scientific hazard and risk analysis with 'user' needs and demands.

A fundamental issue in the representation of landslide hazard and risk involves the linking of methods, concepts and investigations that have been developed at different scales in order to provide an answer at a specified scale. A challenge to landslide hazard and risk research, however, is to not only upscale information, but also to downscale boundary condition information (Figure 26.1) (e.g. Dehn and Buma, 1999; Schmidt and Glade, 2003). Although information extrapolation across different scales occurs in other disciplines (e.g. climatology), the complexities of the geosystem and human system that make up the hazard–risk system ensure that this is a particularly difficult task.

Irrespective of scale, the concepts and approaches to landslide hazard and risk analysis outlined in this book allow a reproducible and, in most cases, objective assessment of the potential consequences of a landslide event. As well as comprehending the ultimate statement of a level of risk, decision makers and planners should also be aware of the concepts, assumptions, methods and limitations involved in its computation. As with any modelling procedure, the limitations of the approach have to be appreciated when using the information for subsequent decision making in the areas of policy and management:

- With any spatial landslide information uncertainties are inherent, which are difficult to evaluate (e.g. Ardizzone *et al.*, 2002; Carrara *et al.*, 1992).
- The resolution and quality of the socio-economic data influence the accuracy of the resulting risk estimate.

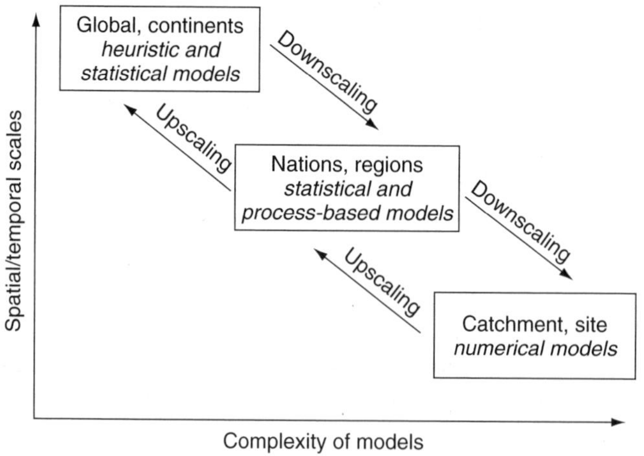


**Figure 26.1** *Bridging the gaps between scale-dependent analysis and the complexity of models*

- In most cases, the vulnerability of structures and of societies can only be roughly estimated or approximated (e.g. Glade, 2003).
- The risk model is always a generalization of reality, and the model performance is strongly dependent on data constraints.
- The calculated landslide risk is a static expression of reality at the time of analysis.

On the other hand, there are many advantages of landslide risk assessments (e.g. Petrascheck and Kienholz, 2003). These are, in particular:

- Risk values and information are transparent and comprehensible.
- Scenarios allow assessment of the consequences of future developments.
- If the reliability of the model performance is strongly dependent on data quantity and quality, then with increasing data availability and quality, the reliability of the risk estimate increases.
- Most models of landslide risk can be adapted to allow for significant changes in the environment, such as vegetation changes or changes in land use (e.g. suburban developments). Therefore the potential exists to regularly update the static risk information.
- The conceptual approach and established methods allow a comparison not only of risk from different landslide types, but also from other natural hazards.

The wide range of perspectives presented in this book show that there are several solutions and approaches to analysing, understanding, managing and reducing landslide risk. At present, the level of understanding of the landslide risk system appears to be more advanced than society's ability to translate this information into risk reduction sytems. There are, however, notable exceptions where the hazard–risk management system is treated in a holistic manner, and the human element is adequately acknowledged. Many developed societies have built into their hazard and resource management legislation requirements to publicly notify and consult with affected parties and in some cases have regard to cultural values. The 'buy-in' of the affected parties and their participation in decisions on the acceptability of risk and risk reduction measures is essential for sustainable management of risk. The understanding of risk perceptions and the evaluation of risk, however, are critical components of risk management that still remain largely within the research arena.

As populations expand and urbanization increases, infrastructure, assets and human behaviour continue to change and thus affect the level of risk. Probably one of the most important research challenges is to understand how risk evolves through time. It is only when this is appreciated that rational land planning and development adjustments can be made to reduce it.

# References

Ardizzone, F., Cardinali, M., Carrara, A., Guzzetti, F. and Reichenbach, P., 2002, Uncertainty and errors in landslide mapping and landslide hazard assessment, *Natural Hazard and Earth System Science*, **2**, 3–14.

Blaikie, P., Cannon, T., Davis, I. and Wisner, B., 1994, *At risk – Natural Hazards, People's Vulnerability, and Disasters* (London: Routledge).

Carrara, A., Cardinali, M. and Guzzetti, F., 1992, Uncertainty in assessing landslide hazard and risk, *ITC Journal*, The Netherlands, **2**, 172–183.

Chung, C.-J.F. and Fabbri, A.G., 1999, Probabilistic prediction models for landslide hazard mapping, *Photogrammetric Engineering & Remote Sensing*, **65**, 1389–1399.

Dehn, M. and Buma, J., 1999, Modelling future landslide activity based on general circulation models, *Geomorphology*, **30**, 175–187.

Glade, T., 2003, Vulnerability assessment in landslide risk analysis, *Die Erde*, **134**, 121–138.

Petrascheck, A. and Kienholtz, H., 2003, Hazard assessment and mapping of mountain risks in Switzerland, in D. Rickenmann and C.-L. Chen (eds), *Debris-flow Hazards Mitigation: Mechanics, Prediction, and Assessment*, 10–12 September 2003, Davos, Switzerland (Rotterdam: Mill Press), 25–38.

Schmidt, M. and Glade, T., 2003, Linking global circulation model outputs to regional geomorphic models: a case study of landslide activity in New Zealand, *Climate Research*, **25**, 135–150.

Winchester, P., 1992, *Power, Choice and Vulnerability* (London: James & James Science Publishers Ltd).

# Glossary

The glossary aims to define terms used throughout the book and in other commonly referred to landslide hazard and risk investigations. The terminology is in accordance with [1]Alexander (2002), [2]Australian Geomechanics Society (2000), [3]Cruden and Fell (1997), [4,5]International Strategy of Disaster Reduction ISDR (http://www.unisdr.org/ => terminology), and [6]International Decade of Natural Disaster Reduction IDNDR (http://www.unisdr.org/ => terminology).

Further explanations and definitions may be obtained from the following:

Aitchison, J. (coordinator), 1989, *International Thesaurus of Refugee Terminology* (Dordrecht: Martinus Nijhoff).
Disaster Management Centre, 1991, *Glossary*, University of Wisconsin.
Disaster Planning, 1985, *A Selected Bibliography: Disaster Preparedness Update Bibliography Series*, No. 5, PAHO, Washington DC.
*Glossary of International Disaster Assistance Terms, 1984: Natural Hazards*, Appendix I, OFDA, Washington DC, 134–150.
*Glossary of Terms for Probabilistic Seismic Risk and Hazard Analysis* (1984): US Geological Survey, Open-File report 84–760.
Gunn, S.W.A., 1990, *Multilingual Dictionary of Disaster Medicine and International Relief* (Dordrecht: Kluwer).
Hagman, G., de Allwood, A., Cutler, P., Kassaye, E., Kessely, L., Kourmayev, G.A. and Tolstopiatov, B. I., 1988, *From Disaster Relief to Development*, Henri Dunant Institute Studies on Development, No. 1, Geneva, 171–185.
*International Glossary of Hydrology*, 1991 (Geneva: WMO).
*International Meteorological Vocabulary*, 1990 (Geneva: WMO).
*Mitigating Natural Disaster Phenomena, Effects and Options – A Manual*, UNDRO Geneva.
*Natural Disasters and Vulnerability Analysis*, 1979 (Geneva: UNDRO).
Ockwell, R., 1990, *Disaster-Related Terminology, Some Observations and Suggestions*, A report to UNDRO.
Steinbrugge, K.V., 1982, *Earthquakes, Volcanoes and Tsunamis – An Anatomy of Hazards*, Skandia America Group, Appendix 2, New York.
UNESCO Division of Earth Sciences, 1990, *Glossary*.

**acceleration**: A change in velocity with time; in seismology and in earthquake engineering, it is expressed as a fraction of gravity (g), with reference to vibrations of the ground or of a structure.[5]

**accelerograph**: Instrument for recording acceleration, particularly in respect of earthquake shaking.[6]

**acceptable risk**: A level of risk that a given society is prepared to accept without any specifically designated management programme. Risk management may aim to reduce all risks to this level.[5]

**advisory**: A message to say that a hazard is in the early stages of approaching, and warnings may follow.[1]

**aftershock**: A smaller earthquake that follows the main shock and originates close to the focus of the main shock. Aftershocks generally decrease in number and magnitude over time.[6]

**alarm**: A visible or audible signal giving warning of danger.

**alert**: Advisory that hazard is approaching and that places emergency forces on standby, pending mobilization, but is less imminent than implied by warning message (see *warning*).

**all-hazards planning**: Most hazardous places are threatened by more than one set of hazards and associated risks. Besides the presence of several classes of extreme phenomena, there may be secondary hazards (e.g. earthquake-induced landslides). It is often more efficient to plan for all the major hazards to be expected in a region rather than separately for single ones. This allows economies of scale to be achieved and risks to be tackled comprehensively.[1]

**antecedent precipitation index**: (see also *antecedent soil moisture*) Weighted summation of past daily precipitation amounts, used as an index of soil moisture.[6]

**anthropogenic hazard, anthropogenic disaster**: See *social disaster*.

**anticyclone** (an atmospheric cell of high pressure): A region where barometric pressure is high relative to that in the surrounding regions at the same atmospheric level.[6]

**areal precipitation**: The average amount of precipitation which has fallen over a specific area.[6]

**ash flow**: Pyroclastic flow including a liquid phase and a solid phase composed mainly of ash.[6]

**assessment (of disaster risk)**: Survey of a real or potential disaster to estimate the actual or expected damage in order to make recommendations for prevention, preparedness and response.[6]

**avalanche**: The rapid and sudden sliding and flowage of masses of usually incoherent and unsorted material such as snow/ice/rock or mixtures of these materials. (Office of US Disaster Assistance)[6]

**background levels (of risk)**: In risk analysis, inherent natural or normal levels of risk in addition to risk from any specific factor.[1]

**benefit–cost analysis**: See *cost–benefit analysis*.

**black-box model**: A model with internal workings that are not specified. The modelling process therefore relates exclusively to the connection between inputs and outputs.[1]

**casualty**: Death or injury (mortality or morbidity) in disaster. Injury can be divided into physical trauma (e.g. fractured bones) and psychological trauma (e.g. post-traumatic stress disorder).[1]

**catastrophe**: In this book the terms *disaster* and catastrophe are used synonymously, because definitions of the two terms are not sufficiently well developed, precise or subject to consensus to be rigorously distinguished from one another. However, some authors regard a catastrophe to be more cataclysmic than a disaster and to affect a larger area. Jurisdictions affected by catastrophe, or so it is argued, are more thoroughly overwhelmed by it that they would be in the case of a mere disaster.[1]

**civil defence**: The progenitor of civil protection. The system of measures, usually run by a government agency, to protect the civilian population in wartime, to respond to disasters, and to prevent and mitigate the consequences of major emergencies in peacetime. The term 'civil defence' is now often referred to as emergency management.

**civil protection**: The process of protecting the general public, organizations, institutions, commerce and industry against disaster, by creating an operational structure for mitigation, preparedness, response and recovery. Military forces do not play a central role in civil protection, which is in the hands of administrative authorities, such as municipal, provincial, state or national governments.[1]

**climatic change**: Change observed in the climate on a global, regional or sub-regional scale caused by natural processes and/or human activity. Climate change refers to a divergence it long-term trend rather than short-term variability.[5]

**consequence**: The outcome of an event expressed qualitatively or quantitatively, being a loss, injury, disadvantage or gain. There may be a range of possible outcomes associated with an event.[2]

**consequence analysis**: Sometimes referred to as *risk identification*, the identification of *elements at risk*, their *vulnerability* and the type of impact or loss expected from a given hazard or hazards and dependent initially on *hazard identification*.

**consequential landslide hazard**: A *hazard (Type I or Type II)* resulting not directly from the landslide itself but as a result of a consequential process set in train by the landslide. For example, a wave set up by a catastrophic landslide entering a water body would be considered a consequential hazard in this context.

**contributions in kind**: Non-cash assistance in materials or services offered or provided in case of disaster.[6]

**control**: The direction of management and rescue activities in an emergency. Authority for control is specified in legislation and emergency plans. It is allied with the processes of directing and assigning tasks to emergency workers, and assuming responsibilities for failures. Many control functions affect multiple organizations.[1]

**control area**: A site with a perimeter around it to prevent the entrance of unauthorized personnel or equipment.[1]

**cost (of event, situation or activities)**: The negative impacts: these may extend beyond damage and losses and may or may not be quantifiable, both direct and indirect, including damage, time, labour, disruption, goodwill, political and intangible losses etc. Whereas losses refer to negative effects to existing resources, costs refer to adverse effects that go beyond existing use.[2]

**cost-effectiveness**: A measure of efficacy obtained by quantifying the costs and benefits of an activity and comparing them in a *cost–benefit ratio*. Cost-effectiveness does not necessarily require a predominance of benefits over costs, nor can all benefits always be quantified as readily as costs can.[1]

**cost–benefit analysis**: Assessment and comparison of the costs and benefits associated with an activity or proposed activity. For example, a comparison of the costs of establishing slope stability measures compared to the accrued economic benefits of being able to occupy that place as a result of the achieved reduction of landslide risk.

**crisis**: In disasters, a point at which normal mechanisms for coping (personal, organizational or institutional) suddenly cease to function as a result of the seriousness of the impact.[1]

**cyclone**: A large-scale closed circulation system in the atmosphere with low barometric pressure and strong winds that rotate counterclockwise in the northern hemisphere and clockwise in the southern hemisphere.[6]

**dam**: (also barrage; barrier; weir) Barrier constructed, or naturally emplaced, across a valley that impounds water or creates a reservoir.[6]

**damage assessment**: Investigation of damaged property and quantification of the value of losses. Monetary estimates usually depend on the cost of repairing the damage, which in turn depends on the adoption of particular techniques and components whose values are known.[1]

**damage classification**: Evaluation and recording of damage to structures, facilities, or objects according to different categories.[6]

**damage ratio**: The cost of damage expressed as a ratio of the total value of the object damaged in an event. When referred to a given magnitude of event, this ratio represents *vulnerabilty.*

**data collection platform (DCP)**: Automatic measuring facility with a radio transmitter to provide contact and transmission of data via satellite.[6]

**database**: A collection of information (in numerical or textual form or as images) that has been classified as and stored in such a way that it can be accessed selectively by potential users. Databases can be stored on cards or as other hard-copy archives, but increasingly they are kept as digital records accessible via computer networks or the Internet.[1]

**debris flow**: A high-density, rapid, downslope flowage of surficial slope-forming material with abundant coarse-grained materials such as rocks, tree trunks, etc.[6]

**decision support system**: A system (commonly a computer model) that uses predetermined logic and strategies to help the decision-making process to become explicit and more rational. Information can be supplied from a databank in order to inform the decision, and rules can be set to guide it (see *expert system*).[1]

**decision tree**: A breakdown of a series of events into smaller, simpler, more manageable, independent segments. These segments are represented as branches of a tree. Each event is given a probability with respect to two decision alternatives for each event.[2]

**declaration of disaster**: Official issuance of a state of emergency upon the occurrence of a large-scale calamity, in order to activate measures aimed at the reduction of the disaster's impact.[5]

**deforestation**: The clearing or destruction of vegetation in a previously forested area.[6]

**Delphi technique**: The use of a group of knowledgeable individuals to arrive independently at an estimate of the outcome of an uncertain situation.[2]

**disaster**: A serious disruption of the functioning of society, causing widespread human, material or environmental losses which exceed the ability of affected society to cope using only its own resources. Disasters are often classified according to their cause (natural or man-made).[5]

**disaster area survey team (DAST)**: A group that is deployed in an area after a disaster to ascertain the extent of damage to population and property and to recommend appropriate responses.[6]

**disaster insurance**: Government-sponsored or private insurance policies for protection against economic losses resulting from disaster.[6]

**disaster legislation**: The body of laws and regulations that governs and designates responsibility for disaster management concerning the various phases of disaster.[6]

**disaster management**: The body of policy and administrative decisions and operational activities that pertains to the various stages of a disaster at all levels (see *emergency management*).[6]

**disaster mitigation**: See *mitigation*.[6]

**disaster phases (syn. disaster cycle, disaster continuum)**: Pre- and post-disaster periods subdivided into particular actions. (Office of US Disaster Assistance)[6]

**disaster relief**: See *relief*.

**disaster response**: A sum of decisions and actions taken during and after disaster, including immediate relief, rehabilitation and reconstruction.[6]

**disaster team**: Multidisciplinary, multisectoral group of persons qualified to evaluate a disaster and to bring the necessary relief.[6]

**displaced person**: Persons who, for different reasons or circumstances, have been compelled to leave their homes. They may or may not reside in their country of origin, but are not legally regarded as refugees.[6]

**drainage basin**: (syn. catchment, river basin, watershed) Area having a common outlet for its runoff.[6]

**dynamic testing**: Analysis of the response of structures under simulated loads of the type imposed by natural hazards.[6]

**earthflow**: A mass movement characterized by downslope translation by flowage of fine surficial material, such as regolith and soil.[6]

**earthquake**: The energy release from a sudden break within the upper layers of the earth, sometimes breaking the surface, resulting in the vibration of the ground, which where strong enough will cause the collapse of buildings and destruction of life and property.[6]

**earthquake forecasting**: See *forecast*.[6]

**earthquake hypocentre (focus)**: The place inside the earth where the faulting which is associated with the earthquake originated.[6]

**earthquake intensity**: A measure of the perceived shaking effects of earthquakes and of the damage they cause to the built and natural environments. Intensity scales are not perfectly correlated with the magnitude of seismic waves and energy expenditure of earthquakes. Although it partly depends on these factors, seismic intensity is also a function of local and site-specific geological conditions, and the resistance of buildings to dynamic displacement. In the Americas the modified Mercalli scale is used; in Europe, the Mercalli–Cancani–Sieberg (MCS) and Medvedev–Sponheuer–Kárník (MSK) scales are used; in Russia the scale is GOST, and in Japan the Japan Meteorological Agency scale is current. Apart from the seven-point JMA scale, the others have 12 divisions, which vary incrementally from imperceptible shaking to cataclysmic damage. Different scales are used, and scales are periodically revised, because building typologies differ from place to place and over time. As the most important feature of the scales is damage to buildings, they must reflect the propensity to damage of each construction type.[1]

**earthquake magnitude**: A standardized measure of the size of the largest seismic waves created by the movement of a geological fault in the earth's brittle crust. Charles Richter's 1935 magnitude scale is now known as local magnitude ($M_L$), and is used in conjunction with other scales such as those for the body waves ($M_B$) and surface waves ($M_S$). As the scales are logarithmic, earthquakes with high magnitude are very much larger than those with moderate or small magnitude. Seismic magnitude is an indirect measure of the amount of physical energy released by an earthquake.[1]

**earthquake swarm**: A series of minor earth tremors (none of which may be identified as the main shock) that occurs within a limited area and time.[6]

**ecosystem**: Basic ecological unit formed by the living environment of the animal and vegetable organisms interacting as a single functional entity.[6]

**elements at risk**: The people, buildings and structures, infrastructure, economic activities, public services, or any other defined values exposed to hazards in a given area.

**emergency**: An imminent or actual event that threatens people, property or the environment and which requires a coordinated and rapid response to minimize its adverse consequences. Emergencies are usually unforeseen and unanticipated, even though they can, and should, be planned for. It is implicit that the consequences of ignoring an emergency or not dealing with it properly are avoidable casualties or damage.

**emergency management**: (see also *civil defence*) Short-term measures taken to respond to particular hazards, risks, incidents, or disasters. Resources and manpower pertaining to government, voluntary and private agencies are organized and directed on the basis of a plan that anticipates needs and coordinates efforts by assigning to particular responders, organizations or field units.[1]

**emergency mapping**: The cartographic depiction of selected aspects of disaster impact or the subsequent emergency relief effort. As the situation can change rapidly because of increases in the knowledge of damage and casualties, or to developments in the relief effort, there is a certain imperative to emergency mapping. It is therefore best carried out using computers with programs, such as geographical information systems, that produce refresher graphics.[1]

**emergency planning**: Planning of actions for the case of a disaster, training of special teams and of population, contingency planning, testing of disaster scenario.[5]

**ENSO**: (El Niño Southern Oscillation) A medium-term, global-scale surface atmospheric oscillation linked to sea–surface temperatures, particularly in the Pacific Ocean. It has three phases; El Niño (negative), neutral, and La Niña (positive). It has a strong influence on global precipitation patterns.

**environmental degradation**: Unfavourable modification of the ecological state and environment through natural processes and/or human activities.[6]

**epicentre**: The point on the earth's surface directly above the place of origin (i.e. focus or hypocentre) of an earthquake. (Office of US Disaster Assistance)[6]

**erosion**: Localized removal of rock or soil as a result of the action of water, ice, wind, coastal processes, or mass movement.

**evacuation**: Precautionary, temporary, planned removal of people or moveable items that, if left in place, would result in avoidable casualties or damage.[1]

**evaluation**: (post-disaster) Appraisal of all aspects of the disaster and its effects.[6]

**evapotranspiration**: The combined loss of water from a given area, by evaporation from the soil and by transpiration from plants.[6]

**event**: An incident or situation, which occurs in a particular place during a particular interval of time.[2]

**event tree analysis**: A technique which describes the possible range and sequence of the outcomes which may arise from an initiating event.[2]

**exceedance probability**: Probability that a given magnitude of an event will be equalled or exceeded.[6]

**expert system**: A computer-based logic system in which formal logic and a database provide answers to particular sequences of questions. It can be used to help make decisions in complex situations such as emergencies.[1]

**exposure**: The length or proportion of time that a person, building or other entity runs a risk. Fixed capital (e.g. houses, bridges, factories) permanently varies the proportion and intensity of exposure. Routine behaviour problems also cause exposure to vary. For example, if the principal risk arises from the collapse of bridges during earthquakes, people will be most exposed to it during daily commuting to or from work or on other forms of regular journey.[1]

**exposure time**: The time period of interest for risk calculations, hazard calculations, or design of structures. For structures, the exposure time is often chosen to be equal to the design lifetime of the structure.[6]

**externalization of risk**: The process by which risk divested by or shifted from an individual to other parties. See *internalization of risk*.

**factor of safety**: The ratio of shear strength to shear stress mobilized along a potential shear plane.

**fault**: A planar or gently curved fracture in the earth's upper layers across which displacement occurs. When this displacement is abrupt it gives rise to an earthquake.[6]

**fault tree**: A flowchart used in engineering to define possible failure patterns in equipment. The fault tree represents combinations of system states and causes of failure that can contribute to a specified major failure (called the top event).[1]

**fault tree analysis**: A systems engineering method for representing the logical combinations of various systems and possible causes which can contribute to a specified event (called the top event).[2]

**first aid**: The immediate but temporary care given on site to the victims of an accident or sudden illness in order to avert complications, lessen suffering, and sustain life until competent services or a physician can be obtained.[6]

**focal depth**: Vertical distance from the earth's surface to the place of origin (hypocentre, focus) of an earthquake.[6]

**focus (earthquake)**: The point beneath the earth's surface where an earthquake rupture starts and from which waves radiate.[6]

**forecast**: Statement or statistical estimate of the occurrence of a future event. This term is used with different meanings in different disciplines (see *prediction*). A forecast is often considered to represent the most rigorous form of prediction by stating, for the occurrence, what will occur, as well as the time and place of occurrence. Lesser forms of prediction may simply provide a probability of occurrence.

**foreshock**: Earthquake which is often part of a distinctive sequence which precedes and originates close to the focus of a large earthquake (main shock).[6]

**frequency**: With respect to natural hazards, the number of events of a given or minimum size per unit time (e.g. the number of earthquakes of magnitude greater than 5.9 per

100 years). The reciprocal of frequency is period, the average interval between events of a given size. For example the return period (or recurrence interval) of an earthquake of magnitude $M_w = 6$ may be 75 years (i.e. the frequency is 1.33 per 100 years). This represents an annual probability of 1.3%. As average values are used, the measure takes no account of irregularity of occurrence. Highly irregular events do not easily yield an accurate measure of frequency and period.[1]

**Global Observing System (GOS)**: The coordinated system of methods, techniques and facilities for making observations on a worldwide scale within the framework of the World Weather Watch.[6]

**grey-box model**: A model whose internal workings are specified by certain simplified processes or procedures that enable the modeller to gain an understanding of how real processes work and thereby to produce a satisfactory output response to input data.[1]

**ground motion**: Seismic vibration of the ground at a particular point, recorded by accelerograph or seismograph in order to determine the vibrational characteristics of an earthquake or explosion.[6]

**groundwater level**: The level at which soil and porous rock begins to be saturated with water. Usually measured as a depth below the ground surface or as a depth above a defined level, such as a shear surface.[6]

**ham radio network**: The international amateur radio network, frequently a valuable contribution by the community to disaster response.[6]

**hazard**: In natural hazard usage, there are two accepted definitions of 'hazard'. The first (*hazard I*) refers to an actual physical entity (process or situation) that has the potential to cause damage (e.g. a large rockslide or a long runout debris flow). This is the common non-technical understanding of hazard. However, this use of the term is also found in some legal and statuary documents, with statements of the form: 'It is council policy to record the date and location of hazards. These include, landslide, debris flow, surface flooding, subsidence etc. . . .' The second definition *(hazard II)* is more technical and refers not to a process but rather a threatening condition resulting from the behaviour of that process, expressed as the probability of occurrence of a damaging landslide.

**hazard I**: A potentially damaging process or situation (the landslide), e.g. an earthquake above a certain intensity or a landslide of sufficient size, depth, or displacement to cause damage or disruption or, as an example of a situation, the presence of weak foundation material.

**hazard II**: The probability of a potentially damaging event (a landslide) occurring in a unit of time. This probability varies with the magnitude of the event (generally small landslides occur more frequently than large landslides). Consequently hazard is often expressed as the probability of occurrence of a given magnitude of event. Defined in this way, hazard represents a state or condition and is assessed and applied to a particular place, e.g. site, unit area of land surface, region or object, e.g. lifelines, hydro dams, etc.

**hazard estimation**: The process of identifying the probability of occurrence of a damaging event.

**hazard identification**: The process of recognizing and accounting for all possible hazards that might occur within the place and time period of interest. For landslides this involves identifying landslide type, *landslide impact characteristics* and *consequential* *landslide hazards*. The process needs to consider the types of element at risk as well as the relationship in time and space between landslides and *elements at risk*. In the overall process of *risk assessment*, hazard identification and *consequence analysis* are interdependent should be carried out simultaneously.

**hydrological forecast**: Statement of expected hydrological conditions for a specified period.[6]

**hydrological warning**: Emergency information on an expected hydrological phenomenon which is considered to be dangerous.[6]

**hypocentre**: See *focus*.[6]

**incidence**: The number of specified occurrences in a given place in a period of time.

**impact**: The effects resulting from a hazard event.

**incident**: In emergency management terms, a sudden event, usually resulting in an emergency, that requires a response from one or more agencies. Incidents are more restricted in scope and consequences than are disasters.[1]

**individual risk**: Total risk (see *societal risk*) divided by the population at risk. For example, if a region with a population of 1 million experiences on average five deaths from landslides per year, the individual risk of being killed, in a year, by a landslide in that region, is 51 000 000, usually expressed in orders of magnitude as $5 \times 10^{-6}$.

**induced seismicity**: Earthquake activity resulting from man-made activities such as mining, large explosions, or forcing large quantities of liquid deep into the ground, e.g. oil-fields, waste disposal or reservoir filling.[6]

**information report**: Report with the same content as that of situation report but, in some instances, issued by an agency in the event that international assistance has not been subject of an official request by the government.[6]

**intensity of disaster impact**: In broad terms, number of deaths, physical and psychological injuries, and people rendered homeless by the disaster; scope of destruction and damage; effect on industrial and commercial productivity, and on employment; effect on human activities, and scale of donations, extra taxation and other financial reparations.[1]

**internalization of risk**: The process by which responsibility for risk is borne or assumed by the individual. See *externalization of risk*.

**involuntary risk**: A risk that is borne because there is no reasonable alternative, and because the bearer is unable or reluctant to forgo the benefits associated with tasking the risk.[1]

**lahar**: A term originating in Indonesia, designating a debris flow or mudflow over the flank of a volcano and composed of water and volcanic deposits. Primary lahars occur during, or as a result of, eruptions; secondary lahars have other causes (e.g. the collapse of a crater wall and sudden drainage of a crater lake).

**land degradation**: Progressive deterioration of land quality or land forms resulting from natural phenomena or human activity.[6]

**landslide**: One of a group of geomorphological processes, also referred to as *mass movement*. Mass movement involves the outward or downward movement of a mass of slope-forming material, under the influence of gravity. Although water and ice may influence this process, these substances do not act as primary transportational agents. Landslides are discrete mass movement features and are distinguishable from other forms of mass movement by the presence of distinct boundaries and rates of movement perceptibly higher than any movement experienced on the adjoining slopes.

Thus this group of processes includes falls, topples, slides, lateral spreads, flow and complex movements as classified by Cruden and Varnes (1996) and Dikau *et al.* (1996). Widespread diffuse forms of mass movement such as creep, subsidence, rebound and sagging are generally not treated as landslides. The criteria used to distinguish different types of landslide generally include: movement mechanism (e.g. slide, flow), nature of the slope material involved (rock, debris, earth), form of the surface of rupture (curved or planar), degree of disruption of the displaced mass, and rate of movement.

**landslide impact characteristics**: Characteristics of the landslide that may control the potential impact including: degree of disruption of the displaced mass, areal extent and distance of runout, depth, area affected, and velocity, discharge per unit width, and kinetic energy per unit area. Collectively these have been considered to represent **'landslide intensity'**.

**landslide stabilization**: Measures to prevent movement and increase the stability of a pre-existing landslide.[6]

**land use control**: A process used in urban and regional planning that imposes restrictions and interdictions on the uses to which particular plots of land are put. It is one of the principal instruments of non-structural mitigation, because it can be used to prevent development, or require structural mitigation, at sites where there is a demonstrably strong risk of damage and casualties. However, it is often less effective where development has already taken place.[1]

**lava flow**: Molten rock which flows downslope from a volcanic vent, typically moving at between a few metres to several tens of kilometres per hour.[6]

**lead time**: Period of a particular hazard between its announcement and arrival, also used for the mobilization of resources needed in relief operations.[6]

**levee (syn. bund, dike, embankment, stop bank)**: Water-retaining earthwork used to confine streamflow within a specified area along the stream or to prevent flooding due to waves or tides; or a natural embankment formed by deposition alongside a stream or debris-flow pathway.

**lifelines**: Vital communications and essential services that are liable to be compromised in disaster. They include the transportation networks along which emergency vehicles and evacuees will travel, main utility corridors for the distribution of electricity and water, and the medical assistance infrastructure.

**likelihood**: Used as a qualitative description of probability or frequency.[2]

**liquefaction**: Loss of resistance to *shear stress* of a water-saturated, silty–sandy soil as a consequence of earth shaking, to the extent that the ground behaves as a liquid rather than a solid.[6]

**logistics:** The range of operational activities concerned with supply, handling, transportation and distribution of materials. Also applicable to the transportation of people.[6]

**long term:** In the aftermath of disaster, the period of reconstruction and post-disaster development, or often persistent effects. A major disaster may create effects that last for 10, 25 or, in a very few cases, 50 years.[1]

**loss:** Any negative consequence, financial or otherwise; a reduction in the value of pre-existing resources (part of the costs experienced as the result of a hazard occurrence (see *costs*).[2]

**macroseismic field:** The area that experiences the visible effects of earthquakes in terms of casualties, damage and disruption. It is usually defined as the area enclosed by the rn/IV isoseismal.[1]

**macrozonation:** The division of a region or large area into zones that express future hazards, vulnerabilities or risks on the basis of accumulated information about where disasters have struck in the past and what distributed spatial effects they have had. Appropriate scales are 1:50 000 to 1:250 000. See also *microzonation*.[1]

**main shock:** The biggest of a particular sequence of earthquakes.[6]

**mass movement:** A general term for the outward and downward movement of slope-forming material under the influence of gravity, without the assistance of water as a transportational agent. Mass movement includes abrupt movements such as *landslides*, as well as slower, more widespread movements such as creep and subsidence.

**mass wasting:** A general term for the reduction of the mass of landforms by downslope transport of soil and rock material brought about by slope processes including mass movement, fluvial, pluvial and wind action.

**mean return period:** The average time between occurrences of a particular hazardous event.[6]

**Mercalli scale:** See *earthquake intensity*.

**microzonation (microzoning):** Subdivision of a region into areas where similar hazard-related effects can be expected. (OFDA). Seismic microzonation is the mapping of a local seismic hazard using a large scale (order of magnitude from 1:5000 to 1:10 000).[6]

**mitigation:** Medium- to long-term measures taken in advance of a disaster aimed at decreasing or eliminating its impact on society and environment. Methods are divided into structural mitigation (e.g. building levees along a river to reduce flooding), semi-structural (e.g. allowing floodable areas to exist along a river floodplain in order to contain flood waves) and non-structural (e.g. flood-damage insurance). Most modern mitigation strategies involve combinations of methods. Mitigation after the event is referred to as alleviation.

**Mobile Satellite Communication System (SATCOM):** Used after breakdown of other communication facilities in disaster-affected areas by disaster aid teams to perform, via satellite, exchange of detailed information by telex, phone or fax with their headquarters concerning detailed requirements for effective delivery of appropriate relief supplies.[6]

**monitor:** To check, supervise, observe critically, or record the progress of an activity, action or system on a regular basis in order to identify change.[2]

**monsoon:** Seasonally heavy rains and wind the direction of which varies from one season to another. They occur particularly in the Indian Ocean and South Asian areas.[6]

**MSK scale:** See *earthquake intensity*.[6]

**mudflow:** The rapid and localized downslope transfer of fine earth material mixed with water.[6]

**natural disaster:** The impact of an extreme natural phenomenon on the human system (lives, livelihoods and activities). For disaster to occur, the impact must exceed the ability of the human system to absorb or reflect it without suffering considerable, albeit temporary, disruption and losses.[1]

**natural hazard:** A hazard resulting form natural processes (see *hazard*).

**non-structural elements:** Those parts of a building (e.g. partitions, ceilings, etc.) that do not belong to the load-bearing system.[6]

**non-governmental organization (NGO):** Non-profit-making organization operating at the local, national or international level. Distinct from a governmental organization, having no statutory ties with a national government.[6]

**non-structural measures:** Actions taken to reduce risk that do not directly involve physical, engineering, or technical measures. For example, land use and building regulations, disaster legislation, public education and information, disaster insurance.[5]

**non-structural mitigation:** Measures used to reduce the impact of future disasters without the use of engineering or architectural techniques. Non-structural mitigation includes insurance coverage, land use and planning measures, and emergency management.[1]

**nowcast:** A description of a current situation, e.g. weather over a short-period (0–2 hours) forecast.[6]

***nuée ardente***: See *pyroclastic flow*.[6]

***N*-year event:** (see also *return period*) Magnitude of an event, the mean return period of which is *N* years.[6]

**organizations (in disaster):** Discrete social systems characterized by high degrees of internal interaction. Sociologists divide them into *adapting* (changing function to meet the needs of the disaster), *emerging* (newly created to meet the needs), *expanding* (absorbing volunteers or conscripted or convoked members), *extending* (enlarging their brief to meet needs created by the disaster), and *redundant* (not useful to the relief effort). Sudden, unexpected and intense impacts can put organizations into crisis, although usually temporarily.[1]

**peak discharge (syn. maximum discharge, peak flow):** Maximum discharge for a given flood hydrograph.[6]

**period:** See *frequency*.

**permafrost:** Layer of soil or rock in which the temperature has been continuously below 0E °C for at least some years.[6]

**planning map:** A map (often in the form of a statutory district plan) on which prescribed or suggested land uses or planned processes are shown. It serves a prescriptive, rather than a representative, function.[1]

**plate tectonics:** The dynamics and associated forms resulting from the earth's upper layers being made up of several mobile large rigid plates whose boundaries are fault zones along which slippage takes place.[6]

**polar front:** Quasi-permanent atmospheric front of great extent, in middle latitudes, which separates polar air from tropical air.[6]

**polder:** A mostly low-lying area artificially protected from surrounding water and within which the water table can be controlled.[6]

**population at risk:** A well-defined population whose lives, property and livelihoods are threatened by given hazards.

**potable water (drinking water):** Water that satisfies health standards, with respect to its chemical and bacteriological composition, and is agreeable to drink.[6]

**precedence approach (stability):** The process by which stability of an area is determined with reference to the stability behaviour observed in other localities with similar characteristics.

**precipitation duration:** Length of precipitation period.

**precipitation gauge:** General term for any device that measures the amount of precipitation; principally a rain gauge or snow gauge.[6]

**precipitation intensity (rainfall intensity):** The rate of precipitation. Amount of precipitation collected in unit time interval. Usually expressed as mm/hour.

**precondition factors (stability):** Inherent factors (usually static) that are necessary but not sufficient for causing a slope to fail.

**prediction:** A statement to the effect that a particular impact will occur in a particular area, with a particular magnitude and set of effects, during a particular time interval. Predictions are the responsibility of bona fide scientists and are usually given in probabilistic terms. The word is virtually synonymous but not as definitive as *forecast*.

**predictor:** A variable or index compiled from several elements, which is known (often empirically) to be highly correlated with a quantity which is to be forecast and is used to forecast it.[6]

**preparatory factors (stability):** Factors (usually dynamic) that reduce the margin of stability of a slope without actually causing failure.

**preparedness (syn. readiness):** Activities designed to minimize loss of life and damage, to organize the temporary removal of people and property from a threatened location and facilitate timely and effective rescue, relief and rehabilitation. See also *prevention*.[6]

**prevention:** Encompasses activities designed to provide permanent protection from disasters. Includes engineering and other physical protective measures, and also legislative measures controlling land use and urban planning. See also *preparedness*.[6]

**probability:** The likelihood of a specific event or outcome, measured by the ratio of specific events or outcomes to the total number of possible events or outcomes. Probability is expressed as a number between 0 and 1, with 0 indicating an impossible event or outcome and 1 indicating an event or outcome is certain.[2]

**probable maximum precipitation (PMP):** The greatest quantity of precipitation per unit time that is expected to occur in a given area with a given recurrence interval (e.g. 500 years). It forms the basis of civil engineering calculations concerning runoff and water yield.[1]

**protocol:** A set of standard procedures for carrying out a task or achieving a well-defined goal. The term is widely used in emergency medicine and in computing.[1]

**public awareness:** The state of understanding in the community as to the nature of the hazard and actions needed to save lives and property before and in the event of disaster.[5]

**pyroclastic flow:** High-density flow of solid volcanic fragments suspended in gas which flows downslope from a volcanic vent (at speeds up to 200 km/h) which may also develop from partial collapse of a vertical eruption cone, subdivided according to fragment composition and nature of flowage into: ash flow, glowing avalanche, *nuée ardente*, pumice flow.[6]

**qualitative approach:** A method by which information is analysed and evaluated verbally, with argument based on judgement and logic.

**quantitative approach:** A method by which information is recorded, analysed and evaluated using numerical scales and techniques.

**quicksand:** Saturated sandy deposits which, under the influence of hydrostatic pressures, are buoyant and are able to flow.[6]

**rain gauge:** See *precipitation gauge*.[6]

**reconstruction (almost syn. rehabilitation):** Actions taken to re-establish a community after a period of rehabilitation subsequent to a disaster. Actions would include construction of permanent housing, full restoration of all services, and complete resumption of the pre-disaster state. (Office of US Disaster Assistance)[6]

**recurrence interval**: See *frequency.*

**rehabilitation**: The operations and decisions taken after a disaster with a view to restoring a stricken community to its former living conditions, while encouraging and facilitating the necessary adjustments to the changes caused by the disaster.[6]

**release**: In risk analysis, the rate at which a hazard strikes, which is usually expressed in terms of frequency and return period, or with respect to the trend of cumulative impacts.[1]

**relief**: Assistance and/or intervention during or after disaster to meet the life preservation and basic subsistence needs. It can be of emergency or protracted duration.[5]

**remote sensing**: The observation and/or study of an area, object or phenomenon from an aerial distance, frequently using data collected by satellite.[6]

**rescue**: See *search and rescue.*[6]

**resettlement**: Actions necessary for the permanent settlement of persons displaced or otherwise affected by a disaster to an area different from their last place of habitation.[6]

**residual risk**: The remaining level of risk after risk treatment measures have been taken.[2]

**response**: The immediate and short-term reactions of the disaster relief community to an emergency situation. It includes search-and-rescue operations, medical assistance, and the early provision of food and shelter for survivors.[1]

**response-generated demands**: Emergency needs arising from the damaged social fabric, rather than directly from the agent that caused the disaster (e.g. the need for shelter for homeless survivors). Agent-generated demands, on the other hand, are measures related specifically to the nature of the hazard agent (e.g. the need for sandbags during a flood).

**return period (recurrence interval)**: See *frequency.*

**Richter scale**: See *earthquake magnitude.*[6]

**risk**: Expected losses (i.e. the probability of specified negative consequence to life, well-being, property, economic activity and other specified values) due to a particular hazard for a given area and reference period. Based on mathematical calculations, risk is the product of hazard and vulnerability.[6]

**risk acceptance**: An informed decision to accept the consequences and the likelihood of a particular risk.[2]

**risk analysis**: The initial stage of *risk assessment* involving *hazard identification, risk identification* and *risk estimation.*

**risk assessment**: The overall process involving hazard identification, estimation of the consequential risk and evaluation of that risk to a point where personal judgements and management decisions can be rationally made.

**risk avoidance**: An informed decision not to become involved in a risk situation.[2]

**risk–benefit analysis**: In terms of disasters, the benefits of inhabiting areas at risk, of carrying out various activities in them and of putting oneself at risk, set off against the costs of damage and losses in disaster (including estimated future costs of the event) and of mitigation works.[1]

**risk communication**: The communication of information about particular risks to the public (individuals, groups, organizations, institutions) and monitoring of the public's response.[1]

**risk control**: That part of risk management which involves the implementation of policies, standards, procedures and physical changes to eliminate or minimize adverse risks.[2]

**risk engineering**: The application of engineering principles and methods to risk management.[2]

**risk estimation**: The process used to produce a measure of the level of risks being analysed. Risk estimation consists of the following steps: frequency analysis, consequence analysis and their integration.[1]

**risk evaluation**: The process of determining the importance and relevance (significance) of the results of *risk analysis* with reference to the social and physical context within which they occur. This process determines whether risk is tolerable or acceptable. Risk evaluation may involve considerations of risk perception, risk communication and risk comparison with the aim of developing some appropriate level or form of response. It generally implicitly or explicitly balances the risk with the benefits associated with exposure to that risk.

**risk financing**: The methods applied to fund risk treatment and the financial consequences of risk. *Note*: In same industries risk financing only relates to funding the financial consequences of risk.[2]

**risk identification**: The process of determining what can happen, why and how.[2]

**risk level**: The level of risk calculated as a function of likelihood and consequence.[2]

**risk management process**: The systematic application of management policies, procedures and practices to the tasks of establishing the context, identifying, analysing, evaluating, treating, monitoring and communicating risk.[2]

**risk perception**: Intuitive understanding of risk based on an individual's own experience and judgement.

**risk reduction**: A selective application of appropriate techniques and management principles to reduce either likelihood of an occurrence or its consequences, or both.[2]

**risk retention**: Intentionally or unintentionally retaining the responsibility for loss or financial burden of loss within the organization.[2]

**risk tolerance**: A decision on the level of the tolerable residual risk after protection measures have been implemented.

**risk transfer**: Shifting the responsibility or burden for loss to another party through legislation, contract, insurance or other means. Risk transfer can also refer to shifting a physical risk or part thereof elsewhere.[2]

**risk treatment**: Selection and implementation of appropriate options for dealing with risk.[2]

**rockfall**: Free-falling or precipitous movement of a detached segment of bedrock of any size from a cliff or other very steep slope.[6]

**rockslide**: A downward, usually sudden and rapid movement of newly detached segments of bedrock over an inclined surface or over pre-existing features. (Office of US Disaster Assistance)[6]

**rupture zone**: Area of fault breakage corresponding to a particular earthquake sequence. Alternatively, with respect to landslides, an area representing the parting of the displaced mass from the *in situ* slope.[6]

**satellite applications**: The use of satellite technology for the purpose of communications or data transmission for monitoring, warning and dissemination of information pertinent to emergency response and/or disaster management.[6]

**scenario**: A hypothetical sequence of events, based closely on observation of real circumstances and on logical projection of their consequences. It is used to develop a detailed picture of how hazards, impacts, relief efforts and so on will probably develop over time. It is thus an important basis for planning.[1]

**sea surge**: A rise in sea level that results in the inundation of areas along coastlines. These phenomena are caused by the movement of ocean and sea currents, low-pressure cells, winds and major storms. (Office of US Disaster Assistance)[6]

**search and rescue**: The process of locating and recovering disaster victims and the application of first aid and basic medical assistance as may be required.[6]

**secondary hazard**: (see *consequential landslide hazard*) Those hazards that occur as a result of another hazard or disaster, i.e. fires or landslides following earthquakes, epidemics following famines, food shortages following drought or floods.[6]

**seismic activity rate**: The mean number per unit time of earthquakes with specific characteristics (e.g. magnitude 6) originating on a selected fault or in a selected area.[6]

**seismic isolation**: Systems used to limit the transfer of strong ground motion to a structure.[6]

**seismic zone**: An area within which ground motion and seismic design requirements for structures are similar.[6]

**seismicity**: The distribution of earthquakes in space and time. (UNDRO)[6]

**seismograph**: An instrument for recording vibratory motion of the ground. (OFDA)[6]

**sensitivity analysis**: The process of systematically varying the numerical values input into calculations or digital models in order to gauge the relative influence of each input value on the scope and nature of output responses.

**severe weather threat index – SWEAT index (threat score)**: An index used to predict thunderstorms and tornadoes.[6]

**shear strength**: The internal resistance (force per unit area) offered to *shear stress*. It is measured by the maximum shear stress that can be sustained without failure.

**shear stress**: The force per unit area within a body of material which tends to cause two adjacent parts of the body to rupture and slide past each other.

**shear wall**: A structural element which resists lateral forces.[6]

**shelter**: Physical protection requirements of disaster victims who no longer have access to normal habitation facilities. Immediate post-disaster needs are met by the use of tents. Alternatives may include polypropylene houses, plastic sheeting, geodesic domes and other similar types of temporary housing.[6]

**short term**: In disasters, a period of a few days to a few weeks that stretches from the impact to the point at which repair and recovery are firmly established processes. It is centred on the initial post-disaster emergency.[1]

**simulation exercise**: Decision-making exercise and disaster drills within threatened communities in order to represent disaster situations to promote more effective coordination of response from relevant authorities and the population.[6]

**situation report**: A brief report that is published and updated periodically during a relief effort and which outlines the details of the emergency, the needs generated and the responses undertaken by all donors as they become known. Situation reports are issued by UNDRO, UNHCR, ICRC and LRCS.[6]

**slide**: See *landslide*.[6]

**SMS/GOES (Synchronous Meteorological Satellites/Geostationary Operational Environmental Satellites)**: Satellites orbiting over the equator at the same rate as Earth's rotation and providing images of visible and infrared portions of the spectrum for the same area every 30 minutes. The satellites can collect and distribute environmental data from remote unattended data collection platforms on land, in water or in the atmosphere and rapidly transmit these data to ground receiving stations.[6]

**social disaster**: The impact on the human system (lives, livelihoods, activities and property) of riots, crowd crushes, terrorist outrages, etc. Together with technological disasters, this category falls under the umbrella heading of *anthropogenic disaster*.[1]

**societal risk**: The total risk attributed to the society responsible for bearing that risk (see *individual risk*).

**soil conditions**: The conditions of earth (moisture content, disaggregation density, etc.) that may mitigate or intensify disaster agents, such as drought, flooding or seismic movement. (Office of US Disaster Assistance)[6]

**soil creep**: The gradual and steady movement of soil and loose rock material down a slope that may be gentle but is usually steep; it is also called surficial creep. (Office of US Disaster Assistance)[6]

**soil moisture**: Content of water in the portion of the soil which is above the table water including water vapour present in the soil pores. In some cases refers strictly to moisture within the root zone of plants.[6]

**specific risk**: Hazard probability × vulnerability for a given element at risk, i.e. the expected degree of loss due to a particular element at risk and type of process.

**storm**: An atmospheric disturbance involving perturbations of the prevailing pressure and wind fields, on scales ranging from tornadoes (1 km across) to extratropical cyclones (2000–3000 km across).[6]

**structural measures**: Disaster-resistant or protective structures (banks, flood walls, dykes, dams).[5]

**structural mitigation**: Engineering and architectural measures to reduce the impact of disasters. Examples include river levees, anti-seismic buildings and avalanche barriers on slopes.[1]

**subsidence**: Collapse of a considerable area of land surface, due to the removal of underlying liquid or solid, or removal of soluble material by means of water.[6]

**susceptibility (landslides)**: The propensity of an area to undergo landsliding. It is a function of the degree of inherent stability of the slope (as indicated by the factor of safety or excess strength) together with the presence of factors capable of reducing excess strength and ultimately triggering movement.

**synoptic chart (weather chart, weather map)**: Geographical map on which meteorological data, analysed or forecast for a specific time, are presented to describe the atmospheric conditions at the synoptic scale.[6]

**technology transfer (or cooperation)**: Information and equipment provided by one country or area to another, along with the responsibility of training individuals in the use of that information, technology and/or equipment. (Office of US Disaster Assistance)[6]

**telemetry**: The use of data communications devices from the sensors *in situ* to a receiving station.[6]

**temporary housing**: See *shelter*.[6]

**tephra**: Solid, fragmentary material ejected during a volcanic eruption such as ash, lapilli, blocks and bombs, except lava.

**terracing**: Horizontal cuts, benches or embankments made along hillsides to reduce erosion, improve cropping, hold back runoff, improve infiltration of rain, or carry out some other conservation function. (Office of US Disaster Assistance)[6]

**thematic map**: A map of the distribution of a particular phenomenon or characteristic (e.g. solid geology, landslides, coasts undergoing erosion).[1]

**threshold**: A state reached (level) where there is an abrupt change in the nature of response to a variable or set of variables.

**tolerable risk**: A level of risk that a society is prepared to live with because there are net benefits in doing so, as long as that risk is monitored and controlled and action is taken to reduce it: see also *risk tolerance*.

**trauma**: Injury of any nature.[6]

**tremor**: A shaking movement of the ground associated with an earthquake or explosion. (Office of US Disaster Assistance)[6]

**triggering factors (stability)**: Factors that are sufficient to instigate slope failure.

**tropical cyclone**: Generic term for a non-frontal synoptic-scale cyclone originating over tropical or sub-tropical waters with organized convection and definite cyclonic surface wind circulation, with winds generally over 33 knots. Referred to as typhoons in Asia and hurricanes in the Atlantic.[6]

**tropical depression**: Cell of low pressure originating in the tropics with wind speed upto 33 knots.[6]

**tropical disturbance**: Light surface winds with indications of cyclonic circulation.[6]

**tsunami**: A series of large waves generated by sudden displacement of seawater (caused by earthquake, volcanic eruption or submarine landslide); capable of propagation over large distances and causing a destructive surge on reaching land. The Japanese term for this phenomenon, which is observed mainly in the Pacific, has been adopted for general usage.[6]

**vector control**: Measures taken to reduce the number of disease-carrying organisms (vectors) and to diminish the risk of their spreading infectious diseases.[6]

**volcanic eruption**: The discharge (aerially explosive) of fragmentary ejecta (see *tephra*), lava and gases from a volcanic vent.[6]

**volcano**: A vent from which a *volcanic eruption* occurs which may be associated with a mountain or other landform such as a caldera.

**voluntary agencies**: Non-governmental agencies or organizations that exist in many countries throughout the world. Some possess personnel trained to assist when disaster strikes. Some have capabilities that extend from the local to national and international levels. (Office of US Disaster Assistance)[6]

**voluntary risk**: A risk that is taken despite full awareness of the consequences.

**vulnerability assessment**: The systematic investigation of vulnerability for all *elements at risk*.

**vulnerability**: The expected degree of loss to a given element or set of elements at risk resulting from the occurrence of a natural phenomenon of a given magnitude. It is directly but inversely related to the resilient qualities of the *element at risk* (e.g. building standards). It may be indirectly related to poverty, lack of mitigation, lack of political power, and marginalization, but it cannot be predicted completely by any of these factors. Hence its origins tend to be complex. It is expressed on a scale from 0 (no damage) to 1 (total loss), or 0% damage to 100% damage respectively, referred to as the damage ratio in cases of quantifiable loss.

**warning**: (1) An alarm signal or message coupled with a recommendation or order to take action (such as mobilize or evacuate). The warning is a more advanced stage of mobilization than the hazard watch, and pertains to conditions in which there is a high probability, or virtual certainty, that disaster will occur relatively soon (within minutes, hours, or at the most a few days). Warnings can be subdivided into the technological processes of conveying the message (i.e. communications systems) and the social process of informing the people and ensuring that they understand and act upon the warning message. Warnings are separate from predictions. Whereas the latter are the responsibility of scientists, warnings are the preserve of civil administrators.[1] (2) Dissemination of message signalling imminent hazard which may include advice on protective measures. See also *alert*.[6]

**warning and evacuation**: Development of disaster prediction methods and establishment of warning systems including communication facilities, shelter and evacuation planning.[5]

**weathering**: The break-up and/or transformation of rocks by mechanical, chemical and biological processes, mainly under the influence of the atmosphere.[6]

**World Weather Watch (WWW)**: The worldwide, coordinated, developing system of meteorological facilities and services provided by WMO members for the purpose of ensuring that all members obtain the meteorological information required both for operational work and research. The essential elements of the WWW are: The Global Observing System, The Global Data-processing System and the Global Telecommunication System (used also for transmission of seismic information in the Far East).[6]

**zonation**: The division of a geographical area of land (e.g. a valley, town or region) into homogeneous sectors with respect to particular criteria (e.g. intensity of hazard, degree of vulnerability or risk, dependence on a particular operations centre).[1]


# References


Alexander, D.E., 2002, *Principles of Emergency Planning and Management* (New York: Oxford University Press).

Australian Geomechanics Society, 2000, Landslide risk management concepts and guidelines, *Australian Geomechanics*, **35**(1), 49–92.

Cruden, D.M. and Varnes, D.J., 1996, Landslide types and processes, in A.K. Turner and R.L. Schuster (eds), *Landslides: Investigation and Mitigation*, Special Report 147 (Washington, DC: National Academy Press), 36–75.

Dikau, R., Brunsden, D., Schrott, L. and Ibsen, M. (eds), 1996, *Landslide Recognition Identification, Movement and Causes* (Chichester: John Wiley & Sons Ltd).

# Thematic Index

# Locations/regions

Adige basin 717–19, 728
Alaska 175, 313, 500, 501, 553
Alcántara 2
Ancona 415, 700
Andorra la Vella 675, 676, 678, 679,
  696, 697
Australia 281, 334, 621–41, 665

Baglung 598–600, 610–17
Bhutan 597–8, 602, 606–7, 616–17

Cairns 68, 621–41
Calabria 700, 705–14, 723–6, 728
California 180, 185–6, 312–13, 318, 320–1,
  327, 344, 354, 357–61, 363–6, 434, 493,
  495–6, 498–501, 503, 505, 509, 511–12
Campania 429, 700, 707, 720, 722, 726, 728
Colorado 83, 215, 313, 354, 356, 359
Conca basin 714
Cuyocuyo 181

East Coast Region 533

Falli-Hölli 301–2, 305
Fei Tsui Road 663–4
Folkestone Warren 585, 586, 588
France 146, 180, 299, 395, 404, 675

Germany 77, 102, 112, 147, 251, 261, 276,
  381, 384, 387–9, 394–6
Gisborne 18, 51, 54, 62, 229, 528, 532,
  538, 540

Glacier National Park 201–15, 218

Hawke's Bay 60, 227, 748
Highlands Court 590
Hong Kong 30, 68, 146, 178–9, 191, 514,
  557, 559, 643–73, 733, 746–8

Isle Of Wight 19, 79, 405–9, 412–14, 418,
  422–4, 564, 588, 589
Italy 23, 53, 102, 140, 142, 148, 180, 183,
  261, 276, 415, 429–64, 470, 553–5, 557,
  563, 572, 699–730

Kent 585
Kwun Lung Lau 647, 657–8, 672

La Baie 165–6, 168, 172
Lazio 700, 712, 726
Lei Yue Mun 652–3, 660–1, 669
Los Angeles 185–6, 359, 494–5, 509,
  512, 649

Madrid 23
Marecchia 714–15
Micronesia 220, 222–33
Montana 201–15

Nepal 597–617
New Zealand 15–18, 23, 25, 29, 45–6, 48,
  51–2, 54–5, 57–8, 60–2, 65–6, 79–80,
  220–2, 225–30, 232, 244–6, 333, 335–6,
  342, 345–9, 517, 522–6, 528–9, 531–3,
  535, 537, 539, 541–2, 747–8, 770